PROGRESS IN BRAIN RESEARCH

VOLUME 164

FROM ACTION TO COGNITION

Other volumes in PROGRESS IN BRAIN RESEARCH

Volume 127: Neural Transplantation II. Novel Cell Therapies for CNS Disorders, by S.B. Dunnett and A. Björklund (Eds.) – 2000, ISBN 0-444-50109-6.
Volume 128: Neural Plasticity and Regeneration, by F.J. Seil (Ed.) – 2000, ISBN 0-444-50209-2.
Volume 129: Nervous System Plasticity and Chronic Pain, by J. Sandkühler, B. Bromm and G.F. Gebhart (Eds.) – 2000, ISBN 0-444-50509-1.
Volume 130: Advances in Neural Population Coding, by M.A.L. Nicolelis (Ed.) – 2001, ISBN 0-444-50110-X.
Volume 131: Concepts and Challenges in Retinal Biology, by H. Kolb, H. Ripps and S. Wu (Eds.) – 2001, ISBN 0-444-50677-2.
Volume 132: Glial Cell Function, by B. Castellano López and M. Nieto-Sampedro (Eds.) – 2001, ISBN 0-444-50508-3.
Volume 133: The Maternal Brain. Neurobiological and Neuroendocrine Adaptation and Disorders in Pregnancy and Post Partum, by J.A. Russell, A.J. Douglas, R.J. Windle and C.D. Ingram (Eds.) – 2001, ISBN 0-444-50548-2.
Volume 134: Vision: From Neurons to Cognition, by C. Casanova and M. Ptito (Eds.) – 2001, ISBN 0-444-50586-5.
Volume 135: Do Seizures Damage the Brain, by A. Pitkänen and T. Sutula (Eds.) – 2002, ISBN 0-444-50814-7.
Volume 136: Changing Views of Cajal's Neuron, by E.C. Azmitia, J. DeFelipe, E.G. Jones, P. Rakic and C.E. Ribak (Eds.) – 2002, ISBN 0-444-50815-5.
Volume 137: Spinal Cord Trauma: Regeneration, Neural Repair and Functional Recovery, by L. McKerracher, G. Doucet and S. Rossignol (Eds.) – 2002, ISBN 0-444-50817-1.
Volume 138: Plasticity in the Adult Brain: From Genes to Neurotherapy, by M.A. Hofman, G.J. Boer, A.J.G.D. Holtmaat, E.J.W. Van Someren, J. Verhaagen and D.F. Swaab (Eds.) – 2002, ISBN 0-444-50981-X.
Volume 139: Vasopressin and Oxytocin: From Genes to Clinical Applications, by D. Poulain, S. Oliet and D. Theodosis (Eds.) – 2002, ISBN 0-444-50982-8.
Volume 140: The Brain's Eye, by J. Hyönä, D.P. Munoz, W. Heide and R. Radach (Eds.) – 2002, ISBN 0-444-51097-4.
Volume 141: Gonadotropin-Releasing Hormone: Molecules and Receptors, by I.S. Parhar (Ed.) – 2002, ISBN 0-444-50979-8.
Volume 142: Neural Control of Space Coding, and Action Production, by C. Prablanc, D. Pélisson and Y. Rossetti (Eds.) – 2003, ISBN 0-444-509771.
Volume 143: Brain Mechanisms for the Integration of Posture and Movement, by S. Mori, D.G. Stuart and M. Wiesendanger (Eds.) – 2004, ISBN 0-444-513892.
Volume 144: The Roots of Visual Awareness, by C.A. Heywood, A.D. Milner and C. Blakemore (Eds.) – 2004, ISBN 0-444-50978-X.
Volume 145: Acetylcholine in the Cerebral Cortex, by L. Descarries, K. Krnjević and M. Steriade (Eds.) – 2004, ISBN 0-444-51125-3.
Volume 146: NGF and Related Molecules in Health and Disease, by L. Aloe and L. Calzà (Eds.) – 2004, ISBN 0-444-51472-4.
Volume 147: Development, Dynamics and Pathology of Neuronal Networks: From Molecules to Functional Circuits, by J. Van Pelt, M. Kamermans, C.N. Levelt, A. Van Ooyen, G.J.A. Ramakers and P.R. Roelfsema (Eds.) – 2005, ISBN 0-444-51663-8.
Volume 148: Creating Coordination in the Cerebellum, by C.I. De Zeeuw and F. Cicirata (Eds.) – 2005, ISBN 0-444-51754-5.
Volume 149: Cortical Function: A View from the Thalamus, by V.A. Casagrande, R.W. Guillery and S.M. Sherman (Eds.) – 2005, ISBN 0-444-51679-4.
Volume 150: The Boundaries of Consciousness: Neurobiology and Neuropathology, by Steven Laureys (Ed.) – 2005, ISBN 0-444-51851-7.
Volume 151: Neuroanatomy of the Oculomotor System, by J.A. Büttner-Ennever (Ed.) – 2006, ISBN 0-444-51696-4.
Volume 152: Autonomic Dysfunction after Spinal Cord Injury, by L.C. Weaver and C. Polosa (Eds.) – 2006, ISBN 0-444-51925-4.
Volume 153: Hypothalamic Integration of Energy Metabolism, by A. Kalsbeek, E. Fliers, M.A. Hofman, D.F. Swaab, E.J.W. Van Someren and R. M. Buijs (Eds.) – 2006, ISBN 978-0-444-52261-0.
Volume 154: Visual Perception, Part 1, Fundamentals of Vision: Low and Mid-Level Processes in Perception, by S. Martinez-Conde, S.L. Macknik, L.M. Martinez, J.M. Alonso and P.U. Tse (Eds.) – 2006, ISBN 978-0-444-52966-4.
Volume 155: Visual Perception, Part 2, Fundamentals of Awareness, Multi-Sensory Integration and High-Order Perception, by S. Martinez-Conde, S.L. Macknik, L.M. Martinez, J.M. Alonso and P.U. Tse (Eds.) – 2006, ISBN 978-0-444-51927-6.
Volume 156: Understanding Emotions, by S. Anders, G. Ende, M. Junghofer, J. Kissler and D. Wildgruber (Eds.) – 2006, ISBN 978-0-444-52182-8.
Volume 157: Reprogramming of the Brain, by A.R. Møller (Ed.) – 2006, ISBN 978-0-444-51602-2.
Volume 158: Functional Genomics and Proteomics in the Clinical Neurosciences, by S.E. Hemby and S. Bahn (Eds.) – 2006, ISBN 978-0-444-51853-8.
Volume 159: Event-Related Dynamics of Brain Oscillations, by C. Neuper and W. Klimesch (Eds.) – 2006, ISBN 978-0-444-52183-5.
Volume 160: GABA and the Basal Ganglia: From Molecules to Systems, by J.M. Tepper, E.D. Abercrombie and J.P. Bolam (Eds.) – 2007, ISBN 978-0-444-52184-2.
Volume 161: Neurotrauma: New Insights into Pathology and Treatment, by J.T. Weber and A.I.R. Maas (Eds.) – 2007, ISBN 978-0-444-53017-2.
Volume 162: Neurobiology of Hyperthermia, by H.S. Sharma (Ed.) – 2007, ISBN 978-0-444-51926-9.
Volume 163: The Dentate Gyrus: A Comprehensive Guide to Structure, Function, and Clinical Implications, by H. E. Scharfman (Ed.) – 2007, ISBN 978-0-444-53015-8

PROGRESS IN BRAIN RESEARCH

VOLUME 164

FROM ACTION TO COGNITION

EDITED BY

C. VON HOFSTEN
K. ROSANDER
Department of Psychology,
Uppsala University,
S-75142 Uppsala,
Sweden

ELSEVIER

AMSTERDAM – BOSTON – HEIDELBERG – LONDON – NEW YORK – OXFORD
PARIS – SAN DIEGO – SAN FRANCISCO – SINGAPORE – SYDNEY – TOKYO

Elsevier
Radarweg 29, PO Box 211, 1000 AE Amsterdam, The Netherlands
Linacre House, Jordan Hill, Oxford OX2 8DP, UK

First edition 2007

Library of Congress Cataloging-in-Publication Data
A catalog record for this book is available from the Library of Congress

British Library Cataloguing in Publication Data
A catalogue record for this book is available from the British Library

ISBN: 978-0-444-53016-5 (this volume)
ISSN: 0079-6123 (Series)

Printed and bound in The Netherlands

07 08 09 10 11 10 9 8 7 6 5 4 3 2 1

List of contributors

H. Ábrahám, Central Electron Microscopic Laboratory, Faculty of Medicine, University of Pécs, Szigeti u. 12, 7643 Pécs, Hungary

K.E. Adloph, Department of Psychology, New York University, 6 Washington Place, Room 410, New York, NY 10003, USA

M. Adriani, Laboratory of Cognitive Neuroscience (LNCO), Ecole Polytechnique Fédérale de Lausanne (EPFL), EPFL-SV-BMI, Station 15, CH 1015 Lausanne, Switzerland

J. Atkinson, Visual Development Unit, Department of Psychology, University College London, Gower Street London WC1E 6BT, London, WC1E 6BT, UK

S.E. Berger, Department of Psychology, College of Staten Island, Graduate Center of the City University of New York, 2800 Victory Boulevard, 4S-221A, Staten Island, NY 10314, USA

A.G. Billard, Learning Algorithms and Systems Laboratory, Ecole Polytechnique Fédérale de Lausanne, EPFL-STI-I2S-LASA, Station 9, CH 1015 Lausanne, Switzerland

S. Biro, Cognitive Psychology Section, Leiden University, Wassenaarseweg 52, 2333 AK Leiden, The Netherlands

O. Blanke, Laboratory of Cognitive Neuroscience (LNCO), Ecole Polytechnique Fédérale de Lausanne (EPFL), EPFL-SV-BMI, Station 15, CH 1015 Lausanne, Switzerland

O. Braddick, Department of Experimental Psychology, University of Oxford, South Parks Road, Oxford, OX1 3UD, UK

S. Caljouw, Institute for Fundamental and Clinical Human Movement Sciences, VU University Amsterdam, Van der Boechorststraat 9, 1081 GG Amsterdam, The Netherlands

J. Call, Max Planck Institute for Evolutionary Anthropology, Deutscher Platz 6, D-04103 Leipzig, Germany

L. Craighero, Department of Biomedical Sciences and Advanced Therapies, Section of Human Physiology, University of Ferrara, via Fossato di Mortara 17/19, 44100-Ferrara, Italy

G. Csibra, School of Psychology, Birbeck College, Malet Street, London, WC1E 7HX, UK

B. Dalla Barba, Dipartimento di Pediatria, Università degli studi di Padova, Via Giustiniani 3, 35100 Padova, Italy

M. Elsabbagh, Centre for Brain and Cognitive Development, Birbeck College, University of London, Henry Wellcome Building, 32 Torrington Square, London, WC1E 7HX, UK

L. Fadiga, Department of Biomedical Sciences and Advanced Therapies, Section of Human Physiology, University of Ferrara, via Fossato di Mortara 17/19, 44100-Ferrara, Italy

G. Gergely, Institute for Psychology of the Hungarian Academy of Sciences, 18-22 Victor Hugo Street, 1132 Budapest, Hungary

T. Gliga, Centre for Brain and Cognitive Development, School of Psychology, Birbeck College, University of London, Malet Street, London, WC1E 7HX, UK

E. Gömöri, Department of Pathology, Faculty of Medicine, University of Pécs, Szigeti u.12, 7643 Pécs, Hungary

G. Gredebäck, Department of Psychology, University of Oslo, Postboks 1094 Blindern, 0317 Oslo, Norway

P. Hauf, Department of Psychology, St. Francis Xavier Universtiy, P.O. Box 5000, Antigonish, NS B2G 2W5, Canada

S. Hunnius, Nijmegen Institute for Cognition and Information, Radboud University Nijmegen, P.O. Box 9104, 6500 HE Nijmegen, The Netherlands

G.M. Innocenti, Department of Neuroscience, Karolinska Institutet, Retzius väg 8, S-17177 Stockholm, Sweden

M.H. Johnson, Centre for Brain and Cognitive Development, Birbeck College, University of London, Henry Wellcome Building, 32 Torrington Square, London, WC1E 7HX, UK

R. Keen, Department of Psychology, Tobin Hall, University of Massachusetts, Amhert, MA 01003, USA

K.D. Kinzler, Department of Psychology, Harvard University, 1130 William James Hall, Cambridge, MA 02138, USA

K. Kovács, Department of Pathology, Faculty of Medicine, University of Pécs, Szigeti u.12, 7643 Pécs, Hungary

H. Kozima, National Institute of Information and Communications Technology, Hikaridai 3-5, Seika, Soraku, Kyoto, 619-0289, Japan

A. Kravják, Department of Pathology, Faculty of Medicine, University of Pécs, Szigeti u.12, 7643 Pécs, Hungary

Y. Kuniyoshi, Department of Mechano-Informatics, School of Information Science and Technology, The University of Tokyo, 7-3-1 Hongo, Bunkyo-ku, Tokyo, 113-8656, Japan

I. Leo, Dipartimento di Psicologia dello Sviluppo e della Socializzazione, Università degli studi di Padova, Via Venezia 8, 35131 Padova, Italy

T.L. Lewis, Department of Psychology, Neuroscience Behaviour, McMaster University, Hamilton, ON, L8S 4K1, Canada

D. Maurer, Department of Psychology, Neuroscience Behaviour, McMaster University, Hamilton, ON, L8S 4K1, Canada

G. Metta, Department of Communication, Computer and Systems Science, University of Genoa, Lira Lab, Viale F. Causa, 13, 16145 Genoa, Italy

C.J. Mondloch, Department of Psychology, Brock Universtiy, 500 Glenridge Avenue, St. Catharines, ON, L2S 3A1, Canada

A. Nagakubo, Intelligent Systems Division, National Institute for Advanced Industrial Science and Technology, 1-1-1 Umezono, Tsukuba, Ibaraki, 305-8568, Japan

C. Nakagawa, National Institute of Information and Communications Technology, Hikaridai 3-5, Seika, Soraku, Kyoto, 619-0289, Japan

L. Natale, Italian Institute of Technology, Via Morego 30, 16163, Genoa, Italy

Y. Ohmura, Department of Mechano-Informatics, School of Information Science and Technology, The University of Tokyo, 7-3-1 Hongo, Bunkyo-ku, Tokyo, 113-8656, Japan

F. Orabona, Lira-Lab, DIST, University of Genoa, Viale Causa 13, 16145, Genoa, Italy

B. Petreska, Learning Algorithms and Systems Laboratory, Ecole Polytechnique Fédérale de Lausanne, EPFL-STI-IS-LASA, Station 9, CH 1015 Lausanne, Switzerland

K. Rosander, Department of Psychology, Uppsala University, Box 1225, Trädgårdsg. 20, S-75142 Uppsala, Sweden

G. Sandini, Italian Institute of Technology, Via Morego 30, 16163 Genoa, Italy

S. Sangawa, Department of Mechano-Informatics, School of Information Science and Technology, The University of Tokyo, 7-3-1 Hongo, Bunkyo-ku, Tokyo, 113-8656, Japan

G. Savelsbergh, Institute for Fundamental and Clinical Human Movement Sciences, VU University Amsterdam, Van der Boechorststraat 9, 1081 GG Amsterdam, The Netherlands; Institute for Biophysical and Clinical Research into Human Movement, Manchester Metropolitan University, Cheshire, England and Faculty of Human Movement Sciences, VU University Amsterdam, Van der Boechorststraat 9, 1081 BT Amsterdam, The Netherlands

L. Seress, Central Electron Microscopic Laboratory, Faculty of Medicine, University of Pécs, Szigeti u. 12, 7643 Pécs, Hungary

K. Shutts, Department of Psychology, William James Hall, Harvard University, Cambridge, MA, USA

F. Simion, Dipartimento di Psicologia dello Sviluppo e della Socializzazione, Università degli studi di Padova, Via Venezia 8, 35131 Padova, Italy

E.S. Spelke, Department of Psychology, Harvard University, 1130 William James Hall, Cambridge, MA 02138, USA

S. Suzuki, Department of Mechano-Informatics, School of Information Science and Technology, The University of Tokyo, 7-3-1 Hongo, Bunkyo-ku, Tokyo, 113-8656, Japan

K. Terada, Department of Mechano-Informatics, School of Information Science and Technology, The University of Tokyo, 7-3-1 Hongo, Bunkyo-ku, Tokyo, 113-8656, Japan

C. Turati, Dipartimento di Psicologia dello Sviluppo e della Socializzazione, Università degli studi di Padova, Via Venezia 8, 35131 Padova, Italy

E. Valenza, Dipartimento di Psicologia dello Sviluppo e della Socializzazione, Università degli studi di Padova, Via Venezia 8, 35131 Padova, Italy

J. van der Kamp, Institute for Fundamental and Clinical Human Movement Sciences, VU University Amsterdam, Van der Boechorststraat 9, 1081 GG Amsterdam, The Netherlands

P. van Hof, Institute for Fundamental and Clinical Human Movement Sciences, VU University Amsterdam, Van der Boechorststraat 9, 1081 GG Amsterdam, The Netherlands

B. Veszprémi, Department of Obstetrics and Gynecology, Faculty of Medicine, University of Pécs, Édesanyák u.7, 7643 Pécs, Hungary

C. von Hofsten, Department of Psychology, Uppsala University, Box 1225, Trädgårdsg. 20, 75142 Uppsala, Sweden

Y. Yasuda, Omihachiman-City Day-Care Center for Children with Special Needs, Tsuchida 1313, Omihachiman, Shiga, 523-0082, Japan

Y. Yorozu, Department of Mechano-Informatics, School of Information Science and Technology, The University of Tokyo, 7-3-1 Hongo, Bunkyo-ku, Tokyo, 113-8656, Japan

Preface

This volume is based on topics discussed during an ESF conference in the series on "Brain Development and Cognition Human Infants" held during the first week of October 2005 at Maratea di Aquafredda in Southern Italy. The idea behind this series of conferences is to explore the phylogenetic and ontogenetic roots of development from a unified perspective, showing how perception, cognition, and action are interrelated in the early development of the child. This problem is examined in both non-human primates and different populations of human children. Development in atypical circumstances can shed new light on typical development, and on the evolutionary process that lead to such advanced functioning as language and culture in humans. Within this theoretical framework, the present conference brought together research on the maturation of cortical processes, mirror neurons, the development of tracking, reaching and locomotion, core knowledge, face perception, social understanding. A special part of the conference examined the work done on developmental robotics. This part was sponsored by EU-project "Robotcub".

Activity is at the very core of development, from the principles that underlie the early structuring of the brain to the processes that enable us to represent and reflect on the world and plan our actions on it. In development, actions bring about change to the neural system and the neural system brings about change to action. Perception, cognition, and motivation develop at the interface between neural processes and action and are crucially dependent on having a body with the experiences that such a body affords. In other words, cognition is a product of the ways in which the child moves through the world and interacts with it.

Thus, perception, cognition, and social functioning are all anchored in the actions of the child. Actions reflect the motives, the problems to be solved, and the constraints and possibilities of the child's body and sensory-motor system. The planning of actions also requires knowledge of the affordances of objects and events that are discovered through the explorative actions of the child. Furthermore, actions are directed into the future and their control is based on knowledge of what is going to happen next. Such knowledge is available because events are governed by rules and regularities. The most basic rules are laws of nature, like gravity and inertia. Others are task specific, like those making it possible to drive a car or ride a bike. Finally, some rules are a function of our innate dispositions and social conventions. They are necessary for the understanding of our own actions and for facilitating social interaction. Children's understanding of other people's actions and intentions emerges from the understanding of their own.

Motor actions are present before birth and, in normally developing children, they evolve in interaction with the evolving perceptual and cognitive systems during the early years of life. In other words, cognition is a product of the ways in which the child moves through the world and interacts with it. Extensive neurophysiological and neuropsychological evidence show that, perception, action, and cognition are closely related in the brain and develop in parallel. We have divided up the sections in the book in terms of how the brain gets structured in development, how perception and cognition evolves in the context of action, how social competence gets structured by these basic abilities and how the principles of embodied cognition can be applied to the structuring of artificial systems. The book discusses both the normal structuring of action, perception, and cognition and how it is expressed in deviant action control, prematurely born children and children with autism.

Claes von Hofsten and Kerstin Rosander
Uppsala University

Acknowledgment

The Maratea conference was sponsored by the European Science Foundation (ESF) and the EU integrated project "Robotcub" (FP6 – 004370).

Contents

III. The Development of Action and Cognition

IV. The Development of Action and Social Cognition

V. The Development of Artificial Systems

SECTION I

The Structuring of the Brain

C. von Hofsten & K. Rosander (Eds.)
Progress in Brain Research, Vol. 164
ISSN 0079-6123

CHAPTER 1

Unaltered development of the archi- and neocortex in prematurely born infants: genetic control dominates in proliferation, differentiation and maturation of cortical neurons

Hajnalka Ábrahám[1,*], Béla Veszprémi[2], Éva Gömöri[3], Krisztina Kovács[3], András Kravják[3] and László Seress[1]

[1]*Central Electron Microscopic Laboratory, University of Pécs, Szigeti u. 12, 7643 Pécs, Hungary*
[2]*Department of Obstetrics and Gynecology, University of Pécs, Pécs, Hungary*
[3]*Department of Pathology, University of Pécs, Pécs, Hungary*

Abstract: The development of cerebral cortex includes highly organized, elaborate and long-lasting series of events, which do not come to an end by the time of birth. Indeed, many developmental events continue after the 40th postconceptual week resulting in a long morphological, behavioral and cognitive development of children. Premature birth causes an untimely dramatic change in the environment of the human fetus and often results in serious threats for life. Cognitive abilities of prematurely born children vary, but a correlation between cognitive impairment and the time of birth is evident. In this study we review the morphological evidence of cortical maturation in preterm and full-term infants. Various aspects of postnatal cortical development including cell proliferation and maturation of neurons in the temporal archi- and neocortex are discussed and compared in preterm infants and age-matched full-term controls.

Our results suggest that cell proliferation and maturation are not influenced by the preterm delivery. In contrast, the perinatal decrease of the number of Cajal–Retzius cells might be regulated by a mechanism that is affected by preterm birth.

We demonstrate that cognitive deficiencies of the prematurely born infants cannot be explained with light microscopically observed alteration of proliferation and maturation of neurons.

Keywords: hippocampus; dentate gyrus; neurogenesis; interneurons; preterm; human cortical development; temporal cortex

In recent decades, primarily through advances in neonatal medicine, there has been an increase in survival rates for prematurely born and low birth-weight infants (Katz and Bose, 1993; Morrison and Rennie, 1997). The fall in perinatal mortality has been accompanied by an increase in short-term morbidity and long-term physical and mental disability of infant survivors of very preterm birth. Therefore, the better chance of survival of preterms is coupled with a growing concern about their neurodevelopmental outcomes (Hack and

*Corresponding author. Tel.: +36 72 536 001 ext.1511; Fax: +36 72 536324; E-mail: hajnalka.abraham@aok.pte.hu

DOI: 10.1016/S0079-6123(07)64001-1

Fanaroff, 1999; Vohr et al., 2000). In early childhood, global cognitive delay, cerebral palsy, blindness and deafness have been reported as the most common signs of the poor outcome. The prematurely born children especially those with lower birth weight have worse academic achievement and lower intelligence scores compared to those born at term and with normal birth weight, and the educational disadvantage persists into adulthood (Hack and Fanaroff, 1999; Vohr et al., 2000; Hack et al., 2002). On neuropsychological testing they perform poorly on measures of attention, executive function, memory, spatial skills, fine and gross motor functions. Problems with these skills are evident in children with no overt neurosensory abnormalities, and even those who have uncomplicated neonatal history, frequently resulting in serious cognitive and educational difficulties by school age (Sykes et al., 1997; Olsen et al., 1998; Breslau et al., 2000). The neurobehavioral outcome of preterm infants has been reported to worsen with younger gestational age at birth and with the lower birth weight (McCormick et al., 1996; Hack et al., 2002) and it was proposed that a premature transition from intrauterine to extrauterine life profoundly disrupts fetal brain development (Peterson et al., 2000). As a support, magnetic resonance imaging (MRI) measurements showed that the extremely preterm infants (gestational age <30 weeks) had reduced cerebral cortical surface with decreased complexity compared to infants born at term (Ajayi-Obe et al., 2000). In addition, smaller volume of cortical gray matter was observed in the amygdala and hippocampus, and the reduction was larger than expected from the overall reduction of the brain volume (Peterson et al., 2000). Adolescents, who were born before the 30th week of gestation, had significantly smaller hippocampal volumes bilaterally than the age-matched full-term controls, despite their equivalent head size (Isaacs et al., 2000). They showed specific deficits in memory, both on tests and according to the parental questionnaires.

The archicortical hippocampal formation plays an essential role in learning and memory functions and has highly organized connections with the associational system of the temporal neocortex. Development of these structures does not come to an end by the 40th postconceptual week, and their postnatal maturation results in the long-lasting cognitive development of infants.

Major subdivisions of the hippocampal formation, including pyramidal cell layer of Ammon's horn are already distinguishable at the 15th gestational week, while layers of the dentate gyrus, as it is observed in the adult, still cannot be identified (Arnold and Trojanowski, 1996). Development of granule cell layer of the dentate gyrus shows a significant delay. Even at term and in the early postnatal period, cell formation and migration of the newly formed neurons have been reported in the human dentate gyrus (Seress et al., 2001). Peri- and postnatal proliferation and maturation of granule cells are accompanied by a prolonged establishment of their connections, as it has been already described for the human mossy cells (Seress and Mrzljak, 1992). According to these evidences, protracted development of neuronal circuits of the hippocampal formation was suggested (Seress, 2001). It is highly possible that prolonged maturation of archicortical areas is accompanied by a similarly long-lasting maturation of the temporal neocortical circuitry.

MRI findings in prematurely born children, such as the reduced size of the hippocampal formation and the reduction in convolution of the cortex may indicate impairments of the archi- and neocortical developmental processes. Therefore our aim was to determine the alterations that may occur at cellular level, and might be responsible for the smaller size of cortical areas and, consequently, for the cognitive and educational difficulties of prematurely born infants. The most plausible assumption behind the lag in growth of the hippocampal formation is the reduced cell proliferation in preterms. In addition, similar probable explanations of the reductions of hippocampal volume and convolution of temporal neocortex are the altered neuronal circuits due to a delayed maturation of nerve cells in the preterms. In the present study an overview is given based on analysis of various aspects of peri- and postnatal brain development in postmortem human brains. Results found in preterms are compared to those obtained from age-matched infants born at term.

In this study, brains of only infants and fetuses were used, whose death was not related to genetic disorders, head injury or neurological diseases including intraventricular, cerebral hemorrhage or periventricular leukomalacia. The gestational age, birth weight, gender, life time, age at death, cause of death and the staining used are summarized in Table 1. Preterm and full-term infants were compared based on the gestational age, and postconceptual time was used for age matching. Postconceptual ages were calculated from the gestational age plus the postnatal life time. For the collection, storage and processing of brain tissues, regulations of the Hungarian Ministry of Health

Table 1. Individual data and clinical diagnosis verified by autopsy of fetuses and preterm and full-term children included in this study

Cases	Gender	Birth weight	Gestational age (weeks) at birth	Postnatal life time	Postconceptual age at death	Cause of death
1.	Female	NA	14	...	14 weeks	Legal abortion (maternal disorder)
2.	Male	NA	15	...	15 weeks	Spontaneous abortion
3.	Male	NA	16	...	16 weeks	Legal abortion (maternal disorder)
4.	Female	NA	16	...	16 weeks	Legal abortion (maternal disorder)
5.	Female	NA	17	...	17 weeks	Spontaneous abortion
6.	Female	NA	18	...	18 weeks	Spontaneous abortion
7.	Female	NA	20	...	20 weeks	Legal abortion (maternal disorder)
8.	Female	700g	22	...	22 weeks	Spontaneous abortion
9.	Female	590g	24	2 days	24 weeks	IRDS
10.	Female	930g	26	1 day	26 weeks	IRDS
11.	Female	830g	27	2 days	27 weeks	IRDS
12.	Male	1100g	28	3 days	28 weeks	IRDS, CHD
13.	Female	1250g	29	8 days	30 weeks	IRDS
14.	Male	1415g	32	5 days	33 weeks	Sepsis, pneumonia
15.	Female	2360g	34	1 day	34 weeks	Pneumonia, IRDS
16.	Female	1890g	35	3 days	35 weeks	IRDS, CHD
17.	Female	2650g	38	2 days	38 weeks	CRI
18.	Male		37	8 days	38 weeks	Esophageal atresia
19.	Male	1620g	34	4 weeks	38 weeks	CRI
20.	Male	2610g	36	2 weeks	38 weeks	CRI
21.	Male	2800g	39	1 day	39 weeks	CRI
22.	Male	3050g	41	...	41 weeks	CRI
23.	Male	780g	27	14 weeks	41 weeks	CRI, IRDS
24.	Male	720g	24	21 weeks	45 weeks	CRI, IRDS
25.	Female	3050g	40	12 weeks	52 weeks	CI
26.	Female	1890g	35	16 weeks	51 weeks	CRI
27.	Male	3050g	37	12 weeks	49 weeks	CRI
28.	Male	2300g	38	22 weeks	60 weeks	CRI
29.	Female	2700g	40	22 weeks	62 weeks	SIDS
30.	Male	990g	27	39 weeks	66 weeks	CI, IRDS
31.	Male	980g	27	32 weeks	59 weeks	CRI, IRDS
32.	Male	2940g	40	55 weeks	1 year	CRI
33.	Female	3100g	41	2 years	2 years	Sepsis, CRI
34.	Female	2810g	40	8 years	8 years	ALL
35.	Female	3000g	40	11 years	11 years	ALL

Note: Shaded rows indicate prematurely born infants.

as well as the policy of Declaration of Helsinki were followed.

The histological techniques and the methods of quantitative evaluation are well-known, generally used, and were described in details in our previous publications (Seress et al., 2001; Meyer et al., 2002; Ábrahám and Meyer, 2003; Ábrahám et al., 2004b). The immunocytochemical markers including the primary antibodies are all commercially available, and were all already included in peer-reviewed publications.

Cytoarchitectonics and cell proliferation in the hippocampal formation

At the age of the 24th gestational week that is considered the earliest time of live birth, the hippocampal formation is still developing and significantly smaller in size than at full-term (Fig. 1). In contrast to Ammon's horn, where pyramidal cells are already formed, the dentate gyrus is still immature and substantial amount of cells with elongated nuclei resembling migrating neurons occurs in the hilus. Proliferating cells in the hippocampal formation were labeled with immunocytochemistry using antibody against Ki-67 protein that is present in the nuclei of cells during the cell cycle except for the G_0 phase (Verheijen et al., 1989a, b).

At the 24th gestational week, numerous proliferating cells expressing Ki-67 protein were seen in all layers of the hippocampal formation including the dentate gyrus (Seress et al., 2001). The hilus and the subgranular layer, which are the sites of divisions of granule cells of the dentate gyrus contained large amount of labeled cells (Fig. 2A). The number of proliferating cells in the hilus decreased with the age and only a limited amount of Ki-67 immunoreactive cells was found at term and afterwards (Fig. 2).

Comparing the size of the hippocampal formation and the cytoarchitectonics of dentate gyrus and Ammon's horn, no difference was seen in pretem infants and in their full-term age matched controls (Fig. 3). Ammon's horn contained similarly low rate of proliferating cells (<0.1%) in both the preterm children and their full-term controls in all examined age groups. Similarly, in the dentate gyrus of preterms no decrease in the rate of cell formation was detected (Table 2). In contrast, the proportion of proliferating cells was slightly higher in all layers of the dentate gyrus in preterms than in their full-term age-matched controls. Ki-67-positive cells were mainly found in the hilar region that is considered to be the secondary germinal matrix of granule cells of the dentate gyrus. Fewer immunoreactive cells were present in the molecular and granule cell layers. Even in extremely preterm infants born at the 24th or 27th gestational weeks

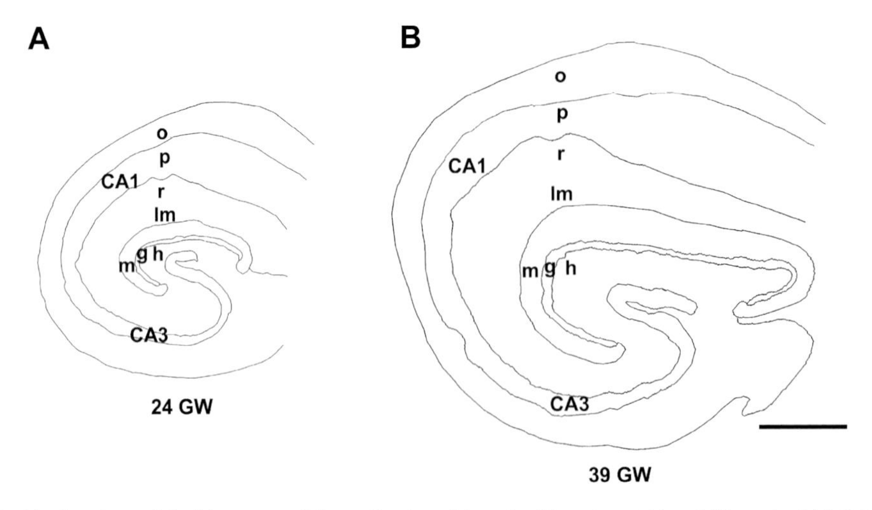

Fig. 1. Camera lucida drawings of the hippocampal formation in a 24-week-old preterm (A) and 39-week-old full-term newborn (B). Scale bar = 1 mm.

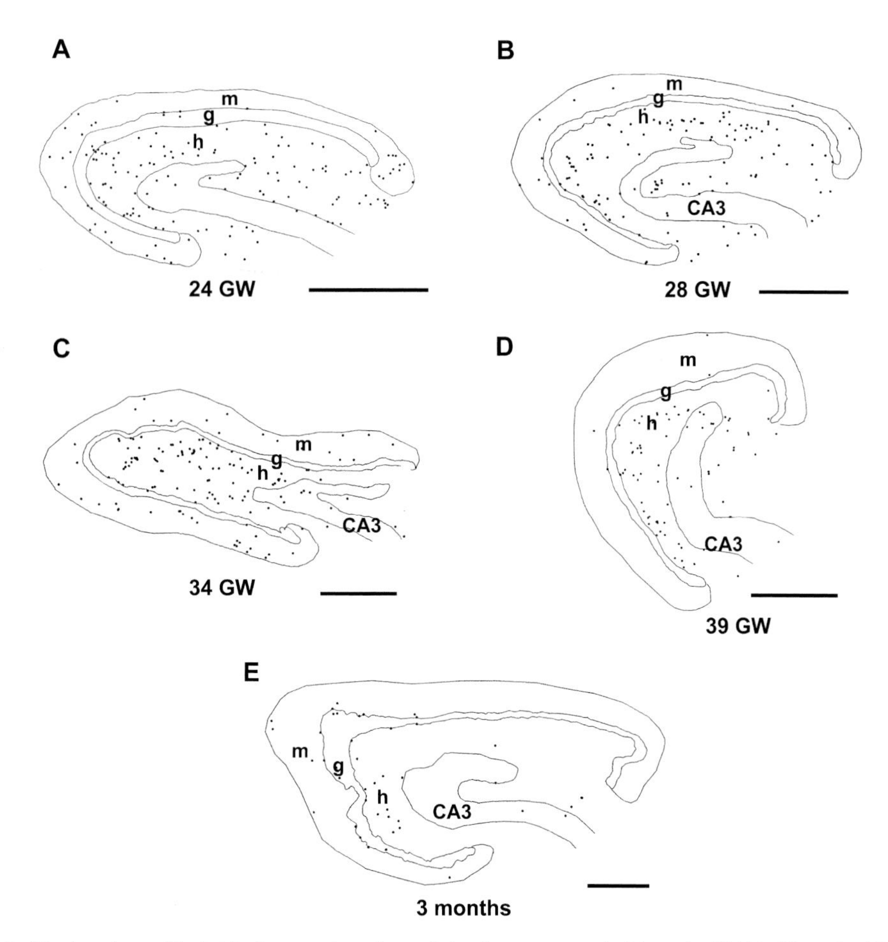

Fig. 2. Camera lucida drawings of individual coronal sections of the dentate gyrus in 24-week-old (A), 28-week-old (B), 34-week-old (C) preterm, in 39-week-old full-term newborn (D) and 3-month-old infant born at term (E). Ki-67-immunoreactive cells are plotted and each dot represents one labeled cell. Scale bars = 0.5 mm.

with extremely low birth weight and lived several weeks, all layers of the dentate gyrus displayed higher rate of cell proliferation than in the age-matched full-term newborn controls (Table 2). In fact, the difference found in the proportion of Ki-67-immunoreactive cells of the dentate gyrus shows parallel correlations with the degree of immaturity at birth. However, the difference in the rate of cell proliferation between preterms and their full-term controls is diminished when the survival time of the very or extremely early born infants is longer. These data may indicate deceleration of postnatal developmental events in preterms, such as the continuously decreasing rate of postnatal cell proliferation. However, at least two evidences contradict this argument. First, the morphology (e.g. size and cytoarchitectonics) of the dentate gyrus is very similar in preterms and in their full-term age-matched controls (Fig. 3), and does not differ as it would be expected on the basis of altered degree of cell proliferation. Second, the percentage of proliferating cells of the dentate gyrus in extremely preterm infants born at the 24th and 27th gestational weeks and lived several weeks is even higher than that found in those fetuses who died at the ages of 24th–28th weeks of gestation (Table 2). Thus, we propose that an increased rate of cell proliferation occurs in the dentate gyrus of preterms shortly after birth. The increased proportion of cell proliferation is found in all layers of the dentate gyrus, and not only in the hilus and subgranular zone, which are the sites of granule cell formation.

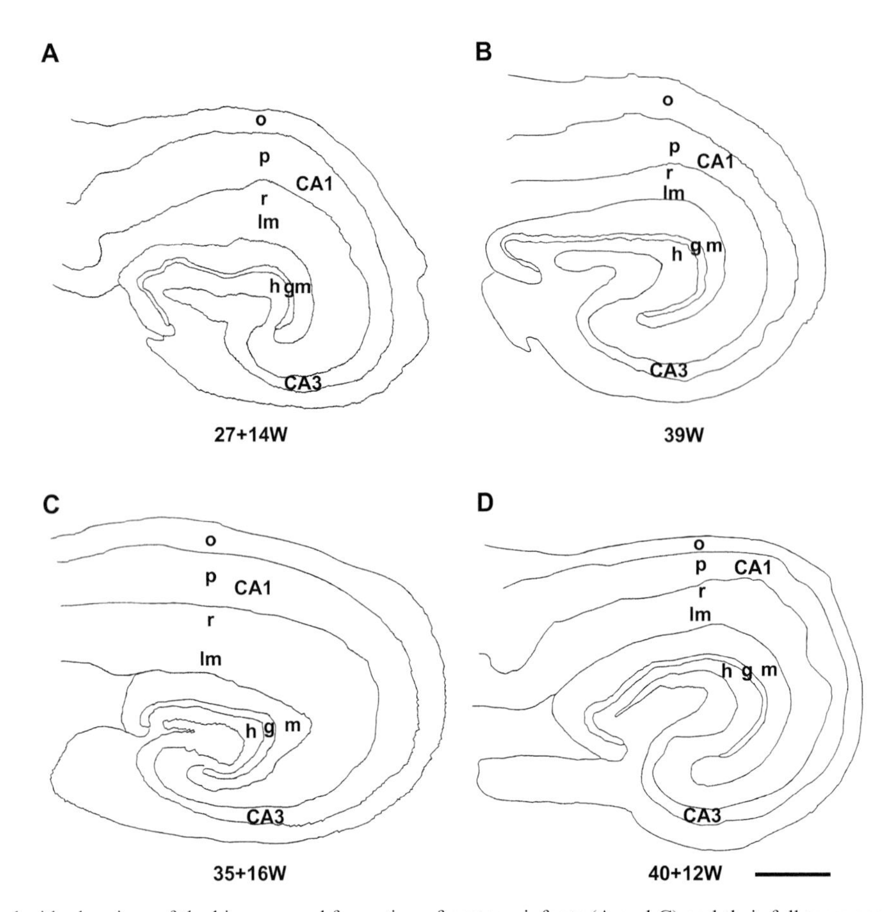

Fig. 3. Camera lucida drawings of the hippocampal formation of preterm infants (A and C) and their full-term age-matched controls (B and D). (A) Hippocampal formation of an infant born at the 27th week of gestation with extremely low birth weight and lived 14 weeks (case no. 23). (B) An infant born at the 39th gestational week (case no. 21). (C) Hippocampal formation of an infant born at the 35th week of gestation and lived 16 weeks (case no. 26). (D) Hippocampal formation of a 3-month-old infant who was born at the age of 40 weeks and lived 12 weeks (case no. 25). Scale bar = 1 mm.

Table 2. Labeling index in different layers of the dentate gyrus in preterm infants and in their age-matched controls

Age in weeks (gestational + extrauterine)	24	28	34	39	27+14	35+16	37+12	38+22	27+39
Case no.	9	12	15	21	23	26	27	28	30
LI in molecular layer(%)	6.83	2.84	2.92	3.1	7.8	3.1	1.27	1.1	1.35
LI in granule cell layer(%)	0.92	0.43	0.62	0.44	2.1	2.1	0.83	0.28	0.31
LI in hilus(%)	4.91	2.32	3.63	1.73	5.1	2.5	0.55	0.51	1.71

Note: Shaded columns indicate preterm infants.

Therefore, divisions of glial and endothelial cells are also expected. In order to prove this, we have identified the proliferating cells studying colocalization of Ki-67 nuclear protein and immunocytochemical markers of glial, endothelial cells and neurons (Ábrahám et al., 2004b).

Vimentin, the marker of undifferentiated and reactive astroglial cells was frequently found in Ki-67 labeled proliferating cells both in preterms and in their full-term age-matched controls. Proliferating glial cells were present in all layers of the dentate gyrus, Ammon's horn and the temporal necortex

(Figs. 4A and B). Although, the proportion of Ki-67/vimentin double-labeled cells indicated a marked variation between the individual cases both in the archi- and neocortex, the rate of proliferating glial cells was higher in preterms than in full-term age-matched controls (Table 3). Glial fibrillary acidic protein (GFAP) found in matured astrocytes colocalized rarely with Ki-67, indicating that mostly undifferentiated glial cells are capable of proliferating. However, this finding supports previous experimental observations that mature glial cells retain their ability to proliferate (Basco et al., 1977).

CD31, an endothelial cell marker labeled substantial amount of Ki-67-positive cells in the dentate gyrus and Ammon's horn (Fig. 4C). However, the intensity of immunostaining varied extremely from case to case due to the sensitivity of CD31 antigen to postmortem delay and, therefore, quantification would not give adequate results. In order to avoid this, Ki-67 immunocytochemistry was combined with periodic-acid Schiff (PAS) reaction that reveals the basal membrane of the blood vessels (Fig. 4D). The advantage of this combination is that PAS staining is even in all sections and, therefore, the results are suitable for quantification. In addition, using this method, proliferating cells in the brain parenchyma can be separated from those that are surrounded by the basal membrane. This latter group of cells contains mainly endothelial cells together with perivascular glial elements. Both in prematurely born infants and in their full-term age-matched controls the proportion of Ki-67 labeled cells was high at the inner side of basal membrane of blood vessels (Table 4). The highest number of proliferating endothelial cells was observed in the molecular layer of the dentate gyrus and in the strata radiatum and lacunosum-moleculare of Ammon's horn that are the places of the largest intensity of neuronal dendritic growth. The values did not differ remarkably but were slightly higher in preterm infants than in controls. In contrast, in the hilus, where cell proliferation is the largest the ratio of labeled endothelial cells was relatively low and slightly decreased in prematurely born infants whereas the proportion of dividing astrocytes was increased in premature infants (Tables 3 and 4). At the same time, there was no sign of an abnormal glial or neuronal accumulation in any area of the hippocampal formation.

Neuronal markers, such as Neu-N or neuron-specific enolase did not show colocalization with Ki-67. These neuronal-specific proteins are found in the differentiated nerve cells that exit from the cell cycle, and they are not expressed by proliferating cells. An earlier marker of postmitotic young migrating neuronal precursors is human β-III isoform of tubulin (Ferreira and Caceres, 1992). Therefore, colocalization of human β-III isoform of tubulin was also investigated using immunohistochemistry. In our material, no colocalization was observed with Ki-67 and human β-III isoform of tubulin either in the hilus or in the granule cell layer of the dentate gyrus, although, Ki-67 labeled cells were frequent in both areas (Figs. 4E and F). Tubulin-immunoreactive granule cells were numerous at the hilar border of the granule cell layer, and a few of them was observed deeper in the cell layer; however, none of them was seen in the hilus.

Thus, our aim to identify the proliferating cells in the postnatal hippocampal formation of preterms and full-term infants was only partially achieved, because of methodological difficulties of working with human brain tissue. The antibody against the Ki-67 protein is widely used by pathologists to detect proliferating cells in tumors, and it was found to be one of the most sensitive markers of cell formation (Verheijen et al., 1989a, b). Ki-67 protein is expressed in the nuclei of cells in all phases of the cell cycle except G_0 indicating that, similarly as we observed, differentiated cells do not contain Ki-67. Therefore, only part of the proliferating cells was identified as glial or endothelial cells, which after reaching a certain level of differentiation retain their ability for mitosis. According to our knowledge, to date no neuron-specific marker has been discovered in human that would label precursor cells in G_1, S, G_2 and M phases. Human β-III isoform of tubulin is found in young migrating neurons (Thomas et al., 1996) and, according to our observation, it is expressed only by postmitotic cells and therefore do not colocalize with the cell proliferation marker Ki-67. Therefore, when compared to other methods used to detect cell proliferation such as ^{3}H-thymidine autoradiography or bromodeoxyuridine

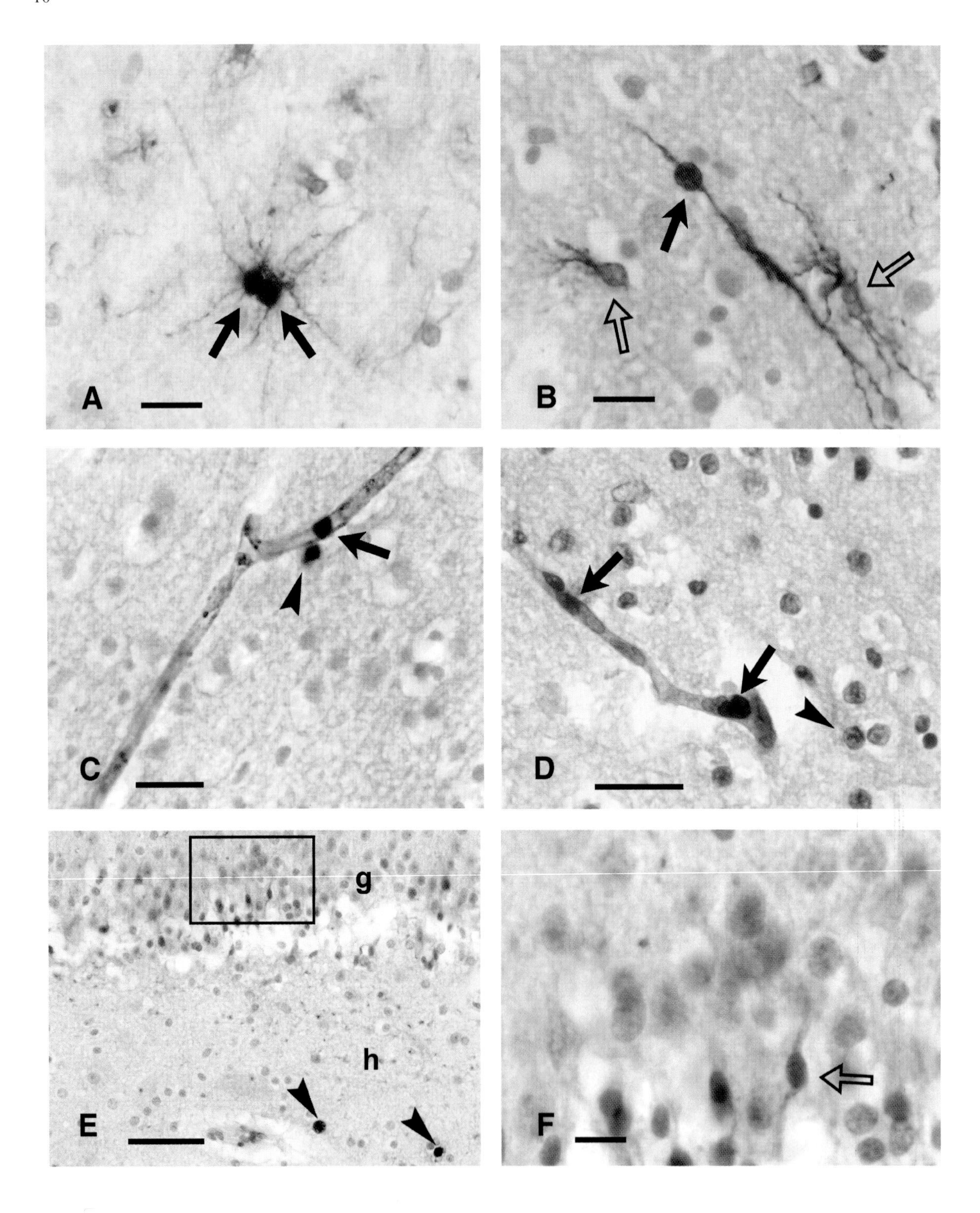
A
B
C
D
g
h
E
F

Table 3. Proportion of the Ki-67-1/vimentin-immunoreactive cells among all Ki-67 labeled cells

Age in weeks (gestational + extrauterine)	39	27+14	38+22	27+39
Case no.	21	23	28	30
Percentage of proliferating identified astrocytes				
Hilus	18	33.3	26	31.9
Molecular layer	10	26	12.5	14.4
CA1 str. radiatum+lacunosum-moleculare	9.7	32	2	11.9
Temporal neocortex	9.5	37	12	5.2

Note: Shaded columns indicate preterm infants.

Table 4. Proportion of the Ki-67-positive cells at the inner side of the PAS-positive basal membrane of the blood vessels among the total number of proliferating cells

Age in weeks (gestational + extrauterine)	39	27+14	38+22	27+39
Case no.	21	23	28	30
Percentage of proliferating identified endothelial cells				
Hilus	21.83	10.9	19	15
Molecular layer	30.7	40.5	47.9	52.9
CA1str.radiatum + lacunosum-moleculare	42	36.8	35.2	46.5
Temporal neocortex	16.6	29	62	42.4

Note: Shaded columns indicate preterm infants.

(BrdU)-labeling, Ki-67 immunohistochemistry has a disadvantage. Namely, it is not suitable for tracing the fate of dividing cells and the cells cannot be labeled once they are differentiated. Although the application of ^{3}H-thymidine autoradiography or BrdU-labeling is theoretically possible in humans, as indeed they were already used (Rakic and Sidman, 1970; Eriksson et al., 1998), the administration of BrdU or radioactive isotopes to human infants should be excluded for ethical reasons and possible health hazards. Therefore, we have to conclude on the basis of available data, which allow the estimation of the proportion of actually proliferating cell types. It can be concluded that approximately two thirds of the dividing cells are identified as glial or endothelial cells in the hippocampal formation and, according to this, neurons may compose only one third of the proliferating cells. However, taking into account that among the proliferating cells undifferentiated glial and endothelial precursors may exist, which still do not express the markers used for identification, the proportion of dividing neuronal elements is less than one third. The low proportion of dividing neurons is supported by the fact that large numbers of proliferating cells are seen also in layers I–VI of the temporal neocortex. It has been shown both in rodent and primate experiments that neocortical neurons are born exclusively in the ventricular and subventricular zones of the

Fig. 4. Photomicrographs showing double-immunoreaction of Ki-67 with vimentin (A and B), CD31 (C), PAS-reaction (D) and beta-III isoform of tubulin (E and F). (A) Dividing vimentin-positive astrocytes containing two Ki-67-labeled nuclei (arrows) in the hilus of an infant born at the age of 27th gestational week and lived 39 weeks (case no. 30). (B) Colocalization of Ki-67 and vimentin (arrow) in the neocortex of an infant born at the 27th gestational week and lived 14 weeks (case no. 23). Open arrows point to vimentin-positive Ki-67-negative astrocytes. (C) Colocalization of Ki-67 with the endothelial cell marker CD31 (arrow) in the neocortex of the infant born at the age of 27th gestational week and lived 39 weeks (case no. 30). Arrowhead shows Ki-67-positive CD31-negative cell. (D) Dividing Ki-67-labeled cells in a blood vessel surrounded by PAS-positive basal membrane (arrows) in the neocortex of the infant born at the age of 27th gestational week and lived 39 weeks (case no. 30). Arrowhead points to a Ki-67-immunoreactive cell in brain parenchyma. (E) Ki-67-positive proliferating cells (arrowheads) in the dentate gyrus of an infant born at the age of 27th gestational week and lived 39 weeks (case no. 30). The Ki-67-immunoreactive proliferating cells do not express beta-III isoform of tubulin that is shown with higher magnification (open arrow) in F. Scale bars = 20 μm for A, B and C, 50 μm for D and E, 10 μm for F.

telencephalon and ganglionic eminences, therefore the possibility of mitotic neurons in cortical layers can be entirely excluded (Anderson et al., 1999; Kornack and Rakic, 2001; Letinic et al., 2002). Furthermore, in the temporal neocortex the proportion of glial and endothelial cells is only 25–75%, suggesting that substantial amount of proliferating non-neuronal cells was not identified. In addition to immature astrocytes, other type of glial cells such as oligodendroglial and microglial cells are among the non-identified proliferating cells. We propose that similarly to the neocortex the dividing cells are centainly not neurons in Ammon's horn and in molecular layer of dentate gyrus, but they are endothelial and glial cells. It is in harmony with the results observed by Boekhoorn et al. (2006) showing that in adult, the proliferating cells of the hippocampal formation are mostly non-neuronal glial and endothelial elements. According to the probability of large amount of non-identified proliferating non-neuronal cells, we propose that the differences between preterms and full-term controls in the proportion of proliferating cells of the dentate gyrus may not reflect higher rate of mitotic neuronal precursors in preterms. The external granule cell layer of the cerebellar cortex contains very similar rate of cell proliferation in prematurely born children and their age-matched full-term controls (Ábrahám et al., 2001, 2004b). Since the external granule cell layer is considered to produce exclusively granule cells, we proposed that postnatal neuronal cell proliferation in the cerebellum is a genetically controlled event and is not altered by premature delivery (Ábrahám et al., 2004b). Similarly to the hippocampal formation, the cerebellum is among those areas of the central nervous system which display reduced volume in prematurely born adolescents demonstrated with MRI (Ajayi-Obe et al., 2000; Isaacs et al., 2000; Allin et al., 2001). If we assume that the cause of the reduced size of the cerebellum in preterms is not the reduced rate of postnatal neuronal proliferation, the same may hold true for the hippocampal formation. The higher degree of cell formation in the hippocampus in prematurely born infants when compared with the age-matched controls strengthens this argument. The unaltered size and the similar convolution of the granule cell layer of the dentate gyrus also support that large differences in neuronal production may not exist in prematurely born infants when compared with controls.

We must take it into account that postmortem detection of cell formation might underestimate the rate of cell proliferation, because clinical data showed severe hypoxic/ischemic episodes including reanimation. However, a possible inaccuracy in the absolute values equally affects the estimation of cell proliferation in preterm infants and their full-term age-matched controls.

What may cause the decrease in brain volume and what can be responsible for the behavioral and cognitive impairments? An altered development of the individual neurons and their connectivity may result in negative functional consequences. It is known from the morphological studies that axonal and dendritic development start early in intrauterine life; however, neuronal maturation is a long process and developing and mature neurons occur simultaneously in the postnatal human and monkey neo- and archicortex (Purpura, 1975; Seress, 1992, 2001). Experimental studies have shown that hypoxia in the developing perinatal hippocampus or cerebellum impairs the dendritic arborization and neuronal connections, while the number of neurons was unchanged (Pokorny and Trojan, 1986; Rees et al., 1998). Such alterations in the dendritic and axonal arborization of neurons may cause reduced volume of cortical white matter in humans. Indeed, several clinical reports showed reduced areas of white matter indicating the disturbance of axonal connections (Murphy et al., 1996; Olsen et al., 1998; Stewart et al., 1999; Peterson et al., 2000).

Development and maturation of hippocampal and neocortical local circuit neurons and hippocampal granule cells

Local circuit neurons

Development of the inhibitory neurons is essential in formation of archi- and neocortical neuronal circuits. Therefore, we examined maturation of three types of interneurons containing calcium-binding proteins, calretinin, calbindin and parvalbumin.

Calretinin is a neurochemical marker of a group of axo-dendritic GABAergic interneurons that provide predominantly interneuron-specific innervation, but although to a lesser extent, also inhibit principal cells (Seress et al., 1993b; Freund and Buzsáki, 1996). Both in the hippocampal formation and in the temporal neocortex calretinin-immunoreactive neurons are mostly bitufted cells oriented perpendicular to hippocampal fissure and to the pial surface of the brain, respectively. In the hippocampus they are found at the border of strata radiatum and lacunosum-moleculare, whereas in the neocortex they are located in all layers, with a higher density in layers II–III. During human ontogenesis, calretinin-immunoreactive interneurons are recognizable in all layers of Ammon's horn at the 16th gestational week. By the 24th week of gestation their adult-like location is prominent at the border of the strata lacusonum-moleculare and radiatum. In addition, their morphology corresponds to that found in adults although at this age cells are smaller and display fewer and shorter dendritic branches. When compared the preterms to their full-term age-matched controls, no significant differences were detected either in the location, or in the dendritic maturation of the cells.

In the neocortex, calretinin-immunoreactive interneurons appear slightly later than in the hippocampus, at the 18th gestational week. At birth, they are found in layers II and III (Fig. 5A); however, they are also numerous in layers IV–VI. In layers II and III calretinin-positive cells mostly display immature morphology that persists during the first few postnatal months (Fig. 5C). In preterms, neither the location nor their density differed from that found in the age-matched full-term controls (Figs. 5A–D).

Another calcium-binding protein, calbindin is the neurochemical marker of axo-dendritic local circuit neurons, which inhibit the glutamatergic pyramidal cells of the cerebral cortex. In addition to interneurons, calbindin is expressed by granule cells of the dentate gyrus and CA1 pyramidal cells of Ammon's horn. Calbindin-containing interneurons are found in the hilus and molecular layer of the dentate gyrus and in strata radiatum and oriens of Ammon's horn (Seress et al., 1993a, 1994; Freund and Buzsáki, 1996). In the temporal neocortex they are prominent in layers II and III, whereas they are less numerous in layers IV–VI.

Cabindin-positive inhibitory neurons of the hippocampal formation initially appear in the fetal period, although slightly later than the calretinin-positive interneurons. At the 16th gestational week only a few very immature-looking immunoreactive cells can be observed. The first calbindin-immunoreactive cells with morphology and location identical to matured hippocampal interneurons are seen only at the 20th or 22nd gestational weeks, both in Ammon's horn and in the dentate gyrus. However, at birth, large numbers of calbindin-positive interneurons still have an immature morphology with short and rarely branched dendritic trees. The distribution and the morphology of calbindin-immunoreactive neurons of the hippocampal formation reveal no difference in preterms when comparing with the age-matched full-term infants.

In the temporal neocortex, calbindin-immunoreactive neurons appear few weeks later than in the archicortical hippocampal formation. At birth, calbindin-positive neurons are abundant in deeper layers of the neocortex whereas the number of calbindin-positive cells in superficial layers (layers II and III) is relatively low (Fig. 5E). In the next few months, the density of the cells is increased in layers II and III, and their number is decreased in deeper layers. Comparing the density and distribution of calbindin-immunoreactive interneurons in the temporal neocortex of prematurely born infants with that of those born at term, no significant difference could be observed (Figs. 5E and F).

The third calcium-binding protein parvalbumin is found in the local circuit neurons known as basket and chandelier cells, which provide perisomatic inhibition to glutamatergic principal cells (Freund and Buzsáki, 1996). Parvalbumin-positive cells are usually large and multipolar and they are localized in close vicinity to the principal cell layers, which are the pyramidal cell layer of Ammon's horn and the granule cell layer of the dentate gyrus (Seress et al., 1993a). In the neocortex,

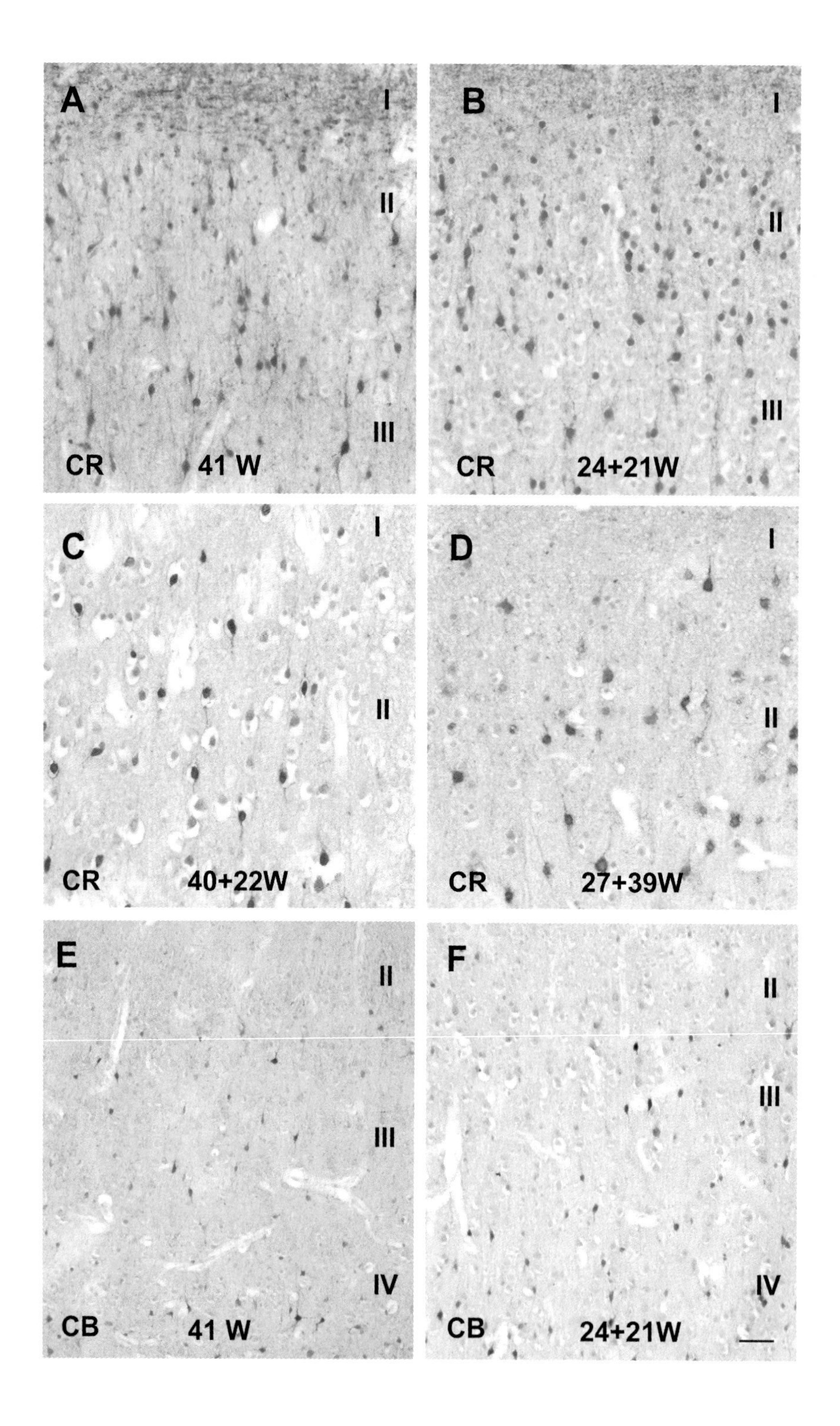
A
I
II
III
CR
41 W
B
I
II
III
CR
24+21W
C
I
II
CR
40+22W
D
I
II
CR
27+39W
E
II
III
IV
CB
41 W
F
II
III
IV
CB
24+21W

parvalbumin-containing cells are abundant in all layers with the highest density in the layers IV–V.

In contrast to the previous two populations of local circuit neurons, parvalbumin expression starts late in development. In the fetal period, parvalbumin-positive cells could not be found either in the hippocampal formation or in the temporal neocortex. At birth, only a few, small parvalbumin-positive somata with short, rarely branching dendrites could be seen in Ammon's horn and in the temporal archi- and neocortex. In the dentate gyrus no parvalbumin-positive cells were visible at birth. In the next few months (before the second year of age) more and more immunoreactive dendrites and axons appeared in both the archi- and neocortical areas (Figs. 6A and C); however, the developmental delay between Ammon's horn and the dentate gyrus is still visible. Even in a 2-year-old child, the morphology of the dendrites and axons of the parvalbumin-immunoreactive cells was still immature and the number of terminal boutons was lower than in 8- or 10-year-old children. In our material, the adult-like morphology of parvalbumin-containing cells and the adult-like pattern of parvalbumin-containing axonal network in the principal cell layers of Ammon's horn and the dentate gyrus appeared in the 8-year-old child. However, we have no specimen between the 2nd and 8th year, therefore, we suggest that maturation of these cells is completed during this period, similar to the development of hilar mossy cells (Seress and Mrzljak, 1992). The parvalbumin expression and the dendritic maturation of these interneurons are similar in the prematurely born children and their full-term age matched controls (Figs. 6A–D).

It has to be emphasized that the lack of parvalbumin expression in inhibitory neurons does not mean that perisomatic inhibitory terminals would be completely missing in the hippocampal formation at birth. Our unpublished electron microscopic observations revealed that perisomatic inhibitory synapses exist both on somata of granule and pyramidal cells in neonates, although those axon terminals are not parvalbumin-immunoreactive.

Interneurons play a crucial role in the processes of memory formation through their role in the γ-oscillation of the hippocampal neuronal network induced by the synchronized activity of GABA-ergic local circuit neurons (Wang and Buzsaki, 1996; Wallenstein and Hasselmo, 1997). Recent investigations showed that parvalbumin-containing local circuit neurons are critical in this process: hippocampal parvalbumin-positive neurons form a syntitium through their dendrodendritic gap junctions (Fukuda and Kosaka, 2003) and supposedly mediate inhibition-based coherent γ-rythms (Tamás et al., 2000). In the hippocampal formation of adult parvalbumin-deficient mice, in which perisomatic inhibitory neurons exist, but they do not express parvalbumin, inhibition-based γ-oscillation increases, resulting in a lower threshold for the development of epileptiform activity (Schwaller et al., 2004). In addition, the lack of parvalbumin in interneurons may affect the higher cognitive functions associated with γ-oscillation (Vreugdenhil et al., 2003).

We propose that the long-lasting postnatal maturation of parvalbumin-containing axo-somatic inhibitory cells in humans might increase the susceptibility of the hippocampal formation in newborns or in infants for stimuli that later cannot induce similar effect. This might clarify why a high fever may cause febril seizures in young infants and may also explain why such early generated seizures disrupt the normal development of the inhibitory circuitry resulting in manifestation

Fig. 5. Photomicrographs showing calretinin- (A–D) and calbindin-immunoreactive (E and F) local circuit neurons in the temporal neocortex of prematurely born infants (B, D and F) and in their age-matched full-term controls (A, C and E). (A) Calretinin-positive interneurons are abundant in layer II and III of the temporal neocortex at birth. (B) Similarly to the newborns, in an infant born extremely preterm with extremely low birth weight (case no. 24), the highest density of calretinin-immunoreactive local circuit neurons is observed in layer II and III. (C) Calretinin-containing interneurons in 5-month-old infant born at term. (D) In infant born at the 27th gestational week and lived 39 weeks (case no. 30), the distribution of calretinin-immunoreactive local circuit neurons corresponds to that found in the 5-month-old infant born at term (C). (E) Calbindin-positive local circuit neurons are less numerous in layers II and III in newborn (case no. 22) than in adults. (F) The number and the distribution of calbindin-immunoreactive interneurons in the preterm infant (case no. 24) show a similarly low density as found in the baby born at term (E). Scale bar = 50 µm for A and B, 30 µm for C and D, 75 µm for E and F.

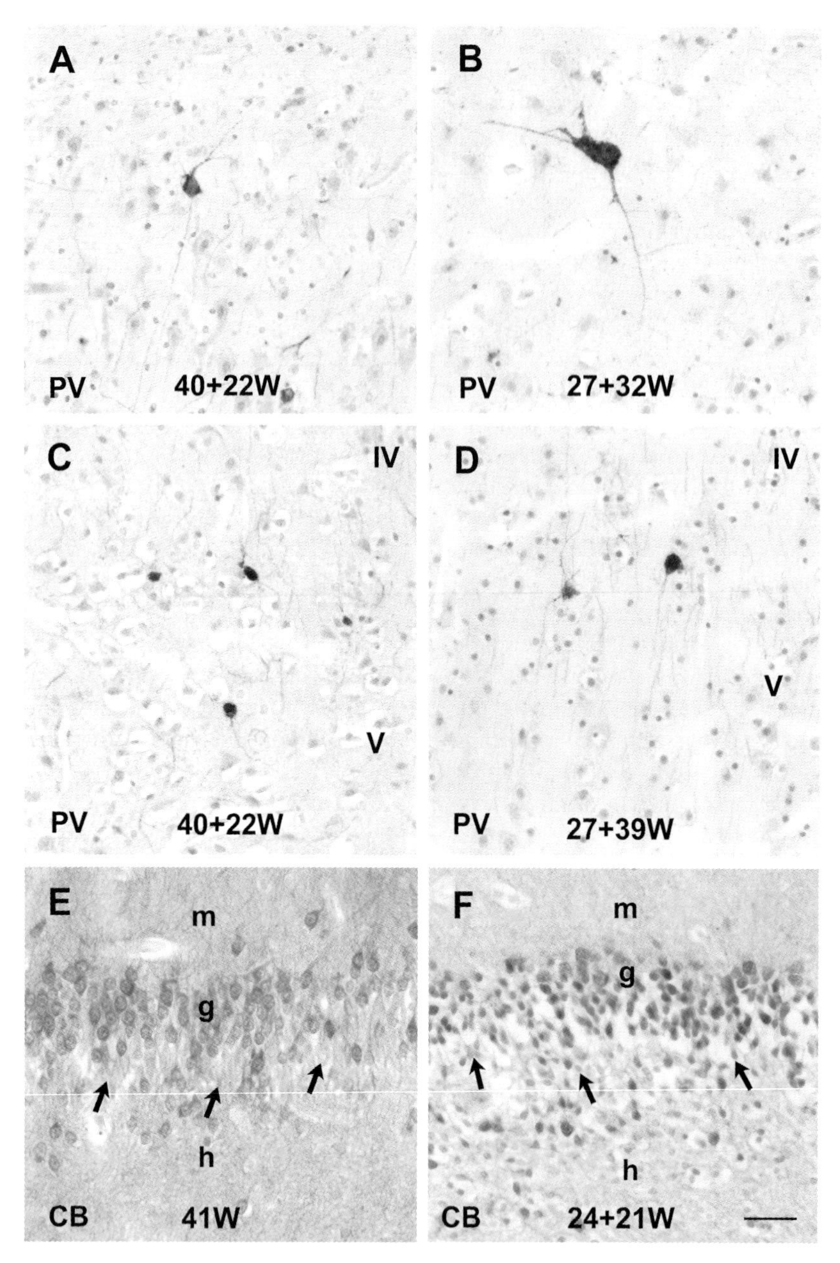

Fig. 6. Photomicrographs showing parvalbumin-immunoreactive local circuit neurons (A–D) and calbindin-containing granule cells of the dentate gyrus (E and F) in prematurely born infants (B, D and F) and in their age-matched full-term controls (A, C and E). (A) Parvalbumin-immunoreactive local circuit neuron in the CA1 pyramidal layer of the hippocampus displaying immature morphology with a few short dendrites in 5-month-old child. (B) In a preterm, born at the 27th week of gestation and lived for 32 weeks (case no. 31) parvalbumin-positive local circuit neuron is similarly immature as observed in infants born at term. (C) In the temporal neocortex of 5-month-old infant, parvalbumin-containing local circuit neurons are located mostly in layers IV and V displaying immature dendritic tree. (D) In preterm born at the 27th week of gestation and lived 39 weeks (case no. 30) the distribution and morphology of parvalbumin-positive local circuit neurons in the temporal neocortex correspond to that found in full-term infants. (E) Calbindin-immunoreactive granule cells of the dentate gyrus occupy mostly the part of the granule cell layer (g) adjacent to the molecular layer (m). At term, numerous calbindin-negative granule cells (arrows) are observed at the border of granule cell layer (g) and hilus (h). (F) Distribution of calbindin-immunoreactive granule cells in the granule cell layer (g) is similar in preterm (case no. 24) as in full-term controls, revealing numerous calbindin-negative neurons in the vicinity of the hilus (h). Scale bar = 25 μm in A and B, 50 μm in C–F.

of epileptic activity and epileptic morphological changes in the hippocampal formation later in life. The developmental delay of parvalbumin-positive interneurons is especially expressed in the dentate gyrus that displays significant neuropathological sings of cellular development in febril-seizure related epilepsy (Houser, 1990).

Granule cells of the dentate gyrus

In the dentate gyrus, granule cells start to express calbindin relatively early. At the 22nd gestational week a few granule cells of the dentate gyrus are already calbindin-positive; however, at birth a large number of neurons in the infrapyramidal blade of the granule cell layer is still calbindin-negative (Fig. 6E). Granule cells form the granule cell layer in an outside-in migrational pattern and older granule cells locate more superficially than the younger ones. Correspondingly, the granule cells locating closer to the molecular layer are calbindin-positive, whereas the deeper locating cells at the hilar border are calbindin-negative (Fig. 6E). The proportion of calbindin-immunoreactive granule cells is increasing with the time; however, as late as the 5th postnatal month substantial amount of the granule cells at the hilar border is still calbindin-negative, indicating arrival of new granule cells from the hilus. Comparing the calbindin-immunoreactivity in granule cells, no significant changes were observed. Even in those who were born extremely prematurely with an extremely low birth weight, calbindin-immunoreactivity revealed the same pattern that observed in the age-matched full-term newborns (Figs. 6E and F).

The adult-like function of the hippocampal formation cannot be expected without an established mature trisynaptic circuitry among the principal cells of the dentate gyrus and Ammon's horn. The perinatal proliferation and migration of a subset of granule cells evidently result in their long-lasting neurochemical maturation. The prolonged development of granule cells determines the postnatal maturation of their target cell such as mossy cells (Seress and Mrzljak, 1992). On the other hand, postnatal maturation of interneurons, especially the late expression of parvalbumin in the perisomatic inhibitory cells coincides with the prolonged maturation of the principal cells (Seress, 1992; Seress and Mrzljak, 1992). Morphological and neurochemical development of principal and GABAergic neurons suggests that neuronal connectivity in the human hippocampus reached and adult-like complexity between the 2nd and 8th years. Therefore, hippocampus may be involved in memory formation in young infants (de Haan et al., 2006), but hippocampus-related adult-like memory formation in humans may not be expected earlier than in early childhood (3–5 years of age).

Persisting Cajal–Retzius cells in the preterm neo- and archicortex

In addition to interneurons, Cajal–Retzius cells are also calretinin-immunoreactive. This peculiar cell-type of cortical layer I was first described by Cajal (1891) and Retzius (1893) and was found in highest number during fetal period of ontogenesis (Meyer and Goffinet, 1998). They express reelin, an extracellular matrix protein that plays crucial role in the migration of the radially migrating cortical neurons (D'Arcangelo et al., 1995). In addition, they contain p73, a nuclear transcription factor that belong to the p53 protein family (Yang et al., 2000; Meyer et al., 2002), and the calcium-binding protein calretinin (Weisenhorn et al., 1994). After cortical migration period comes to an end, the number of Cajal–Retzius cells decreases and only limited number can be found in the cerebral neocortex around term and even in adulthood (Meyer et al., 1999, 2002; Ábrahám and Meyer, 2003). Although higher number of cells is present at around birth in the archicortical hippocampal formation than in the neocortex, the gradual decrease with the age is evident even in the archicortex (Ábrahám et al., 2003, 2004a). Another change that is obvious in the cortical layer I during the third trimester of gestation is the gradual disappearance of the subpial granular layer that is located between the pia mater and Cajal–Retzius cells. Although, either the role of the cells of subpial granular layer during the fetal period or the process as they disappear is not clear, at the 40th week of gestation they are no more visible in full terms.

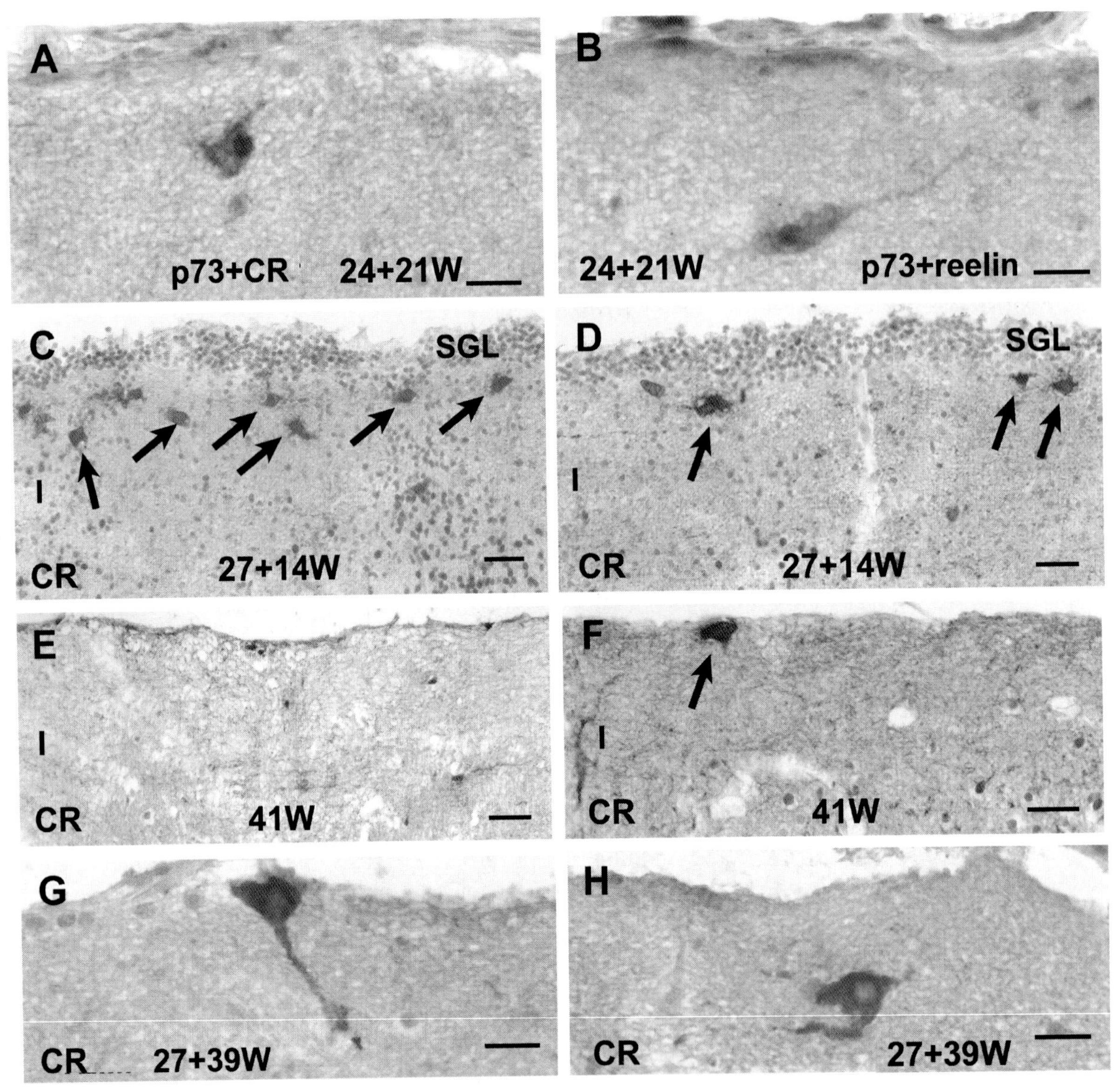

Fig. 7. Cajal–Retzius cells in layer I of the archi- (C and D) and temporal neocortex (A, B, E and F). (A) Cajal–Retzius cell coexpressing p73 (black reaction product) and calretinin (brown) in the temporal neocortical layer I of an infant born at the 24th week of gestation and survived 21 weeks (case no. 24). (B) P73 and reelin-immunoreactive Cajal–Retzius cell in layer I of the temporal neocortex of a prematurely born infant (case no. 24). (C) Calretinin-immunopositive Cajal–Retzius cells (arrows) in layer I below the persisting subpial granular layer (SGL) in the subiculum in an infant born at the 27th gestation weeks and lived for 14 weeks (case no. 23). (D) Calretinin-containing Cajal–Retzius cells (arrows) and the subpial granular layer (SGL) in the temporal neocortex of a preterm (case no. 23). (E) Lack of subpial granular layer and the Cajal–Retzius cells in the subiculum of an infant born at term (case no. 22). (F) One calretinin-immunoreactive Cajal–Retzius cell (arrow) in layer I of the temporal neocortex that lacks the subpial granular layer in full-term newborn (case no. 22). (G, H) Cajal–Retzius cells containing calretinin in layer I of the temporal neocortex in an infant born at the 27th week of gestation and lived 39 weeks (case no. 30). In full-term age-matched controls Cajal–Retzius cells with a similar morphology are only very rarely found. Scale bars = 25 μm in A, B, G and H, 50 μm in C–F.

Table 5. Cajal–Retzius cells in the molecular layer of the temporal archi- and neocortical areas

Case no.	Age in weeks (gestational + extrauterine)	Temporal neocortex	Subiculum	Entorhinal cortex
19.	34+4	2.1±0.13	7.3±2.12	3.1±1.21
20.	36+2	2.9±1.13	NA	NA
21.	39	1.2±0.17	5.3±0.74	3.7±1.19
23.	27+14	6.0±2.15	9.2±1.44	8.2±0.69
24.	24+21	2.3±1.17	NA	NA
28.	38+22	0.6±0.38	5.1±0.34	2.4±0.54
29.	40+22	1.0±0.25	2.37±0.8	1.86±0.37
30.	27+39	1.4±0.53	NA	NA
31.	27+32	1.53±0.44	8.6±1.81	6.52±1.8

Note: Number of Cajal-Retzius cells was counted in five non-consecutive sections and density of the cells was calculated to 1 mm length of the pial surface both in the neo- and archicortex. Data indicate averages ± standard deviation. Shaded rows indicate preterm infants.

The number of Cajal–Retzius cells that are immunoreactive for calretinin or reelin appeared to be higher along the hippocampal fissure in preterm infants than in controls especially along the subiculum and entorhinal cortex as well as in the temporal neocortex (Fig. 7, Table 5). In an extremely preterm infant born at the 27th gestational week and who survived 14 weeks, in addition to Cajal–Retzius cells, the presence of the subpial granular layer was obvious in the subiculum as well as in temporal cortex (Figs. 7C and D). Thus, failure of disappearance of predominantly fetal structures such as Cajal–Retzius cells and the transient subpial granular layer of the archi- and neocortex indicate the arrest of cortical maturation. This is in harmony with the findings Ajayi-Obe et al. (2000) showing that extremely preterm infants, when measured at 38–42 postconceptual weeks of age, had smaller cortical surface that was less complex than in the normal infants born at term.

Cajal–Retzius cells and their protein product reelin are responsible for the radial organization of the mammalian cortex (Bar et al., 2000) and they serve as a stop signal for the radially migrating neurons (Frotscher, 1997). The prolonged existence of stop signal indicated by the large number of Cajal–Retzius cells in preterms means that neuronal migration might differ in preterms than in full-term controls. Differences in timing of cell migration may disturb the formation of neuronal connections that can result in functional alterations. Although the mechanism is not clear, it can be suggested that reelin and Cajal–Retzius cells may play a role in gyrification of the cerebral cortex.

Conclusion

The unaltered neuronal proliferation as well as the morphological similarities of the examined principal cells and local circuit neurons in preterm infants and in their age-matched full-term controls suggest that cell proliferation and maturation are genetically controlled events and are not influenced by the untimely changes of the environment caused by premature birth. Therefore, the smaller volume of brain observed in preterms cannot be explained by lower number of neurons or by their deceleration of maturation. Consequently, neither the altered cell proliferation nor the abnormal maturation of the examined neuronal population can be responsible for the cognitive and behavioral abnormalities described in prematurely born infants.

However, we have no information about the postnatal morphological developmental differences of other interneuronal populations, e.g. somatostatin, neuropeptide Y or nitric-oxide-synthase containing interneurons of the temporal archi- and neocortex in preterm and in full-term control children. Moreover no data are available about the dendritic arborization and spine development the pyramidal cells in preterms. In addition, using light microscopy, very limited information can be obtained about the axonal arborization of the different neuronal populations.

As a consequence, electron microscopic examinations would be needed to determine alteration in the number of synaptic connections between the various types of neurons. Regarding neuronal functions required for the normal learning processes, the normal number, localization and efficacy of the synapses are unquestionable requirements. Since genetic control appears to be strong enough to guide developmental events even after premature birth, we suggest that proper perinatal care may successfully minimize the effects of environmental factors, which may lead to cognitive and functional abnormalities. The alterations of cognitive functions probably have a biochemical and ultrastructural but not light microscopic neuropathological background.

Abbreviations

ALL	acute lymphoid leukemia
CA1, CA3	regions of Ammon's horn
CB	calbindin
CD31	clusters of differentiation 31
CHD	congenital heart disease
CI	cardial insufficiency
CR	calretinin
CRI	cardiorespiratory insufficiency
g	granule cell layer
GW	gestational week
h	hilus of the dentate gyrus
IRDS	infant respiratory distress syndrome
LI	labeling index
lm	stratum lacunosum-moleculare
m	molecular layer
NA	not available
o	stratum oriens
p	stratum pyramidale
PAS	periodic-acid Schiff reaction
PV	parvalbumin
r	stratum radiatum
SIDS	sudden infant death syndrome
SGL	subpial granular layer
W	week

Acknowledgement

The authors acknowledge Mrs. Emese Papp for her excellent technical assistance in the histological preparations of the tissue. This work was supported by the Hungarian National Science Fund (OTKA) with grant no. T047109.

References

Ábrahám, H. and Meyer, G. (2003) Reelin-expressing neurons in the postnatal and adult human hippocampal formation. Hippocampus, 13: 715–727.

Ábrahám, H., Perez-Garcia, C.G. and Meyer, G. (2004a) p73 and Reelin in Cajal-Retzius cells of the developing human hippocampal formation. Cereb. Cortex, 14: 484–495.

Ábrahám, H., Tornóczky, T., Kosztolányi, G. and Seress, L. (2001) Cell formation in the cortical layers of the developing human cerebellum. Int. J. Dev. Neurosci., 19: 53–62.

Ábrahám, H., Tornóczky, T., Kosztolányi, G. and Seress, L. (2004b) Cell proliferation correlates with the postconceptual and not with the postnatal age in the hippocampal dentate gyrus, temporal neocortex and cerebellar cortex of preterm infants. Early Hum. Dev., 78: 29–43.

Ajayi-Obe, M., Saeed, N., Cowan, F.M., Rutherford, M.A. and Edwards, A.D. (2000) Reduced development of cerebral cortex in extremly preterm infants. Lancet, 365: 1162–1163.

Allin, M., Matsumoto, H., Santhouse, A.M., Nosarti, C., AlAsady, M.H., Stewart, A.L., Rifkin, L. and Murray, R.M. (2001) Cognitive and motor function and the size of the cerebellum in adolescents born very pre-term. Brain, 124: 60–66.

Anderson, S., Mione, M., Yun, K. and Rubenstein, J.L. (1999) Differential origins of neocortical projection and local circuit neurons: role of Dlx genes in neocortical interneuronogenesis. Cereb. Cortex, 9: 646–654.

Arnold, S.E. and Trojanowski, J.Q. (1996) Human fetal hippocampal development: I. Cytoarchitecture, myeloarchitecture, and neuronal morphologic features. J. Comp. Neurol., 367: 274–292.

Bar, I., Lambert de Rouvroit, C. and Goffinet, A.M. (2000) The evolution of cortical development. A hypothesis based on the role of Reelin signaling pathway. Trends Neurosci., 23: 633–638.

Basco, E., Hajos, F. and Fulop, Z. (1977) Proliferation of Bergmann-glia in the developing rat cerebellum. Anat. Embryol. (Berl.), 151: 219–222.

Boekhoorn, K., Joels, M. and Lucassen, P.J. (2006) Increased proliferation reflects glial and vascular-associated changes, but not neurogenesis in the presenile Alzheimer hippocampus. Neurobiol. Dis., 24: 1–14.

Breslau, N., Chilcoat, H.D., Johnson, E.O., Andreski, P. and Lucia, V.C. (2000) Neurologic soft signs and low birthweight: their association and neuropsychiatric implications. Biol. Psychiatry, 47: 71–79.

Cajal, S.R. (1891) Sur la structure de l'ecorce cérébrale de quelques mammiféres. La Cellule, 7: 123–176 [English translation. In: DeFelipe J. and Jones E.G. (Eds.), Cajal on the Cerebral Cortex, 1988. Oxford University Press, Oxford, pp. 23–54.].

D'Arcangelo, G., Miao, G.G., Chen, S.C., Soares, H.D., Morgan, J.I. and Curran, T. (1995) A protein related to extracellular matrix proteins deleted in the mouse mutant reeler. Nature, 374: 719–723.

De Haan, M., Mishkin, M., Baldeweg, T. and Vargha-Khadem, F. (2006) Human memory development and its dysfunction after early hippocampal injury. Trends Neurosci., 29: 374–381.

Eriksson, P.S., Perfilieva, E., Bjork-Eriksson, T., Alborn, A.M., Nordborg, C., Peterson, D.A. and Gage, F.H. (1998) Neurogenesis in the adult human hippocampus. Nat. Med., 4: 1313–1317.

Ferreira, A. and Caceres, A. (1992) Expression of the class III beta-tubulin isotype in developing neurons in culture. J. Neurosci. Res., 13: 516–529.

Freund, T.F. and Buzsáki, G. (1996) Interneurons of the hippocampus. Hippocampus, 6: 345–470.

Frotscher, M. (1997) Dual role of Cajal-Retzius cells and reelin in cortical development. Cell Tissue Res., 290: 315–322.

Fukuda, T. and Kosaka, T. (2003) Ultrastructural study of gap junctions between dendrites of parvalbumin-containing GABAergic neurons in various neocortical areas of the adult rat. Neuroscience, 120: 5–20.

Hack, M. and Fanaroff, A.A. (1999) Outcomes of children of extremly low birthweight and gestational age in the 1990s. Early Hum. Dev., 53: 193–218.

Hack, M., Flannery, D.J., Schluchter, M., Cartar, L., Borawski, E. and Klein, N. (2002) Outcomes in young adulthood for very-low-birth-weight infants. N. Engl. J. Med., 346: 149–157.

Houser, C.R. (1990) Granule cell dispersion in the dentate gyrus of humans with temporal lobe epilepsy. Brain Res., 535: 195–204.

Isaacs, E.B., Lucas, A., Chong, W.K., Wood, S.J., Johnson, C.L., Marshall, C., Vargha-Khadem, F. and Gadian, D.G. (2000) Hippocampal volume and everyday memory in children of very low birth weight. Pediatr. Res., 47: 713–720.

Katz, V.L. and Bose, C.L. (1993) Improving survival of the very premature infant. J. Perinatol., 13: 261–265.

Kornack, D.R. and Rakic, P. (2001) Cell proliferation without neurogenesis in adult primate neocortex. Science, 294: 2127–2130.

Letinic, K., Zoncu, R. and Rakic, P. (2002) Origin of GABAergic neurons in the human neocortex. Nature, 417: 645–649.

McCormick, M., Workman-Daniels, K. and Brooks-Gunn, J. (1996) The behavioral and emotional well-being of school-age children with different birthweight. Pediatrics, 97: 18–25.

Meyer, G. and Goffinet, A.M. (1998) Prenatal development of reelin-immunoreactive neurons in the human neocortex. J. Comp. Neurol., 397: 29–40.

Meyer, G., Goffinet, A.M. and Fairen, A. (1999) What is a Cajal-Retzius cell? A reassessment of a classical cell type based on recent observations in the developing neocortex. Cereb. Cortex, 9: 765–775.

Meyer, G., Perez-Garcia, C.G., Ábrahám, H. and Caput, D. (2002) Expression of p73 and Reelin in the developing human cortex. J. Neurosci., 22: 4973–4986.

Morrison, J.J. and Rennie, J.M. (1997) Clinical, scientific and ethical aspects of fetal and neonatal care at extremely preterm periods of gestation. Br. J. Obstet. Gynaecol., 104: 1341–1350.

Murphy, D.J., Squier, M.V., Hope, P.L., Sellers, S. and Johnson, A. (1996) Clinical associations and time of onset of cerebral white matter damage in very preterm babies. Arch. Dis. Child Fetal Neonatal Ed., 75: F27–F32.

Olsen, P., Vainionpaa, L., Paakko, E., Korkman, M., Pyhtinen, J. and Jarvelin, M.R. (1998) Psychological findings in preterm children related to neurologic status and magnetic resonance imaging. Pediatrics, 102: 329–336.

Peterson, B.S., Vohr, B., Staib, L.H., Cannistraci, C.J., Dolberg, A., Schneider, K.C., Katz, K.H., Westerveld, M., Sparrow, S., Anderson, A.W., Duncan, C.C., Makuch, R.W., Gore, J.C. and Ment, L.R. (2000) Regional brain volume abnormalities and long-term cognitive outcome in preterm infants. JAMA, 284: 1939–1947.

Pokorny, J. and Trojan, S. (1986) The development of hippocampal structure and how it is influenced by hypoxia. Acta Univ. Carol. Med. Monogr., 113: 1–79.

Purpura, D.S. (1975) Normal and aberrant neuronal development in the cerebral cortex of human fetus and young infant. In: Brain Mechanisms in Mental Retardation, UCLA Forum in Medical Sciences. Academic Press, New York, pp. 141–169.

Rakic, P. and Sidman, R.L. (1970) Histogenesis of cortical layers in human cerebellum, particularly the lamina dissecans. J. Comp. Neurol., 139: 473–500.

Rees, S., Mallard, C., Breen, S., Stringer, M., Cock, M. and Harding, R. (1998) Fetal brain injury following prolonged hypoxemia and placental insufficiency: a review. Comp. Biochem. Physiol. A. Mol. Interg. Physiol., 119: 653–660.

Retzius, G. (1893) Die Cajal'schen Zellen der Grosshirnrinde beim Menschen und bei Säugethieren. Biologische Untersuchungen von prof. Dr Gustav Retzius. Neue Folge, Vols. 1–8.

Schwaller, B., Tetko, I.V., Tandon, P., Silveira, D.C., Vreugdenhil, M., Henzi, T., Potier, M.C., Celio, M.R. and Villa, A.E. (2004) Parvalbumin deficiency affects network properties resulting in increased susceptibility to epileptic seizures. Mol. Cell Neurosci., 25: 650–663.

Seress, L. (1992) Morphological variability and developmental aspects of monkey and human granule cells: differences between the rodent and primate dentate gyrus. Epilepsy Res. Suppl., 7: 3–28.

Seress, L. (2001) Morphological changes of the human hippocampal formation from midgestation to early childhood. In: Nelson C.A. and Luciana M. (Eds.), Handbook of Developmental Cognitive Neuroscience. MIT Press, Cambridge, MA, pp. 45–58.

Seress, L., Ábrahám, H., Tornóczky, T. and Kosztolányi, G. (2001) Cell formation in the human hippocampal formation from mid-gestation to the late postnatal period. Neuroscience, 105: 831–843.

Seress, L., Gulyas, A.I., Ferrer, I., Tunon, T., Soriano, E. and Freund, T.F. (1993a) Distribution, morphological features, and synaptic connections of parvalbumin- and calbindin D28k-immunoreactive neurons in the human hippocampal formation. J. Comp. Neurol., 337: 208–230.

Seress, L., Leranth, C. and Frotscher, M. (1994) Distribution of calbindin D28k immunoreactive cells and fibers in the monkey hippocampus, subicular complex and entorhinal cortex. A light and electron microscopic study. J. Hirnforsch., 35: 473–486.

Seress, L. and Mrzljak, L. (1992) Postnatal development of mossy cells in the human dentate gyrus: a light microscopic Golgi study. Hippocampus, 2: 127–141.

Seress, L., Nitsch, R. and Leranth, C. (1993b) Calretinin immunoreactivity in the monkey hippocampal formation — I. Light and electron microscopic characteristics and co-localization with other calcium-binding proteins. Neuroscience, 55: 775–796.

Stewart, A.L., Rifkin, L., Amess, P.N., Kirkbride, V., Townsend, J.P., Miller, D.H., Lewis, S.W., Kingsley, D.P., Moseley, I.F., Foster, O. and Murray, R.M. (1999) Brain structure and neurocognitive and behavioral function in adolescnet who were born very preterm. Lancet, 353: 1653–1657.

Sykes, D.H., Hoy, E.A., Bill, J.M., McClure, B.G., Halliday, H.L. and Reid, M.M. (1997) Behavioral adjustment in school of very low birthweight children. J. Child Psychol. Psychiatry, 38: 315–325.

Tamás, G., Buhl, E., Lorincz, A. and Somogyi, P. (2000) Proximally targeted GABAergic synapses and gap junctions synchronize cortical interneurons. Nat. Neurosci., 3: 366–371.

Thomas, L.B., Gates, M.A. and Steindler, D.A. (1996) Young neurons from the adult subependymal zone proliferate and migrate along an astrocyte, extracellular matrix-rich pathway. Glia, 17: 1–14.

Verheijen, R., Kuijpers, H.J., van Driel, R., Beck, J.L., van Dierendonck, J.H., Brakenhoff, G.J. and Ramaekers, F.C. (1989b) Ki-67 detects a nuclear matrix-associated proliferation-related antigen. II. Localization in mitotic cells and association with chromosomes. J. Cell Sci., 92: 131–140.

Verheijen, R., Kuijpers, H.J., Schlingemann, R.O., Boehmer, A.L., van Driel, R., Brakenhoff, G.J. and Ramaekers, F.C. (1989a) Ki-67 detects a nuclear matrix-associated proliferation-related antigen. I. Intracellular localization during interphase. J. Cell Sci., 92: 123–130.

Vohr, B.R., Wright, L.L., Dusick, A.M., Mele, L., Verter, J., Steichen, J.J., Simon, N.P., Wilson, D.C., Broyles, S., Bauer, C.R., Delaney-Black, V., Yolton, K.A., Fleisher, B.E., Papile, L.A. and Kaplan, M.D. (2000) Neurodevelopmental and functional outcomes of extremely low birth weight infants in the National Institute of Child Health and Human Development Neonatal Research Network, 1993–1994. Pediatrics, 105: 1216–1226.

Vreugdenhil, M., Jefferys, J.G., Celio, M.R. and Schwaller, B. (2003) Parvalbumin-deficiency facilitates repetitive IPSCs and gamma oscillations in the hippocampus. J. Neurophysiol., 89: 1414–1422.

Wallenstein, G.V. and Hasselmo, M.E. (1997) GABAergic modulation of hippocampal population activity: sequence learning, place field development, and the phase precession effect. J. Neurophysiol., 78: 393–408.

Wang, X.J. and Buzsaki, G. (1996) Gamma oscillation by synaptic inhibition in a hippocampal interneuronal network model. J. Neurosci., 16: 6402–6413.

Weisenhorn, D.M., Prieto, E.W. and Celio, M.R. (1994) Localization of calretinin in cells of layer I (Cajal-Retzius cells) of the developing cortex of the rat. Brain Res. Dev. Brain Res., 82: 293–297.

Yang, A., Walker, N., Bronson, R., Kaghad, M., Oosterwegel, M., Bonnin, J., Vagner, C., Bonnet, H., Dikkes, P., Sharpe, A., McKeon, F. and Caput, D. (2000) p73-deficient mice have neurological, pheromonal and inflammatory defects but lack spontaneous tumours. Nature, 404: 99–103.

C. von Hofsten & K. Rosander (Eds.)
Progress in Brain Research, Vol. 164
ISSN 0079-6123

CHAPTER 2

Subcortical regulation of cortical development: some effects of early, selective deprivations

Giorgio M. Innocenti*

Department of Neuroscience, Karolinska Institutet, Retzius väg 8, S-17177 Stockholm, Sweden

Abstract: Selective deprivations, such as the lack of sensory input, of social contacts and of language during the critical (sensitive) period of brain development have profound consequences for the structure and function of the adult brain. The field is largely uncharted since only the consequences of the most severe forms of deprivation are known, and that too only in a few systems. It is similarly unknown if the opposite of deprivation, selective over-stimulation in development, which appears to enhance the acquisition of certain skills, for example musical skills, has collateral deprivation-like effects in other domains. In spite of these uncertainties, I propose that the common mechanism underlying the effects of deprivation may be the altered stabilization of neuronal morphologies, particularly connectivity, in the period when their exuberant development is down regulated.

Keywords: visual cortex; corpus callosum; binocular deprivation; critical periods; language; developmental exuberance

The interest in the consequences of selective deprivation in early life goes far back in history. Classical examples are the "experiments" credited to the pharaoh Psammetichus and to the Emperor Fredrick II who deprived children of contact with other humans to study the development of language, the hypothesis being that they would have spoken the primeval human language. As everybody knows, both failed.

Reports of children who were raised by animals or in other ways deprived of human contacts go back to the 14th century and until 2006, 134 cases were listed in the website http://www.feralchildren.com. Equally classical are the studies on the consequences of early social deprivation by Harlow, demonstrating their disruptive effects on social interactions in adulthood (Harlow and Harlow, 1962).

More recently Gregory and Wallace reported a detailed case study of early blindness and its partial recovery after the operation of congenital cataract (Gregory and Wallace, 1963). The classical studies of Wiesel and Hubel on the consequences of visual deprivation in cats and monkeys summarized in Wiesel's Nobel lecture (Wiesel, 1982) greatly enhanced the interest in deprivation studies for at least three reasons. They provided ways of understanding the consequences of visual deprivation, caused by congenital cataract in children. They established a direct link between the functional and the structural levels in visual system organization. Finally they created a rigorous experimental framework within which the old question of the relation between nature and nurture on brain development could be studied.

*Corresponding author. Tel.: +468 52487862;
Fax: +468 315782; E-mail: giorgio.innocenti@ki.se

DOI: 10.1016/S0079-6123(07)64002-3

This short review intends to revisit the consequences of some types of selective deprivation with the intent to bridge the gap, often by way of hypothesis, between animal studies and human development, as well as between functional and structural levels of analysis in different systems. We shall also focus on the ethical issues that deprivation studies raise in relation to what should be considered as one fundamental right of the child, i.e., to develop his/her brain at the best of its genetic potentials.

Deprivation in development causes structural changes in the cerebral cortex

A recent study (Noppeney et al., 2005) of 11 early blind subjects (at <2 years of age), reported decreased volume of the gray matter of the primary visual cortex and of several fiber tracts afferent to it, including the optic radiation, and the splenium of the corpus callosum; in contrast, white matter tracts related to the somatosensory and motor areas were enlarged. These findings are important but their interpretation at the cellular level is not straightforward. Therefore they shall be taken as a starting point for discussing some of the structural changes that might be caused by complete visual deprivation.

The decrease in the gray matter volume could be due to the loss of any one of the cellular populations, neurons or glia, which constitute the cerebral cortex and might undergo programmed cell death in development. This seems to be unlikely though, since no clear evidence of developmental neuronal death was found in the visual cortex of man (Leuba and Kraftsik, 1994) and neuronal death in the human cortex, if it occurs, may be a late gestational phenomenon, preceding the reported effects of visual deprivation (Rabinowicz et al., 1992). Therefore, the decreased gray matter volumes observed by Noppeney et al. (2005) more likely represent a decrease in some elements of the neuropile. This interpretation is strengthened by studies in dark reared kittens, reporting decreased volume of areas 17 and 18, with increased neuronal density and absence of neuronal death, (Takács et al., 1992). Among the components of the neuropile, dendritic and axonal processes and synapses are known to undergo exuberant growth and regression in development (Innocenti and Price, 2005; Fig. 1). They appear, therefore, to be good candidates for the shrinkage of the visual areas caused by early visual deprivation. In addition, decreased myelination could also play a role.

The loss of dendritic processes is a major event in development, leading to the emergence of specific neuronal phenotypes, such as the spiny stellate neuron, in the cat (Vercelli et al., 1992) or layer 5 neurons with short apical dendrite in the rat (Koester and O'Leary, 1992). In addition, it is well known that neurons in area 17 of the monkey undergo loss of dendritic spines in development (Boothe et al., 1979). Dendritic development is regulated by activity (reviewed by Wong and Ghosh, 2002). The development of dendritic pattern is altered by deprivation, at least in the somatosensory representation of the mystacial vibrissae (the barrel field); deprivation does not simply cause regression of the dendritic arbor but rather more complex redistribution of the dendritic branches (Maravall et al., 2004). Surprisingly, sensory deprivation in the primary somatosensory cortex prevents rather than enhance the developmental loss of spines (Zuo et al., 2005).

Geniculo-cortical axons are unlikely to contribute to shrinkage of the neuropile observed in the cases of pure visual deprivation; although they undergo some degree of exuberant growth and remodeling in development, they are not significantly affected by early binocular deprivation of vision (Antonini and Stryker, 1998) or by binocular blockade of retinal activity (Antonini and Stryker, 1993). However, this cannot be excluded after retinal degeneration or injury, which was the cause of early blindness in Noppeney's et al. patients. Indeed, in both, that study and in comparable cases by Shimony et al., 2006, the optic radiation was atrophied.

One observation in the study of Noppeney et al. (2005) seems to establish a direct link between the structural alterations caused by early visual deprivation in man and in animals. In that study a decrease in the volume of the splenium was observed (see Fig. 1), which is known to contain fibers interconnecting the occipital visual areas of the

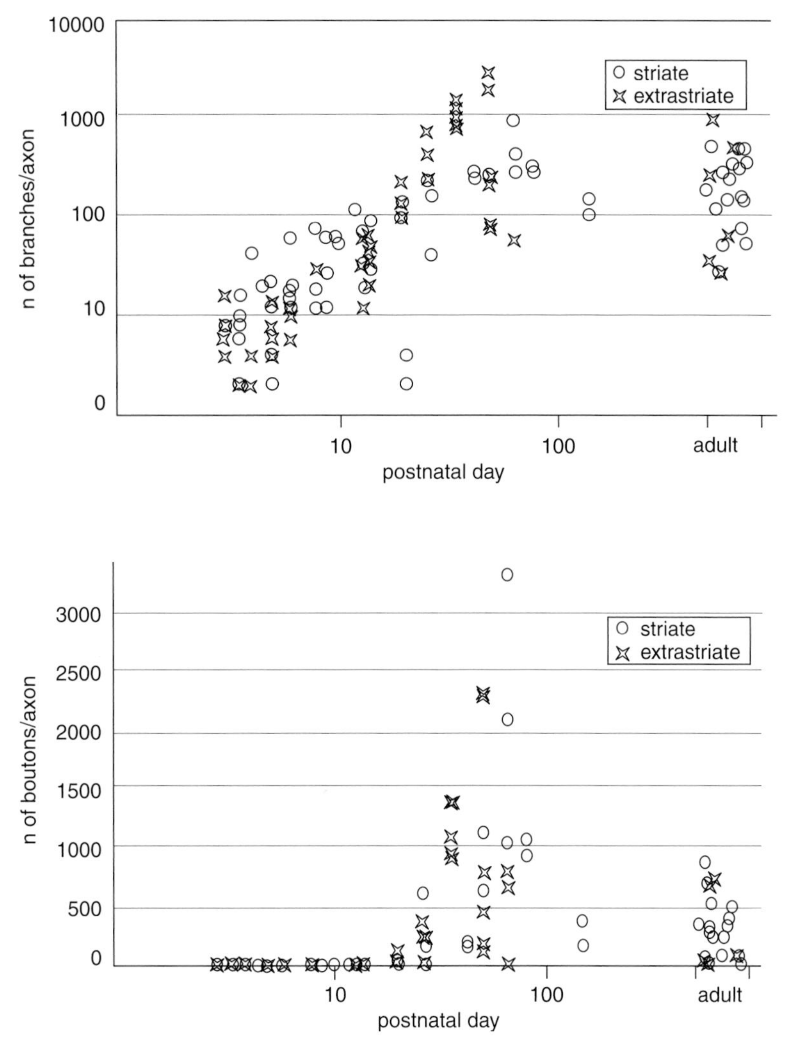

Fig. 1. Developmental changes in the number of branches and boutons in individual callosal axons originating near the border between visual areas 17 and 18 and terminating in areas 17 and 18 (striate) or in extrastriate visual areas. Axons were anterogradely filled with biocytin and reconstructed from serial sections in kittens of different ages by Aggoun-Zouaoui et al. (1996) and Bressoud and Innocenti (1999).

two hemispheres (Hofer and Frahm, 2006). Decreased anisotropy in the splenium was also observed by Shimony et al. (2006). Callosal axons undergo massive numerical reduction in development in both cat and monkey (Berbel and Innocenti, 1988; LaMantia and Rakic, 1990), and probably in man (Clarke et al., 1989). It is well established that binocular visual deprivation by eyelid suture (or by eye enucleation) in the kitten enhances the normal developmental loss of visual callosal axons, as judged by the number of retrogradely labeled callosally projecting neurons, the majority of which are in the supragranular layers (Innocenti and Frost, 1979, 1980; Innocenti et al., 1985; Fig. 2). Interestingly, callosal connections originating from infragranular layers are not affected (Innocenti et al., 1985; Fig. 2) and at least some extrastriate visual areas are also unaffected (Olavarria, 1995; Innocenti, unpublished). The analysis of individual axons has also shown that

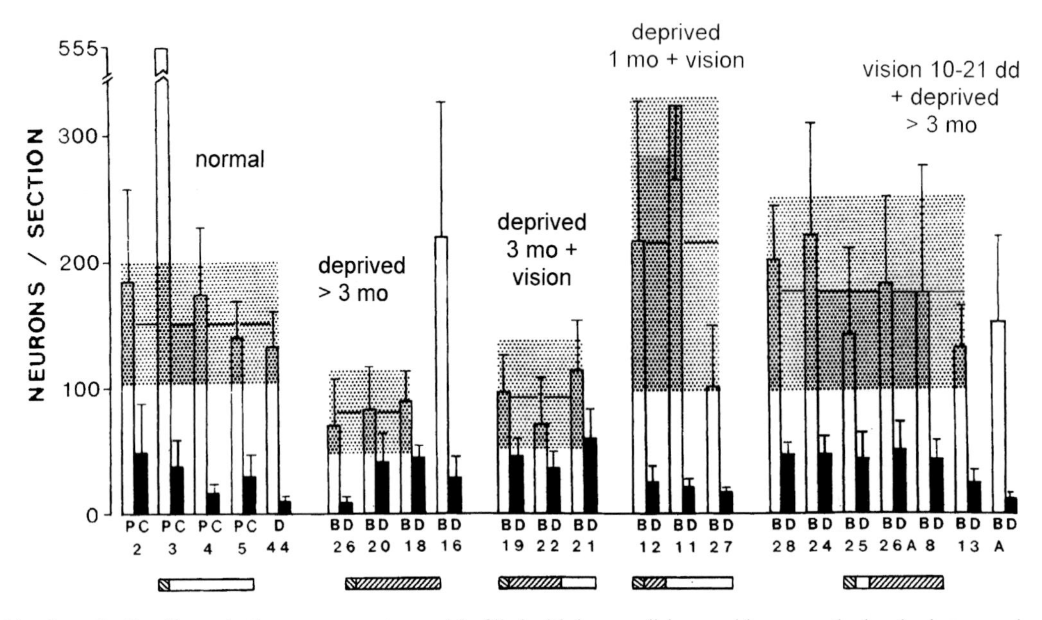

Fig. 2. Number of callosally projecting neurons, retrogradely filled with horseradish peroxidase near the border between visual areas 17 and 18 in normal cats and in cats deprived of vision according to different rearing paradigms. Notice that long-lasting visual deprivation decreases the number of callosally projecting neurons, apparently irreversibly, but only in the supragranular layers (open bars) not in the infragranular layers (filled bars). The horizontal bars are average values across individual experimental animals (letters and number codes) and the shaded areas are standard deviations. Adapted with permission from Innocenti et al. (1985).

the callosal axons, which are maintained in spite of the binocular deprivation, are thinner than normal and have dramatically stunted terminal arbors, i.e., decreased number of branches and of synaptic boutons (Zufferey et al., 1999; Figs. 3 and 6). The local intracortical axons between and within areas 17 and 18 are similarly stunted (Fig. 4) and, in addition, they remain diffuse, instead of acquiring their characteristic patchy distribution (Callaway and Katz, 1991), probably in addition to the loss of connections (Price and Blakemore, 1985; Luhmann et al., 1986, 1990; Fig. 5).

In conclusion, the atrophy of cortico-cortical connections, caused by loss of both axonal and/or dendritic structures in development, provides an attractive explanation for the volumetric decrease of area 17 caused by early visual deprivation (Takács et al., 1992; Noppeney et al., 2005), and is particularly attractive in explaining the volumetric decrease of the splenium of the corpus callosum observed by Noppeney and collaborators.

Deprivation studies in the auditory system have thus far emphasized changes in the lower levels of the auditory pathways. Thus experience contributes to the refinement of topographic specificity in the projection of the auditory nerve to the cochlear nucleus in the cat (Leake et al., 2006 and references therein).

Interestingly, reduced connectivity among limbic structures, the bed nucleus of the stria terminalis, central nucleus of the amygdala and medial prefrontal cortex, but not in the paraventricular nucleus of the hypothalamus and in the insular cortex, were found after injections of the transneuronal tracer pseudorabies virus in the stomach wall in rat pups subjected to handling and maternal separation (Card et al., 2005). This suggests that other forms of early deprivation or abnormal experience may also enhance loss of connections in circuits responsible for visceral/emotional responses.

Some other aspects of animal deprivation studies may be relevant to understanding the consequences

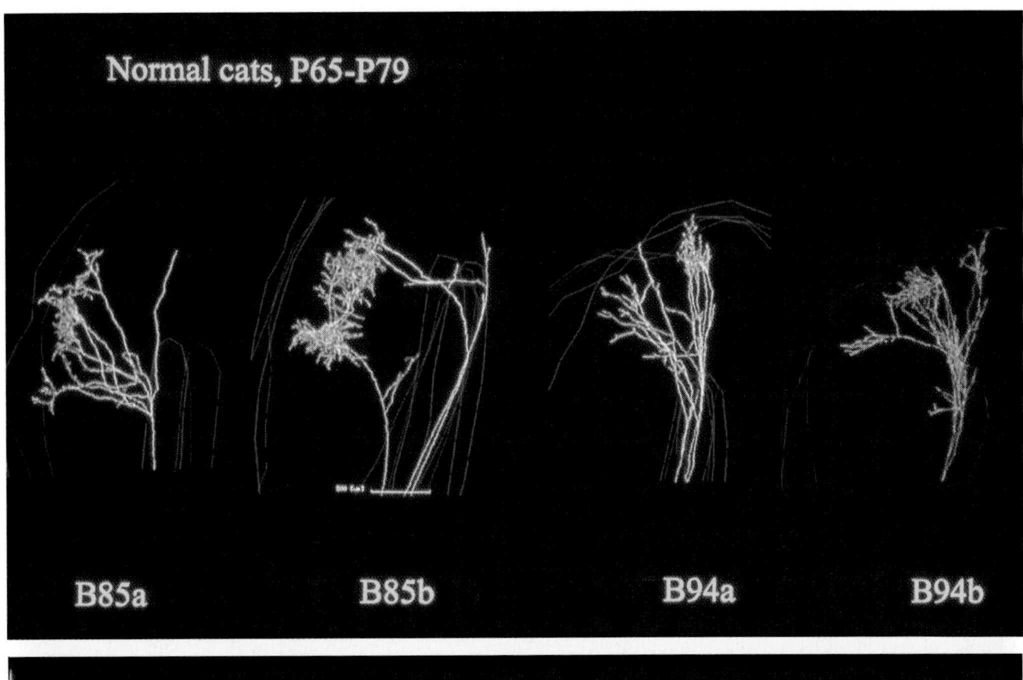

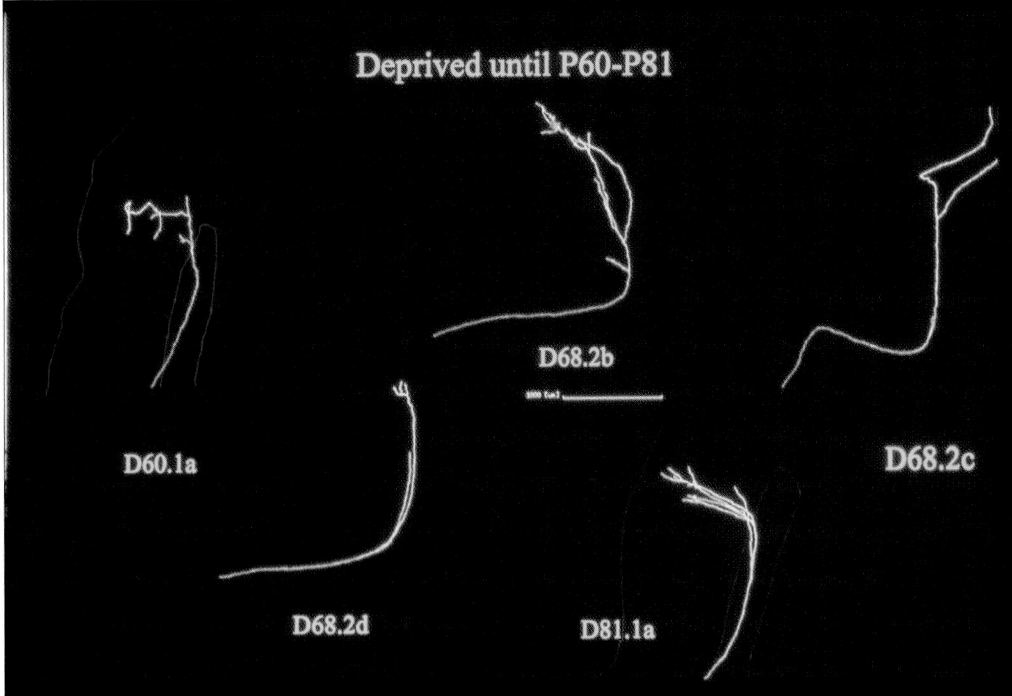

Fig. 3. Examples of individual callosal axons originating and terminating near the border between visual areas 17 and 18, anterogradely filled with biocytin and reconstructed from serial sections in normal adult cats 65–79 days old and in cats which had been deprived of vision by binocular eyelid suture until postnatal days 60–80. The pink dots represent light microscopically identified synaptic boutons. Notice that visual deprivation leads to extremely stunted axonal arbors quantified in Fig. 6. Data from Zufferey et al. (1999).

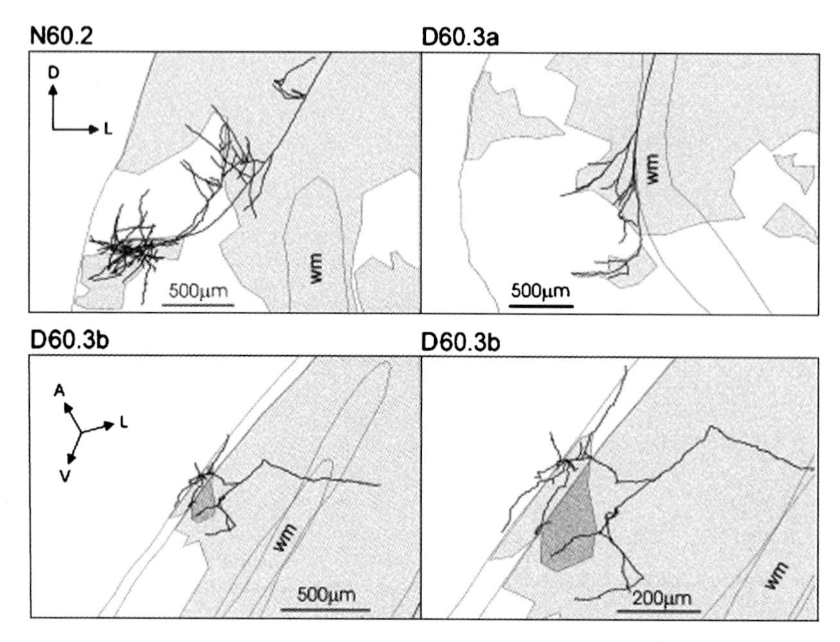

Fig. 4. Examples of individual local axons terminating in visual area 17, anterogradely filled with biocytin and reconstructed from serial sections in a normal cat of 60 days of age (N60.2) and in a cat binocularly deprived of vision by binocular eyelid suture until postnatal day 60 (D60.3a, and b). The gray areas are regions of diffuse labeling; wm is the white matter. Adapted with permission from Zufferey et al. (1999).

of early visual deprivation, and perhaps of other forms of deprivation, in particular language, in humans (discussed below). One intriguing finding was that periods of normal visual experience as short as 10 days, beginning at the time of natural eye opening, had protective effect against a subsequent period of binocular deprivation of several months. In contrast, between 1 and 3 months of binocular deprivation were necessary to cause the loss of callosal projections (Innocenti et al., 1985; Fig. 2). In addition, 8 days of normal visual experience, starting at the time of natural eye opening, and terminated at an age when the terminal arbors of the visual callosal axons are still very scantily branched and are only beginning to form the first synapses is sufficient to trigger a considerable development of callosal and of intracortical terminal arbors (Zufferey et al., 1999; Fig. 6). These findings encourage one to believe that the maturation of at least some fundamental cortical structures may be somewhat protected against the disruptive consequences of deprivation. This may be particularly important for some aspects of human development since, as it will be recalled below, the acquisition of language, which is also impaired in children raised in social isolation, seems to be protected by relatively short periods of exposure, prior to deprivation.

On the other hand, the precise functional consequences of the alterations in axonal morphology are unknown. An attractive interpretation of axonal morphology is that by virtue of their geometry, axons perform transformations (computations) in signal amplification and in the spatial and temporal domains (Innocenti et al., 1994; Innocenti, 1995). Therefore even subtle alteration in axonal geometry in development may have important functional consequences.

Abnormal functional responses after early deprivation

Wiesel and Hubel (1965) described the results of their recordings in area 17 of kittens binocularly deprived of vision before natural eye opening or shortly after (8 days) in these terms: "The cortex of

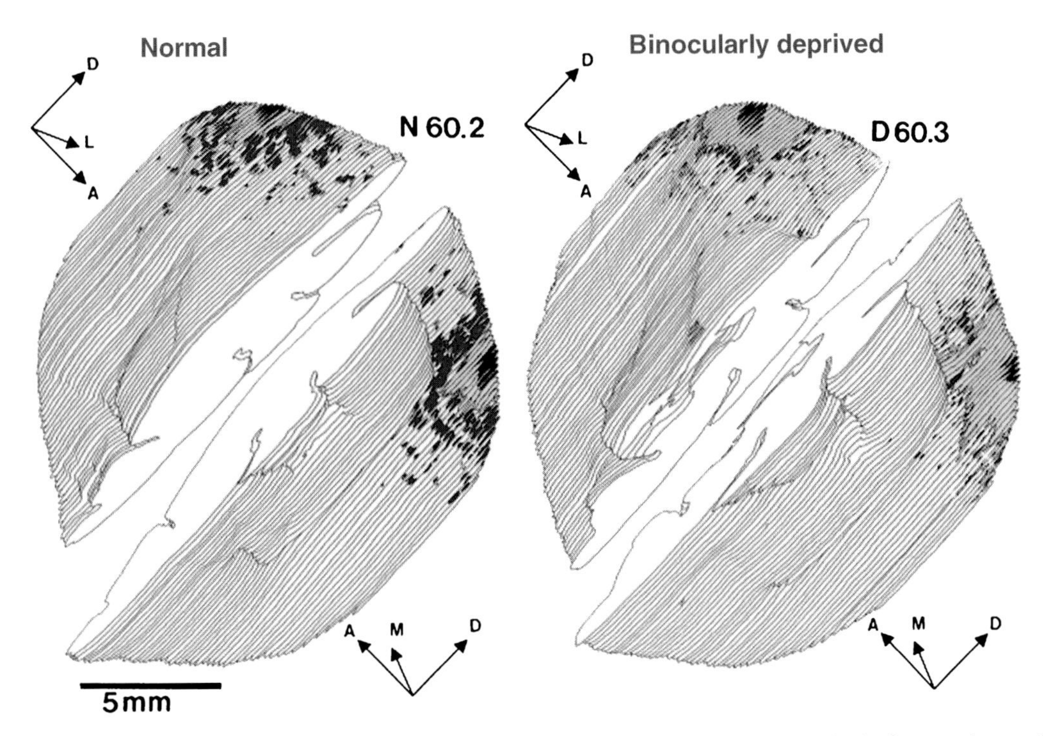

Fig. 5. Three-dimensional reconstruction of local connectivity at the occipital pole (visual areas 17 and 18) of a normal cat and of a cat binocularly deprived of vision by eyelid suture, both aged 60 days. The concentric regions of the dorsal surface of the brains represent the core of a biocytin injection (black) and the regions of diffusely distributed (yellow) and clustered (blue) terminating axons. Adapted with permission from Zufferey et al. (1999).

these animals was... by no means normal. Numerous sluggishly responding unpredictable cells with vaguely defined receptive field properties made the penetrations difficult and frustrating. Of 139 cells encountered 45 (35%) were classed as abnormal. ... Thirty-seven cells (27%) could not be driven by visual stimulation at all and were recognized only by their maintained firing."

The behavioral analysis of these kittens, albeit "*crude*" suggested that "for practical purposes the animal was blind."

Since Wiesel and Hubel's report, it was unequivocally shown that the development of characteristic response properties of visual cortical neurons, including orientation (Blakemore and Cooper, 1970) and direction (Li et al., 2006) specificity requires visual experience and is impaired by deprivation (for review, see also Frégnac and Imbert, 1984). Even more dramatic consequences of early visual deprivation were described for the parietal cortex of the monkey where visual responsiveness was completely lost, in favor of somatosensory responsiveness, while tactile exploration of the environment replaced visual exploration (Hyvärinen et al., 1981).

It seems likely that the deteriorated response properties of cortical neurons are the cause for the permanent deficits in visual performance after early congenital cataract nicely reviewed by Maurer et al. (2005). These deficits include acuity for letters and stripes, sensitivity to peripheral stimuli, sensitivity to global forms and recognition of complex stimuli such as faces.

Many of the deficits reported above may well be caused by the alterations in the connectivity responsible for binding individual neurons into functionally cooperative-neuronal assemblies, i.e., the intra-areal, inter-areal and inter-hemispheric connectivity.

Contrasting with the repeatedly demonstrated poor visual responsiveness of visual areas after early deprivation, an intriguing body of recent literature showed that in blind subjects the primary and secondary visual areas are activated by tactile visual stimuli, notably during Braille reading, and during auditorily driven language tasks (reviewed in Burton, 2003). The effects are larger in early blind individuals than in those who acquired blindness later (Fig. 7; from Burton, 2003) presumably after the period of elimination of exuberant projections and synapses. Most interestingly, the activation of the visual areas seems to contribute to Braille reading in early blind subjects (Cohen et al., 1997).

The pathways conveying tactile and auditory information to the visual areas are unclear. One attractive possibility is the involvement of the parietal areas, which send descending projections to the visual areas and appear to be robustly activated during Braille reading in blind subjects (Sadato et al., 1998; Melzer et al., 2001). As mentioned above, parietal areas were shown to lose visual responsiveness while increasing somatosensory responsiveness in early visually deprived monkeys (Hyvärinen et al., 1981). It was argued elsewhere that the parietal areas, one of the most recent phylogenetic acquisitions of the primate cortex, may be particularly plastic in development as also suggested by the fact that they maintain connections from contralateral somatosensory areas when deprived of cortico-cortical visual input by early lesions of the lower order visual areas (Restrepo et al., 2003). Another possibility is the maintenance of transient connections from auditory and somatosensory areas to visual areas (Innocenti and Clarke, 1984; Dehay et al., 1988; Innocenti et al., 1988) a small fraction of which is

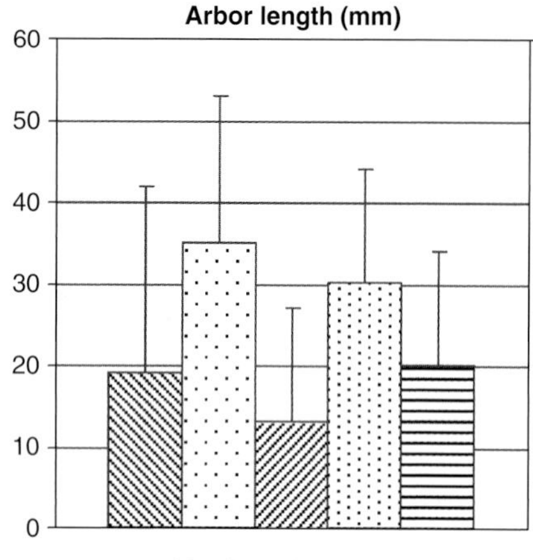

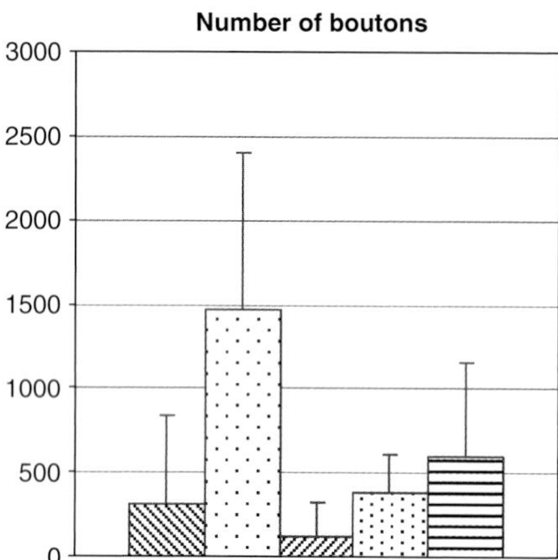

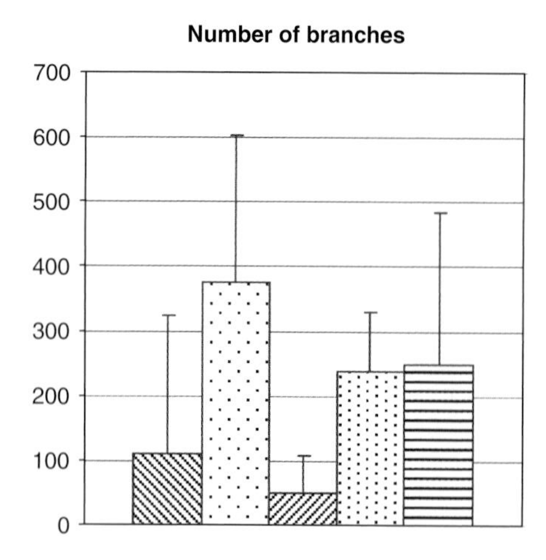

Fig. 6. Total length, number of branches and of boutons per arbor in individual callosal axons originating and terminating near the border between visual areas 17 and 18, anterogradely filled with biocytin and reconstructed from serial sections in normal and in deprived cats. The D68 columns show values from 9 axons in kittens binocularly deprived of vision by eyelid suture until 60–81 days of age. N68 columns show values from 7 axons in normal kittens of 50–80 days of age. The D78 columns show values from 8 axons in one kitten of 78 days of age whose eyelids had been sutured after 8 days of normal vision. The D28 columns show data from 8 axons in kittens deprived of vision as above until postnatal day 29. The N29 columns show data from 3 axons in normal kittens at the age of 27–29 days. Notice that visual deprivation severely affects all the axon parameters already at the end of the first postnatal month. However, a short and early period of normal vision prevents, to some extent the deprivation effect. For statistical evaluations and examples of axons see Zufferey et al. (1999).

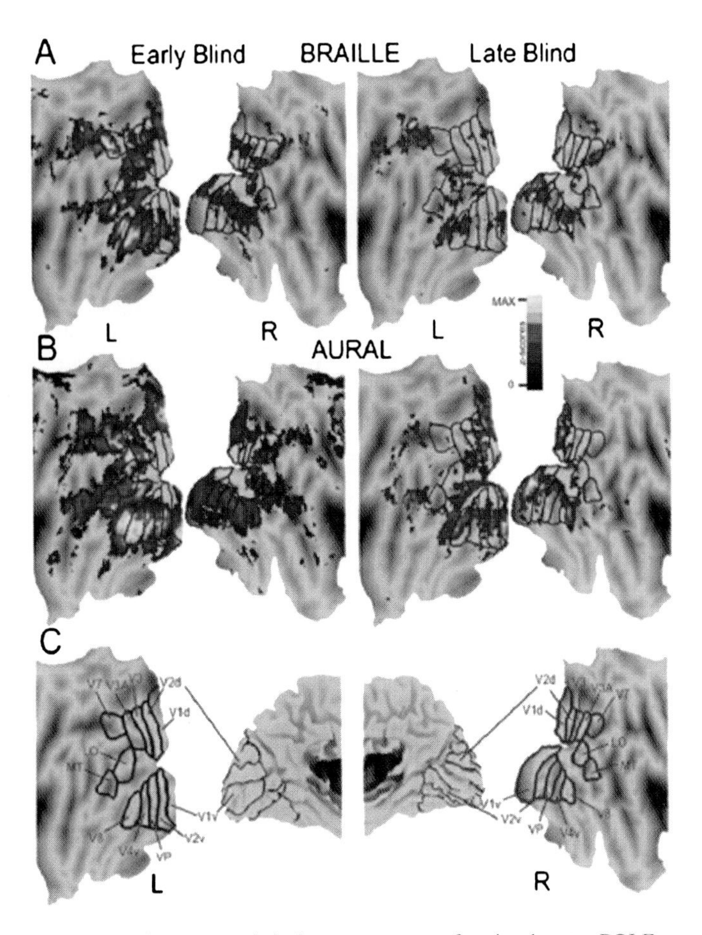

Fig. 7. Two-dimensional, flattened views of z-score statistical parameter maps for visual cortex BOLD responses in early and late blind subjects. Maps in A and B were obtained, respectively, when subjects generated verbs to Braille read or heard nouns. A minimum z-score threshold for all maps was 4.5; $p \leq 0.05$. Maximum z-scores for results from early blind subjects were 30 and 35, respectively, in A and B, and 22 for all results from late blind subjects. (C) Visuotopic borders drawn and labeled according to previous results obtained in sighted subjects. Adapted with permission from Burton (2003).

normally maintained into adulthood (Innocenti et al., 1988; Falchier et al., 2002). Earlier blindness, such as that induced by binocular enucleation prior to the establishment of retinogeniculate and geniculocortical projections in the opossum can also lead to the maintenance of projections to area 17 from auditory, somatosensory and polymodal cortex (Karlen et al., 2006).

Wiesel and Hubel's experiments of visual deprivation in cat and monkey (reviewed in Wiesel, 1982) led to a large and fruitful series of studies many of which capitalized on the paradigm of monocular deprivation. It was repeatedly shown that this rearing condition, if applied during a sensitive period in development, leads to loss of visual responses to the deprived eye to the advantage of the open eye. The structural correlate of this ocular dominance shift was a moderate expansion of the cortical territory occupied by geniculocortical axons pertaining to the open eye, and a severe retraction of those pertaining to the deprived eye. Thus the effects of monocular deprivation on eye dominance were explained as the result of competition between the geniculo-cortical axons carrying information from different eyes. Interestingly though, the competition involves local inhibitory interneurons (reviewed in Hensch, 2005). Along similar lines Dong et al. (2004)

reported that in mouse visual cortex, monocular deprivation causes an increase in the inhibitory postsynaptic potential generated by the activation of inter-areal feedback circuits, while leaving the strength of the feed-forward inhibition unaffected.

The consequences of deprivation are not restricted to the visual modality. Following the seminal work of Van der Loos and Woolsey (1973) showing that the perinatal lesion of the vibrissae impairs the development of the corresponding barrels in the somatosensory cortex of rodents, numerous studies have analyzed the consequences of early deprivation in the same system. Recently, Shoykhet et al. (2005) reported a weakening of the excitatory receptive fields in the barrel field of rats whose whiskers had been trimmed within a critical period extending at least as far as postnatal day (pd) 12. The inhibitory components of the receptive fields were also decreased and so were the inhibitory interactions within adjacent whiskers. These effects may be due to the combined loss of thalamo-cortical and inter-neuronal synapses (Sadaka et al., 2003) and be the cause for the impaired roughness discrimination, using the whiskers, in the adult (Carvell and Simons, 1996).

Alterations in functional connectivity may be caused by isolation in early life, a type of deprivation whose behavioral consequences in primates have been known for many years (Harlow and Harlow, 1962). Bartesaghi et al. (2006) reported decreased field potentials evoked by the stimulation of the commissural pathway in the entorhinal cortex as well as in the dentate gyrus, and the CA3–CA1 region of the hippocampus of guinea pigs raised in isolation from pd 6–7 until pd 90–100. The effects were predominantly in males, absent or less severe in females. The same group reported several deprivation-induced morphological alterations in the hippocampus (Bartesaghi et al., 2003 and references therein).

It is attractive to speculate that structural–functional alterations, similar to those that have been described in the case of visual deprivation, may underlie the deficit in some higher functions of the human brain. Language may be one of these functions although several others are probably still uncharted. This conclusion is strongly supported by the work of Patricia Kuhl and collaborators on the experience-dependent establishment of phonemic boundaries during the early postnatal months (reviewed in Kuhl, 2004). Other hints come from the amazingly large number of reports of children raised in isolation from other humans (Newton, 2002; see also http://www.feralchildren.com). Although some of these reports are controversial, common traits in the best-documented cases are the avoidance of humans, which could have been predicted from Harlow's work (quoted). Less expected is that these children had not acquired the upright posture. Most interesting is the absence of language, which remains at best rudimentary. In this respect the two best-documented cases are those of Victor, the boy of Aveyron, caught in the wild in 1789 at the age of 11 and that of Genie kept in isolation by her father and discovered in 1970 at the age of 13. Neither child acquired proper language in spite of the intensive training they received from professionals. Genie, apparently, never acquired proper syntax. According to Curtiss (1977) "Genie's vocabulary grew by leaps and bounds, but she was still not able to string words together into meaningful sentences. Normal children begin by learning to say simple sentences, like 'No have toy.' Soon they are able to say 'I not have toy.' Eventually they will learn to say, 'I do not have the toy.' Later they will refine the sentence to say, 'I don't have the toy.' Genie seemed to be stuck at the first stage." Although the case of Genie is somewhat controversial, it is important to notice that language involves inter-hemispheric interactions leading to the formation of cooperative-neuronal assemblies in the two hemispheres (Mohr et al., 1994, 2000; Pulvermüller and Mohr, 1996; Friederici and Alter, 2004). Furthermore, the competent areas of the human brain are organized into discrete, interconnected modules (Galuske et al., 2000) in striking similarity to what is found for the visual areas. It is very plausible then, that, as in the case of visual deprivation (discussed above), deprivation of language leads to the loss of inter-hemispheric and local connections with severe consequences for the formation of the neuronal assemblies required for grammatically correct language.

Across systems, the impaired connectivity at the root of the deprivation effects may be due to the altered expression of neurotrophins. In particular,

Gianfreschi et al. (2003) reported that the consequences of dark rearing in the visual cortex of the rat can be counteracted by over-expression of BDNF. Roceri et al. (2004) reported an early (pd 17), transient increase in BDNF expression in prefrontal cortex and hippocampus of maternally deprived rats, and a decrease in the expression of BDNF in the adult prefrontal cortex; interestingly, maternally deprived rats, unlike normally reared rats, did not reduce BDNF levels in the prefrontal cortex and striatum, when subjected to stress.

It seems to be proven beyond doubts that the mechanism responsible for several types of early deprivation is the competition between different sets of developing axons. This applies to the consequences of monocular deprivation (discussed above) as well as to those of sensory substitution in multi-modal areas (Rauschecker, 1995). Nevertheless it can be questioned if competition fully explains the consequences of all kinds of deprivation, in particular those of complete visual deprivation. The volumetric decrease in the visually deprived visual areas observed both in animals and in man (discussed above) speaks against this interpretation since no volumetric decrease would be expected if the visual input were replaced by another, competing input. Instead, the common trait at the roots of all the conditions of early deprivation is probably the fact that connections are initially formed in excess and in a labile form (reviewed in Innocenti and Price, 2005). It seems reasonable to suggest that functional demands should control the subsequent elimination of the superfluous and/or unused connections. However, one may wonder if the elimination of the unused connections really has an adaptive role. In the case of blind subjects whose visual areas appear to be reallocated to processing information from other modalities, for example in Braille reading (discussed above) this appears to be true. On the other hand, the energy-sparing principle that probably underlies the design of brain anatomy, the usually forgotten "*loi de l'economie de protoplasma nerveux transmetteur et des temps de transmission*" (Ramon y Cajal, 1972; Innocenti, 1990) may also dictate the elimination of unused structures, in particular the energy-expensive long axons.

Some problems: structural–structural, functional–functional and structural–functional correlations — species differences; enrichment vs. deprivation; permissive vs. constructive (instructive or selective) role of experience

These days, in spite of the unprecedented, welcome mass of information on the brain several crucial questions remain unanswered. Some of them originate from the difficulty in re-mapping findings across different levels of analysis, methodological approaches and species. In this short review some of these difficulties have already emerged. In particular, the alterations in cortical volumes revealed by structural imaging in man cannot readily be interpreted in terms of changes of one or the other cellular compartments, i.e., neurons, glia, axons, dendrites, synaptic contacts etc., or even blood vessels. Even in the case of structures containing mainly axons, such as the corpus callosum, and without considering the volumes occupied by blood vessels, astrocytes, oligodendrocytes and neurons (the latter are a normal component of the corpus callosum, particularly in development, see Riederer et al., 2004) the volumetric changes could be due to either the number or the size of axons. These in turn are strongly influenced by the number of myelinated axons and by the thickness of their myelin sheath. Tract-tracing based on diffusion anisotropies may, in some cases, reveal interesting peculiarities of white matter tracts although its low resolution usually prevents direct comparisons with experimental tract-tracing using intra-axonal transport or diffusion.

The changes in activity revealed by fMRI or PET, not only lack the spatial and temporal resolutions attained by single cell recordings of action potentials, but also probably reveal a different kind of activity (see Logothetis and Pfeuffer, 2004, for review). As electrically recorded field potentials, they appear to relate to membrane potential shifts that do not necessarily reach threshold for neuronal firing. The relevance of sub-threshold activation for cortical processing remains to be explored. Since in the absence of action potentials the communication between neurons is limited both in space and in time, one rule of thumb may be that the sub-threshold depolarization fields represent region not necessarily involved in processing, but ready to do so.

The relation between brain morphology and function can be surprisingly ambiguous. Thus, subtle and often elusive morphological changes are sometimes associated with extremely severe functional impairments, as in schizophrenia (discussed in Innocenti et al., 2003). In contrast, severe malformation of the visual cortex such as microgyria can allow for rather normal electrophysiological and functional responses in animals and in man (Innocenti et al., 2001; Zesiger et al., 2002), although changes in fast auditory processing have been described in microgyria in male (but not in female) rats (Herman et al., 1997).

Finally, the extrapolations from animal models of developmental pathologies to humans can be challenging. Some developmental processes are constant, albeit not identical, across species and systems. One of them is the appearance of exuberant and labile juvenile connections, a part of which will undergo experience-controlled transition to stabile adult connections (Innocenti and Price, 2005). However, other processes can vary across species or systems, for example the origin of inhibitory interneurons in the cerebral cortex (Letinic et al., 2002) or the adult cortical neuronogenesis (Bhardwaj et al., 2006).

The results of some exceptional experience in development appear to blur the boundary between deprivation and enrichment in early life. Thus, recent studies have reported that early musical training can increase cortical thickness as well as cortical activation at selective locations, and also the size of the corpus callosum (Schlaug et al., 1995), and improves cognitive development. That musical training precedes brain and cognitive differences, rather than the reverse, is suggested by an ongoing longitudinal study (Schlaug et al., 2005). The results published thus far describe one cohort of 5–7 years old children who underwent musical training for 1 year and a cross-sectional comparison of 9–11 years old children with 4-year musical training. In both studies, an increase was observed in cortical thickness and in cortical activation during rhythmic and melodic discrimination tasks. Interestingly, the younger cohort also showed a tendency to increased performance in tasks not directly related to music training such as phonemic awareness, Raven progressive matrices and Key math test.

In neuroanatomical and neurophysiological terms the question seems to be whether early musical training leads to the selective stabilization (Changeux et al., 1973) of synapses, which would have otherwise been eliminated or to the formation of new synapses. This in turns evokes the old debate on the selective vs. instructive role of experience in brain development. Based on the development of visual callosal axons, we argued elsewhere that a more fundamental dichotomy might be whether experience plays a constructive or a permissive role in brain development (Zufferey et al., 1999). Both seem to be true. The permissive role of vision is demonstrated by the finding that, as already mentioned, short periods of visual experience are sufficient to validate the process of axonal growth and differentiation which can then proceed further, in the absence of vision. On the other hand the fact that already at pd 29 axons in binocularly deprived kittens are less developed than those of normal kittens suggests that in normal development experience also plays a constructive role. This constructive role could be the expression of two different mechanisms. The most likely mechanism is the selective validation of individual, newly formed branches and synapses. However, an instructive mechanism cannot be excluded, i.e., that experience induces specific aspects of connectivity.

The considerations above highlight how much crucial and exciting work is ahead for system-oriented studies aiming at clarifying structural–functional relations in normal and abnormal developments with broadly minded animal to human translational approaches.

Neuroethical issues

The evidence that brain development can be influenced by environment raises a number of issues that are beginning to be considered by current debates in the new field of neuroethics (Illes et al., 2005). It should be obvious and universally accepted that the child has the right to experience the conditions that ensure the optimal development of his/her brain, within the limits of its genetic endowment. Yet, this right, which is of paramount importance for the individual and for societies, seems to be explicitly and forcefully written nowhere.

Although in this chapter I have referenced some of the conditions that impair the development of cerebral cortex in a dramatic, well-documented way, we must admit that in general we do not know what these conditions are. More subtle deprivation-induced alterations in cortical and subcortical structures and/or function are a mostly uncharted territory. This is particularly worrying at the time when fast cultural changes in our societies are profoundly modifying the environmental input to the child.

Thus one ought to acknowledge that there are corollaries to the right of the child to the optimal development of his/her brain. One corollary is that research should be supported and encouraged with high priority, aimed at charting all the environmental conditions that affect brain development either adversely or positively. The environmental influences on brain development should be understood in their causal relations, from the cellular, molecular and system levels up to behavior and cognition. Techniques to improve normal development in view of the acquisition of better skills should similarly be investigated to understand if the acquisition of supernormal skill in one domain will be paid by loss of other skills. This includes conditions of early learning that, if applied during the critical period of brain development, may irreversibly modify brain structure and function.

Charting the field, i.e., identifying the ethical issues raised by interventions on the developing brain, if society allows doing so, is where the responsibility of the neuroscientist will end. The decision of what shall be ethically acceptable falls clearly beyond the responsibility of the neuroscientist, although it remains within his/her reach as a human being and moral subject.

References

Aggoun-Zouaoui, D., Kiper, D.C. and Innocenti, G.M. (1996) Growth of callosal terminal arbors in primary visual areas of the cat. Eur. J. Neurosci., 8: 1132–1148.

Antonini, A. and Stryker, M.P. (1993) Development of individual geniculocortical arbors in cat striate cortex and effects of binocular impulse blockade. J. Neurosci., 13: 3549–3573.

Antonini, A. and Stryker, M.P. (1998) Effects of sensory disuse on geniculate afferents to cat visual cortex. Vis. Neurosci., 15: 401–409.

Bartesaghi, R., Raffi, M. and Ciani, E. (2006) Effect of early isolation on signal transfer in the entorhinal cortex-dentate-hippocampal system. Neuroscience, 137: 875–890.

Bartesaghi, R., Raffi, M. and Severi, S. (2003) Effects of early isolation on layer II neurons in the entorhinal cortex of the guinea pig. Neuroscience, 120: 721–732.

Berbel, P. and Innocenti, G.M. (1988) The development of the corpus callosum in cats: a light- and electron-microscopic study. J. Comp. Neurol., 276: 132–156.

Bhardwaj, R.D., Curtis, M.A., Spalding, K.L., Buchholz, B.A., Fink, D., Björk-Eriksson, T., Nordborg, C., Gage, F.H., Druid, H., Eriksson, P.S. and Frisén, J. (2006) Neocortical neurogenesis in humans is restricted to development. Proc. Natl. Acad. Sci. U.S.A., 103: 12564–12568.

Blakemore, C. and Cooper, G.F. (1970) Development of the brain depends on the visual environment. Nature, 228: 477–478.

Boothe, R.G., Greenough, W.T., Lund, J.S. and Wrege, K. (1979) A quantitative investigation of spine and dendrite development of neurons in visual cortex (area 17) of *Macaca nemestrina* monkeys. J. Comp. Neurol., 186: 473–490.

Bressoud, R. and Innocenti, G.M. (1999) Topology, early differentiation and exuberant growth of a set of cortical axons. J. Comp. Neurol., 406: 87–108.

Burton, H. (2003) Visual cortex activity in early and late blind people. J. Neurosci., 23: 4005–4011.

Callaway, E.M. and Katz, L.C. (1991) Effects of binocular deprivation on the development of clustered horizontal connections in cat striate cortex. Proc. Natl. Acad. Sci. U.S.A., 88: 745–749.

Card, J.P., Levitt, P., Gluhowski, M. and Rinamann, L. (2005) Early experience modifies the postnatal assembly of autonomic emotional motor circuits in rats. J. Neurosci., 25: 9102–9111.

Carvell, G.E. and Simons, D.J. (1996) Abnormal tactile experience early in life disrupts active touch. J. Neurosci., 16: 2750–2757.

Changeux, J.-P., Courrège, P. and Danchin, A. (1973) A theory of the epigenesis of neuronal networks by selective stabilization of synapses. Proc. Natl. Acad. Sci. U.S.A., 70: 2974–2978.

Clarke, S., Kraftsik, R., Van der Loos, H. and Innocenti, G.M. (1989) Forms and measures of adult and developing human corpus callosum: is there sexual dimorphism? J. Comp. Neurol., 280: 213–230.

Cohen, L.G., Celnik, P., Pascual-Leone, A., Cornwell, B., Faiz, L., Dambrosia, J., Honda, M., Sadato, N., Gerloff, C., Catalá, M.D. and Hallett, M. (1997) Functional relevance of cross-modal plasticity in blind humans. Nature, 389: 180–183.

Curtiss, S. (1977) Genie: A Psycholinguistic Study of a Modern-Day Wild Child (Perspectives in Neurolinguistics and Psycholinguistics). Academic Press, New York.

Dehay, C., Kennedy, H. and Bullier, J. (1988) Characterization of transient cortical projections from auditory, somatosensory, and motor cortices to visual areas 17, 18, and 19 in the kitten. J. Comp. Neurol., 272: 68–89.

Dong, H., Wang, Q., Valkova, K., Gonchar, Y. and Burkhalter, A. (2004) Experience-dependent development of feedforward and feedback circuits between lower and higher areas of mouse visual cortex. Vision Res., 44: 3389–3400.

Falchier, A., Clavagnier, S., Barone, P. and Kennedy, H. (2002) Anatomical evidence of multimodal integration in primate striate cortex. J. Neurosci., 22: 5749–5759.

Li, Y., Fitzpatrick, D. and White, L.E. (2006) The development of direction selectivity in ferret visual cortex requires early visual experience. Nat. Neurosci., 9: 676–681.

Frégnac, Y. and Imbert, M. (1984) Development of neuronal selectivity in primary visual cortex of cat. Physiol. Rev., 64: 325–434.

Friederici, A.D. and Alter, K. (2004) Lateralization of auditory language functions: a dynamic dual pathway model. Brain Lang., 89: 267–276.

Galuske, R.A.W., Schlote, W., Bratzke, H. and Singer, W. (2000) Interhemispheric asymmetries of the modular structure in the human temporal cortex. Science, 289: 1946–1949.

Gianfreschi, L., Siciliano, R., Walls, J., Morales, B., Kirkwood, A., Huang, Z.J., Tonegawa, S. and Maffei, L. (2003) Visual cortex is rescued from the effects of dark rearing by overexpression of BDNF. Proc. Natl. Acad. Sci. U.S.A., 100: 12486–12491.

Gregory, R.L. and Wallace, J.G. (1963) Recovery from early blindness: a case study. Quartl. J. Exp. Psychol. Monogr. Suppl. 2.

Harlow, H.F. and Harlow, M.K. (1962) Social deprivation in monkeys. Sci. Am., 207: 136–146.

Hensch, T.K. (2005) Critical period plasticity in local cortical neurons. Nat. Rev. Neurosci., 6: 877–887.

Herman, A.E., Galaburda, A.M., Fitch, R.H., Carter, A.R. and Rosen, G.D. (1997) Cerebral microgyria, thalamic cell size and auditory temporal processing in male and female rats. Cereb. Cortex, 7: 453–464.

Hofer, S. and Frahm, J. (2006) Topography of the human corpus callosum revisited: comprehensive fiber tractography using diffusion tensor magnetic resonance imaging. Neuroimage, 32: 989–994.

Hyvärinen, J., Hyvärinen, L. and Linnankoski, I. (1981) Modification of parietal association cortex and functional blindness after binocular deprivation in young monkeys. Exp. Brain Res., 42: 1–8.

Illes, J., Blakemore, C., Hansson, M., Hensch, T.K., Leshner, A., Maestre, G., Magistretti, P., Quirion, R. and Strata, P. (2005) International perspectives on engaging the public in neuroethics. Nat. Rev. Neurosci., 6: 977–982.

Innocenti, G.M. (1990) Pathways between development and evolution. In: Finlay B.L., Innocenti G. and Scheich H. (Eds.), The Neocortex. Plenum Press, New York, pp. 43–52.

Innocenti, G.M. (1995) Exuberant development of connections, and its possible permissive role in cortical evolution. Trends Neurosci., 18: 397–402.

Innocenti, G. and Clarke, S. (1984) Bilateral transitory projection to visual areas from the auditory cortex in kittens. Dev. Brain Res., 14: 143–148.

Innocenti, G.M. and Frost, D.O. (1979) Effects of visual experience on the maturation of the efferent system to the corpus callosum. Nature, 280: 231–234.

Innocenti, G.M., Ansermet, F. and Parnas, J. (2003) Schizophrenia, neurodevelopment and corpus callosum. Mol. Psychiatry, 8: 261–274.

Innocenti, G.M., Berbel, P. and Clarke, S. (1988) Development of projections from auditory to visual areas in the cat. J. Comp. Neurol., 272: 242–259.

Innocenti, G.M. and Frost, D.O. (1980) The postnatal development of visual callosal connections in the absence of visual experience or of the eyes. Exp. Brain Res., 39: 365–375.

Innocenti, G.M., Frost, D.O. and Illes, J. (1985) Maturation of visual callosal connections in visually deprived kittens: a challenging critical period. J. Neurosci., 5: 255–267.

Innocenti, G.M., Lehmann, P. and Houzel, J.-C. (1994) Computational structure of visual callosal axons. Eur. J. Neurosci., 6: 918–935.

Innocenti, G.M., Maeder, P., Knyazeva, M.G., Fornari, E. and Deonna, T. (2001) Functional activation of a microgyric visual cortex in a human. Ann. Neurol., 50: 672–676.

Innocenti, G.M. and Price, D.J. (2005) Exuberance in the development of cortical networks. Nat. Rev. Neurosci., 6: 955–965.

Karlen, S.J., Kahn, D.M. and Krubitzer, L. (2006) Early blindness results in abnormal corticocortical and thalamocortical connections. Neuroscience, 142: 643–658.

Koester, S.E. and O'Leary, D.D.M. (1992) Functional classes of cortical projection neurons develop dendritic distinctions by class-specific sculpting of an early common pattern. J. Neurosci., 12: 1382–1393.

Kuhl, P.K. (2004) Early language acquisition: cracking the speech code. Nat. Rev. Neurosci., 5: 831–843.

LaMantia, A.-S. and Rakic, P. (1990) Axon overproduction and elimination in the corpus callosum of the developing rhesus monkey. J. Neurosci., 10: 2156–2175.

Leake, P.A., Hradek, G.T., Chair, L. and Snyder, R.L. (2006) Neonatal deafness results in degraded topographic specificity of auditory nerve projections to the cochlear nucleus in cats. J. Comp. Neurol., 497: 13–31.

Letinic, K., Zoncu, R. and Rakic, P. (2002) Origin of GABAergic neurons in the human neocortex. Nature, 417: 645–649.

Leuba, G. and Kraftsik, R. (1994) Changes in volume, surface estimate, three-dimensional shape and total number of neurons of the human primary visual cortex from midgestation until old age. Anat. Embryol., 190: 351–366.

Logothetis, N.K. and Pfeuffer, J. (2004) On the nature of the BOLD fMRI contrast mechanism. Magn. Reson. Imaging, 22: 1517–1531.

Luhmann, H.J., Millán, L.M. and Singer, W. (1986) Development of horizontal intrinsic connections in cat striate cortex. Exp. Brain Res., 63: 443–448.

Luhmann, H.J., Singer, W. and Martínez-Millán, L. (1990) Horizontal interactions in cat striate cortex: I. Anatomical substrate and postnatal development. Eur. J. Neurosci., 2: 344–357.

Maravall, M., Koh, I.Y.Y., Lindquist, W.B. and Svoboda, K. (2004) Experience-dependent changes in basal dendritic branching of layer 2/3 pyramidal neurons during a critical period for developmental plasticity in the rat barrel cortex. Cereb. Cortex, 14: 655–664.

Maurer, D., Lewis, T.L. and Mondloch, C.J. (2005) Missing sight: consequences for visual cognitive development. Trends Cogn. Neurosci., 9: 144–151.

Melzer, P., Morgan, V.L., Pickens, D.R., Price, R.R., Wall, R.S. and Ebner, F.F. (2001) Cortical activation during Braille readings is influenced by early visual experience in subjects with severe visual disability: a correlational fMRI study. Hum. Brain Mapp., 14: 186–195.

Mohr, B., Pulvermüller, F., Cohen, R. and Rockstroh, B. (2000) Interhemispheric cooperation during word processing: evidence for callosal transfer dysfunction in schizophrenic patients. Schizophr. Res., 46: 231–239.

Mohr, B., Pulvermüller, F. and Zaidel, E. (1994) Lexical decision after left, right and bilateral presentation of function words, content words and non-words: evidence for interhemispheric interaction. Neuropsychologia, 32: 105–124.

Newton, M. (2002) Savage girls and wild boys, a history of feral children. Faber and Faber, London, pp. 1–184.

Noppeney, U., Friston, K.J., Ashburner, J., Frackowiak, R. and Price, C.J. (2005) Early visual deprivation induces structural plasticity in gray and white matter. Curr. Biol., 15: R488–R490.

Olavarria, J.F. (1995) The effect of visual deprivation on the number of callosal cells in the cat is less pronounced in extrastriate cortex than in the 17/18 border region. Neurosci. Lett., 195: 147–150.

Price, D.J. and Blakemore, C. (1985) The postnatal development of the association projection from visual cortical area 17 to area 18 in the cat. J. Neurosci., 5: 2443–2452.

Pulvermüller, F. and Mohr, B. (1996) The concept of transcortical cell assemblies: a key to the understanding of cortical lateralization and interhemispheric interaction. Neurosci. Biobehav. Rev., 20: 557–566.

Rabinowicz, T., de Courten-Myers, G.M., Petetot, J.M., Xi, G. and de los Reyes, E. (1992) Effect of dark rearing on the volume of visual cortex (areas 17 and 18) and number of visual cortical cells in young kittens. J. Neurosci. Res., 32: 449–459.

Ramon y Cajal, S. (1972) Histologie du système nerveux de l'homme et des vertébrés, Vol. I. Raycar S.A., Madrid, pp. 1–986.

Rauschecker, J.P. (1995) Compensatory plasticity and sensory substitution in the cerebral cortex. Trends Neurosci., 18: 36–43.

Restrepo, C.E., Manger, P.R., Spenger, C. and Innocenti, G.M. (2003) Immature cortex lesions alter retinotopic maps and interhemispheric connections. Ann. Neurol., 54: 51–65.

Riederer, B.M., Berbel, P. and Innocenti, G.M. (2004) Neurons in the corpus callosum of the cat during postnatal development. Eur. J. Neurosci., 19: 2039–2046.

Roceri, M., Cirulli, F., Pessina, C., Peretto, P., Racagni, G. and Riva, M.A. (2004) Postnatal repeated maternal deprivation produces age-dependent changes of brain derived neurotrophic factor expression in selected rat brain regions. Biol. Psychiatry, 55: 708–714.

Sadaka, Y., Weinfeld, E., Lev, D.L. and White, E.L. (2003) Changes in mouse barrel synapses consequent to sensory deprivation from birth. J. Comp. Neurol., 475: 75–86.

Sadato, N., Pascual-Leone, A., Grafman, J., Deiber, M.-P., Ibañez, V. and Hallett, M. (1998) Neural networks for Braille reading by the blind. Brain, 121: 1213–1229.

Schlaug, G., Jäncke, L., Huang, Y., Staiger, J.F. and Steinmetz, H. (1995) Increased corpus callosum size in musicians. Neuropsychologia, 33: 1047–1055.

Schlaug, G., Norton, A., Overy, K. and Winner, E. (2005) Effects of music training on the child's brain and cognitive development. Ann. N.Y. Acad. Sci., 1060: 219–230.

Shimony, J.S., Burton, H., Epstein, A.A., McLaren, D.G., Sun, S.W. and Snyder, A.Z. (2006) Diffusor tensor imaging reveals white matter reorganization in early blind humans. Cereb. Cortex, 16: 1653–1661.

Shoykhet, M., Land, P.W. and Simons, D.J. (2005) Whisker trimming began at birth or on postnatal day 12 affects excitatory and inhibitory receptive fields of layer IV barrel neurons. J. Neurophysiol., 94: 3987–3995.

Takács, J., Saillour, P., Imbert, M., Bogner, M. and Hamori, J. (1992) Effect of dark rearing on the volume of visual cortex (areas 17 and 18) and number of visual cortical cells in young kittens. J. Neurosci. Res., 32: 449–459.

Van der Loos, H. and Woolsey, T.A. (1973) Somatosensory cortex: structural alterations following early injury to sense organs. Science, 179: 395–398.

Vercelli, A., Assal, F. and Innocenti, G.M. (1992) Emergence of callosally projecting neurons with stellate morphology in the visual cortex of the kitten. Exp. Brain Res., 90: 346–358.

Wiesel, T.N. (1982) Postnatal development of the visual cortex and the influence of environment. Nature, 299: 583–591.

Wiesel, T.N. and Hubel, D.H. (1965) Comparison of the effects of unilateral and bilateral eye closure on cortical unit responses in kittens. J. Neurophysiol., 28: 1029–1040.

Wong, R.O.L. and Ghosh, A. (2002) Activity-dependent regulation of dendritic growth and patterning. Nat. Rev. Neurosci., 3: 803–812.

Zesiger, P., Kiper, D., Deonna, T. and Innocenti, G.M. (2002) Preserved visual function in a case of occipito-parietal microgyria. Ann. Neurol., 52: 492–498.

Zufferey, P.D., Jin, F., Nakamura, H., Tettoni, L. and Innocenti, G.M. (1999) The role of pattern vision in the development of cortico-cortical connections. Eur. J. Neurosci., 11: 2669–2688.

Zuo, Y., Yang, G., Kwon, E. and Gan, W.-B. (2005) Long-term sensory deprivation prevents dendritic spine loss in primary somatosensory cortex. Nature, 436: 261–265.

C. von Hofsten & K. Rosander (Eds.)
Progress in Brain Research, Vol. 164
ISSN 0079-6123

CHAPTER 3

The mirror-neurons system: data and models

Laila Craighero[1], Giorgio Metta[2,3], Giulio Sandini[2,3] and Luciano Fadiga[1,3,*]

[1]*D.S.B.T.A., Section of Human Physiology, University of Ferrara, Ferrara, Italy*
[2]*D.I.S.T., LiraLab, University of Genoa, Genoa, Italy*
[3]*The Italian Institute of Technology, Morego, Genoa, Italy*

Abstract: In this chapter we discuss the mirror-neurons system, a cortical network of areas that enables individuals to understand the meaning of actions performed by others through the activation of internal representations, which motorically code for the observed actions. We review evidence indicating that this capability does not depend on the amount of visual stimulation relative to the observed action, or on the sensory modality specifically addressed (visual, acoustical). Any sensorial cue that can evoke the "idea" of a meaningful action activates the vocabulary of motor representations stored in the ventral premotor cortex and, in humans, especially in Broca's area. This is true also for phonoarticulatory actions, which determine speech production. We present also a model of the mirror-neurons system and its partial implementation in a set of two experiments. The results, according to our model, show that motor information plays a significant role in the interpretation of actions and that a mirror-like representation can be developed autonomously as a result of the interaction between the individual and the environment.

Keywords: area F5; mirror-neurons system; canonical neurons; Broca's area; action recognition; speech; single neuron recordings; transcranial magnetic stimulation; brain imaging

Introduction

Since our discovery of mirror neurons we suggested that they might have a role in action recognition and understanding (Di Pellegrino et al., 1992; Gallese et al., 1996; see Rizzolatti and Craighero, 2004). The core of this proposal is the following: when an individual acts, the motor consequences of her action are known by her brain. Mirror neurons allow this knowledge to be extended to actions performed by others. Every time an individual observes an action performed by another individual, neurons that represent that action are activated in the premotor cortex. The observer "understands" someone else's actions because the evoked motor representation corresponds to that generated internally during action execution (see Rizzolatti et al., 2001).

In order to better understand how this mechanism works it is necessary to clarify the functional properties of the monkey's cortical region where mirror neurons have firstly been recorded (named F5 after Matelli et al., 1985).

Premotor mirror neurons: functional properties of monkey area F5

The monkey's area F5 is a premotor area cytoarchitectonically non-homogeneous. Indeed, its part lying on the cortical convexity, that located in the caudal bank of the arcuate sulcus and that in

*Corresponding author. E-mail: fdl@unife.it

DOI: 10.1016/S0079-6123(07)64003-5

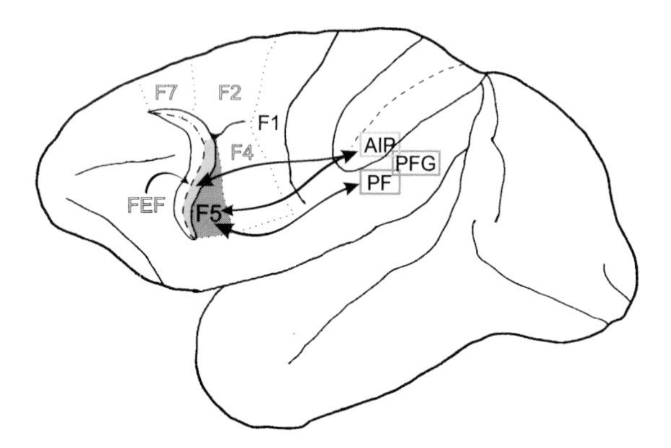

Fig. 1. Lateral view of monkey left hemisphere. Area F5 occupies the caudal bank of the arcuate sulcus (green) and the convexity immediately posterior to it (orange). Area F5 is bidirectionally connected with the inferior parietal lobule (areas AIP-anterior intra-parietal, PF and PFG). Within the frontal lobe, area F5 is connected with hand/mouth representations of primary motor cortex (area F1, labeled in bold in the figure).

the fundus of the arcuate sulcus — which defines the anterior border of area F5 — differ as far as cytoarchitectonics is concerned (Nelissen et al., 2005; Petrides, 2006). While the convexity and the caudal bank are mainly agranular, the fundus is dysgranular. The caudal bank and the convexity of area F5 differ also for their connections with the parietal lobe. While the bank is mainly connected with parietal area AIP (buried inside the intraparietal sulcus), the convexity is mainly connected with the exposed part of the inferior parietal lobule (areas PF and PFG of Barbas and Pandya, 1987; see Fig. 1).

Area F5 contains three types of neurons: motor neurons, "object observation-related" visuomotor neurons (also called canonical neurons), and "action observation-related" visuomotor neurons (mirror neurons). While motor neurons are distributed in the whole area, mirror neurons are mainly located in F5 convexity, and canonical neurons are mainly located in its bank.

Motor neurons

Motor neurons selectively discharge during execution of goal-directed hand/mouth actions (Rizzolatti et al., 1988). The specificity of the goal seems to be an essential prerequisite in activating these neurons. The same neurons that discharge during grasping, holding, tearing, and manipulating are silent when the monkey performs actions that involve a similar muscular pattern but a different goal (e.g., grasping to put away, scratching, and grooming). Further evidence in favor of such a goal-directed representation is given by F5 neurons that discharge when the monkey grasps an object with either the right, the left hand or with the mouth. This observation suggests that some F5 premotor neurons can generalize the goal of the action, independently from the effector. F5 neurons can be sub-divided into several classes on the basis of the action that triggers the neural discharge. The most common types are "grasping," "holding," "tearing," and "manipulating" neurons. Grasping neurons form the most represented class in area F5. Many of them are selective for a particular type of prehension such as precision grip, finger prehension, or whole-hand prehension. In addition, some neurons show specificity for different finger configurations, even within the same grip type. Thus, grasping a large spherical object (whole-hand prehension, requiring the opposition of all fingers) is coded by neurons different from those coding the prehension of a cylinder (also a type of whole-hand prehension but performed with the opposition of the four last fingers and the palm of the hand). Typically, F5 premotor neurons begin to discharge before the contact between the hand and the object. Some of them stop firing immediately after contact, whereas others keep firing for a while after the contact. The temporal relation between grasping movement and neuron discharge varies from neuron to neuron. A group of neurons become active during the initial phase of the movement (opening of the hand), some discharge during hand closure, and others discharge during the entire movement from the opening of the hand until their contact with the object. Taken together, the functional properties of motor F5 neurons suggest that this area stores a set of motor schemata (Arbib, 1997) or, as proposed earlier (Rizzolatti and Gentilucci, 1988), contains a "vocabulary" of motor acts. The "words" of this vocabulary are constituted by populations of neurons. Some of them indicate the general category of an action (hold, grasp, tear, and manipulate), yet others specify the effectors that are appropriate for that action. Finally, a third group is concerned with the temporal segmentation of the actions.

The motor vocabulary of actions of area F5 can also be addressed without explicit action execution. Recent experiments have shown that several F5 neurons discharge at the mere presentation of objects whose shape and size is congruent with the type of grip that is coded motorically by the same neurons (object observation visuomotor neurons) (Murata et al., 1997) or during observation of another monkey or the experimenter making a goal-directed action similar to that coded by the same neurons (action observation visuomotor neurons) (di Pellegrino et al., 1992; Gallese et al., 1996).

Object observation-related visuomotor neurons (canonical neurons)

Object observation visuomotor neurons are active when manipulating an object and when fixating the same object. These neurons discharge at the mere presentation of objects whose shape and size is congruent with the type of grasp coded motorically by the same neurons.

The visual responses of object observation-related F5 neurons have been formally studied by Murata and colleagues (Murata et al., 1997) using a behavioral paradigm, which allowed to test separately the neurons' response to object observation, to the waiting phase between object presentation and movements onset, and during movement execution. The results showed that the majority of these canonical visuomotor neurons are selective to one or at most a few specific objects. Moreover, there is a strict congruence between their visual and motor properties: Neurons that become active when the monkey observes small objects discharge also during precision grip. On the contrary, neurons selectively active when the monkey looks at a large object discharge also during actions directed toward large objects (e.g., whole-hand prehension). The most likely interpretation for the visual discharge of these visuomotor neurons is that there is a close link between the most common three-dimensional stimuli and the actions required to interact with them. Thus, every time a graspable object is presented visually, the corresponding F5 neurons are activated and the action is "automatically" evoked. Under certain circumstances this neural activity guides the execution of the movement directly; under others, it remains an unexecuted representation of the action that might be used for semantic knowledge.

Action observation-related visuomotor neurons (mirror neurons)

Action observation-related visuomotor neurons are active when manipulating an object and when watching someone else performing the same action on the same object.

To be triggered by visual stimuli, action observation visuomotor neurons require an interaction between a biologic effector (hand or mouth) and an object. The sight of the object alone, that of an agent mimicking an action, or an individual making intransitive (non-object-directed) gestures are all ineffective. The object significance for the monkey has no obvious influence on the mirror-neuron response: Grasping a piece of food or a geometric solid produces responses of the same intensity.

Mirror neurons show a large degree of generalization. Very different visual stimuli, but representing the same action, are equally effective. For example, the same mirror neuron that responds to a human hand grasping an object responds also when the grasping hand is that of a monkey. Similarly, the response is, typically, not affected if the action is done near or far from the monkey, despite the fact that the size of the observed hand is obviously different in the two conditions. It is also of little importance for neuron activation if the observed action is eventually rewarded. The discharge is of the same intensity if the experimenter grasps the food and gives it to the recorded monkey or to another monkey introduced in the experimental room.

Typically, mirror neurons show congruence between the observed and executed action. This congruence can be extremely faithful, i.e., the effective motor action (e.g., precision grip) coincides with the action that, when seen, triggers the neurons (e.g., precision grip). For other neurons, the congruence is broader and the motor requirements (e.g., precision grip) are usually stricter than the visual ones (any type of hand grasping).

The most likely interpretation for visual discharge in mirror neurons is that it evokes an internal representation of the observed action. In other terms, the observed action selects, in the F5's motor vocabulary, a congruent "motor word," a potential action.

Actions do not generate only visual consequences and in fact action-generated sound and noise are also very common in nature. One could expect, therefore, that also this sensory information, related to a particular action, can determine a motor activation specific for that same action. Kohler et al. (2002) addressed this point by investigating F5 neurons that discharge when the monkey makes a specific hand action but also when it *hears* the corresponding action-related sounds. Auditory properties of the neurons were studied by using sounds produced by the experimenter's actions and non-action-related sounds. Neurons were not activated by non-action-related sounds, while they responded specifically to the sound of an object breaking and of paper ripping, which are the hand actions more frequently executed by the monkey. Neurons were studied in an experimental design in which two hand actions were randomly presented in vision-and-sound, sound-only, vision-only, and motor conditions (monkeys performing object-directed actions). The authors (Kohler et al., 2002) found that 13% of the investigated neurons discharge both when the monkey performs a hand action and when it hears the action-related sound. Moreover, most of these neurons discharge also when the monkey observed the same action, demonstrating that these "audiovisual mirror neurons" represent actions independently of whether they are *performed*, *heard*, or *seen*.

A typical property of F5 mirror neurons is that their response is quite independent from the observer's point of sight. In other words, the same grasping action activates a given mirror neuron also if it is observed from different points of view (see Fig. 2).

How can the brain achieve an invariant description of a given action by using so different visual information? One possibility is that the system recognizes others' actions by using the same mechanisms it uses to visually control the execution of its own actions. In other terms, the point-of-view dependent visual information could be generalized by the invariance of the motor command driving action execution. To test this hypothesis, we manipulated the amount of visual information on the monkey own acting hand during grasping execution. Results showed that a significant percentage of F5 purely motor neurons are modulated by the vision of the own hand in action and that this modulation is mainly negative (less discharge) when the hand is not visible. These F5 (visuo)-motor neurons may have formed the original nucleus from which mirror neurons may have developed, possibly during the ontogenesis (Gesierich et al., in preparation). Figure 3 depicts a simplified schema of this model, which will be more formally described in the last section of this chapter.

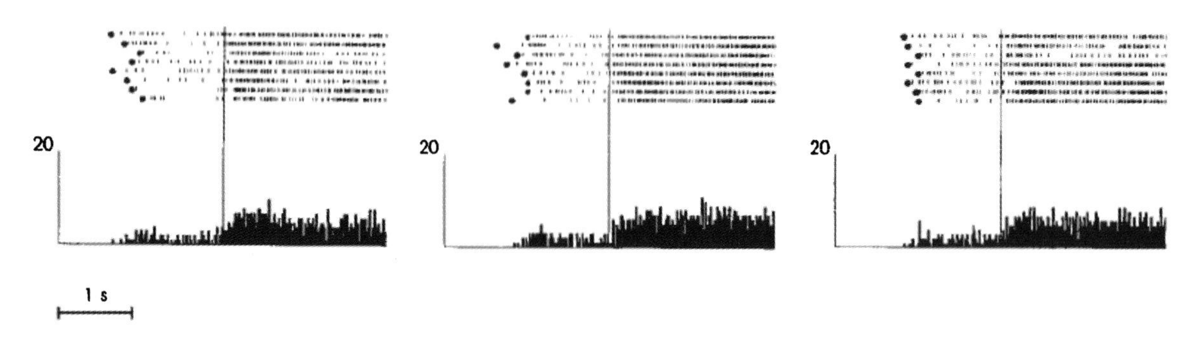

Fig. 2. A typical F5 mirror neuron discharging during the visual presentation of the same grasping movement, in the left visual hemifield (leftmost panel), centrally (central panel) or in the right visual hemifield (rightmost panel). Note the substantial equivalence of the responses. The vertical bars across rasters and histograms indicate the instant at which the experimenter touched the object. Ordinates, spikes per second. Histograms bins, 20 ms. Adapted from Gallese et al. (1996).

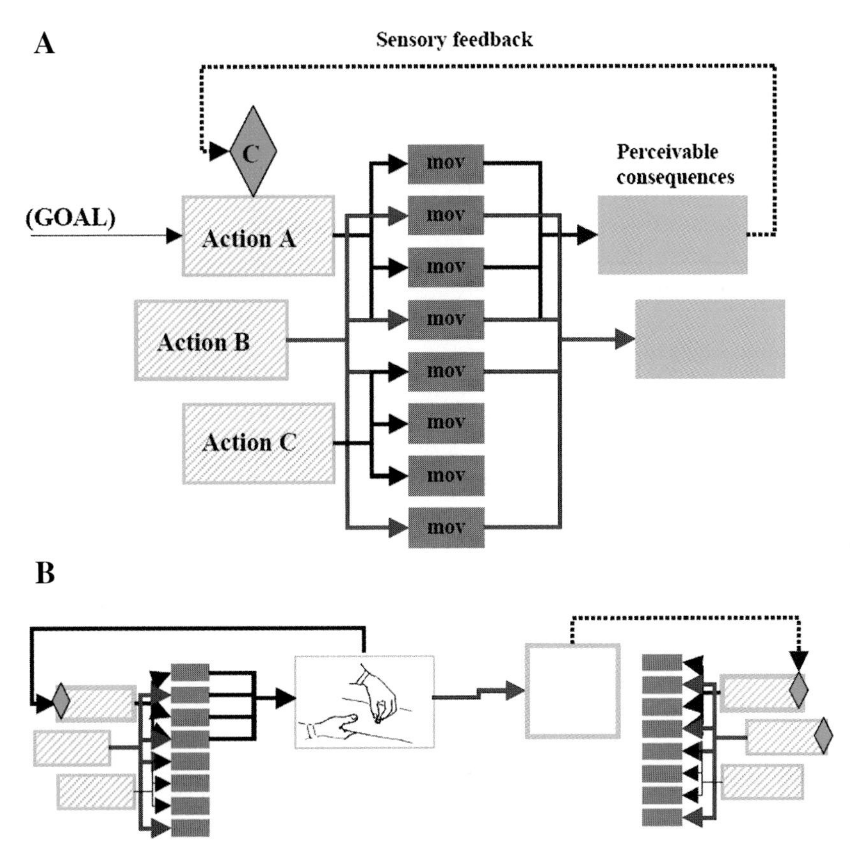

Fig. 3. (A) Simplified model of action representation. The actions level (area F5) is represented in the leftmost part of the figure. Action representations, if activated, activate in turn a set of motor synergies, here depicted in orange (mov, F1 level). Action execution does not produce consequences only on the external environment. Indeed, a series of afferent signals come back, from the periphery to the brain. These proprioceptive, visual, auditory signals (perceivable consequences, in the figure), are constantly monitored by the brain and used to control the development of the ongoing action, signaling also the goal achievement. The hypothesis we suggest is that proprioceptive and motor information, biologically invariant by definition during the actuation of a same motor command, are used by the brain to generalize (and to validate) the visual inputs related to the ongoing action. These visual inputs, that continuously vary depending on the position of the head with respect to the acting hand, are forcedly considered as homologs because they are generated by the same (or very similar) motor program. (B) Simplified model of action recognition. Two individual "brains" are shown, each one organized according to the scheme of A. When the individual on the left grasps an object her motor system receives a visual description of the ongoing movement that could be used to control its correct execution. At the same time, however, the observer's "brain" on the right sees the same scene (with some changes of perspective). Due to the visuomotor coupling she created to control her own movements through the process previously described, this visual representation of the seen action gains the access to the correspondent motor representation (following the dotted line). This is, in our view, the "recognition" played by mirror neurons.

The mirror-neuron system and action recognition: prospective or reactive mechanisms?

The functional properties of F5 neurons indicate that in primates the action representations are addressed not only for motor execution, but also during observation of graspable objects, and perception (visual, acoustical, other?) of actions performed by others. The presence of such a "vocabulary" of actions has important functional implications. Firstly, the execution of motor commands is strongly facilitated. The existence of preformed motor schemata, which are anatomically linked (hardwired) with cortical (primary motor cortex) and sub-cortical motor centers, facilitates the selection of the most appropriate combination of movements simply by addressing the general idea of an action. Thus, the number of variables that the motor system (at the premotor level) has to control to achieve the action goal is reduced. Secondly, it simplifies the association between a given stimulus (i.e., a visually presented object) and the appropriate motor response. This is the case of object observation-related visuomotor responses. Thirdly, it gives the brain a store of "ideas of action" that could be activated whenever visual or acoustic stimuli suggesting that another person is executing an action are perceived. This is the case of action perception-related sensorimotor responses.

On the basis of these functional properties that characterize not only mirror neurons but all the neurons in F5 region, it can be hypothesized that mirror neurons are at the basis of action recognition/understanding, and that this capability is not strictly dependent from the amount of stimulation perceived by the individual.

Evidence in monkeys

The hypothesis that complete visual information about the perceived action is not necessary to determine mirror neurons activation was directly tested by Umiltà and colleagues (Umiltà et al., 2001). The experimental paradigm consisted of two sessions. In the first session, the monkey was shown with a fully visible action directed toward an object or with the mimicry of the same action in the absence of the object. From previous studies it was known that mirror neurons do not discharge when the object is absent. In the second session, the monkey saw exactly the same experimental conditions but with the final part of the action hidden by a screen. Before each trial the experimenter could choose whether to place a piece of food behind the screen so that also the monkey knew whether a target for the action was present. The main result of the experiment was that several neurons discharged in the "hidden" condition, but only when the animal knew that the food was present. This evidence was interpreted as a good demonstration that mirror neurons fire also during the reaching/grasping of an object placed out of sight, as long as the intention and the plausibility of the reaching/grasping action are clear.

The conclusion is that understanding of the action is not fully based on the visual description of the scene but it refers also to the motor representation of the action goal, shared by both the agent and the observer, and triggered by the context in which the action is performed (i.e., the presence or the absence of the food on the table behind the screen).

Evidence in humans

In recent years, a series of brain imaging studies demonstrated that a mirror-neuron system is also present in the human brain. When an individual observes an action, or executes it, a network of cortical areas is activated, including the ventral premotor cortex, the inferior frontal gyrus, the inferior parietal lobule and the superior temporal cortex (see for review Rizzolatti and Craighero, 2004). Furthermore, transcranial magnetic stimulation (TMS) was used to directly investigate the involvement of the motor system in humans during observation of others' actions. TMS is an alternative technique to the single neuron recordings that can be used in humans to obtain good temporal resolution. Single or paired-pulse TMS allows to measure cortical excitability during different phases of an observed action. Moreover, this technique can help to verify the specific involvement of the motor system by discriminating the muscles that are involved in the motor replica. A series of TMS experiments showed that also in humans the mirror system is not strictly dependent on the visual stimulation but it is active whenever a motor representation is addressed (Gangitano et al., 2004; Borroni et al., 2005).

Gangitano and colleagues (Gangitano et al., 2001) in a TMS experiment evoked motor-evoked potentials (MEPs) in the first dorsal interosseus muscle at different time intervals, while subjects were watching a video clip of a hand approaching and grasping a ball and demonstrated that the specific activation of the observer's muscles is temporally coupled to the dynamics of the observed action. In a further experiment Gangitano et al.

(2004) investigated whether this pattern of modulation was the consequence of a "resonant plan" evoked at the beginning of the observation phase or whether the plan was fractioned in different phases sequentially recruited during the course of the ongoing action. The authors therefore used the same procedure as in Gangitano et al. (2001) with the following exception: Subjects were shown video clips representing an unnatural movement, in which the temporal coupling between reaching and grasping components was disrupted, either by changing the time of appearance of maximal finger aperture, or by substituting it with an unpredictable closure. In the first case, the observation of the uncommon movements did not exert any modulation in motor excitability. In the second case, the modulation was limited to the first time of stimulation. Modulation of motor excitability was clearly suppressed by the appearance of the sudden finger closure and was not substituted by any other pattern of modulation. This finding suggests that a motor plan, which includes the temporal features of the natural movement, is activated immediately after the observed movement onset and is discarded when these features cease to match the visual properties of the observed movement. Thus, the human mirror system seems to be able to infer the goal and the probability of an action during the development of its ongoing features.

Recently, Borroni et al. (2005) aimed at verifying the degree of correspondence, especially with respect to a fine temporal resolution, between the observation of prolonged movements and its modulatory effects in the observer. For this purpose the authors asked subjects to watch a cyclic flexion-extension movement of the wrist. The same sinusoidal function was used to fit both observed wrist oscillation and motor resonance effects on the observer's wrist motor circuits. In this way the authors could describe a continuous time course of the two events and precisely determine their phase relation. MEPs were elicited in the right forearm muscle of subjects who were observing a 1 Hz cyclic oscillation of the right hand executed by another person. The results indicated that movement observation elicited a parallel cyclic excitability modulation of the observer's MEP responses following the same period as the observed movement. Interestingly, the MEP modulation preceded the observed movement, being related to time course of muscular activation of the demonstrator and not to the visually perceived movement. This finding indicates that the mirror-neuron system anticipates the movement execution, rather than simply reacting to it.

Thus, the involvement of observer's motor system is not necessarily consequent to the explicit visual description of the complete action but, rather, it may intervene in filling gaps because it gives to the observer an implicit motor knowledge about the observed action. In other words, the mirror system seems to possess the capability to predict the action outcome.

In this line are the data by Kilner and colleagues (Kilner et al., 2004). In an event-related potentials experiment, these authors showed that the readiness potential (RP), a well-known electrophysiological marker of motor preparation, is detectable also during action observation. Furthermore, when the upcoming action is predictable, the rise of the RP precedes the observed movement onset. They recorded electroencephalograms from subjects while they watched a series of short video clips showing an actor's right hand and a colored object. In half of the videos the hand moved, while in the other half it remained stationary. At the beginning of each video the color of the object indicated whether the hand would subsequently move or not. Thus, the observed movements were entirely predictable from the color of object in the video. The results revealed a significant negative gradient that started 500 ms before the onset of the observed hand movement. This activity was comparable with the onset of the movement-related RP produced when subjects actually executed a movement. These results suggest an active role of the mirror system in setting up an anticipatory model of another person's action, endowing our brain with the ability to predict his or her intentions ahead of their realization.

All the reported experiments, however, investigated the involvement of the mirror system during observation of real hands executing goal directed actions. In a very recent experiment our group (Fadiga et al., 2006) wanted to verify if the vision of a real hand is a necessary prerequisite to activate the mirror system, or if any cue suggesting the presence of a hand performing meaningful movements is a sufficient stimulus. To this purpose we submitted human subjects to an fMRI scanning while they were observing a particular category of hand gestures: Hand shadows representing animals opening their mouths. Hand shadows only implicitly "contain" the hand creating them (i.e., hands are not visible but subjects are aware of the fact that the presented animals are produced by the hands). Therefore, they are interesting stimuli that might be used to answer the question of how much and what details of a hand gesture activate the mirror-neuron system. During the fMRI scan, healthy volunteers ($n = 10$) observed videos representing (i) the shadows of human hands depicting animals opening and closing their mouths, (ii) human hands executing sequences of meaningless finger movements, or (iii) real animals opening their mouths. Each condition was contrasted with a "static" condition, where the same stimuli presented in the movie were shown as static pictures (e.g., stills of animals presented for the same amount of time as the corresponding videos). In addition, to emphasize the action component of the gesture, brain activations were further compared between pairs of conditions in a block design.

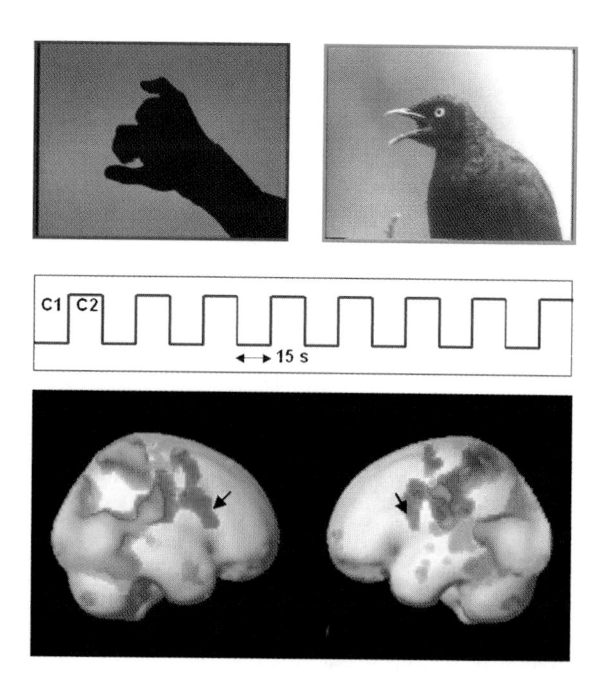

Fig. 4. Cortical activation pattern during observation of animal hand shadows and real animals. Significantly activated voxels ($P<0.001$) in the moving animal shadows (red clusters) and moving real animals (green clusters) conditions after subtraction of the static controls. In the middle part of the figure, the experimental time-course for each contrast is shown (i.e., C1, moving; C2, static). Note the almost complete absence of frontal activation for real animals in comparison to animal shadows, which bilaterally activate the inferior frontal gyrus (arrows). Adapted from Fadiga et al. (2006).

Figure 4 shows, superimposed, the results of the moving vs. static contrasts for animal hand shadows and real animals conditions (red and green spots, respectively). In addition to largely overlapping occipito-parietal activations, a specific differential activation emerged in the anterior part of the brain. Animal hand shadows strongly activated left parietal cortex, pre- and post-central gyri (bilaterally), and bilateral inferior frontal gyrus (BA 44 and 45). Conversely, the only frontal activation reaching significance in the moving vs. static contrast for real animals was located in bilateral BA 6, close to the premotor activation shown in an fMRI experiment by Buccino and colleagues (Buccino et al., 2004) where subjects observed mouth actions performed by monkeys and dogs. This location may therefore correspond to a premotor region where a mirror-neuron system for mouth actions is present in humans. The results shown in Fig. 4 indicate that the shadows of animals opening their mouths, although clearly depicting animals and not hands, convey implicit information about the human being moving her hand in creating them. Indeed, they evoke an activation pattern, which can be superimposed to that evoked by hand action observation (Grafton et al., 1996; Buccino et al., 2001; Grezes et al., 2003). Thus, the results demonstrate that the mirror-neuron system becomes active even if the pictorial details of the moving hand are not explicitly visible: In the case of our stimuli, the brain "sees" the performing hand also behind the visual appearance. Consequently, the human mirror system (or at least part of it) can be seen more as an active interpreter than as a passive perceiver (or resonator).

The possibility to be an active interpreter is based on the knowledge of the context in which the observed action is performed. Different cues coming from the environment activate the representation of the most probable action: in the Umiltà experiment (Umiltà et al., 2001) the knowledge of the presence of the food behind the screen gives plausibility to the hidden action; in the Kilner experiment (Kilner et al., 2004) the color of the object at the beginning of the trial prompts the belief that a grasping towards that object is going to be executed; in the Fadiga experiment (Fadiga et al., 2006) the old memory of playing with shadows on the wall during childhood links the observed animal shadows with hand movements. Recently, we tested the possibility that the mirror-neurons system could be modulated by the canonical neurons activation determined by the vision of the to-be-grasped object (Craighero et al., in press). To this purpose we asked subject to detect the instant at which the demonstrator's hand touched the object. Two different types of grasping on the same object were presented, differing for the type of fingers opposition space: In one case the type of grasping was the one more commonly chosen to grasp the presented object, in the other case it was a less appropriate one. This experimental manipulation created a situation of conflict in terms of motor representations determining two main conditions: A congruent one, in which the motor program evoked by object observation coincides with that executed by the experimenter, and an incongruent one, where the two motor programs differ. Our results showed that subjects' response times are well below those commonly found in simple reaction times tasks (usually around 120–150 ms), indicating that, to accomplish the task, subjects indeed use a predictive model of the seen action. Moreover, response times were shorter for suitable grasping trials than for not suitable ones. This indicates that action prediction is based on the internal motor representation of the seen action, and that whenever incongruence is present between the action evoked in the observer by the to-be-grasped object and the observed action, actually executed on it, the ability to predict the action outcome decreases.

Considering both the functional properties of the neurons of the ventral premotor cortex of the monkey and those of the human mirror-neurons system, as described by TMS and brain imaging experiments, we can argue that the ventral premotor cortex is automatically activated whenever the "idea" of an action is even suggested. This suggestion can derive from the sight of a graspable object, as in the case of canonical neurons (Murata et al., 1997), from the visual (di Pellegrino et al., 1992) or acoustical (Kohler et al., 2002) perception of a transitive action performed by another individual, as in the case of mirror neurons in monkeys and of mirror-neurons system in humans (see Rizzolatti and Craighero, 2004). Moreover, experimental evidence indicates that the idea of an action does not necessarily require a complete perceptual stimulation to be elicited (Umiltà et al., 2001; Fadiga et al., 2006).

In conclusion, the properties of the mirror-neurons system not only are in favor of its role in action understanding, but clearly suggest that mirror neurons are fundamental in interpreting others' intentions and in anticipating the outcome of others people's actions, providing a key mechanism to successfully interact in a social environment.

Broca's area is the core center of the human mirror-neurons system

As discussed in the previous section, experimental evidence demonstrates that a mirror-neurons system is also present in the human brain. The first evidence of the existence of a mirror-like visuomotor activation in the human brain has been provided by Fadiga et al. (1995) by a TMS experiment. The motor cortex of normal human participants was magnetically stimulated and MEPs were recorded from intrinsic and extrinsic hand muscles. It was reasoned that, if the observation of a hand movement activates the premotor cortex, this should, in turn, induce an enhancement of MEPs elicited by the magnetic stimulation of the hand representation of the motor cortex. The results confirmed this hypothesis showing a pattern of muscle facilitation during action observation that strictly resembles that occurring during the actual execution of the observed movements. In other words, looking at a hand closing onto an object evokes a facilitation of the observer's flexors muscles. Strafella and Paus (2000), by using the double stimulus TMS technique, demonstrated the cortical origin of this facilitation. They showed that the interstimulus interval between two close stimulations, which evoked the larger motor facilitation during action observation, was compatible with cortico-cortical facilitating connections.

Further evidence that cortical motor areas are activated during movement observation comes from MEG experiments. Hari and colleagues (Hari et al., 1998) recorded neuromagnetic oscillatory activity of the human precentral cortex elicited by median nerve stimulation in healthy volunteers during rest (i), manipulation of a small object with their right hand (ii), and observation of another individual performing the same task (iii). The cortical 15–25 Hz rhythmical activity was measured. In agreement with previous data (Salmelin and Hari, 1994), this activity was suppressed during movement execution. Most interestingly, the rhythm was also significantly diminished during movement observation. Control experiments confirmed the specificity of the suppression effect. Because the recorded 15–25 Hz activity originates mostly in the anterior bank of the central sulcus, it appears that the human primary motor cortex desynchronizes (and therefore becomes more active) during movement observation in the absence of any active movement. Similar results were obtained also by Cochin and colleagues (Cochin et al., 1998), who recorded EEG from subjects observing videos where human movements were displayed. As a control, moving objects, moving animals, and still objects were presented. The data showed that the observation of human movements, but not that of objects or animals, desynchronizes the EEG pattern of the precentral cortex.

A series of brain imaging experiments were carried out in order to assess which cortical area could be the homolog of the monkey F5 mirror system. Hand grasping movements (Grafton et al., 1996; Rizzolatti et al., 1996) as well as, more recently, more complex hand/arm movements were used as visual stimuli (Decety et al., 1997; Grezes et al., 1998). The results of the first experiments showed that during the observation of hand grasping, among the activation of other areas, there was an activation of the left inferior frontal cortex, in correspondence of the Broca's region, a region historically known to be involved in language production (Broca, 1861). In studies carried out by the Lyon group (Decety et al., 1997; Grezes et al., 1998) the involvement of Broca's area during observation of hand/arm actions was further confirmed. The authors instructed subjects to observe meaningful (with a goal) and meaningless movements. The main result when subjects observed meaningless arm movements was the bilateral activation of the parietal lobe, the activation of the left precentral gyrus and that of the right side of the cerebellum (Grezes et al., 1998). On the contrary, the observation of meaningful hand actions, in addition to the already mentioned frontal and parietal areas, activated the left inferior frontal gyrus (Broca's region). More recently, two additional studies have shown that a meaningful hand-object interaction, more than pure movement observation, is effective in triggering Broca's area activation (Hamzei et al., 2003; Johnson-Frey et al., 2003). Similar conclusions have been reached also for the observation of mouth movements (Campbell et al., 2001). These results, together with comparative cytoarchitectonical data (see Petrides and Pandya, 1994; Nelissen et al., 2005; Petrides, 2006), and fMRI data from Binkofsky and colleagues (Binkofski et al., 1999) demonstrating that Broca's region becomes active also during manipulation of complex objects, suggest that Broca's region has the putative role of human homolog of area F5 in the monkey.

Mirror-neurons system and speech recognition

The presence of an audio-motor resonance (Kohler et al., 2002) in a region that, in humans, is classically considered a speech-related area, immediately evokes the Liberman's motor theory of speech perception (Liberman et al., 1967; Liberman and Mattingly, 1985; Liberman and Whalen, 2000). This theory maintains that the ultimate constituents of speech are not sounds but articulatory gestures that have evolved exclusively at the service of language. Consequently, speech perception and speech production processes can share a common repertoire of motor primitives that, during speech production, are at the basis of the generation of articulatory gestures, and during speech perception are activated in the listener as the result of an acoustically evoked motor "resonance." According to Liberman's theory, the listener understands the speaker when his/her articulatory gestural representations are activated by listening to verbal sounds. Although this theory is not unanimously accepted, it offers a plausible model of an action/perception cycle in the frame of speech processing.

To investigate if speech listening activates listener's motor representations, our group (Fadiga et al., 2002), in a TMS experiment, tested for the presence in humans of a system that motorically "resonates" when an individual listen to verbal stimuli. Healthy subjects were requested to attend to an acoustically presented randomized sequence of disyllabic words, disyllabic pseudowords and bitonal sounds of equivalent intensity and duration. Words and pseudowords were selected according to a consonant-vowel-consonant-consonant-vowel (cvccv) scheme. The embedded consonants in the middle of words and of pseudowords were either a double "f" (labiodental fricative consonant that, when pronounced, requires slight tongue tip mobilization) or a double "r" (lingua-palatal fricative consonant that, when pronounced, requires strong tongue tip mobilization). Bitonal sounds, lasting about the same time as verbal stimuli and replicating their intonation pattern, were used as a control. The excitability of the motor cortex in correspondence of the representation of tongue movements was assessed by using single pulse TMS and by recording MEPs from the anterior tongue muscles. The TMS stimuli were applied synchronously with the double consonant of the presented verbal stimuli (words and pseudowords) and in the middle of the bitonal sounds. Results showed that during speech listening there is an increase of the MEPs recorded from the listeners' tongue muscles when the word strongly involves tongue movements. This indicates that when an individual listens to verbal stimuli his/her speech-related motor centers are specifically activated. Moreover, words-related facilitation was significantly larger than pseudowords related one. These results indicate that the passive listening to words that would involve tongue mobilization (when pronounced) induces an automatic facilitation of the listener's motor cortex. Furthermore, the effect is stronger in the case of words than in the case of pseudowords suggesting a possible unspecific facilitation of the motor speech center due to recognition that the presented material belongs to an extant word.

Similar results were obtained by Watkins and colleagues (Watkins et al., 2003). By using TMS technique they recorded MEPs from a lip (*orbicularis oris*) and a hand muscle (first dorsal *interosseus*) in four conditions: listening to continuous prose, listening to non-verbal sounds, viewing speech-related lip movements, and viewing eye and brow movements. Compared to control conditions, listening to speech enhanced the MEPs recorded from the *orbicularis oris* muscle. This increase was observed only in response to the stimulation of the left hemisphere. No changes of the MEPs in any condition were observed following the stimulation of the right hemisphere. Finally, the size of MEPs elicited in the first *interosseus* muscle did not differ in any condition.

Taken together these experiments show that when an individual listen to verbal stimuli there is an activation of the speech-related motor centers. It is however unclear if this activation could be interpreted in terms of an involvement of motor representations in speech processing and, perhaps, in perception.

In order to investigate the perceptual role of Broca's area and considering that this area has been classically considered specifically involved in phonological processing (at least in production) we decided to use a phonological paradigm in a new experiment: the "phonological priming" task. Phonological priming effect refers to the fact that a target word is recognized faster when it is preceded by a prime word, sharing with it the last syllable (rhyming effect, Emmorey, 1989). In a single pulse TMS experiment (see Fadiga et al., in press) we therefore stimulated participants' inferior frontal cortex while they were performing a phonological priming task. Subjects were instructed to carefully listen to a sequence of acoustically presented pairs of verbal stimuli (dysillabic "cvcv" or "cvccv" words and pseudowords) in which the final phonological overlap was present (rhyme prime) or, conversely, not present. The task of the subjects was to make a lexical decision on the second stimulus (target) by pressing with either the index or the middle finger one of two buttons whether the target was a word or a pseudoword. The pairs of verbal stimuli belonged to four categories, which differed for their lexical content in the prime and in the target (prime-word/target-word (W–W), prime-word/target-pseudoword (W–PW), prime-pseudoword/target-word (PW–W), prime-pseudoword/target-pseudoword (PW–PW)). Each category contained both rhyming and non-rhyming pairs. In some randomly selected trials, we administered single pulse TMS in correspondence of left BA44 (Broca's region, localized by using "Neurocompass," a frameless stereotactic system built in our laboratory) during the interval (20 ms) between prime and target stimuli.

In trials without TMS, there were three main results (Fig. 5): (i) strong and statistically significant facilitation (phonological priming effect) when W–W, W–PW, PW–W pairs are presented; (ii) no phonological priming effect when the PW–PW pair is presented; (iii) faster responses when the target is a word rather than a pseudoword (both in W–W and PW–W).

An interesting finding emerges from the analysis of these results: The presence or absence of lexical content modulates the phonological priming effect. When neither the target nor the prime has access to the lexicon (PW–PW pair) the presence of the rhyme does not facilitate the recognition of the target. In other words, in order to have a phonological effect it is necessary to have access to the lexicon.

In trials with TMS delivery, only W–PW pairs were affected by brain stimulation, the W–PW pair behaving exactly as the PW–PW one. This finding suggests that the stimulation of the Broca's region might have affected the lexical property of the prime (i.e., the meaningfulness of the stimulus). As consequence, the impossibility to access the lexicon determines the absence of the phonological effect. According to our interpretation, the TMS-related effect is absent in the W–W and PW–W pairs because of the presence of a meaningful (W) target. The finding that TMS administered on Broca's region during phonological priming paradigm influences the rhyming effect only in the case of W–PW pairs poses a theoretical problem. In our previous TMS experiment on motor facilitation during speech listening (Fadiga et al., 2002) we have found cortical facilitation during listening of both words and pseudowords. This discrepancy suggests that a cortical area different from Broca's one should be involved in such a "low level" motor resonance. Its localization will be the argument of our future experimental work.

By summarizing the experimental evidence we presented here, we can claim that the activation of

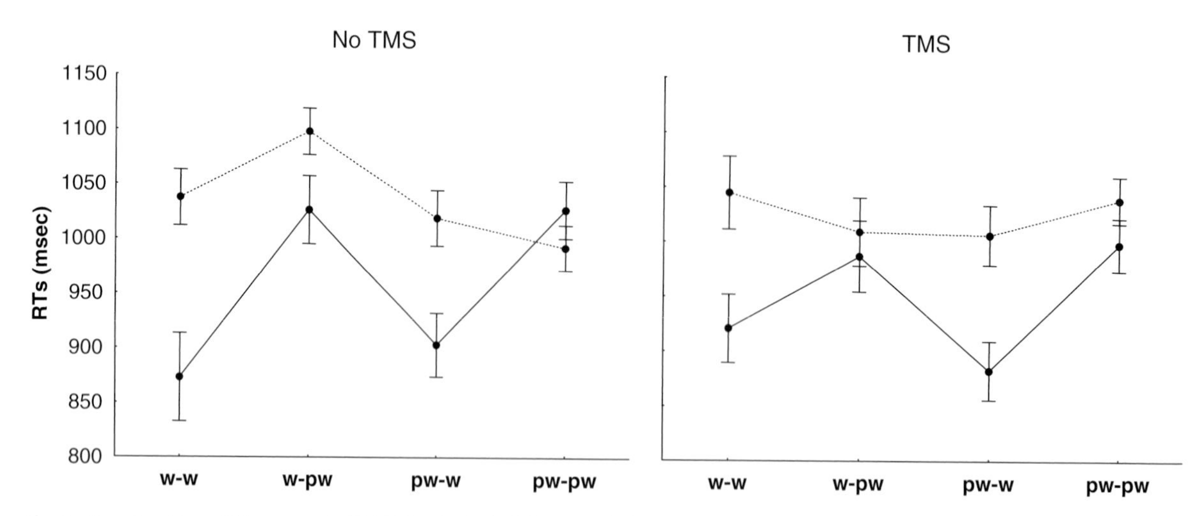

Fig. 5. Reaction times (RTs, msec ± SEM) for the lexical decision during the phonological priming task with and without transcranial magnetic stimulation (TMS) between prime and target. Solid line: conditions in which prime and target share a rhyme. Dashed line: no rhyme. W–W, prime-word/target-word; W–PW, prime-word/target-pseudoword; PW–W, prime-pseudoword/target-word; PW–PW, prime-pseudoword/target-pseudoword.

Broca's region during speech processing, more than indicating a specific role of this area, may reflect its general involvement in meaningful action recognition. This possibility is based on the observation that, in addition to speech-related activation, this area is activated during observation of meaningful hand or mouth actions. Speech represents a particular case of this general framework: among meaningful actions, phonoarticulatory gestures are meaningful actions conveying words. The consideration that Broca's area is the human homolog of the monkey mirror neurons area opens the possibility that human language may have evolved from an ancient ability to recognize visually or acoustically perceived actions performed by others (Rizzolatti and Arbib, 1998).

The motor representation of hand/mouth actions present in Broca's area, which derives from the execution/observation (hearing) matching system already present in monkeys, may have given to this area the capability to deal with verbal communication because of its twofold involvement with motor goals: during execution of own actions and during perception of others' ones. Our hypothesis is that the original role played by this region in generating/extracting action meanings might have been generalized during evolution giving to this area the capability to deal with meanings (and rules) sharing with the motor system similar hierarchical and sequential structures. Recent data from our laboratory on frontal aphasic patients are in line with this idea (see Fadiga et al., in press).

A model of area F5 and of the mirror-neurons system

This section proposes a model of the mirror-neurons system, whose components are in general agreement with the functional properties of area F5 and with the knowledge on the connections that this area maintains with other cortical regions, which describes how the mirror-neurons system intervenes in action recognition (Metta et al., 2006).

It is known that F5 is part of a larger circuit comprising various areas in the parietal lobe (a large reciprocal connection with anterior intraparietal area, AIP), indirectly from superior temporal sulcus (STS), and other premotor and frontal areas. Moreover, it is strongly involved in the generation and control of action indirectly through primary motor cortex (F1), and directly by projecting to motor and medullar interneurons in the spinal cord (see Luppino and Rizzolatti, 2000).

Our model of area F5 revolves around two concepts that are certainly related to the evolution and development of this unique area of the brain. Firstly, we posit that the mirror-neurons system did not appear brand new in the brain but likely evolved from a pre-existing structure devoted solely to the control of grasping action. The reasons for this claim are to be found in the large percentage of motor neurons in F5 (70%) compared to those that have also visual responses. Secondly, we attribute a fundamental role to canonical neurons — and in general that of contextual information specifying the action goal — in the development of the mirror neurons. Since purely motor, canonical, and mirror neurons are found together in F5, it is very plausible that local connections determine at least in part the activation of F5.

Our model follows a forward-inverse approach that has been also proposed in computational motor control theory (Kawato et al., 1987; Wolpert and Miall, 1996) and for explanatory purpose it can be divided into two parts. The first part describes what happens in the actor's brain, the second what happens in the observer's brain when watching another acting individual. As we will see the same structures are used both when acting and when observing an action.

The agent's point of view

We shall consider first what happens from the actor's point of view (see Fig. 6). In her perspective, decision to undertake a particular action is attained by the convergence in area F5 of many factors including the contextual- (by signals from parietal and frontal areas) and object-related information (canonical neurons). Object and context bias the activation of a specific motor plan, which specifies the goal of the motor system in motor

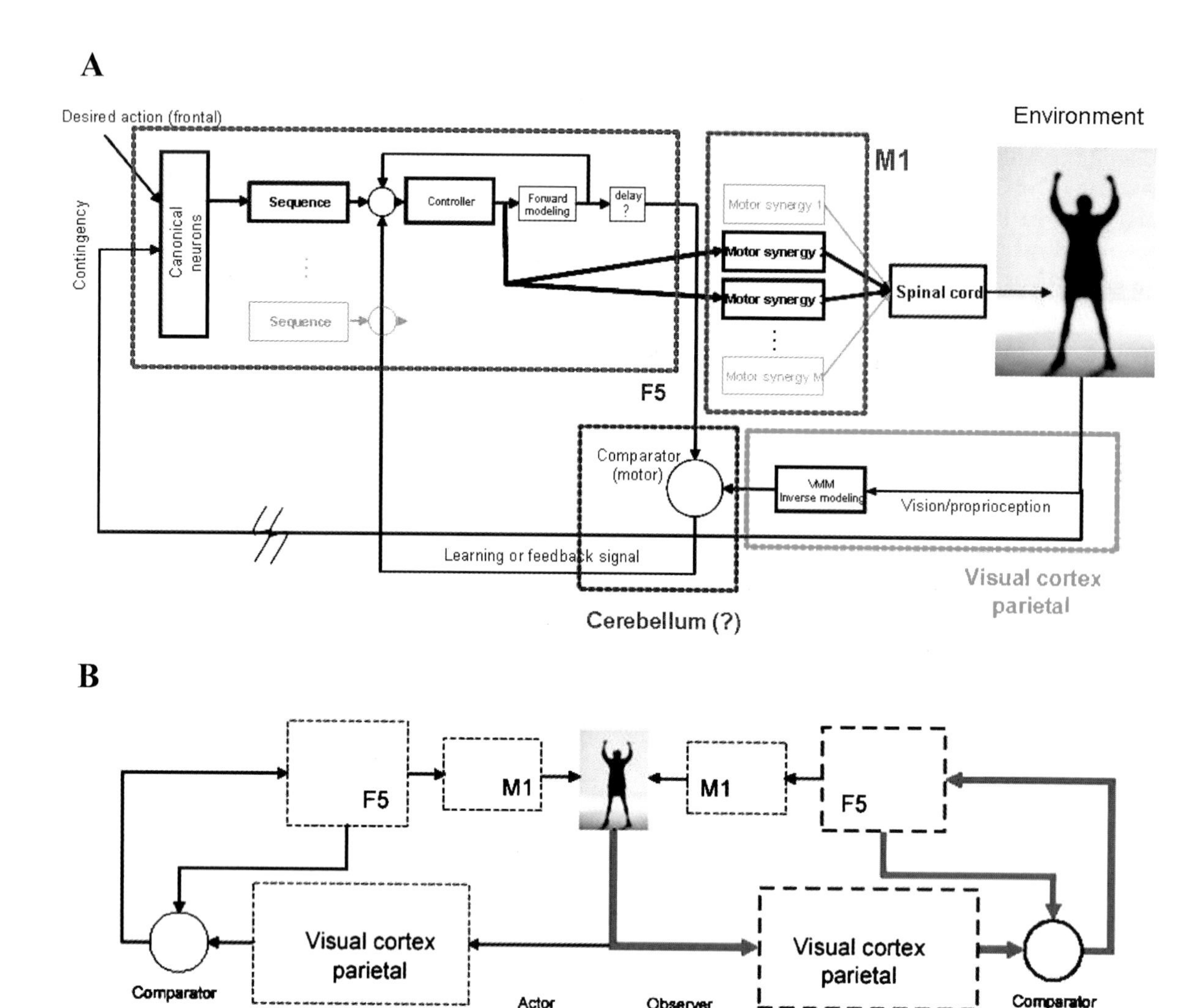

Fig. 6. Model schematics (see text for explanations). (A) Organization of the motor system during execution. (B) The same model working during action observation.

terms and, we generally suppose, it includes temporal information. Our model hypothesize that action specification is initially "described" in frontal areas in some internal reference frame and then transformed into the motor plan by an appropriate controller in F5.

The action plan unfolds mostly open loop (i.e., without employing feedback). A form of feedback (closed loop) is required though to counteract disturbances and to learn from mistakes. This is obtained by relying on a forward or direct model that predicts the outcome of the action as it unfolds in real time. The output of the forward model can be compared with the signals derived from sensory feedback, and differences accounted for (the cerebellum is believed to have a role in this) (Miall et al., 1993; Wolpert and Miall, 1996). A delay module is included to take into account the different propagation times of the neural pathways carrying the predicted and actual outcome of the action. Note that the forward model is relatively simple, predicting only the motor output in advance: Since motor commands are generated internally it is relatively easy to imagine a predictor for this signals.

The inverse model (Visuo-Motor Map, VMM) is much more complicated since it maps sensory feedback (vision mainly) back into motor terms. Visual feedback clearly includes both the hand-related information (STS) and the object information (AIP, IT, and canonical neurons). Finally the predicted and the sensory signals arising from action execution are compared and the feedback error sent back to the controller. The mismatch between the planned and actual action can either be used to compensate on the fly by means of a feedback controller, or to adjust over longer periods of time through learning (Kawato et al., 1987).

The output of area F5, finally activates the motor neurons in the spinal cord (directly or indirectly through motor synergies) to produce the action. This is indicated by a connection to appropriate muscular synergies representing the spinal cord circuits.

Learning of the direct and inverse models can be carried out during ontogenesis by a procedure of self-observation and exploration of the state space of the system: grossly speaking, simply by "detecting" the sensorial consequences of motor commands.

Learning of the affordances of objects (the canonical neurons response) with respect to grasping can also be achieved autonomously by a trial and error procedure, which explores the consequences of trying many different actions of the agent's motor repertoire (different grasp types) to different objects. This includes things such as discovering that small objects are optimally grasped by a pinch or precision grip, while big and heavy objects require a power grasp.

The observer's point of view

In the observer situation motor and proprioceptive information is not directly available. The only readily available information is vision or sound. The central assumption of our model is that the structure of F5 could be co-opted in recognizing the observed actions by transforming visual cues into motor information as before. In practice the inverse model is accessed by visual information, since the observer is not acting herself, visual information directly reaches in parallel the sensorimotor primitives in F5. Only some of them are actually activated because of the "filtering" effect of the canonical neurons and of other contextual information (possibly at a higher level, knowledge of the actor, plausibility of the hand posture, etc.). This procedure could be used then to recognize the action by measuring the most active motor primitive. Thus, in our model, many factors, including the affordances of the target object, determine the recognition and interpretation of the observed action.

Ontogenesis of mirror neurons

Our model gives us also the possibility to hypothesize the ontogenesis of mirror neurons. First of all, the inverse model, the VMM, can be learned through a procedure of self-exploration. Motor commands and correlated visual information are readily available to the developing infant. It is easy to imagine a procedure that learns the inverse model on the basis of this information.

On the top of the VMM, it is plausible that the canonical representation is acquired via the manipulation of a large set of different objects. F5 canonical neurons represent an association between objects' physical properties and the actions they afford: e.g., a small object affords a precision grip, or a coffee mug affords being grasped by the handle. The understanding of object properties and the goal of actions is subsequently fundamental for disambiguating visual information. In fact, certain actions are more likely to be applied to a particular object, and certain objects are more likely to be used in certain actions. A link between action and effects can be traced backward on the basis of our experience: To obtain the effects we have to apply the same action that earlier led to those effects.

Bearing this is mind, when observing some other individual's actions; our understanding can be framed in the same terms: When I see someone grasping a mug full of coffee by the handle, I know that that precise affordance is the most suitable for drinking. If the same mug is empty and I see the agent grasping it by inserting the fingers into it, I can hypothesize that she is going to wash it. Consequently, I can recognize and also predict the outcome of the action on the basis of the link between the contextual information, the type of affordance, and the knowledge of object's properties. It is plausible that a mirror representation derives from the association between the visual information of others' actions and the action-effects link. To obtain this association, however, it is necessary that the observed consequences of an action are recognized as similar in the self or the other individual acting. Only if this happens, the association between the canonical response and the mirror one can then be made.

Role of motor information in the action recognition process

The simplest way of confirming the hypothesis that motor representations are the basis of action recognition is to equip a computer with means of "acting" on objects, collect visual and motor data and build a recognition system that embeds some of the principles of operation that we identified in our model (see Fig. 6). In particular, the hypothesis we would like to test is whether the extra information available during learning (e.g., kinesthetic and tactile) can improve and simplify the recognition of the same actions when they are just observed: i.e., when only visual information is available. Given the current limitations of robotic systems the simplest way to provide "motor awareness" to a machine is by recording grasping actions of human subjects from multiple sources of information including joint angles, spatial position of the hand/fingers, vision, and touch.

For this purpose we assembled a computerized system composed of a data glove (CyberGlove by Immersion), a pair of CCD cameras (Watek 202D), a magnetic tracker (Flock of bird, Ascension), and two touch sensors (FSR). Data was sampled at frame rate, synchronized, and stored to disk by a Pentium class PC. The cyber glove has 22 sensors and allows recording the kinematics of the hand at up to 112 Hz. The tracker was mounted on the wrist and provides the position and the orientation of the hand in space with respect to a base frame. The two touch sensors were mounted on the thumb and index finger to detect the moment of contact with the object. Cameras were mounted at appropriate distance with respect to their focal length to acquire the execution of the whole grasping action with maximum possible resolution.

The glove was lightweight and did not limit any way the movement of the arm and hand. Data recording was carried out with the subject sitting comfortably in front of a table and performing grasping actions naturally toward objects approximately at the center of the table. Data recording and storage was carried out through a custom-designed application; Matlab was employed for post-processing.

We collected a large data set and processing was then performed off-line. The selected grasping types were: power grasp-cylindrical, power grasp-spherical, and precision grasp. Since the goal was to investigate to what extent the system could learn invariances across different grasping types by employing motor information for classification, the experiment included gathering data from a multiplicity of viewpoints. The database contained objects, which afford several grasp types to assure that recognition cannot simply rely on exclusively extracting object features. Rather, according to our model, this is supposed to be a confluence of object recognition (canonical neurons) with hand visual analysis (STS).

A set of three objects was employed: a small glass ball, a parallelepiped, and a large sphere. Each grasping action was recorded from six different subjects (right handed, age 23–29, male/female equally distributed), and moving the cameras to 12 different locations around the subject including two different elevations with respect to the table top which amounts to 168 sequences per subject. Each sequence contained images of the scene from the two cameras synchronized with the cyber glove and the magnetic tracker data.

The visual features were extracted from preprocessed image data. The hand was segmented from the images through a simple color segmentation algorithm. The bounding box of the segmented region was then used as a reference frame to map the view of the hand to a standard size. The orientation of the color blob in the image was also used to rotate the hand to a standard orientation. This data set was then filtered through Principal

Component Analysis (PCA) by maintaining only a limited set of eigenvectors corresponding to the first 2 to 15 largest eigenvalues.

One possibility to test the influence of motor information in learning action recognition is to contrast the situation where motor-kinesthetic information is available in addition to visual information with the control situation where only visual information is available.

In the "motor space" session we used the output of the VMM (see the model schematics) and thus employed motor features for classification. The VMM was approximated from data by using a simple feedforward neural network with sigmoidal units trained with backpropagation. The input of the VMM was the vector of the mapping of the images onto the space spanned by the first N PCA vectors; the output was the vector of joint angles acquired from the data glove. In the "visual space" session the classification was performed in visual space directly.

Classification was always performed by training a Bayesian classifier. In this formulation we identified the likelihood term with the activity of F5 motor neurons (probability of seeing certain features given the performed action and target object) and the priors with the canonical neurons (probability of using a certain action on a given object). The classifier then applies a MAP criterion (maximum a posteriori) by computing the unnormalized posterior probability and taking the maximum over the possible actions.

The results of the experiment are reported in Table 1.

Different sequences were used during the training and the testing phases. During the training phase, 24 sequences from only one point of view were used in the motor space session, while 64 sequences from all the four different points of view were used in the visual space session. Thus, the classifier was trained with the maximum available data only in the latter session. During the testing phase, 96 sequences from four points of view were used in the motor space session, and 32 sequences from four points of view in the visual space one.

The clearest result of this experiment is that the classification in motor space is easier and thus the classifier performs better on the test set. Moreover,

Table 1. Summary of the results of the experiment testing the model

	Motor space session	Visual space session
Training		
No. of sequences	24 (+VMM)	64
No. of points of view	1	4
Classification rate (on the training set) (%)	98	97
Test		
No. of sequences	96	32
No. of points of view	4	4
Classification rate (%)	97	80

the distribution of the data is more "regular" (the likelihood term is simpler) in motor space than in visual space. This is to be expected since the variation of the visual appearance of the hand is larger and depends strongly on the point of view, while the sequence of joint angles tends to be the same across repetitions of the same action. It is also clear that in the experiment the classifier is much less concerned with the variation of the data since this variation has been taken out by the VMM.

Overall, our interpretation of these results is that by mapping in motor space first we are allowing the classifier to choose features that are much better suited for performing optimally, which in turn facilitates generalization. The same is not true in visual space.

The interaction with the environment and the development of mirror-like representation

In order to show that, according to our model, a mirror-neuron-like representation could be acquired by simply relying on the information exchanged during the interaction with the environment, we set forth to the implementation of a complete experiment on a humanoid robot called Cog (Brooks et al., 1999). This was an upper-torso human shaped robot with 22 degrees of freedom distributed along the head, arms, and torso. It lacked hands, it had instead simple flippers that could use to push and prod objects. It could not move from its stand so that the objects it interacted with had to be presented to the robot by a human experimenter. The robot was controlled by a distributed parallel control system based on a real-time operating system and running on a set of Pentium based computers. The robot was equipped with cameras (for vision), gyroscopes simulating the human vestibular system, and joint sensors providing information about the position and torque exerted at each joint.

Since the robot did not have hands, it could not really grasp objects from the table. Nonetheless there are other actions that can be employed in exploring the physical properties of objects, such as touching, poking, prodding, and sweeping. Moreover, since the interaction of the robot's flipper with objects was limited, we employed rolling objects that show a characteristic behavior depending on how they are approached: a toy car, an orange juice bottle, a ball, and a colored toy cube. The robot's motor repertoire besides reaching consisted of four different stereotyped approach movements covering a range of directions of about 180° around the object. The sequence of images acquired during reaching for the object, the moment of impact, and the effects of the action were measured following the approach in Fitzpatrick (2003) and Metta and Fitzpatrick (2003).

The experiment consisted in presenting repetitively each of the four objects mentioned above to the robot. During this stage also other objects were presented at random; the experiment ran for several days and sometimes people walked by the robot and managed to make it poke the most disparate objects. For each successful trial, the robot "stored" the result of the segmentation of the object from the background, the object's principal axis which was selected as representative shape parameter, the action — initially selected randomly from the set of four approach directions — and the movement of the center of mass of the object for some hundreds milliseconds after the impact was detected. We grouped data belonging to the same object by employing a color-based clustering technique. In fact in our experiments the toy car was mostly yellow in color, the ball violet, the bottle orange, etc.

It is possible to describe object behavior in visual terms by estimating the probability of observing object motion relative to the object's own principal axis. Intuitively, this gives information about the rolling properties of the different objects: e.g., the car tends to roll along its principal axis, the bottle at right angle with respect to the axis. For the purpose of generating actions a description of the geometry of poking is required and has to go with the description of the object rolling behavior. This can be easily obtained by collecting many samples of generic poking actions and estimating the average direction of displacement of the object.

Having a visual and a "pragmatic" description of objects, it is now possible to test whether this information can be re-used to make the robot "optimally" poke (i.e., selecting an action that causes maximum displacement) a known object. In practice the same color clustering procedure is used for localizing and recognizing the object, to determine its orientation on the table, its affordance, and finally to select the action that it is most likely to elicit the principal affordance (roll).

A simple qualitative test of the performance determined that out of 100 trials the robot made 15 mistakes. Further analysis showed that 12 of the 15 mistakes were due to poor control of reaching (e.g., the flipper touched the object too early bringing it outside the field of view), and only three to a wrong estimate of the orientation.

Although crude, this implementation shows that with little pre-existing structure the robot could acquire the crucial elements for building knowledge of objects in terms of their affordances. Given a sufficient level of abstraction, our implementation is close to the response of canonical neurons in F5 and their interaction with neurons observed in AIP that respond to object orientation (Sakata et al., 1997).

At this point, we can test whether knowledge about object-directed actions can be reused in interpreting observed actions performed perhaps by a human experimenter. According to our model, whereas the robot identified the motion of the object because of a certain action applied to it, during action observation it could backtrack and derive the type of action from the observed motion of the object. In fact, the same segmentation procedure could visually interpret poking actions generated by a human as well as those generated by the robot.

Thus, observations can be converted into interpreted actions. The action whose effects are closest to the observed consequences on the object (which we might translate into the goal of the action) is selected as the most plausible interpretation given the observation. Most importantly, the interpretation reduces to the interpretation of the "simple" kinematics of the goal and consequences of the action rather than to understanding the "complex" kinematics of the human manipulator. The robot understands only to the extent it has learned to act.

In order to test this possibility we verified whether the robot could imitate the "goal" of a poking action. The step is indeed small since most of the work is actually in interpreting observations. Imitation was generated in the following by replicating the latest observed human movement with respect to the object and irrespective of its orientation. For example, in case the experimenter poked the toy car sideways, the robot imitated him/her by pushing the car sideways (for further details, see Metta and Fitzpatrick, 2003).

In summary, the results from our experiments seem to confirm two facts of the proposed model: first, that motor information plays a role in the recognition process — as would be following the hypothesis of the implication of feedback signals into recognition — and, second, that a mirror-like representation can be developed autonomously on the basis of the interaction between an individual and the environment.

Conclusions

In this paper, starting from known experiments in the monkey, we reviewed the evidence for the existence of a mirror-neurons system in humans. We highlighted the fact that the mirror system is not a passive observer that only "resonates" with the incoming sensory stimulation but rather it works in predicting the future course of action, in filling gaps, and in merging the available evidence for the plausibility of the ongoing observed action. This last aspect includes contextual information which in turn represents the goal of the action and eventually its meaning. We also reviewed the link between the mirror system and speech drawing a parallel between vision and sound but ultimately showing that both impinge on the motor system. Finally, we covered, although briefly, some computation modeling of the mirror system which can be used to clarify certain aspects of the functioning of the biological counterpart. Although still partial, this implementation shows that, in principle, the acquisition of the mirror neurons structure is the almost natural outcome of the development of a control system for grasping. Also, we have put forward a plausible sequence of learning phases involving the interaction between canonical and mirror neurons. This, we believe, is well in accordance with the evidence gathered by neurophysiology. In conclusion, we have embarked in an investigation that is somewhat similar to Liberman's artificial speech recognition attempts. Perhaps, also this time, the mutual rapprochement of neural and engineering sciences might lead to a better understanding of brain functions.

Acknowledgments

This work has been supported by European Commission grants ROBOT-CUB, NEUROBOTICS and CONTACT, by Italian Ministry of Education grants and by Fondazione Cassa di Risparmio of Ferrara.

References

Arbib, M.A. (1997) From visual affordances in monkey parietal cortex to hippocampo-parietal interactions underlying rat navigation. Philos. Trans. R. Soc. Lond. B Biol. Sci., 352: 1429–1436.

Barbas, H. and Pandya, D.N. (1987) Architecture and frontal cortical connections of the premotor cortex (area 6) in the rhesus monkey. J. Comp. Neurol., 256: 211–228.

Binkofski, F., Buccino, G., Posse, S., Seitz, R.J., Rizzolatti, G. and Freund, H. (1999) A fronto-parietal circuit for object manipulation in man: evidence from an fMRI-study. Eur. J. Neurosci., 11: 3276–3286.

Borroni, P., Montagna, M., Cerri, G. and Baldissera, F. (2005) Cyclic time course of motor excitability modulation during the observation of a cyclic hand movement. Brain Res., 1065: 115–124.

Broca, P. (1861) Remarques sur le Siége de la Faculté du Langage Articulé, Suivies d'une Observation d'aphemie (Perte de la Parole). Bulletin de la Société Anatomique de Paris, 6: 330–357.

Brooks, R.A., Breazeal, C., Marjanovi, M., Scassellati, B. and Williamson, M.M. (1999) The Cog project: Building a humanoid robot. In: Nehaniv C.L. (Ed.), Computation for Metaphors, Analogy and Agents. Springer-Verlag, pp. 52–87.

Buccino, G., Binkofski, F., Fink, G.R., Fadiga, L., Fogassi, L., Gallese, V., Seitz, R.J., Zilles, K., Rizzolatti, G. and Freund, H.J. (2001) Action observation activates premotor and parietal areas in a somatotopic manner: an fMRI study. Eur. J. Neurosci., 13: 400–404.

Buccino, G., Lui, F., Canessa, N., Patteri, I., Lagravinese, G., Benuzzi, F., Porro, C.A. and Rizzolatti, G. (2004) Neural circuits involved in the recognition of actions performed by nonconspecifics: an FMRI study. J. Cogn. Neurosci., 16: 114–126.

Campbell, R., MacSweeney, M., Surguladze, S., Calvert, G., McGuire, P., Suckling, J., Brammer, M.J. and David, A.S. (2001) Cortical substrates for the perception of face actions: an fMRI study of the specificity of activation for seen speech and for meaningless lower-face acts (gurning). Brain Res. Cogn. Brain Res., 12: 233–243.

Cochin, S., Barthelemy, C., Lejeune, B., Roux, S. and Martineau, J. (1998) Perception of motion and qEEG activity in human adults. Electroencephalogr. Clin. Neurophysiol., 107: 287–295.

Craighero, L., Bonetti, F., Massarenti, L., Canto, R., Fabbri Destro M. and Fadiga L. (in press) Temporal prediction of touch instant during observation of human and robot grasping. Brain Res. Bull.

Decety, J., Grezes, J., Costes, N., Perani, D., Procyk, E., Grassi, F., Jeannerod, M. and Fazio, F. (1997) Brain activity during observation of actions. Influence of action content and subject's strategy. Brain, 120: 1763–1777.

Emmorey, K.D. (1989) Auditory morphological priming in the lexicon. Lang. Cognitive Proc., 4: 73–92.

Fadiga, L., Craighero, L., Buccino, G. and Rizzolatti, G. (2002) Speech listening specifically modulates the excitability of tongue muscles: a TMS study. Eur. J. Neurosci., 15: 399–402.

Fadiga, L., Craighero, L., Fabbri Destro, M., Finos, L., Cotillon-Williams, N., Smith, A.T. and Castiello, U. (2006) Language in shadow. Soc. Neurosci., 1: 77–89.

Fadiga, L., Fogassi, L., Pavesi, G. and Rizzolatti, G. (1995) Motor facilitation during action observation: a magnetic stimulation study. J. Neurophysiol., 73: 2608–2611.

Fadiga, L., Roy, A.C., Fazio, P. and Craighero, L. (in press) From hand actions to speech: evidence and speculations. In: Haggard P., Rossetti Y. and Kawato M. (Eds.), Sensorimotor Foundations of Higher Cognition. Attention and Performance. Oxford University Press, Oxford.

Fitzpatrick, P. (2003) From First Contact to Close Encounters: A developmentally deep perceptual system for a humanoid robot. Unpublished PhD thesis, MIT, Cambridge, MA.

Gallese, V., Fadiga, L., Fogassi, L. and Rizzolatti, G. (1996) Action recognition in the premotor cortex. Brain, 119: 593–609.

Gangitano, M., Mottaghy, F.M. and Pascual-Leone, A. (2001) Phase-specific modulation of cortical motor output during movement observation. Neuroreport, 12: 1489–1492.

Gangitano, M., Mottaghy, F.M. and Pascual-Leone, A. (2004) Modulation of premotor mirror neuron activity during observation of unpredictable grasping movements. Eur. J. Neurosci., 20: 2193–2202.

Grafton, S.T., Arbib, M.A., Fadiga, L. and Rizzolatti, G. (1996) Localization of grasp representations in humans by positron emission tomography. 2. Observation compared with imagination. Exp. Brain Res., 112: 103–111.

Grezes, J., Armony, J.L., Rowe, J. and Passingham, R.E. (2003) Activations related to "mirror" and "canonical" neurones in the human brain: an fMRI study. Neuroimage, 18: 928–937.

Grezes, J., Costes, N. and Decety, J. (1998) Top down effect of the strategy on the perception of human biological motion: a PET investigation. Cognitive Neuropsychol., 15: 553–582.

Hamzei, F., Rijntjes, M., Dettmers, C., Glauche, V., Weiller, C. and Buchel, C. (2003) The human action recognition system and its relationship to Broca's area: an fMRI study. Neuroimage, 19: 637–644.

Hari, R., Forss, N., Avikainen, S., Kirveskari, E., Salenius, S. and Rizzolatti, G. (1998) Activation of human primary motor cortex during action observation: a neuromagnetic study. Proc. Natl. Acad. Sci. U.S.A., 95: 15061–15065.

Johnson-Frey, S.H., Maloof, F.R., Newman-Norlund, R., Farrer, C., Inati, S. and Grafton, S.T. (2003) Actions or hand-object interactions? Human inferior frontal cortex and action observation. Neuron, 39: 1053–1058.

Kawato, M., Furukawa, K. and Suzuki, R. (1987) A hierarchical neural-network model for control and learning of voluntary movement. Biol. Cybern., 57: 169–185.

Kilner, J.M., Vargas, C., Duval, S., Blakemore, S.J. and Sirigu, A. (2004) Motor activation prior to observation of a predicted movement. Nat. Neurosci., 7: 1299–1301.

Kohler, E., Keysers, C., Umilta, M.A., Fogassi, L., Gallese, V. and Rizzolatti, G. (2002) Hearing sounds, understanding actions: action representation in mirror neurons. Science, 297: 846–848.

Liberman, A.M., Cooper, F.S., Shankweiler, D.P. and Studdert-Kennedy, M. (1967) Perception of the speech code. Psychol. Rev., 74: 431–461.

Liberman, A.M. and Mattingly, I.G. (1985) The motor theory of speech perception revised. Cognition, 21: 1–36.

Liberman, A.M. and Whalen, D.H. (2000) On the relation of speech to language. Trends Cogn. Sci., 4: 187–196.

Luppino, G. and Rizzolatti, G. (2000) The organization of the frontal motor cortex. News Physiol. Sci., 15: 219–224.

Matelli, M., Luppino, G. and Rizzolatti, G. (1985) Patterns of cytochrome oxidase activity in the frontal agranular cortex of the macaque monkey. Behav. Brain Res., 18: 125–136.

Metta, G. and Fitzpatrick, P. (2003) Early integration of vision and manipulation. Adapt. Behav., 11: 109–128.

Metta, G., Sandini, G., Natale, L., Craighero, L. and Fadiga, L. (2006) Understanding mirror neurons: A bio-robotic approach. Interact. Stud., 7: 197–232.

Miall, R.C., Weir, D.J., Wolpert, D.M. and Stein, J.F. (1993) Is the cerebellum a Smith predictor? J. Motor Behav., 25: 203–216.

Murata, A., Fadiga, L., Fogassi, L., Gallese, V., Raos, V. and Rizzolatti, G. (1997) Object representation in the ventral premotor cortex (area F5) of the monkey. J. Neurophysiol., 78: 2226–2230.

Nelissen, K., Luppino, G., Vanduffel, W., Rizzolatti, G. and Orban, G.A. (2005) Observing others: multiple action representation in the frontal lobe. Science, 310: 332–336.

di Pellegrino, G., Fadiga, L., Fogassi, L., Gallese, V. and Rizzolatti, G. (1992) Understanding motor events: a neurophysiological study. Exp. Brain Res., 91: 176–180.

Petrides, M. (2006) Broca's area in the human and the nonhuman primate brain. In: Grodzinsky Y. and Amunts K. (Eds.), Broca's Region. Oxford University Press, New York, pp. 31–46.

Petrides, M. and Pandya, D.N. (1994) Comparative architectonic analysis of the human and the macaque frontal cortex. In: Boller F., Spinnler H. and Hendler J.A. (Eds.), Handbook of Neuropsychology. Elsevier, Amsterdam.

Rizzolatti, G. and Arbib, M.A. (1998) Language within our grasp. Trends Neurosci., 21: 188–194.

Rizzolatti, G., Camarda, R., Fogassi, L., Gentilucci, M., Luppino, G. and Matelli, M. (1988) Functional organization of inferior area 6 in the macaque monkey. II. Area F5 and the control of distal movements. Exp. Brain Res., 71: 491–507.

Rizzolatti, G. and Craighero, L. (2004) The mirror-neuron system. Annu. Rev. Neurosci., 27: 169–192.

Rizzolatti, G., Fadiga, L., Matelli, M., Bettinardi, V., Paulesu, E., Perani, D. and Fazio, F. (1996) Localization of grasp representations in humans by PET: 1. Observation versus execution. Exp. Brain Res., 111: 246–252.

Rizzolatti, G., Fogassi, L. and Gallese, V. (2001) Neurophysiological mechanisms underlying the understanding and imitation of action. Nat. Rev. Neurosci., 2: 661–670.

Rizzolatti, G. and Gentilucci, M. (1988) Motor and visual-motor functions of the premotor cortex. In: Rakic P. and Singer W. (Eds.), Neurobiology of Neocortex. Wiley, Chichester, pp. 269–284.

Sakata, H., Taira, M., Kusunoki, M., Murata, A. and Tanaka, Y. (1997) The TINS Lecture. The parietal association cortex in depth perception and visual control of hand action. Trends Neurosci., 20: 350–357.

Salmelin, R. and Hari, R. (1994) Spatiotemporal characteristics of sensorimotor neuromagnetic rhythms related to thumb movement. Neuroscience, 60: 537–550.

Strafella, A.P. and Paus, T. (2000) Modulation of cortical excitability during action observation: a transcranial magnetic stimulation study. Neuroreport, 11: 2289–2292.

Umiltà, M.A., Kohler, E., Gallese, V., Fogassi, L., Fadiga, L., Keysers, C. and Rizzolatti, G. (2001) I know what you are doing: a neurophysiological study. Neuron, 31: 155–165.

Watkins, K.E., Strafella, A.P. and Paus, T. (2003) Seeing and hearing speech excites the motor system involved in speech production. Neuropsychologia, 41: 989–994.

Wolpert, D.M. and Miall, R.C. (1996) Forward models for physiological motor control. Neural Netw., 9: 1265–1279.

C. von Hofsten & K. Rosander (Eds.)
Progress in Brain Research, Vol. 164
ISSN 0079-6123

CHAPTER 4

Apraxia: a review

Biljana Petreska[1,*], Michela Adriani[2], Olaf Blanke[2] and Aude G. Billard[1]

[1]*Learning Algorithms and Systems Laboratory (LASA), Ecole Polytechnique Fédérale de Lausanne (EPFL), EPFL-STI-I2S-LASA, Station 9, CH 1015 Lausanne, Switzerland*
[2]*Laboratory of Cognitive Neuroscience (LNCO), Ecole Polytechnique Fédérale de Lausanne (EPFL), EPFL-SV-BMI, Station 15, CH 1015 Lausanne, Switzerland*

Abstract: Praxic functions are frequently altered following brain lesion, giving rise to apraxia — a complex pattern of impairments that is difficult to assess or interpret. In this chapter, we review the current taxonomies of apraxia and related cognitive and neuropsychological models. We also address the questions of the neuroanatomical correlates of apraxia, the relation between apraxia and aphasia and the analysis of apraxic errors. We provide a possible explanation for the difficulties encountered in investigating apraxia and also several approaches to overcome them, such as systematic investigation and modeling studies. Finally, we argue for a multidisciplinary approach. For example, apraxia should be studied in consideration with and could contribute to other fields such as normal motor control, neuroimaging and neurophysiology.

Keywords: apraxia; brain lesion; neuropsychological models of apraxia; kinematic studies; computational neuroscience; multidisciplinary approach

Introduction

Apraxia is generally defined as "a disorder of skilled movement not caused by weakness, akinesia, deafferentation, abnormal tone or posture, movement disorders such as tremor or chorea, intellectual deterioration, poor comprehension, or uncooperativeness" (Heilman and Rothi, 1993). Apraxia is thus negatively defined, in terms of what it is not, as a higher order disorder of movement that is *not* due to elementary sensory and/or motor deficits. This definition implies that there are situations where the effector is moved with normal skill (Hermsdörfer et al., 1996). Puzzling parts of apraxia are the *voluntary-automatic dissociation* and *context-dependence*. On the one hand, apraxic patients may spontaneously perform gestures that they cannot perform on command (Schnider et al., 1997). This voluntary-automatic dissociation can be illustrated by an apraxic patient who could use his left hand to shave and comb himself, but could not execute a specific motor action such as opening the hand so as to let go of an object (Lausberg et al., 1999). In this particular case, focusing on the target of the movement rather than on the movement itself increased his chances of a successful execution. On the other hand, the execution of the movement depends heavily on the context of testing (De Renzi et al., 1982). It may be well preserved in a natural context, with a deficit that appears in the clinical setting only, where the patient has to explicitly represent the content of the action outside of the situational props (Jeannerod and Decety, 1995; Leiguarda and Marsden, 2000).

Several authors agree that although apraxia is easy to demonstrate, it has proven difficult to understand. Research on apraxia is filled with

*Corresponding author. Tel.: +41 21 693 54 65;
Fax: +41 21 693 78 50; E-mail: biljana.petreska@a3.epfl.ch

DOI: 10.1016/S0079-6123(07)64004-7

confusing terminology, contradictory results and doubts that need to be resolved (De Renzi et al., 1982; Goldenberg et al., 1996; Graham et al., 1999; Koski et al., 2002; Laeng, 2006). Inconsistencies between similar studies may be explained by differences in the methodological and statistical approaches for the apraxia assessment (i.e., types of gestures used and scoring criteria), chronicity and aetiology of damage and brain lesion localization tools (Haaland et al., 2000). Therefore, it still stands that our understanding of the neural and cognitive systems underlying human praxis is not well established.

The chapter is structured as follows. We first review existing types of apraxia as well as important current and historical models of the apraxic deficit. We then consider the inter- and intra-hemispheric lesion correlates of apraxia. Two other sections are dedicated to the relationship between praxis and language and to the analysis of apraxic errors. We finally discuss the current state-of-the-art in apraxia, and argue for a multidisciplinary approach that encompasses evidence from various fields such as neuroimaging or neurophysiology.

Types of apraxia

This section reviews the current taxonomies of apraxia. Some of the frequently observed types of apraxia have inspired the apraxia models described in the following section, others still challenge them.

Ideational apraxia: was historically defined as a disturbance in the conceptual organization of actions. It was first assessed by performing purposive sequences of actions that require the use of various objects in the correct order (e.g., preparing a cup of coffee) (Poeck, 1983). It was later accepted that ideational apraxia is not necessarily associated to complex actions, but is a larger deficit that also concerns the evocation of single actions. In this view, complex sequences of multiple objects are simply more suitable to reveal the deficit, possibly because of the heavier load placed on memory and attentional resources (De Renzi and Lucchelli, 1988). Nonetheless, the term *conceptual apraxia* was introduced to designate content errors in single actions, excluding sequence errors in multi-staged actions with tools[1] (Ochipa et al., 1992; Heilman et al., 1997). In theoretical models, ideational and conceptual apraxia correspond to a disruption of the conceptual component of the praxis system, i.e., action semantics memory, described in more detail in the "Models of apraxia" section (De Renzi and Lucchelli, 1988; Graham et al., 1999). Patients with ideational apraxia are not impaired in the action execution *per se*, but demonstrate inappropriate use of objects and may fail in gesture discrimination and matching tasks. For example, a patient was reported to eat with a toothbrush and brush his teeth with a spoon and a comb. His inability to use tools could not be explained by a motor production deficit that would characterize ideomotor apraxia (defined below). Interestingly, although he was able to name the tools and point to them on command, he could not match the tools with the objects, hence suggesting a loss of knowledge related to the use of tools.

Ideomotor apraxia: is considered to be a disorder of the production component of the praxis system, i.e., sensorimotor action programs that are concerned with the generation and control of motor activity (Rapcsak et al., 1995; Graham et al., 1999). It is characterized by errors in the timing, sequencing and spatial organization of gestural movements (Leiguarda, 2001). Since the conceptual part of the praxis system is assumed to be intact, patients with ideomotor apraxia should not use objects and tools in a conceptually inappropriate fashion and should not have difficulty with the serial organization of an action (De Renzi et al., 1982). Ideational and ideomotor apraxia have been assessed by testing the execution of various types of gestures: transitive and intransitive (i.e., with or without the use of tools or objects), meaningless non-representational (e.g., hand postures relative to head) and meaningful representational (e.g., waving good-bye), complex sequences with multiple objects, repetitive movements, distal and proximal gestures (e.g., imitation of finger and hand configurations), reaching in peri-personal and body-centered space (e.g., targets in near space or on

[1]Conceptual apraxia is often observed in Alzheimer's disease.

the patient's body), novel movements (i.e., skill acquisition) or imagined movements. These gestures can also be executed under different modalities such as: verbal command, imitation, pantomime and tactile or visual presentation of objects.

The use of various gestures and different modalities to assess apraxia has helped to uncover many interesting *functional dissociations* that are listed below. For example, apraxia was shown to be modality-specific, i.e., the same type of gesture was differentially impaired according to the modality of testing (De Renzi et al., 1982). One dissociation, named *conduction apraxia*, is the syndrome of superior performance on verbal command than on imitation (Ochipa et al., 1994). The opposite pattern has also been observed: very poor performance on verbal command that improved on imitation or when seeing the object (Heilman, 1973; Merians et al., 1997). The extreme occurrence of conduction apraxia, namely the selective inability to imitate with normal performance on verbal command was termed *visuo-imitative apraxia* (Merians et al., 1997). In some cases of visuo-imitative apraxia, defective imitation of meaningless gestures (e.g., fist under chin) contrasts with preserved imitation of meaningful gestures (e.g., hitchhiking) (Goldenberg and Hagmann, 1997; Salter et al., 2004). A surprising case of double dissociation from this kind of visuo-imitative apraxia was described in Bartolo et al. (2001), where the patient showed impairment in meaningful gesture production (both on imitation and verbal command) and normal performance in imitation of meaningless gestures, suggesting that the patient was able to reproduce only movements he did not identify or recognize as familiar. Similarly, the apraxic patients in Buxbaum et al. (2003) responded abnormally to familiar objects (e.g., a key, a hammer or a pen) but normally in recognizing the hand postures appropriate for novel objects (e.g., parallelepipeds differing in size and depth). These two studies argue that the reproduction of a gesture may be constrained by its degree of familiarity, indicating that current models of apraxia would need some refinement.

Furthermore, the representation of transitive and intransitive actions may be dissociable. In Watson et al. (1986), bilateral apraxia was observed only for transitive (e.g., hammering) but not intransitive (e.g., hitchhiking, waving goodbye) movements.[2] Whereas transitive gestures are constrained by the shape, size and function of objects, intransitive actions are related to socio-cultural contexts (Cubelli et al., 2000; Heath et al., 2001). The isolated disturbance of transitive hand movements for use of, recognition and interaction with an object, in the presence of preserved intransitive movements, was named tactile apraxia and usually appears in the hand contralateral to the lesion (Binkofski et al., 2001).

As mentioned in the "Introduction", contextual cues strongly influence the execution of actions. Some studies have systematically manipulated the contextual cues in order to assess their relative importance. For example, patients with impaired pantomime of motor actions showed no deficit in the comprehension of the use of tools or in manipulating the tools (Halsband et al., 2001). Graham et al. (1999) also observed dramatic facilitation in the demonstration of tool use when the patient was given the appropriate or a neutral tool to manipulate.[3] Interestingly, the patient could not prevent himself from performing the action appropriate to the tool he was holding, rather than the requested action. In another study however, gesture execution improved when the object of the action, but not the tool, was given (Clark et al., 1994). Hence, the addition of visual and somaesthetic cues may improve certain aspects of apraxic movements, since it provides mechanical constraints and supplementary information that facilitates the selection of an adequate motor program (Hermsdörfer et al., 2006). Nonetheless, there is the case of a patient that performed much worse when he was actually manipulating the tool than on verbal command[4] (Merians et al., 1999).

Dissociations that concern the nature of the target were also observed. For example, the left brain damaged patients in Hermsdörfer et al. (2003) had prolonged movement times and reduced maximum velocities when the movements were directed

[2]These patients had lesions in the left supplementary motor area (SMA).

[3]The subject had clinically diagnosed corticobasal degeneration.

[4]Ibid.

toward an allocentric target without visual feedback, but performed normally when the target was their own nose. Also, a clear dissociation was found in Ietswaart et al. (2006) between impaired gesture imitation and intact motor programming of goal-directed movements, hence arguing against the interpretation of impaired imitation as a purely executional deficit (see "Models of apraxia").

A particular type of apraxia is *constructional apraxia*, originally described by Kleist as "the inability to do a construction" and defined by Benton as "the impairment in combinatory or organizing activity in which details must be clearly perceived and in which the relationship among the component parts of the entity must be apprehended" (Laeng, 2006). Constructional apraxic patients are unable to spontaneously draw objects, copy figures and build blocks or patterns with sticks, following damage not only to the dominant but also non-dominant hemisphere. Hence, constructional apraxia appears to reflect the loss of bilaterally distributed components for constructive planning and the perceptual processing of categorical and coordinate spatial relations (Platz and Mauritz, 1995; Laeng, 2006).

Apraxia can also be observed in *mental motor imagery tasks*. Motor imagery is considered as a means of accessing the mechanisms of action preparation and imitation, by sharing a common neural basis (Jeannerod and Decety, 1995). Apraxic patients were deficient in simulating hand actions mentally and in imagining the temporal properties of movements[5] (Sirigu et al., 1999). Other apraxic patients showed a deficit in generating and maintaining internal models for planning object-related actions (Buxbaum et al., 2005a). These findings support the notion that the motor impairments observed in apraxic patients result from a specific alteration in their ability to mentally evoke actions, or to use stored motor representations for forming mental images of actions.

Apraxia may also be appropriate to reveal the role of *feedback* during the execution of a movement. Some apraxic patients were impaired in reaching and aiming movements only in the condition without visual feedback (Ietswaart et al., 2001, 2006) and performed worse during pointing with closed eyes (Jacobs et al., 1999; Hermsdörfer et al., 2003). Interestingly, the patients in Haaland et al. (1999) overshot the target when feedback of the hand was removed and undershot the target when the feedback of the target was unavailable. Importantly, these patients continued to rely on visual feedback during the secondary adjustment phase of the movement and never achieved normal end-point accuracy when visual feedback of the hand position or target location was unavailable. These findings also suggest that ideomotor limb apraxia may be associated with the disruption of the neural representations for the extra-personal (spatial location) and intra-personal (hand position) features of movement (Haaland et al., 1999).

The importance of feedback signals was demonstrated in one of our own apraxic patients (unpublished data). We reproduced a seminal study of imitation of meaningless gestures[6] by Goldenberg et al. (2001) on an apraxic patient with left-parietal ischemic lesion. We observed that the patient relied heavily on visual and tactile feedback. He often needed to bring his hand in the field of vision and corrected the hand posture by directly comparing it with the displayed stimulus to imitate. He also used tactile exploration when searching for the correct spatial position on his face. He showed many hesitations and extensive searching which led to highly disturbed kinematic profiles of the gesture (shown in Fig. 3c, d), but often correct final postures.

Apraxia can also be defined in relation to the selectively affected effectors: *orofacial apraxia* or *buccofacial apraxia, oral apraxia, upper and lower face apraxia, lid apraxia, limb apraxia, leg apraxia, trunk apraxia, etc.* Oral apraxia, for example, is defined as the inability to perform mouth actions such as sucking from a straw or blowing a kiss. It should not be confounded with *apraxia of speech* (also called *verbal apraxia*), which is a selective disturbance of the articulation of words (Bizzozero et al., 2000). Motor planning disorders in children are denominated *developmental dyspraxia* (Cermak, 1985). Apraxia can also designate a praxic ability impaired in an isolated manner such as: *gait*

[5]These patients had posterior parietal lesions.

[6]Hand postures relative to the head, an example is shown in Fig. 3a.

apraxia, apraxic agraphia, dressing apraxia, orienting apraxia and *mirror apraxia* (i.e., inability to reach to objects in a mirror (Binkofski et al., 2003)). When the side of brain lesion and affected hand are considered, the terms *sympathetic* and *crossed apraxia* are used. Apraxia can sometimes be related to the specific neural substrate that causes the disorder, for example following subcortical lesions in corticobasal degeneration (Pramstaller and Marsden, 1996; Jacobs et al., 1999; Merians et al., 1999; Hanna-Pladdy et al., 2001; Leiguarda, 2001) or following lesions of the corpus callosum (Watson and Heilman, 1983; Lausberg et al., 1999, 2000; Goldenberg et al., 2001; Lausberg and Cruz, 2004). *Callosal apraxia* for example is particularly appropriate for disentangling the specific hemispheric contributions to praxis.

An extensive list of the types of apraxia and their definitions, including types that were not mentioned above, can be found in Table 1.

Models of apraxia

Contemporary neuropsychological views of apraxia arise from Liepmann's influential work that dates from more than a hundred years ago. Liepmann proposed the existence of an idea of the movement, "movement formulae", that contains the "time-space-form picture" of the action (Rothi et al., 1991). He believed that in right-handers, these movement formulae are stored in the left-parietal lobe, endorsing the view of a left hemispheric dominance for praxis (Faglioni and Basso, 1985; Leiguarda and Marsden, 2000). To execute a movement, the spatiotemporal image of the movement is transformed into "innervatory patterns" that yield "positioning of the limbs according to directional ideas" (Jacobs et al., 1999). Liepmann distinguished between three types of apraxia that correspond to disruptions of specific components of his model (Faglioni and Basso, 1985; Goldenberg, 2003). First, a damaged movement formula (i.e., faulty integration of the elements of an action) would characterize "ideational apraxia". Second, failure of the transition from the movement formula to motor innervation (i.e., inability to translate a correct idea of the movement into a correct act) is defined as "ideomotor apraxia". According to Liepmann, faulty imitation of movements is a purely executional deficit and proves the separation between the idea and execution of a movement, since in imitation the movement formula is defined by the demonstration (Goldenberg, 1995, 2003; Goldenberg and Hagmann, 1997). Finally, loss of purely kinematic (kinaesthetic or innervatory) inherent memories of an extremity is the "limb-kinetic" variant of apraxia.

Another historically influential model is the disconnection model of apraxia proposed by Geschwind (1965). According to this model the verbal command for the movement is comprehended in Wernicke's area and is transferred to the ipsilateral motor and premotor areas that control the movement of the right hand (Clark et al., 1994; Leiguarda and Marsden, 2000). For a left-hand movement, the information needs to be further transmitted to the right association cortex via the corpus callosum. The model postulates that the apraxic disorder follows from a lesion in the left and right motor association cortices, or a disruption in their communication pathways. However this model cannot explain impaired imitation and impaired object use since these tasks do not require a verbal command (Rothi et al., 1991).

Heilman and Rothi (1993) proposed an alternative representational model of apraxia, according to which apraxia is a gesture production deficit that may result from the destruction of the spatiotemporal representations of learned movements stored in the left inferior-parietal lobule. They proposed to distinguish between dysfunction caused by destruction of the parietal areas (where the spatiotemporal representations of movements would be encoded), and the deficit which would result from the disconnection of these parietal areas from the frontal motor areas (Heilman et al., 1982). In the first case, posterior lesions would cause a degraded memory trace of the movement and patients would not be able to correctly recognize and discriminate gestures. In the second case, anterior lesions or disconnections would only provoke a memory egress disorder. Therefore patients with a gesture production deficit with anterior and posterior lesions should perform differently on tasks of gesture discrimination, gesture recognition and novel gesture learning.

Table 1. Taxonomy of apraxia

Type of apraxia	Definition
Ideational apraxia	Initially used to refer to impairment in the conceptual organization of actions, assessed with sequential use of multiple objects. Later defined as conceptual apraxia.
Conceptual apraxia	Impairment in the concept of a single action, characterized by content errors and the inability to use tools.
Ideomotor apraxia	Impairment in the performance of skilled movements, characterized by spatial or temporal errors in the execution of movements.
Limb-kinetic apraxia	Slowness and stiffness of movements with a loss of fine, precise and independent movement of the fingers.
Constructional apraxia	Difficulty in drawing and constructing objects. Impairment in the combinatory or organizing activity in which details and relationship among the component parts of the entity must be clearly perceived.
Developmental dyspraxia	Disorders affecting the initiation, organization and performance of actions in children.
Modality-specific apraxias	*Localized within one sensory system.*
Pantomime agnosia	Normal performance in gesture production tests both on imitation and on verbal command, but poor performance in gesture discrimination and comprehension. Patients with pantomime agnosia can imitate pantomimes they cannot recognize.
Conduction apraxia	Superior performance on pantomime to verbal command than on pantomime imitation.
Visuo-imitative apraxia	Normal performance on verbal command with selectively impaired imitation of gestures. Also used to designate the defective imitation of meaningless gestures combined with preserved imitation of meaningful gestures.
Optical (or visuomotor) apraxia	Disruptions to actions calling upon underlying visual support.
Tactile apraxia	Disturbance of transitive hand movements for use of, recognition and interaction with an object, in the presence of preserved intransitive movements.
Effector-specific apraxias	
Upper/lower face apraxia	Impairment in performing actions with parts of the face.
Oral apraxia	Inability to perform skilled movements with the lips, cheeks and tongue.
Orofacial (or buccofacial) apraxia	Difficulties with performing intentional movements with facial structures including the cheeks, lips, tongue and eyebrows.
Lid apraxia	Difficulty with opening the eyelids.
Ocular apraxia	Impairment in performing saccadic eye movements on command.
Limb apraxia	Used to refer to ideomotor apraxia of the limbs frequently including the hands and fingers.
Trunk (or axial) apraxia	Difficulty with generating body postures.
Leg apraxia	Difficulty with performing intentional movements with the lower limbs.
Task-specific apraxias	
Gait apraxia	Impaired ability to execute the highly practised, co-ordinated movements of the lower legs required for walking.
Gaze apraxia	Difficulty in directing gaze.
Apraxia of speech (or verbal apraxia)	Disturbances of word articulation.
Apraxic agraphia	A condition in which motor writing is impaired but limb praxis and non-motor writing (typing, anagram letters) are preserved.
Dressing apraxia	Inability to perform the relatively complex task of dressing.
Dyssynchronous apraxia	Failure to combine simultaneous preprogrammed movements.
Orienting apraxia	Difficulty in orienting one's body with reference to other objects.
Mirror apraxia	A deficit in reaching to objects presented in a mirror.
Lesion-specific apraxias	
Callosal apraxia	Apraxia caused by damage to the anterior corpus callosum that usually affects the left limb.
Sympathetic apraxia	Apraxia of the left limb due to damage to the anterior left hemisphere (the right hand being partially or fully paralyzed).
Crossed apraxia	The unexpected pattern of apraxia of the right limb following damage to the right-hemisphere.

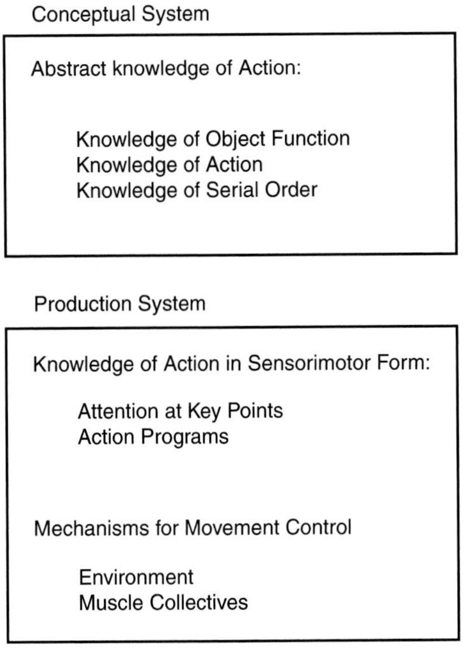

Conceptual System

Abstract knowledge of Action:

Knowledge of Object Function
Knowledge of Action
Knowledge of Serial Order

Production System

Knowledge of Action in Sensorimotor Form:

Attention at Key Points
Action Programs

Mechanisms for Movement Control

Environment
Muscle Collectives

Fig. 1. Roy and Square's cognitive model of limb praxis. Adapted with permission from Roy and Square (1985).

Roy and Square (1985) proposed a cognitive model of limb praxis that involves two systems, i.e., a conceptual system and a production system (illustrated in Fig. 1). The "conceptual system" provides an abstract representation of the action and comprises three kinds of knowledge: (1) knowledge of the functions of tools and objects, (2) knowledge of actions independent of tools and objects and (3) knowledge about the organization of single actions into sequences. The "production system" incorporates a sensorimotor representation of the action and mechanisms for movement control. Empirical support for the division of the praxis system into a conceptual and a production component is provided by a patient who could comprehend and discriminate transitive gestures she was unable to perform (Rapcsak et al., 1995). This model predicts three patterns of impairment (Heath et al., 2001). First, a deficit in pantomime but not in imitation would reflect damage to the selection and/or evocation of actions from long-term memory. Second, a deficit in imitation alone would indicate a disruption of the visual gestural analysis or translation of visual information into movement. Finally, concurrent impairment in pantomime and imitation is thought to reflect a disturbance at the latter, executive stage of gesture production and was the most frequent deficit pattern observed in Roy et al. (2000) and Parakh et al. (2004).

None of these models predict a number of modality-specific dissociations observed in neurologically impaired patients, such as preserved gesture execution on verbal command that is impaired in the visual modality when imitating (Ochipa et al., 1994; Goldenberg and Hagmann, 1997). To account for these dissociations, Rothi et al. (1991) proposed a cognitive neuropsychological model of limb praxis, which reflects more appropriately the complexity of human praxis (illustrated in Fig. 2a). This multi-modular model has input that is selective according to the modality, a specific "action semantics system" dissociable from other semantics systems, an "action reception lexicon" that communicates with an "action production lexicon" and a separate "nonlexical route" for the imitation of novel and meaningless gestures[7] (Rothi et al., 1997).

Although this model is widely used to explain data from multiple neurological studies, it has difficulties concerning several aspects. First, it does not consider the existence of a selective tactile route to transitive actions (Graham et al., 1999). For example, the model fails to explain data from a patient profoundly impaired in gesturing in the verbal and visual modalities, but not with the tool in hand (Buxbaum et al., 2000). Second, imitation of meaningless gestures is assumed to test the integrity of a direct route from visual perception to motor control. However, Goldenberg et al. (1996) have shown that this route is far from direct and involves complex intermediate processing steps. For example, apraxic patients that are impaired in reproducing gestures on their own bodies are also impaired in replicating the gestures on a life-sized manikin (Goldenberg, 1995). Hence, general conceptual knowledge about the human body and the spatial configuration of body parts seems necessary for performing an imitation task (Goldenberg, 1995; Goldenberg et al., 1996;

[7]The vocabulary was borrowed from the literature of language processing.

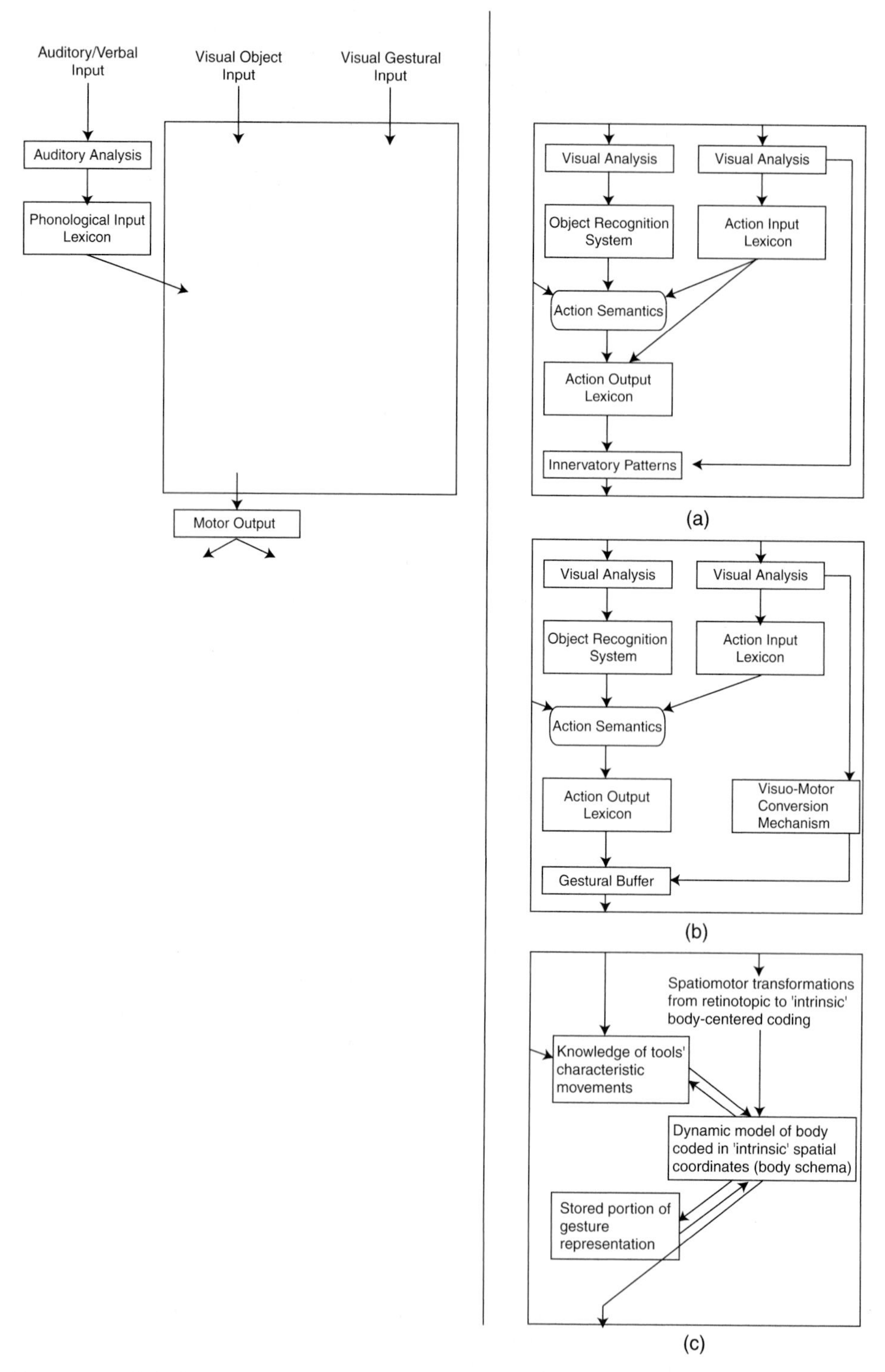

Fig. 2. A cognitive neuropsychological model of limb praxis. The three components on the right are interchangeable with the empty box in the complete model on the left. Under (a) Rothi et al.'s original model of limb praxis. Under (b) the previous model revised by Cubelli et al. and under (c) the model extended by Buxbaum et al. For a detailed description see the text. Adapted respectively with permission from Rothi et al. (1997), Cubelli et al. (2000) and Buxbaum et al. (2000).

Goldenberg and Hagmann, 1997). The belief that imitation is a rather simple and straightforward visuomotor process is misleading as one would have to resolve the "body correspondence problem"[8] to transpose movements from bodies with different sizes and different owners, which are in addition represented in different perspectives (Goldenberg, 1995).

To account for the last observation, Cubelli et al. (2000) have revised Rothi et al.'s cognitive neuropsychological model of limb praxis (illustrated in Fig. 2b). They have added "a visuomotor conversion mechanism" devoted to transcoding the visual input into appropriate motor programs. They have also suppressed the direct link between the "input" and "output action lexicon", leaving only an indirect link through the "action semantics system", as no empirical evidence was found of a patient able to reproduce familiar gestures with obscure meaning, but not unfamiliar gestures (see Fig. 2a, b). Finally, they have also added a "gestural buffer" aimed at holding a short-term representation of the whole action. The model predicts five different clinical pictures (for definitions of the different apraxic disorders please refer to Table 1): (1) a deficit of the "action input lexicon": *pantomime agnosia* (i.e., a difficulty in the discrimination and comprehension of gestures), (2) a deficit of the "action semantics system": conceptual apraxia without ideomotor apraxia, (3) a deficit of the "action output lexicon": conceptual apraxia with spared gesture-meaning associations, (4) a deficit of the "visuomotor conversion mechanism": conduction apraxia (not observed in their study) and (5) a deficit of the "gestural buffer": both ideomotor and ideational apraxia (i.e., impairment in all execution tasks with preserved ability to perform judgment and categorization tasks).

Buxbaum et al. (2000) further extended Rothi et al.'s cognitive neuropsychological model of limb praxis, based on their observation of a patient who performed particularly poorly on tasks that required a spatial transformation of the body. According to their model (illustrated in Fig. 2c), a unitary set of representations named "body schema" calculates and updates the dynamic positions of the body parts relative to one another. Importantly, this dynamic body-centered representation of actions is a common processing stage between the "lexical" and "nonlexical route" and hence subserves both meaningful and meaningless actions. Note that at the level of the "lexical route", there is an additional interaction with the stored representations of learned actions.

Existing models of apraxia still fail to account for additional empirical evidence such as, for example, the differential performance in imitation of hand postures and imitation of finger configurations shown in Goldenberg and Hagmann (1997) and Goldenberg and Karnath (2006). Furthermore, in a study of ideomotor apraxia, Buxbaum et al. (2005b) provided data which is compatible with the influential "mirror neuron hypothesis". Apraxia models cannot easily be reconciled with this hypothesis which, based upon neurophysiological observations from the monkey brain, postulates a "mirror neuron system" underlying both action recognition and action execution (Rizzolatti and Craighero, 2004). Mirror neurons are a special class of visuomotor neurons, initially discovered in area F5 of the monkey premotor cortex (see Fig. 4), which discharge both when the monkey does a particular action and when it observes another individual doing a similar action (Gallese et al., 1996; Rizzolatti and Luppino, 2001; Rizzolatti et al., 2002). Hence, the "mirror neuron system" is believed to map observed actions onto the same neural substrate used to execute these actions. As the same representations appear to subserve both action recognition and action production tasks, it would not be surprising if the perception of a movement is constrained by its executional knowledge. Related to apraxia, the "mirror neuron hypothesis" questions the separation of the "input" and "output lexicon" (Koski et al., 2002).

Contributions of the left- and right-brain hemispheres

Although most apraxia studies show a left-brain hemisphere dominance for praxis, the studies

[8]Here we give a shortened version of the informal statement of the body correspondence problem. Given an observed behavior of the model, i.e., a sequence (or hierarchy) of subgoals, find and execute a sequence of actions using one's own (possibly dissimilar) embodiment which leads through the corresponding subgoals (Nehaniv and Dautenhahn, 2002).

arguing for a significant involvement of the right hemisphere are numerous. Left-brain damage usually affects both hands, whereas right-brain damage affects only the left hand, suggesting that the left hemisphere is not only fully competent for processing movement concepts but also contributes to the generation of movements in the right hemisphere. Apraxic deficits following left hemisphere lesions are also more frequent (De Renzi et al., 1980; Weiss et al., 2001); however, in some rare cases, severe apraxia was observed following right hemisphere lesions (Marchetti and Della Sala, 1997; Raymer et al., 1999). The concept of crossed apraxia was introduced to describe patients with this opposite pattern of limb apraxia that cannot be explained by handedness. Callosal lesions are most suitable for investigating the issues of hemispheric specialization of praxis. For example, split-brain patients were apraxic with their left hands, also suggesting a left hemisphere dominance for processing skilled movement (Watson and Heilman, 1983; Lausberg et al., 1999, 2003), but both hemispheres appeared to contain concepts for skill acquisition (Lausberg et al., 1999) and object use (Lausberg et al., 2003).

In kinematic studies (described in more detail in "The analysis of apraxic errors"), only left-brain damaged patients were impaired in imitation of meaningless movements (Hermsdörfer et al., 1996; Weiss et al., 2001), as well as in pointing movements (Hermsdörfer et al., 2003); whereas right-brain damaged patients had deficits in slow-paced tapping and initiation of aiming movements (Haaland and Harrington, 1996). Hence, the left hemisphere was associated with movement trajectory control (Haaland et al., 2004), sequencing and ballistic movements (Hermsdörfer et al., 2003) and the right hemisphere was related to on-line control of the movement (Hermsdörfer et al., 2003) and closed-loop processing (Haaland and Harrington, 1996).

A left–right dichotomy was also observed for imitation and matching of hand and finger configurations (Goldenberg, 1999). Left-brain damaged patients had more difficulties with imitation than matching and vice versa. In addition, the left hemisphere seemed fully competent for processing hand postures, but needed the right hemisphere's contribution for processing finger postures (Goldenberg et al., 2001; Della Sala et al., 2006a). It was concluded that the left hemisphere mediates conceptual knowledge about the structure of the human body and that the right hemisphere is specialized for visually analyzing the gesture (Goldenberg, 2001; Goldenberg et al., 2001).

Finally, several studies observed similar impairment scores following left- and right-brain lesions, arguing for a bi-hemispheric representation of skilled movement (Haaland and Flaherty, 1984; Kertesz and Ferro, 1984; Roy et al., 1992, 2000; Heath et al., 2001). The less frequent, nevertheless well-detected incidence of limb apraxia following right-brain lesion, was attributed to the sensitivity and precision of the assessment methodology. In addition, right-hemisphere lesions often led to severe face apraxia (Bizzozero et al., 2000; Della Sala et al., 2006b). Hence, a model of widespread praxis, distributed across both hemispheres, may be more appropriate than the unique left-lateralized center previously hypothesized. Moreover, it seems that the degree of left-hemisphere dominance varies within subjects and with the type of movement (Haaland et al., 2004), raising the issue of overlap between the contributions of the right and left hemispheres to specialized praxic functions.

Intra-hemispheric lesion location: a distributed representation of praxis?

Several studies have failed to find a consistent association between the locus of the lesion within a hemisphere and the severity of apraxia (Basso et al., 1980; Kertesz and Ferro, 1984; Alexander et al., 1992; Schnider et al., 1997; Hermsdörfer et al., 2003). Moreover, areas involved in apraxia can also be damaged in non-apraxic patients (Haaland et al., 1999; Buxbaum et al., 2003). However, apraxic deficits are most frequent following parietal and frontal lesions, but were also observed in patients with temporal, occipital and subcortical damages (De Renzi and Lucchelli, 1988; Goldenberg, 1995; Hermsdörfer et al., 1996; Bizzozero et al., 2000).

More specifically, ideomotor apraxia and motor imagery deficits were observed following lesions in the left inferior parietal and the left dorsolateral frontal lobes (Haaland et al., 2000; Buxbaum et al.,

2005a). For example, several studies suggested that Brodmann areas 39 and 40 (i.e., angular and supramarginal gyri of the inferior-parietal lobule) are critical in visuo-imitative apraxia (Goldenberg and Hagmann, 1997; Goldenberg, 2001) and ideomotor limb apraxia (Haaland et al., 1999; Buxbaum et al., 2003). In addition, the superior-parietal lobe appeared crucial in integrating external visual and intra-personal somaesthisic information (Heilman et al., 1986; Haaland et al., 1999). Goldenberg and Karnath (2006), subtracted the lesion overlay of unimpaired from impaired patients and associated disturbed imitation of hand postures with lesions in the inferior-parietal lobe and temporo–parieto–occipital junction, whereas disturbed imitation of finger postures could be related to lesions in the inferior frontal gyrus. Interestingly, parts of the middle and inferior frontal gyri, in the vicinity of Brodmann areas 6, 8 and 46, were involved in all of the ideomotor apraxics in Haaland et al. (1999). Furthermore, premotor lesions (including lesions to the supplementary motor area) particularly affected bimanual actions in Halsband et al. (2001) and transitive actions in Watson et al. (1986).

It has been difficult to disentangle between the specific contributions of the parietal and the frontal cortices, as lesions in these areas lead to similar deficits (Haaland et al., 1999, 2000). For example, target and spatial errors were related to posterior lesions only (Haaland et al., 2000; Halsband et al., 2001; Weiss et al., 2001; Goldenberg and Karnath, 2006), but internal hand configuration errors were present in patients with anterior and posterior lesions (Haaland et al., 2000; Goldenberg and Karnath, 2006). Importantly, only patients with posterior lesions, and not anterior lesions, had difficulties in discriminating between correctly and incorrectly performed actions and in recognizing pantomimes or appropriate hand postures (Halsband et al., 2001; Buxbaum et al., 2005b).

Apraxia can also develop following subcortical lesions (Pramstaller and Marsden, 1996; Graham et al., 1999; Jacobs et al., 1999; Merians et al., 1999; Hanna-Pladdy et al., 2001). In this case, it is not clear whether the apraxia originates from lesions in the basal ganglia, which are extensively connected to the superior-parietal lobe and premotor and supplementary motor areas (Jacobs et al., 1999; Merians et al., 1999), or from the surrounding white matter (i.e., fronto-parietal connections) (Pramstaller and Marsden, 1996).

Failure to find clear correlations between specific lesion loci and different apraxic deficits argues for a widespread cortical and subcortical representation of praxis, distributed across specialized neural systems working in concert (Leiguarda and Marsden, 2000; Hermsdörfer et al., 2003). However, we believe that a selective damage to one of these systems may produce a particular pattern of errors tightly related to a subtype of apraxia.

Praxis and language?

Apraxia is most often seen in association with aphasia (i.e., loss of the ability to speak or understand speech), which renders the assessment of apraxia very difficult. Indeed, one has to provide evidence that the patient has understood the commands so that the motor deficit cannot be attributed to aphasia (De Renzi et al., 1980). Historically, gestural disturbance in aphasics was considered to be a manifestation of damaged abstract knowledge. This idea of a common impaired symbolic function underlying aphasia and apraxia was supported for a long time (Kertesz and Hooper, 1982). However, several large-scale studies failed to find correlations between subtypes of apraxia and aphasia (Goodglass and Kaplan, 1963; Lehmkuhl et al., 1983; Buxbaum et al., 2005b). Moreover, clear evidence of a double dissociation between apraxia and aphasia was presented in Papagno et al. (1993). For example, some patients were able to verbalize a desired movement but could not perform it (Goodglass and Kaplan, 1963), whereas other patients were able to pantomime actions they were unable to name (Rothi et al., 1991). Hence, it seems that many aspects of language and praxis are subserved by independent, possibly contiguous neuronal processes, but concomitant deficits may also appear because of shared neuroanatomical substrates (Kertesz and Hooper, 1982). Nevertheless, the question of how language is related to praxis is a fascinating one and needs further study, as it can give some insight into the existence of a supramodal representation of knowledge, or alternatively shed light onto the communication mechanisms between the praxic- and language-specific representations of knowledge.[9]

The analysis of apraxic errors

There are extensive quantitative analyses of the severity of apraxic errors in single case studies and in large samples of brain-damaged patients. Qualitative analyses however are less numerous and non-standardized, but nonetheless essential for precisely understanding the nature of apraxia. Performances are usually classified in a limited number of response categories such as:[10] temporal errors, spatial errors, content errors, substitutive errors, augmentative errors, fragmentary errors, associative errors (i.e., the correct movement is replaced by another movement that shares one feature), parapraxic errors (i.e., correct execution of a wrong movement), wrong body part errors (e.g., patients that execute a correct movement with the head instead of the hand), body part as tool errors (i.e., a body part is used to represent the imagined tool) and perseveration errors (Lehmkuhl et al., 1983; Poeck, 1983; De Renzi and Lucchelli, 1988; Platz and Mauritz, 1995; Lausberg et al., 1999, 2003; Halsband et al., 2001; Weiss et al., 2001). Perseveration and body parts as tool errors should be accorded some special interest in future studies, as they are prominent in apraxia and their occurrence is far from being elucidated (Poeck, 1983; Raymer et al., 1997; Lausberg et al., 2003). For example, even though normal subjects also commit body part as tool errors,[11] only subjects with brain lesion cannot correct their error after re-instruction (Raymer et al., 1997).

A significant step forward in the analysis of apraxic errors was the use of quantitative 3D kinematic motion analysis. These techniques allowed to show many abnormalities in the kinematic features of apraxic movements such as: deficits in spatial accuracy, irregular velocity profiles, reduced maximum velocities, reduced movement amplitudes, de-coupling of the relationship between instantaneous wrist velocity and trajectory curvature, improper linearity of the movement, wrong orientation of the movement in space and/or deficient joint coordination (Poizner et al., 1990, 1995, 1997; Clark et al., 1994; Platz and Mauritz, 1995; Rapcsak et al., 1995; Merians et al., 1997, 1999; Haaland et al., 1999; Binkofski et al., 2001; Hermsdörfer et al., 2006). An example of an apraxic movement with abnormal kinematics is shown in Fig. 3. Based on kinematic studies it could be concluded that ideomotor limb apraxia impaired the response implementation but not the preprogramming of the movement (Haaland et al., 1999) and decoupled the spatial and temporal representations of the movement (Poizner et al., 1990, 1995). Importantly, the kinematic abnormalities observed were often spatial and not temporal, the longer movement times in the apraxic group could be interpreted as an artifact of the longer distance traveled (Haaland et al., 1999; Hermsdörfer et al., 2006). However, several authors have advised against systematically interpreting the irregular kinematics as an indicator for deficient motor programming or deficient motor implementation (Platz and Mauritz, 1995; Haaland et al., 1999). For example, no correlation could be found between the kinematic abnormalities and apraxic errors in Hermsdörfer et al. (1996). Indeed, movements with degraded kinematics frequently reached a correct final position, while, on the contrary, kinematically normal movements often led to apraxic errors. The abnormal kinematic profile of the gesture probably arose from several corrective and compensatory strategies that the patient used to cope with the apraxic deficit (Goldenberg et al., 1996; Hermsdörfer et al., 1996). For example, hesitant and on-line controlled movements generated multi-peaked velocity profiles in our study (see Fig. 3d). Hence, according to the authors, the basic deficit underlying apraxia may concern the mental representation of the target position. Consistently with this hypothesis, it was found that apraxic

[9]Some authors have posited that an action–recognition mechanism might be at the basis of language development (Rizzolatti and Arbib, 1998).

[10]This list is not extensive. Terminologies can vary a lot across different authors.

[11]There is a hierarchical organization in the performance of actions with increasing difficulty. Children first acquire the ability to actually use objects, then to demonstrate the action with similar substitute objects, then with dissimilar substitute objects, then to use body parts as substitutes, and finally to perform pantomimes with holding imagined objects. This note was taken from Lausberg et al. (2003).

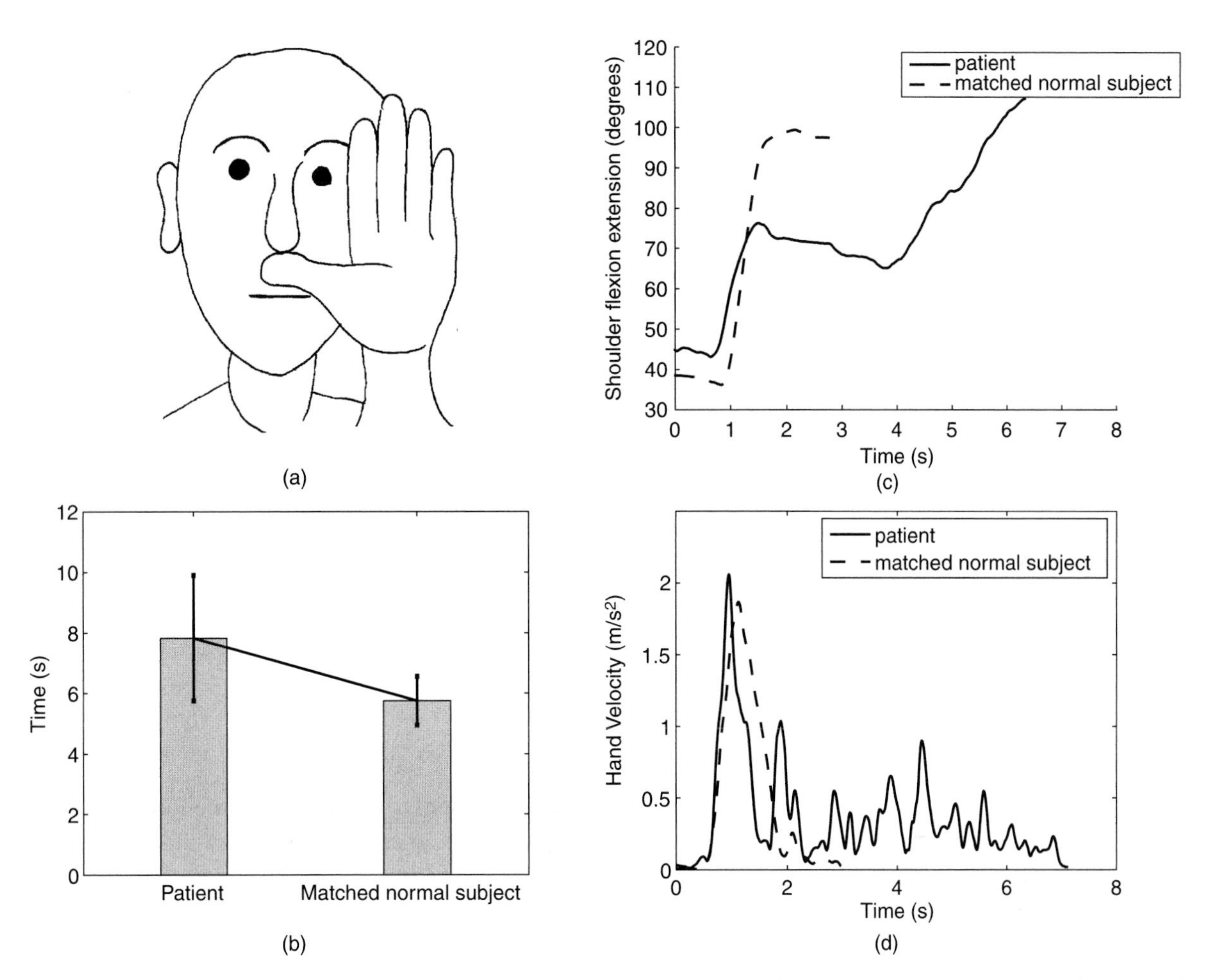

Fig. 3. An example of the abnormal kinematics of an apraxic movement. A patient with left ischemic lesions was tested in a study of imitation of meaningless gestures. The stimulus to imitate for this movement is shown under (a) and represents a hand posture relative to the head. Under (b), the movement times of the patient are longer than those of a matched normal subject (including replacement of the hand in the initial condition). Under (c), the trajectory of the shoulder flexion-extension joint angle of the patient (shown in solid line) contains several irregularities, which are the result from multiple hesitations and changes of directions, whereas the matched normal subject shoulder flexion-extension trajectory (dashed line) is smooth. The speed profile of the patient (solid line) is shown under (d) and contains multiple peaks with reduced maximum velocities that contrast with the simple bell-shaped velocity profile of the matched normal subject (dashed line).

patients relied more than normal subjects on on-line visual information in aiming movements (Ietswaart et al., 2006).

Discussion

We have shown in the preceding sections that apraxia has proven very difficult to assess and understand. Here we will try to provide some hypotheses on why these difficulties might arise and we propose several ways to overcome these.

The complex nature of apraxia

Apraxia designates the impairment of the human praxis system following brain lesion and has to deal with the high complexity and wide range of human praxic functions. Therefore studies of apraxia have separately tackled the faulty execution of many types of gestures (e.g., transitive and intransitive, meaningful and meaningless, peripersonal and body-centered, etc.) of various end-effectors (e.g., mouth, face, leg, limb) in different types of modalities (e.g., visual, auditive, tactile presentation and imitation). The high dimensionality of varying parameters has led to a lack of systematicity in the apraxia assessment and terminologies used. This has also rendered the coherent interpretation of the disorder rather arduous.

It follows that there is a great need to discriminate between different types of actions, as they appear to be differentially impaired in apraxia and hence may involve distinct underlying mechanisms (see "Types of apraxia"). Indeed, it is very likely that the mechanisms of imitation and execution of movements vary according to the type of action that is imitated or executed (Schnider et al., 1997; Goldenberg, 1999; Goldenberg and Karnath, 2006). This suggests that different categories of actions require the use of separate systems at some stage of processing, but the level of separation between the representations underlying actions of different types, or even different actions of the same type, is not at all clear yet.

We will principally argue that it is important to better understand what a particular gesture or execution modality implies in terms of brain resources and brain processes when compared to another gesture/execution modality. For example, a transitive action, i.e., an action that involves an object, is very different from an intransitive action in the sense that it provides supplementary tactile input as a result from the interaction with the object. This tactile sensory input then needs to be integrated to the representation of the action that relies also on other types of sensory inputs such as visual and proprioceptive. Moreover, executing a transitive action in a pantomime condition is also different from executing it with the object in hand, since the action has to be retrieved without the help of tactile input produced by the object. Indeed the movement is somehow modified, for example movement amplitudes in normal subjects were larger in the pantomime condition when compared to actual sawing (Hermsdörfer et al., 2006).

The distinction between meaningful and meaningless gestures would also need some clarification. The reproduction of a recognized meaningful gesture on the one hand, appears entirely based on the internal representation of the gesture. Indeed, the knowledge of a learned skilled act is preferably retrieved from motor memory rather than being constructed *de novo* (Halsband et al., 2001). On the other hand, the reproduction of a meaningless gesture involves a close visual tracking of the imitatee's body configuration and was modeled by a "visuomotor conversion mechanism" or a "body schema" (see Fig. 2b, c). To summarize, a meaningful gesture seems to be, to a certain extent, assimilated to a goal that guides the action from memory, whereas a meaningless gesture is defined as a particular configuration of the body in space and time, with no external referents (Goldenberg, 2001). Hence, imitation of meaningless gestures might be used to test the comprehension and replication of changing relationships between the multiple parts and subdivisions of the refined and complex mechanical device, which is the human body (Goldenberg, 2001). Furthermore, a preserved imitation of meaningless gestures is crucial for the apraxic patient as it might be useful for relearning motor skills. The double dissociation observed between imitation of meaningless and meaningful gestures argues for completely separate processing systems, and is still not accounted for by any of the existing apraxia models previously described. However, meaningless actions involve novel motor sequences that must be analyzed and constructed from the existing movements (Koski et al., 2002) and both meaningless and meaningful gestures appear to engage the body schema, i.e., a dynamic model for coding the body (Buxbaum et al., 2000). Hence, meaningless and meaningful actions may also share some overlapping conceptual representations.

These examples show that there are some common and some distinct processes involved in the different types of movements and modalities used for testing apraxia. Identifying the overlap of these processes would provide a clearer framework for interpreting the patient's performance and would simplify the analysis of the lesion correlates. The choice of the testing condition is crucial, as well as identifying the processes inherent to the chosen condition. However this is a difficult task, since correlations can be found between some very different and even dissociated types of movements.[12] For example, kinematic measures of pointing movements were correlated to gesture imitation, suggesting that the kinematic deficits observed during pointing movements are generalized to more global aiming movements, including movements for imitating hand gestures (Hermsdörfer et al., 2003). Accordingly, gesture imitation is believed to depend upon some of the same cognitive mechanisms as reaching and grasping (Haaland et al., 2000), however the level and extent of interplay is not clear. To make the picture even more complex, the underlying representations may be componential, for example with separate hand posture representations for transitive gestures (Buxbaum et al., 2005b). This leads us to two questions that urge to be answered: (1) What are the basic motor primitives from which all movements are constructed? and (2) Which are the motor components that are related to specific movements?

Beyond the complex nature of apraxia

One way to cope with the complex nature of apraxia is to be even more *precise* and *systematic* in assessing the apraxic disorder. Ideally, the full range of praxic functions, related to different effectors, including mouth, face and foot should be tested in a complete set of modalities (Koski et al., 2002). Moreover, we find it unfortunate that qualitative measures of the errors, such as kinematic measures of the movement trajectory (refer to "The analysis of apraxic errors"), are frequently missing or given in a purely statistical fashion (e.g., 25% of errors in Condition A). As such, these measures do not suffice to understand why the patient succeeds at the execution of some actions, but not other similar actions. For example, in one study the patient was able to evoke some actions (using a razor and a comb) fairly consistently, yet others (hammering and writing) were never produced (Graham et al., 1999). In another study, the same gestures were not always congruently disturbed across the different modes of execution, namely on imitation and on verbal command (Jacobs et al., 1999). We believe that it is this inability to distinguish between different types of errors related to different types of gestures that has prevented us so far from discovering the precise neuroanatomical correlates of apraxia, on top of the difficulty to accurately identify the brain lesion. Hence, the typology and analysis of apraxic errors need to be improved. We encourage extensive categorization of the errors and their characterization via kinematic methods. In addition, the errors should be reported *in relation to* the exact movement and not only specific condition tested.

We also suggest that studies that assess apraxia should more often integrate tasks of motor learning, as patients with apraxia may also be deficient in learning new motor tasks (Heilman et al., 1975; Rothi and Heilman, 1984; Platz and Mauritz, 1995; Lausberg et al., 1999). The main motivation in understanding apraxia is to help the apraxic patients in their everyday lives through the development of efficient rehabilitation methods and training programs.[13] Assessing the exact expression of the apraxic deficit, and especially the patient's motor learning abilities, would help to choose an appropriate therapy for the patient. Efficiently targeting the movements and praxis components specifically affected in each patient would accelerate the process of improving his or her praxic faculties. For the moment, apraxia in relation to motor learning is an under-investigated line of research.

Furthermore, we believe that modeling research may prove very helpful to gain some insight into the details and potential implementation of the processes underlying human praxis. When a roboticist searches for an algorithm for his robot to manipulate objects, he or she has to provide with all the different input signals and implement in practice all the necessary computations and processing resources. For example, the differences

[12] Surprisingly, single finger tapping was a better predictor of the severity of apraxia than goal-directed grasping and aiming (Ietswaart et al., 2006). Single finger tapping is almost never used to assess apraxia.

[13] According to Platz and Mauritz (1995), only patients with ideomotor apraxia and not ideational and constructional apraxia could benefit from a task-specific sensorimotor training.

and similarities between reaching to body-centered versus peripersonal cues would become evident through the development of corresponding algorithms, as they would be explicitly computed. According to Schaal and Schweighofer (2005), computational models of motor control in humans and robots often provide solid foundations that can help us to ground the vast amount of neuroscientific data that is collected today. Thus, biologically inspired modeling studies such as Sauser and Billard (2006) and Hersch and Billard (2006) seem to be very promising approaches in the understanding of the nature of gestures and in emphasizing the differences and similarities of their underlying processes.

Although neuropsychological models are essential for the understanding of apraxia, they do not address the question of the precise neural representation of the action and how this representation can be accessed. In a neurocomputational model, one has to take into account the computational principles of movement that reproduce the behavioral and kinematic results of the patient, as well as propose a biologically plausible implementation of the black-box components of apraxia models. In this view, we have a developed a simple neurocomputational model described in Petreska and Billard (2006), that accounts for the callosal apraxic deficit observed in a seminal experimental study of imitation of meaningless gestures (Goldenberg et al., 2001). Our model combines two computational methods for unsupervised learning applied to a series of artificial neural networks. The biologically inspired and distributed representations of sensory inputs self-organize according to Kohonen's algorithm (Kohonen, 2001) and associate with antihebbian learning (Gerstner and Kistler, 2002). The appropriate transformations between sensory inputs needed to reproduce certain gestures are thus learned within a biologically plausible framework. It is also possible to impair the networks in a way that accounts for the performance of Goldenberg et al.'s apraxic patient in all of the conditions of the study. The model also suggests potential neuroanatomical substrates for this task. We believe that the development of neurocomputational models is a good way to probe our understanding of apraxia and is compatible with the view of integrating knowledge from different lines of research, a point that we will defend in the following section.

Toward a multidisciplinary approach

We believe that apraxia can be best dismantled by adopting a multidisciplinary approach. Future models of apraxia will need to encompass knowledge and data from studies of *normal human motor control*, human brain imaging and monkey brain neurophysiology. Fortunately, several authors have already attempted to combine different sources of evidence: by considering apraxia in the neurophysiological framework (e.g., Leiguarda and Marsden, 2000) or by validating a model of apraxia using neuroimaging methods (e.g., Hermsdörfer et al., 2001; Peigneux et al., 2004; Chaminade et al., 2005; Mühlau et al., 2005).

Normal human motor control: has been extensively studied via behavioral, psychophysical, kinematic or computational methods for decades, giving rise to several principles of movement, such as: spatial control of arm movements (Morasso, 1981), maps of convergent force fields (Bizzi et al., 1991), uncontrolled manifold concepts (Scholz and Schöner, 1999), τ-coupling in the perceptual guidance of movements (Lee et al., 1999) and inverse and forward internal models (Wolpert and Ghahramani, 2000). Studies of motor control have also inspired several models for reaching like: minimum jerk trajectory control (Flash and Hogan, 1985), vector-integration-to-endpoint model (Bullock and Grossberg, 1988), minimum torque change model (Uno et al., 1989) and stochastic optimal feedback control (Todorov and Jordan, 2002) (for a review refer to Desmurget et al. (1998)). Proposed models for grasping (e.g., schema design (Oztop and Arbib, 2002)) are reviewed in Jeannerod et al. (1995) and models for sensorimotor learning (such as the modular selection and identification for control model (Haruno et al., 2001)) in Wolpert et al. (2001). In addition, it was also shown that the amplitude and direction of pointing movements may be independently processed (Vindras et al., 2005) or that the kinematics and dynamics for reaching may be separately learned (Krakauer et al., 1999). Investigation of apraxia can only benefit from taking into account the rich knowledge of the computational processes of movement used by the brain; and obviously, apraxia models would need to be compatible with the current general theories of movement control.

Progress in describing the contribution of specific brain regions to human praxis through the study of brain-damaged patients has been limited by the variability in the size, location and structures affected by the lesion (Koski et al., 2002). *Human brain imaging studies*, particularly positron emission tomography (PET) and functional magnetic resonance (fMRI) overcome this difficulty to a certain extent and have an essential role in resolving the neuroanatomical correlates of human functions. Despite the evident difficulties and limitations to study movements with neuroimaging, numerous studies have addressed the question of the representation of human praxis, making significant contributions to the understanding of the neural substrates underlying visuomotor control (for a review see Culham et al. (2006)). In order to give an idea of the number of praxis functions that have been addressed with brain imaging technologies, we will mention some of them: observation of meaningful and meaningless actions with the intent to recognize or imitate (Decety et al., 1997), hand imitation (Krams et al., 1998), visually guided reaching (Kertzman et al., 1997; Desmurget et al., 1999; Grefkes et al., 2004), object manipulation and tool-use (Binkofski et al., 1999; Johnson-Frey et al., 2005), real and/or imagined pantomimes (Moll et al., 2000; Choi et al., 2001; Rumiati et al., 2004) and sequential organization of actions (Ruby et al., 2002). The areas specialized for the perception of body parts and postures have been consistently identified[14] (Peigneux et al., 2000; Downing et al., 2001). Most importantly, several brain imaging studies have been conducted in relation to apraxia (Hermsdörfer et al., 2001; Peigneux et al., 2004; Chaminade et al., 2005; Mühlau et al., 2005) with the intent to test the neuroanatomical hypothesis of the neuropsychological models previously described.

Neurophysiological studies: allow the investigation of brain processes at the neuronal level and are essential to the understanding of the principles of neural computation. Certainly the monkey brain differs from the human brain, however this discrepancy can be overcome to some extent through the search of homologies (Rizzolatti et al., 2002; Arbib and Bota, 2003; Orban et al., 2004; Sereno and Tootell, 2005). Sensorimotor processes such as reaching and grasping for example, have been extensively studied: several parallel parietofrontal circuits were identified, each subserving a particular sensorimotor transformation (Kalaska et al., 1997; Wise et al., 1997; Matelli and Luppino, 2001; Battaglia-Mayer et al., 2003). Without going into the details of the representations used in each of these functionally distinct parietal and frontal areas (illustrated in Fig. 4), we will mention those that seem relevant for understanding apraxia. For example, LIP-FEF neurons discharge in relation with eye movements and are sensitive to the direction and amplitude of eye saccades (Platt and Glimcher, 1998), VIP-F4 neurons construct a representation of the "peripersonal space" confined to the head (Duhamel et al., 1998), MIP-F2 neurons have a crucial role in the planning, execution and monitoring of reaching movements (Eskandar et Assad, 1999; Simon et al., 2002; Raos et al., 2004) and finally AIP-F5 neurons mediate motor responses selective for hand manipulation and grasping movements (Cohen and Andersen, 2002). Furthermore, multiple space representations appear to coexist in the brain that integrate multisensory inputs (e.g., visual, somatosensory, auditory and vestibular inputs) (Graziano and Gross, 1998). For example, neurons in area 5 appear to combine visual and somatosensory signals in order to monitor the configuration of the limbs (Graziano et al., 2000) and the receptive fields of VIP neurons respond congruently (i.e., with matching receptive fields) to tactile and visual stimulation (Duhamel et al., 1998). It is very interesting that the modality-specific activities are spatially aligned: the visual receptive field corresponding to the arm or the face may shift along with that body part when it is passively

[14]Interestingly, these occipital and visually specialized areas are not only modulated by the visual presentation of body configurations, but also when the person executes a limb movement (Astafiev et al., 2004), indicating a bi-directional flow of the information.

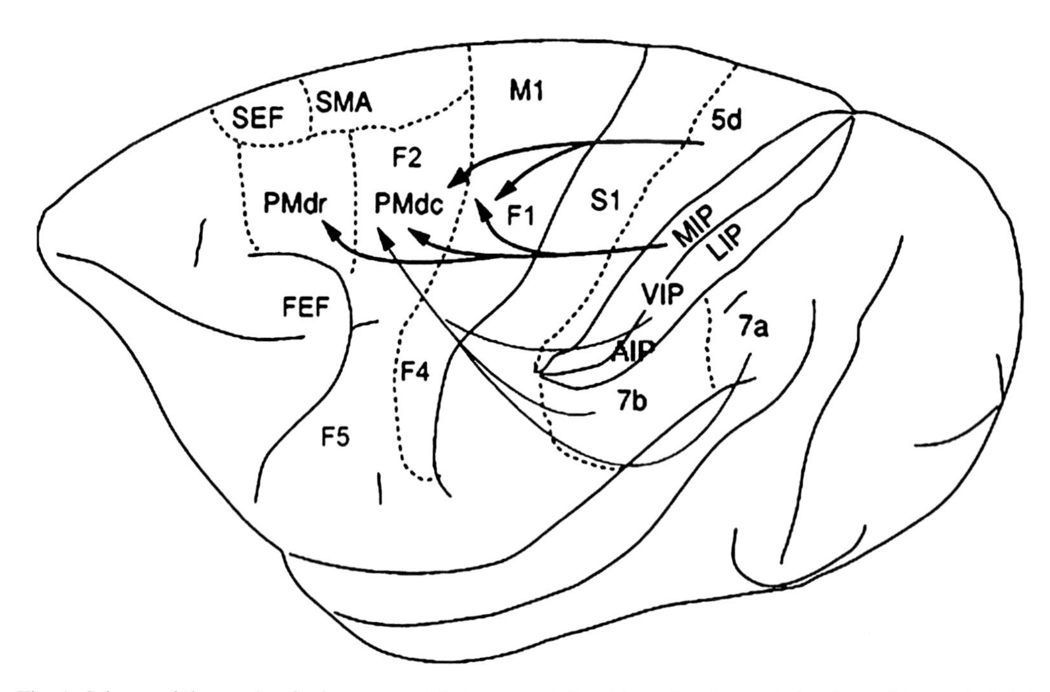

Fig. 4. Schema of the monkey brain areas and their connectivity. Adapted with permission from Wise et al. (1997).

moved (Graziano et al., 1997). In addition, neurophysiological data can give us insight into how the arm posture modulates the activity of somatosensory neurons (Helms Tillery et al., 1996) and how it affects the neurons that compute the trajectory of the hand (Scott et al., 1997). It should be noted that several sensorimotor transformations are needed in order to grasp an object, the motor command being in hand coordinates and the object's location in gaze coordinates. To compute these transformations, the brain appears to use multiple body-centered frames of references (Graziano and Gross, 1998): the frames of references underlying VIP area neurons appear to be organized along a continuum from eye to head coordinates (Duhamel et al., 1997; Avillac et al., 2005) and direct transformations from head to body-centered representations are possible in the posterior-parietal cortex (Buneo et al., 2002; Buneo and Andersen, 2006) with an error estimate of the target position computed in a common eye reference frame (Batista et al., 2002; Cohen and Andersen, 2002). Finally, it was also shown that tools may be integrated into the "body schema" at the neuronal level (Iriki et al., 1996; Maravita et al., 2003).

To conclude, we strongly believe that this multidisciplinary approach should be *bidirectional*. Not only apraxia can be interpreted in the neuropsychological and neurophysiological frameworks, but these research domains would also benefit from taking into consideration observations from apraxia. For example, one could learn enormously on how the normal human praxis system functions by looking at how it is affected by apraxia.

Acknowledgments

This work is supported in part by the Swiss National Science Foundation, through grant 620-066127 of the SFN Professorships Program, by the Sport and Rehabilitation Engineering Program at EPFL and by the Robotcub Project.

References

Alexander, M.P., Baker, E., Naeser, M.A., Kaplan, E. and Palumbo, C. (1992) Neuropsychological and neuroanatomical dimensions of ideomotor apraxia. Brain, 115: 87–107.

Arbib, M. and Bota, M. (2003) Language evolution: neural homologies and neuroinformatics. Neural Netw., 16: 1237–1260.

Astafiev, S.V., Stanley, C.M., Shulman, G.L. and Corbetta, M. (2004) Extrastriate body area in human occipital cortex responds to the performance of motor actions. Nat. Neurosci., 7(5): 542–548.

Avillac, M., Denève, S., Olivier, E., Pouget, A. and Duhamel, J.-R. (2005) Reference frames for representing visual and tactile locations in parietal cortex. Nat. Neurosci., 8(7): 941–949.

Bartolo, A., Cubelli, R., Della Sala, S., Drei, S. and Marchetti, C. (2001) Double dissociation between meaningful and meaningless gesture reproduction in apraxia. Cortex, 37: 696–699.

Basso, A., Luzzatti, C. and Spinnler, H. (1980) Is ideomotor apraxia the outcome of damage to well-defined regions of the left hemisphere? Neuropsychological study of CAT correlation. J. Neurol. Neurosurg. Psychiatry, 43: 118–126.

Batista, A.P., Buneo, C.A., Snyder, L.H. and Andersen, R.A. (1999) Reach plans in eye-centered coordinates. Science, 285(5425): 257–260.

Battaglia-Mayer, A., Caminiti, R., Lacquiniti, F. and Zago, M. (2003) Multiple levels of representation of reaching in the parieto-frontal network. Cereb. Cortex, 13: 1009–1022.

Binkofski, F., Buccino, G., Stephan, K.M., Rizzolatti, G., Seitz, R.J. and Freund, H.-J. (1999) A parieto-premotor network for object manipulation: evidence from neuroimaging. Exp. Brain Res., 128: 210–213.

Binkofski, F., Butler, A., Buccino, G., Heide, W., Fink, G., Freund, H.-J. and Seitz, R.J. (2003) Mirror apraxia affects the peripersonal mirror space. A combined lesion and cerebral activation study. Exp. Brain Res., 153: 210–219.

Binkofski, F., Kunesch, E., Classen, J., Seitz, R.J. and Freund, H.-J. (2001) Tactile apraxia: Unimodal apractic disorder of tactile object exploration associated with parietal lobe lesions. Brain, 124: 132–144.

Bizzi, E., Mussa-Ivaldi, F.A. and Giszter, S. (1991) Computations underlying the execution of movement: a biological perspective. Science, 253: 287–291.

Bizzozero, I., Costato, D., Della Sala, S., Papagno, C., Spinnler, H. and Venneri, A. (2000) Upper and lower face apraxia: role of the right hemisphere. Brain, 123: 2213–2230.

Bullock, D. and Grossberg, S. (1988) Neural dynamics of planned arm movements: emergent invariants and speed-accuracy properties during trajectory formation. Psychol. Rev., 95(1): 49–90.

Buneo, C.A. and Andersen, R.A. (2006) The posterior parietal cortex: sensorimotor interface for the planning and online control of visually guided movements. Neuropsychologia, 44(13): 2594–2606.

Buneo, C.A., Jarvis, M.R., Batista, A.P. and Andersen, R.A. (2002) Direct visuomotor transformations for reaching. Nature, 416: 632–635.

Buxbaum, L.J., Giovannetti, T. and Libon, D. (2000) The role of dynamic body schema in praxis: evidence from primary progressive apraxia. Brain Cogn., 44: 166–191.

Buxbaum, L.J., Johnson-Frey, S.H. and Bartlett-Williams, M. (2005a) Deficient internal models for planning hand–object interactions in apraxia. Neuropsychologia, 43: 917–929.

Buxbaum, L.J., Kyle, K.M. and Menon, R. (2005b) On beyond mirror neurons: internal representations subserving imitation and recognition of skilled object-related actions in humans. Brain Res. Cogn. Brain Res., 25: 226–239.

Buxbaum, L.J., Sirigu, A., Schwartz, M.F. and Klatzky, R. (2003) Cognitive representations of hand posture in ideomotor apraxia. Neuropsychologia, 41: 1091–1113.

Cermak, S. (1985) Developmental dyspraxia. In: Roy E.A. (Ed.), Neuropsychological Studies of Apraxia and Related Disorders. North-Holland, Amsterdam, pp. 225–248.

Chaminade, T., Meltzoff, A.N. and Decety, J. (2005) An fMRI study of imitation: action representation and body schema. Neuropsychologia, 43(2): 115–127.

Choi, S.H., Na, D.L., Kang, E., Lee, K.M., Lee, S.W. and Na, D.G. (2001) Functional magnetic resonance imaging during pantomiming tool-use gestures. Exp. Brain Res., 139(3): 311–317.

Clark, M.A., Merians, A.S., Kothari, A., Poizner, H., Macauley, B., Rothi, L.J.G. and Heilman, K.M. (1994) Spatial planning deficits in limb apraxia. Brain, 117: 1093–1106.

Cohen, Y.E. and Andersen, R.A. (2002) A common reference frame for movement plans in the posterior parietal cortex. Nat. Rev. Neurosci., 3: 553–562.

Cubelli, R., Marchetti, C., Boscolo, G. and Della Sala, S. (2000) Cognition in action: testing a model of limb apraxia. Brain Cogn., 44: 144–165.

Culham, J.C., Cavina-Pratesi, C. and Singhal, A. (2006) The role of parietal cortex in visuomotor control: what have we learned from neuroimaging? Neuropsychologia, 44(13): 2668–2684.

Decety, J., Grèzes, J., Costes, N., Jeannerod, M., Procyk, E., Grassi, E. and Fazio, F. (1997) Brain activity during observation of actions. Influence of action content and subject's strategy. Brain, 120: 1763–1777.

Della Sala, S., Faglioni, P., Motto, C. and Spinnler, H. (2006a) Hemisphere asymmetry for imitation of hand and finger movements, Goldenberg's hypothesis reworked. Neuropsychologia, 44(8): 1496–1500.

Della Sala, S., Maistrello, B., Motto, C. and Spinnler, H. (2006b) A new account of face apraxia based on a longitudinal study. Neuropsychologia, 44(7): 1159–1165.

De Renzi, E., Faglioni, P. and Sorgato, P. (1982) Modality specific and supramodal mechanisms of apraxia. Brain, 105: 301–312.

De Renzi, E. and Lucchelli, F. (1988) Ideational apraxia. Brain, 111: 1173–1185.

De Renzi, E., Motti, F. and Nichelli, P. (1980) Imitating gestures. A quantitative approach to ideomotor apraxia. Arch. Neurol., 37(1): 6–10.

Desmurget, M., Epstein, C.M., Turner, R.S., Prablanc, C., Alexander, G.E. and Grafton, S.T. (1999) Role of the posterior parietal cortex in updating reaching movements to a visual target. Nat. Neurosci., 2(6): 563–567.

Desmurget, M., Pélisson, D., Rossetti, Y. and Prablanc, C. (1998) From eye to hand: planning goal-directed movements. Neurosci. Biobehav. Rev., 22(6): 761–788.

Downing, P.E., Jiang, Y., Shuman, M. and Kanwisher, N. (2001) A cortical area selective for visual processing of the human body. Science, 293: 2470–2473.

Duhamel, J.-R., Bremmer, F., BenHamed, S. and Graf, W. (1997) Spatial invariance of visual receptive fields in parietal cortex neurons. Nature, 389: 845–848.

Duhamel, J.-R., Colby, C.L. and Goldberg, M.E. (1998) Ventral intraparietal area of the macaque: congruent visual and somatic response properties. J. Neurophysiol., 79: 126–136.

Eskandar, E.N. and Assad, J.A. (1999) Dissociation of visual, motor and predictive signals in parietal cortex during visual guidance. Nat. Neurosci., 2(1): 88–93.

Faglioni, P. and Basso, A. (1985) Historical perspectives on neuroanatomical correlates of limb apraxia. In: Roy E.A. (Ed.), Neuropsychological studies of apraxia and related disorders. North-Holland, Amsterdam, pp. 3–44.

Flash, T. and Hogan, N. (1985) The coordination of arm movements: an experimentally confirmed mathematical model. J. Neurosci., 5: 1688–1703.

Gallese, V., Fadiga, L., Fogassi, L. and Rizzolatti, G. (1996) Action recognition in the premotor cortex. Brain, 119: 593–609.

Gerstner, W. and Kistler, W. (2002) Spiking neuron models: single neurons, populations, plasticity. Cambridge University Press.

Geschwind, N. (1965) Disconnexion syndromes in animals and man. Part I. Brain, 88: 237–294.

Goldenberg, G. (1995) Imitating gestures and manipulating a mannikin — the representation of the human body in ideomotor apraxia. Neuropsychologia, 33(1): 63–72.

Goldenberg, G. (1999) Matching and imitation of hand and finger postures in patients with damage in the left or right hemispheres. Neuropsychologia, 37(5): 559–566.

Goldenberg, G. (2001) Imitation and matching of hand and finger postures. Neuroimage, 14: S132–S136.

Goldenberg, G. (2003) Apraxia and beyond: life and work of Hugo Liepmann. Cortex, 39(3): 509–524.

Goldenberg, G. and Hagmann, S. (1997) The meaning of meaningless gestures: a study of visuo-imitative apraxia. Neuropsychologia, 35(3): 333–341.

Goldenberg, G., Hermsdörfer, J. and Spatt, J. (1996) Ideomotor apraxia and cerebral dominance for motor control. Cogn. Brain Res., 3: 95–100.

Goldenberg, G. and Karnath, H.-O. (2006) The neural basis of imitation is body part specific. J. Neurosci., 26(23): 6282–6287.

Goldenberg, G., Laimgruber, K. and Hermsdörfer, J. (2001) Imitation of gestures by disconnected hemispheres. Neuropsychologia, 39: 1432–1443.

Goodglass, H. and Kaplan, E. (1963) Disturbance of gesture and pantomime in aphasia. Brain, 86: 703–720.

Graham, N.L., Zeman, A., Young, A.W., Patterson, K. and Hodges, J.R. (1999) Dyspraxia in a patient with corticobasal degeneration: the role of visual and tactile inputs to action. J. Neurol. Neurosurg. Psychiatry, 67: 334–344.

Graziano, M.S.A., Cooke, D.F. and Taylor, C.S. (2000) Coding the location of the arm by sight. Science, 290: 1782–1786.

Graziano, M.S.A. and Gross, C.G. (1998) Spatial maps for t he control of movement. Curr. Opin. Neurobiol., 8(2): 195–201.

Graziano, M.S.A., Hu, X.T. and Gross, C.G. (1997) Visuospatial properties of ventral premotor cortex. J. Neurophysiol., 77: 2268–2292.

Grefkes, C., Ritzl, A., Zilles, K. and Fink, G.R. (2004) Human medial intraparietal cortex subserves visuomotor coordinate transformation. Neuroimage, 23: 1494–1506.

Haaland, K.Y. and Flaherty, D. (1984) The different types of limb apraxia errors made by patients with left vs. right hemisphere damage. Brain Cogn., 3(4): 370–384.

Haaland, K.Y. and Harrington, D.L. (1996) Hemispheric asymmetry of movement. Curr. Opin. Neurobiol., 6: 796–800.

Haaland, K.Y., Harrington, D.L. and Knight, R.T. (1999) Spatial deficits in ideomotor limb apraxia. A kinematic analysis of aiming movements. Brain, 122: 1169–1182.

Haaland, K.Y., Harrington, D.L. and Knight, R.T. (2000) Neural represantations of skilled movement. Brain, 123: 2306–2313.

Haaland, K.Y., Prestopnik, J.L., Knight, R.T. and Lee, R.R. (2004) Hemispheric asymmetries for kinematic and positional aspects of reaching. Brain, 127: 1145–1158.

Halsband, U., Schmitt, J., Weyers, M., Binkofski, F., Grützner, G. and Freund, H.-J. (2001) Recognition and imitation of pantomimed motor acts after unilateral parietal and premotor lesions: a perspective on apraxia. Neuropsychologia, 39: 200–216.

Hanna-Pladdy, B., Heilman, K.M. and Foundas, A.L. (2001) Cortical and subcortical contributions to ideomotor apraxia: analysis of task demands and error types. Brain, 124: 2513–2527.

Haruno, M., Wolpert, D.M. and Kawato, M. (2001) MOSAIC model for sensorimotor learning and control. Neural Comput., 13: 2201–2220.

Heath, M., Roy, E.A., Black, S.E. and Westwood, D.A. (2001) Intransitive limb gestures and apraxia following unilateral stroke. J. Clin. Exp. Neuropsychol., 23(5): 628–642.

Heilman, K.M. (1973) Ideational apraxia — a re-definition. Brain, 96: 861–864.

Heilman, K.M., Maher, L.M., Greenwald, M.L. and Rothi, L.J. (1997) Conceptual apraxia from lateralized lesions. Neurology, 49(2): 457–464.

Heilman, K.M. and Rothi, L.J. (1993) Apraxia. In: Heilman K.M. and Valenstein E. (Eds.), Clinical Neuropsychology (3rd ed.). Oxford University Press, New York, pp. 141–163.

Heilman, K.M., Rothi, L.G., Mack, L., Feinberg, T. and Watson, R.T. (1986) Apraxia after a superior parietal lesion. Cortex, 22(1): 141–150.

Heilman, K.M., Rothi, L.J. and Valenstein, E. (1982) Two forms of ideomotor apraxia. Neurology, 32: 342–346.

Heilman, K.M., Schwartz, H.D. and Geschwind, N. (1975) Defective motor learning in ideomotor apraxia. Neurology, 25(11): 1018–1020.

Helms Tillery, S.I., Soechting, J.F. and Ebner, T.J. (1996) Somatosensory cortical activity in relation to arm posture: nonuniform spatial tuning. J. Neurophysiol., 76(4): 2423–2438.

Hermsdörfer, J., Blankenfeld, H. and Goldenberg, G. (2003) The dependence of ipsilesional aiming deficits on task demands, lesioned hemisphere, and apraxia. Neuropsychologia, 41: 1628–1643.

Hermsdörfer, J., Goldenberg, G., Wachsmuth, C., Conrad, B., Ceballos-Baumann, A.O., Bartenstein, P., Schwaiger, M. and Boecker, H. (2001) Cortical correlates of gesture processing: clues to the cerebral mechanisms underlying apraxia during the imitation of meaningless gestures. Neuroimage, 14: 149–161.

Hermsdörfer, J., Hentze, S. and Goldenberg, G. (2006) Spatial and kinematic features of apraxic movement depend on the mode of execution. Neuropsychologia, 44: 1642–1652.

Hermsdörfer, J., Mai, N., Spatt, J., Marquardt, C., Veltkamp, R. and Goldenberg, G. (1996) Kinematic analysis of movement imitation in apraxia. Brain, 119: 1575–1586.

Hersch, M. and Billard, A.G. (2006) A biologically-inspired model of reaching movements. In Proceedings of the 2006 IEEE/RAS-EMBS International Conference on Biomedical Robotics and Biomechatronics, Pisa, pp. 1067–1072.

Ietswaart, M., Carey, D.P. and Della Sala, S. (2006) Tapping, grasping and aiming in ideomotor apraxia. Neuropsychologia, 44: 1175–1184.

Ietswaart, M., Carey, D.P., Della Sala, S. and Dijkhuizen, R.S. (2001) Memory-driven movements in limb apraxia: is there evidence for impaired communication between the dorsal and the ventral streams? Neuropsychologia, 39: 950–961.

Iriki, A., Tanaka, M. and Iwamura, Y. (1996) Coding of modified body schema during tool use by macaque postcentral neurons. Neuroreport, 7: 2325–2330.

Jacobs, D.H., Adair, J.C., Macauley, B., Gold, M., Gonzalez Rothi, L.J. and Heilman, K.M. (1999) Apraxia in corticobasal degeneration. Brain Cogn., 40: 336–354.

Jeannerod, M., Arbib, M., Rizzolatti, G. and Sakata, H. (1995) Grasping objects: the cortical mechanisms of visuomotor transformation. Trends Neurosci., 18: 314–320.

Jeannerod, M. and Decety, J. (1995) Mental motor imagery: a window into the representational stages of action. Curr. Opin. Neurobiol., 5: 727–732.

Johnson-Frey, S.H., Newman-Norlund, R. and Grafton, S.T. (2005) A distributed left hemisphere network active during planning of everyday tool use skills. Cereb. Cortex, 15: 681–695.

Kalaska, J.F., Scott, S.H., Cisek, P. and Sergio, L.E. (1997) Cortical control of reaching movements. Curr. Opin. Neurobiol., 7: 849–859.

Kertesz, A. and Ferro, J.M. (1984) Lesion size and location in ideomotor apraxia. Brain, 107: 921–933.

Kertesz, A. and Hooper, P. (1982) Praxis and language: the extent and variety of apraxia in aphasia. Neuropsychologia, 20(3): 275–286.

Kertzman, C., Schwarz, U., Zeffiro, T.A. and Hallett, M. (1997) The role of posterior parietal cortex in visually guided reaching movements in humans. Exp. Brain. Res., 114: 170–183.

Kohonen, T. (2001) Self-Organizing Maps. 3rd ed., Springer Series in Information Sciences. Vol. 30. Springer, Berlin.

Koski, L., Iacoboni, M. and Mazziotta, J.C. (2002) Deconstructing apraxia: understanding disorders of intentional movement after stroke. Curr. Opin. Neurol., 15: 71–77.

Krakauer, J.W., Ghilardi, M.-F. and Ghez, C. (1999) Independent learning of internal models for kinematic and dynamic control of reaching. Nat. Neurosci., 2(11): 1026–1031.

Krams, M., Rushworth, M.F., Deiber, M.-P., Frackowiak, R.S. and Passingham, R.E. (1998) The preparation, execution and suppression of copied movements in the human brain. Exp. Brain Res., 120(3): 386–398.

Laeng, B. (2006) Constructional apraxia after left or right unilateral stroke. Neuropsychologia, 44(9): 1595–1606.

Lausberg, H. and Cruz, R.F. (2004) Hemispheric specialisation for imitation of hand-head positions and finger configurations: a controlled study in patients with complete callosotomy. Neuropsychologia, 42: 320–334.

Lausberg, H., Cruz, R.F., Kita, S., Zaidel, E. and Ptito, A. (2003) Pantomime to visual presentation of objects: left hand dyspraxia in patients with complete callosotomy. Brain, 126: 343–360.

Lausberg, H., Davis, M. and Rothenhäusler, A. (2000) Hemispheric specialization in spontaneous gesticulation in a patient with callosal disconnection. Neuropsychologia, 38: 1654–1663.

Lausberg, H., Göttert, R., Münssinger, U., Boegner, F. and Marx, P. (1999) Callosal disconnection syndrome in a left-handed patient due to infarction of the total length of the corpus callosum. Neuropsychologia, 37: 253–265.

Lee, D.N., Craig, C.M. and Grealy, M.A. (1999) Sensory and intrinsic coordination of movement. Proc. R. Soc. Lond. B, 266: 2029–2035.

Lehmkuhl, G., Poeck, K. and Willmes, K. (1983) Ideomotor apraxia and aphasia: an examination of types and manifestations of apraxic symptoms. Neuropsychologia, 21(3): 199–212.

Leiguarda, R. (2001) Limb apraxia: cortical or subcortical. Neuroimage, 14: S137–S141.

Leiguarda, R.C. and Marsden, C.D. (2000) Limb apraxias: higher order disorders of sensorimotor integration. Brain, 123: 860–879.

Maravita, A., Spence, C. and Driver, J. (2003) Multisensory integration and the body schema: close to hand and within reach. Curr. Biol., 13: R531–R539.

Marchetti, C. and Della Sala, S. (1997) On crossed apraxia. Description of a right-handed apraxic patient with right supplementary motor area damage. Cortex, 33(2): 341–354.

Matelli, M. and Luppino, G. (2001) Parietofrontal circuits for action and space perception in the macaque monkey. Neuroimage, 14: S27–S32.

Merians, A.S., Clark, M., Poizner, H., Jacobs, D.H., Adair, J.C., Macauley, B., Rothi, L.J.G. and Heilman, K.M. (1999) Apraxia differs in corticobasal degeneration and left-parietal stroke: a case study. Brain Cogn., 40: 314–335.

Merians, A.S., Clark, M., Poizner, H., Macauley, B., Gonzalez Rothi, L.J. and Heilman, K.M. (1997) Visual-imitative dissociation apraxia. Neuropsychologia, 35(11): 1483–1490.

Moll, J., de Oliveira-Souza, R., Passman, L.J., Souza-Lima, F. and Andreiuolo, P.A. (2000) Functional MRI correlates of real and imagined tool-use pantomime. Neurology, 54: 1331–1336.

Morasso, P. (1981) Spatial control of arm movements. Exp. Brain Res., 42: 223–227.

Mühlau, M., Hermsdörfer, J., Goldenberg, G., Wohlschläger, A.M., Castrop, F., Stahl, R., Röttinger, M., Erhard, P., Haslinger, B., Ceballos-Baumann, A.O., Conrad, B. and Boecker, H. (2005) Left inferior parietal dominance in gesture imitation: an fMRI study. Neuropsychologia, 43: 1086–1098.

Nehaniv, C.L. and Dautenhahn, K. (2002) The correspondence problem. In: Dautenhahn K. and Nehaniv C.L. (Eds.), Imitation in Animals and Artifacts. MIT Press, London, pp. 41–61.

Ochipa, C., Rothi, L.J.G. and Heilman, K.M. (1992) Conceptual apraxia in Alzheimer's disease. Brain, 115: 1061–1071.

Ochipa, C., Rothi, L.J.G. and Heilman, K.M. (1994) Conduction apraxia. J. Neurol. Neurosurg. Psychiatry, 57: 1241–1244.

Orban, G.A., Van Essen, D. and Vanduffel, W. (2004) Comparative mapping of higher visual areas in monkeys and humans. Trends Cogn. Sci., 8(7): 315–324.

Oztop, E. and Arbib, M.A. (2002) Schema design and implementation of the grasp-related mirror neuron system. Biol. Cybern., 78: 116–140.

Papagno, C., Della Sala, S. and Basso, A. (1993) Ideomotor apraxia without aphasia and aphasia without apraxia: the anatomical support for a double dissociation. J. Neurol. Neurosurg. Psychiatry, 56: 286–289.

Parakh, R., Roy, E., Koo, E. and Black, S. (2004) Pantomime and imitation of limb gestures in relation to the severity of Alzheimer's disease. Brain Cogn., 55: 272–274.

Peigneux, P., Van der Linden, M., Garraux, G., Laureys, S., Degueldre, C., Aerts, J., Del Fiore, G., Moonen, G., Luxen, A. and Salmon, E. (2004) Imaging a cognitive model of apraxia: the neural substrate of gesture-specific cognitive processes. Hum. Brain Mapp., 21: 119–142.

Peigneux, P., Salmon, E., Van der Linden, M., Garraux, G., Aerts, J., Delfiore, G., Deguel-dre, C., Luxen, A., Orban, G. and Franck, G. (2000) The role of lateral occipi-totemporal junction and area MT/V5 in the visual analysis of upper-limb postures. Neuroimage, 11: 644–655.

Petreska, B. and Billard, A.G. (2006) A neurocomputational model of an imitation deficit following brain lesion, in Proceedings of 16th International Conference on Artificial Neural Networks (ICANN 2006). Lecture Notes in Computer Science, LNCS 4131: 770–779.

Platt, M.L. and Glimcher, P.W. (1998) Response fields of intraparietal neurons quantified with multiple saccadic targets. Exp. Brain Res., 121: 65–75.

Platz, T. and Mauritz, K.-H. (1995) Human motor planning, motor programming, and use of new task-relevant information with different apraxic syndromes. Eur. J. Neurosci., 7: 1536–1547.

Poeck, K. (1983) Ideational apraxia. J. Neurol., 230: 1–5.

Poizner, H., Clark, M.A., Merians, A.S., Macauley, B., Rothi, L.J.G. and Heilman, K.M. (1995) Joint coordination deficits in limb apraxia. Brain, 118: 227–242.

Poizner, H., Mack, L., Verfaellie, M., Rothi, L.J.G. and Heilman, K.M. (1990) Three-dimensional computergraphic analysis of apraxia. Neural representations of learned movement. Brain, 113(1): 85–101.

Poizner, H., Merians, A.S., Clark, M.A., Rothi, L.J.G. and Heilman, K.M. (1997) Kinematic approaches to the study of apraxic disorders. In: Rothi L.J.G. and Heilman K.M. (Eds.), Apraxia: The neuropsychology of Action. Psychology Press, Hove, UK, pp. 93–109.

Pramstaller, P.P. and Marsden, C.D. (1996) The basal ganglia and apraxia. Brain, 119: 319–340.

Raos, V., Umiltá, M.-A., Gallese, V. and Fogassi, L. (2004) Functional properties of grasping-related neurons in the dorsal premotor area F2 of the macaque monkey. J. Neurophysiol., 92: 1990–2002.

Rapcsak, S.Z., Ochipa, C., Anderson, K.C. and Poizner, H. (1995) Progressive ideomotor apraxia: evidence for a selective impairment of the action production system. Brain Cogn., 27: 213–236.

Raymer, A.M., Maher, L.M., Foundas, A.L., Heilman, K.M. and Rothi, L.J.G. (1997) The significance of body part as tool errors in limb apraxia. Brain Cogn., 34: 287–292.

Raymer, A.M., Merians, A.S., Adair, J.C., Schwartz, R.L., Williamson, D.J.G., Rothi, L.J.G., Poizner, H. and Heilman, K.M. (1999) Crossed apraxia: Implications for handedness. Cortex, 35(2): 183–199.

Rizzolatti, G. and Arbib, M.A. (1998) Language within our grasp. Trends Neurosci., 21(5): 188–194.

Rizzolatti, G. and Craighero, L. (2004) The mirror-neuron system. Annu. Rev. Neurosci., 27: 169–192.

Rizzolatti, G., Fogassi, L. and Gallese, V. (2002) Motor and cognitive functions of the ventral premotor cortex. Curr. Opin. Neurobiol., 12: 149–154.

Rizzolatti, G. and Luppino, G. (2001) The cortical motor system. Neuron, 31: 889–901.

Rothi, L.J.G. and Heilman, K.M. (1984) Acquisition and retention of gestures by apraxic patients. Brain Cogn., 3(4): 426–437.

Rothi, L.J.G., Ochipa, C. and Heilman, K.M. (1991) A cognitive neuropsychological model of limb praxis. Cogn. Neuropsychol., 8(6): 443–458.

Rothi, L.J.G., Ochipa, C. and Heilman, K.M. (1997) A cognitive neuropsychological model of limb praxis and apraxia. In: Rothi L.J.G. and Heilman K.M. (Eds.), Apraxia: The neuropsychology of Action. Psychology Press, Hove, UK, pp. 29–49.

Roy, E.A., Black, S.E., Winchester, T.R. and Barbour, K.L. (1992) Gestural imitation following stroke. Brain Cogn., 30(3): 343–346.

Roy, E.A., Heath, M., Westwood, D., Schweizer, T.A., Dixon, M.J., Black, S.E., Kalbfleisch, L., Barbour, K. and Square, P.A. (2000) Task demands and limb apraxia in stroke. Brain Cogn., 44: 253–279.

Roy, E.A., Square, P.A. (1985) Common considerations in the study of limb, verbal and oral apraxia. In: Roy, E.A. (Ed.), Neuropsychological Studies of Apraxia and Related Disorders. North-Holland, Amsterdam, Series Advances in Psychology, Vol. 23, pp. 111–161.

Ruby, P., Sirigu, A. and Decety, J. (2002) Distinct areas in parietal cortex involved in long-term and short-term action planning: a PET investigation. Cortex, 38: 321–339.

Rumiati, R.I., Weiss, P.H., Shallice, T., Ottoboni, G., Noth, J., Zilles, K. and Fink, G.R. (2004) Neural basis of pantomiming the use of visually presented objects. Neuroimage, 21(4): 1224–1231.

Salter, J.E., Roy, E.A., Black, S.E., Joshi, A. and Almeida, Q.J. (2004) Gestural imitation and limb apraxia in corticobasal degeneration. Brain Cogn., 55(2): 400–402.

Sauser, E.L. and Billard, A.G. (2006) Parallel and distributed neural models of the ideomotor principle: an investigation of imitative cortical pathways. Neural Netw., 19(3): 285–298.

Schaal, S. and Schweighofer, N. (2005) Computational motor control in humans and robots. Curr. Opin. Neurobiol., 6: 675–682.

Schnider, A., Hanlon, R.E., Alexander, D.N. and Benson, D.F. (1997) Ideomotor apraxia: behavioral dimensions and neuroanatomical basis. Brain Lang., 57: 125–136.

Scholz, J.P. and Schöner, G. (1999) The uncontrolled manifold concept: identifying control variables for a functional task. Exp. Brain Res., 126(3): 289–306.

Scott, S.H., Sergio, L.E. and Kalaska, J.F. (1997) Reaching movements with similar handpaths but different arm orientations. II. Activity of individual cells in dorsal premotor cortex and parietal area 5. J. Neurophysiol., 78: 2413–2416.

Sereno, M.I. and Tootell, R.B. (2005) From monkeys to humans: what do we now know about brain homologies? Curr. Opin. Neurobiol., 15(2): 135–144.

Simon, O., Mangin, J.-F., Cohen, L., Le Bihan, D. and Dehaene, S. (2002) Topographical layout of hand, eye, calculation, and language-related areas in the human parietal lobe. Neuron, 33(3): 475–487.

Sirigu, A., Daprati, E., Pradat-Diehl, P., Franck, N. and Jeannerod, M. (1999) Perception of self-generated movement following left parietal lesion. Brain, 122: 1867–1874.

Todorov, E. and Jordan, M.I. (2002) Optimal feedback control as a theory of motor coordination. Nat. Neurosci., 5(11): 1226–1235.

Uno, Y., Kawato, M. and Suzuki, R. (1989) Formation and control of optimal trajectory in human multijoint arm movement. Minimum torque-change model. Biol. Cybern., 61: 89–101.

Vindras, P., Desmurget, M. and Viviani, P. (2005) Error parsing in visuomotor pointing reveals independent processing of amplitude and direction. J. Neurophysiol., 94: 1212–1224.

Watson, R.T., Fleet, W.S., Rothi, L.J.G. and Heilman, K.M. (1986) Apraxia and the supplementary motor area. Arch. Neurol., 43(8): 787–792.

Watson, R.T. and Heilman, K.M. (1983) Callosal apraxia. Brain, 106: 391–403.

Weiss, P.H., Dohle, C., Binkofski, F., Schnitzler, A., Freund, H.-J. and Hefter, H. (2001) Motor impairment in patients with parietal lesions: disturbances of meaningless arm movement sequences. Neuropsychologia, 39: 397–405.

Wise, S.P., Boussaoud, D., Johnson, P.B. and Caminiti, R. (1997) Premotor and parietal cortex: corticocortical connectivity and combinatorial computations. Annu. Rev. Neurosci., 20: 25–42.

Wolpert, D.M. and Ghahramani, Z. (2000) Computational principles of movement neuroscience. Nat. Neurosci., 3: 1212–1217.

Wolpert, D.M., Ghahramani, Z. and Flanagan, J.R. (2001) Perspectives and problems in motor learning. Trends Cogn. Sci., 5(11): 487–494.

SECTION II

The Early Development of Perception and Action

C. von Hofsten & K. Rosander (Eds.)
Progress in Brain Research, Vol. 164
ISSN 0079-6123

CHAPTER 5

Effects of early visual deprivation on perceptual and cognitive development

Daphne Maurer[1,*], Catherine J. Mondloch[2] and Terri L. Lewis[1,3,4]

[1]Department of Psychology, Neuroscience & Behaviour, McMaster University, Hamilton, ON, L8S 4K1, Canada
[2]Department of Psychology, Brock University, 500 Glenridge Avenue, St. Catharines, ON, L2S 3A1, Canada
[3]Department of Ophthalmology, The Hospital for Sick Children, Toronto, ON, M5G 1X8, Canada
[4]Department of Ophthalmology and Vision Sciences, University of Toronto, Toronto, ON, M5S 1A8, Canada

Abstract: During early infancy, visual capabilities are quite limited. Nevertheless, patterned visual input during this period is necessary for the later development of normal vision for some, but not all, aspects of visual perception. The evidence comes from studies of children who missed early visual input because it was blocked by dense, central cataracts in both eyes. In this article, we review the effects of bilateral congenital cataracts on two aspects of low-level vision – acuity and contrast sensitivity, and on three aspects of higher-level processing of faces. We end by discussing the implications for understanding the developmental mechanisms underlying normal perceptual and cognitive development.

Keywords: visual deprivation; perceptual development; plasticity; acuity; face processing; contrast sensitivity

Newborns can see, but there are serious limitations on their visual perception: their visual acuity is 30–40 times worse than that of adults (Atkinson et al., 1977; Banks and Salapatek, 1978; Brown and Yamamoto, 1986; van Hof-van Duin and Mohn, 1986; Courage and Adams, 1990; reviewed in Maurer and Lewis, 2001a, b) and their processing of faces is very limited (e.g., de Haan et al., 2001; Cashon and Cohen, 2003, 2004; Bertin and Bhatt, 2004; Bhatt et al., 2005). There is rapid progress during infancy, such that by 6–8 months of age, visual acuity is only 6 times worse than that of adults (Mayer et al., 1995; reviewed in Maurer and Lewis, 2001a, b) and most types of face processing have emerged. By 4–6 years of age, acuity is adult-like (Mayer and Dobson, 1982; Ellemberg t al., 1999a), but some types of face processing continue to improve into adolescence (Carey et al., 1980; Bruce et al., 2000; Mondloch et al., 2002, 2003b). In this chapter, we will evaluate the role of visual input in driving the postnatal changes in acuity and in face processing by contrasting the perceptual development of children with normal eyes to that of children who were deprived of patterned visual input at birth because they were born with dense, central cataracts in both eyes.

*Corresponding author. Tel.: +1 905 525 9140 ext. 23030; Fax: +1 905 529 6225; E-mail: maurer@mcmaster.ca

Children treated for congenital cataract

In the patients we selected for our studies, the cataracts were central and so dense that they blocked all patterned visual input to the retina. Inclusion criteria included that the infant did not

DOI: 10.1016/S0079-6123(07)64005-9

fixate a light or follow it when it moved, that the cataract blocked completely an ophthalmologist's view of the retina, and/or that the ophthalmologist described the cataract as dense and central. In all cases, the cataracts were present from the time of the first eye examination, which always occurred before 6 months of age. Because it is unlikely that complete cataracts develop rapidly during infancy, we have assumed that the deprivation was always present from birth. Treatment involved surgical removal of the cataractous lens and fitting the eyes with compensatory contact lenses. We have monitored the patients' visual development from the time when the visual deprivation ended, that is, from the time the contact lenses allowed the first focused patterned input to the retina.

Children treated for bilateral congenital cataract afford an opportunity to evaluate the role of visual input in driving the postnatal perceptual changes seen in normal development. However, there are limitations to this natural experiment that must be kept in mind when interpreting any deficits. First, before treatment, the retina may not be completely deprived of visual input because changes from bright light to complete darkness may be transmitted through the cataractous lens sufficiently well to cause small changes in the illumination of the retina. Second, after treatment, the contact lenses provide a fixed focus, such that objects at some specific distance are in perfect focus, but objects closer to the child, or farther away, are increasingly out of focus. Typically, the contact lenses are fit to give the child perfect focus at arms' length until the child begins to walk, at which point the contact lens for one eye is changed to focus perfectly farther from the child. In addition, children treated for bilateral congenital cataract often develop secondary eye problems such as nystagmus (small, jiggly eye movements) or strabismus (misaligned eyes). As a result of the fixed focus and such secondary eye problems, children treated for bilateral congenital cataract do not receive completely normal visual input at any point in their lives. Thus, any deficits may arise from the initial complete deprivation of patterned visual input and/or from the continuing milder alterations of visual input. In fact, however, control experiments suggest that the deficits we have found arise from the original deprivation caused by the cataracts and not from later perturbations.

A final limitation is that the conclusions are limited to the variability in the duration of deprivation found in our natural sample. No child in the studies we will report here had visual deprivation lasting less than the first month of life and, in most of our studies, none had deprivation lasting more than the first year of life. Thus, we do not know if the outcome would be better with shorter deprivation, nor if it would be as good after longer deprivation.

Visual acuity

Visual acuity in adults and children old enough to read typically is measured by having them read letters on an eye chart containing letters of decreasing size. The smallest letters that can be read accurately provide a measure of visual acuity. Visual acuity in infants typically is measured by determining the narrowest stripe width that the infant can see. One method — Teller Acuity cards — takes advantage of infants' natural preference to look at something patterned, like stripes, in preference to a plain gray. Infants are shown cards with a patch of stripes to one side (to the left or right of center) on a gray background of matched luminance. A peephole in the middle of the card allows a tester, unaware of the size and side of the stripes, to watch the infant's reaction to each card to determine if the child prefers looking to the right or left. The tester then inverts the card 180° to see if the infant's looking preference switches to the other side (e.g., from the left side of center to the right side), and decides, based on the infant's reaction, where the stripes are located on the card and, thus, whether the baby can see them. Over trials, the size of stripe is decreased until the tester observes that the baby is responding randomly. The estimate of the baby's grating acuity is the smallest stripe size for which the baby shows a preference. Based on this method, Mayer and colleagues (Mayer et al., 1995) provided normative data for babies from birth to 4 years of age (see Fig. 1). Adults with normal eyes have a grating acuity slightly better than 1 min of arc (one-sixtieth of a degree of visual angle). As shown in Fig. 1,

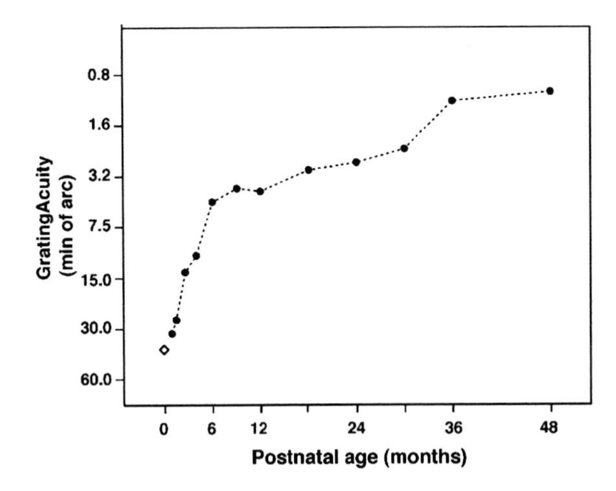

Fig. 1. Grating acuity from birth to 4 years of age. Shown are the normative values from the Teller Acuity Card procedure described in the text (Mayer et al., 1995). Each dot represents the smallest stripe size, in minutes of arc, for which infants at a particular age showed a reliable looking preference. Adapted with permission from Maurer and Lewis (2001b).

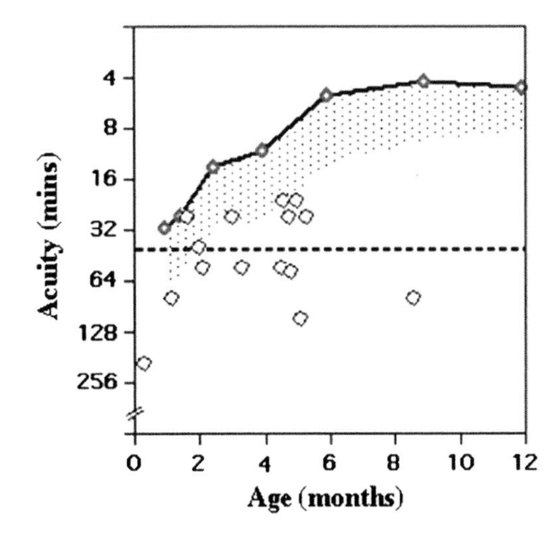

Fig. 2. Grating acuity of the right eye on the immediate test of children treated for bilateral congenital cataract. Each dot represents the acuity of a treated eye plotted at the age when contact lenses first allowed patterned visual input after surgery. The solid line and stippled area represent the mean acuity and lower 95% prediction limit (95% confidence that acuity will be at least as good as this value) for the normative group tested on Teller Acuity cards. The dotted line represents the geometric mean of the patients' acuity values. The data from the left eye are similar. Adapted with permission from Maurer et al. (1999).

newborns' acuity is more than 40 times worse that that of adults. Acuity improves rapidly over the first 6 months, followed by more gradual improvements over the next 4 years. Not until 4–6 years of age does grating acuity reach adult levels (Mayer and Dobson, 1982; Ellemberg et al., 1999a).

We have studied the acuity of children longitudinally after treatment for bilateral congenital cataracts (Lewis et al., 1995; Ellemberg et al., 1999b; Maurer et al., 1999; reviewed in Maurer and Lewis, 2001a, b). In one study, we measured the acuity of 12 patients on the day when they could first see — the day when they received their first contact lenses, which provided the first focused patterned visual input to the retina after removal of the cataracts (Maurer et al., 1999). The 1-week delay between surgery and the fitting of the first contact lenses was sufficient for the eyes to heal from the surgeries. On the first measurement, which occurred within 10 min of the end of visual deprivation, acuity in each eye was like that of newborns, regardless of the patient's age, which ranged from 1 to 9 months. As a result, those treated later had a larger deficit in acuity compared to children with normal eyes (see Fig. 2). These results indicate that patterned visual input drives the rapid improvement in acuity evident during the first 6 months of normal development (see Fig. 1). However, the visual system of the cataract patients was not dormant during the period of visual deprivation. This was evident when we retested their visual acuity after just 1 h of visual input: there was an improvement in almost every case (see Fig. 3), such that the mean acuity increased over the hour of visual experience to that of a typical 6-week-old with normal eyes. No such change occurred in an age-matched control group. Patients continued to improve faster than normal over the next week and month (see Fig. 4).

A control experiment with six additional infants treated for bilateral congenital cataract confirmed that the rapid improvement after treatment was driven by visual input and not by non-visual factors such as adjusting to the contact lens. For the control experiment, the immediate test occurred, as before, within 10 min of the infant receiving the contact lenses. Then, one eye was patched while the other eye received 1 h of visual input, after which both eyes received the usual retest of acuity.

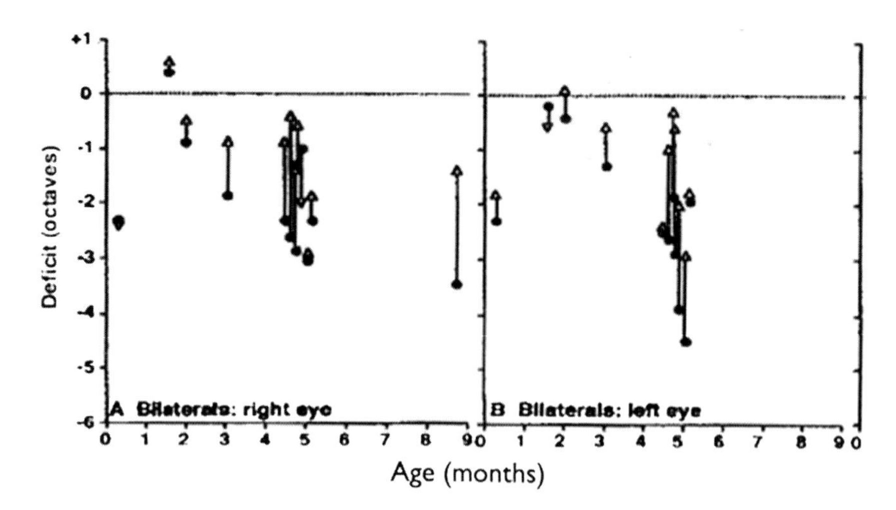

Fig. 3. Change in grating acuity after the first hour of visual input for children treated for bilateral congenital cataract. Each connected set of points represents the acuity deficit in octaves for one eye on the immediate test and the test after 1 h of visual input, plotted at the age when contact lenses first allowed patterned visual input after surgery. The dotted line at zero represents no deficit, and negative values represent deficits, with larger numbers representing larger deficits. Data for the right eyes are shown in Panel A and data for the left eyes in Panel B. Adapted with permission from Maurer et al. (1999).

Acuity improved in the eye that received visual input but not in the fellow patched eye that did not (see Fig. 5), a result indicating that visual input caused the rapid improvement in acuity. The amount of improvement in the experienced eye was similar to that observed in the main experiment. Combined, the results indicate that patterned visual input during infancy not only drives the initial rapid improvement seen in infants with normal eyes, but that it also allows accelerated recovery after visual deprivation.

Despite the accelerated recovery immediately after treatment, children treated for bilateral congenital cataract do not develop normal visual acuity. The developmental progression is well illustrated by our longitudinal studies of contrast sensitivity (Maurer et al., 2006). The contrast sensitivity function represents the amount of contrast needed to see stripes of various size, or spatial frequency — the greater the sensitivity, the less contrast needed to see the stripes. As shown in Fig. 6, adults are most sensitive to mid spatial frequencies (i.e., 3–5 cycles per degree of visual angle): for those frequencies they can still see the stripes when the contrast is low. There is a decrease in sensitivity for higher spatial frequencies up to the acuity cutoff, above which adults cannot see stripes even of maximum contrast (black and white). There is a smaller drop-off for low spatial frequencies (wide stripes). As shown in Fig. 6, contrast sensitivity is quite good by 4 years of age and reaches adult levels by age 7 (Ellemberg et al., 1999a).

The pattern was quite different in children treated during infancy for bilateral congenital cataract whose contrast sensitivity we measured longitudinally beginning between 5 and 8 years of age. As shown in Figs. 7 and 8, 1–2 years after the initial test, contrast sensitivity for low spatial frequencies (wide stripes) had improved more than in the control group so that an initial deficit had vanished or decreased dramatically. Some of the improvement occurred after 7 years of age, that is, after the age at which development is usually complete. Contrast sensitivity for mid spatial frequencies did not change between tests, while that of the control group increased, leading the patients to have an increased deficit. Contrast sensitivity for high spatial frequencies (10–20 cycles per degree) was not measurable because the patients could not see such thin stripes at any age. Thus, visual input during middle childhood allows partial recovery from the effects of early visual deprivation, but only at the low spatial frequencies where vision began to recover rapidly immediately after treatment. The asymptotic

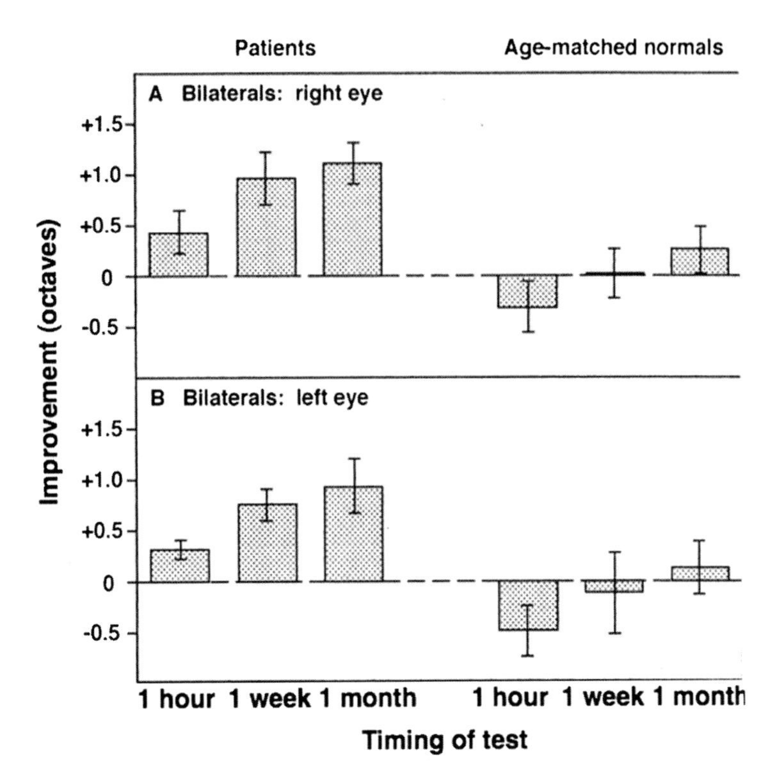

Fig. 4. Mean improvement in acuity in octaves (±1 s.e.) for children treated for bilateral congenital cataract from the immediate test to the test after the first hour of visual input, 1 week later, and 1 month later (left side) and for children in the age-matched control group (right side). Data for the right eye are shown in Panel A and data for the left eye in Panel B. Adapted with permission from Maurer et al. (1999).

sensitivity leaves the patient treated for bilateral congenital cataract with the contrast sensitivity of a typical toddler with normal eyes (Gwiazda et al., 1997). Complete recovery at higher spatial frequencies may be possible if the deprivation is especially short: a few patients treated at 6–8 days of age have achieved normal letter acuity (Kugelberg, 1992), as did 1 of 13 cases treated before 7 weeks of age in another cohort (Magnusson et al., 2002).

In sum, visual deprivation prevents the normal development of spatial vision. Although sensitivity to low spatial frequencies (wide stripes) can recover to normal levels, beginning with rapid improvement immediately after treatment, sensitivity to mid and high spatial frequencies does not. For those high spatial frequencies, there is a sleeper effect: visual deprivation during the first few months of life prevents the development of normal sensitivity to high spatial frequencies (10–20 cycles per degree) that infants with normal eyes typically do not begin to perceive until 2 years of age (Mayer et al., 1995). Studies of binocularly deprived monkeys suggest that the deficits are likely to arise at the level of the primary visual cortex, V1, where cells are sluggish, have abnormally large receptive fields, and reduced acuity, unlike cells in the retina and lateral geniculate nucleus, which respond normally (Crawford et al., 1975; Hendrickson and Boothe, 1976; Blakemore and Vital-Durand, 1983, 1986; Crawford et al., 1991).

Face processing

The poor contrast sensitivity of infants with normal eyes limits the information that they can perceive in faces: they can readily see the oval contour, the hair, and the basic layout of features but not the details of the internal features. Nevertheless, our

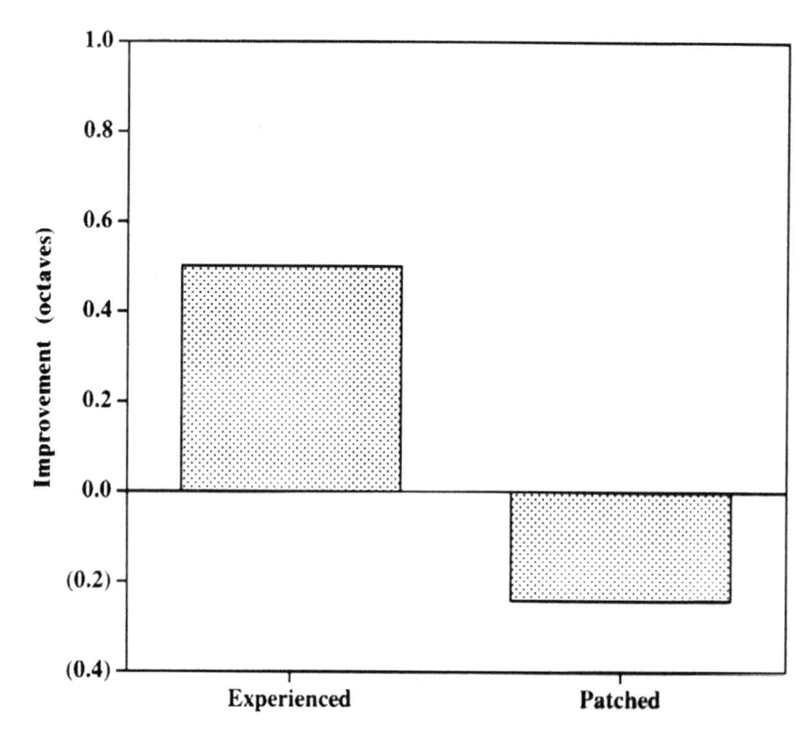

Fig. 5. Mean improvement in acuity in octaves between the immediate and 1 h tests for the six children treated for bilateral congenital cataract in the control experiment. Shown are the results for the eye that received visual experience (left side) and the fellow eye that was patched (right side).

studies of children treated for bilateral congenital cataract indicate that visual deprivation during this period when input is normally so limited prevents the later development of some, but not all, aspects of face processing (reviewed in Mondloch et al., 2003a). Here we report the ultimate deficit or ability in such patients: what they were able to achieve after removal of the cataract and contact lens fitting, followed by many years of (nearly) normal visual input (range 9 years to more than 20 years). In each case, the results from patients ($n = 11$–17 depending on task) were compared to those from controls matched on age, sex, handedness, and race/ethnic group.

Face detection

Adults can readily detect that a stimulus is a face based on its first-order relations (the ordinal relations that position two eyes above a nose, which is in turn above a mouth) (Diamond and Carey, 1986). They do so rapidly even when some of the individual features are missing (e.g., a line drawing with eyes and nose but no mouth) and even when the normal facial features are replaced by an arrangement of fruit or vegetables forming the correct first-order relations for a face (Moscovitch et al., 1997). Similarly, they can detect a face in an upright two-tone Mooney face (see Fig. 9) in which the perception of individual features has been degraded by transforming all luminance values to black or white (Kanwisher et al., 1998).

Patients treated for bilateral congenital cataract develop normal face detection (Mondloch et al., 2003a). To test face detection, we gave them a task consisting of brief presentations (100 ms) of either a Mooney face or a scrambled Mooney face (see Fig. 9 for examples) and asked them to indicate whether the stimulus was a face or nonface. We chose Mooney faces because they cannot be classified as faces based on individual features. As shown in Fig. 10, patients' accuracy and reactions times were normal. Thus, early visual deprivation does not prevent the later development of normal

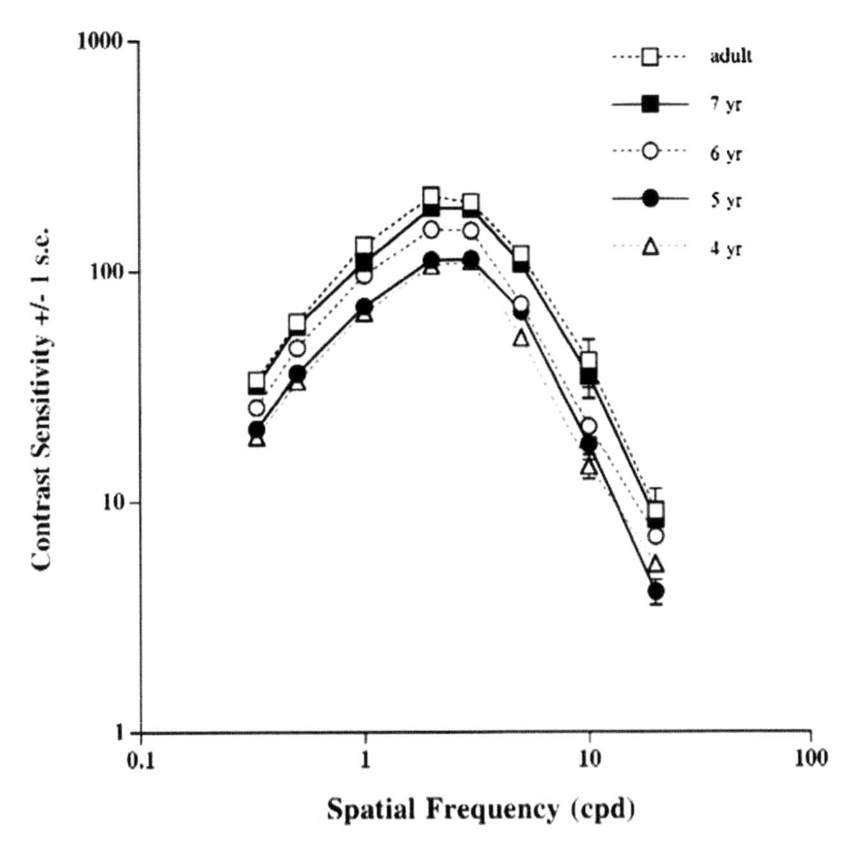

Fig. 6. Development of normal contrast sensitivity. Shown is the mean contrast sensitivity (±1 s.e.) of adults and four age groups of children for various spatial frequencies. When not shown, standard error bars are smaller than the symbols. Adapted with permission from Ellemberg et al. (1999a).

sensitivity to the first-order relations that underlie face detection. Preliminary data from longitudinal studies using a simpler task during infancy suggest that the ultimately normal performance may represent recovery from an earlier deficit (Mondloch et al., 1998). We are currently using ERP and fMRI to determine whether patients treated for bilateral congenital cataract use the normal neural networks for face detection or whether the plasticity extends to the recruitment of a different system that can, nevertheless, achieve normal accuracy and reaction time.

Recognition of facial identity

Adults can recognize thousands of individual faces rapidly and accurately and do so despite changes in the appearance of individual features caused by alterations in other cues that they must monitor such as head and eye orientation, emotional expression, or sound being spoken (Bahrick et al., 1975; see Bruce and Young, 1998, for a review). The reliable cues to individual identity (those that do not change with a trip to the hairdresser) are the shape of the head contour, the shape of individual internal features (e.g., eyes, nose, mouth, eyebrows), and the metric distances among the features, a configural cue called second-order relations. Although adults use all of these cues to decode facial identity, there is considerable evidence that their expertise in face recognition comes primarily from exquisite sensitivity to second-order relations (reviewed in Maurer et al., 2002). For example, adults are better at recognizing individual upright faces than at recognizing individual objects, but the superiority diminishes if

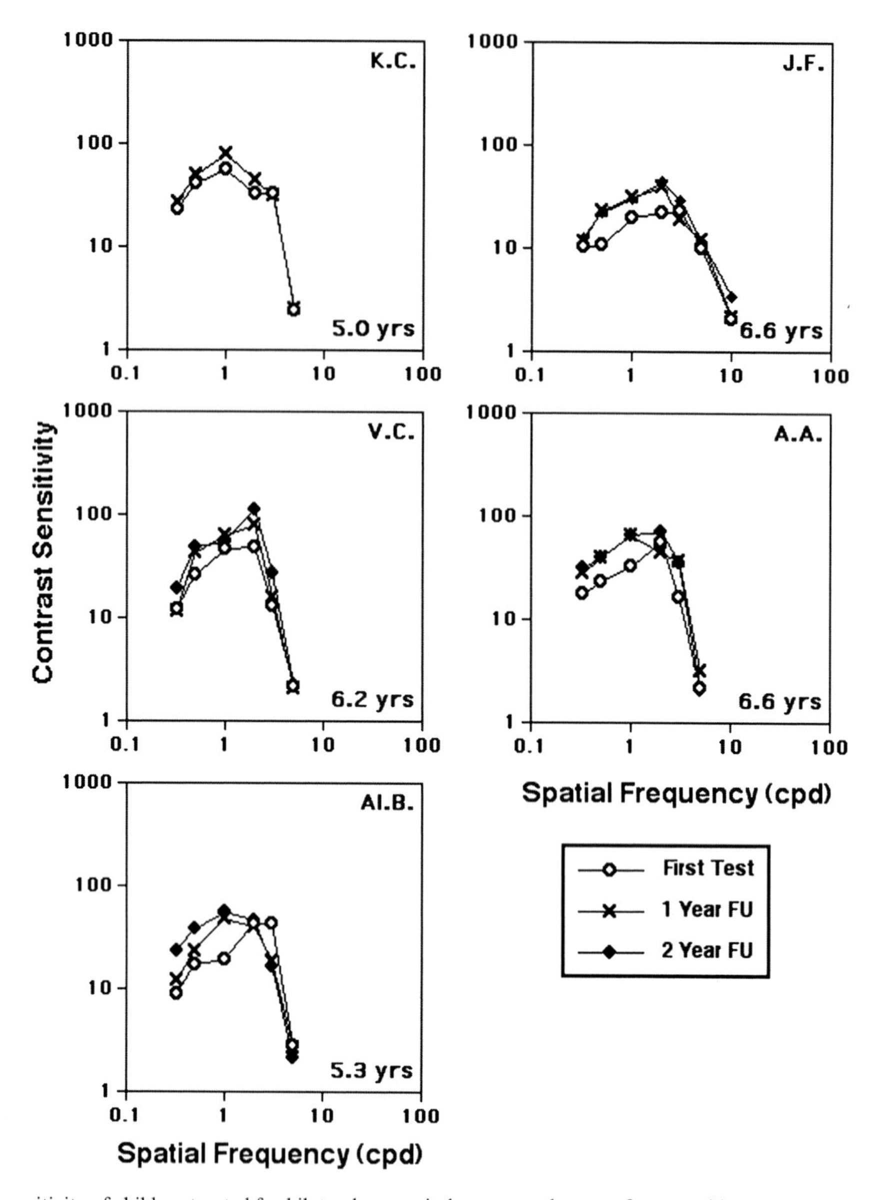

Fig. 7. Contrast sensitivity of children treated for bilateral congenital cataract who were first tested between age 5 and 8 years (○) and then retested 1 year (×) and/or 2 years (◇) later. Repeat tests from two older children are also shown. The age at the first test is indicated in the bottom right corner of the box for each patient. Adapted with permission from Maurer et al. (2006).

the stimuli are inverted (Yin, 1969), just as their sensitivity to second-order relations in faces plummets with inversion, much more than their sensitivity to facial features (Leder and Bruce, 1998; Freire et al., 2000; Mondloch et al., 2002; Malcolm et al., 2005; Leder and Carbon, 2006; Rhodes et al. 2006; but see Riesenhuber et al., 2004; Sekuler et al., 2004; Yovel and Kanwisher, 2004, 2005; see also Collishaw and Hole, 2000).

Our first study of face processing in children treated for bilateral congenital cataract indicated that they have deficits in recognizing the identity of a face they saw about one-half second earlier if the orientation of the head changed (e.g., from

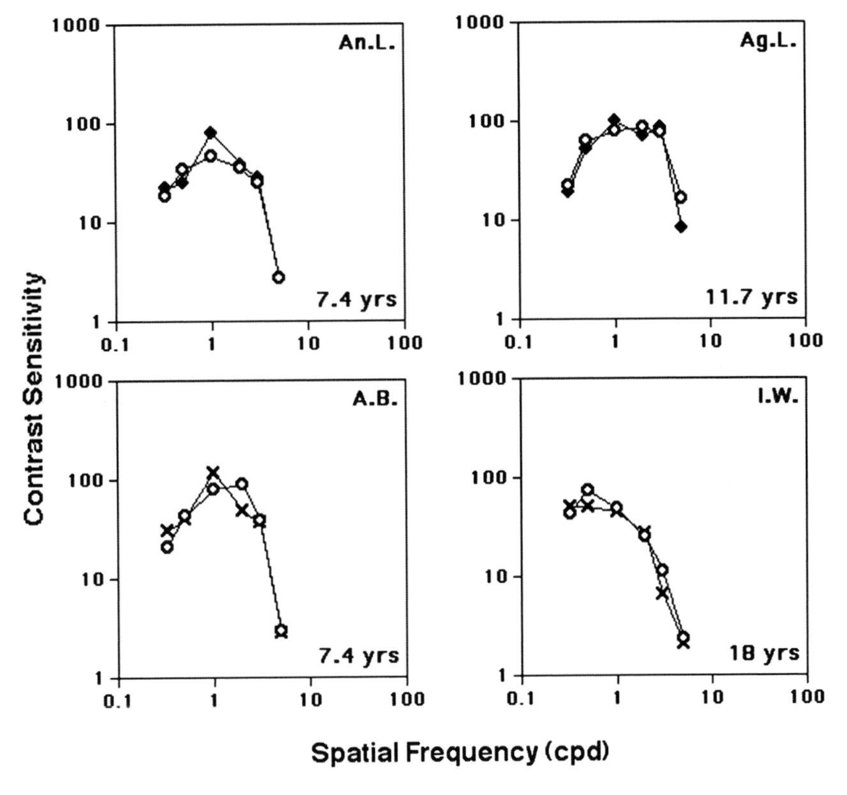

Fig. 7. *Continued.*

looking up to turned 45° toward the side) (Geldart et al., 2002). Good performance on this task depends on sensitivity to second-order relations because as the head is rotated, the shape of features and the external contour appear to change or are occluded, but the basic layout of the face determined by bone structure — the second-order relations — remains constant. As would be expected, the accuracy of adults with normal eyes on this task drops dramatically if the stimuli are inverted (Mondloch et al., 2003b). Therefore, we suspected that early visual deprivation might interfere with the development of sensitivity to second-order relations. Because the patients were normal on our measures of lip-reading, matching emotional expression, and matching direction of eye gaze (Geldart et al., 2002) — all of which could be solved by attending to specific features and none of which were impaired by inversion (Mondloch et al., 2003b) — we suspected that the patients might have normal featural processing.

To test these predictions directly, we created a task (which has come to be called the "Jane" test of facial identity) in which subjects make same/different judgments about pairs of faces presented sequentially that differ only in the shape of the external contour, only in the shape of the eyes and mouth, or only in the spacing between the eyes and between the eyes and mouth (Mondloch et al., 2002). Figure 11 illustrates the faces used. Changes of each type are presented in separate blocks in order to encourage reliance on contour processing, featural processing, and processing of second-order relations, respectively. Control experiments with normal adults confirmed that, as expected if it is a valid measure of sensitivity to second-order relations, inversion decreased accuracy for the spacing set much more than it did for the other two sets (Mondloch et al., 2002). Studies of children with normal eyes indicate that accuracy for the feature and contour sets is (nearly) adult-like by 6 years of age but that accuracy for the spacing

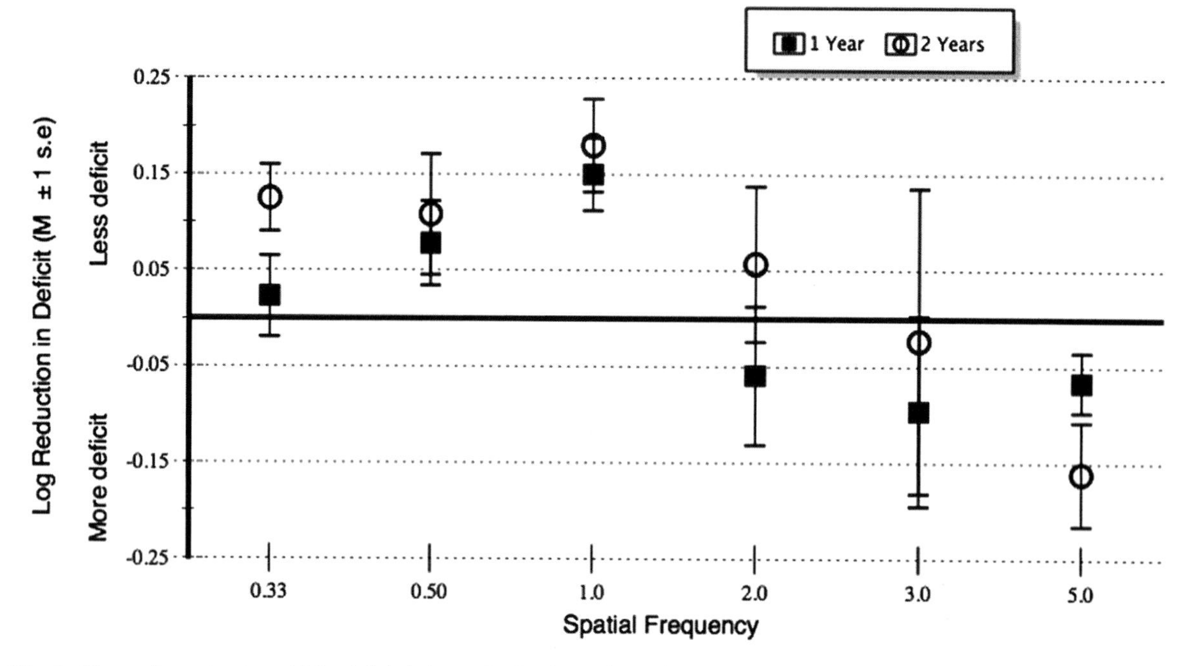

Fig. 8. Change in contrast sensitivity deficit in log units for the patients shown in Fig. 7 who were first tested before 8 years of age. Each point represents the mean change in deficit (±1 s.e.) after 1 year (closed symbols) or after 2 years (open symbols). Positive values represent a reduction in the size of the deficit (i.e., normalization) and negative values represent an increase in the size of the deficit (i.e., greater deviation from normal). Adapted with permission from Maurer et al. (2006).

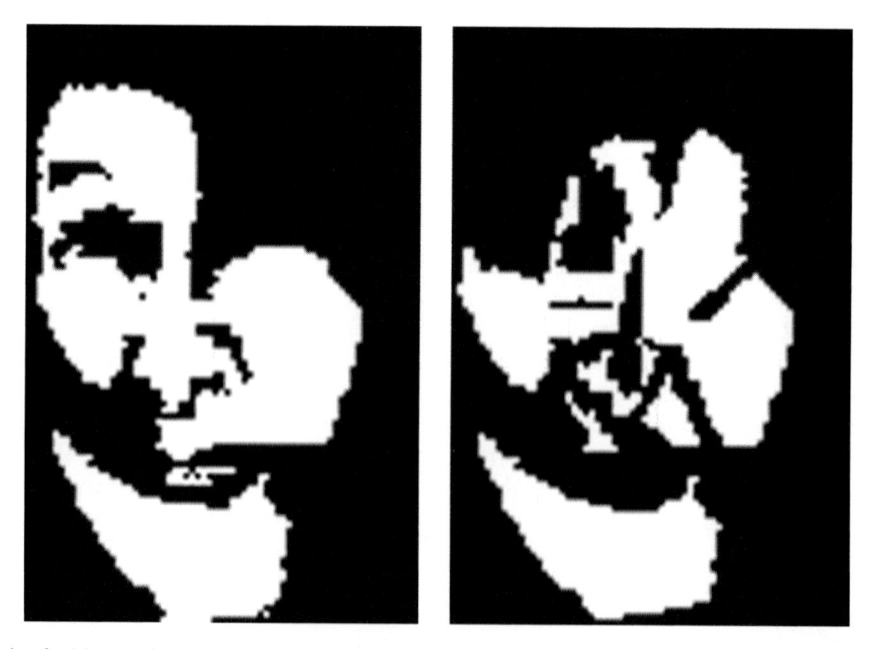

Fig. 9. An example of a Mooney face (A) and a scrambled Mooney face (B) of the type used to test face detection in children treated for bilateral congenital cataract. All luminance values are set to white and black, a manipulation that eliminates veridical facial features. Nevertheless, adults can detect the first-order relations that define a face when the stimuli are upright.

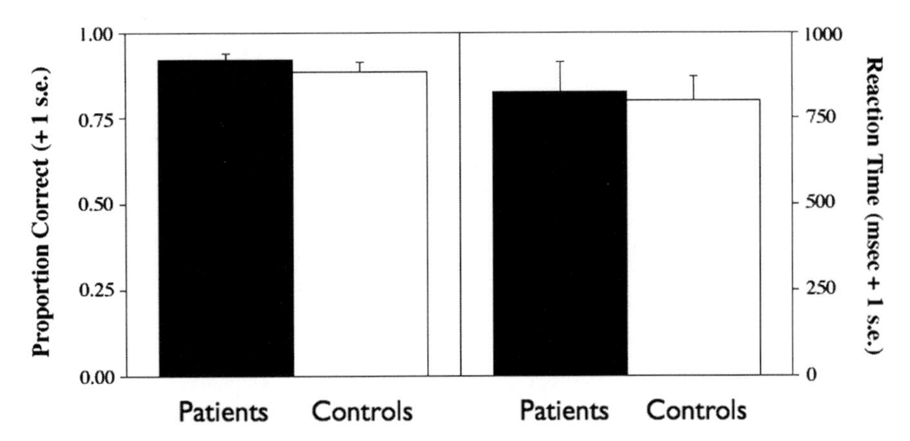

Fig. 10. Mean accuracy (±1 s.e.) (left panel) and mean reaction time (right panel) for determining whether a Mooney image was a face or non-face in patients treated for bilateral congenital cataract and in the age-matched control group.

set improves even after 14 years of age (Mondloch et al., 2002, 2003a). This is despite the fact that accuracy for the contour and spacing sets is identical in adults with normal eyes, that adults' accuracy for the featural set is in the typical range that we have found with a much larger set of faces (Mondloch et al., unpublished data), and that the spacing differences cover most of the variance in the normal population of adult Caucasian female faces (Farkas, 1981).

Sensitivity to second-order relations not only develops more slowly than sensitivity to other cues to facial identity, it also emerges later in development. A simpler version of the Jane task indicated that 4-year-olds are able to recognize the faces of children they learned from a storybook and a picture of their own face when tested with foils with contour or feature differences. In contrast, they perform at chance when tested with foils with spacing differences, even though the spacing differences captured most of the variability among children's faces (Mondloch et al., 2006b; but see McKone and Boyer, 2006, for evidence of earlier sensitivity to spacing differences as a cue to typicality). Note, however, that when the spacing differences exceed natural limits, sensitivity to spacing differences in faces is apparent in infants as young as 5 months (but not 3 months) of age (Bertin and Bhatt, 2004; Bhatt et al., 2005).

As indicated in Fig. 12, patients treated for bilateral congenital cataract performed normally on the contour and featural sets but had significant impairments on the spacing set, even when the initial deprivation had ended by 2 months of age (Le Grand et al., 2001, 2003; Mondloch et al., 2003a). Subsequent studies with children treated for unilateral cataract suggested that it is specifically input to the right hemisphere during early infancy that is necessary to set up the system so that it can later gain expertise in recognizing faces based on second-order relations (Le Grand et al., 2003). Thus, visual input during early infancy, at a time when the infant demonstrates no sensitivity to second-order relations, is necessary to set up the neural substrate — presumably in the right hemisphere — that will allow the later development of normal sensitivity to second-order relations.

Holistic face processing

One reason for the patients' deficit in processing second-order relations might be that they never learned to process faces holistically. Unlike objects, adults process faces as a holistic Gestalt, gluing the features together into a whole that is difficult to parse into individual features. One measure of holistic processing is the composite face effect (e.g., Young et al., 1987; Hole, 1994). When adults are asked to judge the identity of faces from just the top half, they have difficulty doing so if the top half is aligned with the bottom half of another person's

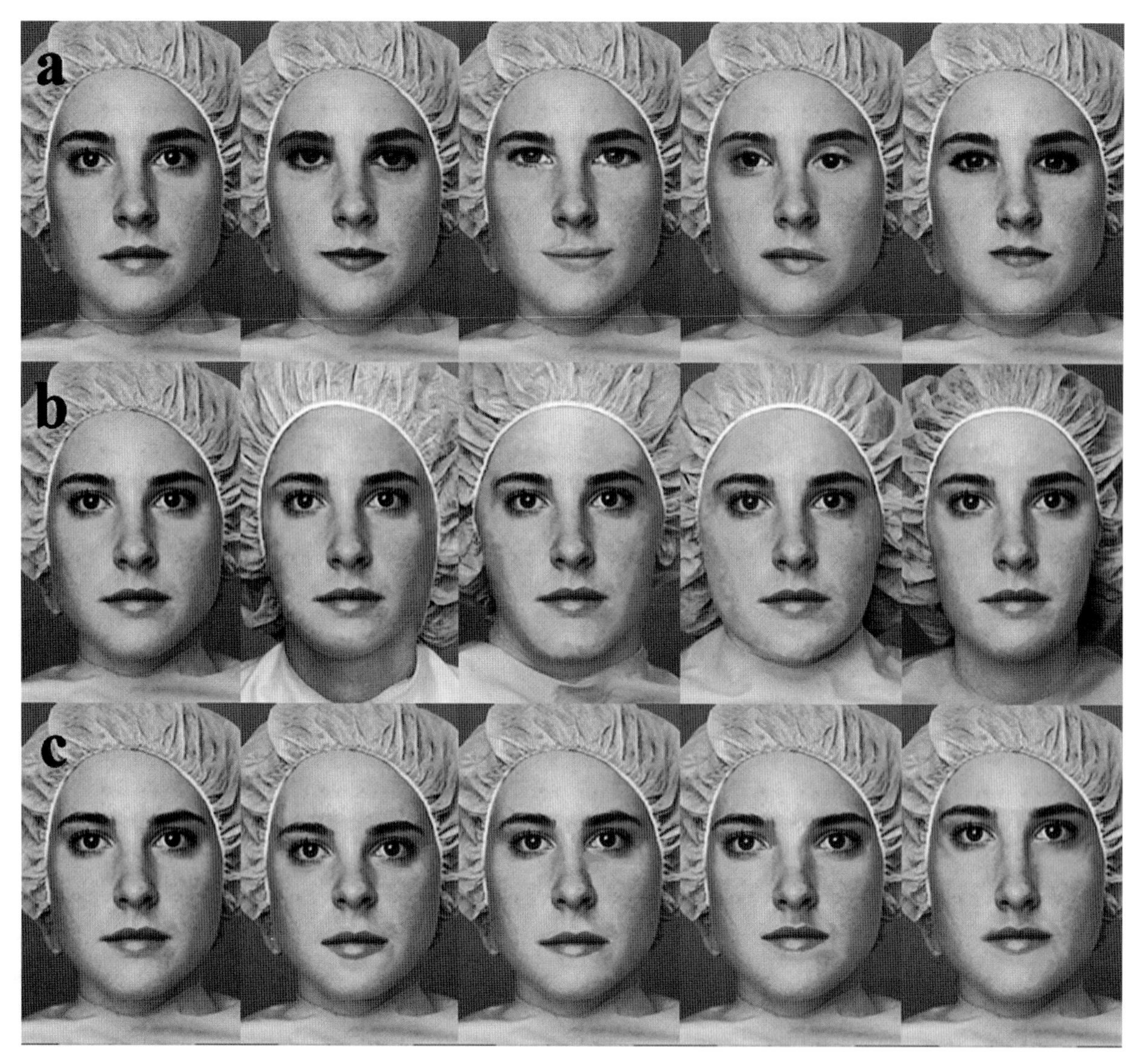

Fig. 11. Faces from the Jane task. Jane is shown as the left-most image in each row. Faces in the top row (the Feature Set) differ from Jane only in the shape of the eyes and mouth. Faces in the middle row (the Contour Set) differ from Jane only in the shape of the external contour. Faces in the bottom row (the Spacing Set) differ from Jane only in the spacing between the eyes and between the eyes and mouth. Adapted with permission from Le Grand et al. (2003).

face, presumably because holistic processing glues the features in the top and bottom halves together so tightly that it makes it difficult to attend to just one half. Misaligning the two halves to break holistic processing, or inverting the stimuli, makes the task much easier (see Fig. 13). Children as young as 4–6 years of age show an adult-like composite face effect: just like adults, they are 20–25% less accurate in seeing that the top halves of two unfamiliar faces are the same when they are aligned with the confusing bottom halves of other faces than when the two halves are misaligned (de Heering et al., 2007; Mondloch et al., in press). Such early development of holistic face processing may facilitate the development of sensitivity to second-order relations by forcing the child to pay attention to the proportions of the face and to relate them to the proportions of an average face at the center of an

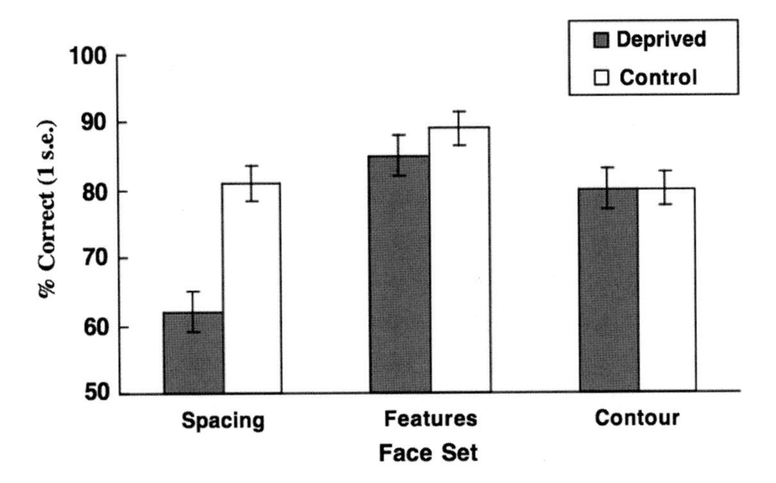

Fig. 12. Mean accuracy (±1 s.e.) for the Spacing, Feature, and Contour Sets of the Jane task (see Fig. 11). Shown are the results for children treated for bilateral congenital cataracts and for age-matched controls.

n-dimensional face space (Rhodes et al., 1987, 2003; Valentine, 1991; Rhodes and Jeffery, 2006).

Patients treated for bilateral congenital cataract do not show normal holistic processing of faces, even when the deprivation ended by 3 months of age (Le Grand et al., 2004). Importantly, they demonstrated this deficit by *superior* performance on the composite face task. Unlike the control group, their accuracy in judging that the top halves of the two sequential faces were the same was as high when the two tops were aligned with different bottom halves as when they were misaligned. In fact, their accuracy in the critical condition (same/aligned) where holistic processing impairs normal performance was significantly *higher* than that of the control group.

Cashon and Cohen (2003, 2004) have tested for the first signs of holistic processing during infancy by testing whether babies treat a switched face with the internal features of one familiar face and the external features of another familiar face like a novel face (as it would be if the internal and external features are integrated holistically) or as a familiar face (as it would be if the features are processed separately). The results indicate that 4-month-olds, but not 3-month-olds, process the internal and external features holistically. Combined, the results indicate that visual input during the first 3 months of life — before the first manifestations of holistic processing — is necessary to set up the system for its later development.

Summary and developmental implications

In summary, early visual deprivation from congenital cataract prevents the later development of normal visual acuity, contrast sensitivity for mid and high spatial frequencies, and two aspects of configural face processing: holistic face processing and decoding of identity based on second-order relations. It does not prevent the development (or more, likely, allows recovery) of normal contrast sensitivity for low spatial frequencies, normal face detection, and normal featural processing. The deficits described here are all examples of sleeper effects: visual deprivation during a period in normal infancy before the first manifestations of functional ability prevents their later development (Maurer et al., 2007).

Because the cataracts blocked all patterned visual input to the retina, we do not know if specific types of input are necessary for each visual capability; for example, whether it is specifically input from faces that is necessary for the normal development of face processing. However, the fact that holistic processing and sensitivity to second-order relations later become refined for the types of faces

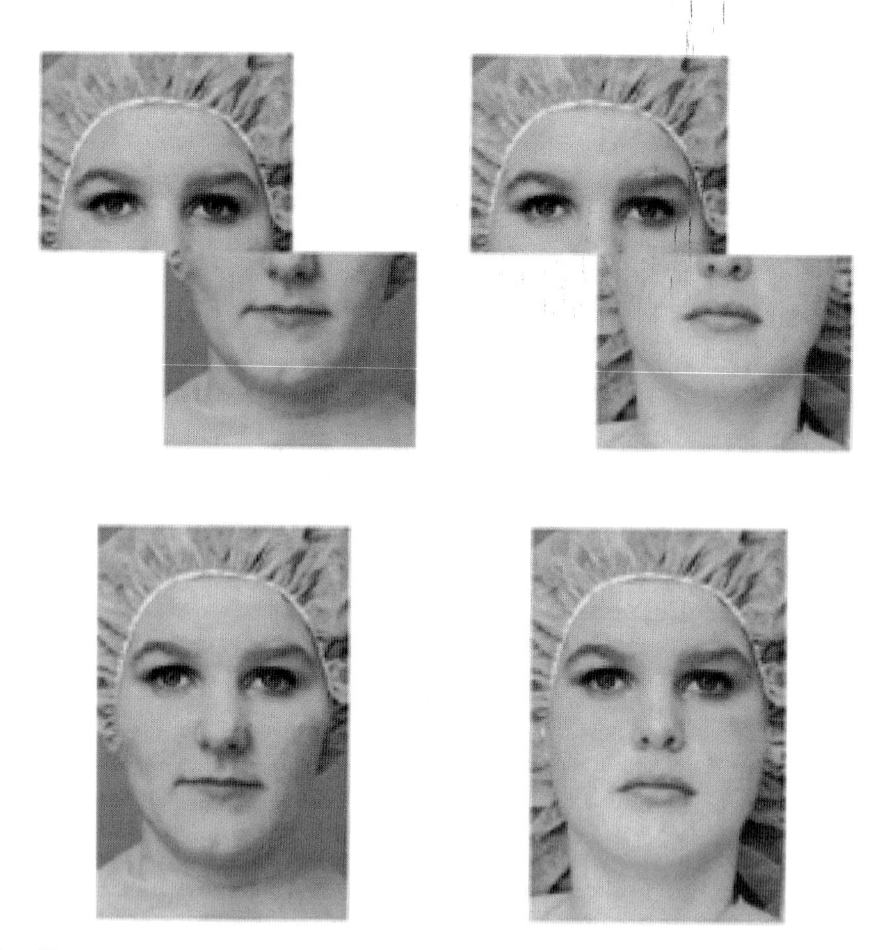

Fig. 13. Examples of faces used on same trials in the composite face task. Subjects are asked to indicate whether the top halves of the two faces are the same when the bottom half is different and misaligned (top panel) or aligned (bottom panel). Holistic processing makes the top halves of the aligned faces look different. Adapted with permission from Le Grand et al. (2004).

the individual typically observes (own race; own species) indicates that it is likely that face input *per se* plays some role (Michel et al., 2006; Mondloch et al., 2006a; Rhodes et al., 2006).

One possible explanation of such sleeper effects is that visual input during early infancy is necessary to set up or preserve the optimal neural architecture for the visual capability. In the absence of visual input, the requisite cells and/or connections may fail to develop or be lost through competitive interactions from inputs from other sensory modalities, as suggested by the specialization of the visual cortex, including the primary visual cortex, for touch, hearing, and perhaps even language in the congenitally blind (Kujala et al., 1995; Cohen et al., 1999; Röder et al., 2000; Bavelier and Neville, 2002; Burton et al., 2002a, b; Sadato et al., 2002; Amedi et al., 2003; Burton et al., 2003; Gizewski et al., 2003; reviewed in Maurer et al., 2005). By this account, the visual capability cannot develop normally at a later point in development because the optimal neural architecture to support it is not available.

Alternatively, the optimal architecture may be preserved and the needed connections formed, but those connections may be silenced or visual neuronal responses may be actively inhibited because of stronger input from other modalities during the initial deprivation and perhaps even subsequently. This possibility is suggested by evidence that the visual cortex can become responsive to tactile and auditory inputs even when blindness begins as late as adolescence (Cohen et al., 1999; Sadato et al.,

2002) and, to a lesser extent, even in adulthood (Burton et al., 2002a, 2004). Further support comes from evidence that a blind adult, who had been born without natural lenses, slowly became able to perceive unified objects when he was first given compensatory glasses at age 29 (Mandavilli, 2006; but see Fine et al., 2003). Such evidence suggests that there are multimodal connections to the visual cortex that can develop or be preserved even in the absence of sensory input and that can be revealed in adulthood.

A third — and not mutually exclusive — possibility is that early visual deprivation leads to the recruitment of alternative pathways to support vision that bypass the primary visual cortex and that send input to higher visual centers via the superior colliculus, pretectum, and pulvinar. That possibility is suggested by evidence from cats that were deprived of early visual input by hoods covering their heads who later are able to learn to make visual discriminations, although it takes much more than the normal number of trials (Zablocka et al., 1976, 1980; Zernicki, 1979; Zablocka and Zernicki, 1996). Subsequent selective lesions indicate that the deprived cats use an alternative pathway to perform the task: lesions to the primary visual cortex impaired the performance of the normal cats in the control group but not the deprived group, whereas lesions of the pretectum or superior colliculus impaired performance of the deprived group but not the normal group. If children treated for bilateral congenital cataract use such an alternative pathway, then the pattern of recovery and deficit may simply reflect the limits on the functions that the alternative pathway can support.

Future research using neuroimaging techniques may help to distinguish among these hypotheses. It is also possible that training with feedback in the areas of deficit might lead to improved or even normal visual capabilities, as it has for contrast sensitivity deficits in adults with anisometropic amblyopia, a reduction in vision caused by unequal refractive errors in the two eyes during early childhood (Zhou et al., 2006). Additional evidence for the likely benefit of training comes from studies indicating that the vision of adults with normal eyes improves after playing action video games (Green and Bavelier, 2003, 2006, 2007) and from studies that have effectively trained adults to reduce the other-race disadvantage in face recognition (Elliott et al., 1973; Goldstein and Chance, 1985). Training studies with adults and children with a history of early visual deprivation from bilateral cataracts may help to elucidate whether the deficits reflect permanent changes in the neural architecture or whether there is sufficient residual plasticity to allow additional recovery. Whatever the outcome, our results indicate that early visual input shapes the nervous system of the infant with normal eyes in ways that permit the child to later develop the acute sensitivity to the details of pattern that is needed for reading and sensitivity to the configural properties of faces that is vital to social interactions.

Acknowledgments

The testing of cataract patients was supported by grants to DM from the Social Sciences and Humanities Research Council (Canada), Natural Sciences and Engineering Council (Canada), the Canadian Institutes of Health (Canada), the National Institutes of Health (U.S.), and the Human Frontiers Science programme. We thank Drs. Henry Brent and Alex Levin from The Hospital for Sick Children for their cooperation and support and the cataract patients for their generous assistance.

References

Amedi, A., Raz, N., Pianka, P., Malach, R. and Zohary, E. (2003) Early 'visual' cortex activation correlates with superior verbal memory performance in the blind. Nat. Neurosci., 6(7): 758–766.

Atkinson, J., Braddick, O. and Moar, K. (1977) Contrast sensitivity of the human infant for moving and static patterns. Vis. Res., 17(9): 1045–1047.

Bahrick, H.P., Bahrick, P.O. and Wittlinger, R.P. (1975) Fifty years of memory for names and faces: a cross-sectional approach. J. Exp. Psychol. Gen., 104(1): 54–75.

Banks, M.S. and Salapatek, P. (1978) Acuity and contrast sensitivity in 1-, 2-, and 3-month-old human infants. Invest. Ophthalmol. Vis. Sci., 17(4): 361–365.

Bavelier, D., and Neville, H. (2002) Cross-modal plasticity: where and how? Nat. Rev. Neurosci., 3 (6): 443–452.

Bertin, E. and Bhatt, R.S. (2004) The Thatcher illusion and face processing in infancy. Dev. Sci., 7(4): 431–436.

Bhatt, R.S., Bertin, E., Hayden, A. and Reed, A. (2005) Face processing in infancy: developmental changes in the use of different kinds of relational information. Child Dev., 76(1): 169–181.

Blakemore, B. and Vital-Durand, F. (1983) Visual deprivation prevents the postnatal maturation of spatial contrast sensitivity neurons of the monkey's striate cortex. J. Physiol., 345: 45.

Blakemore, B. and Vital-Durand, F. (1986) Effects of visual deprivation on the development of the monkey's lateral geniculate nucleus. J. Physiol., 380(1): 493–511.

Brown, A.M. and Yamamoto, M. (1986) Visual acuity in newborn and preterm infants measured with grating acuity cards. Am. J. Opthamol., 102(2): 245–253.

Bruce, V., Campbell, R.N., Doherty-Sneddon, G., Import, A., Langton, S., McAuley, S. and Wright, R. (2000) Testing face processing skills in children. Br. J. Dev. Psychol., 18(3): 319–333.

Bruce, V. and Young, A.W. (1998) In the Eye of the Beholder: The Science of Face Perception. Oxford University Press, Oxford.

Burton, H., Diamond, J. and McDermott, K. (2003) Dissociating cortical regions activated by semantic and phonological tasks: a fMRI study in blind and sighted people. J. Neurophysiol., 90(3): 1965–1982.

Burton, H., Sinclair, R. and McLaren, D. (2004) Cortical activity to vibrotactile stimulation: An fMRI study in blind and sighted individuals. Hum. Brain Mapp., 23(4): 210–238.

Burton, H., Snyder, A., Conturo, T., Akbudak, E., Ollinger, J. and Raichle, M. (2002a) Adaptive changes in early and late blind: a fMRI study of Braille reading. J. Neurophysiol., 87(1): 589–607.

Burton, H., Snyder, A., Diamond, J. and Raichle, M. (2002b) Adaptive changes in early and late blind: an fMRI study of verb generation to heard nouns. J. Neurophysiol., 88(6): 3359–3371.

Carey, S., Diamond, R. and Woods, B. (1980) Development of face recognition: a maturational component? Dev. Psychol., 16(4): 257–269.

Cashon, C.H. and Cohen, L.B. (2003) The construction, deconstruction, and reconstruction of infant face perception. In: Pascalis O. and Slater A. (Eds.), The Development of Face Processing in Infancy and Early Childhood: Current Perspectives. NOVA Science Publishers, New York, pp. 55–68.

Cashon, C.H. and Cohen, L.B. (2004) Beyond U-shaped development in infants' processing of faces: an information-processing account. J. Cogn. Dev., 5(1): 59–80.

Cohen, L., Weeks, R., Sadato, N., Celnik, P., Ishii, K. and Hallett, M. (1999) Period of susceptibility for cross-modal plasticity in the blind. Ann. Neurol., 45(4): 451–460.

Collishaw, S.M. and Hole, G.J. (2000) Featural and configural processes in the recognition of faces of different familiarity. Perception, 29(8): 893–909.

Courage, M. and Adams, R. (1990) Visual acuity assessment from birth to three years using the acuity card procedure: cross-sectional and longitudinal samples. Optom. Vis. Sci., 67(9): 713–718.

Crawford, M.L., Blake, R., Cool, S. and von Noorden, G. (1975) Physiological consequences of unilateral and bilateral eye closure in macaque: some further observations. Brain Res., 85(1): 150–154.

Crawford, M.L.J., Pesch, T.W., von Noorden, G.K., Harwerth, R.S. and Smith, E.L. (1991) Bilateral form deprivation in monkeys: electrophysiologic and anatomic consequences. Invest. Ophthalmol. Vis. Sci., 32(8): 2328–2336.

De Haan, M., Johnson, M., Maurer, D. and Perrett, D. (2001) Recognition of individual faces and average face prototypes by 1- and 3-month-old infants. Cogn. Dev., 16(2): 659–678.

De Heering, A., Houthuys, S. and Rossion, B. (2007) Holistic face processing is mature at 4 years of age: evidence from the composite face effect. J. Exp. Child Psychol., 96(1): 57–70.

Diamond, R. and Carey, S. (1986) Why faces are and are not special: an effect of expertise. J. Exp. Psychol. Gen., 115(2): 107–117.

Ellemberg, D., Lewis, T.L., Liu, C.H. and Maurer, D. (1999a) The development of spatial and temporal vision during childhood. Vis. Res., 39(14): 2325–2333.

Ellemberg, D., Lewis, T.L., Maurer, D., Liu, C.H. and Brent, H.P. (1999b) Spatial and temporal vision in patients treated for bilateral congenital cataracts. Vis. Res., 39(20): 3480–3489.

Elliott, E., Wills, E. and Goldstein, A. (1973) The effects of discrimination training on the recognition of white and oriental faces. Bull. Psychon. Soc., 2: 71–73.

Farkas, L. (1981) Anthropometry of the Head and Face in Medicine. Elsevier, New York.

Fine, I., Wade, A.R., May, M.G., Goodman, D.F., Boynton, G.M. and Wandell, B.A. (2003) Long-term visual deprivation affects visual perception and cortex. Nat. Neurosci., 6(9): 915–916.

Freire, A., Lee, K. and Symons, L.A. (2000) The face-inversion effect as a deficit in the encoding of configural information: direct evidence. Perception, 29(2): 159–170.

Geldart, S., Mondloch, C., Maurer, D., de Schonen, S. and Brent, H. (2002) The effects of early visual deprivation on the development of face processing. Dev. Sci., 5(4): 490–501.

Gizewski, E., Gasser, T., de Greiff, A., Boehm, A. and Forsting, M. (2003) Cross-modal plasticity for sensory and motor activation patterns in blind subjects. Neuroimage, 19(3): 968–975.

Goldstein, A. and Chance, J. (1985) Effects of training on Japanese face recognition: reduction of the other-race effect. Bull. Psychon. Soc., 23: 211–214.

Green, C. and Bavelier, D. (2003) Action video game modifies visual selective attention. Nature, 423(6939): 534–537.

Green, C.S. and Bavelier, D. (2006) Effect of action video games on the spatial distribution of visuo-spatial attention. J. Exp. Psychol. Hum. Percept. Perform., 32(6): 1465–1478.

Green, C.S. and Bavelier, D. (2007) Action video game experience alters the spatial resolution of vision. Psychol. Sci., 18(1): 88–94.

Gwiazda, J., Bauer, J., Thorn, F. and Held, R. (1997) Development of spatial contrast sensitivity from infancy to adulthood: psychophysical data. Optom. Vis. Sci., 74(10): 785–789.

Hendrickson, A. and Boothe, R. (1976) Morphology of the retinal and dorsal lateral geniculate nucleus in dark-reared monkeys (*Maccaca nemestrina*). Vis. Res., 16(5): 517–521.

Hole, G. (1994) Configurational factors in the perception of unfamiliar faces. Perception, 23(1): 65–74.

Kanwisher, N., Tong, F. and Nakayama, K. (1998) The effect of face inversion on the human fusiform face area. Cognition, 68(1): B1–B11.

Kugelberg, U. (1992) Visual acuity following treatment of bilateral congenital cataracts. Doc. Ophthalmol., 82(3): 211–215.

Kujala, T., Huotilainen, M., Sinkkonen, J., Ahonen, A., Alho, K., Hamalainen, M., Ilmoniemo, R.S., Kajola, M., Knuutila, J.E., Lavikainen, J., Salonen, O., Simola, J., Standerskjold-Nordenstam, C.-G., Tiitinen, H., Tissari, S.O. and Naatanen, R. (1995) Visual cortex activation in blind humans during sound discrimination. Neurosci. Lett., 183(1–2): 143–146.

Leder, H. and Bruce, V. (1998) Local and relational aspects of face distinctiveness. Q. J. Exp. Psychol., 51(3): 449–473.

Leder, H. and Carbon, C. (2006) Face-specific configural processing of relational information. Br. J. Psychol., 97(part 1): 19–29.

Le Grand, R., Mondloch, C., Maurer, D. and Brent, H.P. (2001) Early visual experience and face processing. Nature, 410(6831): 890 Correction: 2001, 412 (6849) p. 786.

Le Grand, R., Mondloch, C., Maurer, D. and Brent, H. (2003) Expert face processing requires visual input to the right hemisphere during infancy. Nat. Neurosci., 6(10): 1108–1112 Erratum: 2003, 6 (12) 1329.

Le Grand, R., Mondloch, C., Maurer, D. and Brent, H. (2004) Impairment in holistic face processing following early visual deprivation. Psychol. Sci., 15(11): 762–768.

Lewis, T.L., Maurer, D. and Brent, H.P. (1995) The development of grating acuity in children treated for unilateral or bilateral congenital cataract. Invest. Ophthalmol. Vis. Sci., 36(10): 2080–2095.

Magnusson, G., Abrahamsson, M. and Sjöstrand, J. (2002) Changes in visual acuity from 4 to 12 years of age in children operated for bilateral congenital cataracts. Br. J. Ophthalmol., 86(12): 1385–1398.

Malcolm, G., Leung, C. and Barton, J.J.S. (2005) Regional variations in the inversion effect for faces: differential effects for feature shape, spatial relations, and external contour. Perception, 33(10): 1221–1231.

Mandavilli, A. (2006) Look and learn. Nature, 441(7091): 271–272 (News feature).

Maurer, D., Ellemberg, D. and Lewis, T.L. (2006) Repeated measures of contrast sensitivity reveal limits to visual plasticity after early binocular deprivation in humans. Neuropsychologica, 44(11): 2104–2112.

Maurer, D., Le Grand, R. and Mondloch, C. (2002) The many faces of configural processing. Trends Cong. Sci., 6(6): 255–260.

Maurer, D. and Lewis, T. (2001a) Visual acuity and spatial contrast sensitivity: normal development and underlying mechanisms. In: Nelson C. and Luciana M. (Eds.), Handbook of Developmental Cognitive Neuroscience. MIT Press, Cambridge, MA, pp. 237–250.

Maurer, D. and Lewis, T.L. (2001b) Visual acuity: the role of visual input in inducing postnatal change. Clin. Neurosci. Res., 1(1): 239–247.

Maurer, D., Lewis, T.L., Brent, H.P. and Levin, A.V. (1999) Rapid improvement in the acuity of infants after visual input. Science, 286(5437): 108–110.

Maurer, D., Lewis, T.L. and Mondloch, C. (2005) Missing sights: consequences for visual cognitive development. Trends Cogn. Sci., 9(3): 144–151.

Maurer, D., Mondloch, C.J. and Lewis, T.L. (2007) Sleeper effects. Dev. Sci., 10(1): 40–47.

Mayer, D.L., Beiser, A.S., Warner, A.F., Pratt, E.M., Raye, K.N. and Lang, J.M. (1995) Monocular acuity norms for the Teller acuity cards between ages one month and four years. Invest. Ophthalmol. Vis. Sci., 36(3): 671–685.

Mayer, D.L. and Dobson, V. (1982) Visual acuity development in infants and young children as assessed by operant preferential looking. Vis. Res., 22(9): 1141–1151.

McKone, E. and Boyer, B. (2006) Sensitivity of 4-year-olds to featural and second-order relational changes in face distinctiveness. J. Exp. Child Psychol., 94(2): 134–162.

Michel, C., Rossion, B., Han, J., Chung, C.-S. and Caldara, R. (2006) Holistic processing is finely tuned for faces of our own race. Psychol. Sci., 17(7): 608–615.

Mondloch, C.J., Geldart, S., Maurer, D. and Le Grand, R. (2003b) Developmental changes in face processing skills. J. Exp. Child Psychol., 86(1): 67–84.

Mondloch, C.J., Le Grand, R. and Maurer, D. (2002) Configural face processing develops more slowly than featural face processing. Perception, 31(5): 553–566.

Mondloch, C.J., Le Grand, R. and Maurer, D. (2003a) Early visual experience is necessary for the development of some–but not all–aspects of face processing. In: Pascalis O. and Slater A. (Eds.), The Development of Face Processing in Infancy and Early Childhood: Current Perspectives. Nova Science Publishers, New York, pp. 99–117.

Mondloch, C.J., Leis, A. and Maurer, D. (2006b) Recognizing the face of Johnny, Suzy, and me: insensitivity to the spacing among features at four years of age. Child Dev., 77(1): 234–243.

Mondloch, C.J., Lewis, T.L., Maurer, D. and Levin, A.V. (1998) The effects of visual experience on face preferences during infancy. Dev. Cognit. Neurosci., Tech. Rep. No. 98.4.

Mondloch, C.J., Maurer, D. and Ahola, S. (2006a) Becoming a face expert. Psychol. Sci., 17(11): 930–934.

Mondloch, C.J., Pathman, T., Maurer, D., Le Grand, R. and De Schonen, S. (in press) The composite face effect in six-year-old children: Evidence of adultlike holistic face processing. Vis. Cogn.

Moscovitch, M., Winocur, G. and Behrman, M. (1997) What is special about face recognition? Nineteen experiments on a person with visual object agnosia and dyslexia but normal face recognition. J. Cogn. Neurosci., 9(5): 555–604.

Rhodes, G., Brennan, S. and Carey, S. (1987) Identification and ratings of caricatures: Implications for mental representations of faces. Cogn. Psychol., 19(4): 473–497.

Rhodes, G., Hayward, W. and Winkler, C. (2006) Expert face coding: configural and component coding of own-race and other-race faces. Psychon. Bull. Rev., 13(3): 499–505.

Rhodes, G. and Jeffery, L. (2006) Adaptive norm-based coding of facial identity. Vis. Res., 46(18): 2977–2987.

Rhodes, G., Jeffery, L., Watson, T.L., Clifford, C.W.G. and Nakayama, K. (2003) Fitting the mind to the world: face adaptation and attractiveness aftereffects. Psychol. Sci., 14(6): 558–566.

Riesenhuber, M., Jarudi, I., Gilad, S. and Sinha, P. (2004) Face processing in humans is compatible with a simple shape-based model of vision. Proc. Biol. Sci., 271(supplement 6): S448–S450.

Röder, B., Rösler, F. and Neville, H. (2000) Event-related potentials during auditory language processing in congenitally blind and sighted people. Neuropsychologia, 38(11): 1482–1502.

Sadato, N., Okada, T., Honda, M. and Yonekura, Y. (2002) Critical period for cross-modal plasticity in blind humans: a functional MRI study. Neuroimage, 16(2): 389–400.

Sekuler, A., Gaspar, C., Gold, J. and Bennett, P. (2004) Inversion leads to quantitative not qualitative changes in face processing. Curr. Biol., 14(5): 391–396.

Valentine, T. (1991) A unified account of the effects of distinctiveness, inversion, and race in face recognition. Q. J. Exp. Psychol. A, 43(2): 161–204.

Van Hof-van Duin, J. and Mohn, G. (1986) The development of visual acuity in normal fullterm and preterm infants. Vis. Res., 26(6): 909–916.

Yin, R.K. (1969) Looking at upside down faces. J. Exp. Psychol., 81: 141–145.

Young, A.W., Hellawell, D. and Hay, D.C. (1987) Configurational information in face perception. Perception, 16(6): 747–759.

Yovel, G. and Kanwisher, N. (2004) Face perception: domain specific, not process specific. Neuron, 44(5): 889–898.

Yovel, G. and Kanwisher, N. (2005) The neural basis of the behavioral face-inversion effect. Curr. Biol., 15(24): 2256–2262.

Zablocka, T. and Zernicki, B. (1996) Discrimination learning of grating orientation in visually deprived cats and the role of the superior colliculi. Behav. Neurosci., 110(3): 621–625.

Zablocka, T., Zernicki, B. and Kosmal, A. (1976) Visual cortex role in object discrimination in cats deprived of pattern vision from birth. Acta Neurobiol. Exp. (Wars), 36(1–2): 157–168.

Zablocka, T., Zernicki, B. and Kosmal, A. (1980) Loss of object discrimination after ablation of the superior colliculus-pretectum in binocularly deprived cats. Behav. Brain Res., 1(6): 521–531.

Zernicki, B. (1979) Effects of binocular deprivation and specific experience in cats: behavioral, electrophysiological, and biochemical analyses. In: Brazier M. (Ed.), Brain Mechanisms in Memory and Learning: From the Single Neuron to Man. Raven Press, New York, pp. 179–195.

Zhou, Y., Huang, C., Xu, P., Tao, L., Qiu, Z., Li, X. and Lu, Z.-L. (2006) Perceptual learning improves contrast sensitivity and visual acuity in adults with anisometropic amblyopia. Vis. Res., 46(5): 739–750.

C. von Hofsten & K. Rosander (Eds.)
Progress in Brain Research, Vol. 164
ISSN 0079-6123

CHAPTER 6

Visual tracking and its relationship to cortical development

Kerstin Rosander*

Department of Psychology, Uppsala University, Box 1225, S-75142 Uppsala, Sweden

Abstract: Measurements of visual tracking in infants have been performed from 2 weeks of age. Although directed appropriately, the eye movements are saccadic at this age. Over the first 4 months of life, a rapid transition to successively smoother eye movements takes place. Timing develops first and at 7 weeks of age the smooth pursuit is well timed to a sinusoidal motion of 0.25 Hz. From this age, the gain of the smooth pursuit improves rapidly and from 4 months of age, smooth pursuit dominates visual tracking in combination with head movements. This development reflects massive cortical and cerebellar changes. The coordination between eyes–head–body and the external events to be tracked presumes predictive control. One common type of model for explaining the acquisition of such control focuses on the maturation of the cerebellar circuits. A problem with such models, however, is that although Purkinje cells and climbing fibers are present in the newborn, the parallel and mossy fibers, essential for predictive control, grow and mature at 4–7 months postnatally. Therefore, an alternative model that also includes the prefrontal cerebral cortex might better explain the early development of predictive control. The prefrontal cortex functions by 3–4 months of age and provides a site for prediction of eye movements as a part of cerebro-cerebellar nets.

Keywords: visual development; vor; cerebellum; infant brain; smooth pursuit; eye–head coordination; predictive models

Introduction

Adult humans track a moving object with eye movements that keep the gaze fixated on the object. Such smooth pursuit (SP) stabilizes the moving object on the retina. When the eye movement is not optimal, saccades quickly adjust the retinal image. The background to this seemingly simple behavior is quite complex. In a natural situation the object is tracked with the head as well as the eyes. This presumes that visual and vestibular information are combined simultaneously to direct the gaze on the object. When the head moves, gaze will still be on target. This is also the case if the body suddenly turns or moves: the moving object is visually tracked and gaze is on the target. In this situation both the vestibular ocular reflex (VOR) that counter-rotates the eyes when the head moves and the vestibulo-collic reflex (VCR) counter-rotating the head relative to the body (Peterson and Richmond, 1988) must be suppressed. To keep the projection of the moving object on the fovea over changes in its velocity and direction of motion requires that the control of eye rotation compensates for all delays in the motor system. Thus, the oculomotor system that controls the eye muscles must be driven predictively (Pavel, 1990; Leigh and Zee, 1999).

*Corresponding author. Tel.: +46 18 4712139;
Fax: +4618 4712123; E-mail: kerstin.rosander@psyk.uu.se

DOI: 10.1016/S0079-6123(07)64006-0

That process must, in addition to the velocity and position of the tracked object and the delays in the sensorimotor system, involve all sensory information: visual, vestibular, and proprioceptive. In an experimental situation, when the lag of the eyes relative to the object motion is less than 120 ms, we consider the eye movements to be predictive (Robinson, 1965). The earliest form of visual tracking is saccadic and it is not until ~7 weeks of age that smooth eye movements (SEMs) are consistently observed. However, already in neonates, vestibularly controlled eye movements are smooth and evidently the eye muscles are ready to control such gradual movements (Rosander and von Hofsten, 2000). Thus, it is the neural system for visual control of SEMs that is the limiting factor and not the motor system itself.

The control of SP eye movements and the complexity of the associated predictive processes are expressions of cognitive abilities (Kowler, 1990). This gives them an important role for understanding early cognitive development. Successful tracking is realized through a motor command that, similarly to reaching, is regulated with a predictive model of external and internal factors. In the literature, the learning of such internal models in adults has been extensively discussed (i.e., Wolpert et al., 1998). There is however no such data for how and when they appear in development. The purpose of this chapter is to relate behavioral data from visual and visual-vestibular gaze adjustments to neuro-anatomical data relevant for predictive models.

A procedure suitable for investigating gaze control in young infants

We have constructed a safe and versatile "mini-room" that has been used extensively in infants. The advantage is that the infant can concentrate on the task, and much data is obtained during a few minutes. This apparatus has been described earlier (von Hofsten and Rosander, 1996, 1997) and is shown in Fig. 1. The infant is placed in a chair at the centre of a cylinder that constitutes the visual field. The inside surface is homogeneously white or patterned in specific ways and an object is placed on its inner surface, in front of the infant's eyes. The object, in most studies an orange-coloured happy face, has a small video-camera at its centre, thus allowing continuous monitoring of the infant's face during the experiment from the point of view of the stimulus. Other types of objects have been constructed to replace the routinely used "happy face." During the experimental trials, the object is moved by a motor giving an oscillation according to a sinusoidal or triangular velocity function. Another motor controls the oscillation of the chair, thus providing a condition for testing the vestibular function (VVOR condition). Finally, the cylinder and chair can be oscillated in synchrony. The last condition gives rise to a conflict between visual and vestibular information — the former induces visual tracking and the later a counter-rotation of the eyes. In order to stabilize gaze on the object in this condition, the VOR must be completely inhibited (vestibular inhibition (VINHIB) condition).

Measurements of head, object, and chair motions are performed using a motion analysis system with reflective markers (Qualisys). Small (4 mm diameter) markers are placed on the object, the head, and on the upper part of the chair (Rosander and von Hofsten, 2000). Eye movements are registered with pre-amplified EOG (Westling, 1992) and this signal is synchronized with other measurements. In the analysis, the Euclidian coordinates of all markers and the EOG measurements are transformed to angular coordinates. The eye motion data file is then stripped from saccades $>40^{\circ}$/s for identifying the smooth tracking component (SEM). Gain is estimated with Fourier analysis and timing with cross correlations. Some examples of eye and head measurements of a 4-week-old infant are shown in Figs. 2(SEM) and 3(VVOR).

The development of visual tracking

Visual tracking of an object is based on the engaging of attention on it and the commands to voluntarily track it. The ability to engage and disengage attention on targets is present at birth. However, as infants get better at stabilizing gaze

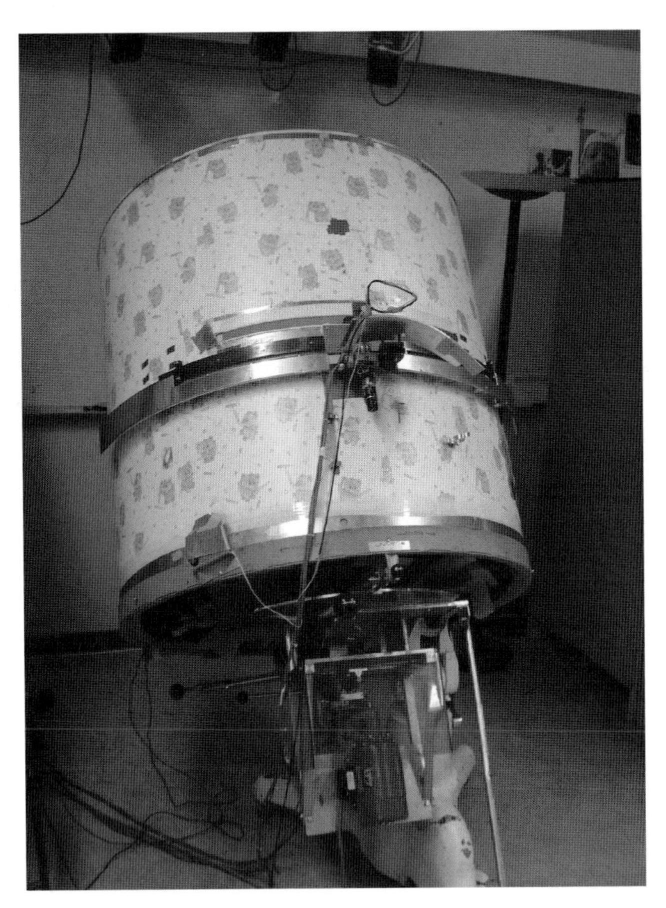

Fig. 1. The drum at the centre of which an infant chair is placed. The surface of the drum is manually rotated 180° when the infant has been safely fastened with attached EOG electrodes and motion markers. An object, usually a 7° happy "face," is oscillating in a slit in front of the infant. The slit is seen in the middle of the photo.

on an attractive target, they become less able to look away from it. This phenomenon has been termed "obligatory attention" or "sticky fixation" (Stechler and Latz, 1966; Mayes and Kessen, 1989). It has been suggested that it reflects the early maturation of a basal-ganglia/nigral pathway that induces a nonspecific inhibition of the superior colliculus (SC) (Johnson, 1990). The phenomenon of obligatory attention expresses itself rather dramatically in object tracking. von Hofsten and Rosander (1996, 1997) found that 2- and 3-month-old infants almost never looked away from the moving object they were tracking smoothly even though each trial had a duration of 20–30 s.

Newborns have a very minute ability to track a moving object with SEM. However, it has been shown that if an object with bright contrast is gently moved in front of the infant within reach, he or she will turn the head and eyes towards it and occasionally touch the object (von Hofsten, 1982). Dayton and Jones (1964) observed smooth following of a 15° object in neonates up to velocities of ~15°/s and Kremenitzer et al. (1979) observed smooth tracking in newborns for a large, 12°, stimulus moving at 9°–30°/s. Bloch and Carchon (1992) used a small boll, 4° in diameter, in their study of visual tracking in neonates of 3 days and 2–4-weeks-old. Although the velocity was low (8.7°/s), mostly saccadic tracking and head movements were observed. Phillips et al. (1997) measured 1- to 4-month-old infants' eye movement when they tracked a small (1.7°) red lamp in hold–ramp–hold

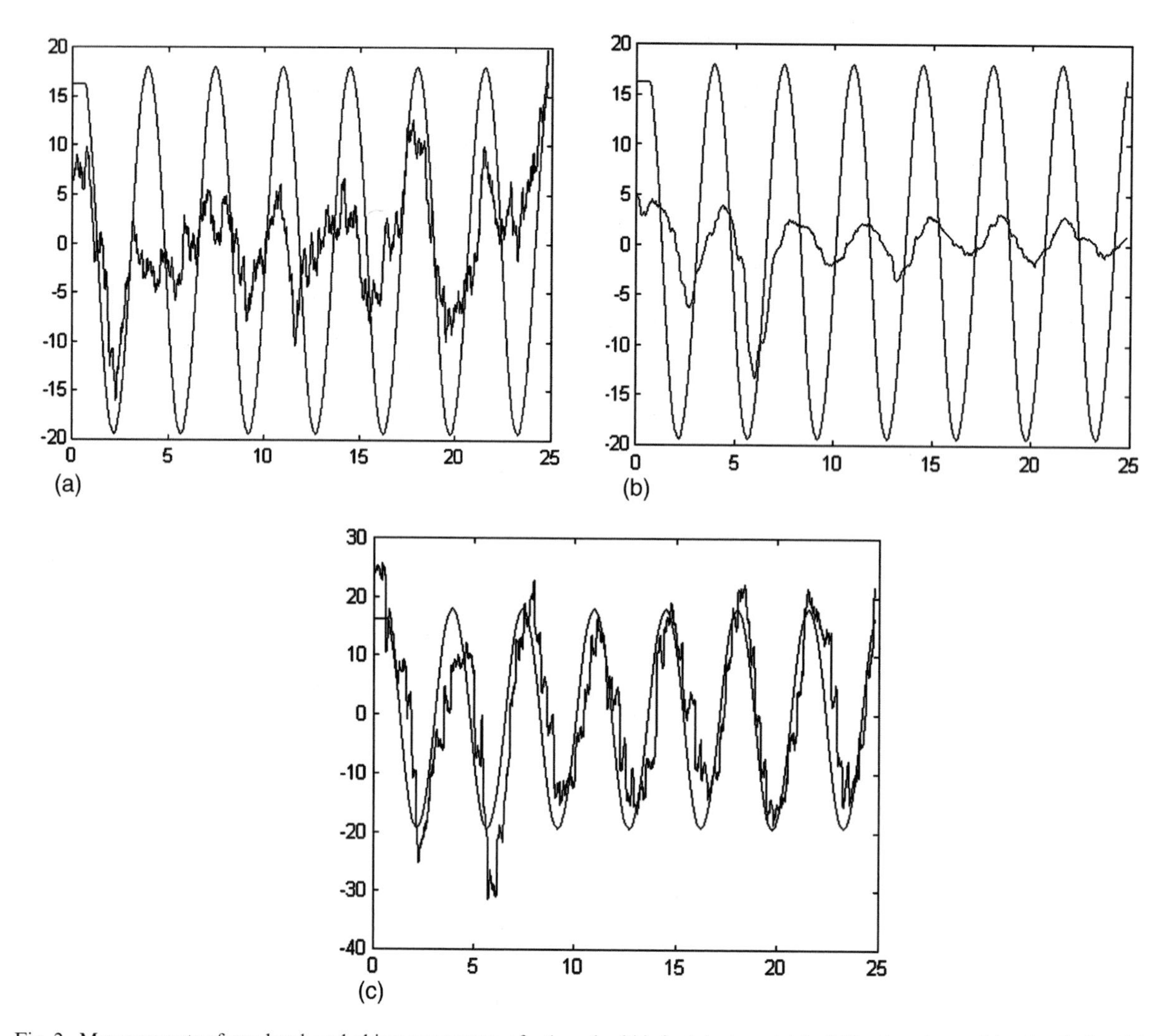

Fig. 2. Measurements of eye, head, and object movements of a 4-week-old infant during a trial of 25 s. Angular position is calculated (y-axis) as a function of time (x-axis); (a) eye-and object movements, (b) head-and object movements, (c) calculated gaze and object motion.

motion pattern. All infants showed a small amount of SP for the velocity interval 8°–32°/s. With increasing age, the number of saccades decreased and the intervals of smooth tracking became longer. At 3 months of age the smooth tracking dominated. A similar result was obtained by Aslin (1981) who used an object, a vertical rod 2° × 8°, moving sinusoidally with 10°/s. He found saccadic tracking up to 6 weeks of age after which SP was observed. We measured eye and head movements for object velocities between 0.1 and 0.4 Hz (von Hofsten and Rosander, 1996, 1997). In one experiment (von Hofsten and Rosander, 1996), an object (size 8°) moved sinusoidally at 0.1, 0.2, and 0.3 Hz, corresponding to 8.9°, 17.8°, and 26.7°/s, against a red-and-white vertically striped background. Values similar to Phillips et al. (1997) were found for eye gain (saccades plus SP), i.e., close to 0.6 at 1 and 2 months and 0.7–0.8 at 3 months of age.

Smooth pursuit

von Hofsten and Rosander (1997) separated the saccadic and smooth tracking components in a longitudinal study of infants from 2 to 5 months of

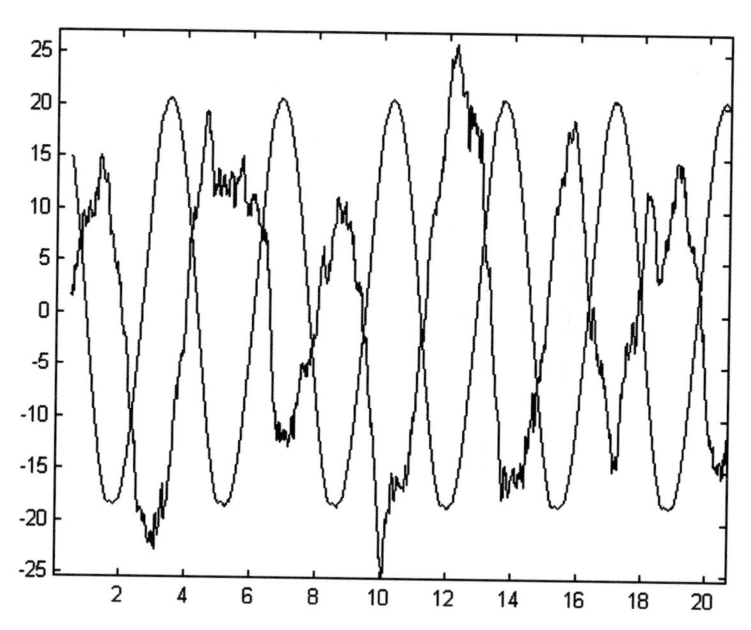

Fig. 3. Measurements of eye and body movements of a 4-week-old infant for a trial of 20 s. The object is stationary (VVOR). Axes as in Fig. 2.

age. The infants were shown sinusoidal and triangular oscillations of 0.2 and 0.4 Hz with amplitudes of 10° and 20° corresponding to average speeds of 10, 20, and 40°/s. They found that SP primarily developed between 2 and 3 months of age (von Hofsten, 2004). All the infants studied showed a similar increase in gain between these two ages. Gain was also dependent on object velocity. At 2 months it was 0.45, on the average, for the 40°/s and 0.80 for the 10°/s motion. The lag of the SP was less than 100 ms already at 2 months and did not change much over age. At 5 months of age, infants tracked periodic motions up to 0.6 Hz but at higher velocities the gain and phase deteriorated similarly to adults.

Stimulus size

In general, the younger the infant, the larger is the object needed for attentive tracking. Rosander and von Hofsten (2002) measured tracking of objects varying in size between 2.5° and 35°. The infants were studied longitudinally from 6.5–14 weeks. The objects moved sinusoidally at 0.25 Hz. The object size had no effect on SP gain at any of the ages studied. The gain increased with age, independently of object size. At 6.5 weeks of age, it was ~0.4 and at 14 weeks it was 0.8. There was an effect of object size in both the 6.5- and 9-week-old infants: smaller objects elicited more saccades. When the SP gain was low, the saccades compensated rather precisely and optimally. Furthermore, while the SP lag was minimal for all the ages and sizes studied, the lag of the raw eye movement, SP + saccades, improved substantially from 6.5 weeks of age where it was 200–300 ms to 12.5 weeks where it was less than 100 ms. There was no effect of object size. This indicates that at 12.5 weeks of age, both these forms of tracking are predictively controlled.

If the background around the object moves with it, this could either enhance or suppress the tracking. A greater extent of motion argues for enhanced tracking and the diminished motion contrast between the fixation object and the background in this condition argues for a suppression of tracking. Rosander and von Hofsten (2000) measured visual tracking for an object moving sinusoidally at 0.25 Hz. The background was a striped pattern and moved with the object. The gain and lag of the eyes relative to the object (alternatively the head

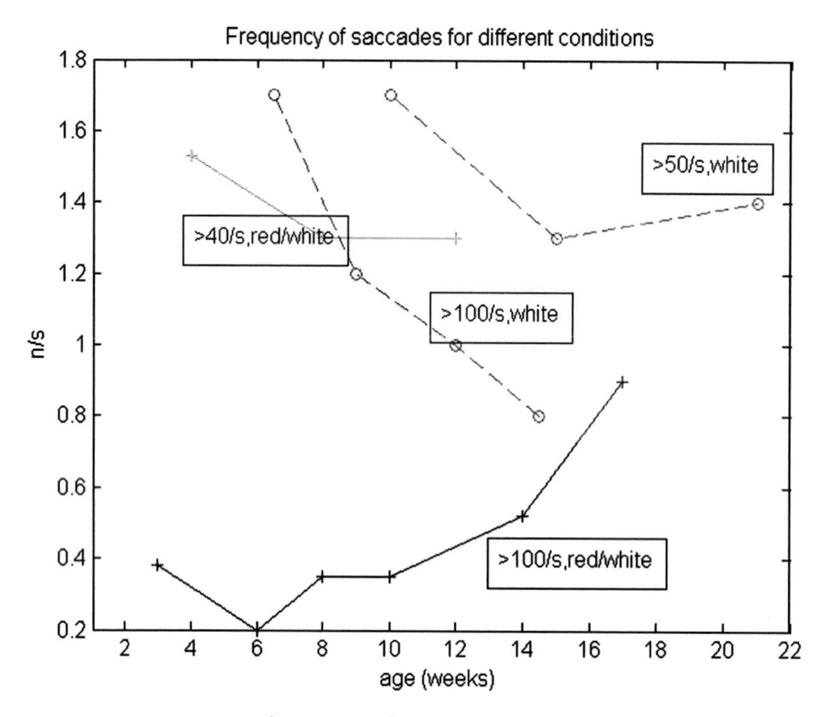

Fig. 4. The number of saccades (y-axis), $>100\,s^{-1}$ and $>40\,s^{-1}$ for visual tracking of a 0.25 Hz moving object if the background is striped ("red/white") or white.

slip) was similar for this study and other studies either using no background or a patterned stationary background (von Hofsten and Rosander 1996, 1997). However, the saccades (>100°/s) were fewer compared to conditions with no moving patterned background (Fig. 4). At 2 weeks of age the eye lagged almost 1.0 s, at 4 weeks 0.2 s, and at 10 weeks of age 0.05 s (Rosander and von Hofsten, 2000). Thus, the timing of the eye is rather precise when the gain of the SEMs start to function.

In the literature, two kinds of pursuit eye movements have been distinguished: optokinetic response (OKR) and SP. The function of OKR is to enable stabilization of gaze during ego motion while the function of SP is to enable gaze stabilization on a small moving object in a complex visual scene. The OKR is a phylogenetically older system and is present in all mammals while the SP system is only present in primates (Buttner and Buttner-Ennever, 1988; Paigie, 1994). It has been argued that OKR and SP are distinct systems. However, the systems have related neural pathways and both systems are predictive (Krauzlis and Stone, 1999). Our results suggest that both systems are indeed closely connected. There are almost no effects of size. Furthermore, it has been proposed that OKR develops before SP. Our results indicate that there is no shift in control over the age period studied. One might argue then that even the smallest object used activates the OKR system. But a small 2.5° stimulus can hardly be considered to control gaze during ego motion. The question is then whether there is any sense in distinguishing between OKR and SP. Also experimental data from adults suggest that SP and OKR are parts of a common system (Heinen and Watamaniuk, 1998).

Motion trajectory

In adults several experiments have been performed with unpredictable motion patterns (Barnes and Lawson, 1989), challenging the predictive oculomotor processing. Developmental studies have shown that "biological motion" is detected better than other types of motions early in life (Fox and

McDaniel, 1982). One common type of biological motion is the sinus motion, and combinations of such functions. We performed an experiment comparing the sine function with constant object motion that abruptly reversed at the end points of the trajectory (triangular motion). The drum set up was used, and the object was moved horizontally with the two motion patterns and with two amplitudes, 10° and 20°. Two-, 3-, and 5-month-old infants were measured when they were looking at an object that moved either according to a sinusoidal or a triangular function. Significant differences between motion types were obtained for eye gain, number of saccades, and SP timing. At 0.4 Hz the constant velocity motion gave higher gaze and SP gains than the sinusoidal one but very bad timing (around 250 ms). This indicates that because the triangular motion did not give an indication of when the motion would reverse, the tracking continued for a short time in the previous direction after which the eyes caught up with a reactive saccade. Because of this there were also more saccades for the triangular motion as compared to the sinusoidal one. Furthermore, the higher oscillation frequency and/or the higher amplitude induced more saccades. Drastic effects were seen for the timing. At 2 and 3 months of age, the lag was ~75 ms for sinusoidal motion, and 250 ms for the triangular motion. At 5 months of age all sinusoidal motions were anticipated predictively and the lag for the triangular motion was much reduced. The fact that the reversal of the triangular motion is not predictable from what happens just before (motion is constant), suggests that the reduced lag at 5 months is due to an ability to predict the periodicity of the motion.

Another question concerns the form of the trajectory. Gronqvist et al. (2006) studied visual tracking of circular, vertical, and horizontal motions in 5-, 7-, and 9-month-olds. They measured gaze movements with an infrared corneal reflection technique (ASL) camera and simultaneously head movements with "Flock of Birds." Similar to adults, infants tracked horizontal motions better than vertical ones. Learning was also found to improve within trials. Gredeback et al. (2005) found no consistent predictive tracking of a circular path until 8 months of age.

Head and eye coordination

In addition to eye movements, head movements are also used to stabilize gaze on a moving object. In 1-month-old infants, Bloch and Carchon (1992) showed that head movements were used more than in newborns. Also Daniel and Lee (1990) found that between 11 and 28 weeks, the head was increasingly used in tracking and some infants used more head than eye movements. von Hofsten and Rosander (1996) found that head gain increased marginally with age from 1 to 3 months for the tracking of an object moving at 0.1, 0.2, or 0.3 Hz. It was below 0.2 for all velocities. This result is similar to those found by von Hofsten and Rosander (1997). We found that the head made up around 10% of the gain at 2 and 3.5 months of age. However, at 5 months of age, almost 50% of the gain was accounted for by head movements. This can be seen in Fig. 5 that shows how the change in head movement gain is balanced with a change in gain for the eye movements. The resulting gaze is still on target. We found that it was convenient to relate eye position and eye velocity to the difference between object and head, i.e., the head slip. During tracking with both eye and head it means that the eye movement gain should optimally be equal to the head slip. von Hofsten and Rosander (1997) measured the lag of the head movements and found it to be considerable (~0.4 s). However, the lagging head did not deteriorate gaze tracking. Instead the eyes were compensating for the lagging head in such a way that the resulting gaze lag was close to 0. This means that the head movements are taken into account in the programming of the eye movements. However, for high oscillation frequencies, the lag of the head and the lead of the eyes tended to counteract each other. For instance, at 0.6 Hz a 0.4 s lag corresponds to almost a 90° phase lag. Thus the combined gain was less than the gain of either single head or eyes. In such cases the subject would fill in the smooth tracking with saccades.

Although the head contributes to the tracking of moving objects, it also moves for other reasons. One of the most important functions of the vestibular system is to compensate head movements unrelated to the tracking task by counter-rotating the eyes in equal amount in synchrony with the

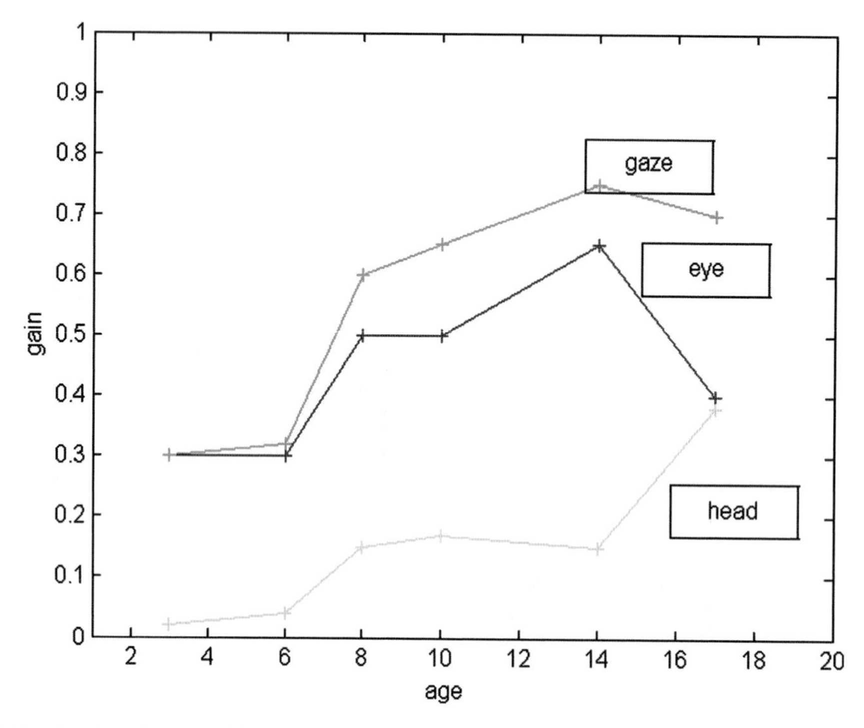

Fig. 5. Eye, head, and gaze position gain for tracking a 0.25 Hz moving object as a function of age in weeks.

head movements (VOR). This keeps the image stable on the retina in spite of the fact that head movements always include some fast head oscillations (Skavenski et al., 1979; Gresty, 1992). In adults, VOR is very effective for high frequencies in the interval 1–6 Hz. We found that this kind of compensatory eye movements appear very early in development (von Hofsten and Rosander, 1996). They are present in 1-month-old infants but the amplitude of the compensating eye movements is generally much higher than the inducing head movements. By 3 months of age the head and eye movements are much better scaled to each other and the compensation is much more effective.

Before 5–6 weeks of age essentially all head movements are compensated, even those that may contribute to the tracking. The problem that has to be solved in the development of eye–head coordination is how to separate the head movements that contribute to gaze tracking from the head movements that have to be compensated. It has been proposed that such smooth tracking with eye and head requires a suppression of the VOR (cf. Fukushima, 2003a). At the age when SP appears, the oculomotor system begins to utilize head movements in the tracking task, but sometimes the head movements involved in the tracking are compensated giving the paradoxical effect that the head tracking is totally ineffective. The VOR for the head movements involved in the tracking must not be compensated but a general suppression of VOR includes the compensations of head movements unrelated to the tracking task and they still need to be compensated if gaze is to be stabilized on the moving target. Separating the head movements that contribute to gaze tracking from those that need to be compensated is a problem that is solved at around 3 months of age (Fig. 6).

Apart from the vestibular system, several other systems are involved in the stabilization of body and gaze in space and the coordination between eyes and head. The vestibulo-collic response (VCR) will induce counter-rotation of the head to keep it stable in space in response to body rotation. The collic-ocular response, guided by proprioceptive information from the neck will initiate a counter-rotation of

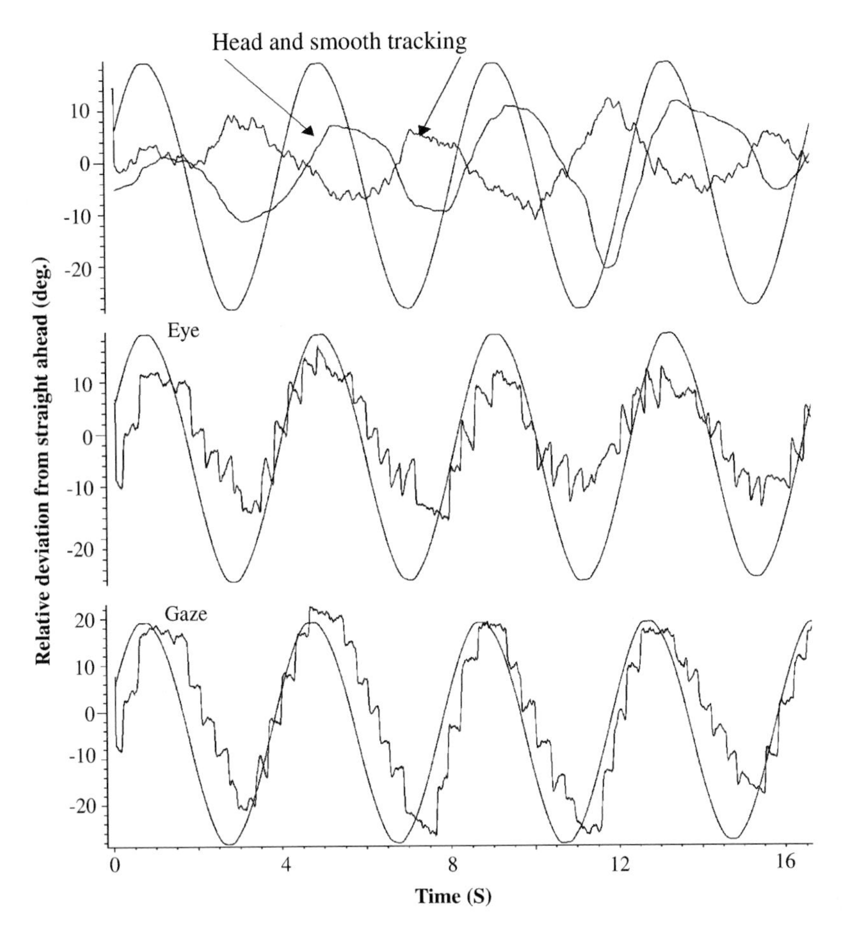

Fig. 6. Measurements of eye movements for an infant (6.5 weeks of age) that was not able to suppress VOR during the trial. The upper figure shows measured head and object movements together with calculated smooth eye movement component. The middle figure shows measured eye and object movements, and the lower figure the calculated gaze.

the eyes to compensate for neck-rotation. The final solution of the problem of coordinating head and eye movements needs to take all these different systems into account and has to be based on predictive control of intentional movements.

Brain development from the perspective of visual tracking

The visual pathways

In the human neonate two neural pathways have been suggested to process visual motion information. The primary pathway from the retina via the lateral geniculate nucleus (LGN) to visual cortex and MT+, and the secondary one via the SC and pulvinar to MT (Atkinson, 2000). The visual cortex is the cornerstone of visual processing. In primary visual area v17 (V1) Layer 4, most visual input comes from the LGN. The extrastriate areas, primary visual area v18 (V2), primary visual area v19 (V3), V4, and V5 (MT), process and distribute visual information further. For example, it has been suggested that the V3 processes large pattern coherent motions, as the striped pattern described earlier.

There are a few studies in humans on the development of neural growth between layers in the visual cortex. Prenatally, the intracolumnar connections between layers 2/3 and 5 develop, and the first intercolumnar connections are observed within layers 4 and 5 (Burkhalter et al., 1993). Burkhalter et al. (1993) also conclude that the intracolumnar connections develop before the intercolumnar ones reflect the processing of local features in the visual field. The magnocellular pathway characterized by the fibers from the LGN reaching Layer 4C in the V1, continue to Layer 4B and further towards the MT area. Sandersen (1991) (referring to Rakic, 1976) reported that the neuro-genesis in primate area 18 (V2) is slightly ahead of area 17 (V1), and that area 19 (V3) is ahead of area 18. Becker et al. (1984) studied the dendritic development in the visual cortex and found that at 4 months of age the dendritic length in Layer 5 had reached its maximum. Burkhalter (1993) reports that at 4 months of age forward connections from V1 to V2 are well matured. Taken together, at 4 months the maturation of the visual cortex is well on its way.

The maturation of the LGN in humans was studied by Garey and de Courten (1983). Their study concerns the dendritic and somatic spines, both common at 4 months of age. At 9 months of age, such spines have disappeared and the cells look like those in adults (Garey, 1984). Furthermore, the volume of the LGN doubles between birth and 6 months of age, after which no changes in volume are observed. In macaque infants Movshon et al. (2005) concluded that the maturation of neurons in the LGN is driven by the retinal cells. From 1 week of age, the spatial and temporal resolutions increase (Movshon et al., 2005).

The visual cortex increases 4-fold in size between birth and 4 months of age (see Garey, 1984). Huttenlocher et al. (1982) investigated the synaptogenesis in the primary visual cortex (V1) and found that it increases rapidly between 2 and 4 months of age reaching a peak at 8 months of age. Such high production of synapses during the 2–4 month period is rather general and is found in visual, prefrontal, somatosensory, and motor parts of the cerebral cortex (Rakic et al., 1986). After 1 year of age, the synaptic density decreases to adult values at 11 years of age, a process that permits plasticity or individual differences (Sandersen, 1991). For the development of SEM and for perception of motion direction, the V5/MT+ area, downstream from the primary visual cortex is essential. It has been suggested that the development of visual motion perception and SP between 2 and 5 months of age reflects the onset of functioning of the MT+ area. Kiorpes and Movshon (2004) measured receptive-field properties in infant macaques and found direction-sensitivity in the majority of MT cells at 1 week of age, (corresponding to 1 month in humans) although the neuronal dynamics was not adult-like. According to a histology study by Flechsig (1905), the MT+ area becomes myelinized during the first month in human infants. Rosander et al. (2007) reported that the evoked response potential (ERP) for motion stimulus in this area was discernable at 2, distinct at 3, and massive at 5 months of age. The maturation of the MT+ structure may be compared to the development of motion perception and SP gain.

An alternative visual pathway from the retina via SC has by some authors been suggested to be exuberant (Sandersen, 1991). However, for motion stimuli, this pathway that bypasses V1 still remains in adults and is active for non-conscious fear (Morris et al., 1999) and fast moving stimuli (ffytche et al., 1995; Buchner et al., 1997; Sincich et al., 2004). The subcortical, collicular pathway is considered to dominate motion processing until around 2 months of age (Dubowitz et al., 1986; Snyder et al., 1990; Martin et al., 1999; Atkinson, 2000).

The neural pathways for visual motion are illustrated in Fig. 7. As observed in Fig. 7, one important connection is MT+ to frontal eye field (FEF). In adults, O'Driscoll (2000) found that the gain of SP is set in the FEF area. At 4 months of age the prefrontal cortex is functioning (Csibra et al., 1997), as shown by the ERP for anticipatory or reactive saccades. In their discussion of such prospective control Canfield and Kirkham (2001) suggest that FEF may function even earlier, i.e., at 3 months of age. For language perception, Dehaene-Lambertz et al. (2002) obtained MRI response in the right prefrontal area of 3-month-old infants. The fact that the prefrontal and FEF areas may function at such young age is supported by the reported neural development by Mrzljak et al. (1990). During the postnatal period of 0–6 months

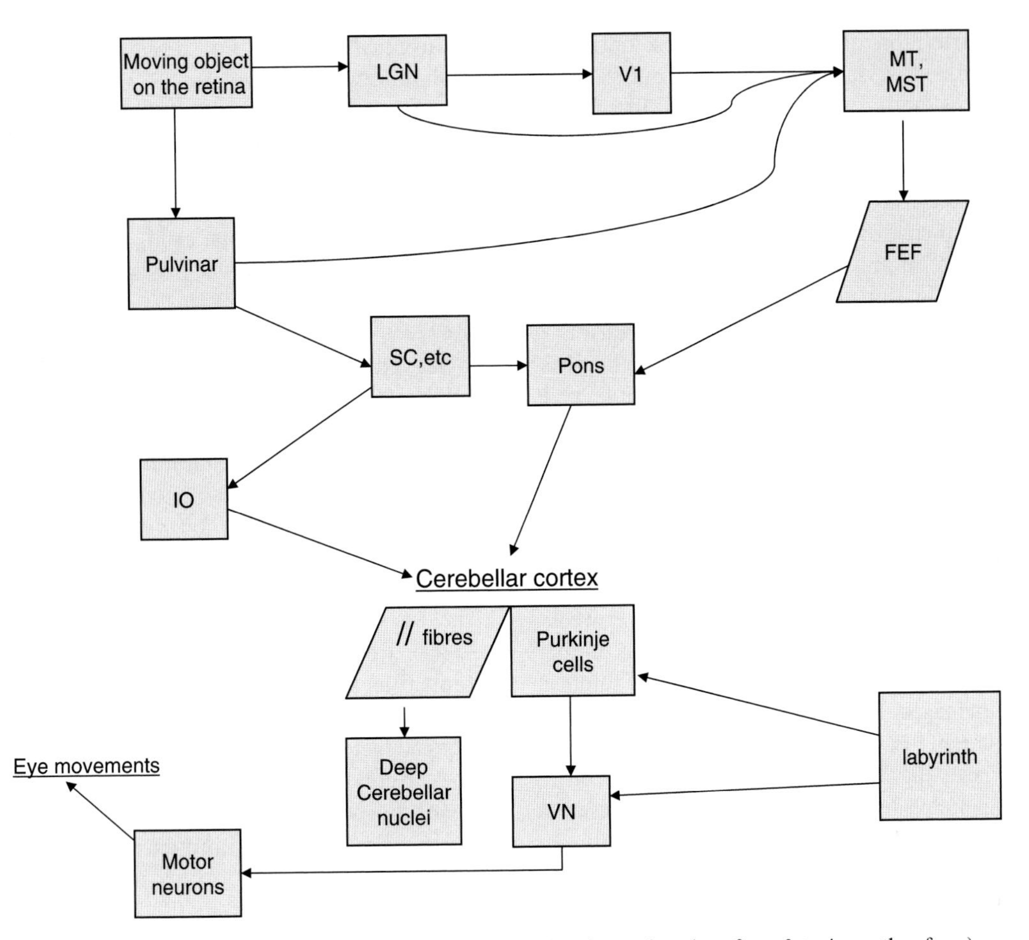

Fig. 7. An illustration of main visual motion pathways (rhombus = functions from 3 to 4 months of age).

most types of projections and local circuits are developed. Successive functioning may then take place to process SP gain for oculomotor functioning according to the compartment model in Fig. 7. Fukushima (2003b) suggests that caudal neurons in the FEF set the SP gain as well as maintaining a prediction of the eye movement. Furthermore, also head movements are involved in gaze estimates (see Fig. 1 in Fukushima, 2003b).

Cerebellum

The structure of cerebellum differs from that of cerebrum, consisting of three lobes and an underlying deep cerebellar nuclei (the fastigial, interpositus, and dentate). One lobe, the flocculus in the vestibulo-cerebellum is of special interest as it is essential for VOR and SP. This is a phylogenetic old part of cerebellum. There are two major inputs to cerebellum: one via the mossy fibers (MFs) and one via the climbing fibers (CFs) from the inferior olive (IO). MFs also branch to the deep nuclei.

The cerebellum has a central role in motor regulation, learning, memory, and action planning. The key loop includes the IO, the CFs, the Purkinje cells (PCs), plus the connecting MFs and parallel fibers (PF), the signals of which constitute the Purkinje output to the cerebellar nuclei (Miall, 1998). The MFs provide, from the pontine nuclei in the brainstem, information about the external world and intended and actual movements. The CFs from the IO to cerebellar cortex provide error signals related to movement production. All output from the cerebellum go via PCs to deep nuclei and VN for further distribution over the brain. The output to the cerebral cortex via the thalamus is massive.

There is some data on the maturation of CFs, MFs, and PFs. At birth, the CFs from the IO are connected to the PCs (Gudovic and Milutinovic, 1996). These cells, that form the cerebellar output, are then in their 3rd stage of maturation (Zecevic and Rakic, 1976) and especially their dendritic tree expands slowly during the first year of life. As pointed out (Zecevic and Rakic, 1976) this is a complex process, as the PFs connect to the PCs. These fibers are formed when the granule cells migrate from the external granular layer (EGL) inwards, causing this layer to get thinner with age. In their study of the histogenesis of the human cerebellum, Rakic and Sidman (1970) and Lavezzi et al. (2006) found that the EGL is formed at ~10–11 weeks after conception. The thickness decreases from 25 to 30 μ during the first 6 weeks of life. At 2 months, the EGL starts looking thinner and between 5 and 7 months of age the involution process increases. The layer is not visible at 12 months of age. Friede (1973) found a similar pattern although at a slightly younger age. Abraham et al. (2001) concluded that the EGL disappears between 8 and 11 months of age. The MFs are immature in the newborn (Seress and Mrzljak, 2004).

In summary, it can be argued that the PF–PC connection (glomerulus) begins to function at 1.5–2 months of age and has matured considerably at 5 months of age. The MFs follow the same course, but the CFs are connected to the PCs at birth.

Cerebellar functioning during development

An important function of cerebellum is to regulate sensory-motor coordination. Miall and Reckess (2002) stated that "one of the fundamental functions of the cerebellum is to act as a sensory predictor" (p. 212). This is manifested, at least to some extent in adaptive learning (Ito, 2001). What is learnt is the dynamics, not the individual part, of a movement (Ito, 1993). The cerebellar pathways create networks that serve sensory-motor functions. Most studies of cerebellar function reveal that predictive models have their site in loops that include the IO, PC, PF, MF units. However, if this neuroanatomical basis is not developed, the cerebellar functioning is incomplete.

Two types of responses have been studied in infants that illustrate the development of cerebellar learning. The first one is the conditioned eye blink response. It develops gradually during the first 5 months of life (Ivkovich et al., 1999; Klaflin et al., 2002). Little et al. (1984) measured the eye-blink response in infants aged 10, 20, and 30 days. For the first 10 days, no learning occurred, but it was observed for the 20- and 30-days groups. Klaflin et al. (2002) found a dramatic increase in the response at 5 months of age that they relate to the maturation of the cerebellar circuits. The interpretation of this response is that the MF carries the conditioned stimulus information, and the CF the unconditioned one. Then, the learning of eye blink is explained by the long-term depression (LTD) at the PF synapse on the PC (Ekerot and Jörntell, 2003). For this response, the PFs play an important role and the dramatic increase in learning at 5 months of age is in agreement with the neurodevelopment of the PFs.

A second example of cerebellar learning is the vestibulo-ocular reflex. The vestibulo-cerebellar cortex receives several inputs: from MFs originating in the labyrinths, from the vestibular nuclei, from CFs originating in the IO, and from MFs from the LGN and/or the SC. In the flocculonodular region, a translation of the degree of head motion relative to eye rotation is performed, so that the eye movements completely compensate the head movements. The VOR shows adaptable learning, and it has been suggested (Ito, 1993) that the glomerulus with the PF signal is a model for the desired eye trajectory resulting from the VOR. According to Blazquez et al. (2003), however, one might alternatively consider the important role of the brainstem structures. For rotations at frequencies of 0.2–0.5 Hz, the VOR does not function perfectly in neonates. The phase is, up to 1 month of age, not perfectly compensatory (Cioni, 1984; Weissman et al., 1989) and the eyes move too fast relative to the head. From then on, the VOR gain is close to 1.0 (Eviatar et al., 1974; Weissman et al., 1989; Finocchio et al., 1991). Rosander and von Hofsten (2000) showed that the gain is close to 1.0 at 8 weeks of age. In Fig. 8 the gains of the

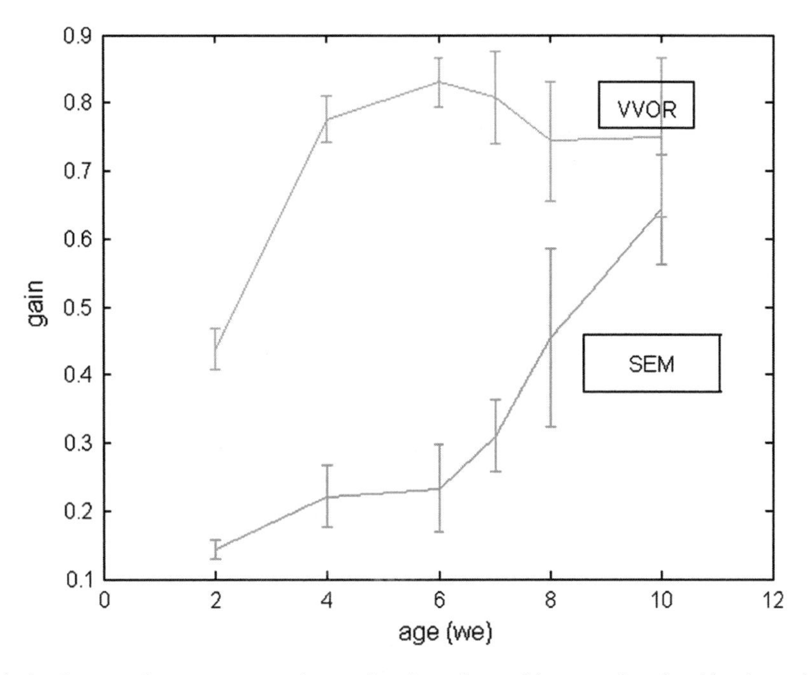

Fig. 8. Calculated gain for the smooth eye component in two situations: the tracking one when the object is moving (lower curve) and the vestibular one when the object is stationary (upper curve) as a function of age.

estimated SEM and VOR gain are shown. At 2 weeks of age the VOR gain is low but it increases rapidly up to 4–6 weeks of age. At 0.25 Hz the timing of the eye is ahead of the head with 25–75 ms at 2 weeks of age. The gain is around 0.45. From these data one can conclude that essential parts of the VOR loop have matured at 1–1.5 months of age. This is confusing considering that if there are no MF input (which is the main sensory input) and no PFs, the VOR should appear to be jerky and nonprecise, as the flocculus (Ito model) estimates the gain.

Taken together, the eyeblink response and the VOR develop in synchrony with the expanding PFs that constitute a critical part of the motor learning and predictive ability properties of the cerebellum. Maturation of these functions appears to take place in synchrony with neuronal growth and differentiation being established at 4 months of age.

Visual– vestibular interaction

When a moving object is tracked with the head and eye, visual and vestibular information must be utilized in an optimal way. All brain areas that are involved in the production of SEMs are also receiving vestibular information and a great deal of the integration of vestibular and visual information takes place in the cerebral cortex (Fukushima, 2003a). In adults it is assumed that the areas of VN, flocculus, vermis, and the brainstem nuclei are reached by visual information of moving objects, as well as by vestibular information. Lesions in the flocculus area impair the ability to suppress vestibular signals in a visual-vestibular interaction situation (Zee et al., 1981). We have performed two studies on how the visual-vestibular interaction is challenged when the infant and the environment (the drum) rotate in synchrony during light conditions (VINHIB described above). In the first study (2000), suppression of the gain of VOR was observed from 3 weeks of age. The counter-rotation of the eyes were not fully compensatory, the mean phase between eye and chair being ~140°. The VCR was not suppressed for the age period studied. On the contrary, it increased up to 18 weeks of age. In the experiment, the stimulus was a small happy face on a striped background. The second experiment was similar to the first one but the background was white. SEMs, saccades, and gaze were compared for three conditions: SEM, VVOR, and VINHIB. If no eye movements are observed during the last condition, this would imply a perfect inhibition of the VOR. Similary, a perfect inhibition of the VCR means that no head movements are observed. Some smooth compensatory eye movements were observed in the VINHIB condition from the youngest age level at 2 weeks to the oldest at 16 weeks showing that VOR was never completely suppressed. The gain was constant up to 10 weeks of age and diminished thereafter. The VCR was small until 10 weeks of age when head mobility increased substantially. The saccades were directed so that the eye movement gain was decreased, contrary to the VVOR and SEM conditions where the saccades increased the gain.

When the SEM and VVOR vectors for smooth eye position are added, the "theoretical curve" in Fig. 9 is obtained. It fits well with the "experimental" curve, the SEM gain in the VINHIB case. This result indicates that the neural processing for the inhibition of VOR is related to the VOR and SEM processing, supporting the notion of a common neural site like the flocculus (Buttner and Waespe, 1984). In VINHIB it seems easier to inhibit the vestibular induced eye movements if the background is neutral, and without texture of stripes.

Development and predictive models

Oculomotor control and regulation, as all motor control, is realized through predictive models. Such models provide a desired, predicted external state (the moving object) and how internal factors (delays in muscle activation) shall be regulated in order to fulfil that desired state. The goal of eye tracking is to stabilize gaze on the object of interest, and gaze position is based on eye, head, and body movements. SEMs begin to contribute to tracking from 6 weeks of age and attain an adult-like function at around 12–16 weeks of age. Head movements are never predictive in this age period but the eye movements compensate for the lagging head so that gaze is predictive.

Predictive models are the result of a learning process. One part is to create neural models of the

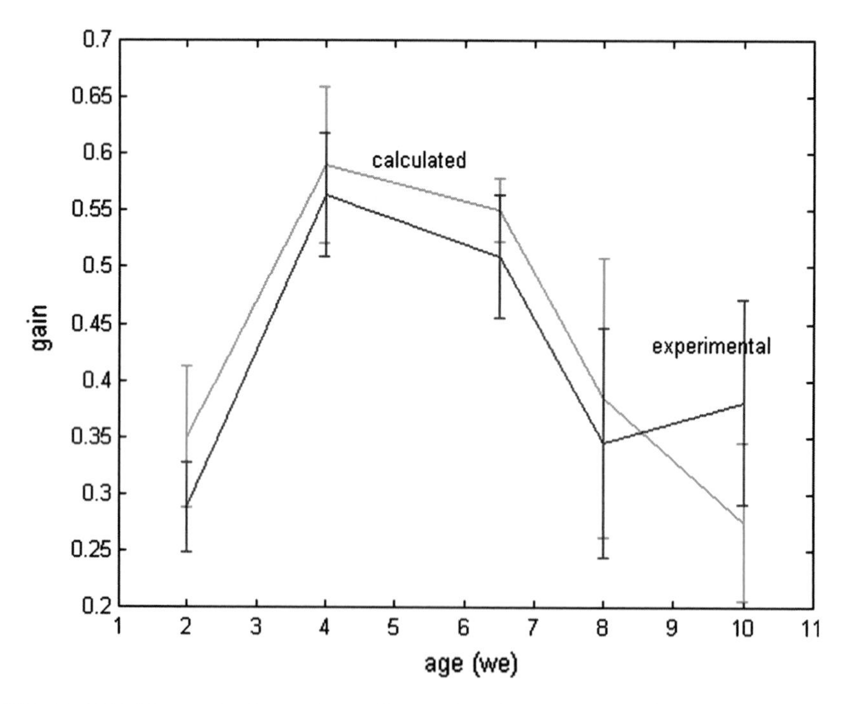

Fig. 9. Gains of SEM in the VINHIB situation ("experimental"), together with vector sums of SEMs in the object tracking and VVOR situations ("theoretical") as a function of age.

motor system in the body (Glickstein, 1993; Miall et al., 1993), another to learn the dynamics of movements (Ito, 1993). There is much evidence that cerebellum is the neural site for such models. They assume an adult functioning cerebellum with the PFs, PCs, and CF as key structures, in which LTD is formed and modified by CF error signals (Kawato, 1999; Ito, 2001). Ito (1993) discusses how parallel processing, not looping, functions for cerebro-cerebellar connections. Feed-forward or inverse control of actions in cerebellum is parallel to feedback control in the cerebral cortex (Ito, 1993). In their discussion of trans-cerebellar loops, Altman and Bayer (1997) suggested two types of loops regulating motor activity: a reactive one that includes error-correcting and negative feedback and a proactive one that has to do with error-avoiding and feedback. Both the upper proactive and the lower reactive loops originate in the deep nuclei. The reactive or lower one is phylogenetically older and has vestibular and spinal input: the cerebello-rubro-olivocerebellar loop. Hilbert and Caston (2000) discussed these models recently in connection with the motor learning in the Lurcher mice. The loops are regulatory and not part of executive function. The reactive regulatory task is to compensate for consequences of ongoing or previous actions. This is realized by feedback. The second type, the proactive one (cerebello-thalamo-cortico-ponto-cerebellar), is to make adjustments in preparation for actions. Recent research on the cerebro-cerebellar loops indicate that the closed loops between sensorimotor cerebral areas and the cerebellum are matched by closed loops with frontal, cingulate, parahippocampal, and occipito-temporal prestriate areas (Miall, 1998). The cerebellar projections to the prefrontal cortex, via thalamus, influence non-motor activities, as planning and spatial working memory utilizes information from the dorsal visual stream (Middleton and Strick, 2001).

The gain of the SP improves dramatically between 2 and 4 months of age, in synchrony with the maturation of PFs, and with the expansion of the visual pathways from the visual area to the MT+ area. However, even though the gain is insufficient at 2 months, the timing is predictive. How can visual tracking of a moving object be predictive at 2 months of age when the PFs in cerebellum are still immature? This means that some other process gives predictive eye movements, as for example a corresponding proactive loop that includes area 8 (FEF). Keeler (1990) has discussed a dynamic system for cerebellar functioning where prediction is established in the cerebral cortex.

Recently, Manto (2006) discussed the cerebello-cortical interactions. Some neurons in the deep cerebellar nuclei project to the IO, and loops are thus formed for cerebellar learning and tuning of temporal patterns. In the young infant (<1 month) such a system may function as a reactive loop, the "Phylogenetic old loop" in Kawato's model (1999). One possibility is that there are alternative systems; error-based brainstem (SC) loops or the more versatile nucleo-olivary pathways suggested by Bengtsson and Hesslow (2006). During the maturation of granule cells and MFs, a proactive loop may be formed that involves visual and vestibular information. Especially, a connection between the cerebellar and the cortical areas, that is MT+ and FEF, is established at ~4 months of age. At this age, behavior data have been presented showing that the visual tracking system is almost at the adult level (high gain and very small lag), with a smart coupling between head and eye movements. It is also possible that the LTD in cerebellum is utilized for tracking over occluders. That ability, built on prediction and an ability to represent object motion, is established at 17–20 weeks of age (von Hofsten et al., 2007). Finally, at this age, visually guided catching of moving objects is established in infants constituting one of the important purposes of visual tracking. This will challenge the perception–action loop demanding learning to establish cerebellar internal models.

Abbreviations

CF	climbing fiber
ERP	evoked response potential
FEF	frontal eye field
IO	inferior olive
LGN	lateral geniculate nucleus
LTD	long term depression

LTP	long term potentiation
MFs	mossy fibers
MT	median temporal area
PFs	parallel fibers
PC	purkinje cell
SC	superior colliculus
SEM	smooth eye movements
SP	smooth pursuit
V1	primary visual area, v17
V2, V3	visual areas v18 and v19, respectively
VCR	vestibulo-collic reflex
VINHIB	vestibular inhibition
VOR	vestibular ocular reflex
VVOR	visual vestibular ocular reflex

Acknowledgments

This study was supported by the Swedish Research Council (421-2003-1508) and the EU integrated project FP6-004370 (Robotcub). The author thanks Claes von Hofsten for valuable comments and discussions.

References

Abraham, H., Tornóczky, T., Kosztolányi, G. and Seress, L. (2001) Cell formation in the cortical layers of the developing human cerebellum. Int. J. Dev. Neurosci., 19: 53–62.

Altman, J. and Bayer, S.A. (1997) Development of the Cerebellar System in Relation to its Evolution, Structure, and Functions. CRC Press, Boca Raton, FL.

Aslin, R. (1981) Development of smooth pursuit in human infants. In: Fischer B., et al. (Eds.), Eye Movements: Cognition and Visual Development. Erlbaum, Hillsdale, NJ, pp. 31–51.

Atkinson, J. (2000) The Developing Visual Brain, Oxford Psychology Series 32. Oxford University Press, UK.

Barnes, G.R. and Lawson, J.F. (1989) Head-free pursuit in the human of a visual target moving in a pseudo-random manner. J. Physiol., 410: 137–155.

Becker, L.E., Armstrong, D.L., Chan, F. and Wood, M. (1984) Dendritic development in human occipital cortical neurons. Dev. Brain Res., 13: 117–124.

Bengtsson, F. and Hesslow, G. (2006) Cerebellar control of the inferior olive. Cerebellum, 5: 7–14.

Blazquez, P.M., Hirata, Y., Heiney, S.A., Green, A.M. and Highstein, S.M. (2003) Cerebellar signatures of vestibule-ocular reflex motor learning. J. Neurosci., 23(30): 9742–9751.

Bloch, H. and Carchon, I. (1992) On the onset of eye-head coordination in infants. Behav. Brain Res., 49: 85–90.

Buchner, H., Gobbelé, R., Wagner, M., Fuchs, M., Waberski, T.D. and Beckmann, R. (1997) Fast visual evoked potential input into human area V5. Neuroreport, 8(11): 2419–2422.

Burkhalter, A. (1993) Development of forward and feedback connections between areas V1 and V2 of human visual cortex. Cereb. Cortex, 3: 476–487.

Burkhalter, A., Bernardo, K.L. and Charles, V. (1993) Development of local circuits in human visual cortex. J. Neurosci., 13(5): 1916–1931.

Buttner, U. and Buttner-Ennever, J.A. (1988) Present concepts of oculomotor organization. In: Buttner-Ennever J.A. (Ed.), Reviews of Oculomotor Research, Vol. 2. Neuroanatomy of the Oculomotor System. Elsevier, Amsterdam, pp. 3–32.

Buttner, U. and Waespe, W. (1984) Purkinje cell activity in the primate flocculus during optokinetic stimulation, smooth pursuit eye movements and VOR suppression. Exp. Brain Res., 55: 97–104.

Canfield, R.L. and Kirkham, N.Z. (2001) Infant cortical development and the prospective control of saccadic eye movements. Infancy, 2(2): 197–211.

Cioni, G. (1984) Development of the dynamic characteristics of the horizontal vestibulo-ocular reflex in infancy. Neuropediatrics, 15: 125–130.

Csibra, G., Johnson, M.H. and Tucker, L.A. (1997) Attention and oculomotor control: a high-density ERP study of the gap effect. Neuropsychologica, 35(6): 855–865.

Daniel, B.M. and Lee, D. (1990) Development of looking with head and eyes. J. Exp. Child Psychol., 50: 200–216.

Dayton, G.O. and Jones, M.H. (1964) Analysis of characteristics of fixation reflex in infants by use of direct current electrooculography. Neurology, 14: 1152–1156.

Dehaene-Lambertz, G., Dehaene, S. and Hertz-Pannier, L. (2002) Functional neuroimaging of speech perception in infants. Science, 298: 2013–2015.

O'Driscoll, G.A. (2000) Functional neuroanatomy of smooth pursuit and predictive saccades. Neuroreport, 11: 1335–1340.

Dubowitz, L.M.S., Mushin, J., De Vries, L. and Arden, G.B. (1986) Visual function in the newborn infant: is it cortically mediated? Lancet, 327(8490): 1139–1141.

Ekerot, C. and Jörntell, H. (2003) Parallel fiber receptive fields: a key to understanding cerebellar operation and learning. Cerebellum, 2: 101–109.

Eviatar, L., Eviatar, A. and Naray, I. (1974) Maturation of neuro-vestibular reponses in infants. Dev. Med. Child Neurol., 16: 435–446.

ffytche, D.H., Guy, C.N. and Zeki, S. (1995) The parallel visual motion inputs into areas V1 and V5 of human cerebral cortex. Brain, 118(Pt6): 1375–1394.

Finocchio, D.V., Preston, K.L. and Fuchs, A. (1991) Infants' eye movements: quantification of the vestibular-ocular reflex and visual-vestibular interactions. Vis. Res., 10: 1717–1730.

Flechsig, P. (1905) Gehirnphysioligie und Willenstheorien. Paper presented at Fifth International Psychology Congress, Rome. (Reprinted in: Some papers on the cerebral cortex (von Bonin G., trans.), Springfield, IL. Thomas, 1960, pp. 181–200.

Fox, R. and McDaniel, C. (1982) The perception of biological motion by human infants. Science, 218: 486–487.
Friede, R.L. (1973) Dating the development of human cerebellum. Acta Neuropathol., 23: 48–58.
Fukushima, K. (2003a) Roles of the cerebellum in pursuit-vestibular interactions. Cerebellum, 2: 223–232.
Fukushima, K. (2003b) Frontal cortical control of smooth pursuit. Curr. Opin. Neurobiol., 13: 647–654.
Garey, L.J. (1984) Structural development of the visual system of man. Hum. Neurol., 3: 75–80.
Garey, L.J. and de Courten, C. (1983) Structural development of the lateral geniculate nucleus and visual cortex in monkey and man. Behav. Brain Res., 10: 3–13.
Glickstein, M. (1993) Motor skills but not cognitive tasks. TINS, 16(11): 450–451.
Gredeback, G., von Hofsten, C., Karlsson, J. and Aus, K. (2005) The development of two-dimensional tracking: a longitudinal study of circular pursuit. Exp. Brain Res., 163(2): 204–213.
Gresty, M. (1992) Disorders of head-eye coordination. Baillière's Clin. Neurol., 1: 317–343.
Gronqvist, H., Gredeback, G. and von Hofsten, C. (2006) Developmental asymmetries between horizontal and vertical tracking. Vis. Res., 46(11): 1754–1761.
Gudovic, R. and Milutinovic, K. (1996) Regression changes in inferior olivary nucleus compared to changes of Purkinje cells during development in humans. J. Brain Res., 37(1): 67–72.
Heinen, S.J. and Watamaniuk, S. (1998) Spatial integration in human smooth pursuit. Vis. Res., 38: 3785–3794.
Hilbert, P. and Caston, J. (2000) Motor learning in lurcher mice during aging. J. Neurosci., 102(3): 615–623.
von Hofsten, C. (1982) Eye-hand coordination in newborns. Dev. Psychol., 18: 450–461.
von Hofsten, C. (2004) An action perspective on motor development. TICS, 8(6): 266–272.
von Hofsten, C., Kochukhova, O. and Rosander, K. (2007) Predictive tracking over occlusions by 4-month-old infants. Dev. Sci., In press.
von Hofsten, C. and Rosander, K. (1996) The development of gaze control and predictive tracking in young infants. Vis. Res., 36: 81–96.
von Hofsten, C. and Rosander, K. (1997) Development of smooth pursuit tracking in young infants. Vis. Res., 37: 1799–1810.
Huttenlocher, P.R., de Courten, C., Garey, L.J. and van der Loos, H. (1982) Synaptogenesis in human visual cortex: evidence for synapse elimination during normal development. Neurosci. Lett., 33: 247–252.
Ito, M. (1993) Movement and thought: identical control mechanisms by the cerebellum. TINS, 16(11): 448–450.
Ito, M. (2001) Cerebellum long-term depression: characteristics, signal transduction, and functional roles. Physiol. Rev., 81(3): 1143–1195.
Ivkovich, D., Collins, K.L., Eckerman, C.O., Krasnegor, N.A. and Stanton, M.E. (1999) Classical delay eyeblink conditioning in 4- and 5-month-old human infants. Psychol. Sci., 10(1): 4–8.
Johnson, M.H. (1990) Cortical maturation and the development of visual attention in early infancy. J. Cogn. Neurosci., 2: 81–95.
Kawato, M. (1999) Internal models for motor control and trajectory planning. Curr. Opin. Neurol., 9: 718–727.
Keeler, J.D. (1990) A dynamical system view of cerebellar function. Physica D, 42: 396–410.
Kiorpes, L. and Movshon, J.A. (2004) Development of sensitivity to visual motion in macaque monkeys. Vis. Neurosci., 21(6): 851–859.
Klaflin, D., Stanton, M., Herbert, J., Greer, J. and Eckerman, C.O. (2002) Effect of delay interval on classical eyeblink conditioning in 5-month-old human infants. Dev. Psychobiol., 41: 329–340.
Kowler, E. (1990) The role of visual and cognitive processes in the control of eye movement. In: Kowler E., (Ed.), Eye Movements and Their Role in Visual and Cognitive Processes. Reviews of Oculomotor Research. Amsterdam, Elsevier, pp. 1–70.
Krauzlis, R.N. and Stone, L.S. (1999) Tracking with the mind's eye. TINS, 22(12): 544–550.
Kremenitzer, J.P., Vaughan, H.G., Kurtzberg, D. and Dowling, K. (1979) Smooth-pursuit eye movements in the newborn infant. Child Dev., 50: 442–448.
Lavezzi, A.M., Ottaviani, G., Terni, L. and Matturri, L. (2006) Histological and biological developmental characterization of the human cerebellar cortex. Int. J. Dev. Neurosci., 24: 365–371.
Leigh, R.J. and Zee, D.S. (1999) The Neurology of Eye Movements. Oxford University Press, New York, USA.
Little, A.H., Lipsitt, L.P. and Rovee- Collier, C. (1984) Classical conditioning and retention of the infant's eyelid reponse: effects of age and interstimulus interval. J. Exp. Child Psychol., 37: 512–524.
Manto, M. (2006) On the cerebello-cerebral interactions. Cerebellum, 5: 286–288.
Martin, E., Joeri, P., Loenneker, T., Ekatodramis, D., Vitacco, D., Hennig, J. and Marcar, V. (1999) Visual processing in infants and children studies using functional MRI. Pediatr. Res., 46(2): 135–140.
Mayes, L.C. and Kessen, W. (1989) Maturational changes in measures of habituation. Infant Behav. Dev., 12: 437–450.
Miall, R.C. (1998) The cerebellum, predictive control and motor coordination. Novartis Found. Symp., 218: 272–284 and discussion, pp. 284–290.
Miall, R.C. and Reckess, C.Z. (2002) The cerebellum and the timing of coordinated eye and hand tracking. Brain Cogn., 48: 212–226.
Miall, R.C., Weir, D.J., Wolpert, D.M. and Stein, J.F. (1993) Is the cerebellum a Smith predictor? J. Mot. Behav., 25(3): 203–216.
Middleton, F.A. and Strick, P.L. (2001) Cerebellar projections to the prefrontal cortex of the primate. J. Neurosci., 21(2): 700–712.
Morris, J.S., Öhman, A. and Dolan, R.J. (1999) A subcortical pathway to the right amygdala mediating "unseen" fear. Proc. Natl. Acad. Sci., 96(4): 1680–1685.

Movshon, J.A., Kiorpes, L., Hawken, M.J. and Cavanagh, J.R. (2005) Functional maturation of the Macaque's lateral geniculate nucleus. J. Neurosci., 25(10): 2712–2722.

Mrzljak, M., Uylings, H.B.M., Eden, C.G. and Judas, M. (1990) Neuronal development in human prefrontal cortex in prenatal and postnatal stages. Prog. Brain Res., 85: 185–222.

Paigie, G.D. (1994) Senescence of human vision-vestibular interactions: smooth pursuit, optokinetic, and vestibular control of eye movements with aging. Exp. Brain Res., 98: 355–372.

Pavel, M. (1990) Predictive control of eye movement. In: Kowler E. (Ed.), Eye Movements and Their Role in Visual and Cognitive Processes. Reviews of Oculomotor Research, Vol. 4. Elsevier, Amsterdam, pp. 71–114.

Peterson, B.W. and Richmond, F.J. (1988) Control of Head Movements. Oxford University Press, New York.

Phillips, J.O., Finocchio, D.V., Ong, L. and Fuchs, A.F. (1997) Smooth pursuit in 1- to 4-months-old infants. Vis. Res., 37: 3009–3020.

Rakic, P. (1976) Differences in the time of origin and in eventual distribution of neurons in areas 17 and 18 of visual cortex in rhesus monkey. Exp. Brain Res., Suppl. 1: 244–248.

Rakic, P., Bourgeois, J., Eckenhoff, M., Zecevic, N. and Goldman-Rokic, P. (1986) Concurrent overproduction of synapses in diverse regions of the primate cerebral cortex. Science, 232: 232–235.

Rakic, P. and Sidman, R.L. (1970) Histogenesis of cortical layers in human cerebellum, particularly the lamina dissecans. J. Comp. Neurol., 139: 473–500.

Robinson, D.A. (1965) The mechanics of human smooth pursuit eye movements. J. Physiol., 180: 569–591.

Rosander, K. and von Hofsten, C. (2000) Visual-vestibular interaction in early infancy. Exp. Brain Res., 133: 321–333.

Rosander, K. and von Hofsten, C. (2002) Development of gaze tracking of small and large objects. Exp. Brain Res., 146: 257–264.

Rosander, K., Nyström, P., Gredebäck, G. and von Hofsten, C. (2007) Cortical processing of visual motion in young infants. Vis. Res., 47(12): 1614–1623.

Sandersen, K. (1991) The ontogenesis of the subcortical and cortical visual centres. In: Dreher B. and Robinson S.R. (Eds.), Vision and Visual dysfunction. Vol. 3: Neuroanatomy of the Visual Pathways and their Development, Chapter 3. MacMillan Press Ltd., London, UK, pp. 360–376.

Seress, L. and Mrzljak, L. (2004) Postnatal development of mossy cells in the human dentate gyrus: a light microscopic Golgi study. Hippocampus, 2(2): 127–141.

Sincich, L.C., Park, K.F., Wohlgemuth, M.J. and Horton, J.C. (2004) Bypassing V1: a direct geniculate input to area MT. Nat. Neurosci., 7(10): 1123–1128.

Skavenski, A.A., Hansen, R.M., Steinman, R.M. and Winterson, B.J. (1979) Quality of retinal image stabilization during small natural and artificial body rotations in man. Vis. Res., 19: 675–683.

Snyder, R.D., Hata, S.K., Brann, B.S. and Mills, R.M. (1990) Subcortical visual function in the newborn. Pediatr. Neurol., 6: 333–336.

Stechler, G. and Latz, E. (1966) Some observations on attention and arousal in the human infant. J. Am. Acad. Child Psychol., 5: 517–525.

Weissman, B.M., DiScenna, A.O. and Leigh, R.J. (1989) Maturation of the vestibule-ocular reflex in normal infants during the first 2 months of life. Neurology, 39: 534–538.

Westling, G. (1992) Wiring diagram of EOG amplifiers. Department of Physiology, Umeå University.

Wolpert, D.M., Miall, R.C. and Kawato, M. (1998) Internal models in the cerebellum. TICS, 2(9): 338–347.

Zecevic, N. and Rakic, P. (1976) Differentiation of Purkinje cells and their relationship to other components of developing cerebellar cortex in man. J. Comp. Neurol., 167: 27–48.

Zee, D.S., Yamazaki, A., Butler, P.H. and Gunduz, G. (1981) Effects of ablation of flocculus and paraflocculus on eye movements in primate. J. Neurophysiol., 46(4): 878–899.

C. von Hofsten & K. Rosander (Eds.)
Progress in Brain Research, Vol. 164
ISSN 0079-6123

CHAPTER 7

Visual and visuocognitive development in children born very prematurely

Janette Atkinson[1,*] and Oliver Braddick[2]

[1]*Visual Development Unit, Department of Psychology, University College London, London, UK*
[2]*Department of Experimental Psychology, University of Oxford, Oxford, UK*

Abstract: Preterm birth is a risk factor for deficits of neurological and cognitive development. Four cohort studies are reported investigating the effects of very premature birth (<32 weeks gestation) on visual, visuocognitive and visuomotor function between birth and 6–7 years of age. The first study used two measures of early visual cortical function, orientation reversal visual event-related potentials (OR-VERP) and fixation shifts under competition. Both these functional measures of visual development correlated with the severity of brain abnormality observed on structural MRI at and before term, and were sensitive predictors of neurodevelopmental outcome at 2 years. The second study compared VERP measures for orientation-reversal and direction-reversal (DR) stimuli, from 2 to 5 months post-term age, in healthy very premature infants compared to infants born at term. The groups did not differ on the development of OR-VERP responses, but the development of the DR-VERP motion responses was delayed in the premature group despite the absence of any brain damage visible on ultrasound, consistent with the developmental vulnerability we have identified in the dorsal cortical stream. The third study used the Atkinson Battery of Child Development for Examining Functional Vision (ABCDEFV) to assess sensory, perceptual, cognitive and spatial visual functions, together with preschool tests of attention and executive function. The premature group showed delays on these tests in line with severity of observed perinatal brain damage on structural MRI at term age. Deficits on certain spatial tasks (e.g. block-construction copying) and executive function tests (e.g. the detour box task) were apparent even in children with minimal damage apparent on MRI. The fourth study tested a large cohort of 6- to 7-year old children born before 32 weeks gestation, across a wide range of cognitive domains, including new tests of spatial cognition and memory. The premature group as a whole showed significant deficits on both auditory and visual tests of attention and attentional control from the TEA-Ch battery, on tests of location memory, block construction and on many visuocognitive and visuomotor tests. Development was generally relatively normal on language tests and on WPPSI scores. Factor analysis showed that while general cognitive ability accounted for the largest part of the variance, significant deficits, and a relationship to MRI results, were primarily in spatial, motor, attention and executive function tests. A model is proposed suggesting that the cluster of deficits seen in children born prematurely may be related to networks involving the cortical dorsal stream and its connections to parietal, frontal and hippocampal areas.

Keywords: prematurity; visual development; visuo-cognitive development ABCDEFV; VERP

*Corresponding author. Tel.: (+44) 207 679 7574;
Fax: (+44) 207 679 7576; E-mail: j.atkinson@ucl.ac.uk

DOI: 10.1016/S0079-6123(07)64007-2

Introduction

Preterm birth is a major cause of childhood neurological and psychiatric impairment, including lifelong physical and mental health problems (e.g. Jones et al., 1998; Larsson et al., 2005). Half of surviving infants born before 26 weeks gestation show neurodevelopmental impairment at 30 months of age (Wood et al., 2000). Among less immature infants, over one-third develop neurocognitive (learning disability), motor or behavioural deficits and disorders of attention (e.g. Bhutta et al., 2002; Marlow et al., 2005). As well as major disability, there is evidence for more widespread but more subtle deficits of cognitive function. Both levels of disability have an enormous impact on quality of life for individuals, their families and society and incur heavy health care costs (Stevenson et al., 1996). Identifying and understanding the developmental pathway linking premature birth to later cognitive development is therefore a major objective, which may yield advances in early diagnosis and therapy.

Studying this pathway should also improve our broader understanding of brain and cognitive development, in understanding the precursors of cognitive achievements and the ways in which the developing human brain shows functional plasticity in the face of early damage.

Links between brain structure and function in early life have been made possible by advances in brain imaging: first, cranial ultrasound which is now a standard clinical assessment for preterm and many term infants, and second, magnetic resonance imaging (MRI). Technical advances have made MRI possible within the neonatal intensive care unit (Maalouf et al., 1999), improving the identification of perinatal cerebral lesions in the preterm infant (e.g. Roelants-van Rijn et al., 2001; Woodward et al., 2006). Brain imaging through the period around term has also identified characteristic anomalies in children born preterm (Rutherford, 2002; Dyet et al., 2006), particularly in white matter. Thus we have an increasingly detailed picture of early structural development of the brain in preterm infants.

However, these anatomical findings are only beginning to be related to the neurocognitive outcome of these children (as in the recent example by Woodward et al. (2005) correlating neonatal cerebral tissue volumes with working memory deficits at 2 years). Standard methods of clinical assessment in the early years of life for these children depend heavily on motor competences and so may not fully reflect accurately their cognitive abilities. There is a need for broader assessments of infants' and toddlers' cognitive and cerebral functions, and for specific information on the early function of specific brain systems that underpin particular domains of cognitive development at different ages. Such measures in infancy could help to indicate the immediate functional consequences of brain damage, related to its severity and cerebral location. They may also be able to serve as early surrogates for later measures of neurological and cognitive outcome in specific psychological domains.

In particular, visual function is an area which develops early and where there are a number of milestones, identified in typical development the first year of life, which signal the maturation of specific cortical systems. Vision also serves as the basis for early developing sensory-motor and cognitive skills, which can be tracked through early childhood.

In this chapter we report four studies that have used tests of visual, visuocognitive and visuomotor development which we have designed to tap specific visual neural networks and their consequences for cognitive development, and related these to structural information from neonatal brain imaging (ultrasound or MRI). Two of these studies concentrate on measures of specific visual cortical functions in the first year; the other two follow-up outcomes in visually related cognitive functions up to early school age.

Model of early visual development

We study development of the preterm infants in terms of the neurobiological models we have devised from previous research for typically developing infants (Atkinson, 1984, 2000; Atkinson and Braddick, 2003). The newborn, starting with very limited visual behaviour, develops many of the

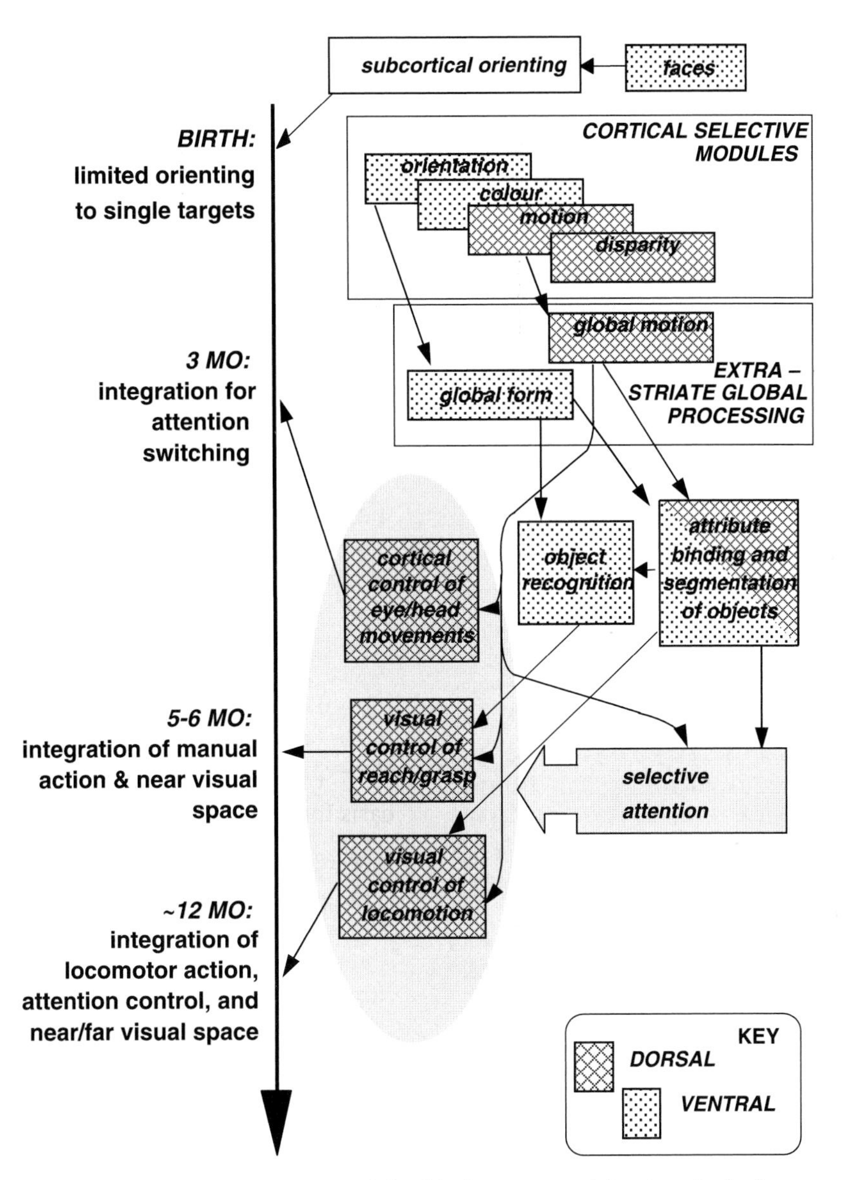

Fig. 1. Model of the developmental sequence of visual brain systems and functions, in the first year of life.

complex visual processes necessary for pattern and depth perception in the first few months of life. Some key points of this developmental sequence, schematized in Fig. 1 are discussed below.

Newborn

At birth the infant can make saccadic eye movements and imprecise, slow head movements to orient towards high contrast targets, a function mediated by a subcortical system involving the superior colliculus. This newborn 'Where?' orienting system only operates well when there is no competition between targets for the newborn's attention. It operates supramodally across sensory modalities, as a non-specific alerting system, as in the ability of newborns to orient, albeit inaccurately and slowly, to a lateral auditory stimulus in the first few hours after birth.

Onset of visual cortical functions

Over the first months of life a set of specific neural channels become functional, sequentially in time, for processing different visual attributes in the cortex. In typically developing infants, cortical orientation selectivity has an onset around 3 weeks to 3 months of post-term age (depending on the spatiotemporal properties of the stimulus); directional motion selectivity around 2–3 months; and binocular interaction underpinning stereoscopic depth perception around 4–5 months. We have devised 'marker tasks' to identify the responses of these cortical mechanisms, specific for particular visual attributes. These tasks use both electrophysiological measures (visual evoked potentials (VEP) or visual event related potential (VERP)) and behavioural methods, such as forced-choice preferential looking (FPL) and infant control habituation–dishabituation paradigms. The studies described below gauge the development of orientation sensitivity from the orientation reversal (OR)-VERP: this is a steady state response to a stripe pattern, which rapidly alternates between two orientations 90 degrees apart (Braddick et al., 1986). A similar measurement can be made for directional motion using random dot patterns that rapidly reverse their direction of motion: the direction-reversal visual event related potential (DR-VERP) (Wattam-Bell, 1991; Braddick, 1993).

Dorsal and ventral streams: the first stages of 'local' processing

Pools of neurons, sensitive to these different visual attributes, provide the first stage of the two main cortical streams of processing, called the 'dorsal' and 'ventral' streams (Mishkin et al., 1983; Milner and Goodale, 1995). The dorsal stream is used for processing information for motion, spatial relations and directing actions (Where?/How?), while the ventral stream processes information for identifying objects and faces (What?/Who?). Our finding that orientation selectivity emerges earlier than direction selectivity in the first months of life implies that at this local level the ventral stream starts to function in normal development earlier than the dorsal stream. This local processing is believed to take place in the striate visual cortex, visual area 1 (V1) (Atkinson, 2000).

Dorsal and ventral streams: a second stage of global processing

A second stage of processing in the dorsal and ventral streams, in extrastriate cortex, integrates information over wide areas to analyse its global structure. The development of global motion processing, a function of dorsal stream processing, can be compared with global processing of static form in the ventral stream, by measuring analogous coherence thresholds for form and motion. Neurons responding to concentrically organized patterns have been found in visual area 4 (V4) in macaques (Gallant et al., 1993)), an extrastriate area at a similar level in the ventral stream to visual area 5/middle temporal area (V5/MT) in the dorsal stream, which has been identified as a key area for global motion perception (Zeki, 1974; Newsome and Paré, 1988). Our current studies of typically developing infants (see Braddick and Atkinson, this volume) suggest at this global level of dorsal and ventral stream processing, global motion processing (dorsal stream) operates at an earlier stage of development (around 3 months) than global form processing (ventral stream) found at 4–6 months of age at the earliest. This is a reversal of the sequence for form and motion processing at the earlier local processing level, discussed above.

Integration within the dorsal and ventral streams

Sensitivity in these channels is followed by the development of integrative processes across channels within a single stream so that the infant can build up internal representations of objects, including discrimination of individuals using cortical face recognition mechanisms. However, faces may be a special case for which there may be an earlier subcortical mechanism operating from birth, which biases visual attention to configurations that are 'face-like' in the newborn (Simion et al., 2006). This may be replaced by a cortical system operating from a few months after birth with a more mature sensitivity to facial features and configurations (Morton and Johnson, 1991).

Selective attention

The cortical channels used for analysing visual attributes are linked into the first functional cortical networks for selective attention. Orienting or switching attention to a peripheral novel stimulus or target, when the infant is already fixating a centrally presented stimulus, requires modulation and disengagement of the subcortical orienting system by cortical processes. We have developed the fixation shift (FS) paradigm, and found that, in normally developing infants, the cortical system for active switches of attention between competing targets starts to function around 3–4 months of age (Atkinson et al., 1992; Hood and Atkinson, 1993). Evidence for the role of the cortex in disengagement comes from studies of infants who have undergone hemispherectomy, surgical removal of one complete hemisphere to relieve intractable epilepsy (Braddick et al., 1992). Post-operatively these infants can shift gaze towards a target appearing in the peripheral field contralateral to the removed hemisphere when an initial central fixation target disappears, but fail to disengage to fixate the peripheral target when the central target remains visible, although they can do so towards a target in the intact visual field, reflecting the need for a cortical 'disengage' mechanism.

Integration across the dorsal and ventral streams

Throughout development there must be interactions and integration between information in the dorsal and ventral streams. For objects to be represented, information about colour, shape and texture must be integrated with motion information at a relatively early stage, so that objects can be segregated from each other in space and separated from their background. This basic figure-ground segmentation, and its relation to dorsal and ventral stream function, is discussed by Braddick and Atkinson elsewhere in this volume.

Onset of function in different modules of the dorsal stream controlling actions

These processes of segmentation provide object representations, which are integrated with dorsal-stream spatial information, to allow the infant to act and respond first with selective attentional eye and head orienting action systems, then later with the emergence of action systems associated with reaching and grasping, and later still with exploratory action systems involving locomotion.

Integration across visuomotor and attentional systems

These action systems require maturation and integration of both visual attentional systems and visuomotor systems, both integral aspects of dorsal stream function. Further discussion of these action systems can be found in chapters by Braddick and Atkinson, and by von Hofsten in this volume.

Measures of visual brain development in the preschool and early school years

The child's broader visual capabilities change a great deal in the preschool years. The changes in visuomotor, visuoperceptual, visuocognitive and spatial behaviour and understanding in this period are underpinned by the development of the massive interconnectivity between different cortical areas and of cortical–subcortical networks. The Atkinson Battery of Child Development for Examining Functional Vision (ABCDEFV) is a battery that we have devised, which brings together tests for developmental milestones for functional vision across these domains (Atkinson et al., 2002a). The battery includes standard procedures for measuring core vision (acuity, accommodation, visual fields and control of eye movements) and additional tests for higher level functions in visuoperceptual and visuocognitive domains (e.g. shape matching, embedded figures tests, spatial tasks such as copying specific block constructions). There are around 20 short subtests in the battery and each has been standardized, so that any individual child between 1 month and

6 years can be given a normal age-equivalence score to compare with their chronological age. Some children with specific developmental disorders show very uneven profiles across subtests and this allows the examiner to identify areas of concern for further testing and remediation.

Alongside the ABCDEFV, we have designed a number of tests to tap particular cortical networks, based on evidence from adult lesion patients and models from behavioural and electrophysiological studies of non-human primates. For example, we have developed a preschool test of executive function called the 'counter-pointing' task. The child has to inhibit a familiar prepotent spatial response (pointing to a target when it appears on one side of a central fixation point), by pointing to the opposite side of the screen. By 4 years of age typically developing children can learn the new association rule and inhibit this prepotent response rapidly and automatically. This is similar in rationale to the anti-saccade task, used to identify frontal impairment in adult patients. Under Study 3 below we describe other executive function tasks, which we have used for preschool children, devised by other researchers (Russell detour box, Hood Tubes).

We have also devised tests to tap development of the ventral and dorsal streams, from comparisons of global form and global motion coherence thresholds. These take to 4 years and beyond the tests of sensitivity to global form and global motion, which we have discussed above for infants in the first 6 months of life. They take the form of computer games ('find the ball in the grass', 'find the road in the snowstorm'). These measures show that ventral stream processing for global form matures to adult levels somewhat earlier (around 7–8 years), than dorsal stream processing for global motion coherence, where adult-like thresholds are achieved at around 10–11 years (Gunn et al., 2002; Atkinson and Braddick, 2005).

The disadvantage of global motion processing compared to global form processing is seen, to a much greater degree, in a number of developmental disorders. We first identified this in children with Williams syndrome (WS) (Atkinson et al., 1997), and tests on WS adults showed that it was a persisting feature of this genetic disorder (Atkinson et al., 2006). However, these motion coherence deficits are also found in children with hemiplegia (Gunn et al., 2002), and a similar relative impairment of global motion sensitivity has also been found in autistic children (Spencer et al., 2000), in studies of developmental dyslexia (Cornelissen et al., 1995; Hansen et al., 2001; Ridder et al., 2001), in fragile-X syndrome (Kogan et al., 2004) and in children who had early visual deprivation due to congenital cataract (compare Ellemberg et al. (2002); Lewis et al. (2002)). As this motion coherence deficit appears to be a common feature across a wide variety of paediatric disorders with different etiologies and neurodevelopmental profiles, it is likely to represent an early vulnerability (dorsal stream vulnerability) in the motion-processing stream at an early stage of development (Braddick et al., 2003). These tests of motion and form coherence have also been used to test at 6–7 years of age children born very prematurely, and are discussed below.

Studies of prematurely born children

In the four studies we report here, the range of tests described above have provided age-appropriate measures to follow the development of children in premature cohorts from a few months of post-term age to 7 years, across a range of different visual and cognitive domains, and to compare these measures with brain imaging data from the neonatal period.

The first, third and fourth cohort studies, described here, have been carried out in collaboration with the neonatal paediatrics and imaging team at the Hammersmith Hospital in London. These children had structural MRI in the neonatal period, and at term, as part of the protocol. The overall plan of these studies is schematized in Fig. 2. Study 1 looks at these infants with measures of cortical function in the first year, and compares this to neurodevelopmental follow-up at age 2 years. Study 3 takes the similar group of children with a range of visual and visuocognitive measures up to age 5 years, and Study 4 up to 6–7 years.

The second study was a collaboration with the neonatal paediatric team at the John Radcliffe Hospital in Oxford. Our testing followed a similar

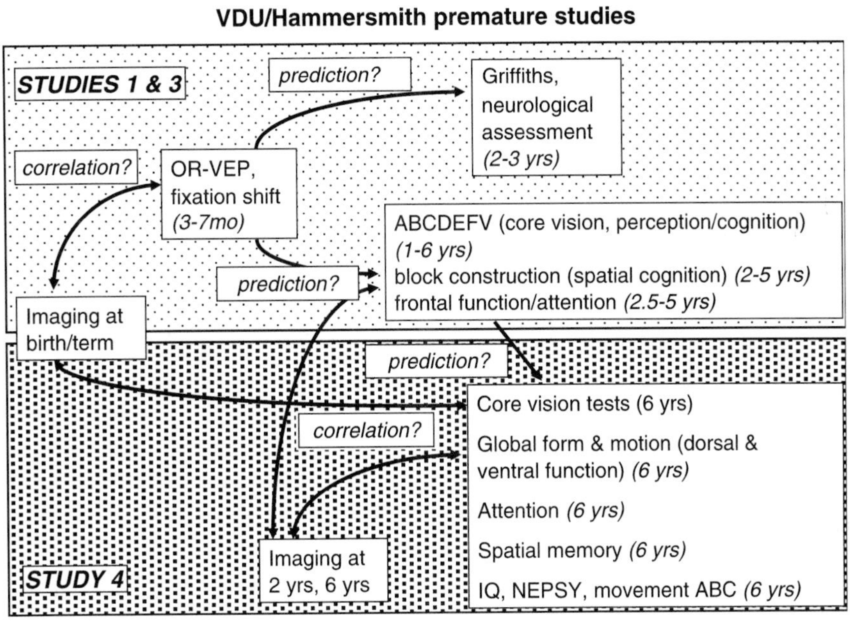

VDU/Hammersmith premature studies

STUDIES 1 & 3
prediction?
Griffiths, neurological assessment (2-3 yrs)
correlation?
OR-VEP, fixation shift (3-7mo)
ABCDEFV (core vision, perception/cognition) (1-6 yrs)
block construction (spatial cognition) (2-5 yrs)
frontal function/attention (2.5-5 yrs)
prediction?
Imaging at birth/term
prediction?
Core vision tests (6 yrs)
correlation?
Global form & motion (dorsal & ventral function) (6 yrs)
Attention (6 yrs)
Spatial memory (6 yrs)
Imaging at 2 yrs, 6 yrs
STUDY 4
IQ, NEPSY, movement ABC (6 yrs)

Fig. 2. Scheme of the follow up studies of prematurely born infants in collaboration with the Hammersmith group (Studies 1, 3 and 4 in this paper).

protocol as the first study with the Hammersmith group, except that the opportunity was taken to compare the results of orientation-reversal (OR) with DR-VERP testing. Brain imaging of the Oxford group was by serial ultrasound rather than MRI, and the focus of this study was on presumed healthy infants who had no cerebral damage visible on any of the serial ultrasound scans.

Study 1: onset of function in the visual cortex in very premature infants

A number of investigators have found visual measures to be effective indicators of premature infants' neurological status (e.g. Cioni et al., 2000; SanGiovanni et al., 2000; Guzzetta et al., 2001). However, measures such as visual acuity are not necessarily specific to the cortical aspects of vision. In this study we focused on the markers of cortical function, OR-VERP and FSs under competition, which have been outlined above. These measures have been well established in our work with both typically developing infants (Braddick et al., 1986, 2005; Atkinson et al., 1992; Braddick, 1993) and clinical groups of children (Hood and Atkinson, 1990; Braddick et al., 1992; Mercuri et al., 1996, 1997a, b, 1998; Atkinson et al., 2002b, 2003). Studies of full term infants with perinatal brain insults (Mercuri et al., 1996, 1997a, 1998) have shown that these measures reflect cerebral damage as assessed from neonatal MRI, although the relationship is not necessarily with the specific brain structures classically considered as involved in vision (Mercuri et al., 1997b). These measures are also effective predictors of later neurodevelopmental status in these groups (Mercuri et al., 1999). The present study aimed to test whether these functional cortical measures are also diagnostic of early brain damage and predictive of later outcome in a group born very prematurely.

Population and methods

Twenty-six preterm infants, born before 32 weeks gestational age (mean 28.1 weeks, s.d. 2.7 weeks) were included in the study. Images from neonatal and term MRI (T1- and T2-weighted scans in transverse and coronal planes) were examined for the presence of parenchymal lesions, intraventricular haemorrhage, ventricular dilatation and, in the term T2 scanes, diffuse excessive high signal intensity (DEHSI) within the white matter (Maalouf et al., 1999; Counsell et al., 2003). Intraventricular haemorrhage (IVH) was considered a significant finding if it was followed by ventricular dilatation at term (Dyet et al., 2006). DEHSI was visually rated as 'none/mild', 'moderate' or 'severe'. Infants were then categorized into three groups on a composite measure derived from the overall MRI data (see Appendix B). Eight children fell into Category 0 (mild), eight into Category 1 (moderate) (one of these did not complete both the cortical function tests) and ten into Category 2 (severe).

These infants had the OR-VERP and FS cortical marker tests, along with core vision assessment from the ABCDEFV battery, first between 2 and 11 months post term. Details of the procedures can be found in Appendix A. Whenever possible, infants who failed one or both cortical marker tests on their first visit early in this period were retested later within this age range. The 'pass' criterion for the OR-VERP was the occurrence of a statistically reliable response on the circular-variance test (Wattam-Bell, 1985) for eight reversals per second by the age of 7 months post term; for the FS test, a difference between competition and non-competition mean latencies below 0.5 s by the same age, with at least 4/5 initial fixations on the correct side in both conditions. The median age of the first visit was 5.0 months post term. By this age, normative data (Braddick et al., 1986, 2005; Atkinson et al., 1992) show positive results in 85% or more of healthy infants born at term.

The children subsequently had a neurodevelopmental assessment (Griffiths Developmental Scales; Huntley, 1996) at 2 years. We took a developmental quotient (DQ) of 80 or below as a criterion of poor neurodevelopmental outcome at 2 years, for evaluating the predictive power of the early visual measures.

Results

Thirteen of 25 children gave significant OR-VERP responses. Figure 3 shows that successful

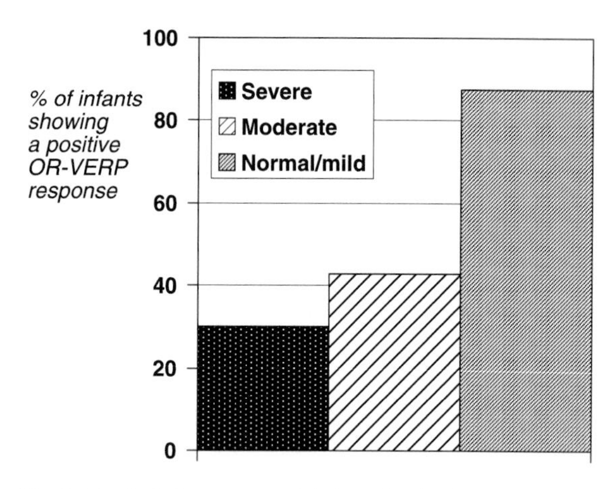

Fig. 3. Study 1: Percentage of infants in the three MRI categories who showed a statistically significant orientation reversal VERP by 7 months of age.

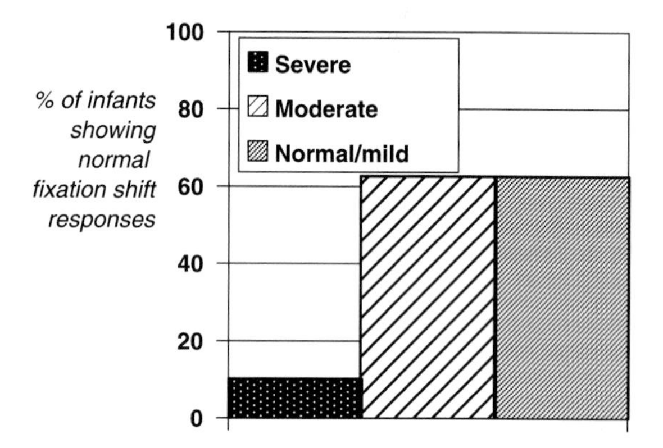

Fig. 4. Study 1: Percentage of infants in the three MRI categories who met norms for fixation shift under competition by 7 months of age.

establishment of the OR-VERP response decreased across the three categories of severity of the MRI findings (statistically significant at the level $p = 0.04$ on Fisher's Exact Probability Test for a 3×2 table, http://faculty.vassar.edu/lowry/VassarStats.html). The group categorized as 'normal/mild' on MRI scan have a high percentage of normal results for this test.

Eleven out of 26 infants met the pass criterion for the FS test. Figure 4 shows the breakdown between MRI categories. As with the OR-VERP responses, the proportion of infants meeting the pass criterion shows a significant decrease with MRI severity ($p = 0.03$ on Fisher's Exact

Probability Test) However, for this measure, unlike the OR-VERP, even the 'normal/mild' group show performance falling well below age norms, to the same level as the 'moderate' group. It is possible that the FS test may be a marker of the longer term, more subtle, neurodevelopmental problems encountered frequently in the preterm population.

As is generally found for a group of children with neurological impairments, there were a number of problems detected in core vision testing. In particular five children were strabismic, one had nystagmus and eight had unilateral or bilateral narrowing of visual fields for their corrected age. Fields were tested by confrontation with a Sheridan Tests for Young Children and Retardates (STYCAR) ball brought in from one side, and so this test represents a less formal version of the measurement made in the FS test. It should be emphasized however that even the children with detected vision problems all passed age norms on preferential looking acuity testing (Salomao and Ventura, 1995) and no child in the group would be considered on the basis of these results to have a major cerebral sensory visual impairment. We do not consider, therefore, that the cortical findings reflect sensory visual impairment alone, either at ocular or cerebral levels. The results imply that conventional vision examination does not necessarily identify children with significant developmental disorders of central visual processing, which are associated with broader problems that are manifest in later neurodevelopmental assessment.

We also examined the relationship between a 'fail' on these functional cortical measures in the first year, and the results of the Griffiths developmental testing at 2 years. The cortical measures were effective predictors. For predicting DQ of 80 or below, the OR-VEP test had sensitivity 86%, specificity 65%, positive predictive value 50% and negative predictive value 92%; the corresponding figures for FS were 100, 67, 54 and 100%.

The overall performance of the preterm group on the cortical tests is considerably worse than would be found in a normal term-born population. In other studies (Atkinson et al., 2002b; Braddick et al., 2005), we have found that between 80 and 85% of such infants show a significant OR-VERP at 12–17 weeks, and our fixation-shift criterion would also be achieved by the great majority of typically developing infants in this age range (Atkinson et al., 1992).

In an earlier study of prematurely born infants from a different centre (Atkinson et al., 2002b), we studied OR-VERP in low birth weight infants born at 24–32 weeks gestation. No MR brain imaging was available, but nine of the infants had abnormal cranial ultrasound scans. In four, the abnormalities were severe (cystic periventricular leucomalacia (PVL) or large intraventricular haemorrhages) and none of these four showed significant OR-VERP by 4.5 months. This group would have been placed in the 'severe' category in the classification of the present study. Among the children thought to have no abnormality on cranial ultrasound, 68% showed significant OR-VERP, a figure very close to 66% in the 'normal/mild' and 'moderate' groups combined from the present study.

The sensitivity and specificity figures, for predicting developmental quotient, are similar to those we have reported for full-term infants with hypoxic-ischaemic encephalopathy (Mercuri et al., 1999).

In conclusion, these tests of visual cortical function can provide an effective indication of functional impact of perinatal brain damage. The high sensitivity implies that children who have a poor developmental outcome have a very high probability of failing these tests; conversely, the high negative predictive values mean that a pass on these tests is a strong predictor of a relatively good developmental outcome. However, the lower specificity implies that a substantial proportion of children who do not have a markedly poor outcome at 2 years nonetheless may fail the cortical tests in infancy. The scope for developmental testing at 2 years is limited and the criterion of $DQ \leq 80$ identifies only a particular group of more seriously impaired children, particularly those with problems in the motor domain. If the Griffiths test at this age is a relatively insensitive measure (Barnett et al., 2004), it may still be the case that deficits of early cortical function relate to later outcome where more sensitive cognitive measures can be taken.

The tests we have used in this study, orientation selectivity and FSs, represent just two aspects of cortical visual development. In the next study we report, we have looked at another aspect of the sequence of cortical development, directional selectivity.

Study 2: development of orientation and motion responses in infants born preterm

We have discussed above the relation between early development of cortical responses to local motion and orientation, as the first stages of information processing in the dorsal and ventral streams respectively. There are reasons to believe that the former might be more sensitive to the developmental problems of infants born preterm. As discussed above, many developmental disorders are characterized by 'dorsal stream vulnerability'. Motion detection relies on the efficient transmission of visual signals with precise timing from different locations and may therefore be susceptible to the abnormalities in fibre organization and myelination (Braddick et al., 2003). Such abnormalities are commonly reported in preterm infants (Hüppi et al., 1998; Arzoumanian et al., 2003; Counsell et al., 2003, 2006).

In this study, therefore, we compared the development of directional motion responses to orientation responses, in a group of infants born preterm and term born controls, to examine whether directional responses revealed greater delays, which might be a precursor of the motion-processing problems seen in older children. We were particularly interested in delays that might be detectable in apparently neurologically healthy babies.

Population and methods

The study group had 17 infants born at 32 weeks gestation or below (median 28.8 weeks), at the John Radcliffe Hospital, Oxford. They all had <72 h mechanical ventilation and normal cranial ultrasound from the period from birth to 36 weeks gestation, in the period between 2004 and 2006. Twenty-six healthy, full-term infants, born within 14 days of their expected date of birth were recruited as controls, were also recruited from the John Radcliffe Hospital and tested in the same way as the preterms.

Children were tested between 9.9 and 17.1 weeks post-term age (median 13.0 weeks). The OR stimulus was identical to that used in the first study above. On the same visit, their VERP responses were tested with the DR stimulus developed from that of Wattam-Bell (1991) and Braddick et al. (2005): random dots reversed their direction of horizontal motion twice per second. One thousand white random dots, each with a 200 ms lifetime, moved horizontally at 8.5°/s on a black background. Both the OR and direction reversal (DR) responses are measured by Fourier analysis of the steady-state signal: evidence for the development of cortical orientation- or direction-selectivity comes from the statistical significance of the response at the reversal frequency (F2, the second harmonic of the sweep frequency). Details of the procedure can be found in the Appendix A.

Results

Figure 5 shows scatter plots of the F2 signal: noise ratio for the two stimuli as a function of their post-term age at test. The figures show that this response measure for the OR stimulus is similar in preterm- and term-born infants, and shows little variation across the age range for either group. However, the DR responses increase with age for both groups, with consistently smaller responses for the preterm group. A one-way analysis of variance (ANOVA) of the DR signal:noise ratio, with age-at-test as a covariate, showed significant main effects of group, $F(1,41) = 8.09$ ($p<0.01$) and age-at-test $F(1,41) = 9.149$ ($p<0.01$). A similar analysis for the OR response indicated no effect of group or age-at-test.

Of the individual infants, only the youngest preterm infant (9.9 weeks corrected age) and only 2/26 of the term born infants had OR responses, which failed to reach significance. In comparison, for the DR stimulus only 40% of the 10 preterm infants aged 2–3 months showed significant responses compared with 67% of the 12 term born controls in this age group. At 3–4 month post

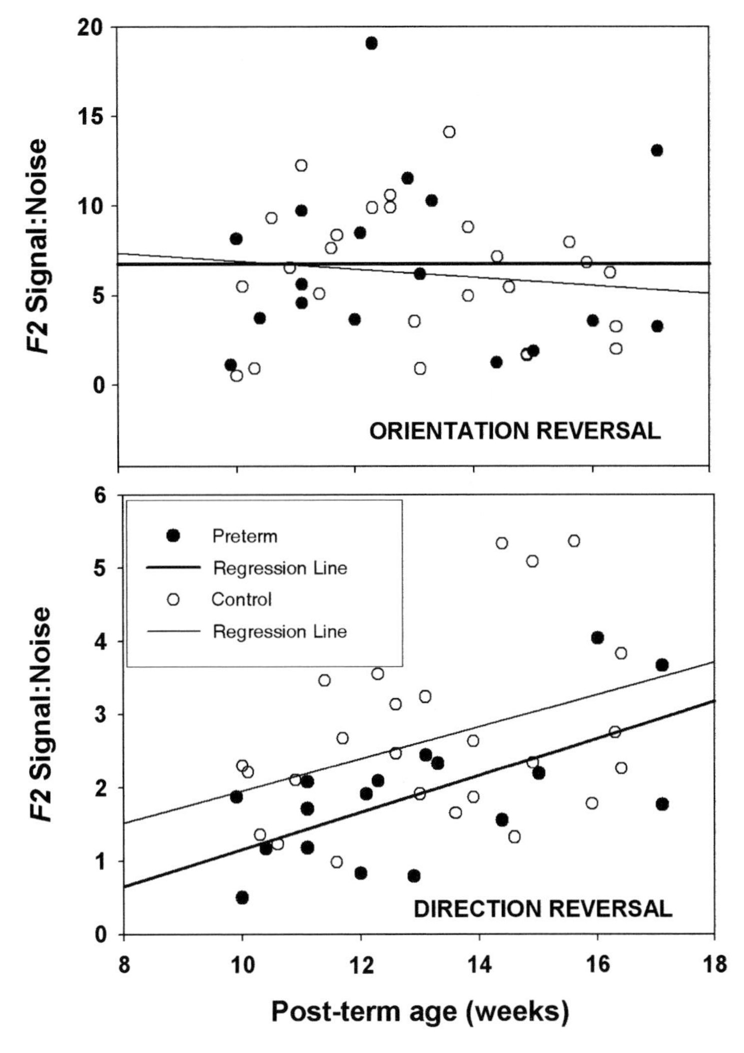

Fig. 5. Study 2: Signal:noise ratio (a measure of the reliability of the VERP response) as a function of post-term age, for orientation-reversal stimulus (upper) and direction reversal (lower). Solid circles = neurologically healthy prematurely born infants; open circles = control infants.

term, 63% of the preterm and 79% of the term-born group gave significant F2 DR responses. However this difference did not reach significance ($\chi^2 = 2.45$; d.f. = 1; $p = 0.12$). Comparing the infants aged 2–3 months and 3–4 months separately, only 40% (4/10) of the preterm infants were showing DR responses but 8/12 (66.7%) of term born controls had DR responses at this age. In the 3- to 4-month olds, 5/8 (62.5%) of the preterm infants and 11/14 (78.6%) of the control infants showed significant F2 DR responses.

The relative delay of DR compared to OR responses in normal development is similar to that found by Braddick et al. (2005). The finding that OR responses in neurologically healthy prematurely born infants are similar to term-born infants, matched for gestational age, is consistent with the results of the infants who were normal on ultrasound in our earlier study (Atkinson et al., 2002b) and with the no/mild damage grouping Study 1. However, even in these infants, there is a delay in maturation of motion responses compared to the term-born group; comparison of the parallel regression lines of Fig. 5 suggests that this delay is equivalent to ~4 weeks.

Discussion

We have suggested above that visual motion processing may be susceptible to anomalies of prematures' brain development that may not be apparent on standard clinical cerebral ultrasound examination. Neonatal MRI studies, notably those using diffusion weighted imaging, have found that at least 50% of preterm infants have white-matter abnormalities, including signal abnormalities such as DEHSI (see Study 1 above), loss of volume, cystic abnormality, enlarged ventricles and thinning of the corpus callosum in addition to delayed myelination (Hüppi et al., 1998; Maalouf et al., 1999; Counsell et al., 2006). Cranial ultrasound has poor sensitivity for the detection of white matter injury (Inder et al., 2003). The delay of responses on the DR stimulus may reflect the consequences of such white matter injury. Future work combining diffusion-weighted MRI with DR measures post term will be needed to test this hypothesis.

Global processing of motion is starting to develop very rapidly at the ages when DR responses are first observed (see Braddick and Atkinson, this volume). It remains to be determined what are the consequences of prematures' delay for the development of global processing (see Study 4 below). But we know, as discussed above, that global motion processing problems are widely seen in later childhood, in children with neurodevelopmental problems (Braddick et al., 2003). Vulnerability for motion processing deficits may be present at a very early stage of dorsal stream development, even in preterm infants who are 'neurologically normal' by conventional clinical measures.

Study 3: development of the visual brain between 1 and 6 years in very premature children

In Studies 1 and 2, we have considered early functioning of the lowest levels of the ventral and dorsal cortical streams. As cortical visual processing is an early-developing aspect of brain function and is functionally significant for the acquisition of many cognitive, motor and social skills, it is important to understand the developmental sequence of visual cognition throughout the preschool years. It is also necessary to devise and apply tests that specifically tap development of various aspects of developmental spatial cognition, where the neural underpinnings have already been suggested from animal models and from adult patients with specific focal lesions.

In this third study we measure functional vision of prematurely born children between birth and 5 years, using the ABCDEFV test battery, which has been briefly described above, for assessing developmental progress in sensory, perceptual, cognitive and spatial vision (Atkinson et al., 2002a). Further details of the ABCDEFV subtests are given in Appendix A.

We have supplemented the ABCDEFV with three tests designed to assess the development of 'control processes' sometimes called 'executive function'; these are attentional processes believed to indicate that the frontal lobes have reached a certain level of maturation (Robbins, 1996; Duncan and Owen, 2000). These tests, the Russell detour box (Biro and Russell, 2001), the Hood Gravity Tubes (Hood, 1995) and counter-pointing (Atkinson et al., 2003), all tap the ability to inhibit a pre-potent response, in modes appropriate to different ages between 2 and 5 years. We have already used them to demonstrate the differential impairment of frontal control functions in children with WS (Atkinson et al., 2003). A fuller description of the procedures is provided in Appendix A.

Population and protocol

Twenty-nine children born before 32 weeks gestational age (mean gestational age 28.6 weeks) were studied in collaboration with the Hammersmith Hospital. This cohort overlapped with those in Study 1 and their neonatal and term MRI results were classified according to the same scheme as described for that study. Seven children fell into the category with the least problems (normal/very mild MRI abnormalities), 9 into the 'moderate' category and 13 into the 'severe' MRI category. They were studied between 5 months and 6 years of age, using age-appropriate components of the ABCDEFV and the frontal executive function tests.

Results

Core vision

Overall, 12 out of 29 children failed at least one test appropriate for their age group in the period between 1 year and 5 years on the ABCDEFV Core Vision Tests. None failed the preferential looking acuity test used at the younger ages with binocular viewing; one child failed the Cambridge Crowded Cards test of acuity at 5 years. Many of the failures (7 out of 29 tested) were on measures of peripheral visual fields on one or both sides. It should be noted that this test, which requires the child to switch fixation from the tester's face to a small white ball moving slowly along an arc inwards from the periphery, has an attentional as well as a sensory component, somewhat similar to the FS test described under Study 1. A more classically visual disorder is convergent strabismus, shown by seven children, all of whom were in the severe MRI group. The percentage of children failing at least one of these core visual tests is correlated with the extent of abnormality seen on MRI, as shown in Fig. 6, a relationship at the $p = 0.007$ level (Fisher's Exact Probability Test).

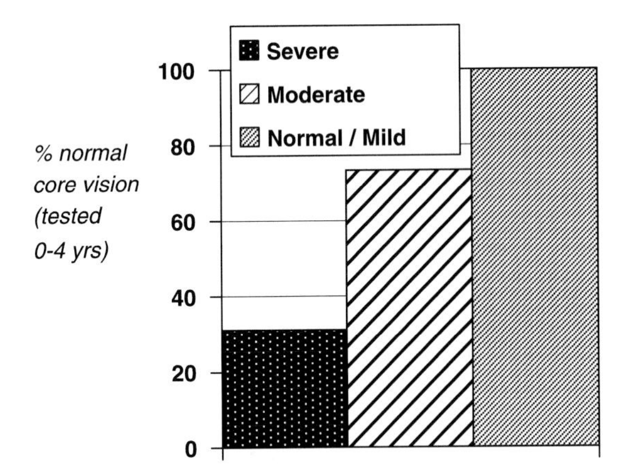

Fig. 6. Study 3: Percentage of infants in the three MRI categories who met age norms for core vision tests of the ABCDEFV battery.

Visuoperceptual and visuocognitive tests

Many children in this premature group failed at least one age-appropriate subtest of the ABCDEFV additional visuoperceptual and visuocognitive tests from the ABCDEFV on at least one occasion (17 out of 21). Figure 7 shows that again there was an orderly decline in performance as a function of MRI severity, which was statistically significant ($p = 0.009$, Fisher's Exact Probability Test). All of the 'severe' groups (10 out of 10 tested) failed at least one age-appropriate test between 1 and 5 years of age, with many failing all the age appropriate tests throughout the age range. In the group with minimal brain imaging abnormality, only one out of four children failed one

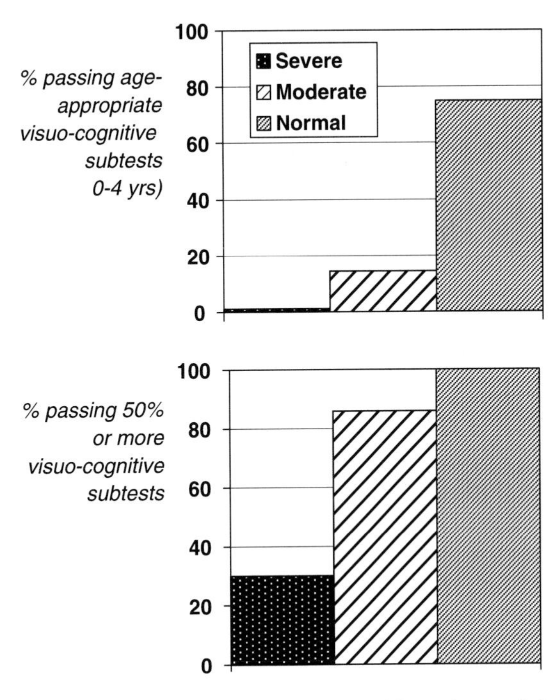

Fig. 7. Study 3: Results on the visuocognitive subtests of the ABCDEFV battery (object permanence, shape matching, embedded figures, Frostig cats, envelope) for infants in the three MRI categories who met age norms for core vision tests of the ABCDEFV battery. Upper chart: percentage of children meeting age norms on all tests taken. Lower chart: percentage of children meeting norms for at least 50% of tests taken.

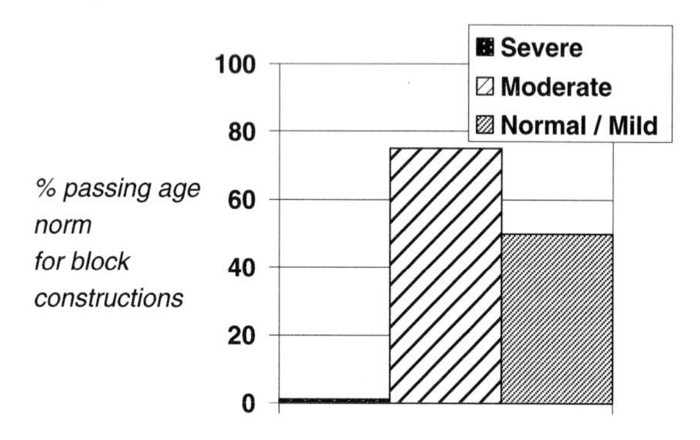

Fig. 8. Study 3: Percentage of children in the three MRI categories who met age norms for the block-construction task.

test, while in the 'moderate group' six out of seven children did so. The subtests with the most failures were the shape-matching test (9 out of 13 children tested on at least one occasion) and the block-construction copying test.

The block-construction copying test, like the other visuocognitive tests summarized above, involves manipulation of visuospatial information, but in this case the graded series of constructions allows the test to be demanding for children up to at least 5 years. All three groups were below normal performance overall on this test (Fig. 8).

Frontal executive function

Figure 9 shows performance on the three executive control tests for the 26 children who were tested in the appropriate age range ($N = 7$, 8, and 11 in the three MRI groups). 25 out of 26 failed at least one of the three tests on one occasion. The very low proportion passing suggests that these tests are highly sensitive to the effects of preterm birth; but because pass rates were so low overall, they showed no significant relationship to the severity of the MRI category ($p = 0.59$, Fisher's Exact Probability Test).

Overall, therefore, performance on a wide range of visual and visuocogntive functions was associated with the level of brain abnormality seen on neonatal MRI. However, this relationship depends on the nature and complexity of the function concerned. For core visual function, in particular binocularity and visual fields, deficits are associated

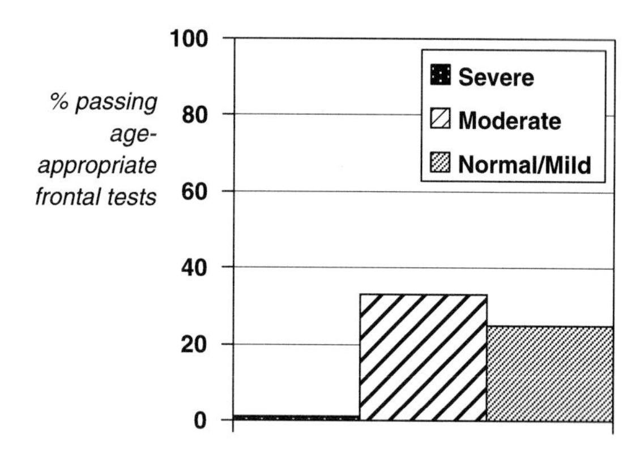

Fig. 9. Study 3: Percentage of children in the three MRI categories who passed age-appropriate frontal executive function tasks (Hood gravity tubes, 3–5 years; Russell detour box, 2–4 years; counter-pointing, 4 years on). See Appendix A for further details of procedures.

with the most severe MRI group. The visuocognitive tests showed poor performance in the 'severe' and 'moderate' groups, but relatively good results for those in the normal/mild group. For the block construction and the frontal tests, especially the latter, even the 'normal/mild' group show some deficit compared to age norms between 1 and 5 years of age. Thus more complex cognitive functions, particularly those involving the frontal lobes, are susceptible to the effects of prematurity, even when these are not qualitatively apparent on neonatal and term MRI. These frontal and attentional deficits are likely to be significant in terms of school age problems such as ADHD. Study 4, described below, enabled us to examine a similar group of children at age 6–7 years, when a wide variety of visual, cognitive and spatial abilities could be tested.

Study 4: development of a cohort of premature infants at 6–7 years

The fourth study aimed to build up a fuller picture of the areas in which premature children were impaired and unimpaired, at an age where many areas of competence could be assessed by tests with norms established in the general population. We examined, at 6–7 years post-term age, a premature cohort from the Hammersmith ($N = 56$, born before 32 weeks gestation), for whom neonatal/term MRI assessment was available, similar to that in Studies 1 and 3. A wide-ranging test battery was used, some being standard cognitive measures and some newly devised to address particular areas of function believed to depend on specific brain systems.

Protocol

The test battery included:

- verbal and performance IQ measured with the Wechsler Intelligence Scale for Children (WISC) (or WPPSI if necessary);
- British Picture Vocabulary Scale (BPVS) (Dunn et al., 1982);
- ABCDEFV core vision tests;
- block-copying task as for ABCDEFV, but timed and related to new timed age norms; and
- Movement Assessment Battery for Children (Movement ABC), a standardized assessment of everyday motor competence within three categories: manual dexterity (e.g. posting coins in a small slit in a box), balance (e.g. walking along a white line on the floor, with one foot closely in front of the other) and ball skills (e.g. kicking a ball into a space between two goal posts) It is normalized for ages 4–12 years (Henderson and Sugden, 1992).
- Four subtests from the Test of Everyday Attention for Children (TEA-Ch) (Manly et al., 1999, 2001): 'Sky Search' (selective attention and visual search task), 'Score' (sustained auditory attention), 'Walk- Don't Walk' (sustained attention and inhibition of an action), 'Opposite Worlds' (switches of attention between two rules and inhibition of a prepotent response). More details are given in Appendix A.
- Coherence threshold measurements for global motion and form coherence (Gunn et al., 2002) to assess function of extra-striate dorsal and ventral streams (further details in the appendix).

  Location memory within different co-ordinate frames (the 'Town Square' test, Nardini et al. (2006), see Appendix A). Children saw a toy hidden under one of the 12 cups on a square board with landmarks, and had to locate it (a) from the same viewpoint, where egocentric cues suffice, (b) after walking partway around it, requiring either updating of egocentric location, or allocentric cues provided by room and board landmarks, (c) when the board is rotated, so a representation relative to local board landmarks is required. The latter, which normally emerges only beyond 5 years of age, is supported by hippocampal processing (O'Keefe and Burgess, 1996).

Results and discussion

The performance of the group as a whole is summarized in Figs. 10–12. To provide a common metric for the diverse tests, the children's scores on the different tests were converted to the equivalent percentile of the norms for the age group. Figure 10 shows the mean percentile score for each test, with the 95% confidence interval on this mean.

In some areas, the children taken as a whole do not show any deficit: in both WPPSI/WISC verbal and performance scales and BPVS vocabulary, the means are very close to the normative 50th percentile, as is the core vision measure of acuity. However, in three of the four attention scores, in both types of global processing (motion and form), all the spatial memory conditions of the 'town square', block constructions and most markedly the Movement ABC battery, the mean percentile score differs significantly from 50%.

It may be more revealing to consider the proportion of children who perform very badly on particular tests, falling below the 15th or even the 5th percentile of the typically developing population. Figures 11 and 12 show the number reaching these criteria of deficit. In general, the pattern broadly follows that shown by the means of Fig. 10 (i.e. the lower bars in Fig. 10 correspond to higher bars in Figs. 11 and 12). This is particularly striking in the motor performance domain, with almost half the ex-premature children falling below the 5th percentile from normative data.

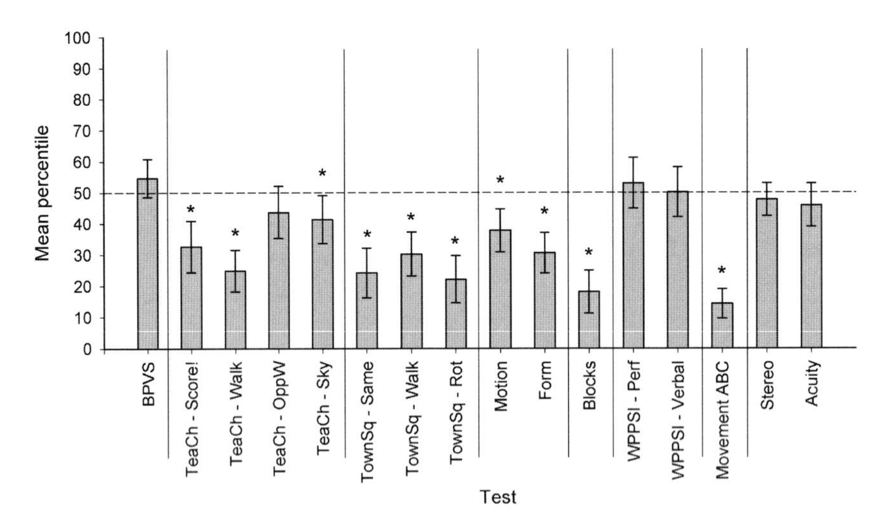

Fig. 10. Study 4: Performance of ex-premature infants at age 6–7 years. Each test is scored in terms of the percentile norms for this age group, and the mean value is plotted (error bars = 95% confidence intervals). Stars indicate test on which the mean score of the ex-premature group differed significantly from 50%. BPVS = British Picture Vocabulary Scale; TEA-Ch = Test of Everyday Attention for Children, with four subtests (see Appendix A for details); TownSq = Town Square test of spatial memory (see Appendix A for details of the three sub-tests); Motion = motion coherence threshold; Form = Form Coherence threshold; Blocks = timed block constructions; WPPSI = Wechsler Preschool and Primary Scale of Intelligence (Performance and Verbal subscales). Fuller accounts of the tests are in the text and Appendix A.

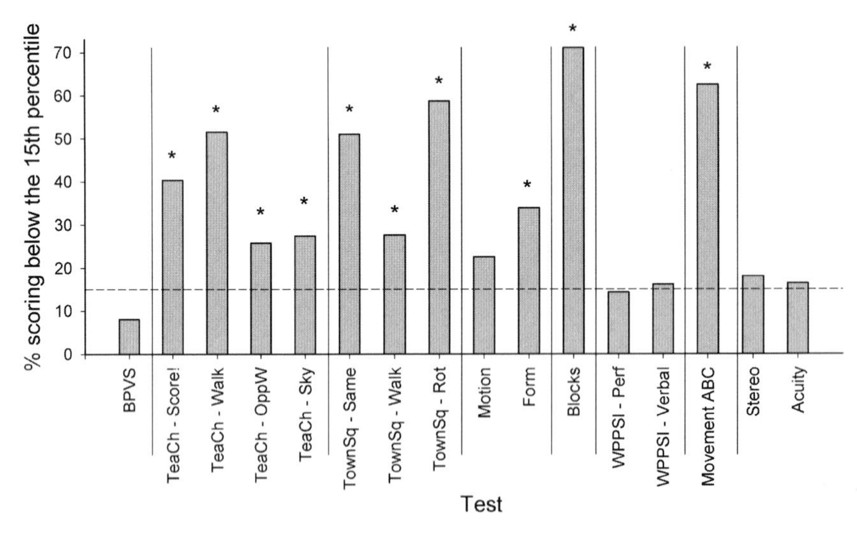

Fig. 11. Study 4: Performance of ex-premature infants at age 6–7 years, plotting the percentage of children in the group who scored below the 15th percentile from the age norms for each test. See Fig. 10 for explanation of test names.

The main differences are that all the components of the TEA-Ch now show significantly more poor performance than would be expected from the general population, and that motion coherence sensitivity shows more very poor performance (<5th percentile) than does form coherence, even though their mean results were similar. This is consistent with the differential deficit for global motion over global form (dorsal stream vulnerability) that we have identified in other clinical

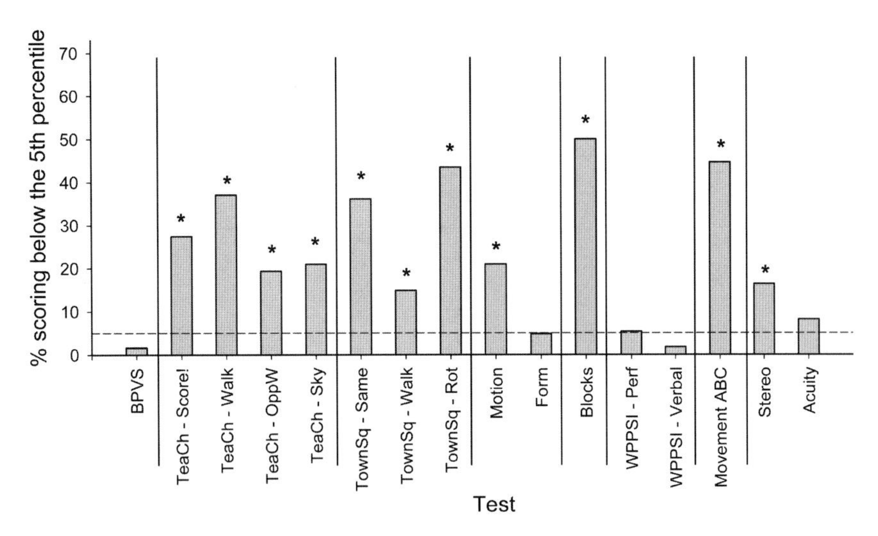

Fig. 12. Study 4: Performance of ex-premature infants at age 6–7 years, plotting the percentage of children in the group who scored below the 15th percentile from the age norms for each test. See Fig. 10 for explanation of test names.

groups, although it should be noted that in the data it is apparent from looking at a relatively small percentage of very poor performers, rather than any differential depression of mean levels of form and motion sensitivity. The significant group of ex-premature children with strabismus, and hence no stereovision, means that stereo is another area where a significant group falls below the 5th percentile level of performance.

The idea of dorsal stream vulnerability can be related to the main areas of deficit: the control of action as indicated by the Movement ABC scores, which of course requires the dorsal stream transformation of visual information, particularly in the parietal lobe, and the spatial manipulations of information required for the block-construction task and the various town-square subtests. Beyond these spatial-parietal functions, there also appears to be a significant area of deficit in the 'attention' tests of the TEA-Ch, some of which (e.g. 'Walk-don't Walk' and 'Opposite Worlds') might better be described as frontal executive control tasks.

Some more understanding of the pattern of results may be gained by looking at the pattern of correlations between these scores, and their relationship to the neonatal/term MRI categories. Table 1 shows the results of a factor analysis (varimax rotation) of the scores on which Figs. 10–12 are based, together with statistical tests of whether scores on the four main factors vary significantly with (a) the difference between the 'severe' MRI category and the other two categories, and (b) the children's gestational age at birth. The four factors account for about two-thirds of the total variance in the scores.

Factor 1 loads heavily on both subscales of the Wechsler intelligence test, on the block-construction test, vocabulary, and moderately on two tests in the TEA-CH attention scale and the Movement ABC. This seems to represent a general cognitive development factor. However, it relates only weakly to the MRI results, and Figs. 10 and 13 show that not all of its component tests are impaired in the premature group. That is, it may account for variance between children, but this variance is not necessarily a consequence of the damage for which premature children are at risk.

Factor 2 loads most heavily on the counter-pointing test, and on the 'score' and 'opposite worlds' test of the TEA-Ch, and so can be interpreted as related to frontal control functions. This is the factor most clearly related to the early MRI condition.

Factor 3 is related to motor ability (Movement ABC score) and the TEA-CH 'score' test. It is also clearly related to form coherence sensitivity. This last is a puzzling result, which requires further investigation.

Table 1. Factor analysis of the data of Study 4, showing all factors with eigenvalue ⩾ 1, varimax rotation

	Component			
	1	2	3	4
BPVS	**0.86**	0.18	−0.03	−0.18
T.score	0.31	**0.44**	**0.48**	0.10
T.walk	0.17	0.25	0.15	**0.77**
T.oppw	**0.51**	**0.54**	0.24	0.30
T.sky	**0.60**	0.33	0.37	0.05
Motion1	0.05	−0.07	−0.15	**0.82**
Form1	−0.04	0.03	**0.78**	−0.17
Blocks	**0.74**	0.26	−0.04	0.19
dpoint	0.08	**0.89**	−0.08	−0.01
cpoint	**0.45**	**0.50**	0.35	0.13
W.Perf	**0.79**	−0.10	0.32	0.16
W.Verbal	**0.72**	0.20	0.16	0.30
MABC.Tot	**0.41**	−0.04	**0.58**	0.22
% of variance	27.5	14.2	12.9	12.6
Groups (0 + 1) vs. 2	*(one-tailed)	*(two-tailed)	ns	ns
Gestational age	ns	ns	*(two-tailed)	ns

Notes: Loadings over 0.4 are shown in bold type. The last two rows shows which factors differed between MRI groups, and as a function of gestational age at birth.
*Indicates a significant effect at the $p<0.05$ level (one- and two-tailed).

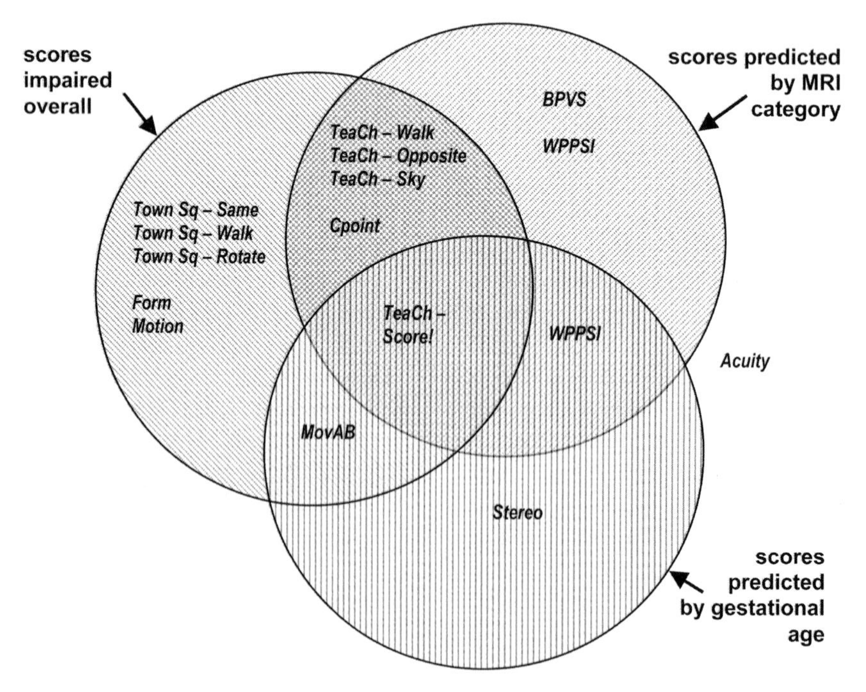

Fig. 13. Study 4: Diagram summarizing the relationship between the sets of test scores which are (a) significantly poorer that the age norms for the ex-premature group as a whole (impaired overall); (b) vary significantly as a function of the neonatal/term MRI category of the child; (c) vary significantly as a function of the child's gestational age at birth.

Finally, Factor 4 is strongly related to motion coherence sensitivity, and more weakly to a range of tests. This appears to be a fairly pure measure of the 'dorsal stream vulnerability' we have discussed above. It is striking that, as in earlier data (Braddick et al., 1998) there is no correlation between form and motion coherence sensitivity, even though these measures might be expected to share some common ability to attend to a demanding visual task and detect a visual pattern in background noise.

Figure 13 shows diagrammatically the relation of individual tests to MRI results and gestational age. It is notable that there are a number of measures, notably the 'town square' tests and both form and motion coherence, on which the premature group as a whole perform poorly but which do not relate to either MRI or degree of prematurity. Clearly there must be features of brain development, which are compromised by premature birth but are not reflected in the MRI categories we have used. An important focus for future research will be to investigate whether more detailed information on brain structure can be related to these patterns of ability and impairment. Such information might relate to the regional distribution of brain damage, or to more quantitative measures, either of brain volume or of white matter organization as derived from diffusion-weighted techniques. Some preliminary analysis has suggested that the different demands on spatial representation placed by different versions of the Town Square test are dissociated by quantitative MRI analysis, with the encoding of allocentric relations between objects and landmarks appearing particularly sensitive to volumetric differences.

Figure 13 also shows the point made from the analysis of Factor 2, that among the performance measures most closely related to the MRI findings are those which reflect the executive control processes of the frontal lobes.

General discussion

These four studies have looked at various stages in the development of infants whose premature birth has created a risk factor for brain development. Study 1 shows that measures of visual cortical function, identifiable in the first months of life, relate to the anomalies of brain structure, especially white matter, that are common in this group and also help to predict subsequent development. However, not all aspects of cortical function are equally affected by prematurity and its associated risks.

The other studies provide evidence that three related areas of function are most affected: spatial function, selective attention and executive control. All of these have links with the dorsal visual cortical stream, although they involve other brain systems as well. The early levels of the dorsal stream, in extra-striate visual areas, are involved in motion processing. Study 2 has shown that the development of the motion system is delayed in prematurity, and Study 4 found that very poor sensitivity to global motion is disproportionately common in these children at 6 years, a result consistent with our earlier findings on hemiplegic children (Gunn et al., 2002) some but not all of whom were ex-premature.

Within the parietal lobe, dorsal stream information feeds into the systems that are used in visuospatial manipulation, such as the block construction found to be impaired in Studies 3 and 4, and into the visual control of action. The Movement ABC results in Study 4 show that action systems are among the aspects of brain function most affected by prematurity. Spatial information is also used in encoding locations, as in the town square task; however, this process involves hippocampal systems which integrate parietal information about self-motion with ventral-stream information to encode landmarks (Burgess, 2002). All these spatial systems appear to be vulnerable; further research is needed to determine whether this involves specific vulnerability of the hippocampus as well as the parietal dorsal-stream systems that provide some of its input.

Apart from motor co-ordination, Studies 3 and 4 found the most vulnerable functions to be those involved in the allocation of attention and the selection and executive control of behaviour. A frontal-parietal network has been identified as involved in a wide variety of tasks of this kind (Duncan and Owen, 2000; Kastner and

Ungerleider, 2000). Again, further research is needed to determine how far this vulnerability is a consequence of the susceptibility to subtle damage of frontal structures themselves, and how far it is a consequence of a developmental cascade from parietal impairment of the kind affecting other dorsal-stream function. Studies 3 and 4 showed that these functions were impaired overall in the group; Study 4 showed that the aspects tapped by the TEA-Ch battery were among the functions that could be related to earlier MRI status.

Table 2 summarizes the functions found to be affected and unaffected, particularly in Study 4, with an indication of the brain areas likely to be implicated.

Overall, these four studies show that there are specific delays in many systems within the visuomotor, spatial and attentional domains in these cohorts of extremely premature born children, which persist through the preschool years into early school age. Although for some of these children these deficits are relatively mild, there must be a concern that even in mild cases a consequence may be subtle learning difficulties (e.g. ADHD), which persist and have cascading effects on later academic abilities.

However, given that most of these children had identifiable brain abnormalities from birth on MRI, brain development shows remarkable plasticity and recovery of almost normal function in many areas. Normal early language development and mild visuomotor problems would seem to characterize the development of many children in this group. However, the additional delays or deficits on tests of attention, seen in some children from a few months of age and persisting into school age is a consistent finding across many in the group. It remains an open question as to whether very early remedial treatment in infancy or the preschool years might reduce these problems later in life.

Table 2. Summary of functions that are impaired and those that are relatively normal in children born prematurely

Normal	Impaired
Language ($F \times P \times T$)	Attention/executive — (F)
Narrative and configural memory ($T \times H$)	Global motion/form ($O \times P \times T$)
Visual acuity (O)	Movement ABC (P)
	Spatial memory — egocentric and sensorimotor updating (P)
	Spatial memory — allocentric ($H \times F$)
	Block copying ($P \times O \times T$)

Note: F: frontal; *P*: parietal; *T*: temporal; *O*: occipital; *H*: hippocampus.

Abbreviations

ABCDEFV	Atkinson Battery of Child Development for Examining Functional Vision
ANOVA	analysis of variance
BPVS	British Picture Vocabulary Scale
DEHSI	diffuse excessive high signal intensity
DQ	developmental quotient
DR	direction reversal
F2	second harmonic frequency
FPL	forced-choice preferential looking
FS	fixation shift
IVH	intraventricular haemorrhage
Movement ABC	Movement Assessment Battery for Children
MRI	magnetic resonance imaging
OR	orientation reversal
PLIC	posterior limb of the internal capsule
PVL	periventricular leucomalacia
STYCAR	Sheridan Tests for Young Children and Retardates
TEA-Ch	Test of Everyday Attention for Children
V1	visual area 1 (= primary visual cortex)
V4	visual area 4
V5/MT	visual area 5/middle temporal area
VEP	visual evoked potential
VERP	visual event-related potentials
WPPSI	Wechsler Primary and Preschool Scale of Intelligence
WISC	Wechsler Intelligence Scale for Children
WS	Williams syndrome

Acknowledgements

This research has been supported by grants from the UK Medical Research Council. It has depended on the collaboration of David Edwards, Mary Rutherford, Frances Cowan, Eugenio Mercuri, Leigh Dyet, Daniella Ricci and Rachel Rathbone at the Hammersmith Hospital, and of Andrew Wilkinson at the John Radcliffe Hospital, Oxford. The research of the Hammersmith group is supported by the Medical Research Council, the Garfield Weston Foundation and Wellbeing.

Many members of the Visual Development Unit in Oxford and London contributed to this work; in particular we thank John Wattam-Bell, Shirley Anker, Dee Birtles, Marko Nardini, Kate Breckenridge and Stephanie Riak Akuei for invaluable help. This work would not be possible without the co-operation of the participating families and their children; we thank them all.

Appendix A. test procedures

Studies 1 and 2

Orientation-reversal VERPs were tested, by steady-state recording of electrical responses to a high contrast sine wave grating pattern, spatial frequency 0.3 c/°, presented on a CRT screen at a 40 cm viewing distance from the infant's eyes. The grating orientation alternated between 45° and 135° at a rate of eight reversals per second, embedded in a sequence of random phase shifts at 24 shifts per second. VEPs were recorded with three gold cup electrodes positioned on the forehead, vertex and 1 cm above the inion. 200 sweeps (two reversals per sweep) were averaged, and the signal at 8 Hz (F2) extracted by Fourier analysis. Statistical significance of the VERP signal at this frequency was assessed by the 'circular-variance' test (Wattam-Bell, 1985). For calculation of signal:noise ratio, noise was estimated from the mean amplitude of frequency components at 0.02 Hz intervals for 1 Hz either side of the stimulus frequency, excluding the stimulus frequency itself.

(Study 2) Direction-reversal VERP

The stimulus consisted of 1000 white random dots, size 0.92°, lifetime 200 ms moving horizontally at 8.5°/s on a black background. The motion direction reversed two times per second. Recording and analysis procedures were similar to the OR-VERP above. Fuller details of both VERP methods can be found in Braddick et al. (2005).

(Study 1) Fixation shifts

The procedure was based on Atkinson et al. (1992). Infants were seated 40 cm in front of a large screen, 51 × 38 cm. They initially fixated a central face-like figure alternating between two formats at 3 Hz. A target of adjacent bright and dark stripes, each 2.9 × 14.7 cm, reversing in contrast at 3 Hz, then appeared 13.5 cm either left or right of the centre. An observer records the time and direction of the child's first fixation.

Within a block of 25 trials, 'non-competition' (central target disappears at onset of peripheral target), 'competition' (central target remains visible) and 'confrontation' (peripheral targets appear on either side of centre simultaneously) trials were mixed in random order.

Atkinson Battery of Child Development for Examining Functional Vision (ABCDEFV)

The tests in this battery have been divided into 'Core vision tests', measures of basic visual function, and 'Additional Tests', measures of visuo-cognitive processing with different areas of spatial cognition differentiated in different sub-tests. The tests have been selected to cover, with suitable adjustments for age, the age range tested in Studies 1 and 3 from birth to 5 years. They are fully described in Atkinson et al. (2002a) which details the procedures and age-appropriate pass-fail criteria which has been determined from normative data for each test.

ABCDEFV core vision tests

These are appropriate for any child capable of making saccadic eye movements. They do not require reaching and grasping nor smooth pursuit eye movements. The following subtests were included in the assessment: (1) orthoptic assessment (eye movements, pupillary response to light, strabismus, visual fields assessed with 'STYCAR' balls; (2) binocular and monocular optokinetic nystagmus; (3) visual acuity, tested by Teller Preferential Looking Acuity Cards for children under 4 years, Cambridge Crowding Cards for children over 4 years mental age; (4) assessment of refractive errors using the VPR-1 videorefractor ability to change focus (accommodation) for targets at different distances; (5) maintain attention on receding target to distance of 2 m or more; (6) Lang stereo test. A child was scored 'pass' if all of these core vision tests were passed at an age appropriate level.

ABCDEFV 'additional tests' of visuocognitive ability

The battery includes a group of subtests appropriate for ages from 12 months to 5 years, which requires the child to perform cognitive and sensory-motor operations on visual information. The tests used require a basic capability of reaching and grasping with at least one arm and hand, but do not require high acuity, contrast sensitivity, a pincer grasp or very fine finger movements.

1. Object permanence: Finding a partially or wholly covered toy, corresponding to Stages 3–4 of Piaget's Object Permanence (9–18 months)
2. Shape matching: The child must select and insert the appropriately shaped wooden form in the correct slot of a specially designed from board of either three or five shapes to be matched (square, circle, star, triangle, rectangle) (2–4 years)
3. Embedded figures: (a) Animals test: the child is required to name or point to as many as they can out of five outline animal shapes that are superimposed on each other (correct naming not required); (b) Frostig cats: the child has to identify as many as possible of four cat silhouettes in a tree drawing. Tree and cats are both depicted by line shading giving a masking effect (2–4 years)
4. Letter in envelope test: This requires the child to place a rectangular card in a rectangular envelope. The child passes the test if appropriate alignment is used on the first attempt (2.5–4 years).
5. Block-construction copying: The child has to build a copy of a geometric three-dimensional construction made up of six to eight small rectangular blocks. There is a range of geometrical constructions (rows, towers, cross, bridge, enclosure) appropriate for different ages (1.5–5 years). In Study 3, the test was administered and scored as described in Atkinson et al. (2002a). In Study 4, to allow an assessment of the 6- to 7-year olds tested, the three most demanding shapes were used (cross, enclosure, bridge), and the scoring based on time taken against norms from a control group of the same age range.

In the presentation of results, pass/fail data tests 1–4 have been presented together (Fig. 7) and the block-construction results are presented separately (Fig. 8).

(Study 3) Frontal executive function tests

Detour box

This apparatus, originally devised by Hughes and Russell (1993) consists of a 'magic box' with an aperture at the front through which the child can see a brightly coloured ball on a platform. The tester explains if he or she reaches through the aperture to pick up the ball, the ball falls through a trapdoor and disappears from view (achieved by a photocell beam) and demonstrates that the ball can be obtained by a less direct method. In the *lever task*, the child can use a lever to push the ball down a chute into a tray it can be picked out; in the less direct *switch task*, the child operating a switch inactivates the photocell allowing the child to retrieve the ball by hand. On each task trials continue until three successive correct responses are made (to a maximum of 15 trials). Performance is measured by the number of errors before this criterion is reached. Normative data (Hughes and Russell, 1993; Biro and Russell, 2001) indicate that the switch task is typically mastered by 3.5 years compared to 2.5 years for the lever task.

Hood gravity tubes

This test, devised by Hood (1995), consists of a set of flexible opaque tubes, each leading to an opaque cup. The child sees a ball dropped into the top of one tube, and is invited to select the cup into which it has fallen. Young children make the 'gravity error' of selecting the cup vertically beneath the release point even when the tube does not connect the two. This can be considered as an inability to inhibit a prepotent response based on a simple gravitational trajectory. Age norms have been established by Hood (1995) between 2 and 4 years for mastering increasingly complex configurations (numbers of tubes and balls).

Counterpointing

The display is identical to the non-competition condition of the fixation shift test described above. In the pointing condition, introduced first, the child has to point to the target as quickly as possible after it appears. In the counterpointing condition, when the target appears, the child is told to point as quickly as possible to the opposite side of the screen from the target ('the computer is trying to catch you out — don't let it trick you, point to the other side from the stripes to make the stripes disappear'). 20 successive trials are run in each condition.

(Study 4) Form and motion coherence

Coherence thresholds were tested as described by Gunn et al. (2002). The form stimulus was a static array of randomly oriented short line segments (white lines on a black background, density 1.3 segments/deg^2) containing a 'target' area on one side of the display where segments were orientated tangentially to form concentric circles. The proportion of tangentially oriented ('coherent') line segments among the randomly oriented 'noise' segments in the target area defined the coherence value for each trial. The motion stimulus comprised two random dot kinematograms (white dots on a black background, density 4 dots/deg^2), one of which was segregated into three horizontal strips, with the direction of the coherent motion of the middle 'target' strip being opposite to that of the two outer strips. The dots on the opposite side of the screen moved uniformly. A variable proportion of the dots oscillated horizontally across each array forming coherent motion (velocity 6°/s), while the remaining dots moved in random directions (incoherent motion). The direction of coherent motion reversed every 240 ms. 'Signal' dots had a limited lifetime of 6 video frames (120 ms). The additional 'noise' created by the disappearance of signal dots at the end of their lifetime was taken into account when calculating coherence levels on this task.

Thresholds were obtained by two-alternative forced-choice: children were asked, 'The snow is falling, can you find the road?' (motion) or 'Can you find the ball hidden in the long grass?' (form). Coherence level was initially set to 100% and after two to six practice trials was varied according to a two-up/one-down staircase procedure. Threshold was defined as the mean coherence level during the last four reversals.

'Town Square'

The procedure is described by Nardini et al. (2006). 12 inverted cups are laid out on a board 82 × 82 cm with distinctive toy houses and animals laid out to provide landmarks along two edges. The child stands near one corner of the board and sees a toy hidden under one cup; he/she has to point to the hiding location after one of the three alternative manipulations: 'same' (the child walks along one edge, then back to the starting point); 'walk'(the child walks along two edges, and so has to locate the cup from a viewpoint at the opposite corner); 'rotate' (the child walks as in 'same', but the board is rotated through 135°). The child's view of the board is obscured while walking. Children completed four randomly ordered trials for each condition. Performance scores are based on the distance of the cup selected from the correct cup, corrected for the chance distribution of distances. Age norms are derived from the results of Nardini et al. (2006).

TEA-Ch

We used four sub-tests of the TEA-Ch test (Manly et al., 2001) to distinguish four attentional domains: selective attention, sustained attention, response inhibition and attentional switching. These sub-tests are:

'Sky Search' (selective attention) — the time and accuracy of the child are scored for identifying 20 specified targets (pairs of identical spaceships) within an array of 108 distracters (pairs of non-matching spaceships). Motor speed is controlled for by a parallel task in which only targets are present. 'Score!' (sustained auditory attention) requires the child to keep count of a string of identical sounds interrupted by long pauses, thus requiring him/her to actively maintain attention in a task with little inherent stimulation. 'Walk don't walk!' (sustained attention and response inhibition) requires the child to mark successive squares along a path in response to a series of tones at an increasing rate. When a different sound occurs, the action must be inhibited. The task is paced by the speed of repetition of the tones and gets faster. The child has to maintain the attention to the sound and not lapse into an automatic response. 'Opposite worlds' (attentional switching) is a timed test using paths of digits (1s and 2s); initially the child has to say the correct name ('one' or 'two') to each successive digit; this is interchanged with a task where the child has to reverse the names, saying 'one' for 2 and vice versa.

Appendix B. MRI categories

Normal/mild [0]: No lesions in neonatal MRI; any neonatal intraventricular haemhorrhage resolved without ventricular dilation at term; no diffuse excessive high signal intensity (DEHSI), or isolated mild DEHSI, visible at term.

Moderate [1]: Any of the following conditions: moderate or severe DEHSI; moderate DEHSI with ventricular dilation; ventricular dilatation at term following germinal layer haemorrhage or intraventricular haemorrhage, but not requiring taps or shunts; neonatal focal punctate lesions; small venous infarct with normal basal ganglia and thalamus; isolated absent myelination in the posterior limb of the internal capsule (PLIC) at term equivalent age; cerebellar atrophy.

Severe [2]: Ventricular dilation requiring intervention, cystic periventricular leucomalacia, large venous infarction, or infarction within basal ganglia or thalamus, with involvement of the PLIC.

References

Arzoumanian, Y., Mirmiran, M., Barnes, P.D., Woolley, K., Ariagno, R.L., Moseley, M.E., Fleisher, B.E. and Atlas, S.W. (2003) Diffusion tensor brain imaging findings at term-equivalent age may predict neurologic abnormalities in low birth weight preterm infants. Am. J. Neuroradiol., 24: 1646–1653.

Atkinson, J. (1984) Human visual development over the first six months of life. A review and a hypothesis. Hum. Neurobiol., 3: 61–74.

Atkinson, J. (2000) The Developing Visual Brain. Oxford University Press, Oxford.

Atkinson, J., Anker, S., Rae, S., Hughes, C. and Braddick, O. (2002a) A test battery of child development for examining functional vision (ABCDEFV). Strabismus, 10(4): 245–269.

Atkinson, J., Anker, S., Rae, S., Weeks, F., Braddick, O. and Rennie, J. (2002b) Cortical visual evoked potentials in very low birth weight premature infants. Arch. Dis. Child. Fetal Neonatal Ed., 86: F28–F31.

Atkinson, J. and Braddick, O. (2003) Neurobiological models of normal and abnormal visual development. In: De Haan M. and Johnson M. (Eds.), The Cognitive Neuroscience of Development. Psychology Press, Hove, pp. 43–71.

Atkinson, J. and Braddick, O. (2005) Dorsal stream vulnerability and autistic disorders: The importance of comparative studies of form and motion coherence in typically developing children and children with developmental disorders (Commentary on Milne, E., Swettenham, J. and Campbell, R., 'Motion perception and autistic spectrum disorder: A review). Cahiers Psychol. Cognit. (Curr. Psychol. Cognit.), 23(1–2): 49–58.

Atkinson, J., Braddick, O., Anker, S., Curran, W. and Andrew, R. (2003) Neurobiological models of visuospatial cognition in children with Williams syndrome: measures of dorsal-stream and frontal function. Dev. Neuropsychol., 23: 141–174.

Atkinson, J., Braddick, O., Rose, F.E., Searcy, Y.M., Wattam-Bell, J. and Bellugi, U. (2006) Dorsal-stream motion processing deficits persist into adulthood in Williams syndrome. Neuropsychologia, 44: 828–833.

Atkinson, J., Hood, B., Wattam-Bell, J. and Braddick, O.J. (1992) Changes in infants' ability to switch visual attention in the first three months of life. Perception, 21: 643–653.

Atkinson, J., King, J., Braddick, O., Nokes, L., Anker, S. and Braddick, F. (1997) A specific deficit of dorsal stream function in Williams' syndrome. Neuroreport, 8: 1919–1922.

Barnett, A.L., Guzzetta, A., Mercuri, E., Henderson, S.E., Haataja, L., Cowan, F. and Dubowitz, L. (2004) Can the Griffith's scales predict neuro-motor and perceptual-motor impairment in full term infants with neonatal encephalopathy? Arch. Dis. Child., 89: 637–643.

Bhutta, A.T., Cleves, M.A., Casey, P.H., Cradock, M.M. and Anand, K.J. (2002) Cognitive and behavioral outcomes of school-aged children who were born preterm: a meta-analysis. JAMA, 288: 728–737.

Biro, S. and Russell, J. (2001) The execution of arbitrary procedures by children with autism. Dev. Psychopathol., 13: 97–110.

Braddick, O., Atkinson, J., Hood, B., Harkness, W., Jackson, G. and Vargha-Khadem, F. (1992) Possible blindsight in babies lacking one cerebral hemisphere. Nature, 360: 461–463.

Braddick, O., Atkinson, J. and Wattam-Bell, J. (2003) Normal and anomalous development of visual motion processing: motion coherence and 'dorsal stream vulnerability'. Neuropsychologia, 41: 1769–1784.

Braddick, O., Birtles, D., Wattam-Bell, J. and Atkinson, J. (2005) Motion- and orientation-specific cortical responses in infancy. Vision Res., 45: 3169–3179.

Braddick, O.J. (1993) Orientation- and motion-selective mechanisms in infants. In: Simons K. (Ed.), Early Visual Development: Normal and Abnormal. Oxford University Press, New York.

Braddick, O.J., Lin, M.-H., Atkinson, J. and Wattam-Bell, J. (1998) Motion coherence thresholds: effect of dot lifetime and comparison with form coherence. Perception, 27: S199.

Braddick, O.J., Wattam-Bell, J. and Atkinson, J. (1986) Orientation-specific cortical responses develop in early infancy. Nature, 320: 617–619.

Burgess, N. (2002) The hippocampus, space and viewpoints in episodic memory. Q. J. Exp. Psychol., 55A: 1057–1080.

Cioni, G., Bertuccelli, B., Boldrini, A., Canapicchi, R., Fazzi, B., Guzzetta, A. and Mercuri, E. (2000) Correlation between visual function, neurodevelopmental outcome, and magnetic resonance imaging findings in infants with periventricular leucomalacia. Arch. Dis. Child. Fetal Neonatal Ed., 82: F134–F140.

Cornelissen, P., Richardson, A., Mason, A., Fowler, S. and Stein, J. (1995) Contrast sensitivity and coherent motion detection measured at photopic luminance levels in dyslexics and controls. Vision Res., 35: 1483–1494.

Counsell, S.J., Allsop, J.M., Harrison, M.C., Larkman, D.J., Kennea, N.L., Kapellou, O., Cowan, F.M., Hajnal, J.V., Edwards, A.D. and Rutherford, M.A. (2003) Diffusion-weighted imaging of the brain in preterm infants with focal and diffuse white matter abnormality. Pediatrics, 112: 1–7.

Counsell, S.J., Shen, Y., Boardman, J.P., Larkman, D.J., Kapellou, O., Ward, P., Allsop, J.M., Cowan, F.M., Hajnal, J.V., Edwards, A.D. and Rutherford, M.A. (2006) Axial and radial diffusivity in preterm infants who have diffuse white matter changes on magnetic resonance imaging at term-equivalent age. Pediatrics, 117: 376–386.

Duncan, J. and Owen, A.M. (2000) Common regions of the human frontal lobe recruited by diverse cognitive demands. Trends Neurosci., 23: 475–483.

Dunn, L., Dunn, P., Whetton, D. and Pintillie, D. (1982) British Picture Vocabulary Scale. NFER, Windsor.

Dyet, L.E., Kennea, N., Counsell, S.J., Maalouf, E.F., Ajayi-Obe, M., Duggan, P.J., Harrison, M., Allsop, J.M., Hajnal, J., Herlihy, A.H., Edwards, B., Laroche, S., Cowan, F.M., Rutherford, M.A. and Edwards, A.D. (2006) Natural history of brain lesions in extremely preterm infants studied with serial magnetic resonance imaging from birth and neurodevelopmental assessment. Pediatrics, 118: 536–548.

Ellemberg, D., Lewis, T.L., Maurer, D., Brar, S. and Brent, H.P. (2002) Better perception of global motion after monocular than after binocular deprivation. Vision Res., 42: 169–179.

Gallant, J.L., Braun, J. and Van Essen, D.C. (1993) Selectivity for polar, hyperbolic, and Cartesian gratings in macaque visual cortex. Science, 259: 100–103.

Gunn, A., Cory, E., Atkinson, J., Braddick, O., Wattam-Bell, J., Guzzetta, A. and Cioni, G. (2002) Dorsal and ventral stream sensitivity in normal development and hemiplegia. Neuroreport, 13: 843–847.

Guzzetta, A., Cioni, G., Cowan, F. and Mercuri, E. (2001) Visual disorders in children with brain lesions: 1. Maturation of visual function in infants with neonatal brain lesions: correlation with neuroimaging. Eur. J. Paediatr. Neurol., 5: 107–114.

Hansen, P.C., Stein, J.F., Orde, S.R., Winter, J.L. and Talcott, J.B. (2001) Are dyslexics' visual deficits limited to measures of dorsal stream function? Neuroreport, 12: 1527–1530.

Henderson, S.E. and Sugden, D.A. (1992) The Movement ABC Manual. The Psychological Corporation, London.

Hood, B. (1995) Gravity rules for 2–4 year olds. Cogn. Dev., 10: 577–598.

Hood, B. and Atkinson, J. (1990) Sensory visual loss and cognitive deficits in the selective attentional system of normal infants and neurologically impaired children. Dev. Med. Child Neurol., 32: 1067–1077.

Hood, B. and Atkinson, J. (1993) Disengaging visual attention in the infant and adult. Infant Behav. Dev., 16: 405–422.

Hughes, C. and Russell, J. (1993) Autistic children's difficulty with mental disengagement from an object: its implications for theories of autism. Dev. Psychol., 29: 498–510.

Huntley, M. (1996) The Griffiths Mental Development Scales: from Birth to 2 Years. Association for Research in Infant and Child Development (ARICD), The Test Agency Ltd., Henley on Thames, UK.

Hüppi, P.S., Maier, S.E., Peled, S.Z., Gary, P., Barnes, P.D., Jolesz, F.A. and Volpe, J.J. (1998) Microstructural development of human newborn cerebral white matter assessed in vivo by diffusion tensor magnetic resonance imaging. Pediatr. Res., 44: 584–590.

Inder, T.E., Anderson, N.J., Spencer, C., Wells, S. and Volpe, J.J. (2003) White matter injury in the premature infant: a comparison between serial cranial sonographic and MR findings at term. Am. J. Neuroradiol., 24: 805–809.

Jones, P.B., Rantakallio, P., Hartikainen, A.-L., Isohanni, M. and Sipila, P. (1998) Schizophrenia as a long-term outcome of pregnancy, delivery, and perinatal complications: a 28-year follow-up of the 1966 North Finland general population birth cohort. Am. J. Psychiatry, 155: 355–364.

Kastner, S. and Ungerleider, L.G. (2000) Mechanisms of visual attention in the human cortex. Annu. Rev. Neurosci., 23: 315–341.

O'Keefe, J. and Burgess, N. (1996) Geometric determinants of the place fields of hippocampal neurones. Nature, 381: 425–428.

Kogan, C.S., Bertone, A., Cornish, K., Boutet, I., Der Kaloustian, V.M., Andermann, E., Faubert, J. and Chaudhuri, A. (2004) Integrative cortical dysfunction and pervasive motion perception deficit in fragile X syndrome. Neurology, 63: 1634–1639.

Larsson, H.J., Eaton, W.W., Madsen, K.M., Vestergaard, M., Olesen, A.V., Agerbo, E., Schendel, D., Thorsen, P. and Mortensen, P.B. (2005) Risk factors for autism: perinatal factors, parental psychiatric history, and socioeconomic status. Am. J. Epidemiol., 161: 916–925.

Lewis, T.L., Ellemberg, D., Maurer, D., Wilkinson, F., Wilson, H.R., Dirks, M. and Brent, H.P. (2002) Sensitivity to global form in glass patterns after early visual deprivation in humans. Vision Res., 42: 939–948.

Maalouf, E.F., Duggan, P.J., Rutherford, M.A., et al. (1999) Magnetic resonance imaging of the brain in a cohort of extremely preterm infants. J. Pediatrics, 135: 351–357.

Manly, T., Nimmo-Smith, I., Watson, P., Anderson, V., Turner, A. and Roberston, I.H. (2001) The differential assessment of children's attention: the test of everyday attention for children (TEA-Ch), normative sample and ADHD performance. J. Child Psychol. Psychiatry, 42: 1065–1081.

Manly, T., Robertson, I.H., Anderson, V. and Nimmo-Smith, I. (1999) The Test of Everyday Attention for Children: TEA-Ch. Thames Valley Test Company, Bury St Edmunds, Suffolk, England.

Marlow, N., Wolke, D., Bracewell, M.A. and Samara, M. (2005) Neurologic and developmental disability at six years of age after extremely preterm birth. N. Engl. J. Med., 352: 9–19.

Mercuri, E., Atkinson, J., Braddick, O., Anker, S., Cowan, F., Rutherford, M., Pennock, J. and Dubowitz, L. (1997a) Visual function in full-term infants with hypoxic-ischaemic encephalopathy. Neuropaediatrics, 28: 155–161.

Mercuri, E., Atkinson, J., Braddick, O., Anker, S., Nokes, L., Cowan, F., Rutherford, M., Pennock, J. and Dubowitz, L. (1996) Visual function and perinatal focal cerebral infarction. Arch. Dis. Child., 75: F76–F81.

Mercuri, E., Atkinson, J., Braddick, O., Anker, S., Nokes, L., Cowan, F., Rutherford, M., Pennock, J. and Dubowitz, L. (1997b) Basal ganglia damage in the newborn infant as a predictor of impaired visual function. Arch. Dis. Child., 77: F111–F114.

Mercuri, E., Braddick, O., Atkinson, J., Cowan, F., Anker, S., Andrew, R., Wattam-Bell, J., Rutherford, M., Counsell, S. and Dubowitz, L. (1998) Orientation-reversal and phase - reversal visual evoked potentials in full-term infants with brain lesions: a longitudinal study. Neuropaediatrics, 29: 1–6.

Mercuri, E., Haataja, L., Guzzetta, A., Anker, S., Cowan, F., Rutherford, M., Andrew, R., Braddick, O., Cioni, G., Dubowitz, L. and Atkinson, J. (1999) Visual function in term infants with hypoxic-ischaemic insults: correlation with neurodevelopment at 2 years of age. Arch. Dis. Child Fetal Neonatal Ed., 80: F99–F104.

Milner, A.D. and Goodale, M.A. (1995) The Visual Brain in Action. Oxford University Press, New York.

Mishkin, M., Ungerleider, L. and Macko, K.A. (1983) Object vision and spatial vision: two critical pathways. Trends Neurosci., 6: 414–417.

Morton, J. and Johnson, M.H. (1991) CONSPEC and CONLERN: a two-process theory of infant face recognition. Psychol. Rev., 98: 164–181.

Nardini, M., Burgess, N., Breckenridge, K. and Atkinson, J. (2006a) Differential developmental trajectories for egocentric, environmental and intrinsic frames of reference in spatial memory. Cognition, 101: 153–172.

Newsome, W.T. and Paré, E.B. (1988) A selective impairment of motion processing following lesions of the middle temporal area (MT). J. Neurosci., 8: 2201–2211.

Ridder, W.H., Borsting, E. and Banton, T. (2001) All developmental dyslexic subtypes display an elevated motion coherence threshold. Optom. Vis. Sci., 78: 510–517.

Robbins, T.W. (1996) Dissociating executive functions of the prefrontal cortex. Philos. Trans. R. Soc. Lond. B., 351: 1463–1471.

Roelants-van Rijn, A.M., Groenendaal, F., Beek, F.J., Eken, P., van Haastert, I.C. and de Vries, L.S. (2001) Parenchymal brain injury in the preterm infant: comparison of cranial ultrasound, MRI and neurodevelopmental outcome. Neuropediatrics, 32: 80–89.

Rutherford, M.A. (2002) MRI of the Neonatal Brain. W B Saunders, London.

Salomao, S.R. and Ventura, D.F. (1995) Large sample population age norms for visual acuities obtained with Vistech-Teller Acuity Cards. Invest. Ophthalmol. Vis. Sci., 36: 657–670.

SanGiovanni, J.P., Allred, E.N., Mayer, D.L., Stewart, J.E., Herrera, M.G. and Leviton, A. (2000) Reduced visual resolution acuity and cerebral white matter damage in very-low-birthweight infants. Dev. Med. Child Neurol., 42: 809–815.

Simion, F., Turati, C., Valenza, E., Leo, I. (2006) The emergence of cognitive specialization in infancy: the case of face preference. In: Munakata, Y. and Johnson, M. (Eds.).

Processes of Change in Brain and Cognitive Development: Attention and Performance, Vol. XXI. Oxford University Press, Oxford, pp. 189–208.

Spencer, J., O'Brien, J., Riggs, K., Braddick, O., Atkinson, J. and Wattam-Bell, J. (2000) Motion processing in autism: evidence for a dorsal stream deficiency. Neuroreport, 11: 2765–2767.

Stevenson, R.C., McCabe, C.J., Pharoah, P.O. and Cooke, R.W. (1996) Cost of care for a geographically determined population of low birthweight infants to age 8–9 years. I. Children without disability. Arch. Dis. Child. Fetal Neonatal Ed., 74: F114–F117.

Wattam-Bell, J. (1985) Analysis of infant visual evoked potentials (VEPs) by a phase-sensitive statistic. Perception, 14: A33.

Wattam-Bell, J. (1991) The development of motion-specific cortical responses in infants. Vision Res., 31: 287–297.

Wood, N.S., Marlow, N., Costeloe, K., Gibson, A.T. and Wilkinson, A.R. (2000) Neurologic and developmental disability after extremely preterm birth. EPICure study group. N. Engl. J. Med., 343: 378–384.

Woodward, L.J., Anderson, P.J., Austin, N.C., Howard, K. and Inder, T.E. (2006) Neonatal MRI to predict neurodevelopmental outcomes in preterm infants. N. Engl. J. Med., 355: 685–694.

Woodward, L.J., Edgin, J.O., Thompson, D. and Inder, T.E. (2005) Object working memory deficits predicted by early brain injury and development in the preterm infant. Brain, 128: 2578–2587.

Zeki, S.M. (1974) Functional organization of a visual area in the posterior bank of the superior temporal sulcus of the rhesus monkey. J. Physiol., 236: 549–573.

C. von Hofsten & K. Rosander (Eds.)
Progress in Brain Research, Vol. 164
ISSN 0079-6123

CHAPTER 8

Development of brain mechanisms for visual global processing and object segmentation

Oliver Braddick[1] and Janette Atkinson[2,*]

[1]*Department of Experimental Psychology, University of Oxford, South Parks Road, Oxford OX1 3UD, UK*
[2]*Visual Development Unit, Department of Psychology, University College London, London, UK*

Abstract: Objects have specific cognitive attributes, elicit particular visuo-motor responses, and require visual processes beyond primary visual cortex to combine information over extended regions as a basis for the segmentation and integration of visual objects. As well as segmentation, the assignment of region boundaries to differentiate figure from ground is a key process whose operation can be observed in infants during the early months. Global organization of both motion and pattern information plays a role in object segmentation and integration. These two types of global processing are associated with different brain systems, have different developmental courses, and are differentially vulnerable in developmental disorders.

In infancy, specific visual attributes determine the selection of a manual response (reach-and-grasp vs. surface exploration) and also the detailed kinematic parameters of each class of response. These are taken to reflect the properties of distinct visuo-motor modules whose properties emerge between 4 and 12 months of age. While these modules are a component of the dorsal cortical stream, they must interact with ventral stream processing in development and in the mature system.

Keywords: infant vision; visual development; object perception; segmentation; figure-ground; global form & motion; reaching

Introduction

From Jean Piaget onwards, the object concept has been a central theme in the study of the development of infant cognition. Work has concentrated on infants' understanding of the basic properties of objects: a single object can have effects through multiple sensory modalities; it remains the same object when its sensory effects change with distance, lighting, etc; it cannot occupy the same space as another object; it remains the same object even though part of it is hidden from view, and it continues to exist even when it disappears from view.

However, the infant's visual input is a continuous stream of information in time and space. All these object properties presuppose that the infant's visual processing can identify a section of this stream as a distinct entity which can be assigned a common label, and to which these properties can be attributed. The object must be visually segmented from its background, and as the opposite side of the same coin, the elements that make it up must be grouped together. In this paper, we first consider the early development of the visual

*Corresponding author. Tel.: +441865 271355; Fax: +441865 271354; E-mail: Oliver.Braddick@psy.ox.ac.uk

DOI: 10.1016/S0079-6123(07)64008-4

processes that make such segmentation and grouping possible.

In the world of the infant, objects are not just external entities to be identified and whose properties have to be understood. They are also the potential targets of the infant's actions. In fact, although recent work on infants' understanding of object properties has concentrated on the perceptual recognition of these properties, Piaget's theory treated the object as a sensory-motor, not just a sensory entity. The fact that a region of the visual field is distinctive and can be segmented does not necessarily make it an object in the sense of a target for action. In the latter part of this paper, we consider what defines an object as something which evokes a reaching and grasping action on the part of an infant. Some visually segmented regions are surfaces rather than objects, and require a different kind of action.

Segmentation by pattern and motion contrast

Segmentation requires first the local analysis of basic visual properties such as motion, texture, and colour which may differentiate an object from its background. Within primary visual cortex, neurons extract such information through receptive fields with selective responses to these local properties. These forms of cortical selectivity first develop in the first 2–3 months of life (Braddick et al., 1989; Braddick, 1993; Atkinson, 2000). However, the ability to respond selectively, e.g. to contour orientation, does not necessarily mean that the infant has a functioning mechanism that can detect the difference in texture orientation between adjoining regions and use it to define an object boundary. We examined this question (Atkinson and Braddick, 1992) by testing infants' preferential looking towards a rectangular region, which contained obliquely oriented texture elements, within a surround whose texture had the opposite oblique orientation. We found such a preference around 4 months of age, but not in 2-month-olds. Furthermore, the preference, while statistically reliable, was relatively weak. Sireteanu and Rieth (1992), in a similar experiment, only found evidence of segmentation in infants of 6 months and older. It appears that while young infants can segment a shape with boundaries defined by texture differences, such segmentation does not form a very salient object for young infants.

It has been suggested that the salience of texture boundaries depends on inhibitory connections between neurons that have similar orientation tuning in adjacent regions of the visual field (Nothdurft, 1990). (Such an arrangement would generate a signal at texture boundaries, by analogy with the way that ganglion cells with centre-surround opponency are activated at boundaries defined by luminance.) Horizontal connections between cortical orientation columns are believed to mediate such 'surround inhibition'. One neurally based hypothesis for the lack of salience of objects defined by such boundaries in infants is that these conditions may be relatively immature in the early months, even if the orientation selectivity of individual neurons is established (Burkhalter et al., 1993).

Segmentation by motion provides an interesting contrast to this relatively weak orientation-based segmentation. Infants' sensitivity to directional motion has been demonstrated, from about 7 weeks of age onwards, through preferential looking towards a region in a random dot motion pattern which is segmented by opposed directions of motion, compared to a region containing the same speed of dot motion but in a uniform direction (Wattam-Bell, 1992, 1994). This preference is sufficiently strong and reliable that it can be used to estimate motion thresholds in individual infants (Wattam-Bell, 1992, 1994; Mason et al., 2003). It might be proposed that the development of underlying directional selectivity occurs even earlier, with segmentation relatively delayed as it is for orientation. However, all attempts to test the development of cortical sensitivity to the direction of uniform motion have shown later-developing and weaker responses, compared to segmentation (Wattam-Bell, 1996). The argument made by Banton and Bertenthal (1996) to the contrary rests on the occurrence of the optokinetic response. However, in very young infants this is a subcortical reflex, with a quite different neural basis from discrimination performance (Braddick et al., 2003; Mason et al., 2003). In addition, independent VEP (visual evoked potential) methods, which do not depend on segmentation, show that the development of responses to uniform direction develops later than that for orientation (Braddick, 1993; Braddick et al., 2005). Thus, it appears that around the time cortical motion sensitivity first develops, it can be used effectively to segment differently moving regions. In fact, this segmentation is one of the most effective ways of tapping young infants' motion processing.

Why is directional motion an effective segmentation cue for infants as soon as we can find evidence for its elementary cortical processing, in contrast to oriented contours which provide only a weak segmentation cue well after development of the mechanisms for their elementary processing? One reason may be their relevance to the visual ecology of the infant. Segmentation by motion can arise either from self-moving objects against a static background, or from motion parallax whenever the observer moves, either actively or passively. In the former case, the self-motion itself is an important property and is associated with significant types of object, notably members of one's own species and one's own body parts. Motion parallax, as well as giving a quite ubiquitous basis for segmentation, has the important property that it is correlated with relative depth. It therefore provides an initial basis for infants to organize their perceptual world in 3-D, and has been shown to provide information that can be used by infants in discriminations of 3-D object shape (Yonas et al., 1987; Arterberry and Yonas, 1988, 2000). Along with stereopsis which develops shortly afterwards, motion parallax may play an important role in identifying and calibrating some of the more complex 'pictorial' sources of depth information which are present in the optic array; sensitivity to these develops over later months of life (Yonas and Granrud, 1985). Work on infants' perception of partially occluded objects has also emphasised the importance of the motion shared by parts of the same object, but not shared by the occluder or background, in grouping these parts as a unity distinct from the occluder (Kellman and Spelke, 1983; Johnson and Aslin, 1998).

In contrast to this dominant role of motion in segmenting the visual world, it is rarer to encounter adjacent regions of the optic array which are primarily distinguished by texture. In particular, common motion is a valid cue for grouping the parts of an object even when they differ in luminance, colour, texture, etc, and contrasting motions between adjacent regions almost always indicates a physical separation. In contrast, differences in texture are not always present between objects, and when present are often accompanied by more obvious luminance or colour boundaries. Thus motion-based segmentation offers environmental information that is considerably more significant and reliable than texture-based segmentation.

An alternative type of account for the developmental advantage of motion-based segmentation is in terms of the maturation of the underlying neural mechanisms. The work of Burkhalter et al. (1993) suggested that while horizontal connections in primary visual cortex (V1) were relatively immature in early infancy, the connections in layer 4B (containing the direction selective cells which project to area V5/MT) are the earliest to develop and start to emerge before term. Thus these authors suggest that the basis for lateral interactions in motion-sensitive mechanisms appears before lateral interactions in the pattern-, colour-, and disparity-sensitive mechanisms of V1. Relative neural maturation may be also an issue in the different extra-striate regions to which these mechanisms project. Little is known about the comparative developmental sequence of specific extra-striate areas. However, the motion-sensitive area V5/MT is known to be one of the most mature cerebral areas at birth, in terms of myelination and cyto-architectonics (Flechsig, 1920; Walters et al., 2003; Bourne and Rosa, 2006). In maturity, V5/MT contains neurons with centre-surround opponency in terms of motion direction (Born and Tootell, 1992), so it may be the early development of connectivity in this extra-striate area which underlies infants' sensitivity to motion segmentation.

Of course, accounts in terms of ecological significance and of neural maturation are not incompatible. Presumably the maturational sequence reflects, at least in part, selection pressure on those aspects of brain development which are most critical for successful development within the infant's visual environment. Segmentation of objects is critical for effective use of vision, and as suggested above, differential motion is a particularly important segmentation cue.

Figure-ground distinction

To define objects it is necessary to segment visually the object from surrounding regions, but this is not sufficient. When an object occludes another object or a background, the bounding contour between the two regions of the field is intrinsic to the occluding object and defines its characteristic shape. This boundary is not useful in characterising the occluded object or surface, since the latter physically extends behind the bounding contour and any contour, which defines its characteristic shape is interrupted or concealed. Thus the bounding contour 'belongs to' the occluding object or figure, but not to the occluded surface or ground. In well-known demonstrations such as Rubin's face/vase figure (see for example http://en.wikipedia.org/wiki/Rubin_vase) the figure-ground relationship is unstable and the bounding contour may be assigned to the region on either side to produce different perceived objects. However, the two forms of border-ownership are incompatible alternatives and only one region may be seen as a figural object at any one time.

A number of studies have shown that infants are sensitive to the cues defining occlusion of one region by another. At such an occlusion edge, contours that are surface markings have aligned terminators which create an 'illusory contour'. We have shown that 2-month infants can preferentially detect such a configuration over non-aligned terminators (Curran et al., 1999). Such illusory contours are best known when, in conjunction with luminance contour segments, they are organized to create figures such as the Kanizsa square. It has been demonstrated (Ghim, 1990; Kavsek and Yonas, 2006) that infants aged 4–6 months respond preferentially to such organization. It is to be noted that in all these experiments, there was an element of common movement added to the static aligned configuration.

For infants to perceive bounded regions as identifiable objects, then, their visual systems must implement this border-ownership, and process the shape defined by the intrinsic contour of the figure differently from that defined by the same contour as a non-intrinsic outline of the visible part of the ground. Most studies of figure-ground relations have depended on observers' subjective reports of which part of the display was seen as 'figure'. Such reports are not available from infants. However a more objective test is possible, since adults also recognise the shape of the figure more accurately than the shape of the ground (Driver and Baylis, 1996).

Subjective and objective methods have identified a number of visual properties which contribute to determining whether a region is seen as 'figure' or ground'. Several of these are listed in the legend of Fig. 1. The figure illustrates the design of a habituation experiment (Atkinson et al., 2004), that we carried out to test whether infants extract shape information from figure regions preferentially from ground regions.

In the habituation phase, the infant was shown a pair of identical patterns (Fig. 1, left) each containing a shape occluding part of a rectangular background region. This was an 'infant-controlled habituation' procedure, i.e. on each trial the stimuli were presented until the infant looked away. Identical presentations were repeated until the average looking time over 3 consecutive trials had dropped to below 50% of the maximum looking time, or 12 trials had been completed. At that point the test phase was entered, in which two shapes were presented for 10 s, corresponding to the 'figure' shape, and to the larger of the two 'ground' shapes from the habituation display (Fig. 1, right). There were two such test trials, in which the left–right positions of the two shapes were exchanged. Looking times towards each shape during the trial were measured. Infants who showed a strong side bias (>75% looking towards one side of the display across all test trials) were excluded from the analysis.

We wished to avoid the possibility that infants' recovery from habituation was determined solely by the familiarity of the light–dark contrast of the figure in the test phase. We therefore used

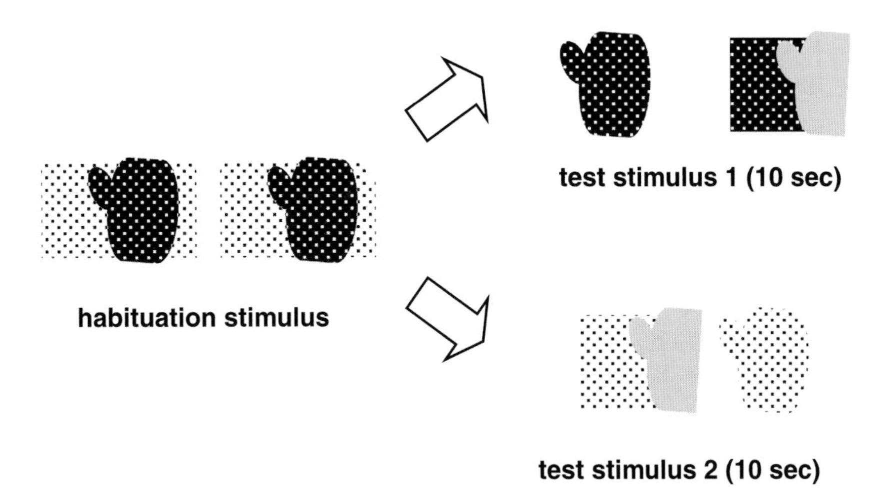

Fig. 1. Stimulus sequence used to test figure-ground assignment in infants. In the habituation stimulus, the curved shape in the middle of each panel is predisposed to be seen as 'figure' by adults because, compared to the surrounding 'ground' regions it has (a) lower luminance; (b) continuity of its bounding contour across T-junctions; (c) contours that are predominantly convex outwards; (d) cusps that are convex inwards (e.g. where the elliptical protrusion joins the main body). In the actual displays the regions were distinguished by colour contrast, not by the texture shown in these diagrams. Mirror-image versions of the shapes, and the left–right ordering of the two shapes in the successive test trials, were counterbalanced across infants.

coloured patterns. In the habituation phase this was either pink or green, with the figure and ground differing in luminance but having the same chromaticity, and a grey outer surround. The isolated shapes in the test phase had the same chromaticity as in the habituation phase, but both shapes had the same luminance and contrast against the grey background. This corresponded either to the figure luminance or the ground luminance from the test phase, with one of these luminance values being used in each of the two test trials. The colour (pink vs. green), the use of dark or light test stimuli on the first test trial, and the use of leftward vs. rightward facing figures, were all counterbalanced across infants.

If the infants preferentially encoded the shape of the figure over that of the ground parts in the habituation display, the 'ground' shape in the test trial should be less familiar, and so would be predicted to have the greater looking time.

The left bar of Fig. 2 shows the mean distribution of looking times between the two patterns in the test phase, for a group of 16 infants aged 10–14 weeks. It is clear that performance is very close to equal looking time (49.7%:50.3%), so close that it seemed fruitless to gather data for a larger group. However, a critical question with any habituation experiment is whether there is any intrinsic preference between the stimuli, independent of any effect of the habituation. We therefore tested a new, larger group of infants over a wider age range (7–21 weeks, mean 14 weeks) with only the test trials from this procedure. The results are shown in the right hand bar of Fig. 2. It is clear there is a modest but highly significant preference for the 'figure shape', in the absence of any habituation. No age trend appeared in this preference.

Taking the group as a whole, the preference experiment showed that infants looked more at the shape with the 'figure' characteristics (although it should be emphasised that in this display, both shapes are figures, in the sense that they appear to adults as objects against a uniform background). We cannot infer from this result that infants encode the shape of a figure region preferentially to that of a ground region. However, we can conclude that some of the stimulus characteristics that bias a region towards 'figure' assignment (e.g. convexity, presence of concave cusps) also attract infants' attention. This sensitivity that attracts attention to 'figure' cues may be a developmental

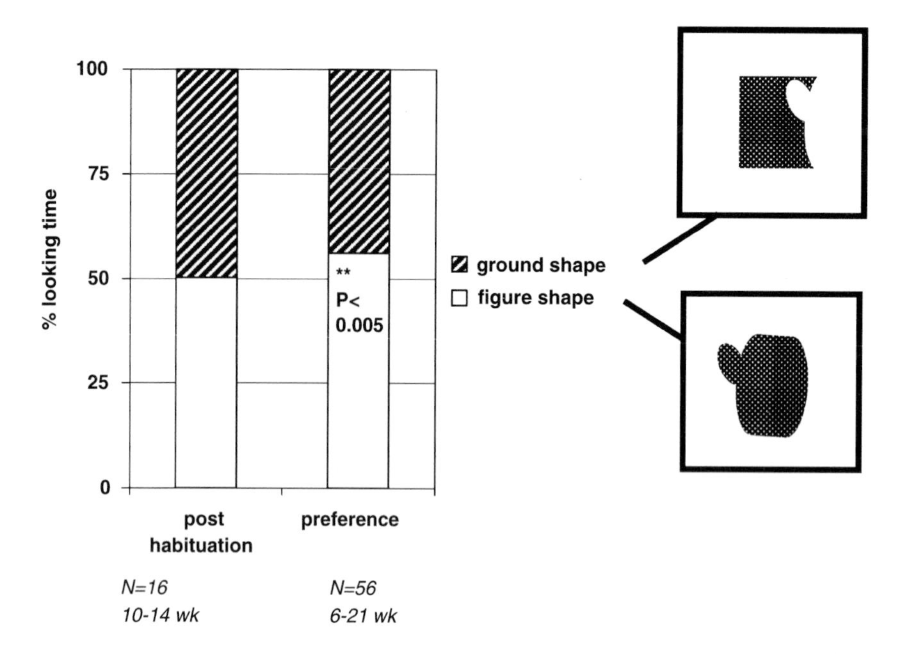

Fig. 2. Results of habituation and preference tests with the figure-ground stimuli illustrated in Fig. 1.

precursor of preferential encoding of the shape of the attended region.

If the preference for the figure following habituation was significantly less than its spontaneous preference, that would be evidence of preferential encoding during habituation, even if the two figures did not show significantly different post-habituation looking times. The data are in this direction. But given the modest preferences involved and the variability, very large numbers of infants would have to be tested to provide sufficient power in a test of this difference between conditions. However, we can say that during the age range up to 4 months studied here, infants are starting to pick up the attributes that distinguish ground shapes from figure shapes.

We cannot yet say at what age this ability can serve to assign border-ownership, the defining characteristic of the figure-ground difference in adults. It is possible that border assignment develops along with sensitivity to the figure cues, but the habituation method is not sufficiently sensitive to pick it up. Alternatively, there may be a developmental mechanism by which some of the figure cues, such as convex boundary, initially attract the infant's attention to the region within that boundary. This selective attention may in turn develop the association between the interior region and the bounding contour that determines the figure-ground distinction, and determines the 'object quality' of the figure.

Global form and motion coherence

The discussion above has emphasised the importance of segmentation, and the boundaries between regions, in defining an object. However, structure and uniformity within a region are also important. Experimentally, the analysis of within-region coherence has two advantages. First, it can be defined quantitatively, as the percentage of elements within the region, which have a common property, such as direction of motion. Secondly, it can be related to levels of neural organization in the visual system.

The motion system can be taken as an example. In primary visual cortex (V1), the neurons that respond selectively to motion direction have small receptive fields, and so represent the local motion of small elements. In higher-tier areas, receptive field sizes increase and so motion information can be integrated over a larger area. Sensitivity to motion coherence is evidence for such integration, or global processing; coherence cannot be identified from the motion of elements taken individually. Neural responses in area V5/MT increase linearly with the percentage of coherently moving dots (Britten et al., 1993; Rees et al., 2000), and damage to this area seriously impairs motion coherence sensitivity, both in humans (Baker et al., 1991) and macaques (Newsome and Paré, 1988). Although most work has concentrated on V5/MT, a number of other extra-striate areas respond differentially to coherent vs. random motion, and so are implicated in global processing, notably (in humans) area V3A (Braddick et al., 2000).

Measures of infant performance indicate that global motion processes operate at an early stage of development. Coherence thresholds can be measured using preferential looking in Wattam-Bell's segmented motion display, and show a marked improvement over the weeks following the age at which direction discrimination first emerges (Wattam-Bell, 1994). A related method varies the coherence in terms of the spread of motion directions about the mean value (Banton et al., 1999). This work shows that infants of 12 weeks and over can integrate directional distributions whose standard deviation is as large as 68°.

These results suggest that very soon after local motion signals are first available in the developing brain, the processes that integrate them into global representations are operating quite efficiently. We have made similar comments above on the early presence of segmentation processes in the motion-processing pathway. Global integration is based on the connectivity between V1 and extra-striate areas including V5. It is possible that these connections are established very early, at least in a crude form, but that their function is latent until local directional selectivity emerges in V1 around age 7 weeks (Wattam-Bell, 1996). This delay might reflect some minimum level of temporal precision in the developing visual system, which is needed to make directional motion responses possible.

Global form processing can be studied in an analogous manner. We have devised a measure of 'form coherence' thresholds: subjects are required to detect the organization of short line segments into concentric circles, with 'noise' introduced by randomizing the orientation of a proportion of the line segments. Neurons responding to concentric organization of this kind have been reported in area V4 in macaques (Gallant et al., 1993) — an extra-striate area occupying a similar position in the ventral stream to that of V5 in the dorsal stream.

We have previously identified, from fMRI studies of normal adults, specific areas involved in our form and motion coherence tasks (Braddick et al., 2000, 2001). This work has shown that anatomically distinct circuits are activated in global processing of form and motion, although each circuit involve parts of both the parietal and temporal lobes, and cannot therefore be said to be strictly 'dorsal' and 'ventral' in the human brain. However, the activated areas do include dorsal stream areas V5 and V3A for motion, and anatomically ventral areas for form. It has also been found that the BOLD response in fMRI studies increases linearly with degree of coherence in an area analogous to V5 (Rees et al., 2000) and we find that areas in the lingual/fusiform gyrus, which may include V4, similarly show a linear response with form coherence (Braddick et al., 2002b).

The global form stimulus can be used in a preferential-looking method with infants, where the partially coherent circular configuration appears on one side of the display and a random array of line segments on the other. This method has demonstrated that 4-month-olds show a significant but weak response to global form, which is absent in 2-month-olds (Braddick et al., 2002a). Thus, although local orientation sensitivity emerges earlier in development than directional selectivity (Braddick, 1993; Braddick et al., 2005), global organization based on orientation is less effective in determining infant behaviour than global organization based on motion.

The form and motion tests described above used somewhat different stimulus configurations, and different groups of infants. Recently, we have compared infants' sensitivity to global form and motion more directly, both by behavioural means and by visual event-related potentials (VERP). In each case, the stimuli were designed to be as comparable as possible in their geometry, and the same infants were tested for form and motion. Behaviourally, we have tested infants between 8 and 18 weeks for forced-choice preferential looking with stimuli of the kind illustrated in Fig. 3. The elements in the display are either short arc segments (form test) or limited-lifetime dots each moving through the same arc at constant speed (motion test). Preference was tested between one side of the display where the arcs were randomly oriented, and the other side where there is a region in which the arcs are arranged with a coherent concentric organization. (Note that, to maintain the same spatio-temporal properties on each side, the dot motions have equal linear rather than angular speed, and so the appearance is of a slow 'whirlpool' effect rather than of rigid rotation.) Each infant was tested over either one or two blocks of 10 trials with each stimulus, the order of form and motion testing being counterbalanced across infants.

Fig. 3. Display used for preferential looking test of global form coherence. The region of coherent concentric organization could appear in either the left or right half of the display. The display for testing motion coherence sensitivity was similar, except that the short arc elements illustrated were replaced by dots moving through the same arc paths at 6.5°/s.

Figure 4 shows the preferential looking results for 40 infants aged between 8 and 18 weeks. For the motion display, the infants show a consistent preference (although not particularly strong) throughout the age range. For the comparable form display, looking behaviour is at chance for the younger infants. Unlike motion, it shows a significant regression with age, with a clear preference for the coherent form only in the infants over 13 weeks.

The same issue can be addressed electrophysiologically. In a sequence of the kind illustrated in Fig. 5, a coherent pattern (similar to the right hand side of Fig. 2) and an incoherent pattern (similar to the left hand side) are presented alternately. A neural response will be expected at each transition, not only because of the change of global organization, but also because local orientations

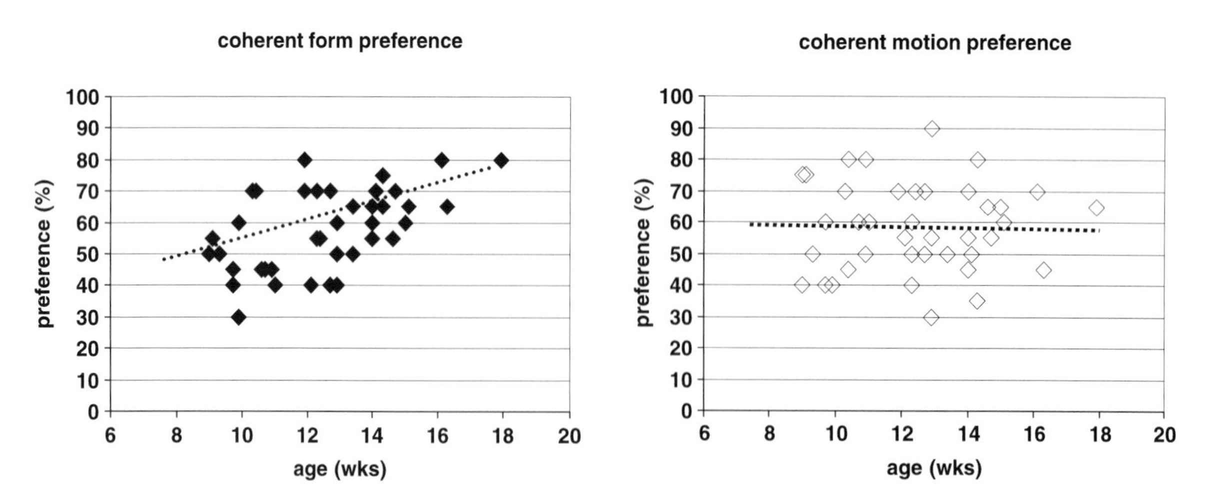

Fig. 4. Preferential looking results for the form coherence display of Fig. 2, and a geometrically similar motion coherence display, on the same group of infants aged 9–18 weeks. Dotted lines show regression of % preference on age for each test

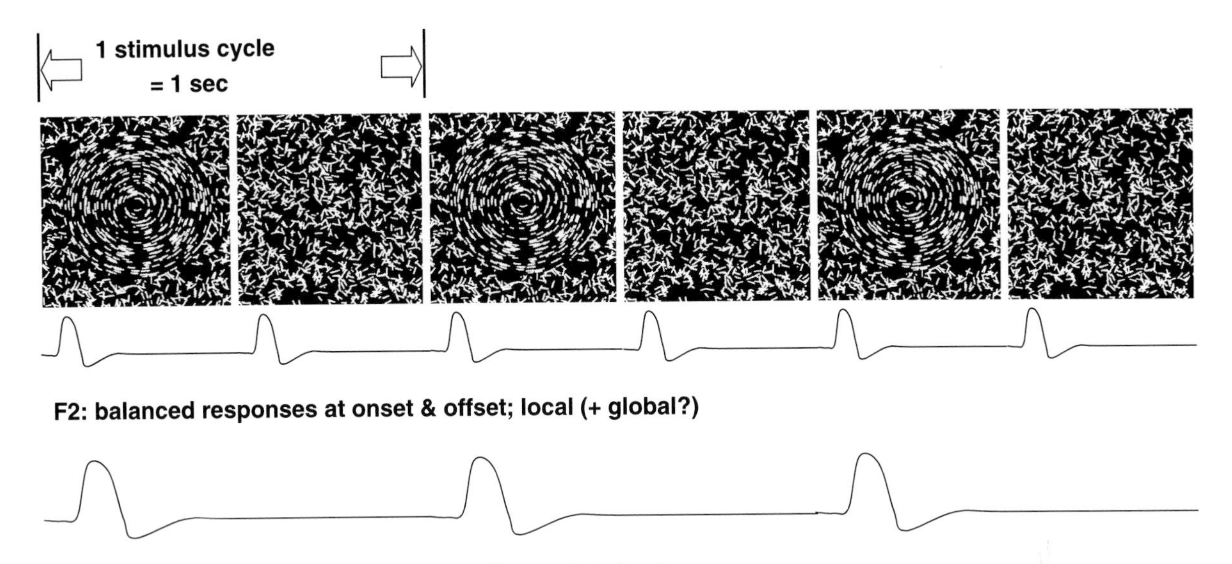

Fig. 5. Stimulus sequence used to elicit global-form VERP responses. The sequence used for global motion responses is similar except that the elements are moving dots moving along short arc trajectories, rather than the static arc segments used for the form test.

or motions will change in a random fashion everywhere. Thus a response at this frequency – the second harmonic of the stimulus cycle, or 'F2' — may arise from mechanisms sensitive to either the local or the global structure of the stimulus; the two contributions cannot readily be separated. However, the local events at the onset and offset of the global pattern are symmetrical in a statistical sense; because the non-coherent pattern is random, the orientation changes at different location will follow a uniform random distribution at both onset and offset. Any differential response at onset and offset must therefore come from processes sensitive to the global structure. Such global responses will appear at the fundamental frequency of the stimulus cycle (F1), as illustrated in Fig. 5

We can verify that F1 does measure a global response, by testing these VERPs in adults with variable coherence levels (Braddick et al., 2006a, b). The amplitude of the F2 form response is independent of coherence; that of the F2 motion response decreases gently with decreasing coherence, but for both motion and form there remains a substantial amplitude at 0% coherence when the sequence consists of a series of random patterns. In contrast, the F1 response varies linearly with coherence, and is absent at 0% coherence when the random sequence contains no distinction between onset and offset. This demonstration confirms that the F1 response depends on the presence of global structure.

Both form and motion responses can be measured in infants. However, they show different developmental courses. Figures 5–6 show the findings from such recordings on three groups of infants, aged 2–3 months, 3–4 months, and 4–5 months, respectively (Braddick et al., 2006a, b). As in adults, the recorded responses to global motion are stronger than those for global form. This difference is most marked in the younger groups, where twice as many children reach the criterion of statistical significance for this response to motion compared to form (Fig. 6). By 4–5 months, the two responses are converging, both in the proportion of infants with a significant response and in the average signal:noise ratio taken over all infants. It is noteworthy to compare the global responses, indicated by the F1 frequency (Fig. 7a), with the F2 responses that have their primary origin in local processing (Fig. 7b). For the latter responses, the developmental pattern is quite different: the signal:noise ratio rises with age for the motion stimulus, but declines for the form stimulus. Thus

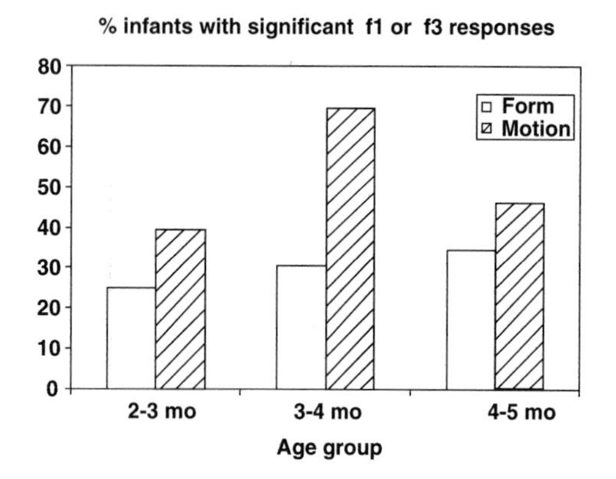

Fig. 6. Percentage of infants in three age groups showing a statistically significant global VEP response (odd-numbered harmonics) to the form sequence shown in Fig. 4, and its global-motion equivalence. Significance is determined by the circular-variance test (Wattam-Bell, 1985) on 200 sweeps of a 1 Hz stimulus. $N = 26, 22, 18$ infants in the 2–3, 3–4, and 4–5 month age groups, respectively.

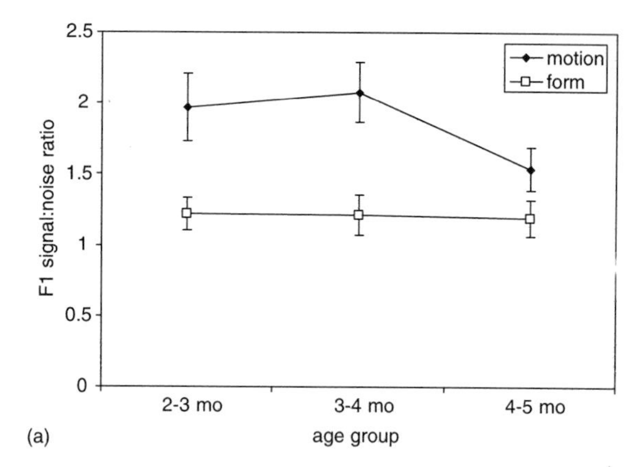

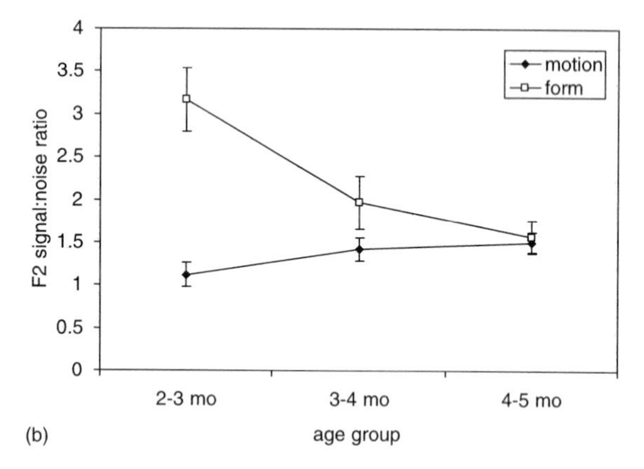

Fig. 7. Signal:noise ratio of (a) the global F1 component and (b) local F2 component in the VEP responses to form and motion sequences, for the infants whose data appear in Fig. 5. Error bars = standard error of the mean. Noise is estimated from the mean amplitude of frequency components at 0.005 Hz intervals for 0.25 Hz either side of the stimulus frequency, excluding the stimulus frequency itself.

the relative changes in the global F1 response do not simply reflect an overall general change in the properties of the brain's electrical responses that can be recorded, but are specific to those that reflect global processing in the two domains.

Behavioural and electrophysiological findings therefore converge on a common developmental account. This provides much more confidence in the findings than relying on either method alone, each with their separate limitations. For motion, the global responses are not quite as consistent as those with a local component (Wattam-Bell, 1996; Braddick et al., 2005), but they are present at an age very soon after the onset of these local signals, supporting the view that global processes can integrate local motion signals almost as soon as the latter become available in cortical development. In contrast, the global form VERPs show a later developmental course that extends over some months of infancy.

Development of global processing through middle childhood: 'dorsal stream vulnerability'

The initially accelerated development of global motion compared to global form processing is only the first stage in the development of these two functions. Our work has used coherence thresholds for global motion and form as indicators of the course of maturation of dorsal and ventral-stream extra-striate visual areas, respectively beyond infancy. Over the period 4–10 years of age, the relative advantage seen for global motion processing is reversed; motion coherence thresholds show more variation, and reach adult values more slowly, than do the corresponding thresholds for form processing (Gunn et al., 2002; Atkinson and Braddick, 2005). Furthermore, this relative immaturity of global motion processing is accentuated in a wide variety of developmental disorders.

We first saw this in children with Williams Syndrome (WS) (Atkinson et al., 1997), and tests on WS adults showed that an impairment of global motion relative to global form was a persisting feature of this genetic disorder (Atkinson et al., 2006). However, deficits specific to global motion processing are not special to WS. Hemiplegic children with IQ performance in the normal range showed significantly poorer global motion performance for age than typically developing children, while their global form sensitivity was unaffected (Gunn et al., 2002), and a similar relative impairment of global motion sensitivity has also been found in autistic children (Spencer et al., 2000), in studies of developmental dyslexia (Cornelissen et al., 1995; Hansen et al., 2001; Ridder et al., 2001), in fragile-X disorder (Kogan et al., 2004) and in children who had early visual deprivation due to congenital cataract (compare Ellemberg et al. (2002) with Lewis et al. (2002)). As this motion coherence deficit appears to be a common feature across a wide variety of paediatric disorders with very different aetiologies and neurodevelopmental profiles, we suggest that it represents an early vulnerability in the motion processing stream of a very basic nature, which we have termed 'dorsal stream vulnerability' (Braddick et al., 2003).

Overall, then, the process of extracting the global structure of the visual array requires processes of segmentation between regions, and processes that detect coherent structure within regions. Segmentation and coherence are global processes that operate, probably at extra-striate level, in the domains of both motion and form. In the delineation and representation of objects, both motion and form carry information and must ultimately be integrated. These two domains, however, are handled at least at the first processing levels of the brain, by substantially separate neural networks, they have distinctive developmental trajectories through infancy and childhood, and are differently vulnerable in neurodevelopmental disorders.

Objects as targets for reaching: the effect of size

As discussed in the introduction, an object is not only the result of sensory organization, but crucially has visual characteristics defining it as a target for action. We turn now to these visuomotor aspects. Infants develop the ability to initiate rapidly a hand movement towards a seen object, move the hand smoothly and accurately toward the target object, and form the hand appropriately. This behaviour emerges towards the middle of the first year of life, and during a period between about 6–15 months of age it is very readily triggered, with infants reaching on a high proportion of occasions when a small object is placed within their reach. It represents a specialized and refined coupling of the visual and motor systems, suggesting the development of specifically tuned visuomotor networks (Atkinson, 2000; Atkinson and Braddick, 2003) which can compute visual object properties that characterise the 'affordance' (Gibson, 1966, 1977) of graspability. With age, grasping behaviour becomes elicited more selectively, and integrated better into action sequences with larger behavioural goals.

Such networks, translating visual information into action, are key components of the dorsal cortical stream (Milner and Goodale, 1995). Single unit neurophysiology in the monkey has demonstrated populations of neurons in the parietal lobe, forming distinct subsystems that encode the information need to control eye movements, reaching, and grasping (Sakata et al., 1997). These distinctions are supported by human neuroimaging and neuropsychology (Culham and Valyear, 2006). In contrast, the analysis of basic visual properties that define 'salience', and so elicit visual attention, is not necessarily a dorsal stream function, as in the case of global form coherence discussed above. Newman et al. (2001) compared the same targets for the way they elicited 'preferential looking' and 'preferential reaching'. This work found a transition, from a stage before 8 months in which reaching and looking were driven by common visual parameters, to a later stage in which the visual properties of objects which specify 'graspability' in preferential reaching (for older infants) were distinct from those determining looking. Thus the targets of the two responses could be dissociated. We have proposed a developmental model of interacting modules (Fig. 8, from Atkinson and Braddick (2003)) describing the

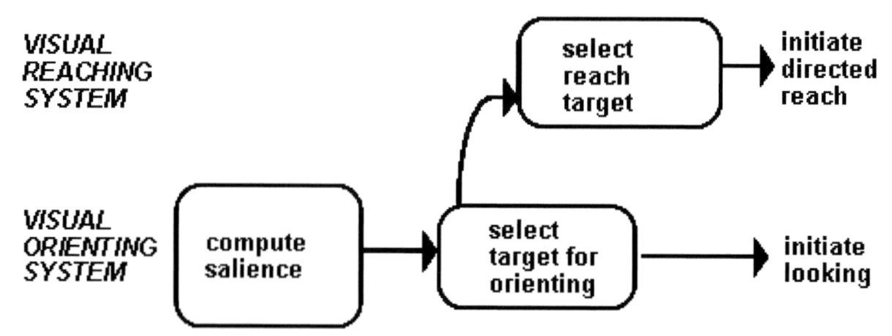

STAGE 1: Reaching develops under the control of general orienting response to salient and novel objects, which also includes head/eye orienting (looking)

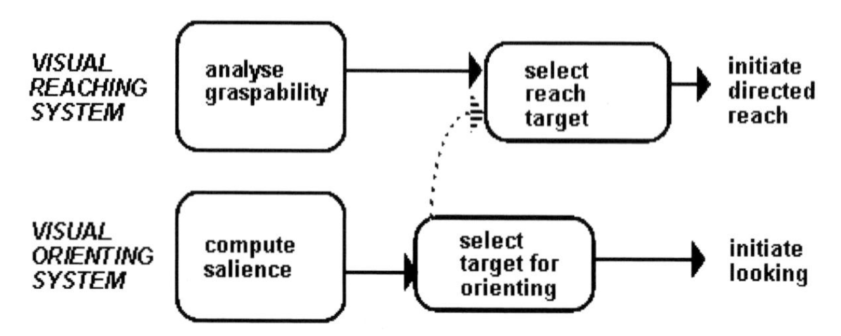

STAGE 2: Visual analysis of object graspability can over-ride salience in selection of reach target: reaching and looking can be dissociated

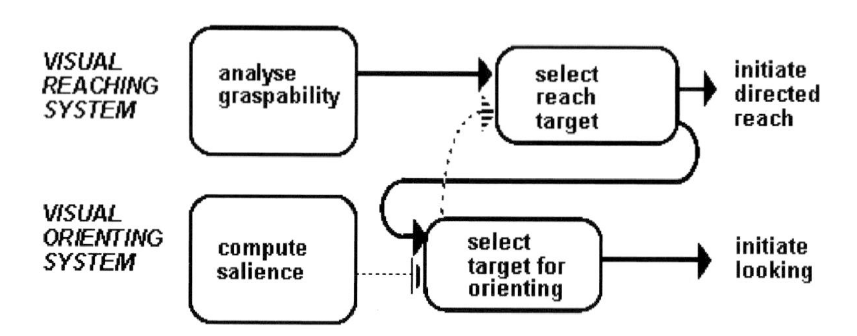

STAGE 3: Reaching target selected, from visual analysis of graspability can entrain head/eye orienting, overriding control by salience: looking is coupled to the demands of reaching

Fig. 8. Model of successive stages in the development and interaction of visuomotor modules controlling looking (orienting) and reaching (Atkinson and Braddick, 2003).

relative maturation and differentiation of the dorsal stream networks controlling eye and hand movements, from a common analysis of specific visual attributes.

During the period of readily-triggered reaching, the class of stimuli that elicit this behaviour must be defined by specific visual parameters. In particular, if an object is sufficiently large, such as a table top, it becomes effectively unbounded and functions as a 'surface', a possible target for manual exploration rather than a potential reaching target for reach to grasp. In the parietal lobe, orientation of extended surfaces is encoded by a distinct neural population form that encoding the shape of graspable objects. A recent study in our laboratory aimed to explore the development of this distinction in infants, by examining both the gross characteristics and the kinematic details of the behaviour elicited by small and large objects.

A number of studies have investigated the sensitivity of reaching behaviour to object size. Many have concentrated on the anticipatory scaling of hand opening to object size, which is first seen around 7–9 months (von Hofsten and Rönnqvist, 1988; Newell et al., 1989). The selection of bimanual reaching for larger objects appears to occur somewhat earlier, at 5–6 months (Clifton et al., 1991; Siddiqui, 1995; Fagard, 2000). However, most studies have used a range of sizes within the range that could be physically grasped. We selected a much wider range of sizes, to explore the range over which the affordances of a graspable object may be lost.

Our study tested 70 infants, divided into three age groups: 4–6 months ($n = 19$); 6–8 months ($n = 28$), and 8–10.3 months ($n = 23$). The target objects were six lightweight cardboard boxes covered in uniform yellow material, varying in width from 1.6 cm to 46.7 cm (full figures are in the caption to Fig. 9). These sizes were selected such that it was possible for the two smallest objects (1 and 2) to be grasped with one hand by all children in this age range. Object 3 could be grasped and lifted with two hands but not effectively with one. The larger objects 4, 5 and 6 were all on occasion grasped and moved bimanually by infants in the study, but could only be lifted off the surface with difficulty, if at all. The objects were presented one at a time in a 'theatre' with a screen that could be raised or lowered to reveal the object. The infants' reaching movements were videotaped and a two-camera Elite motion-tracking system (BTS, Milan) was used to record the kinematics. For this behavioural analysis, three measures were: (i) whether it was a successful reach, i.e. at least one hand contacted the object (reach/non-reach); (ii) whether the reach was bimanual (i.e. the second hand had started its movement towards the object before the first hand made contact); (iii) type of contact following the reach, scored as either 'reach and grasp' (RG) or 'non-grasp contact' (NG) when the hand was not

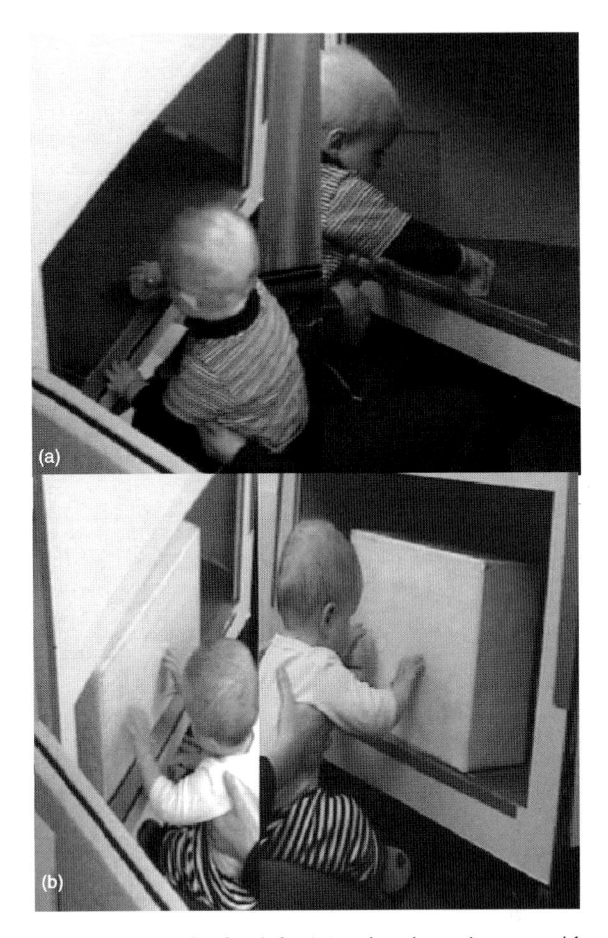

Fig. 9. (a) Example of an infant's 'reach and grasp' contact with object 2. (b) Example of a 'non-grasp' contact with object 5.

grasping or configured for grasping, for example the open hand touched the object but without any closing of the figures or the open fingers were brushed over the surface of the object. Figure 9 shows examples of the two kinds of contact.

Incidence of reaching (Fig. 10)

Our results illustrate how readily the reaching response is visually elicited in infants between 4 and 10 months: none of the objects elicited a reach in less than 78% of valid trials, in any of the age groups (Fig. 10). Only infants in the two older groups showed a systematic variation in reaching frequency with object size. The lack of differential reaching under 6 months is consistent with earlier findings using smaller ranges of object size

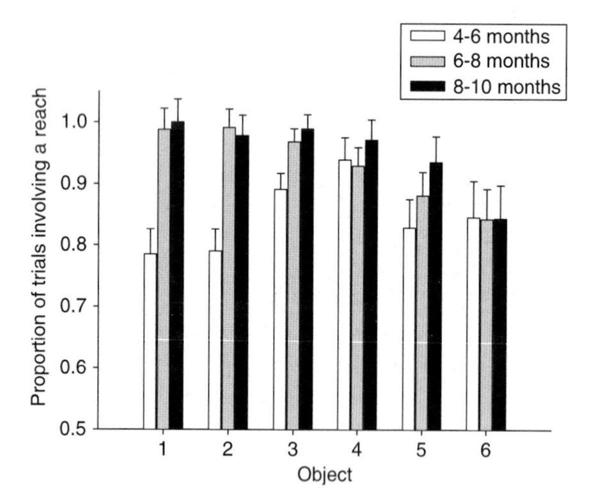

Fig. 10. Proportion of trials with a valid reach in the three age groups. Object sizes (width × height × depth, mm): (1) 16 × 40 × 16; (2) 32 × 42 × 32; (3) 92 × 105 × 95; (4) 270 × 195 × 107; (5) 379 × 260 × 158; (6) 467 × 370 × 195.

(e.g. Newell et al., 1989; Fagard and Jacquet, 1996; Fagard, 2000; Newman et al., 2001). It might be taken to indicate that, in the younger group, processing of visual size information is not effectively used to determine the affordance of reaching and grasping. However, other data discussed below suggest that this insensitivity to object size is not absolute, and that even the youngest age group can show differential behaviour in other respects.

Bimanual reaching (Fig. 11)

The larger objects can be manipulated, if at all, with two hands but not with one. In both 6–8 and 8–10 month age groups the proportion of bimanual reaching increased steadily with object size, suggesting that this aspect of object affordance is visually processed and the information delivered to motor control (Fig. 11). The 4–6 month group did not show this pattern; in particular, bimanual reaches did not increase from objects 1–4 and the largest object showed the lowest proportion of bimanual reaches for any object, in this age range.

Reach and grasp actions (Fig. 12)

The proportion of reaches that were RG, compared to NG contact, decreased systematically

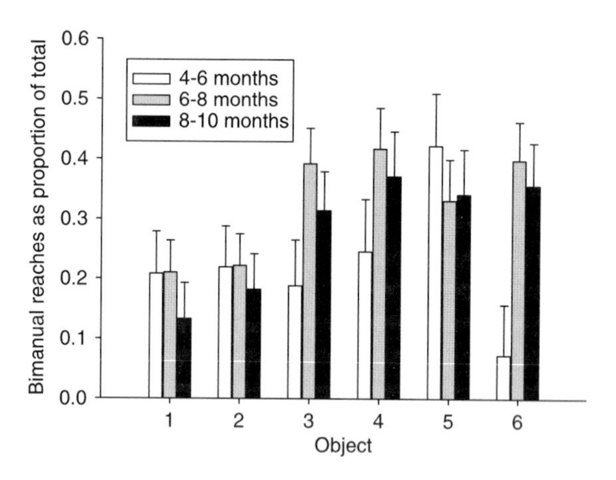

Fig. 11. Proportion of reaches that were bimanual, for each object (see caption for Fig. 9) for each age group. A bimanual reach was defined as one in which both hands contacted the object, and the second hand had started its movement towards the object before the first hand made contact.

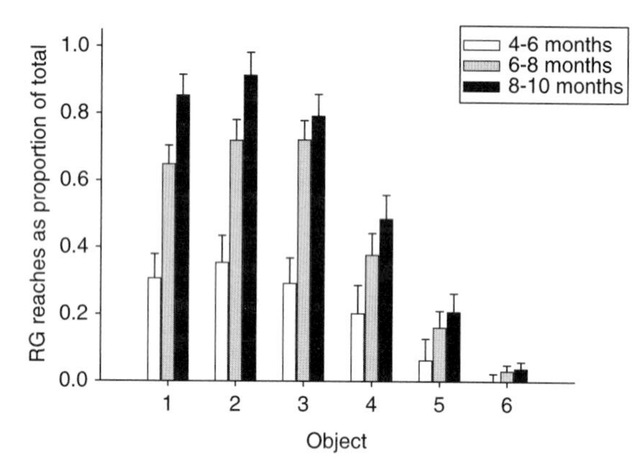

Fig. 12. Proportion of reaches that were categorised as 'reach and grasp', for each age group and object.

with object size (Fig. 12). Although this aspect of RG control is substantially better developed in the two older groups, the relation to size was statistically significant for each of the three age groups. The observational classification of RG and NG reaches was supported by kinematic measures. In particular, as shown in the bars at the right of Fig. 13, hand speed at contact with the object was 50% faster in NG reaches than in RG reaches, consistent with an action which is not planned so as to terminate in a controlled grasp. The finer-grained analysis provided by the kinematic measures also shows the sensitivity of reaching

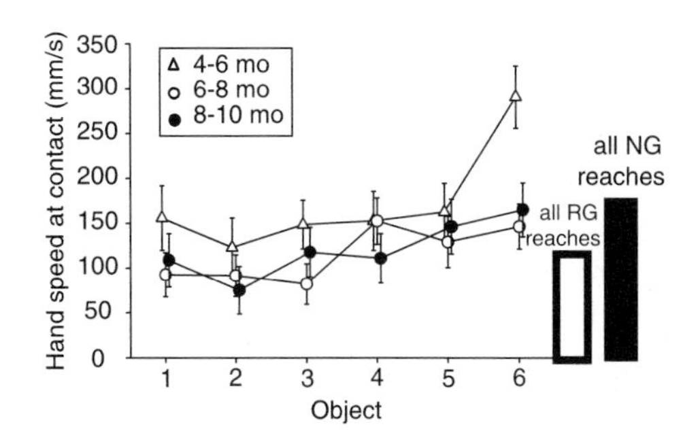

Fig. 13. Kinematic measures before contact are sensitive to object size, and distinguish different types of reach. Means for hand speed at contact as a function object size for infants in the three age groups (△ = 4–6 months, ○ = 6–8 months, ● = 8–10 months). Vertical lines depict standard errors of the means. Bars at right show the mean hand speed at contact (across all ages and objects) for reaches categorised by observation as reach and grasp (RG) or non-grasp (NG).

behaviour to object size. With increasing size, there was (i) a decrease in total reach duration, (ii) an increase in maximum velocity, (iii) a decrease in percentage deceleration time, and (iv) an increase in end velocity with increasing object size. All of these relationships, which also characterise the RG/NG distinction, are consistent with a less carefully calibrated approach to the object when it is larger. As an example, the plots in Fig. 13 show the statistically significant trend of end velocity, a trend that showed no interaction with age group.

Our results show that infants across the whole age range reported here, when presented with the appearance of an object, show manual responses that are sensitive in various ways to object size. The response measures are derived purely from the period up to contact with the object, so this sensitivity must reflect visual information picked up before or during the reach.

In summary, the 6–8 month and 8–10 month infants demonstrated sensitivity to visual size information in terms of which objects would trigger reaching, which objects elicited grasping action, and which elicited bimanual vs. unimanual reaches. The 4–6 month group showed this only in respect of grasping action, even though such an action occurred much less often overall in this youngest group. Thus for the youngest infants, in an age range where successful reaches and grasps are a relatively recent achievement, size sensitivity appears only in the form of the response (reach/grasp vs. non-grasp contact) and not in the frequency with which it is elicited. In both groups over 6 months, the mechanism initiating reaching is more selectively (but also more readily) engaged.

Although infants reached on the majority of trials for objects across the whole size range, the non-grasp hand actions associated with reaching for the larger objects suggested that visual information from these objects was frequently eliciting a response of surface exploration, rather than grasping that could support manipulation, as illustrated in Fig. 9(b).

Within both grasp and non-grasp actions, the kinematics showed a degree of variation with object size that is comparable to the variation between these response categories. Thus it appears that size information acts in two ways for infants: it can determine the choice of a category of action plan (bimanual vs. unimanual, 'reach and grasp' vs. 'surface exploration'), and it can also modulate the detailed kinematics within the scheme of the selected action plan. Both kinds of size sensitivity are most readily demonstrated over a range of sizes that includes very large objects (>30 cm). For infants over 6 months, visually available size information defines a category for which both unimanual and bimanual reaching actions are elicited with high frequency, and so can be considered an 'object' not only for perceptual purposes but also for visuo-motor purposes.

We propose that distinct neural networks underlie the visuo-motor processes that generate grasping of an 'object' and manual exploration of a 'surface'. The classification of parietal neurons which combine different visual stimulus preferences with a motor role (Sakata et al., 1997) suggests the kind of elements which may participate in such networks. Our data illustrate the progressive differentiation and refinement of these processes between 4 and 10 months of age.

Summary and conclusions: dorsal and ventral systems

We have presented evidence on the development of two aspects of the 'visual object'. We considered first the processes of segmentation, boundary assignment, and detection of coherence that allow a region of the visual field to be a candidate object. Second, we considered the developing sensitivity to the properties that define the affordance of object-oriented actions, specifically reaching and grasping. In the second sense, 'objectness' is a characteristic of the visuomotor modules which are part of the dorsal stream, probably in the parietal lobe (Rizzolatti et al., 1997) and connecting into systems for the planning and execution of actions. Within the dorsal-ventral division, the place of the first set of processes is less clear-cut. Segmentation, grouping and boundary assignment define something whose shape and other properties are appropriate for recognition, classification and connecting with the semantic properties of a particular object class — the processes for which the ventral stream is supposed to encode visual information.

However, it would be over-simple to consider that segmentation and coherence are specifically ventral-stream processes. Only a coherent object, defined by segmentation and grouping, is an appropriate target for a reach and grasp action (or for many other actions). It can be argued, therefore, that segmentation and grouping processes play a similar role in both streams and that the same processes may well be common to both streams. On the other hand, as we have discussed above, the two types of coherence, motion and form, activate different brain structures, have different histories in normal and atypical development, and can be associated with dorsal- and ventral-stream systems, respectively.

It is probable, then, that there are considerable interactions between the streams. On the one hand, much information that is relevant to defining and selecting the target of an action is analysed by ventral, form-based recognition processes. On the other hand, modes of information processing, that are closely associated with the dorsal systems that prepare spatial representations for action control, are also important in defining and segmenting a region for recognition as a distinct object. We have concentrated on motion-based segmentation in this article, but segmentation by binocular disparity is another, closely associated, example.

In the future, developmental studies may provide an important route into the study of these interactions. Systems which become closely coupled in the mature system may show more separation when they first become established, and the information used for early reaching, for instance, may be more restricted than that which ultimately becomes integrated into the reaching control system. The dissociation of form and motion deficits in developmental disorders may provide a further route for understanding the relationship between interacting components of visual and visuo-motor systems.

Abbreviations

BOLD	brain oxygen level dependent
F1	fundamental frequency
F2	second harmonic frequency
fMRI	functional magnetic resonance imaging
NG	non-grasp
RG	reach and grasp
V1	Visual area 1 (= primary visual cortex)
V3A	Visual area 3A
V4	visual area 4
V5/MT	Visual area 5/Middle temporal area
VEP	Visual evoked potential
VERP	visual event-related potentials

Acknowledgements

This research has been supported by grants from the UK Medical Research Council. Many members of the Visual Development Unit in Oxford and London contributed to this work; in particular we thank John Wattam-Bell, Dee Birtles, Dorothy Cowie, Shirley Anker, Kate Breckenridge, and Marko Nardini for their collaboration and support in the experiments described here. This work would not have been possible without the co-operation of volunteer families and their children; we thank them all.

References

Arterberry, M.E. and Yonas, A. (1988) Infants' sensitivity to kinetic information for 3-dimensional object shape. Percept. Psychophys., 44: 1–6.

Arterberry, M.E. and Yonas, A. (2000) Perception of three-dimensional shape specified by optic flow by 8-week-old infants. Percept. Psychophys., 62: 550–556.

Atkinson, J. (2000) The developing visual brain. Oxford University Press, Oxford.

Atkinson, J. and Braddick, O. (1992) Visual segmentation of oriented textures by infants. Behav. Brain Res., 49: 123–131.

Atkinson, J. and Braddick, O. (2003) Neurobiological models of normal and abnormal visual development. In: De Haan M. and Johnson M. (Eds.), The Cognitive Neuroscience of Development. Psychology Press, Hove, pp. 43–71.

Atkinson, J. and Braddick, O. (2005) Dorsal stream vulnerability and autistic disorders: the importance of comparative studies of form and motion coherence in typically developing children and children with developmental disorders. Cahiers de psychologie cognitive (Current Psychology of Cognition), 23: 49–58.

Atkinson, J., Braddick, O.J. and Cowie, D. (2004) Infant research on figure-ground and global coherence reveals gaps in knowledge of adult vision. Perception S33: Abstract 9b.

Atkinson, J., King, J., Braddick, O., Nokes, L., Anker, S. and Braddick, F. (1997) A specific deficit of dorsal stream function in Williams' syndrome. Neuroreport, 8: 1919–1922.

Atkinson, J., Braddick, O., Rose, F.E., Searcy, Y.M., Wattam-Bell, J. and Bellugi, U. (2006) Dorsal-stream motion processing deficits persist into adulthood in Williams Syndrome. Neuropsychologia, 44: 828–833.

Baker, C.L., Hess, R.F. and Zihl, J. (1991) Residual motion perception in a "motion-blind" patient, assessed with limited-lifetime random dot stimuli. J. Neurosci., 1(11): 454–481.

Banton, T. and Bertenthal, B.I. (1996) Infants' sensitivity to uniform motion. Vision Res., 36: 1633–1640.

Banton, T., Bertenthal, B.I. and Seaks, J. (1999) Infants' sensitivity to statistical distributions of motion direction and speed. Vision Res., 39: 3417–3430.

Born, R.T. and Tootell, R.B. (1992) Segregation of global and local motion processing in primate middle temporal visual area. Nature, 357: 497–499.

Bourne, J.A. and Rosa, M.G. (2006) Hierarchical development of the primate visual cortex, as revealed by neurofilament immunoreactivity: early maturation of the middle temporal area (MT). Cereb. Cortex, 16: 405–414.

Braddick, O.J. (1993) Orientation- and motion-selective mechanisms in infants. In: Simons K. (Ed.), Early Visual Development: Normal and Abnormal. Oxford University Press, New York.

Braddick, O., Atkinson, J. and Wattam-Bell, J. (2003) Normal and anomalous development of visual motion processing: motion coherence and 'dorsal stream vulnerability'. Neuropsychologia, 41(13): 1769–1784.

Braddick, O., Birtles, D., Wattam-Bell, J. and Atkinson, J. (2005) Motion- and orientation-specific cortical responses in infancy. Vision Res., 45: 3169–3179.

Braddick, O., Curran, W., Atkinson, J., Wattam-Bell, J. and Gunn, A. (2002a) Infants' sensitivity to global form coherence. Invest. Ophthalmol. Vis. Sci. 43: E-Abstract 3995.

Braddick, O.J., Atkinson, J. and Wattam-Bell, J. (1989) Development of visual cortical selectivity: binocularity, orientation, and direction of motion. In: von Euler C. (Ed.), Neurobiology of Early Infant Behaviour. MacMillan, London, pp. 165–172.

Braddick, O.J., Birtles, D., Mills, S., Warshafsky, J., Wattam-Bell, J. and Atkinson, J. (2006a) Brain responses to global perceptual coherence. J. Vision 6: Abstract 426.

Braddick, O.J., Birtles, D.B., Warshafsky, J., Akthar, F., Wattam-Bell, J. and Atkinson, J. (2006b) Evoked potentials specific to global visual coherence in adults and infants. Perception, 35: 3.

Braddick, O.J., O'Brien, J., Rees, G., Wattam-Bell, J., Atkinson, J. and Turner, R. (2002b) Quantitative neural responses to form coherence in human extrastriate cortex. Annual Meeting of the Society for Neuroscience (Program no. 721.9).

Braddick, O.J., O'Brien, J.M.D., Wattam-Bell, J., Atkinson, J., Hartley, T. and Turner, R. (2001) Brain areas sensitive to coherent visual motion. Perception, 30: 61–72.

Braddick, O.J., O'Brien, J.M.D., Wattam-Bell, J., Atkinson, J. and Turner, R. (2000) Form and motion coherence activate independent, but not dorsal/ventral segregated, networks in the human brain. Curr. Biol., 10: 731–734.

Britten, K.H., Shadlen, M.N., Newsome, W.T. and Movshon, J.A. (1993) Responses of neurons in macaque MT to stochastic motion signals. Visual Neurosci., 10: 1157–1169.

Burkhalter, A., Bernardo, K.L. and Charles, V. (1993) Development of local circuits in human visual cortex. J. Neurosci., 13: 1916–1931.

Clifton, R.K., Rochat, P., Litovsky, R.Y. and Perris, E.E. (1991) Object representation guides infants reaching in the dark. J. Exp. Psychol.: Hum. Percept. Perform., 17: 323–329.

Cornelissen, P., Richardson, A., Mason, A., Fowler, S. and Stein, J. (1995) Contrast sensitivity and coherent motion detection measured at photopic luminance levels in dyslexics and controls. Vision Res., 35: 1483–1494.

Culham, J.C. and Valyear, K.F. (2006) Human parietal cortex in action. Curr. Opin. Neurobiol., 16: 205–212.

Curran, W., Braddick, O.J., Atkinson, J., Wattam-Bell, J. and Andrew, R. (1999) Development of illusory-contour perception in infants. Perception, 28: 527–538.

Driver, J. and Baylis, G.C. (1996) Edge-assignment and figure-ground segmentation in short-term visual matching. Cogn. Psychol., 31: 248–306.

Ellemberg, D., Lewis, T.L., Maurer, D., Brar, S. and Brent, H.P. (2002) Better perception of global motion after monocular than after binocular deprivation. Vision Res., 42: 169–179.

Fagard, J. (2000) Linked proximal and distal changes in the reaching behavior of 5- to 12-month-old human infants grasping objects of different sizes. Infant Behav. Dev., 23: 317–329.

Fagard, J. and Jacquet, A.Y. (1996) Changes in reaching and grasping objects of different sizes between 7 and 13 months of age. Br. J. Dev. Psychol., 14: 65–78.

Flechsig, P. (1920) Anatomie des Menschlichen Gehirns und Rückenmarks. Thieme, Leipzig.

Gallant, J.L., Braun, J. and Van Essen, D.C. (1993) Selectivity for polar, hyperbolic, and Cartesian gratings in macaque visual cortex. Science, 259: 100–103.

Ghim, H.R. (1990) Evidence for perceptual organization in infants — perception of subjective contours by young infants. Infant Behav. Dev., 13: 221–248.

Gibson, J.J. (1966) The senses considered as perceptual systems. Houghton Mifflin, Boston.

Gibson, J.J. (1977) The theory of affordances. In: Shaw R. and Bransford J. (Eds.), Perceiving, Acting, and Knowing: Toward an Ecological Psychology. Lawrence Erlbaum Associates, Hillsdale, NJ, pp. 67–82.

Gunn, A., Cory, E., Atkinson, J., Braddick, O., Wattam-Bell, J., Guzzetta, A. and Cioni, G. (2002) Dorsal and ventral stream sensitivity in normal development and hemiplegia. Neuroreport, 13: 843–847.

Hansen, P.C., Stein, J.F., Orde, S.R., Winter, J.L. and Talcott, J.B. (2001) Are dyslexics' visual deficits limited to measures of dorsal stream function? NeuroReport, 12, 1527–30.

von Hofsten, C. and Rönnqvist, L. (1988) Preparation for grasping an object: A developmental study. J. Exp. Psychol.: Hum. Percept. Perform., 14: 610–621.

Johnson, S.P. and Aslin, R.N. (1998) Young infants' perception of illusory contours in dynamic displays. Perception, 27: 341–353.

Kavsek, M. and Yonas, A. (2006) The perception of moving subjective contours by 4-month-old infants. Perception, 35: 215–227.

Kogan, C.S., Bertone, A., Cornish, K., Boutet, I., Der Kaloustian, V.M., Andermann, E., Faubert, J. and Chaudhuri, A. (2004) Integrative cortical dysfunction and pervasive motion perception deficit in fragile X syndrome. Neurology, 63: 1634–1639.

Kellman, P.J. and Spelke, E.S. (1983) Perception of partly occluded objects in infancy. Cogn. Psychol., 15: 483–524.

Lewis, T.L., Ellemberg, D., Maurer, D., Wilkinson, F., Wilson, H.R., Dirks, M. and Brent, H.P. (2002) Sensitivity to global form in glass patterns after early visual deprivation in humans. Vision Res., 42: 939–948.

Mason, A.J.S., Braddick, O. and Wattam-Bell, J. (2003) Motion coherence thresholds in infants — different tasks identify at least two distinct motion systems. Vision Res., 43(10): 1149–1157.

Milner, A.D. and Goodale, M.A. (1995) The Visual Brain in Action. Oxford University Press, Oxford.

Newell, K.M., Scully, D.M., McDonald, P.V. and Baillargeon, R. (1989) Task constraints and infant grip configurations. Dev. Psychobiol., 22: 817–832.

Newman, C., Atkinson, J. and Braddick, O. (2001) The development of reaching and looking preferences in infants to objects of different sizes. Dev. Psychol., 37: 1–12.

Newsome, W.T. and Paré, E.B. (1988) A selective impairment of motion processing following lesions of the middle temporal area (MT). J. Neurosci., 8: 2201–2211.

Nothdurft, H.C. (1990) Texton segregation by associated differences in global and local luminance distribution. Proc. R. Soc. London B, 239: 295–320.

Rees, G., Friston, K. and Koch, C. (2000) A direct quantitative relationship between the functional properties of human and macaque V5. Nat. Neurosci., 3: 716–723.

Ridder, W.H., Borsting, E. and Banton, T. (2001) All developmental dyslexic subtypes display an elevated motion coherence threshold. Optom. Vision Sci., 78: 510–517.

Rizzolatti, G., Fogassi, L. and Gallese, V. (1997) Parietal cortex: from sight to action. Curr. Opin. Neurobiol., 7: 562–567.

Sakata, H., Taira, M., Kusunoki, M., Murata, A. and Tanaka, Y. (1997) The parietal association cortex in depth perception and visual control of hand action. Trends Neurosci., 20: 350–357.

Siddiqui, A. (1995) Object size as a determinant of grasping in infancy. J. Genet. Psychol., 156: 345–358.

Sireteanu, R. and Rieth, C. (1992) Texture segmentation in infants and children. Behav. Brain Res., 49: 133–139.

Spencer, J., O'Brien, J., Riggs, K., Braddick, O., Atkinson, J. and Wattam-Bell, J. (2000) Motion processing in autism: evidence for a dorsal stream deficiency. Neuroreport, 11: 2765–2767.

Walters, N.B., Egan, G.F., Kril, J.J., Kean, M., Waley, P., Jenkinson, M. and Watson, J.D.G. (2003) In vivo identification of human cortical areas using high-resolution MRI: An approach to cerebral structure–function correlation. Proc. Natl. Acad. Sci. U.S.A., 100: 2981–2986.

Wattam-Bell, J. (1985) Analysis of infant visual evoked potentials (VEPs) by a phase-sensitive statistic. Perception, 14: A33.

Wattam-Bell, J. (1992) The development of maximum displacement limits for discrimination of motion direction in infancy. Vision Res., 32: 621–630.

Wattam-Bell, J. (1994) Coherence thresholds for discrimination of motion direction in infants. Vision Res., 34: 877–883.

Wattam-Bell, J. (1996) The development of visual motion processing. In: Vital-Durand F., Braddick O. and Atkinson J. (Eds.), Infant Vision. Oxford University Press, Oxford.

Yonas, A., Arterberry, M.E. and Granrud, C. (1987) Four-month-old infants' sensitivity to binocular and kinetic information for three-dimensional-object shape. Child Dev., 58: 910–917.

Yonas, A. and Granrud, C.E. (1985) Development of visual space perception in infants. In: Mehler J. and Fox R. (Eds.), Neonate Cognition: Beyond the Blooming Buzzing Confusion. Lawrence Erlbaum Associates, Hillsdale, NJ, pp. 45–67.

C. von Hofsten & K. Rosander (Eds.)
Progress in Brain Research, Vol. 164
ISSN 0079-6123

CHAPTER 9

How face specialization emerges in the first months of life

Francesca Simion[1,*], Irene Leo[1], Chiara Turati[1], Eloisa Valenza[1] and Beatrice Dalla Barba[2]

[1]*Dipartimento di Psicologia dello Sviluppo e della Socializzazione, Università degli studi di Padova, Padova, Italy*
[2]*Dipartimento di Pediatria, Università degli studi di Padova, Padova, Italy*

Abstract: The present chapter deals with the topic of the ontogeny and development of face processing in the first months of life and is organized into two sections concerning face detection and face recognition. The first section focuses on the mechanisms underlying infants' visual preference for faces. Evidence is reviewed supporting the contention that newborns' face preferences is due to a set of non-specific constraints that stem from the general characteristics of the human visuo-perceptual system, rather than to a representational bias for faces. It is shown that infants' response to faces becomes more and more tuned to the face category over the first 3 months of life, revealing a gradual progressive specialization of the face-processing system. The second section sought to determine the properties of face recognition at birth. In particular, a series of experiments are presented to examine whether the inner facial part is processed and encoded when newborns recognize a face, and what kind of information — featural or configural — newborns' face recognition rely on. Overall, results are consistent with the existence of general constraints present at birth that tune the system to become specialized for faces later during development.

Keywords: development; perception; specialization; face; cognition; infancy

Introduction

The human face represents a unique, highly salient, and biologically significant visual stimulus, which provides critical cognitive and social information regarding identity (Valentine et al., 1998), direction of attention (Langton et al., 2000), intentions (Baron-Cohen, 1995), and emotions (Ekman, 1982). Adults are experts in processing faces and can recognize thousands of individual faces. In contrast, except with special training (e.g., Gauthier and Tarr, 1997), they do not demonstrate the same expertise with any other stimulus category.

Behavioral studies show that in adults face perception involves cognitive operations that are different from those involved in the recognition of other objects. Face recognition is more disrupted by inversion (turning the face upside down) than is object recognition (i.e., the inversion effect; Yin, 1969). Accuracy at discriminating individual face parts is higher when the entire face is presented than when the parts are presented in isolation, whereas the same holistic advantage is not found for parts of other objects (i.e., part–whole effect; Tanaka and Farah, 1993).

*Corresponding author. Tel: +39 049 8276522;
Fax: +39 049 8276511; E-mail: francesca.simion@unipd.it

DOI: 10.1016/S0079-6123(07)64009-6

Although these findings suggest that faces are special, there is less agreement as to how the term "special" should be defined. According to some authors faces are special because of the existence of specific processing mechanisms that recruit dedicated brain areas and are separated from processing mechanisms used to recognize other objects (Farah et al., 2000; Kanwisher, 2000). In contrast, other authors maintain that the mechanisms utilized for processing faces are not specialized for faces per se, but rather are used for performing fine-grained discrimination between visually similar exemplars of any stimulus category (Gauthier and Tarr, 2002). Thus, face-processing abilities would be the result of general processes devoted to the highly expert identification of within-category exemplars from any object class (e.g., Gauthier and Logothetis, 2000; Tarr and Gauthier, 2000).

In order to examine the nature of the operations that occur to process faces and whether such operations are exclusively involved in face processing, it is important to refer to specific tasks, as different tasks make different demands. Tasks like face detection, face discrimination, and face recognition are known to require different processes (Sergent, 1989). Therefore, certain tasks may require face-specific processes, whereas other tasks may be performed on the basis of processes that are not specific to faces. Face detection involves a decision as to whether a given stimulus is a face, and implies the capacity to detect that all faces share the same first order relational features with two eyes above a nose that is above a mouth. Face discrimination (or simultaneous matching) involves a comparison between two simultaneously presented faces and a decision as to their sameness or difference. Face recognition (or delayed matching) involves a judgment of previous occurrence and, thus, whether a face has been seen earlier.

In the literature on face processing, the distinction between these processes has been originally proposed by Bruce and Young (1986) in one of the most influential models of adults' face processing (Burton et al., 1990). This is a bottom-up hierarchical and branching model, composed by different stages. In the first stage, called "structural encoding," the raw, view-dependent image undergoes those processes that are necessary for later perceptual processing. The representation of the face at this stage depends on both the viewing condition (i.e., angle of profile) and facial configuration (i.e., expression, eye gaze, mouth position). The representation that is produced at the structural encoding stage is then processed further by separate systems that perceive personal identity, expressions, and speech related mouth movements. The model thus proposes that face recognition is a sequential process, which involves several stages and is independent of and parallel to the other processes designed to deal with different kinds of facial information. Human neuro-imaging and evoked potential studies support the existence of distinct aspects of face processing as mediated by distinct neural systems located in the occipito-temporal-visual extra-striate cortex (Haxby et al., 1991; Sergent et al., 1992; Allison et al., 1994; Puce et al., 1996; Hoffman and Haxby, 2000; McCarthy and Huettel, 2002).

The notion that different tasks involving faces depend on different processing mechanisms is also supported by Johnson (2005), according to whom in adults the cortical route for face recognition would be separated from a rapid, low spatial frequency (LSF) subcortical route that involves the superior collicolus, pulvinar, and amygdala and mediates face detection.

This chapter will examine the emergence of the specialized cognitive system to process faces in the first months of life. In so doing, we will discuss separately the processes involved in face detection and face recognition. As for face detection, we will examine the developmental time course of the preference newborns manifest toward faces, with the aim of investigating whether the same mechanisms underlie the phenomenon of face preference at birth and at later stages during development. As for face recognition, we will examine the nature of the perceptual information the human system uses to recognize a familiar face at birth.

Face detection

Mechanisms underlying face preference at birth

Many studies document that newborns, when presented with face and non-face patterns, spontaneously look longer at and orient more frequently toward the configuration that represents a face

(Johnson and Morton, 1991; Valenza et al., 1996; Macchi Cassia et al., 2004). Furthermore, they look longer at attractive than unattractive faces (Slater et al., 1998), appear to be sensitive to the presence of the eyes, and prefer to look at faces that engage them in eye contact (Batki et al., 2000; Farroni et al., 2002). These visual preferences have been interpreted as supporting the hypothesis that faces are special, because they are processed by an innately specified face module that would not require experience to manifest itself. Evidence in favor of this view comes also from an infant, who, at 1 day of age, sustained brain damage that resulted in a profound impairment in face recognition (Farah et al., 2000). This module has been thought to be an experience-independent mechanism dedicated to face processing and selective response of newborns to face stimuli is considered as the direct precursor of the adult cortical face-processing system. In this perspective, the possibility is admitted that the region of the cortex that in adults responds selectively to faces is active from birth. In contrast, other authors argue that the functional and neuroanatomical specialization for faces observed in adults would emerge during development as a result of fine-tuning by experience of parts of the visual system (de Schonen and Mathivet, 1989; Nelson, 2001). In this experience-expectant perspective, the cortical tissue has gained, through evolutionary pressures, the potential to become specialized for face processing. However, specialization for face processing would emerge only if the critical type of input is provided within crucial time windows (Greenough and Black, 1992; Nelson, 2001, 2003; de Schonen, 2002). This approach is contingent on a probabilistic epigenesis of cognitive development that views interactions between genes, structural brain changes, and psychological functions as bidirectional (Black et al., 1998). In particular, the partial functioning of neural pathways would shape subsequent development of neural structures and circuits that are the basis for further functional development. This process would be a progressive tuning of certain cerebral tissues, from a large range of visual information to the specific type of information the faces convey (Johnson, 2000; Johnson and de Haan, 2001; Nelson, 2001, 2003).

Within the experience-expectant perspective, it remains unclear the nature of the constraints that induce face preference at birth, and in particular if these constraints are specific or general. Besides the experience-expectant one, other hypotheses have been proposed. The first, called "sensory hypothesis," explained all infants' visual preferences on the basis of the visibility of the stimuli. According to it, face detection, as well as detection of stimuli other than faces, relies mainly on the psychophysical properties that determine the visibility of the pattern. Newborns prefer to look at faces merely because these stimuli happen to best match the sensitivity of their visual system (Kleiner, 1987).

A second view explains newborns' face preference on the basis of either an early face representation provided by evolutionary pressures (Slater and Quinn, 2001) or of the existence of an LSF face configuration detector (Johnson, 2005). It maintains that face detection is supported by a "quick and dirty" subcortical route sensitive to a surface or area with more dark areas in the upper half (Johnson, 2005). According to this view, a content-specific template is present at birth, which allows the human system to detect a face, providing the bootstrapping starting point for developing a specialization for faces.

Still another view on face preference in newborns assumes that the presence at birth of non-specific attentional or perceptual biases may be sufficient to produce the emergence of the functional and neural specialization for faces observed later in development. In particular, this view is consistent with a theory accounting for the newborns' visual preferences in terms of a number of general constraints, including both low level (i.e., contrasts and spatial frequencies content) and higher-level variables (i.e., the structural properties of a stimulus described by the phase spectrum according to the Fourier's analysis). In other words, it would not be necessary to postulate any content-specific template to explain visual preference for faces in newborns. Face preference at birth would reflect a predilection for a collection of general structural properties that attract newborns' gaze. Two are the general structural properties that would play a role in determining face preference in newborns. The first property, termed *congruency*, is defined by the presence of a congruent spatial relation between the disposition of the inner features and the shape of the outer contour, with more features being located in the widest portion of the configuration (Macchi Cassia et al., 2002; Turati et al., 2005). The second property, termed *up-down asymmetry*, is defined by the presence of more patterning in the upper part of the configuration than in the lower part (Macchi Cassia et al., 2002; Simion et al., 2002; Turati et al., 2002).

In line with this latter possibility, in our laboratory we demonstrated that newborns orient their gaze more frequently to and look longer at geometrical stimuli with a higher density of elements in the upper part. Thus, because newborns show a preference for up-down asymmetrical patterns with more elements in the upper part (Simion et al., 2002), we hypothesized that the presence of an up-down asymmetry in the distribution of the face features, with more features placed in the upper part, could be the reason why newborns prefer a face. This prediction has been supported by studies with face-like configurations and with real faces (Turati et al., 2002; Macchi Cassia et al., 2004; Simion et al, 2006). By manipulating the location of three square elements within a head-shaped contour, we demonstrated that the correct disposition of the eyes in the upper part is not necessary to induce a preference because an upright stimulus with two blobs randomly located in the upper part is always preferred over an upside down stimulus (Fig. 1).

Even more interesting is the result showing a visual preference for a nonface-like arrangement of elements located in the upper portion of the stimulus over a face-like arrangement positioned in the lower portion of the pattern (Turati et al., 2002, Experiment 3) (Fig. 2).

These findings strongly suggest that newborns' preference for faces may be ascribed to a non-specific attentional bias for patterns with a higher density of elements in the upper part (i.e., top-heavy patterns) rather than to a specific face detector. This conclusion has been supported by similar results obtained using real faces and manipulating the position of the inner features within the face (Macchi Cassia et al., 2004; Simion et al., 2006) (Fig. 3).

In these series of studies, the position of the inner features was manipulated so that a scrambled face with more elements in the upper part was created: When a veridical face was contrasted with a scrambled face with more elements in the upper part, newborns showed a preference for the scrambled face.

These findings provide clear evidence that, at birth, the human visual processing system is functionally organized to prefer the configuration that

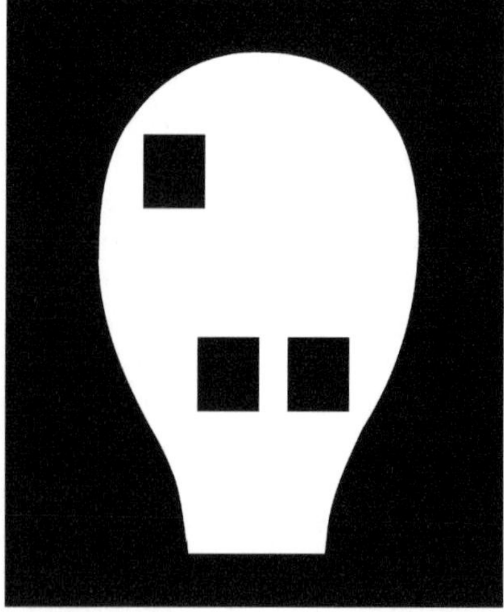

Fig. 1. The face-like configurations used in Turati et al. (2002) in which the position of the three inner elements was manipulated to create upright nonface-like configuration with more elements in the upper part and the upside-down configuration with a reversed position of the inner elements. (Adapted with permission from Turati et al., 2002.)

Fig. 2. The face-like configurations used in Turati et al. (2002) in which the nonface-like configuration with the elements placed in the upper portion of the pattern was contrasted with a configuration with the inner elements positioned in a face-like arrangement, but placed in the lower portion of the pattern. (Adapted with permission from Turati et al., 2002.)

displays a higher stimulus density in the upper part. Although this nonspecific structural property induces a visual preference in newborns even when it is embedded in non-face stimuli, it is crucial in eliciting newborns' preferential response to faces. Therefore, one can conclude that the specific preference for faces at birth is attributable to the cumulative effect of a set of non-specific constraints that stem from the general characteristics of the human visuo-perceptual system rather than by a face detector.

The importance of non-specific properties in face preference has been recently questioned by a study showing a preference for an upright schematic or naturalistic face only under positive polarity condition (i.e., black internal features against a white head shape), whereas no such preference occurred in the negative polarity condition (i.e., white internal features against a black head shape). Because in the negative and positive polarity conditions the same number of elements is present in the upper part of the faces, these results were interpreted as contradicting the explanations of face preference in terms of general biases toward configurations with more elements in the upper part, thus supporting the existence of a mechanism sensitive to the unique form of the human face under natural lighting conditions (Farroni et al., 2005).

However, the absence of significant results (i.e., null results) under the negative contrast polarity condition should be interpreted with caution because they admit alternative explanations. First, a large number of stimulus variables have been proposed as affecting newborns' preferences. In particular, at birth, the attractiveness of a pattern is affected by the amplitude spectra as well as by the phase spectra (Slater et al., 1985). The reversal of contrast polarity can be described in the spatial frequencies domain as 180° shifts in the phase angles of all spatial frequencies and this shift could interfere with newborns' preferences (Dannemiller and Stephens, 1988). Second, the phase spectra of certain patterns cannot be arbitrarily shifted without destroying the discriminability of the pattern (Kemp et al., 1996). Third, a change in polarity might affect the process of figure–ground segregation. So, segmenting white local elements is more difficult than segmenting black local elements. Experiments that either verify if the contrast polarity effect is limited to face-like patterns or if the change in polarity decreases the discriminability of stimuli other than faces are required to test the role of contrast polarity in determining

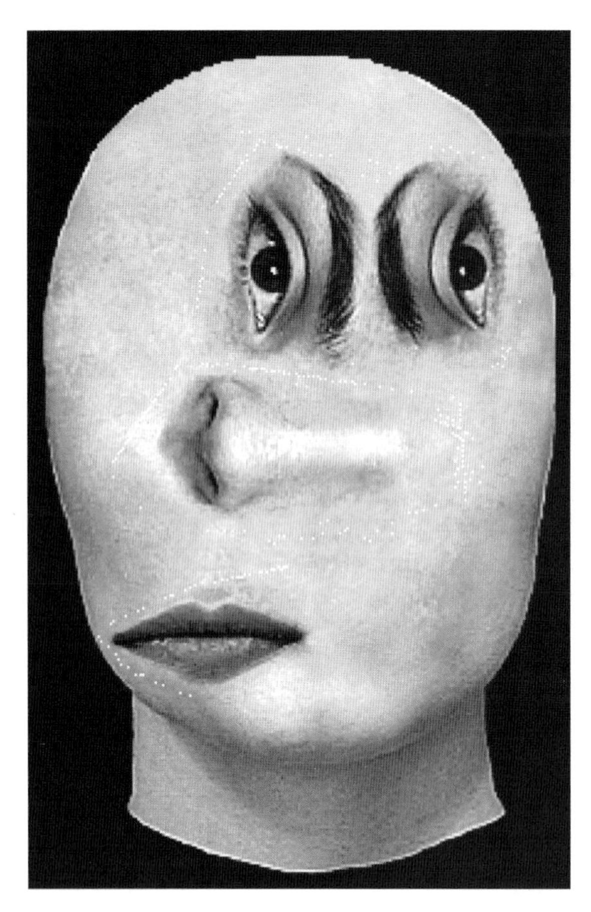

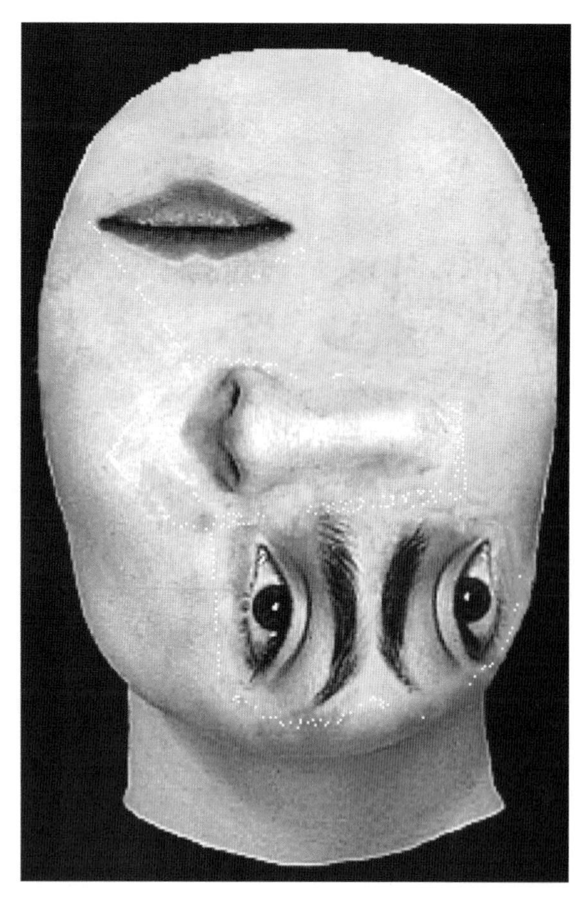

Fig. 3. In this stimulus pairs used in Macchi Cassia et al. (2004) the location of the internal facial features was manipulated to create two asymmetrical, nonface-like configurations that differed exclusively in the up-down positioning of the inner features along the horizontal plane. (Adapted with permission from Macchi Cassia et al., 2004.)

newborns' preferences. Finally, a mechanism underlying face preference, which is more face-related than previously supposed, cannot explain the data demonstrating that an upright stimulus with three blobs randomly located in the upper part is always preferred over a face-like pattern (Turati et al., 2002) and that a scrambled face with more elements in the upper part is always preferred to a real face (Macchi Cassia et al., 2004). To conclude, as for now the notion of a preference for stimuli with more elements in the upper part is well established. In the newborns' visual world, faces are the patterns most likely to display such asymmetry. As a consequence, at birth the preference for face is more parsimoniously attributable to a preference for up-down asymmetrical patterns.

How does face specialization emerge?

Even though it has been demonstrated that at birth general biases constrain the system to attend to faces, a question that is still open concerns the developmental time course of these biases and the time when the system becomes specialized to process this special category of stimuli.

Some recent neuropsychological studies that measured event-related-potential (ERP) (Halit et al., 2003) or performed positron emission tomography (PET) scans (Tzourio-Mazoyer et al., 2002) suggested that, by 2–3 months of age, there are the first signs of cortical specialization for faces. In particular, the existence of an ERP component that differentiates human faces from monkey faces in

3-month-olds supports the notion that some specialization for the human face is already present at this age (Halit et al., 2003). Furthermore, a PET study demonstrated the activation of a distributed network of cortical areas, which largely overlapped the adult face-processing network, in 2-month-old infants when they were looking at unknown women's faces (Tzourio-Mazoyer et al., 2002). However, signs of a gradual process of specialization have never been demonstrated in 3-month-old infants using behavioral, rather than neurophysiological measures. In our laboratory, a series of behavioral studies were carried out to test whether the general structural property of up-down asymmetry that induce face preference at birth still operates at 3 months of age, when a certain degree of cortical specialization for faces begins to emerge.

In a first experiment, we investigated if face preference can be still observed in 3-month-old infants. A clear preference for upright real faces was obtained when contrasting photographs of real faces in the canonical upright orientation and the upside-down version of the same faces (Turati et al., 2005). After having demonstrated a face preference in 3-month-old infants, we investigated whether the bias toward up-down asymmetrical non-face patterns with more elements in the upper part is still present after 3 months (Turati et al., 2005). The same experiments with geometrical figures carried out with newborns (Simion et al., 2002) were replicated with 3-month-old infants. Results indicated that 3-month-olds' responses varied as a function of the type of up-down asymmetrical stimuli that were presented. When infants were shown elements organized into a T, their visual behavior appeared to be radically different from that exhibited by newborns (Simion et al., 2002), in that the visual preference for the top-heavy upright T disappeared when contrasted with an inverted T. However, when the other type of stimuli, with four elements in the upper or lower half of the configuration, was considered, a different pattern of results emerged: that is, 3-month-olds' behavior paralleled that shown by newborns with the top-heavy configuration being preferred over the bottom-heavy configuration. These results suggest that, in 3-month-old infants, the determinants of infants' preference for non-face top-heavy patterns are still active but are less powerful than in newborns. This weak preference for up-down asymmetrical patterns suggests that up-down asymmetry might still play a role in determining 3-month-olds' preference for faces (Turati et al., 2005). To test this possibility, we directly compared a natural face and a top-heavy scrambled face with more elements in the upper part (Simion et al., 2006). If older infants, as newborns, prefer the non-face stimulus with a great number of elements in the upper part, we can conclude that the same basic constraints mediate face preference at birth and later during development. In contrast, if older infants prefer the real face to the top-heavy scrambled face, we can conclude that different processes mediate face preference at different ages. Results strongly support this second hypothesis because 3-month-old infants always prefer the real face demonstrating that at this age the up-down asymmetry in the distribution of the inner features can no longer be considered as a crucial factor able to induce infants' preference for a real face (Simion et al., 2006). Thus, it appears that the bias for up-down asymmetric stimuli at birth acts as an early facilitating factor that leads to an increased specialization for faces later in development, thus disproving the hypothesis that the same general constraints mediate face preference at different stages during infancy (Simion et al., 2006).

Overall, the whole pattern of evidence discussed in the first section of this chapter is consistent with an experience-expectant perspective on face-processing development, which emphasizes the role of both general biases and exposure to certain experiences in driving the emergence of the later-developing face-processing system. To develop in its adult-like expert form, the face-processing system might not require a highly specific a priori template (i.e., a face-specific bias at birth), but rather a number of general biases might work together to provide the minimal information that would be sufficient to bootstrap the system. Combined with the prolonged exposure to human faces typical of the infant's environment, these general biases drive the system toward the emergence of the functional and neural specialization that can be observed later in development.

Face recognition

Recognizing, identifying, and responding appropriately to different faces is a crucial cognitive achievement of our species. This sophisticated competence refers to the ability to discriminate among different exemplars of the face category, recognizing a face as familiar. It rests on recognition memory competencies and differs from that of face detection, which refers to the capacity to perceptually discriminate between a face and a non-face visual object. The complexity of face processing in adults has directed some investigators to studying face recognition in preschool-aged children (e.g., Carey and Diamond, 1977; Diamond and Carey, 1986; Sangrigoli and de Schonen, 2004) and infants (e.g., Pascalis et al., 1998; Blass and Camp, 2004) to elucidate its rules in neurologically and cognitively less complex epochs. The following section of the present chapter focuses on newborns' ability to recognize individual faces.

Literature is consistent in showing that newborns possess face-recognition skills despite their limited visual abilities and immature cortical visual areas. Three-day-old neonates are capable to discriminate their mother's face from a female unfamiliar face (Field et al., 1984; Bushnell et al., 1989; Pascalis et al., 1995; Bushnell, 2001). Such recognition ability can also be generalized to unfamiliar faces. Pascalis and de Schonen (1994) demonstrated that, after being habituated with a photograph of a stranger's face, newborns show a visual preference for a novel face even after a 2 min retention interval (Pascalis and de Schonen, 1994).

Contrary to what happens for face detection, where researchers still diverge about the specificity of the mechanisms involved in newborns' preference for faces (see the previous section of the present chapter), the different models on infants' face processing converge to suggest that newborns' face recognition is mediated by a general-purpose mechanism, which, from birth, allows infants to learn face and non-face visual stimuli (de Schonen and Mathivet, 1989; de Schonen, 2002).

Despite the apparent general agreement on newborns' face recognition capabilities and on the mechanisms that mediate such competences, still little is known about the perceptual cues that drive the system to discriminate and recognize individual faces at birth. In particular, issues concerning newborns' ability to process the internal face representation are still unresolved. In the literature on adults' face processing, a number of studies have indicated that the processing of human faces relies both on outer features and inner features (e.g., O'Donnell and Bruce, 2001), both on local and configural information (e.g., Diamond and Carey, 1986; Hay and Cox, 2000; Leder and Bruce, 2000; Freire and Lee, 2001; Maurer et al., 2002). Although these issues have been extensively investigated in school and preschool-aged children, where studies have used stimuli and procedures similar to those of adults, the same issues have received less attention in infancy and, even less, in newborns. In particular, it is still unknown (a) whether newborns are able to process and encode the inner facial part when they recognize a face; and (b) what kind of visual information (i.e., featural or configural) newborns rely on during face recognition.

Are newborns able to process the inner portion of a face?

The relative importance of the inner and outer features of the face changes with age in relation to the familiarity of the person to be recognized. As for highly familiar or famous faces, there is a transition from a greater reliance on outer features (the hairline) to a greater reliance on inner features (the eyes, nose, and mouth) (Campbell and Tuck, 1995; Campbell et al., 1995, 1999). As for unfamiliar faces, a more stable developmental trend emerges, in that at all ages recognition is easier when it is driven by outer rather than inner facial cues (Newcombe and Lie, 1995; Want et al., 2003).

Some studies seem to suggest that there is an inability of newborns to respond to the internal features of the face. Pascalis et al. (1995) reported that newborns' preference for the mother's face over a female stranger's face disappeared when only the inner features of the two faces were visible (i.e., when both women wore scarves around their heads) and the outer contour of the head and hairline were masked. Based on these findings, the authors concluded that, in order to recognize and prefer the mother, newborns used the outer features of the face. This conclusion is consistent with the observation that newborns' visual scanning patterns of schematic face configurations tend to be focalized toward the well-contrasted area corresponding to the external frame (Salapatek, 1968; Maurer, 1983). Conversely, a study by Slater et al. (2000) on infant perception of face attractiveness demonstrated that newborns fixate longer face photographs judged attractive by adults and that this preference is determined by the inner rather than outer facial features. Thus, available evidence does not depict a consistent scenario concerning whether perceptual information conveyed by the internal features is available and useful for newborns' face recognition.

A series of experiments were carried out in our laboratory to determine what information newborns process and encode when they discriminate, learn, and recognize individual faces (Turati et al., 2006). Newborns' ability to discriminate and recognize experimentally familiarized images of real faces was tested in three different conditions: when the face was fully visible, when only the inner features were visible, and when the outer features alone were visible (Turati et al., 2006, Experiment 1). Our findings demonstrate that inner and outer facial features alone convey sufficient information to allow newborns to recognize a face (Fig. 4).

Furthermore, we investigated whether newborns' face recognition is maintained when the perceptual appearance of the familiar face changed from the habituation to the test phase. The purpose was to test whether newborns, by relying only on a portion of it, are able to recognize a face to which they were previously habituated in its full-face version, and whether habituation to a portion of a face allows newborns to subsequently recognize the full face from which such a portion was extrapolated. More specifically, we hypothesized that, if the outer features are more salient than the inner features, newborns should keep their recognition ability whenever the outer features remain visible, whereas recognition should disappear whenever the outer features are removed (Turati et al., 2006, Experiment 2). Results evidenced that newborns failed to recognize the familiar face when the presence versus absence of the outer features was manipulated between the habituation and the test phase (Fig. 5).

Based on these results, infants' failure to recognize their mother's face when she wore a scarf (Pascalis et al., 1995) might be attributed to infants' difficulty in recognizing a perceptual similarity between two stimuli, which highly differ in their appearance (i.e., such as a fully visible face and a face lacking in its outer parts), rather than to their inability to process the inner facial features per se. However, the presence of a perceptual modification between the familiarized and the test face per se is not sufficient to explain our results (Turati et al., 2006, Experiment 2). Indeed, when the modification related to the presence versus absence of the inner features, with the outer features maintained invariant, as in the outer features condition, newborns' face recognition was preserved. Therefore, newborns were able to recognize a modified version of the familiar face on condition that a salient cue was kept constant. This suggests that, early in life, the outer facial features may deliver more salient information about an individual face, enjoying an advantage over the inner features in driving face recognition.

To summarize, our results showed that, few days after birth, either the inner or the outer features of a face are sufficient cues for face recognition. This leads to the conclusion that the limited resolution capacities of the visual system at birth do not prevent few-day-old infants from detecting and discriminating the information embedded in the inner portion of the face. Yet, our findings also revealed that the outer part of the face enjoys an advantage over the inner part in determining recognition. Although infants at birth can use both the external portion of the head region and the internal facial features to recognize an individual face, the outer features may be processed more readily than the inner features.

What kind of visual information (featural or configural) newborns' rely on when they recognize a face?

Although agreed-upon terminology and definitions of featural and especially configural information are

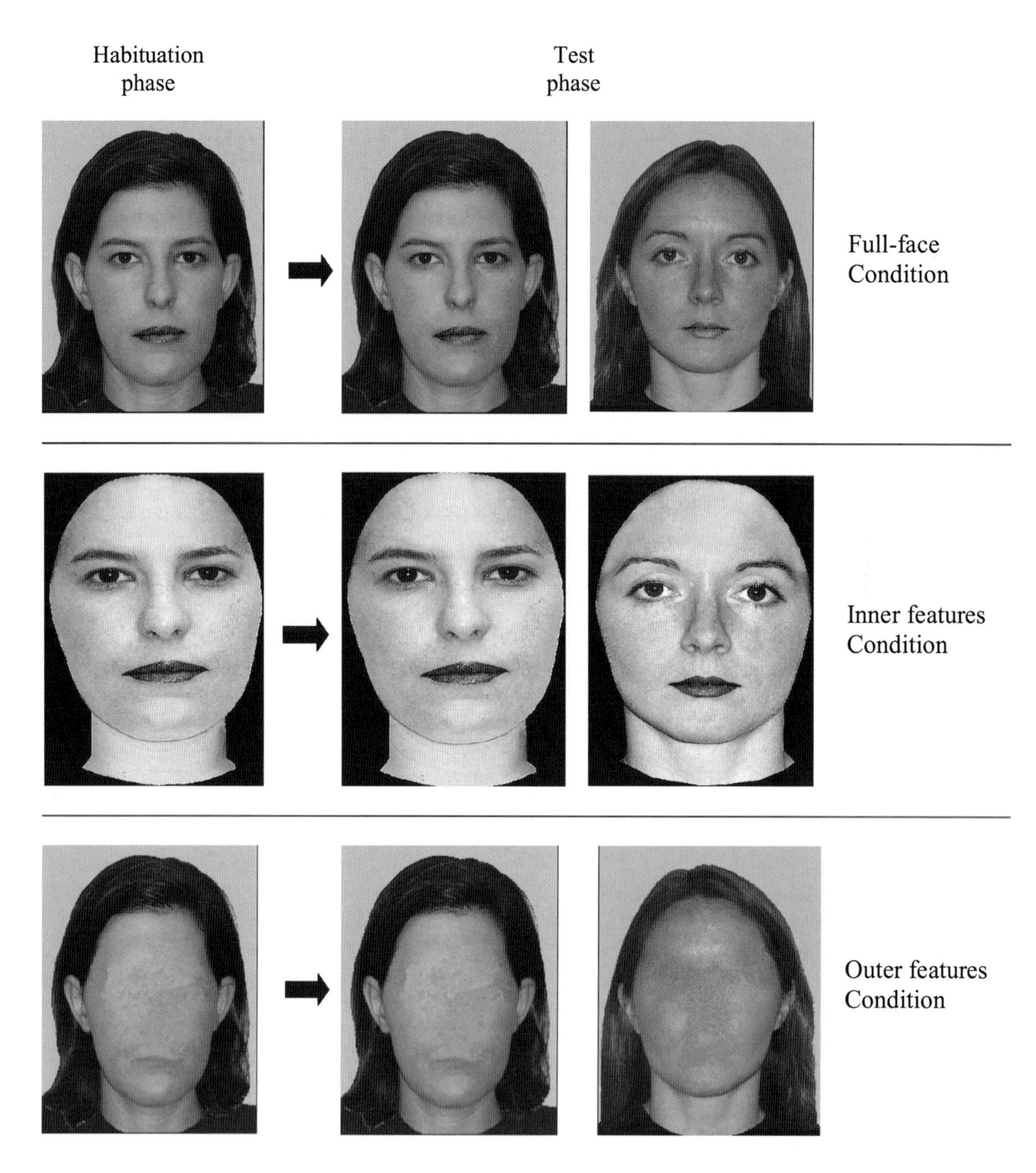

Fig. 4. The stimuli used in Turati et al. (2006) shown in the habituation and test phases in the full-face, inner features, and outer features conditions of Experiment 1. (Adapted with permission from Turati et al., 2006.)

lacking, it is often accepted that the former refers to facial features, such as the nose, mouth, and eyes, and the latter to the arrangement and spatial relation among facial features. Because all faces share the same first-order relations, recognition of individual faces requires the encoding of information about subtle variations in the shape of internal features (e.g., eyes with different shapes). Second-order

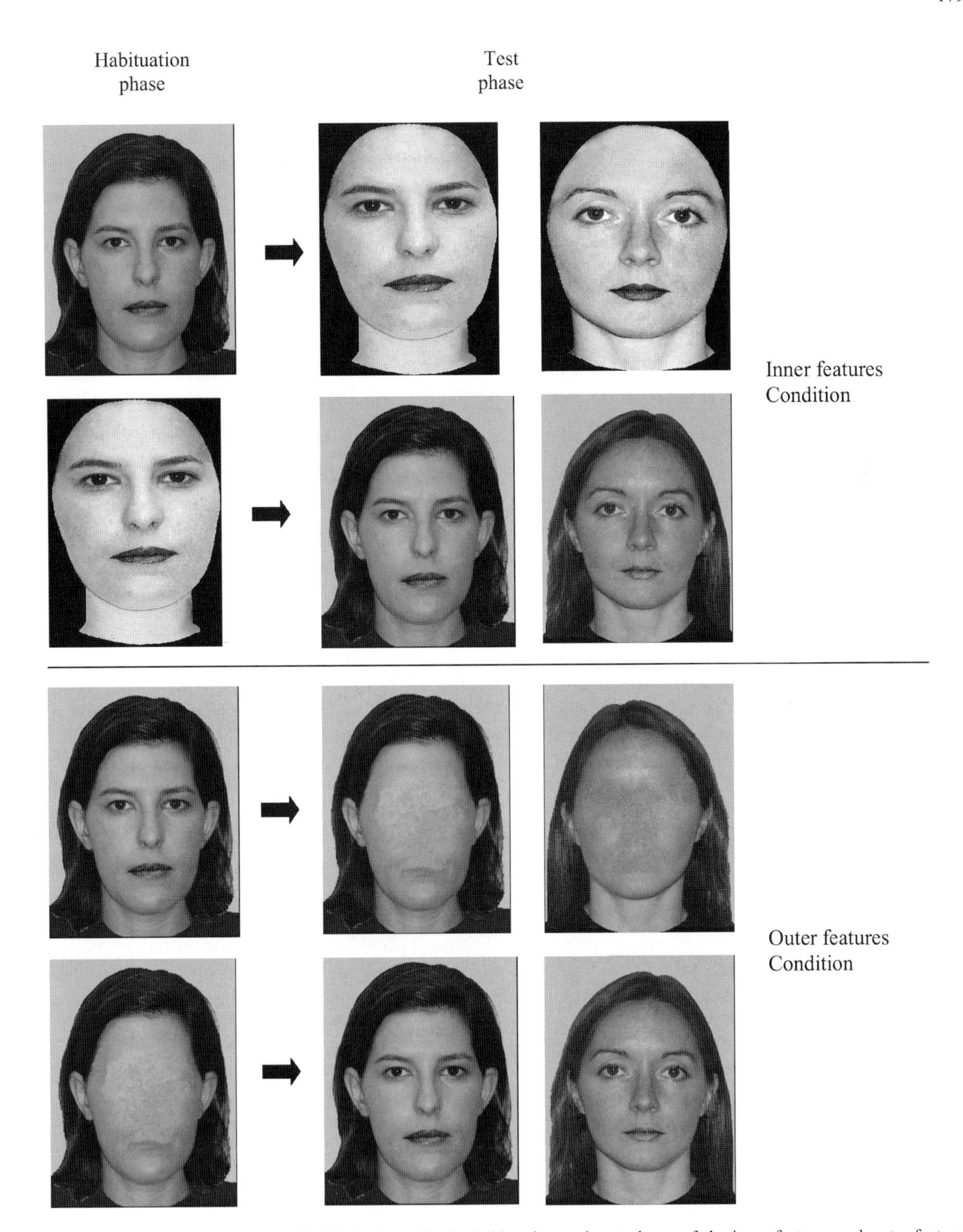

Fig. 5. The stimuli used in Turati et al. (2006) shown in the habituation and test phases of the inner features and outer features conditions of Experiment 2. (Adapted with permission from Turati et al., 2006.)

relations refer to the spatial distances among internal features (configural information; e.g., the distance between the eyes). Research revealed that adults and children are able to identify faces using both featural and configural information, but configural information acquires more and more relevance with development (e.g., Carey and Diamond, 1977; Sergent, 1984; Searcy and Bartlett, 1996; Deruelle and de Schonen, 1998; Leder and Bruce, 2000; Freire and Lee, 2001; Maurer et al., 2002). Many researchers argued that adult-like expertise in recognizing facial identity is not achieved until adolescence (e.g., Carey et al., 1980) and that children may rely less on configural processing than do adults (e.g., Carey et al., 1980).

What kind of visual information do newborns rely on when they recognize the inner portion of a face? Are newborns sensitive to the spatial relation between the local features comprised within the inner region of the face?

This issue has been addressed in our laboratory by three different lines of research. The first explored the spatial frequencies the infants' visual system uses for the identification of faces (de Heering et al., in press). The second line of research sought to determine whether newborns' ability to recognize individual faces is affected by the orientation (Turati et al., 2006, Experiment 3). The third line examined whether sensitivity to the spacing among features is present in the first days of life by testing newborns' response to the so-called "Thatcher illusion" (see below; Leo and Simion, 2007).

Based on what is known about the functional properties of the visual system at birth (i.e., Atkinson, 2000), newborns' visual sensitivity is limited to certain spatial frequency ranges. This limitation is well represented by the newborns' contrast sensitivity function (CSF), which, compared to adults' CSF, appears shifted downward and leftward, underlying the importance of LSF to process and recognize visual stimuli early in life (i.e., Banks and Salapatek, 1981).

In adults, a number of studies emphasize the relevance of spatial frequencies for face processing and have put forward the appealing possibility that the configuration of a visual face stimulus is largely subtended by coarse visual information, as provided by LSF (Morrison and Schyns, 2001; Goffaux et al., 2005; Goffaux and Rossion, 2006).

A study conducted in our laboratory was aimed at investigating what visual information newborn infants use during recognition of the inner portion of a face by systematically manipulating the spatial frequencies of the face stimuli (de Heering et al., in press). Two experiments were carried out using filtered faces and the visual habituation technique with an infant control procedure. In Experiment 1, stimuli were filtered using the cutoff point of 1 cycle per degree (cpd), because newborns' visual acuity does not exceed this limit (Atkinson et al., 1977; Banks and Ginsburg, 1985; Acerra et al., 2002). Results indicate that newborns were able to recognize the image of a real face to which they were habituated in the lowpass condition (LSF Group, <1 cpd filtering), but not in the highpass condition (high spatial frequency (HSF) Group, >1 cpd filtering). In Experiment 2, a 0.5 cpd cutoff was chosen so that both the lowpass (LSF Group, <0.5 cpd filtering) and the highpass (HSF Group, between 0.5 and 1 cpd) versions of the stimuli contained half of the frequencies comprised in the range to which the newborns' visual system is sensitive. Newborns were able to recognize the familiar face only in the lowpass condition (LSF Group).

These findings provide the first direct demonstration that few-day-old infants effectively rely on LSF rather than HSF bands to recognize a face, supporting the idea of an LSF advantage in individual face recognition at birth. Newborns seem to treat in a privileged manner those coarse perceptual cues that mainly convey configural information. This claim is further supported by the second line of research pursued in our laboratory, which tested whether newborns' ability to recognize individual faces is affected by orientation (Turati et al., 2006). It is well known that, in adults, memory and recognition for faces is disproportionately affected by stimulus inversion, as compared with that for other classes of familiar and complex mono-oriented objects, such as houses, airplanes, etc. (e.g., Yin, 1969). Converging behavioral data suggest that this "face inversion effect" results from the disruption of the configural information embedded in faces. When faces are inverted, adults continue to be able to use featural information nearly as well, but are impaired in their ability to use second-order relations (e.g., Bartlett and Searcy, 1993; Leder and Bruce, 2000). An analogue to the "inversion effect" has been documented in 4-month-old infants, who have been shown to process faces in a recognition task differently according to whether the stimuli are presented upright or inverted (Turati et al., 2004). Moreover, it has been shown that stimulus inversion disrupts newborns' preference for attractive over unattractive face ("attractiveness effect;" Slater et al., 2000).

In this vein, we examined whether newborns' ability to recognize a full face or a face in which only the outer or the inner features were visible is preserved when the face is presented inverted, that is rotated 180°. Results showed that the inversion of the face stimuli differentially affected newborns' recognition performance in the three examined conditions. Newborns were still able to recognize a face as familiar only on condition that the external features were present (i.e., full face and external features only). In contrast, recognition was disrupted when only the inner portion of the inverted face was shown.

When presented solely with the inner portion of an inverted face, newborns were unable to recognize the familiarized face, as inferred by the lack of preference for the novel face. Overall, the lack of recognition for upside-down faces when only the inner features were shown suggests that early in life faces are processed on the basis of the specific spatial relation among the features that make each face unique (i.e., the distance between the eyes, nose, and mouth).

The third line of research on visual information newborns rely on when they recognize a face attempted to show in a more direct manner that sensitivity to the spacing among features is present at birth (Leo and Simion, 2007). In particular, we examined newborns' ability to recognize small configural changes within realistic faces by testing whether newborns are sensitive to the "Thatcher illusion." This illusion is created by rotating 180° the eyes and mouth within an image of a smiling face. The resulting image is defined by adults as a grotesque expression when is viewed upright. In contrast, when the entire image is rotated 180°, this bizarre expression disappears. It has been argued that the "Thatcher illusion" is a result of the interfering effects of inversion on the processing of configural information (e.g., Bartlett and Searcy, 1993). Our purpose was to investigate whether newborns were able to discriminate between an unaltered realistic face and the same face with the eyes and the mouth rotated 180° (i.e., "thatcherized"): faces were presented in their canonical upright orientation or upside down. Results showed that newborns discriminated an unaltered and a "thatcherized" face when stimuli were presented in the upright orientation, but failed to discriminate the same stimuli when they were 180° rotated. As noted previously, some researchers have suggested that the only differences between an unaltered and a "thatcherized" face are changes in second order relational information (e.g., Bartlett and Searcy, 1993). Thus, evidence seems to indicate that the ability to process configural information is available at birth.

Collectively, the results of the reported studies suggest that several sources of information are utilized for individual face recognition soon after birth, in that 1- to 3-day-old infants are capable of recognizing a familiar face based on either the inner or outer features, and on either local or configural perceptual properties. Few-day-old babies find it easier to recognize familiar people from the outer parts of their faces, although recognition based solely on the inner features is also possible. Thus, early in development the outer features of the face enjoy an advantage over the inner features for individual face recognition. Also, evidence provided support to the claim that the visual system constrains newborns to attend in a privileged manner to certain properties of the information embedded in faces that are at the basis of the configual processing. However, it is likely that configural processing will gradually become predominant, with respect to local processing, later in development.

Conclusions

The evidence reviewed herein suggests that the development of face processing is an activity-dependent and experience-expectant process

(Nelson, 2003) that stems from general constraints of the human visuo-perceptual system shortly after birth. To develop in its adult-like form, the face-processing system seems not to require a highly specific a priori representational bias. Rather, at birth a number of general biases seem to work together to provide the minimal information that would be sufficient to set the system to become specialized.

As for face detection, results are clear in indicating that some general biases allow newborns to orient toward certain structural properties that faces share with other stimuli. More intriguing are the results showing that the same general biases cannot explain face preferences 3 months later. Thus, evidence reveals a modification in the determinants of 3-month-old infants' face preference, in that, contrary to what happens at birth, the mechanisms at the basis of face preference differ from those involved in visual preferences for non-face patterns. At 3 months of age, face preference appears to be the product of more specific mechanisms that respond more selectively to the perceptual characteristics that distinguish faces from other stimulus categories. As already suggested by some computational models (Acerra et al., 2002; Bednar, 2003), this change in the mechanisms underlying face preference in the first months of life might be viewed as the gradual and progressive emergence of an increasingly complex system for face processing rather than the appearance of a novel and independent mechanism.

Evidence provided on the controversial issue of what cues newborns use to recognize individual faces leads to a similar picture, in that the nature of perceptual information driving face recognition does not seem to undergo qualitative and marked changes early in life. Rather, our findings suggest that several sources of information are used for individual face recognition soon after birth, in that 1- to 3-day-old infants are capable of recognizing a familiar face based on either the inner or outer features, and on either featural or configural perceptual properties. Nevertheless, it is likely that, with development, come changes in the relative importance of one or the other source of information, one being more prominent than the other to act as an effective cue for recognition. Specifically, our data showed that early in life, the outer features deliver more salient information about an individual face than the inner features, although recognition based solely on the inner features is also possible. Similarly, we reported evidence showing that the visuo-perceptual system at birth constrains newborns to process those coarse visual cues of a face that convey configural information, although it is likely that configural processing will gradually become predominant, with respect to featural processing, only later in development.

Overall, the pattern of results is consistent with an experience-expectant perspective on face processing that emphasizes the relevance of both general biases and exposure to certain experiences to drive the system to become functionally specialized to process faces in the first months of life. Most probably by virtue of the prolonged experience with human faces the processes responsible for infants' visual processes impinge on the development of face processing, shifting from broadly tuned to a wide range of visual stimuli to increasingly tuned to human faces.

Acknowledgments

The authors are deeply indebted to Prof. C. Umiltà for his precious comments and suggestions on earlier versions of this work. We thank the nursing staff of the Pediatric Clinic of the University of Padua for their invaluable collaboration and the parents and infants who participated in these studies. We thank also Sandro Bettella for writing the software. The studies reported in the chapter were supported by grants from the Ministero dell'Università e della Ricerca Scientifica e Tecnologica (No. 2003112997_004 and No. 2005119101_003).

References

Acerra, F., Burnod, I. and de Schonen, S. (2002) Modeling aspects of face processing in early infancy. Dev. Sci., 5: 98–117.

Allison, T., Ginter, H., McCarthy, G., Nobre, A.A., Puce, A., Luby, M. and Spencer, D.D. (1994) Face recognition in human extrastriate cortex. J. Neurophysiol., 71: 821–825.

Atkinson, J. (2000) The Developing Visual Brain. Oxford University Press, Oxford, UK.

Atkinson, J., Braddick, O. and Moar, K. (1977) Development of contrast sensitivity over the first 3 months of life in the human infant. Vision Res., 17(9): 1037–1044.

Banks, M.S. and Ginsburg, A.P. (1985) Infant pattern preferences: a review and new theoretical treatment. In: Reese H.W. (Ed.), Advances in Child Development and Behavior, Vol. 19. Academic Press, New York, pp. 207–246.

Banks, M.S. and Salapatek, P. (1981) Infant pattern vision: a new approach based on the contrast sensitivity function. J. Exp. Child Psychol., 31: 1–45.

Baron-Cohen, S. (1995) Mindblindness: An Essay on Autism and Theory of Mind. MIT Press, Cambridge.

Bartlett, J.C. and Searcy, J. (1993) Inversion and configuration of faces. Cogn. Psychol., 25: 281–316.

Batki, A., Baron-Cohen, S., Wheelwright, S., Connellan, J. and Ahluwalia, J. (2000) Is there an innate gaze module? Evidence from human neonates. Infant Behav. Dev., 23(2): 223–229.

Bednar, J.A. (2003) The role of internally generated neural activity in newborn and infant face preferences. In: Pascalis O. and Slater A. (Eds.), The Development of Face Processing in Infancy and Early Childhood: Current Perspectives. Nova Science Publishers, New York, pp. 133–142.

Black, J.E., Jones, T.A., Nelson, C.A. and Greenough, W.T. (1998) Neuronal plasticity and the developing brain. In: Alessi N.E., Coyle J.T., Harrison S.I. and Eth S. (Eds.), Handbook of Child and Adolescent Psychiatry, Vol. 6. Basic Psychiatric Science and Treatment. Wiley, New York, pp. 31–53.

Blass, E.M. and Camp, C.A. (2004) The ontogeny of face identity: I. Eight- to 21-week-old infants use internal and external face features in identity. Cognition, 92(3): 305–327.

Bruce, V. and Young, A. (1986) Understanding face recognition. Br. J. Psychol., 77: 305–327.

Burton, A.M., Bruce, V. and Johnston, R.A. (1990) Understanding face recognition with an interactive activation model. Br. J. Psychol., 81(3): 361–380.

Bushnell, I.W.R. (2001) Mother's face recognition in newborn infants: learning and memory. Infant Child Dev., 10: 67–74.

Bushnell, I.W.R., Sai, F. and Mullin, J.T. (1989) Neonatal recognition of the mother's face. Br. J. Dev. Psychol., 7: 3–15.

Campbell, R., Coleman, M., Walker, J., Benson, P.J., Wallace, S., Michelotti, J. and Baron-Cohen, S. (1999) When does the inner-face advantage in familiar face recognition arise and why? Vis. Cogn., 6: 197–216.

Campbell, R. and Tuck, M. (1995) Recognition of parts of famous-face photographs by children: an experimental note. Perception, 24: 451–456.

Campbell, R., Walker, J. and Baron-Cohen, S. (1995) The use of internal and external face features in the development of familiar face identification. J. Exp. Child Psychol., 59: 196–210.

Carey, S. and Diamond, R. (1977) From piecemeal to configurational representation of faces. Science, 195: 312–314.

Carey, S., Diamond, R. and Woods, B. (1980) Development of face recognition: a maturational component? Dev. Psychol., 16(4): 257–269.

Dannemiller, J.L. and Stephens, B.R. (1988) A critical test of infant pattern preference models. Child Dev., 59(1): 210–216.

Deruelle, C. and de Schonen, S. (1998) Do the right and left hemispheres attend to the same visuo-spatial information within a face in infancy? Dev. Neuropsychol., 14: 535–554.

Diamond, R. and Carey, S. (1986) Why faces are not special: an effect of expertise. J. Exp. Psychol. Gen., 115: 107–117.

O'Donnell, C. and Bruce, V. (2001) Familiarisation with faces selectively enhances sensitivity to changes made to the eyes. Perception, 30: 755–764.

Ekman, P. (1982) Emotion in the Human Face. Cambridge University Press, Cambridge.

Farah, M.J., Rabinowitz, C., Quinn, G.E. and Liu, G.T. (2000) Early commitment of neural substrates for face recognition. Cogn. Neuropsychol., 17: 117–123.

Farroni, T., Csibra, G., Simion, F. and Johnson, H.M. (2002) Eye contact detection in humans from birth. Proc. Natl. Acad. Sci., 99(14): 9602–9605.

Farroni, T., Johnson, H.M., Menon, E., Zulian, L., Faraguna, D. and Csibra, G. (2005) Newborns' preference for face-relevant stimuli: Effects of contrast polarity. Proc. Natl. Acad. Sci., 102(47): 17245–17250.

Field, T.M., Cohen, D., Garcia, R. and Greenberg, R. (1984) Mother-stranger face discrimination by the newborn. Infant Behav. Dev., 7: 19–25.

Freire, A. and Lee, K. (2001) Face recognition in 4- to 7-year-old: processing of configural, featural, and paraphernalia information. J. Exp. Child Psychol., 80: 347–371.

Gauthier, I. and Logothetis, N.K. (2000) Is face recognition not so unique after all? Cogn. Neuropsychol., 17: 125–142.

Gauthier, I. and Tarr, M.J. (1997) Becoming a "Greeble" expert: exploring mechanisms for face recognition. Vision Res., 37(12): 1673–1682.

Gauthier, I. and Tarr, M.J. (2002) Unraveling mechanisms for expert object recognition: bridging brain activity and behavior. J. Exp. Psychol. Hum. Percept. Perform., 28(2): 431–446.

Goffaux, V., Hault, B., Michel, C., Voung, Q.C. and Rossion, B. (2005) The respective role of low and high spatial frequencies in supporting configural and featural processing of faces. Perception, 34: 77–86.

Goffaux, V. and Rossion, B. (2006) Faces are "spatial": holistic face perception is supported by low spatial frequencies. J. Exp. Psychol. Hum. Percept. Perform., 32(4): 1023–1039.

Greenough, W.T. and Black, J.E. (1992) Induction of brain structure by experience: substrates for cognitive development. In: Gunnar M. and Nelson C.A. (Eds.), Behavioral Developmental Neuroscience. Minnesota Symposia on Child Psychology, Vol. 24. Erlbaum, Hillsdale, NJ, pp. 35–52.

Halit, H., de Haan, M. and Johnson, M.H. (2003) Cortical specialization for face processing: face sensitive event-related potential components in 3- and 12-month-old infants. Neuroimage, 19: 1180–1193.

Haxby, J.V., Grady, C.L., Horwitz, B., Ungerleideir, L.G., Mishkin, M., Carson, R.E., Herscovitch, P., Schapiro, M.B. and Rapoport, S.I. (1991) Dissociations of spatial and object visual processing pathways in human extrastriate cortex. Proc. Natl. Acad. Sci., 88: 1621–1625.

Hay, D.C. and Cox, R. (2000) Developmental changes in the recognition of faces and facial features. Infant Child Dev., 9: 199–212.

de Heering, A., Turati, C., Rossion, B., Bulf, H., Goffaux and Simion, F. (in press) Newborns' face recognition is based on spatial frequencies below 0.5 cycles per degree. Cognition, doi: 10.1016/j.cognition.2006.12.012.

Hoffman, E.A. and Haxby, J.V. (2000) Distinct representation of eye gaze and identity in the distributed human neural system for face perception. Nat. Neurosci., 3: 80–84.

Johnson, M.H. (2000) Functional brain development in infants: elements of an interactive specialization framework. Child Dev., 71: 75–81.

Johnson, M.H. (2005) Subcortical face processing. Nat. Rev. Neurosci., 6: 766–774.

Johnson, M.H. and de Haan, M. (2001) Developing cortical specialization for visual-cognitive function: the case of face recognition. In: McClelland J.L. and Seigler R.S. (Eds.), Mechanisms of Cognitive Development: Behavioral and Neural Perspectives. Lawrence Erlbaum Associates, Mahwah, NJ, pp. 253–270.

Johnson, M.H. and Morton, J. (1991) Biology and Cognitive Development: The Case of Face Recognition. Basil Blackwell, Oxford, England.

Kanwisher, N. (2000) Domain specificity in face perception. Nat. Neurosci., 3: 759–763.

Kemp, R., Pike, G., White, P. and Musselman, A. (1996) Perception and recognition of normal negative faces: the role of shape from shading and pigmentation cues. Perception, 25: 37–52.

Kleiner, K.A. (1987) Amplitude and phase spectra as indices of infants' pattern preferences. Infant Behav. Dev., 10: 49–59.

Langton, S.R.H., Watt, R.J. and Bruce, V. (2000) Do the eyes have it? Cues to the direction of social attention. Trends Cogn. Sci., 4(2): 50–59.

Leder, H. and Bruce, V. (2000) When inverted faces are recognized: the role of configural information in face recognition. Q. J. Exp. Psychol. A, 53: 513–536.

Leo, I. and Simion, F. (2007) Face processing at birth: a Thatcher illusion study. Poster presented at the *SRCD Biennial Meetings*, Boston, MA, USA March 29-April 1.

Macchi Cassia, V., Turati, C. and Simion, F. (2004) Can a non specific bias toward top-heavy patterns explain newborns' face preference? Psychol. Sci., 15: 379–383.

Macchi Cassia, V., Valenza, E., Pividori, D. and Simion, F. (2002) Facedness vs. non-specific structural properties: what is crucial in determining face preference at birth. Poster presented at the *International Conference on Infant Studies*. Toronto, Ontario, Canada.

Maurer, D. (1983) The scanning of compound figures by young infants. J. Exp. Child Psychol., 35: 437–448.

Maurer, D., Le Grand, R. and Mondloch, C.J. (2002) The many faces of configural processing. Trends Cogn. Sci., 6: 255–260.

McCarthy, G. and Huettel, S. (2002) A functional brain system for face processing revealed by event-related potentials and functional MRI. Int. Congr. Ser., 1226: 3–16.

Morrison, D.J. and Schyns, P.G. (2001) Usage of spatial scales for the categorization of faces, objects, and scenes. Psychon. Bull. Rev., 8: 454–469.

Nelson, C.A. (2001) The development and neural bases of face recognition. Infant Child Dev., 10: 3–18.

Nelson, C.A. (2003) The development of face recognition reflects an experience-expectant and activity-dependent process. In: Pascalis O. and Slater A. (Eds.), The Development of Face Processing in Infancy and Early Childhood: Current Perspectives. Nova Science Publishers, New York, pp. 79–88.

Newcombe, N. and Lie, E. (1995) Overt and covert recognition of faces in children and adults. Psychol. Sci., 6: 241–245.

Pascalis, O., de Haan, M., Nelson, C.A. and de Schonen, S. (1998) Long-term recognition memory for faces assessed by visual paired comparison in 3- and 6-month-old infants. J. Exp. Psychol. Learn. Mem. Cogn., 24: 249–260.

Pascalis, O. and de Schonen, S. (1994) Recognition memory in 3- to 4-day-old human neonates. Neuroreport, 5: 1721–1724.

Pascalis, O., de Schonen, S., Morton, J., Deruelle, C. and Fabre-Grenet, M. (1995) Mother's face recognition by neonates: a replication and an extension. Infant Behav. Dev., 18: 79–85.

Puce, A., Allison, T., Asgari, M., Gore, J.C. and McCarthy, G. (1996) Differential sensitivity of human visual cortex to faces, letterstrings, and textures: a functional magnetic resonance imaging study. J. Neurosci., 16: 5205–5215.

Salapatek, P. (1968) Visual scanning of geometric figures by the human newborn. J. Comp. Physiol. Psychol., 66: 247–258.

Sangrigoli, S. and de Schonen, S. (2004) Effect of visual experience on face processing: a developmental study of inversion and non-native effects. Dev. Sci., 7: 74–87.

de Schonen, S. (2002) Epigenesis of the cognitive brain: a task for the 21st century. In: Backman L. and von Hofsten C. (Eds.), Psychology at the Turn of the Millennium, Vol. 1. Psychology Press, Hove, UK, pp. 55–88.

de Schonen, S. and Mathivet, E. (1989) First come, first served: a scenario about the development of hemispheric specialization in face recognition during infancy. Eur. Bull. Cogn. Psychol., 9: 3–44.

Searcy, J.H. and Bartlett, J.C. (1996) Inversion and processing of component and spatial-relational information in faces. J. Exp. Psychol. Hum. Percept. Perform., 22(4): 904–915.

Sergent, J. (1984) Configural processing of faces in the left and the right cerebral hemispheres. J. Exp. Psychol. Hum. Percept. Perform., 10(4): 554–572.

Sergent, J. (1989) Structural processing of faces. In: Young A.W. and Ellis H.D. (Eds.), Handbook of Research on Face Processing. North-Holland, Amsterdam, pp. 57–91.

Sergent, J., Ohta, S. and MacDonald, B. (1992) Functional neuroanatomy of face and object processing: a positron emission tomography study. Brain, 115(1): 15–36.

Simion, F., Turati, C., Valenza, E. and Leo, I. (2006) The emergence of cognitive specialization in infancy: the case of face preference. In: Johnson M. and Munakata M. (Eds.), Processes of Change in Brain Development: Attention and Performance XXI. Oxford University Press, New York, pp. 189–208.

Simion, F., Valenza, E., Macchi Cassia, V., Turati, C. and Umiltà, C. (2002) Newborns' preference for up-down asymmetrical configurations. Dev. Sci., 5: 427–434.

Slater, A., Bremner, G., Johnson, S.P., Sherwood, P., Hayes, R. and Brown, E. (2000) Newborn infants' preference for attractive faces: the role of internal and external facial features. Infancy, 1: 265–274.

Slater, A., Earle, D.C., Morison, V. and Rose, D. (1985) Pattern preferences at birth and their interaction with habituation induced novelty preferences. J. Exp. Child Psychol., 39: 37–54.

Slater, A. and Quinn, P.C. (2001) Face recognition in the newborn infant. Infant Child Dev., 10: 21–24.

Slater, A., Von der Schulenburg, C., Brown, E., Badenoch, M., Butterworth, G., Parsons, S. and Samuels, C. (1998) Newborn infants prefer attractive faces. Infant Behav. Dev., 21(2): 345–354.

Tanaka, J.W. and Farah, T. (1993) Parts and wholes in face recognition. Q. J. Exp. Psychol., 46a: 225–245.

Tarr, M.J. and Gauthier, I. (2000) FFA: a flexible fusiform area for subordinate level visual processing automatized by expertise. Nat. Neurosci., 8: 764–769.

Turati, C., Macchi Cassia, V., Simion, F. and Leo, I. (2006) Newborns' face recognition: the role of inner and outer facial features. Child Dev., 77(2): 297–311.

Turati, C., Sangrigoli, S., Ruel, J. and de Schonen, S. (2004) Evidence of the face-inversion effect in 4-month-old infants. Infancy, 6: 275–297.

Turati, C., Simion, F., Milani, I. and Umiltà, C. (2002) Newborns' preference for faces: what is crucial? Dev. Psychol., 38: 875–882.

Turati, C., Valenza, E., Leo, I. and Simion, F. (2005) Three-month-old visual preference for faces and its underlying visual processing mechanisms. J. Exp. Child Psychol., 90(3): 255–273.

Tzourio-Mazoyer, N., de Schonen, S., Crivello, F., Reutter, B., Aujard, Y. and Mazoyer, B. (2002) Neural correlates of woman face processing by 2-month-old infants. Neuroimage, 15: 454–461.

Valentine, D., Edelman, B. and Abdi, H. (1998) Computational and neural network models of face processing. J. Biol. Syst., 6(2): 1–7.

Valenza, E., Simion, F., Macchi Cassia, V. and Umiltà, C. (1996) Face preference at birth. J. Exp. Psychol. Hum. Percept. Perform., 22: 892–903.

Want, S.C., Pascalis, O., Coleman, M. and Blades, M. (2003) Recognizing people from the inner or outer parts of their faces: developmental data concerning "unfamiliar" faces. Br. J. Dev. Psychol., 21: 125–135.

Yin, R.K. (1969) Looking at upside-down faces. J. Exp. Child Psychol., 81: 141–145.

C. von Hofsten & K. Rosander (Eds.)
Progress in Brain Research, Vol. 164
ISSN 0079-6123

CHAPTER 10

The early development of visual attention and its implications for social and cognitive development

Sabine Hunnius*

Department of Pediatric and Developmental Psychology, Tilburg University, Tilburg, The Netherlands

Abstract: Looking behavior plays a crucial role in the daily life of an infant and forms the basis for cognitive and social development. The infant's visual attentional systems undergo rapid development during the first few months of life. During the last decennia, the study of visual attentional development in infants has received increasing interest. Several reliable measures to investigate the early development of attentional processes have been developed, and currently a number of new methods are giving fresh impetus to the field. Research on overt and covert as well as exogenously and endogenously controlled attention shifts is presented. The development of gaze shifts to peripheral targets, covert attention, and visual scanning behavior is treated. Whereas most attentional mechanisms in very young infants are thought to be mediated mainly by subcortical structures, cortical mechanisms become increasingly more functional throughout the first months. Different accounts of the neurophysiological underpinnings of attentional processes and their developmental changes are discussed. Finally, a number of studies investigating the implications of attentional development for early cognitive and social development are presented.

Keywords: visual attention; infant; covert attention; overt attention; orienting; disengagement; scanning; inhibition of return

Introduction

Anyone who has ever observed a few-weeks-old baby look around and examine his environment must have noticed how different his visual behavior is from that of an adult. A young infant tends to move his eyes in a slow and sluggish way, often it seems as if he does not notice interesting things in his peripheral visual field, and sometimes he appears just to gaze into space for long periods of time. The baby likes to look at patterns with high contrast, and from time to time his gaze appears to get "stuck" at an object or location. The infant may keep on staring there, although there may be other equally or more salient things to look at in his environment, for example a colorful toy or his mother's face. Were we to observe the same infant 3 or 4 months later, we would see that his looking behavior has changed dramatically. The infant now examines objects by scanning them quickly and systematically and tends to alternate intense inspections with brief looks away. His eye movements are fast, and he tracks moving objects or persons easily. He seems bright and alert, and complex, colorful, and moving stimuli attract his attention particularly easily.

*Present address: Nijmegen Institute for Cognition and Information, Radboud University, Nijmegen, The Netherlands.
Tel.: +31-24-36 12648; Fax: +31-24-36 16066;
E-mail: s.hunnius@nici.ru.nl

DOI: 10.1016/S0079-6123(07)64010-2

Vision plays a crucial role in the daily life of an infant. As young infants are unable to move around on their own or to reach for, grasp, and manipulate objects easily, they explore their environment and learn about the world mainly by looking. Looking is also one of the most important ways in which infants communicate with their caretakers. During face-to-face interaction, caretakers use their infant's looking behavior as an indicator of attention and adjust their communicative behavior accordingly (Papoušek and Papoušek, 1983). Early face-to-face interaction forms the beginning of social communication and also plays an important role in the development of social cognition.

Attentional processes in early childhood have attracted increasing interest from researchers during the last decades, and as a consequence, the early development of visual attention has become a fruitful field of study. In the early 1960s, infant looking measures were introduced to developmental psychology (see paragraph "Looking measures"), and since then, the study of infant attention has developed from a tool to investigate various aspects of early perceptual and cognitive development into a fully mature field in its own right. So, it has been stated recently that finally infant attention as a field of study has "grown up" (Colombo, 2002).

The research on the development of infant visual attention has produced a large body of well-established findings and well-reasoned models to explain them, and its progress has regularly been documented in review papers and books over the last years (see Atkinson, 2000; Johnson, S.P., 2001, Johnson, M.H., 2005a). To date, however, our knowledge of early attentional development is mainly based on experimental studies carried out in carefully controlled laboratory environments (but see Bornstein and Ludemann, 1989), and we have only very little precise information on young infants' visual and attentional behavior in more natural contexts and on how developmental changes in attentional behavior are connected to the infant's social-emotional and cognitive development.

Fortunately, in addition to attempts to understand the fundamental processes in infant attentional development, a functional view on infant attention is becoming increasingly important. Contextual effects on attentional processes are being studied, and the integration of early attentional development with findings on the development of early social-emotional processes and cognitive skills is receiving more and more attention (see Posner and Rothbart, 1980; Johnson, S.P., 2001; Colombo, 2002; Johnson, M.H., 2005b).

In addition, new research methods have given a special impetus to the field during the recent years. On the one hand, a number of new techniques have become available, particularly in the domain of measuring brain activity. On the other hand, established techniques for research with adults have been adapted and refined to make them suitable for the testing of infants.

This article has several goals. First of all, the most important methods to study infant attention and the advances made in this area will be reviewed. Then, an overview of recent findings on early attentional development will be provided and different accounts of the neurophysiological underpinnings of these developmental changes will be presented. Special attention will be paid to the recent progress made in understanding the intertwining of early attentional development and social-emotional and cognitive development.

Methods of studying the early development of attentional processes

Looking measures

Attention, vision, and human looking patterns have long fascinated researchers (see Bell, 1823; Müller, 1826), and also Wilhelm Preyer (1882), whose book *Die Seele des Kindes* (The Mind of the Child) is often considered as the beginning of infant psychology, describes his careful observations of the developmental changes in his infant son's visual behavior.

The simple observation of gaze was the earliest method of studying eye movements and visual attention, and still is very common (Hainline, 1993; Wade and Tatler, 2005). Two important paradigms in infant psychology are based on the observation of looking behavior: In the preferential looking paradigm infants are presented with two stimuli, and a reliably longer looking duration to one of them is interpreted as evidence that the infant discriminates the two displays (see Fantz, 1961; Fantz et al., 1962; also see Fig. 1). The other influential paradigm, the visual habituation procedure (and the familiarization method), is based on infants' tendency to look at novel rather than familiar stimuli and to show a decrement in looking time with continued exposure to the same stimulus. When after repeated exposure to one stimulus a novel one is presented, look duration to the unfamiliar stimulus will be longer if the infant discriminates the two stimuli (Bornstein, 1985). Both procedures have been extremely important in studying infants' developing vision (for a review see Kellman and Arterberry, 1998), and habituation-based methods have also widely been used to examine the early development of attentional and cognitive processes. However, overall looking measures are very unspecific and global, and the suitability of these methods particularly for the investigation of more complex

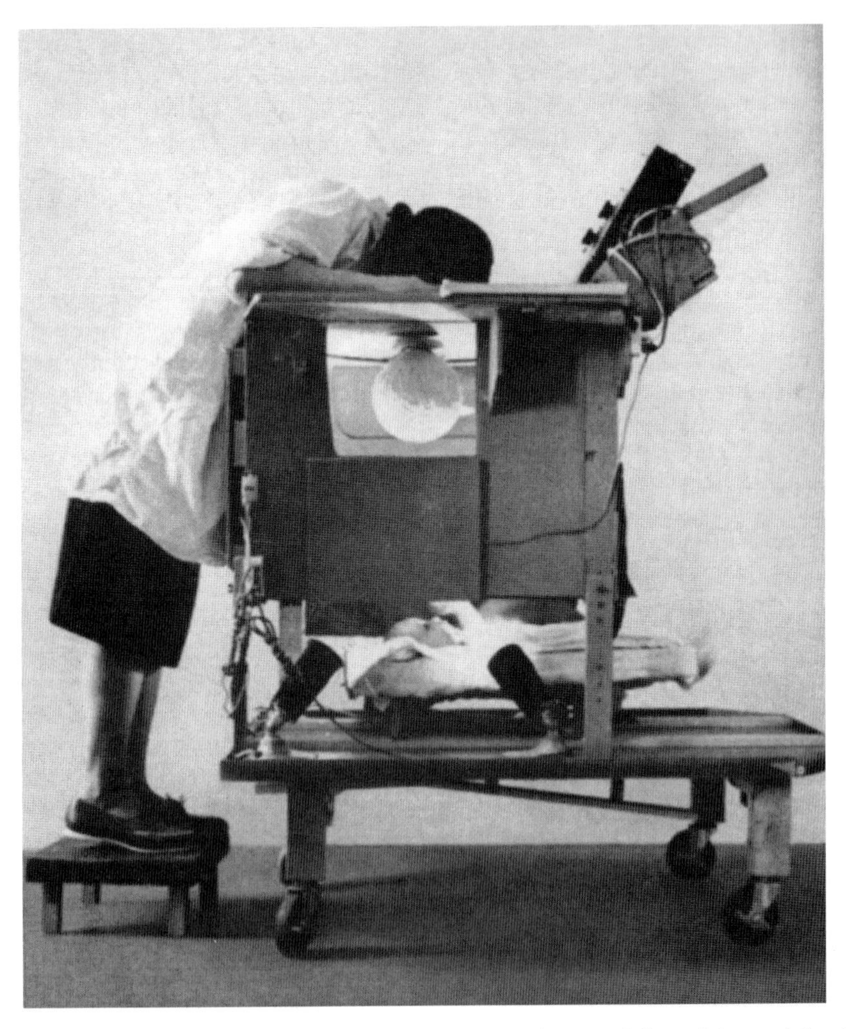

Fig. 1. Looking chamber used in Fantz' (1961) research on visual preferences in young infants. A human infant lies in a crib in the chamber, looking at objects hung from the ceiling. The observer, watching through a peephole, records the attention given to each object (copyright David Linton).

processes must be questioned (cf., Haith, 1998; Aslin and Fiser, 2005).

Measuring eye movements in infants

There are two more precise methods of measuring eye movements and fixations: electro-oculography (EOG) and corneal-reflection photography. EOG is based on measuring the change in electrical potential caused by the rotation of the eye. It was first used in the 1920s (Schott, 1922; Meyers, 1929), and is still applied frequently also in infant research (see Richards and Holley, 1999; Rosander and von Hofsten, 2004). However, the method has several limitations, especially when used with young infants (Finocchio et al., 1990; Aslin and McMurray, 2004). It can be sensitive to artifacts and it requires electrodes to be attached to the subject's face. Furthermore, EOG provides data only on the relative displacement of the eye and not on where the subject is looking.

For corneal reflection eye-tracking, an (infrared) light source is used to create a reflection off the front surface of the eyeball. The reflection is displaced when the subject moves fixation, and the information about the relative position of the corneal reflection with respect to the center of the pupil and its change are used to determine whether an eye movement took place. However, to gather information about the location of fixation, the corneal reflection eye-tracking system has to be individually calibrated before the measurement in order to map the output data onto the field the subject is looking at (Harris et al., 1981; Bronson, 1983).

The technique of infrared corneal photography was first applied with human infants in the 1960s (Salapatek and Kessen, 1966; Haith, 1969) and has been improved in many respects since then (Aslin and McMurray, 2004; Haith, 2004; Hunnius and Geuze, 2004a). It has become substantially more accurate as a consequence of increased sampling rates and custom-built calibration procedures for infants. The new eye-tracking systems also allow a less restricted testing situation with less or no need to restrain the infant's head movements and minimal demands on the infant's postural control. Since very recently, it is thus possible to examine how infants look at different stimuli in a more precise and at the same time more natural way than ever before, and this technique might provide a powerful tool for investigating perceptual and cognitive functions in infants (Hayhoe, 2004). Whereas overall looking measures offer, at best, global information about the processing of stimuli, infants' eye movement behavior can reveal more specific information about how infants encode complex visual stimuli, for example through measures of fixation locations, fixation durations, and the time of occurrence of fixation shifts or anticipatory eye movements (see Johnson et al., 2003; Falck-Ytter et al., 2006). This might help to extend our knowledge of young infants' visual behavior and the processes that underlie it, beyond what can be examined using habituation-based measures alone (Haith, 2004).

Heart rate measures

Heart rate (HR) is controlled directly by the autonomic nervous system, which is closely linked to the cerebral cortex where higher level cognitive processes, including attentional processes, are mediated. As a consequence of the connection between these systems, changes in HR measures occur in association with changes in attentional status and sensory and cognitive processing. HR measures, such as changes in cardiac cycle length or respiratory sinus arrhythmia, have therefore frequently been used to investigate attentional processes during visual tasks in adults (see Coles, 1972; Walker and Sandman, 1979) as well as in infants (for an overview, see Reynolds and Richards, in press). It has been proposed that changes in HR measures can serve as an index for attention phases during visual information processing (Graham, 1979; Porges, 1980; Richards, 1988; Richards and Hunter, 1998).

The HR-based differentiation between sustained attention, when an infant is focusing attention on an object or stimulus and is actively processing it, and attention termination, when the infant is still looking but is no longer encoding the stimulus (Richards and Casey, 1991), has made a particularly important contribution to our understanding of infants' attentional mechanisms and looking behavior during visual tasks (Finlay and Ivinskis, 1984; Richards, 1997, 2005b; Hicks and Richards, 1998; Richards and Holley, 1999; see paragraph "The development of overt shifts of attention requiring disengagement").

Marker tasks

Another way to study the development of attentional processes and to connect changes in performance to brain development is the use of marker tasks (Johnson, 2005a). This method uses behavioral tasks previously used in neurophysiological or brain imaging studies of adults or non-human primates whose neurological basis is thus relatively well established. Investigations of infants' performance on the same tasks at different ages and in various contexts provide insight into the interrelations between developmental changes in observable behavior and brain structures.

The marker task approach is frequently used to investigate neurodevelopmental models of infant visual attention (Atkinson and Braddick, 2003; Johnson, 2005a) and to determine whether behavioral observations are consistent with expectations derived from our current understanding of neurological development (see Clohessy et al., 1991; Hood, 1993; Hunnius et al., 2006b). However, the approach has been criticized as the same behavior might be mediated by different neurological structures at different stages of development (Goldman-Rakic, 1971). Comparisons between and generalizations across different groups of participants, such as infants, adults, and patients, should therefore be handled with care (Hood et al., 1998a). Further, the fact that the marker task approach focuses primarily on neurological underpinnings to explain performance without taking into account other variables such as endogenous states, has been commented on (Hainline, 1993).

Measures of brain activity

The technological progress of the last years has generated new methods for imaging brain activity in infants and has also refined the existing techniques. Many of the methods used with adults involve unpleasant procedures or the administration of radioactive substances (e.g., computed tomography or positron emission tomography (PET)), or are simply too complex and too sensitive to use with delicate subjects, who are indifferent to instructions (e.g., magnetoencephalography). However, a number of techniques can be used successfully with young children, and their application has pushed our insight into the neural underpinnings of attention and attentional development steadily further during the last decade (for an overview, see Thomas and Casey, 2003).

The classic technique of electro-encephalography (EEG; see Brazier, 1961) has been implemented in infants for many years (see Gibbs and Gibbs, 1941). In particular the use of event-related potentials (ERPs), or averaged electrophysiological responses to internal or external stimuli, has contributed greatly to our understanding of early attentional processes (see Csibra et al., 1998, 2001; Richards, 2001, 2005a).

Two relatively new methods, which are being used increasingly frequently in studies with young infants and are promising in the context of research on visual attention, are functional magnetic resonance imaging (fMRI) and near infrared spectroscopy (NIRS; see Meek et al., 1998; Dehaene-Lambertz et al., 2002; Taga et al., 2003; Csibra et al., 2004). Unlike EEG, which measures neural activity directly, these two techniques are based on the cerebral hemodynamic responses correlated with neural activity. The implementation of fMRI allows brain activity to be localized with a high degree of spatial precision. However, the technique has a limited time resolution, requires a very rigid constraint of the infant's head movements, and also exposes the infant to strong magnetic fields and to intense noise. NIRS is less sensitive to the infant's movements, but measurements might not be as precise as with fMRI. Although there are a number of practical issues still to be solved (see Aslin and Mehler, 2005), NIRS can be considered a powerful new approach to the study of the developing brain.

The new approaches offer interesting possibilities and perspectives for future research. However, obtaining evidence of where neural activity is located during a particular task is of limited scientific value on its own (and has been criticized as "neurophrenology" or "neophrenology", see Uttal, 2001; Aslin and Fiser, 2005). Only on the basis of strong theoretical and empirical foundations can brain imaging methods unfold their exciting potential (Hood, 2001).

Developmental changes in visual attention and eye movements

Visual orienting

Eye movements emerge during prenatal development (Prechtl, 1984). They have been observed around 16–18 weeks gestation, but are naturally not associated with visual stimulation before birth (although there might be some light reaching the fetus' eyes in utero). From the first days of their life, when awake and alert, infants selectively attend to different aspects of their visual environment. As newborns, for instance, they preferentially orient to and spend more time looking at a face-like pattern rather than a non-facelike stimulus (see Johnson et al., 1991a; Valenza et al., 1996) or a novel rather than a familiar stimulus (see Slater et al., 1988).

Shifts of gaze and shifts of attention are tightly associated, although they do not necessarily occur conjointly, and orienting — the aligning of attention with a source of sensory input — does not always have to be directly observable (Stelmach et al., 1997). Whereas eye movements that shift gaze from one location to another are named overt orienting, in covert orienting, foveation and visual attention do not coincide, and eyes, head, and body may remain stationary while attention shifts (Posner, 1980; Wright and Ward, 1998). In addition, shifts of attention — overt as well as covert — can be exogenously or endogenously generated. While exogenously controlled shifts of attention are automatic, as when infants orient to a salient stimulus appearing in their visual field, endogenously triggered orienting involves voluntary or strategic attention shifts to locations of interest (Jonides, 1980; Posner and Raichle, 1994). Scanning a visual display or following another person's gaze, for instance, might be primarily under endogenous control. These distinctions, which initially evolved from the literature on adult visual attention (Klein et al., 1992), have guided research on the development of visual orienting in infancy (see Fig. 2; cf., Johnson, 1994; Butcher and Kalverboer, 1997; Klein, 2005). In practice, however, the initiation of most attention shifts has both exogenous and endogenous components (Klein et al., 1992), and this of course also holds for infants' everyday attentional behavior.

Overt orienting to exogenous cues: the localization of stimuli in the peripheral visual field

Extensive research has been carried out to examine the development of reliable gaze shifting in

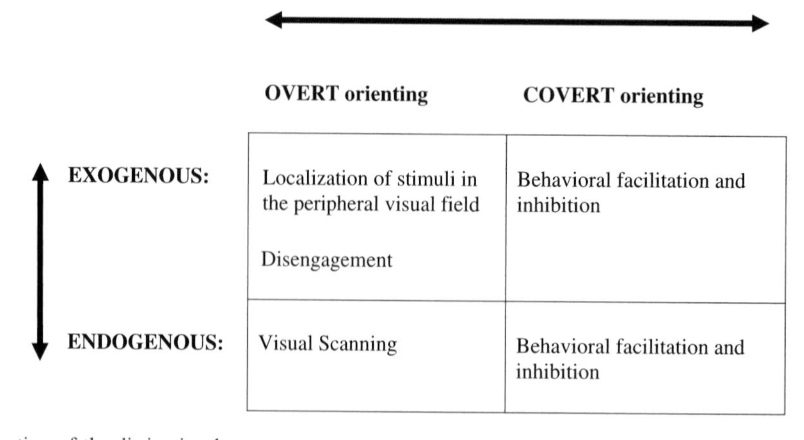

Fig. 2. Illustration of the distinction between overt versus covert orienting and exogenous versus endogenous control.

young infants and the factors that influence it (for reviews, see Mayer and Fulton, 1993; Maurer and Lewis, 1998). The sudden onset of a stimulus in the peripheral visual field triggers relatively automatic saccades, which accordingly have also been named "visual grasp reflex" (Rafal, 1998) or "attention-getting" mechanism (Cohen, 1972).

Infants as young as 1 month of age are able to localize a target, which appears in their peripheral visual field (Aslin and Salapatek, 1975; Lewis and Maurer, 1992), and this reaction has been shown to be present even at birth (Harris and MacFarlane, 1974; Lewis and Maurer, 1992). During the first months of life, overt orienting becomes steadily more efficient and infants look to a peripheral target more frequently and with shorter latencies (see Matsuzawa and Shimojo, 1997; Butcher et al., 2000). The structure of the eye movements has been described to change from a series of small movements even for targets at moderate eccentricity (Aslin and Salapatek, 1975; Salapatek et al., 1980; Richards and Hunter, 1997; Hunter and Richards, 2003; but cf., Hainline, 1993; Hainline et al., 1984) to the highly accurate, single large eye movement, possibly followed by one or two small corrective movements, characteristic of adults (Bartz, 1967; Weber and Daroff, 1971; Prablanc et al., 1978; Viviani and Swensson, 1982).

The eccentricity to which infants move their gaze to locate a target has been found to increase rapidly during the first 4 months of age (Harris and MacFarlane, 1974; Lewis and Maurer, 1992). Despite this fast early development, several studies report that an adult-like performance is not attained before the end of infancy (Dobson et al., 1998) or even the school-age years (Bowering et al., 1996; Tschopp et al., 1998).

A number of studies have demonstrated that the developmental course of the overt orienting response depends on the experimental setup and the specific characteristics of the triggering stimuli. In the usual experimental setup (see Fig. 3 for a schematic representation of the stimulus sequence in a gaze shifting task), simple physical characteristics of the peripheral target — such as its contrast, size, or luminance (Cohen, 1972; Lewis et al., 1985; Atkinson et al., 1992) — affect the probability and latency of its localization most strongly. Also when orienting is studied in a slightly more realistic laboratory situation, in which infants have to shift their gaze to stimuli surrounded by competing visual targets, stimulus salience mainly determines where infants look (Ross and Dannemiller, 1999; Dannemiller, 2002, 2005). More complex stimulus characteristics seem not to influence infants' simple gaze shifting (Hunnius and Geuze, 2004b), which is consistent with the automatic character of the orienting response.

The development of overt shifts of attention requiring disengagement

Being able to shift attention and gaze flexibly is a prerequisite for many behaviors, which play an important role in early development. Young infants explore and monitor their environment by looking from one location to another. They regulate the flow of visual input by alternating intense inspections of a stimulus with short looks away, and they control their arousal by regularly shifting their gaze away from an interaction partner (see paragraph "Visual attention and early social-emotional development").

However, when an infant is already looking at something — for example, a toy or a person's face — the actual gaze shift to a new location is preceded by the disengagement of attention and gaze. Whereas the ability to carry out a simple shift of gaze to a peripheral target has been shown to be functional around birth (see paragraph "Overt orienting to exogenous cues: the localization of stimuli in the peripheral visual field"), infants between approximately 1 and 4 months of age have been reported to have difficulty looking away from a stimulus, once their attention has been engaged. As a consequence, they may exhibit long periods of staring. This phenomenon of difficulty with disengagement has been described frequently in the infant literature and has been referred to with such diverse terms as "obligatory attention" (Stechler and Latz, 1966), "attention tropism" (Caron et al., 1977), and "sticky fixation" (Hood, 1995). It can be observed in a

Disengagement task

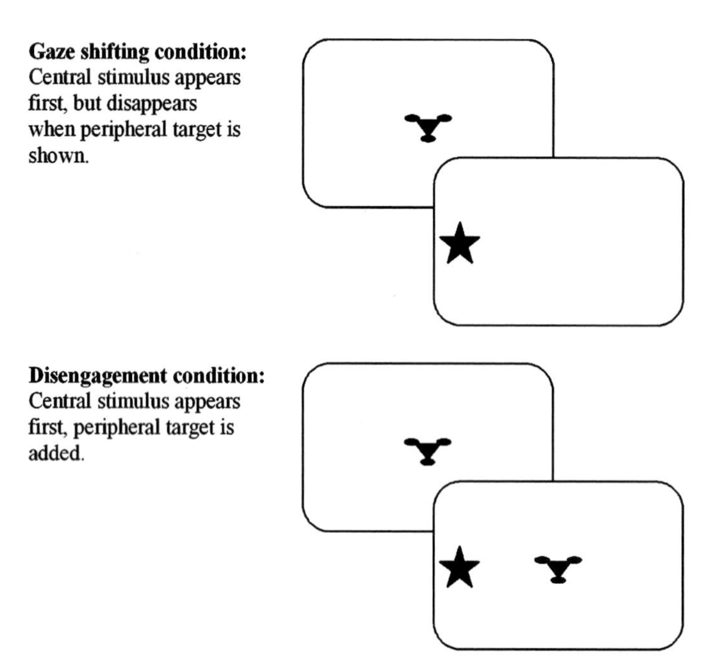

Inhibition of return task

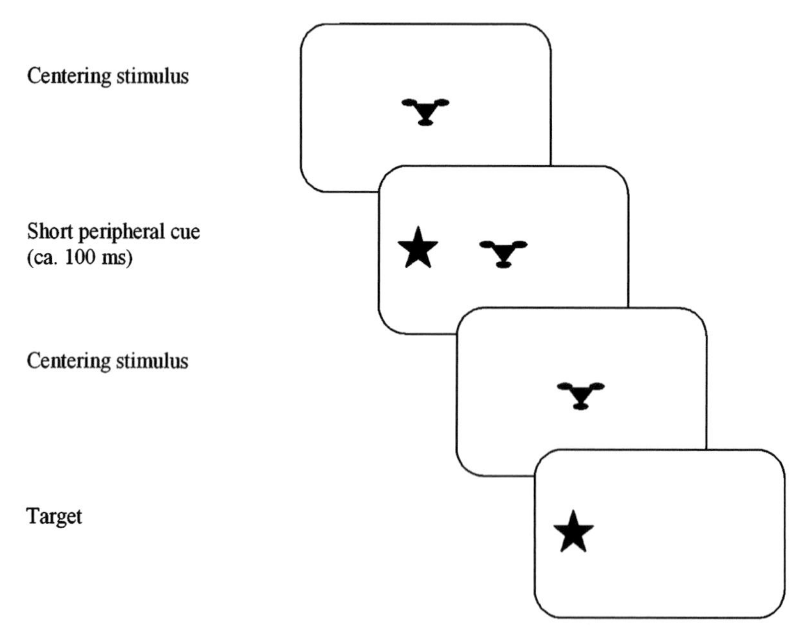

Fig. 3. Stimulus sequences in a disengagement task (simple gaze shifting versus disengagement) and in an IOR task.

laboratory context (see Fig. 3 for a schematic representation of the stimulus sequence in a disengagement task; Harris and MacFarlane, 1974; Aslin and Salapatek, 1975; but cf., Goldberg et al., 1997), in free looking situations (Stechler and Latz, 1966), or during social interaction (Hopkins and van Wulfften Palthe, 1985), and is also reported frequently by mothers and other caretakers as an everyday experience. From HR studies we know that infants are especially unlikely to shift their gaze to a target in the periphery when their attention is actively engaged by the central stimulus (Richards, 1997; Richards and Hunter, 1997; Hicks and Richards, 1998) and that they still detect and process peripheral stimuli, even when their gaze remains on the central stimulus (Finlay and Ivinskis, 1984).

It is not yet entirely clear whether and to what extent the stimulus competition effect is already present in newborns. A number of studies report that newborn infants respond to stimuli less far in the periphery when a central stimulus is present than when it is turned off (Harris and MacFarlane, 1974; MacFarlane et al., 1976), but findings on the latency of gaze shifts are mixed (Hood et al., 1996; Farroni et al., 1999).

Disengagement difficulties seem to increase between birth and 1 month of age (MacFarlane et al., 1976; Schwartz et al., 1987; Farroni et al., 1999) and are greatest in infants of 1–2 months (Johnson et al., 1991b; Hood and Atkinson, 1993; Butcher and Kalverboer, 1999). After 2 months, however, disengaging attention and shifting gaze away from a stimulus become increasingly efficient. By 4 months of age, infants are able to move their attention and gaze easily and rapidly, and staring behavior becomes rare (Hood and Atkinson, 1993; Butcher et al., 2000; Hunnius and Geuze, 2004b). From the age of approximately 6 months, infants' performance when shifting gaze between two stimuli is comparable to that of adults (Hood and Atkinson, 1993; Matsuzawa and Shimojo, 1997). Fixation shifts are faster in adults when two stimuli are presented successively than when they overlap (Saslow, 1967; Fischer and Weber, 1993), and it has been suggested that the phenomenon of obligatory attention is an extreme manifestation of this "gap effect" (Csibra et al., 1998; Hood et al., 1998a).

Infants' visual reactions in a disengagement experiment are modulated by the physical attributes of the stimuli used (Tronick, 1972; Finlay and Ivinskis, 1984; Butcher et al., 2000). Unlike the simple orienting response, the probability and latency of infants' shifts of gaze from a stimulus currently under attention to a target in the periphery are also influenced by higher order characteristics of the two stimuli (Hunnius and Geuze, 2004b). The degree of disengagement difficulty young infants might experience in daily life is therefore likely to vary between different situations.

Various explanations of disengagement difficulties in infants have been put forward. Some authors attribute staring behavior to young infants' inability to break off a fixation (Hood, 1995; Hood et al., 1996). Others argue that young infants have difficulty generating an eye movement while processing a stimulus that is currently in their central visual field (Johnson, 1990). Yet others (Rothbart et al., 1994) suggest that shifts of gaze are preceded by covert shifts of attention, and disengagement problems reflect difficulty shifting attention covertly.

A number of studies have addressed the implications of attentional development for young infants' gaze shifting in social situations. Hood et al. (1998b), for instance, showed that infants as young as 3 months of age are able to follow another person's gaze, if the experimental setup allows them to overcome their disengagement difficulties. The emergence of reliable disengagement corresponds with a change in the infant's looking behavior during natural face-to-face interaction. At this age, the infant starts to shift his gaze away more often during the interaction, either in order to regulate arousal (Stifter and Moyer, 1991) or to explore other locations that are becoming increasingly more interesting to them (Kaye and Fogel, 1980).

Covert orienting: facilitation and inhibition

As mentioned earlier, not all attention shifts occur overtly and together with an eye movement. The focus of attention can also be moved covertly while gaze is maintained at one location. Accordingly, the focus of attention has been compared to a beam of light that can be shifted between different locations independently from movements of the eyes (Posner et al., 1980; Eriksen and St. James, 1986). During fixation of a stimulus, rapid shifts of covert attention take place (Saarinen and Julesz, 1991) in order to select the next location to look at (Posner and Driver, 1992). Recently, it has been suggested that covert attention can also be directed to more than one location at a time, which supports a more complex model of covert attention than the metaphor of a unitary and indivisible attentional spotlight implies (see Hahn and Kramer, 1998; Awh and Pashler, 2000; McMains and Somers, 2004).

Covert shifts of attention to and away from a spatial location have been shown to affect subsequent responses to a target at this location (Posner, 1978). For approximately 300 ms, the perceptual processing of stimuli at the previously attended location is enhanced (Posner, 1980; Posner et al., 1985). If the interval between cue and target stimulus is longer than 300 ms, the response facilitation effect is reversed and the processing of targets in the vicinity of the cue is impaired (Posner and Cohen, 1984; Klein, 2005). This effect, which lasts for 2–3 s after the triggering event (Samuel and Kat, 2003), has been named inhibition of return (IOR; Posner et al., 1985).

In infants, the development of covert attention shifting has been studied by examining the emergence of response facilitation and IOR for cued targets. For this purpose, the spatial cuing paradigm initially developed by Posner (1980; Posner and Cohen, 1984) has been adapted for the use with infants (Hood, 1993; see Fig. 3 for a schematic representation of the stimulus sequence in an infant IOR task): While infants are looking at a central stimulus, a short peripheral cue is presented. As young infants are very unlikely to shift their gaze away from the central stimulus they are currently fixating (cf., paragraph "The development of overt shifts of attention requiring disengagement"), only a covert shift of attention can be carried out to the location of the cue. The central fixation stimulus subsequently disappears, and, following a delay, a target is presented either at the cued location or on the opposite side of the central fixation stimulus. The occurrence of facilitation or inhibition is then interpreted as evidence for the ability to shift attention covertly.

There is broad evidence that infants show facilitation of their reactions to a cued location after the age of 4 months (Hood and Atkinson, 1991; in Hood, 1995; Johnson et al., 1994) and possibly even from approximately 3 months on (Johnson and Tucker, 1996; Richards, 2000a; Richards, 2000b in Richards, 2005b). IOR after covert attention shifts, on the other hand, has been observed consistently only in infants of 4 months and older (Hood and Atkinson, 1991; Hood, 1995; Johnson, 1994; Johnson and Tucker, 1996; Butcher et al., 1999; but see Richards, 2000b in Richards, 2005b). Richards (2000a, 2001, 2005a) combined measures of infant looking behavior with ERPs to explore the development of covert attention and the hypothesized dissociation in the development of facilitation and IOR. The changes in the location of ERP activity he observed suggested a gradual development of the ability to shift attention covertly throughout the first months of infancy. Whereas infants of 3 months seem to show facilitation mainly as a result of automatic saccadic programming, infants from the age of 4 months might show both facilitation and IOR as a result of a shift of covert attention (Richards, 2000a).

At the same time, several studies have also demonstrated inhibition in much younger infants. After overt orienting to a cue, a few-days-old infants tended to look less frequently and more slowly in the direction of the preceding saccade (Valenza et al., 1994; Simion et al., 1995). These findings suggest that one of the mechanisms involved in overt IOR is already functional in neonates and is thus ready long before the ability to shift attention covertly emerges.

Interestingly, IOR, initially identified in terms of location-based coordinates, may also be defined in terms of object-based coordinates, and may move with objects (Tipper et al., 1991, 1994; Gibson and Egeth, 1994; Yi et al., 2003). It has been suggested that looking behavior is initially determined by environment-based IOR before object-based IOR starts to emerge (Harman et al., 1994).

By 8 months, cueing of a part of an object leads to IOR to the whole object, which shows that object-centered IOR is present by this age (Johnson and Gilmore, 1998). These mechanisms of location- and object-centered IOR are thought to play an important role in infants' growing ability to explore their environment in a functional and systematic way. It has been argued that IOR is an adaptive evolutionary mechanism that hinders attention from returning to a location where it has recently been engaged (either overtly or covertly), and thus biases orienting towards novel stimuli (Posner and Cohen, 1984).

During recent years, the significance of the covert visual attentional system in a social context has received increasing interest (see Butcher and Kalverboer, 1997). It has been shown that the processing of an object at which another person is looking is enhanced in adults as well as in infants (Friesen and Kingstone, 1998; Hood et al., 1998b; Driver et al., 1999). Such endogenously triggered shifts of covert attention by a person's eye direction can be elicited from infants as young as 3 months of age as indexed by frequencies of looks, saccadic latencies, and ERPs (Hood et al., 1998b; Farroni et al., 2003; Reid et al., 2004). It has been suggested that this attentional mechanism and the early sensitivity to other person's gaze direction provide the foundation for social interaction and particularly for the establishment of joint visual attention with a caregiver.

Visual scanning

Once very young infants have oriented to a stimulus, they scan it actively. However, their scanning patterns differ from those of older infants and adults. They tend to examine only limited parts of the stimulus (Haith, 1980; Bronson, 1990), to spend long periods fixating a few single locations (Salapatek, 1968; Bronson, 1996), and to ignore other stimuli in their visual field (Salapatek, 1975; Haith, 1980; Bronson, 1996). During the first weeks of life, infants have also been shown to look at the most salient features of a stimulus pattern, such as edges or outer contours, rather than (stationary) inner parts of a stimulus ("externality effect", Salapatek, 1975; Milewski, 1976; "contour salience effect", Bronson, 1991).

From 2–3 months of age, infants start to explore a stimulus under examination more consistently and more extensively. They fixate more locations and various features, exhibit more brief fixations, and scan more rapidly over an array of stimulus figures (Salapatek and Kessen, 1966; Leahy, 1976; Bronson, 1982, 1990, 1994, 1996; Hunnius and Geuze, 2004a). Salient parts of a stimulus still attract the infants' gaze, but they have gained volitional, strategic control over their scanning behavior (Bronson, 1994). Bronson (1994) has described the developmental changes as a gradual transition from a non-flexible, infant-like way of scanning — characterized by extremely long fixations directed to single salient parts of a stimulus — to a more advanced, adult-like scanning mode with brief fixations and extensive fixation patterns.

Adults have been shown to adapt their scanning patterns to the stimuli they are exploring. Different scanning patterns have been reported for stimuli that differ in physical characteristics, such as luminance, texture, or color (von Wartburg et al., 2005), as well as in semantic aspects (Loftus and Mackworth, 1978; Henderson et al., 1999) or familiarity (Althoff and Cohen, 1999). Far less is known about the impact of stimulus characteristics on visual scanning in young infants. Recent research has provided evidence that young infants' ability to tailor their scanning behavior to the characteristics of the stimuli under examination evolves around 3 months of age (Johnson and Johnson, 2000; Hunnius and Geuze, 2004a). In their longitudinal study, Hunnius and Geuze (2004a) showed that, when exploring a complex, abstract stimulus in contrast to the well-known face of their mother, infants started to exhibit a scanning pattern with slightly longer fixation durations from the age of 14 weeks.

However, there are indications that some stimulus properties can influence the degree of maturity of scanning behavior during the first months of life. In infants of about 3 months of age, flickering stimuli have been shown to elicit less advanced scanning with more extremely long fixations around single prominent features compared to displays with continuous luminosity (Bronson,

1990), and it has been suggested that this also holds for complex moving, colorful stimuli compared to relatively simple, achromatic geometric forms (Johnson and Johnson, 2000; Hunnius and Geuze, 2004a). Before mature scanning behavior has become established, scanning patterns might be particularly susceptible to the characteristics of the stimuli.

As infants grow older, they attain increasing intentional control over their eye movements. By 4–5 months of age, they are able to examine their environment in an efficient and flexible way. They can shift gaze rapidly and reliably between and within visual stimuli and are able to direct their gaze to relevant locations. Eye movements are now generated in accordance with the strategic demands of ongoing information processing and when familiar stimuli are scanned, recursive scanning patterns can be observed (Bronson, 1982). Most of these results on the development of scanning, however, stem from studies that examined infants' scanning of rather simple, abstract stimuli. How infants visually explore more natural stimuli, such as complex visual scenes, is largely unknown to date.

Faces are frequent stimuli in the world of an infant, and young infants' perception of faces has been studied extensively (see Simion et al., 2007/this issue; Pascalis and Slater, 2003). A number of studies have examined how infants explore facial stimuli and how their exploratory behaviors change as they grow older. It has been suggested that the limitations that have been found in young infants' scanning of abstract stimuli — such as the so-called externality effect — also hold for their visual exploration of faces. When infants younger than 2 months of age scan faces, they tend to restrict their fixations to the perimeter of the face, whereas infants older than 2–3 months are more likely to also inspect the internal facial features, especially the eyes and, to a much lesser extent, the mouth (Maurer and Salapatek, 1976; Haith et al., 1977; Hainline, 1978). One has to keep in mind, though, that the studies on which these results are based have used mostly photographs or drawings of faces (e.g., Hainline, 1978; Gallay et al., 2006), and when real faces were used, they were usually still faces (e.g., Maurer and Salapatek, 1976; Bronson, 1982). Only very few studies have examined infants' face scanning in more realistic interaction situations (but see Hunnius et al., 2007; Merin et al., 2007). When presented with a face that is talking and moving naturally, even infants as young as 6 weeks of age direct their gaze at the internal features of the face (Hunnius and Geuze, 2004a). Although their scanning patterns are indeed characterized by relatively few fixations, some of them extremely long, they spend most of the time looking at the eye and mouth region rather than the hairline or the perimeter of the face (see Fig. 4; Hunnius and Geuze, 2004a). This is in line with earlier findings that if the internal elements of a pattern are moving or flickering, they are more likely to be looked at even by very young infants (Bushnell, 1979; Girton, 1979).

The studies described imply that even very young infants are able to establish eye contact with their caregivers. Infants at risk for autism, however, show reduced looking at their mother's eyes during social interaction, but increased gaze at her mouth (Merin et al., 2007). From a very early age, healthy infants appear to be able to distinguish between faces with eyes directed at them and faces with averted gaze and to prefer those with eyes directed at them (Farroni et al., 2002; Hains and Muir, 1996). Mutual gaze plays a crucial role in the contact and communication between caregiver and infant, and is thought to be an important factor in the infant's social development and for the quality of the infant–caregiver relationship (Keller and Gauda, 1987; Schölmerich et al., 1995).

Neuropsychological background of visual attention and eye movements and its development

Neurophysiological background of eye movement generation

This overview of the early development of visual attention and eye movements raises the question which neurophysiological processes underlie the functions and developmental changes described above? The following two paragraphs are dedicated to this question.

Fig. 4. Example of a scan path of a 6-week-old infant while looking at a video of her mother's naturally moving face. Dots represent the location and duration of visual fixations.

One of the currently predominant models of eye movement generation is the one developed by Peter Schiller (Schiller, 1985, 1998). His most recent model (Schiller, 1998) is based on adult primate electrophysiological and lesion data and distinguishes between two different, but partly overlapping, neural systems of eye movement control: the anterior and the posterior eye movement control system. The anterior system is responsible for saccades that are voluntary or planned, whereas the posterior system is thought to generate fast, reactive eye movements and orienting responses.

The pathways of the anterior system, originating in retinal ganglion cells that are specialized for the analysis of fine detail and color (Richards and Hunter, 1998), project through the lateral geniculate nucleus to the striate cortex, and run — mainly through the temporal lobe — to the frontal eye fields. From there, they project via the basal ganglia and the superior colliculus to the eye movement centers of the brain stem. However, these brain stem structures also receive direct input from the frontal eye fields within the anterior eye movement control system. The posterior eye movement control system, on the other hand, receives most of its input from retinal ganglion cells that are located in the peripheral retina and are specialized for the detection of sudden changes (Richards and Hunter, 1998). Its pathways project via the lateral geniculate nucleus to the striate cortex and then run, either directly or indirectly via the parietal lobe, through the basal ganglia to the superior colliculus.

Both systems, the anterior as well as the posterior eye movement systems, thus control the activity of the superior colliculus. Their excitatory or inhibitory input plays an important role in the generation of eye movements to interesting locations and at the same time in the inhibition of automatic eye movements in order to ensure the well-organized input of visual information.

Neuropsychological models of attentional development in infants

The visual and attentional behavior of an awake, alert infant is largely determined by the developmental status of the brain structures that form the visual system. Changes observed in behavior and its neural correlates can be due to maturation, but can as well occur as a response to experience (Greenough et al., 1987), and there are also several examples of neural and behavioral changes that are the result of an interaction between intrinsic factors and environmental aspects (Greenough et al., 1987; Johnson and Morton, 1991). Maturation therefore both lays the basis for experience (e.g., by influencing which visual features are salient at a particular age) and depends on experience because the neural processes, which produce structural change, are driven by input.

Anatomical (Conel, 1939–1967) and PET scan studies (Chugani, 1994) have demonstrated that, generally, subcortical brain structures are more mature at birth than cortical structures. During early infancy, the superior colliculus is one of the most mature structures involved in the generation of eye movements and is thought to play a crucial role in the generation of eye movements.

Gordon Bronson was one of the first to propose a model, which applied findings from research on adult neurological systems (e.g., Trevarthen, 1968; Schneider, 1969) to infant visual behavior (Bronson, 1974). According to Bronson's model, the early development of visual attention can be viewed as a shift from subcortical to cortical processing. Visual behavior in the newborn thus is mainly controlled by means of the phylogenetically older visual system. It is only at 2–3 months of age that the locus of control switches to the primary visual system and its predominantly cortical pathways.

Bronson's original model, based on "two visual systems" (Schneider, 1969) and a subcortical–cortical dichotomy, has been criticized as being too simplistic and incomplete (Atkinson, 1984; Johnson, 1990). Further, the early presence of certain perceptual abilities, such as pattern recognition (Slater et al., 1983) or orientation discrimination (Atkinson et al., 1988), has given rise to the notion that there is at least some degree of cortically mediated visual processing at birth. It is now known that several comparatively independent cortical streams of visual processing exist (see Van Essen, 1985) and that they undergo rapid development during infancy, as a result of the generation and pruning of synapses, myelination, and neurotransmitter development (see de Haan and Johnson, 2003).

Bronson's latest model (Bronson, 1994, 1996) is based on the existence of two pathways — the "striate" and the "poststriate" networks (Bronson, 1996) — which are similar to the posterior and anterior eye movement control system proposed by Schiller (1998, 1985) and on the assumption that the changes observed in early visual behavior can be explained by reference to the maturational state of these pathways. During the first few weeks of life, eye movements are mainly controlled by the striate networks. These areas are highly responsive to stimulus salience; accordingly, young infants' visual behavior tends to be mainly salience-guided. Once the fovea is aligned with an area of high salience, fixations are often concentrated around this area because — due to the anatomical structure of the retina — salient areas close to fixation produce higher levels of striate activity than comparable areas further away. As highly salient stimuli produce long-lasting activity, fixations tend to be long in young infants. From about 6 weeks of age, the poststriate networks with their pathways through the parietal and frontal cortex become increasingly effective. This system comprises areas that are able to encode the location and the spatial features of visual stimuli. These pathways project to the superior colliculus and to the brain stem centers, which directly generate eye movements. Older infants thus can draw on these poststriate capacities to override salience effects and move their eyes intentionally to locations of interest.

Mark Johnson (1990, 1995a, 2005; Johnson et al., 1998) has similarly proposed a model of visual and attentional development that builds on Schiller's (1985) model of eye movement control, in particular on four distinct pathways Schiller describes — three cortical and one subcortical. Johnson argues that the characteristics of visually guided behavior mirror the degree of functionality of these four pathways and that the developmental state of the primary visual cortex determines which of these pathways is functional. In correspondence with the inside-out pattern of postnatal development in the cerebral cortex (see paragraph Nowakowski, 1987; Rakic, 1988), he hypothesizes that during early infancy the deeper layers of the cortex tend to be more active than more superficial ones. In newborn infants, only the deeper layers of the primary visual cortex are functional, and visually guided behavior is therefore controlled predominantly by the subcortical pathway.

The fast orienting response to abrupt changes in the peripheral visual field is generally thought to be mediated by subcortical structures (see Johnson, 1995a; Bronson, 1996; Atkinson and Braddick, 2003), notably the superior colliculus (Wurtz and Munoz, 1995). As mentioned above (see paragraph "Overt orienting to exogenous cues: the localization of stimuli in the peripheral visual field"), this simple orienting system, which is especially sensitive to movement, is thought to be relatively mature at birth. The stepwise saccades to peripheral targets observed in young infants and their disappearance during the first few months of life (Aslin and Salapatek, 1975; Richards and Hunter, 1997; Hunter and Richards, 2003), while poorly understood, may be evidence for the ongoing maturation of this system. Also IOR following overt orienting, which has been shown to be present already in newborn infants, is thought to be mediated mainly by subcortical structures, such as the superior colliculus (Valenza et al., 1994; Hood et al., 1998b).

Reliable gaze shifting away from stimuli currently under attention requires cortical control over the superior colliculus, and young infants have been shown to have trouble looking away from a salient stimulus (see paragraph "The development of overt shifts of attention requiring disengagement"). Johnson (1990, 1995a) has attributed the onset and decline of sticky fixation to the maturation of different cortical structures: During the first month, the nigral pathway, providing inhibitory input from the deeper layers of the primary visual cortex to the superior colliculus, becomes increasingly functional. This as yet unregulated tonic inhibition has as a temporary consequence the infant's disengagement difficulties. During the third and the fourth month, pathways through the parietal and frontal areas become functional, ending the tonic inhibition that initially caused the staring behavior and allowing the more differentiated regulation of collicular activity. Tests of an infant whose superior colliculus was intact but whose right cortex had been removed for the treatment of seizures showed sticky fixation after the age of 7 months to stimuli on the left side (Braddick et al., 1992). This supports the notion that modulation of the subcortical orienting system by cortical processes is necessary to end infants' disengagement difficulties. However, it does not confirm Johnson's (1990, 1995a) hypothesis that cortical inhibition of the superior colliculus is a prerequisite for the occurrence of sticky fixation. Other researchers have stressed the role of covert attentional mechanisms in disengagement, attributing the changes in infant orienting behavior to a network of brain structures, known as the "posterior attention system" (PAS; Posner and Petersen, 1990), including the posterior parietal lobe, the pulvinar nuclei of the thalamus, and the superior colliculus (see Johnson et al., 1991b; Rothbart et al., 1994). Studies investigating associations between infants' performance on disengagement and covert attention tasks, however, have been inconclusive on this issue (Johnson et al., 1991b; Butcher, 2000).

Whereas the subcortical components of the PAS, the superior colliculus and the pulvinar, appear to be functional around birth, the posterior parietal lobe is much less mature at birth, undergoing rapid development around 3–4 months of age (Conel, 1939–1967; Chugani et al., 1987; Chugani, 1994). Around this age, facilitation and IOR after covert orienting start to emerge (see paragraph "Covert orienting: facilitation and inhibition"). This is consistent with the view that developmental changes in areas of the PAS such as the parietal cortex play an important role in the ontogeny of covert attention (Johnson et al., 1994; Richards, 2005b).

At approximately the same age, but probably slightly earlier than reliable disengagement of attention emerges (Hunnius et al., 2006a), infants' scanning has improved substantially, having changed from mainly exogenously elicited reflexive to endogenously driven, volitional visual behavior (see paragraph "Visual scanning"). Whereas the phenomenon that young infants do not attend to a stationary pattern within a larger frame or pattern (externality effect) is characteristic of looking behavior that is controlled subcortically, purposive saccades are presumably mediated by a pathway extending to frontal structures, such as the frontal eye fields, that matures at approximately 3 months of age (Johnson, 1995a). The increasingly mature state of this parvocellular pathway is thought not only to lead to well-organized scanning behavior, but also to enhance anticipatory eye movements (Haith et al., 1988; Canfield and Haith, 1991; Csibra et al., 2001) and the inhibition of saccades (Johnson, 1995b).

Functional visual attention

Visual attention and early cognitive development

In an adult's life, attentional processes play an important role in many daily activities. Reading, recognizing a familiar face in a group of people, or walking through a crowded mall are only a few examples of skills which depend on fast, accurate shifts of attention and gaze. For infants, gaining control over attention and gaze shifts is crucial to be able to explore the environment and to learn about the surrounding world.

Several studies have tried to link early cognitive development and looking behavior. We know that attention patterns in early infancy are related to later cognitive functioning (Fagan and McGrath, 1981; Colombo, 1993) and that infants' looking duration during habituation shows continuity with cognition in later childhood (Rose et al., 1986; Bornstein, 1998). Furthermore, longer look durations are related to slower and more variable gaze shifting in a disengagement task (Frick et al., 1999), and it has been suggested that individual differences in the efficiency of disengagement of attention form the basis for the relationship between look duration and cognitive performance in infants (Colombo et al., 2001). Bronson (1991) compared scanning patterns of 12-week-old infants who were fast or slow in processing a stimulus, and showed that the infants who processed the stimulus more slowly also exhibited a less extensive scanning style characterized by frequent prolonged fixations (but cf., Krinsky-McHale, 1993). Differences in scanning patterns are also related to other perceptual and cognitive competences. In a recent study, Johnson et al. (2004) examined infants' ability to perceive object unity in a display of a moving rod partly occluded by a box and their scanning behavior. They showed that the scanning patterns of infants who appeared to perceive object unity were characterized by more fixations of longer duration on the relevant parts and the movement of the figure. Further research combining attentional and cognitive measures is needed and promising, as it might increase our understanding of how basic attentional and perceptual processes are connected to more complex cognitive skills.

Visual attention and early social-emotional development

The ability to shift attention and gaze swiftly and reliably plays a role not only in early cognitive development, but is also closely connected to the development of self-regulatory competences (Posner and Rothbart, 1981; Rueda et al., 2004; Rothbart et al., 2006). Looking away from a stimulus, which is annoying or too intense, is a way of regulating sensation, and it is therefore not surprising that sticky fixation has often been described as causing infants distress (Stechler and Latz, 1966; Tennes et al., 1972). Further, it is a daily observation of many mothers that they can relieve their fussy infant's distress by attracting his attention to something new or interesting, such as a toy, and the effectiveness of distraction as a soothing technique has also been demonstrated in a laboratory context (Harman et al., 1997).

A number of studies have addressed the associations between the development of attention and the development of emotion regulation. Infants' ease of disengagement in an experimental task was shown to be related to their level of distress when faced with limitations and to their soothability (Johnson et al., 1991b; McConnell and Bryson, 2005; but cf., Ruddy, 1993). Stifter and Braungart (1995) found that, orienting away (to the mother or an object) during a stressful situation was associated with a decrease in negative affect in 5-month-old infants. The ability to redirect attention away from distressing stimuli has similarly been shown to be related to lower levels of negative affect in infants of 13.5 months of age (Rothbart et al., 1992).

As mentioned before, early face-to-face interaction also calls on the infants' regulation skills. Infants shift their gaze regularly in order to regulate visual input and to avoid an unpleasant or too intense interaction (Cohn and Tronick, 1983; Stifter and Moyer, 1991; Hunnius et al., 2007). The still-face procedure (Tronick et al., 1978) has been used to assess parent–infant interaction, coping, and the regulation of arousal in a situation of face-to-face interaction. Abelkop and Frick (2003) have investigated infants' looking behavior during a still-face situation as well as their performance on attention tasks, and found that attentional measures showed moderate stability within cognitive and social contexts. Several studies thus have provided support to the notion that attentional processes affect infants' regulatory skills and social behavior, and future research should continue to explore the interrelations of attentional and emotional development in infancy.

Abbreviations

EEG	electro-encephalography
EOG	electro-oculography
ERP	event-related potential
fMRI	functional magnetic resonance imaging
HR	heart rate
IOR	inhibition of return
ms	millisecond(s)
NIRS	near infrared spectroscopy
PAS	posterior attention system
PET	positron emission tomography

Acknowledgment

I would like to thank Phillipa R. Butcher and Reint H. Geuze for their comment on an earlier version of this article.

References

Abelkop, B.S. and Frick, J.E. (2003) Cross-task stability in infant attention: new perspectives using the still-face procedure. Infancy, 4: 567–588.

Althoff, R.R. and Cohen, N.J. (1999) Eye-movement-based memory effect: a reprocessing effect in face perception. J. Exp. Psychol. Learn. Mem. Cogn., 25: 997–1010.

Aslin, R.N. and Fiser, J. (2005) Methodological challenges for understanding cognitive development in infants. Trends Cogn. Sci., 9: 92–98.

Aslin, R.N. and McMurray, B. (2004) Automated corneal-reflection eye-tracking in infancy: methodological developments and applications to cognition. Infancy, 6: 155–163.

Aslin, R.N. and Mehler, J. (2005) Near-infrared spectroscopy for functional studies of brain activity in human infants: promise, prospects, and challenges. J. Biomed. Opt., 10: 011009.

Aslin, R.N. and Salapatek, P. (1975) Saccadic localization of visual targets by the very young human infant. Percept. Psychophys., 17: 293–302.

Atkinson, J. (1984) Human visual development over the first six months of life. Hum. Neurobiol., 3: 61–74.

Atkinson, J. (2000) The Developing Visual Brain. Oxford Psychology Series 32. Oxford University Press, Oxford, UK.

Atkinson, J. and Braddick, O. (2003) Neurobiological models of normal and abnormal visual development. In: de Haan M. and Johnson M.H. (Eds.), The Cognitive Neuroscience of Development. Psychology Press, Hove, UK, pp. 43–71.

Atkinson, J., Hood, B.M., Wattam-Bell, J., Anker, S. and Tricklebank, J. (1988) Development of orientation discrimination in infancy. Perception, 17: 587–595.

Atkinson, J., Hood, B.M., Wattam-Bell, J. and Braddick, O. (1992) Changes in infants' ability to switch visual attention in the first three months of life. Perception, 21: 643–653.

Awh, E. and Pashler, H. (2000) Evidence for split attentional foci. J. Exp. Psychol. Hum. Percept. Perform., 26: 834–846.

Bartz, A.E. (1967) Fixation errors in eye movements to peripheral stimuli. J. Exp. Psychol., 75: 444–446.

Bell, C. (1823) On the motions of the eye, in illustration of the uses of the muscles and nerves of the orbit. Philos. Trans. R. Soc., 113: 166–186.

Bornstein, M.H. (1985) Habituation of attention as a measure of visual processing in human infants: summary, systematization, and synthesis. In: Gottlieb G. and Krasnegor N.A. (Eds.), Measurement of Audition and Vision in the First Year of Postnatal Life: A Methodological Overview. Ablex, Norwood, NJ, pp. 253–300.

Bornstein, M.H. (1998) Stability in mental development from early life: methods, measures, models, meanings, and myths. In: Simion F. and Butterworth G. (Eds.), The Development of Sensory, Motor and Cognitive Capacities in Early Infancy: From Perception to Cognition. Psychology Press, Hove, England, pp. 301–332.

Bornstein, M.H. and Ludemann, P.M. (1989) Habituation at home. Infant Behav. Dev., 12: 525–529.

Bowering, E.R., Maurer, D., Lewis, T.L., Brent, H.P. and Riedel, P. (1996) The visual field in childhood: normal development and the influence of deprivation. Dev. Cogn. Neurosci. Technical Report, No. 96.1.

Braddick, O., Atkinson, J., Hood, B., Harkness, W., Jackson, G. and Vargha-Khadem, F. (1992) Possible blindsight in infants lacking one cerebral hemisphere. Nature, 360: 461–463.

Brazier, M.A.B. (1961) A History of the Electrical Activity of the Brain: The First Half-Century. Macmillan, New York.

Bronson, G.W. (1974) The postnatal growth of visual capacity. Child Dev., 45: 873–890.

Bronson, G.W. (1982) The Scanning Patterns of Human Infants: Implications for Visual Learning. Monographs on Infancy No. 2. Ablex, Norwood, NJ.

Bronson, G.W. (1983) Potential sources of error when applying a corneal reflex eye-monitoring technique to infant subjects. Behav. Res. Methods Instrum., 15: 22–28.

Bronson, G.W. (1990) Changes in infants' visual scanning across the 2- to 14-week age period. J. Exp. Child Psychol., 49: 101–125.

Bronson, G.W. (1991) Infant differences in rate of visual encoding. Child Dev., 62: 44–54.

Bronson, G.W. (1994) Infants' transitions towards adult-like scanning. Child Dev., 65: 1243–1261.

Bronson, G.W. (1996) The growth of visual capacity: evidence from infant scanning patterns. In: Rovee-Collier C. and Lipsitt L.P. (Eds.), Advances in Infancy Research, Vol. 11. Ablex, Norwood, NJ, pp. 109–141.

Bushnell, I.W.R. (1979) Modification of the externality effect in young infants. J. Exp. Child Psychol., 28: 211–229.

Butcher, P.R. (2000) Longitudinal studies of visual attention in infants: The early development of disengagement and inhibition of return. Doctoral dissertation. University of Groningen, Groningen, The Netherlands.

Butcher, P.R. and Kalverboer, A.F. (1997) The early development of covert visuo-spatial attention and its impact on social looking. Early Dev. Parent., 6: 15–26.

Butcher, P.R. and Kalverboer, A.F. (1999) A phase transition in infants' ability to disengage visual attention. In: Savelsbergh G.J.P., van der Maas H.L.J. and van Geert P.L.C. (Eds.), Non-Linear Developmental processes. Royal Netherlands Academy of Arts and Sciences, Amsterdam, pp. 81–92.

Butcher, P.R., Kalverboer, A.F. and Geuze, R.H. (1999) Inhibition of return in very young infants: a longitudinal study. Infant Behav. Dev., 22: 303–319.

Butcher, P.R., Kalverboer, A.F. and Geuze, R.H. (2000) Infants' shifts of gaze from a central to a peripheral stimulus: a longitudinal study of development between 6 and 26 weeks. Infant Behav. Dev., 23: 3–21.

Canfield, R.L. and Haith, M.M. (1991) Young infants' visual expectations for symmetric and asymmetric stimulus sequences. Dev. Psychol., 27: 198–208.

Caron, A., Caron, R., Minichiello, M.D., Weiss, S.J. and Friedman, S.L. (1977) Constraints on the use of the familiarization-novelty method in the assessment of infant discrimination. Child Dev., 48: 747–762.

Chugani, H.T. (1994) Development of regional brain glucose metabolism in relation to behavior and plasticity. In: Dawson G. and Fischer K. (Eds.), Human Behavior and the Developing Brain. Guilford, New York, pp. 153–175.

Chugani, H.T., Phelps, M.E. and Mazziotta, J.C. (1987) Positron emission tomography study of human brain functional development. Ann. Neurol., 22: 487–497.

Clohessy, A.B., Posner, M.I., Rothbart, M.K. and Vecera, S.P. (1991) The development of inhibition of return in early infancy. J. Cogn. Neurosci., 3: 345–350.

Cohen, L.B. (1972) Attention-getting and attention-holding processes of infant visual preferences. Child Dev., 43: 869–879.

Cohn, J.F. and Tronick, E.Z. (1983) Three-month-old infants' reaction to simulated maternal depression. Child Dev., 54: 185–193.

Coles, M.G. (1972) Cardiac and respiratory activity during visual search. J. Exp. Psychol., 96: 371–379.

Colombo, J. (1993) Infant Cognition: Predicting Later Intellectual Functioning. Sage, Newbury Park, CA.

Colombo, J. (2002) Infant attention grows up: the emergence of a developmental cognitive neuroscience perspective. Curr. Direct. Psychol. Sci., 11: 196–199.

Colombo, J., Richman, W.A., Shaddy, D.J., Greenhoot, A.F. and Maikranz, J.M. (2001) Heart rate-defined phases of attention, look duration, and infant performance in the paired-comparison paradigm. Child Dev., 72: 1605–1616.

Conel, J.L. (1939–1967) The Postnatal Development of the Human Cerebral Cortex, Vol. 1–8. Harvard University Press, Cambridge, MA.

Csibra, G., Henty, J., Volein, A., Elwell, C., Tucker, L., Meek, J. and Johnson, M.H. (2004) Near infrared spectroscopy reveals neural activation during face perception in infants and adults. J. Pediatr. Neurol., 2: 85–89.

Csibra, G., Tucker, L.A. and Johnson, M.H. (1998) Neural correlates of saccade planning in infants: a high-density ERP study. Int. J. Psychophysiol., 29: 201–215.

Csibra, G., Tucker, L.A. and Johnson, M.H. (2001) Differential frontal cortex activation before anticipatory and reactive saccades in infants. Infancy, 2: 159–174.

Dannemiller, J.L. (2002) Relative color contrast drives competition in early exogenous orienting. Infancy, 3: 275–302.

Dannemiller, J.L. (2005) Evidence against a maximum response model of exogenous visual orienting during early infancy and support for a dimensional switching model. Dev. Sci., 8: 567–582.

Dehaene-Lambertz, G., Dehaene, S. and Hertz-Pannier, L. (2002) Functional neuroimaging of speech perception in infants. Science, 298: 2013–2015.

Dobson, V., Brown, A.M., Harvey, E.M. and Narter, D.B. (1998) Visual field extent in children 3.5–30 months of age tested with a double-arc LED perimeter. Vision Res., 38: 2743–2760.

Driver, J., Davis, G., Ricciardelli, P., Kidd, P., Maxwell, E. and Baron-Cohen, S. (1999) Gaze perception triggers reflexive visuospatial orienting. Vis. Cogn., 6: 509–540.

Eriksen, C.W. and St. James, J.D. (1986) Visual attention within and around the field of focal attention: a zoom lens model. Percept. Psychophys., 40: 225–240.

Fagan, J.F. and McGrath, S.K. (1981) Infant recognition memory and later intelligence. Intelligence, 5: 121–130.

Falck-Ytter, T., Gredebäck, G. and von Hofsten, C. (2006) Infants predict other people's action goals. Nat. Neurosci., 9: 878–879.

Fantz, R.L. (1961) The origin of form perception. Sci. Am., 204: 66–72.

Fantz, R.L., Ordy, J.M. and Udelf, M.S. (1962) Maturation and pattern vision in infants during the first six months. J. Comp. Physiol. Psychol., 55: 907–917.

Farroni, T., Csibra, G., Simion, F. and Johnson, M.H. (2002) Eye contact detection in humans from birth. Proc. Natl. Acad. Sci. U.S.A., 198: 9602–9605.

Farroni, T., Mansfield, E.M., Lai, C. and Johnson, M.H. (2003) Infants perceiving and acting on the eyes: tests of an evolutionary hypothesis. J. Exp. Child Psychol., 85: 199–212.

Farroni, T., Simion, F., Umiltà, C. and Dalla Barba, B. (1999) The gap effect in newborns. Dev. Sci., 2: 174–186.

Finlay, D. and Ivinskis, A. (1984) Cardiac and visual responses to moving stimuli presented either successively or simultaneously to the central and peripheral visual fields in 4-month-old infants. Dev. Psychol., 20: 29–36.

Finocchio, D.V., Preston, K.L. and Fuchs, A.F. (1990) Obtaining a quantitative measure of eye movements in human infants: a method of calibrating the electrooculogram. Vision Res., 30: 1119–1128.

Fischer, B. and Weber, H. (1993) Express saccades and visual attention. Behav. Brain Sci., 16: 553–610.

Frick, J.E., Colombo, J. and Saxon, T.E. (1999) Individual and developmental differences in disengagement of fixation in early infancy. Child Dev., 70: 537–548.

Friesen, C.K. and Kingstone, A. (1998) The eyes have it! Reflexive orienting is triggered by nonpredictive gaze. Psychon. Bull. Rev., 5: 490–495.

Gallay, M., Baudouin, J.-Y., Durand, K., Lemoine, C. and Lécuyer, R. (2006) Qualitative differences in the exploration of upright and upside-down faces in four-month-old infants: an eye-movement study. Child Dev., 77: 984–996.

Gibbs, F.A. and Gibbs, E.L. (1941) Atlas of Electroencephalography. Addison-Wesley, Cambridge, MA.

Gibson, B.S. and Egeth, H. (1994) Inhibition of return to object-based and environment-based locations. Percept. Psychophys., 55: 323–339.

Girton, M.R. (1979) Infants' attention to intrastimulus motion. J. Exp. Child Psychol., 28: 416–423.

Goldberg, M.C., Maurer, D. and Lewis, T.L. (1997) Influence of a central stimulus on infants' visual fields. Infant Behav. Dev., 20: 359–370.

Goldman-Rakic, P.S. (1971) Functional development of the prefrontal cortex in early life and the problem of neuronal plasticity. Exp. Neurol., 32: 366–387.

Graham, F.K. (1979) Distinguishing among orienting, defense, and startle reflexes. In: Kimmel H.D., van Olst E.H. and Orlebeke J.F. (Eds.), The Orienting Reflex in Humans. Lawrence Erlbaum, Hillsdale, NJ, pp. 137–167.

Greenough, W.T., Black, J.E. and Wallace, C.S. (1987) Experience and brain development. Child Dev., 58: 539–559.

de Haan, M. and Johnson, M.H. (2003) Mechanisms and theories of brain development. In: de Haan M. and Johnson M.H. (Eds.), The Cognitive Neuroscience of Development. Psychology Press, Hove, UK, pp. 1–18.

Hahn, S. and Kramer, A.F. (1998) Further evidence for the division of attention among non-contiguous locations. Vis. Cogn., 5: 217–256.

Hainline, L. (1978) Developmental changes in visual scanning of face and nonface patterns by infants. J. Exp. Child Psychol., 25: 90–115.

Hainline, L. (1993) Conjugate eye movements of infants. In: Simons K. (Ed.), Early Visual Development, Normal and Abnormal. Oxford University Press, New York, pp. 47–79.

Hainline, L., Turkel, J., Abramov, I., Lemerise, E. and Harris, C. (1984) Characteristics of saccades in human infants. Vision Res., 24: 1771–1780.

Hains, S.M.J. and Muir, D.W. (1996) Infant sensitivity to adult eye direction. Child Dev., 67: 1940–1951.

Haith, M.M. (1969) Infrared television recording and measurement of ocular behavior in the human infant. Am. Psychol., 24: 279–282.

Haith, M.M. (1980) Rules that Babies Look By. Lawrence Erlbaum, Hillsdale, NJ.

Haith, M.M. (1998) Who put the cog in infant cognition? Is rich interpretation too costly? Infant Behav. Dev., 21: 167–179.

Haith, M.M. (2004) Progress and standardization in eye movement work with human infants. Infancy, 6: 257–265.

Haith, M.M., Bergman, T. and Moore, M.J. (1977) Eye contact and face scanning in early infancy. Science, 198: 853–855.

Haith, M.M., Hazan, C. and Goodman, G.S. (1988) Expectation and anticipation of dynamic visual events by 3.5-month-old babies. Child Dev., 59: 467–479.

Harman, C., Posner, M.I., Rothbart, M.K. and Thomas-Thrapp, L. (1994) Development of orienting to locations and objects in human infants. Can. J. Exp. Psychol., 48: 301–318.

Harman, C., Rothbart, M.K. and Posner, M.I. (1997) Distress and attention interactions in early infancy. Motiv. Emot., 21: 27–43.

Harris, C.M., Hainline, L. and Abramov, I. (1981) A method for calibrating an eye-monitoring system for use with human infants. Behav. Res. Methods Instrum., 13: 11–20.

Harris, P. and MacFarlane, A. (1974) The growth of the effective visual field from birth to seven weeks. J. Exp. Child Psychol., 18: 340–348.

Hayhoe, M.M. (2004) Advances in relating eye movements and cognition. Infancy, 6: 267–274.

Henderson, J.M., Weeks, P.A. and Hollingworth, A. (1999) The effects of semantic consistency on eye movements during complex scene viewing. J. Exp. Psychol. Hum. Percept. Perform., 25: 210–228.

Hicks, J.M. and Richards, J.E. (1998) The effects of stimulus movement and attention on peripheral stimulus localization by 8- to 26-week-old infants. Infant Behav. Dev., 21: 571–589.

Hood, B.M. (1993) Inhibition of return produced by covert shifts of visual attention in 6-month-old infants. Infant Behav. Dev., 16: 245–254.

Hood, B.M. (1995) Shifts of visual attention in the human infant: a neuroscientific approach. In: Rovee-Collier C. and Lipsitt L.P. (Eds.), Advances in Infancy Research, Vol. 9. Ablex, Norwood, NJ, pp. 163–216.

Hood, B.M. (2001) Combined electrophysiological and behavioral measurement in developmental cognitive neuroscience: some cautionary notes. Infancy, 2: 213–217.

Hood, B.M. and Atkinson, J. (1991, March) Inhibition of return in 3-, 6-month-olds and adults. Paper presented at the Biennial Meeting of the Society for Research in Child Development. Seattle, WA, USA.

Hood, B.M. and Atkinson, J. (1993) Disengaging visual attention in the infant and adult. Infant Behav. Dev., 16: 405–422.

Hood, B.M., Atkinson, J. and Braddick, O.J. (1998a) Selection-for-action and the development of orienting and visual attention. In: Richards J.E. (Ed.), Cognitive Neuroscience of Attention: A Developmental Perspective. Lawrence Erlbaum, Mahwah, NJ, pp. 219–250.

Hood, B.M., Murray, L., King, F., Hooper, R., Atkinson, J. and Braddick, O. (1996) Habituation changes in early infancy: longitudinal measures from birth to 6 months. J. Reprod. Infant Psychol., 14: 177–185.

Hood, B.M., Willen, D.J. and Driver, J. (1998b) Adult's eyes trigger shifts of visual attention in human infants. Psychol. Sci., 9: 131–134.

Hopkins, B. and van Wulfften Palthe, T. (1985) Staring in infancy. Early Hum. Dev., 12: 261–267.

Hunnius, S. and Geuze, R.H. (2004a) Developmental changes in visual scanning of dynamic faces and abstract stimuli in infants: a longitudinal study. Infancy, 6: 231–255.

Hunnius, S. and Geuze, R.H. (2004b) Gaze shifting in infancy: a longitudinal study using dynamic faces and abstract stimuli. Infant Behav. Dev., 27: 397–416.

Hunnius, S., Geuze, R.H. and van Geert, P.L.C. (2006a) Associations between the developmental trajectories of visual scanning and disengagement of attention in infants. Infant Behav. Dev., 29: 108–125.

Hunnius, S., Geuze, R.H., Zweens, M.J. and Bos, A.F. (2006b) Effects of preterm experience on the developing visual system: a longitudinal study of shifts of attention and gaze in early infancy. Manuscript submitted for publication.

Hunnius, S., de Wit, T.C.J. and von Hofsten, C. (2007, May). Young infants' eye movements and scanning of faces during face-to-face interaction. Poster presented at the 37th Annual Meeting of the Jean Piaget Society, Amsterdam, The Netherlands.

Hunter, S.K. and Richards, J.E. (2003) Peripheral stimulus localization by 5- to 14-week old infants during phases of attention. Infancy, 4: 1–25.

Johnson, M.H. (1990) Cortical maturation and the development of visual attention in early infancy. J. Cogn. Neurosc., 2: 81–95.

Johnson, M.H. (1994) Visual attention and the control of eye movements in early infancy. In: Umiltà C. and Moscovitch M. (Eds.), Attention and Performance XV: Conscious and Nonconscious Information Processing. MIT Press, Cambridge, MA, pp. 291–310.

Johnson, M.H. (1995a) The development of visual attention: a cognitive neuroscience perspective. In: Gazzaniga M.S. (Ed.), The Cognitive Neurosciences. MIT Press, Cambridge, MA, pp. 735–747.

Johnson, M.H. (1995b) The inhibition of automatic saccades in early infancy. Dev. Psychobiol., 28: 281–291.

Johnson, M.H. (2005a) Developmental Cognitive Neuroscience: An introduction (2nd ed.). Blackwell Publishers, London.

Johnson, M.H. (2005b) The ontogeny of the social brain. In: Mayr U., Awh E. and Keele S.W. (Eds.), Developing Individuality in the Human Brain: A Tribute to Michael I. Posner. American Psychological Association, Washington, pp. 125–140.

Johnson, M.H., Dziurawiec, S., Ellis, H. and Morton, J. (1991a) Newborns' preferential tracking of face-like stimuli and its subsequent decline. Cognition, 40: 1–19.

Johnson, M.H. and Gilmore, R.O. (1998) Object-centered attention in 8-month-old infants. Dev. Sci., 1: 221–225.

Johnson, M.H., Gilmore, R.O. and Csibra, G. (1998) Toward a computational model of the development of saccade planning. In: Richards J.E. (Ed.), Cognitive Neuroscience of Attention: A Developmental Perspective. Lawrence Erlbaum, Mahwah, NJ, pp. 103–130.

Johnson, M.H. and Morton, J. (1991) Biology and cognitive development: The Case of Face Recognition. Blackwell, Oxford, UK.

Johnson, M.H., Posner, M.I. and Rothbart, M.K. (1991b) Components of visual orienting in early infancy: contingency learning, anticipatory looking, and disengaging. J. Cogn. Neurosci., 3: 335–344.

Johnson, M.H., Posner, M.I. and Rothbart, M.K. (1994) Facilitation of saccades toward a covertly attended location in early infancy. Psychol. Sci., 5: 90–93.

Johnson, M.H. and Tucker, L.A. (1996) The development and temporal dynamics of spatial orienting in infants. J. Exp. Child Psychol., 63: 171–188.

Johnson, S.P. (2001) Neurophysiological and psychophysical approaches to visual development. In: Kalverboer A.F., Gramsbergen A. (Series Eds.) and Hopkins J.B. (Section Ed.), Handbook of Brain and Behaviour in Human Development: IV. Development of perception and cognition. Elsevier, Amsterdam, pp. 653–675.

Johnson, S.P., Amso, D. and Slemmer, J.A. (2003) Development of object concepts in infancy: evidence for early learning in an eye tracking paradigm. Proc. Natl. Acad. Sci. U.S.A., 100: 10568–10573.

Johnson, S.P. and Johnson, K.L. (2000) Early perception-action coupling: eye movements and the development of object perception. Infant Behav. Dev., 23: 461–483.

Johnson, S.P., Slemmer, J.A. and Amso, D. (2004) Where infants look determines how they see: eye movements and object perception performance in 3-month-olds. Infancy, 6: 185–201.

Jonides, J. (1980) Towards a model of the mind's eye's movement. Can. J. Psychol., 34: 103–112.

Kaye, K. and Fogel, A. (1980) The temporal structure of face-to-face communication between mothers and infants. Dev. Psychol., 16: 454–464.

Keller, H. and Gauda, G. (1987) Eye contact in the first months of life and its developmental consequences. In: Rauh H. and Steinhausen H.-Ch. (Eds.), Advances in Psychology, No. 46. Psychobiology and Early Development. Elsevier, Amsterdam, pp. 129–143.

Kellman, P.J. and Arterberry, M.E. (1998) The Cradle of Knowledge: Development of Perception in Infancy. MIT Press, Cambridge, MA.

Klein, R.M. (2005) On the role of endogenous orienting in the inhibitory aftermath of exogenous orienting. In: Mayr U., Awh E. and Keele S.W. (Eds.), Developing Individuality in the Human Brain: A Tribute to Michael I. Posner. American Psychological Association, Washington, pp. 45–64.

Klein, R., Kingstone, A. and Pontefract, A. (1992) Orienting of visual attention. In: Rayner K. (Ed.), Eye Movements and Visual Cognition: Scene Perception and Reading. Springer, New York, pp. 46–65.

Krinsky-McHale, S. (1993, March) Visual scanning during information-processing in infancy. Poster presented at the Biennial Meeting of the Society for Research in Child Development. New Orleans, USA.

Leahy, R.L. (1976) Development of preferences and processes of visual scanning in the human infant during the first 3 months of life. Dev. Psychol., 12: 250–254.

Lewis, T.L. and Maurer, D. (1992) The development of the temporal and nasal visual fields during infancy. Vision Res., 32: 903–911.

Lewis, T.L., Maurer, D. and Blackburn, K. (1985) The development of young infants' ability to detect stimuli in the nasal visual field. Vision Res., 25: 943–950.

Loftus, G. and Mackworth, N.H. (1978) Cognitive determinants of fixation location during picture viewing. J. Exp. Psychol. Hum. Percept. Perform., 4: 565–572.

MacFarlane, A., Harris, P. and Barnes, I. (1976) Central and peripheral vision in early infancy. J. Exp. Child Psychol., 21: 532–538.

Matsuzawa, M. and Shimojo, S. (1997) Infants' fast saccades in the gap paradigm and development of visual attention. Infant Behav. Dev., 20: 449–455.

Maurer, D. and Lewis, T.L. (1998) Overt orienting toward peripheral stimuli: normal development and underlying mechanisms. In: Richards J.E. (Ed.), Cognitive Neuroscience of Attention: A Developmental Perspective. Lawrence Erlbaum, Mahwah, NJ, pp. 51–102.

Maurer, D. and Salapatek, P. (1976) Developmental changes in the scanning of faces by young infants. Child Dev., 47: 523–527.

Mayer, D.L. and Fulton, A.B. (1993) Development of the human visual field. In: Simons K. (Ed.), Early Visual Development, Normal and Abnormal. Oxford University Press, New York, pp. 117–129.

McConnell, B.A. and Bryson, S.E. (2005) Visual attention and temperament: developmental data from the first 6 months of life. Infant Behav. Dev., 28: 537–544.

McMains, S.A. and Somers, D.C. (2004) Multiple spotlights of attentional selection in human visual cortex. Neuron, 42: 677–686.

Meek, J.H., Firbank, M., Elwell, C.E., Atkinson, J., Braddick, O. and Wyatt, J.S. (1998) Regional hemodynamic responses to visual stimulation in awake infants. Pediatr. Res., 43: 840–843.

Merin, N., Young, G.Y., Ozonoff, S. and Rogers, S.J. (2007) Visual fixation patterns during reciprocal social interaction distinguish a subgroup of 6-month-old infants at-risk for autism from comparison infants. J. Autism Dev. Disord., 37: 108–121.

Meyers, I.L. (1929) Electronystagmography. A graphic study of the action currents in nystagmus. Arch. Neurol. Psychiatry, 21: 901–918.

Milewski, A. (1976) Infant's discrimination of internal and external pattern elements. J. Exp. Child Psychol., 22: 229–246.

Müller, J. (1826) Zur vergleichenden Physiologie des Gesichtssinnes des Menschen und der Thiere nebst einem Versuch über die Bewegungen der Augen und über den menschlichen Blick. C. Cnobloch, Leipzig, Germany.

Nowakowski, R.S. (1987) Basic concepts of CNS development. Child Dev., 58: 568–595.

Papoušek, H. and Papoušek, M. (1983) Biological basis of social interactions: implications of research for an understanding of behavioural deviance. J. Child Psychol. Psychiatry, 24: 117–129.

Pascalis, O. and Slater, A. (2003) The Development of Face Processing in Infancy and Early Childhood: Current Perspectives. Nova Science Publishers, New York.

Porges, S.W. (1980) Individual differences in attention: A possible physiological substrate. In: Keogh, B.K. (Ed.), Advances in Special Education, Vol. 2. JAI Press, Greenwich, CT, pp. 111–134.

Posner, M.I. (1978) Chronometric Explorations of Mind. Lawrence Erlbaum, Hillsdale, NJ.

Posner, M.I. (1980) Orienting of attention. Q. J. Exp. Psychol., 32: 3–25.

Posner, M.I. and Cohen, Y. (1984) Components of visual orienting. In: Bouma H. and Bouwhuis D.G. (Eds.), Attention and Performance X. Lawrence Erlbaum, Hillsdale, NJ, pp. 531–556.

Posner, M.I. and Driver, J. (1992) The neurobiology of selective attention. Curr. Opin. Neurobiol., 2: 165–169.

Posner, M.I. and Petersen, S.E. (1990) The attention system of the human brain. Annu. Rev. Neurosci., 13: 25–42.

Posner, M.I., Rafal, R.D., Choate, L.S. and Vaughan, J. (1985) Inhibition of return: neural basis and function. Cogn. Neuropsychol., 2: 211–228.

Posner, M.I. and Raichle, M.E. (1994) Images of Mind. Scientific American Books, New York.

Posner, M.I. and Rothbart, M.K. (1981) The development of attentional mechanisms. In: Howe H.E. and Flowers J.H. (Eds.), Nebraska Symposium on Motivation. University of Nebraska Press, Lincoln, NB, pp. 1–52.

Posner, M.I., Snyder, C.R. and Davidson, B.J. (1980) Attention and the detection of signals. J. Exp. Psychol., 109: 160–174.

Prablanc, C., Massé, D. and Echallier, J.F. (1978) Error-correcting mechanisms in large saccades. Vision Res., 18: 557–560.

Prechtl, H.F.R. (1984) Continuity and change in early neural development. In: Prechtl H.F.R. (Ed.), Continuity of Neural Functions from Prenatal to Postnatal Life. Clinics in Developmental Medicine, Vol. 94. Blackwell, Oxford, UK, pp. 1–15.

Preyer, W. (1882) Die Seele des Kindes. Beobachtungen über die geistige Entwicklung des Menschen in den ersten Lebensjahren. Th. Grieben's Verlag, Leipzig, Germany.

Rafal, R. (1998) The neurology of visual orienting: a pathological disintegration of development. In: Richards J.E. (Ed.), Cognitive Neuroscience of Attention: A Developmental Perspective. Lawrence Erlbaum, Mahwah, NJ, pp. 181–218.

Rakic, P. (1988) Specification of cerebral cortical area. Science, 241: 170–176.

Reid, V.M., Striano, T., Kaufman, J. and Johnson, M.H. (2004) Eye gaze cueing facilitates neural processing of objects in 4-month-old infants. Neuroreport, 15: 2553–2555.

Reynolds, G.D. and Richards, J.E. (in press) Infant heart rate: A developmental psychophysiological perspective. In: Schmidt L.A. and Segalowitz S.J. (Eds.), Developmental Psychophysiology. Cambridge University Press.

Richards, J.E. (1988) Heart rate changes and heart rate rhythms, and infant visual sustained attention. In: Ackles P.K., Jennings J.R. and Coles M.G.H. (Eds.), Advances in Psychophysiology, Vol. 3. JAI Press, Greenwich, CT, pp. 189–221.

Richards, J.E. (1997) Peripheral stimulus localization by infants: attention, age and individual differences in heart rate variability. J. Exp. Psychol. Hum. Percept. Perform., 23: 667–680.

Richards, J.E. (2000a) Localizing the development of covert attention in infants with scalp event-related potentials. Dev. Psychol., 36: 91–108.

Richards, J.E. (2000b) The development of covert attention to peripheral targets and its relation to attention to central visual stimuli. Paper presented at the International Conference on Infant Studies, Brighton, England.

Richards, J.E. (2001a) Cortical indexes of saccade planning following covert orienting in 20-week-old infants. Infancy, 2: 135–157.

Richards, J.E. (2005a) Localizing cortical sources of event-related potentials in infants' covert orienting. Dev. Sci., 8: 255–278.

Richards, J.E. (2005b) Development of covert orienting in young infants. In: Itti L., Rees G. and Tsotsos J.K. (Eds.), Neurobiology of Attention. Elsevier Academic Press, Amsterdam, pp. 82–88.

Richards, J.E. and Casey, B.J. (1991) Heart rate variability during attention phases in young infants. Psychophysiology, 28: 43–53.

Richards, J.E. and Holley, F.B. (1999) Infant attention and the development of smooth pursuit tracking. Dev. Psychol., 35: 856–867.

Richards, J.E. and Hunter, S.K. (1997) Peripheral stimulus localization by infants with eye and head movements during visual attention. Vision Res., 37: 3021–3035.

Richards, J.E. and Hunter, S.K. (1998) Attention and eye movement in young infants: neural control and development. In: Richards J.E. (Ed.), Cognitive Neuroscience of Attention: A Developmental Perspective. Lawrence Erlbaum, Hillsdale, NJ, pp. 131–162.

Rosander, K. and von Hofsten, C. (2004) Infants' emerging ability to represent occluded object motion. Cognition, 91: 1–22.

Rose, D.H., Slater, A. and Perry, H. (1986) Prediction of childhood intelligence from habituation in early infancy. Intelligence, 10: 251–263.

Ross, S.M. and Dannemiller, J.L. (1999) Color contrast, luminance contrast and competition within exogenous orienting in 3.5-month-old infants. Infant Behav. Devel., 22: 383–404.

Rothbart, M.K., Posner, M.I. and Kieras, J. (2006) Temperament, attention, and the development of self-regulation. In: McCartney K. and Phillips D. (Eds.), Blackwell Handbook of Early Childhood Development. Blackwell, Malden, MA, pp. 338–357.

Rothbart, M.K., Posner, M.I. and Rosicky, J. (1994) Orienting in normal and pathological development. Dev. Psychopathol., 6: 635–652.

Rothbart, M.K., Ziaie, H. and O'Boyle, C.G. (1992) Self-regulation and emotion in infancy. In: Eisenberg N. and Fabes R.A. (Eds.), Emotion and Its Regulation. Jossey-Bass, San Francisco, CA, pp. 7–23.

Ruddy, M.G. (1993) Attention shifting and temperament at 5 months. Infant Behav. Dev., 16: 255–259.

Rueda, M.R., Posner, M.I. and Rothbart, M.K. (2004) Attentional control and self-regulation. In: Baumeister R.F. and Vohs K.D. (Eds.), Handbook of Self-Regulation: Research, Theory, and Applications. Guilford, New York, pp. 357–370.

Saarinen, J. and Julesz, B. (1991) The speed of attentional shifts in the visual field. Proc. Natl. Acad. Sci., 88: 1812–1814.

Salapatek, P. (1968) Visual scanning of geometric figures by the human newborn. J. Comp. Physiol. Psychol., 66: 247–258.

Salapatek, P. (1975) Pattern perception in early infancy. In: Cohen L.B. and Salapatek P. (Eds.), Infant Perception: From Sensation to Cognition, Vol. 1. Academic Press, New York, pp. 133–248.

Salapatek, P., Aslin, R.N., Simonson, J. and Pulos, E. (1980) Infant saccadic eye movements to visible and previously visible targets. Child Dev., 51: 1090–1094.

Salapatek, P. and Kessen, W. (1966) Visual scanning of triangles by the human newborn. J. Exp. Child Psychol., 3: 155–167.

Samuel, A.G. and Kat, D. (2003) Inhibition of return: a graphical meta-analysis of its time course and an empirical test of its temporal and spatial properties. Psychon. Bull. Rev., 10: 897–906.

Saslow, M.G. (1967) Latency for saccadic eye movement. J. Opt. Soc. Am., 57: 1030–1033.

Schiller, P. (1985) A model for the generation of visually guided saccadic eye movements. In: Rose D. and Dobson V.G. (Eds.), Models of the Visual Cortex. Wiley, Chichester, UK, pp. 62–70.

Schiller, P. (1998) The neural control of visually guided eye movements. In: Richards J.E. (Ed.), Cognitive Neuroscience of Attention. Lawrence Erlbaum, Mahwah, NJ, pp. 5–50.

Schneider, G.E. (1969) Two visual systems. Science, 163: 895–902.

Schölmerich, A., Leyendecker, B. and Keller, H. (1995) The study of early interaction in a contextual perspective: culture, communication, and eye contact. In: Valsiner J. (Ed.), Child Development within Culturally Structured Environments. Vol. III: Comparative-Cultural and Constructivist Perspectives. Ablex, Norwood, NJ, pp. 29–50.

Schott, E. (1922) Über die Registrierung des Nystagmus und anderer Augenbewegungen vermittels des Saitengalvanometers. Deutsches Archiv für Klinische Medizin, 140: 79–90.

Schwartz, T.L., Dobson, V., Sandstrom, D.J. and Van Hof-van Duin, J. (1987) Kinetic perimetry assessment of binocular visual field shape and size in young infants. Vision Res., 27: 2163–2175.

Simion, F., Leo, I., Turati, C., Valenza, E. and Dalla Barba, B. (2007) How face specialization emerges in the first months of life. Progr. Brain Res., 164: 169–185.

Simion, F., Valenza, E., Umiltà, C. and Dalla Barba, B. (1995) Inhibition of return in newborns is temporo-nasal asymmetrical. Infant Behav. Dev., 18: 189–194.

Slater, A., Morison, V. and Rose, D. (1983) Locus of habituation in the human newborn. Perception, 12: 593–598.

Slater, A., Morison, V. and Somers, M. (1988) Orientation discrimination and cortical function in the human newborn. Perception, 17: 597–602.

Stechler, G. and Latz, E. (1966) Some observations on attention and arousal in the human infant. J. Am. Acad. Child Psychol., 5: 517–525.

Stelmach, L.B., Campsall, J.M. and Herdman, C.M. (1997) Attentional and ocular movements. J. Exp. Psychol. Hum. Percept. Perform., 23: 823–844.

Stifter, C.A. and Braungart, J.M. (1995) The regulation of negative reactivity in infancy: function and development. Dev. Psychol., 31: 448–455.

Stifter, C.A. and Moyer, D. (1991) The regulation of positive affect: gaze aversion activity during mother-infant interaction. Infant Behav. Dev., 14: 111–123.

Taga, G., Asakawa, K., Maki, A., Konishi, Y. and Koizumi, H. (2003) Brain imaging in awake infants by near-infrared optical topography. Proc. Natl. Acad. Sci., 100: 10722–10727.

Tennes, K., Emde, R., Kisley, A. and Metcalf, D. (1972) The stimulus barrier in early infancy: an exploration of some formulations of John Benjamin. In: Holt R.R. and Peterfreund E. (Eds.), Psychoanalysis and Contemporary Science. Macmillan, New York, pp. 206–234.

Thomas, K.M. and Casey, B.J. (2003) Methods for imaging the developing brain. In: de Haan M. and Johnson M.H. (Eds.), The Cognitive Neuroscience of Development. Psychology Press, Hove, UK, pp. 19–41.

Tipper, S.P., Driver, J. and Weaver, B. (1991) Object-centred inhibition of return of visual attention. Q. J. Exp. Psychol. A, 43: 289–298.

Tipper, S.P., Weaver, B., Jerreat, L.M. and Burak, A.L. (1994) Object-based and environment-based inhibition of return of visual attention. J. Exp. Psychol. Hum. Percept. Perform., 20: 478–499.

Trevarthen, C.B. (1968) Two mechanisms of vision in primates. Psychol. Forsch., 31: 299–337.

Tronick, E. (1972) Stimulus control and the growth of the infant's effective visual field. Percept. Psychophys., 11: 373–376.

Tronick, E.Z., Als, H., Adamson, L., Wise, S. and Brazelton, T.B. (1978) The infant's response to entrapment between contradictory messages in face-to-face interaction. J. Am. Acad. Child Psychol., 17: 1–13.

Tschopp, C., Safran, A.B., Viviani, P., Reicherts, M., Bullinger, A. and Mermoud, C. (1998) Automated visual field examination in children aged 5–8 years. Part II: normative values. Vision Res., 38: 2211–2218.

Uttal, W.R. (2001) The New Phrenology: The Limits of Localizing Cognitive Processes in the Brain. MIT Press, Cambridge, MA.

Valenza, E., Simion, F., Cassia, V.M. and Umiltà, C. (1996) Face preference at birth. J. Exp. Psychol. Hum. Percept. Perform., 22: 892–903.

Valenza, E., Simion, F. and Umiltà, C. (1994) Inhibition of return in newborn infants. Infant Behav. Dev., 17: 293–302.

Van Essen, D.C. (1985) Functional organization of primate visual cortex. In: Peters A. and Jones E.G. (Eds.), Cerebral Cortex, Vol. 3. Plenum, New York, pp. 259–329.

Viviani, P. and Swensson, R.G. (1982) Saccadic eye movements to peripherally discriminated visual targets. J. Exp. Psychol. Hum. Percept. Perform., 8: 113–126.

Wade, N.J. and Tatler, B.W. (2005) The Moving Tablet of the Eye: The Origins of Modern Eye Movement Research. Oxford University Press, New York.

Walker, B.B. and Sandman, C.A. (1979) Human visual evoked responses are related to heart rate. J. Comp. Physiol. Psychol., 93: 717–729.

von Wartburg, R., Ouerhani, N., Pflugshaupt, T., Nyffeler, T., Wurtz, P., Hügli, H. and Müri, R.M. (2005) The influence of colour on oculomotor behaviour during image perception. Neuroreport, 16: 1557–1560.

Weber, R.B. and Daroff, R.B. (1971) The metrics of horizontal saccadic eye movements in normal humans. Vision Res., 11: 921–928.

Wright, R.D. and Ward, L.M. (1998) The control of visual attention. In: Wright R.D. (Ed.), Visual Attention. Oxford University Press, New York, pp. 132–186.

Wurtz, R. and Munoz, D. (1995) Role of monkey superior colliculus in control of saccades and fixation. In: Gazzaniga M.S. (Ed.), The Cognitive Neurosciences. MIT Press, Cambridge, MA, pp. 533–548.

Yi, D.-J., Kim, M.-S. and Chun, M.M. (2003) Inhibition of return to occluded objects. Percept. Psychophys., 65: 1222–1230.

SECTION III

The Development of Action and Cognition

C. von Hofsten & K. Rosander (Eds.)
Progress in Brain Research, Vol. 164
ISSN 0079-6123

CHAPTER 11

Visual constraints in the development of action

Geert Savelsbergh[1,2,*], Simone Caljouw[1], Paulion van Hof[1] and John van der Kamp[1,3]

[1]*Institute for Fundamental and Clinical Human Movement Sciences, VU University Amsterdam, Van der Boechorststraat 9, 1081 BT Amsterdam, The Netherlands*
[2]*Institute for Biophysical and Clinical Research into Human Movement, Manchester Metropolitan University, Cheshire, UK*
[3]*Institute for Human Performance, University of Hong Kong, Hong Kong*

Abstract: The chapter's aim is to understand the role of visual information in the control of avoidance and interception behaviors in infancy from the ecological psychology approach to perception and action. We show that during infancy developmental change in action is associated with the use of different information sources and that this process of attunement promotes the perceived action possibilities (affordances). In the final section, we position these findings within Milner and Goodale's two-visual system model, which holds that perception and action are mediated by two functionally and neuron-anatomically separate visual (sub-)systems.

Keywords: action; control; two visual system; constraints; information; affordances; Goodale-Milner model; optical variable; avoidance; interception

Introduction

Infants are actively engaged in perceiving and acting on the world. For instance, they reach for and pick up a toy, mouth the toy, bang it on a surface. These activities are not only fun, but also provide the infants with information about the toy's properties and ways to improve their actions. It are these apparently simple activities that provide the researcher a direct window into young infants' perception and action capabilities, and how they are related to each other during early development. They can show, on the one hand, how infants learn to perceive the environment, in particularly, how they learn to perceive what action is appropriate in a situation. On the other hand, they can demonstrate how infants become better in performing their actions, i.e., how they learn to adjust their movements to the requirements of the situation. And finally, the infants' playful activities may provide a window into the relationship between perception and action during early development.

Development of perception and action is adaptive; developmental changes result in the infant being better suited to act in a particular, personal environment. Infants become more proficient actors; they learn to better tune their movements to the situation at hand. Infants also get better in perceiving what action is appropriate in a particular situation. Therefore, visual processes for perception and action should be linked early in development. However, the majority of infancy research that has examined perception or action either emphasized the development of perception and cognition or the development of action. The

*Corresponding author. Tel.: +31 20 5988461; Fax: +31 20 5982000; E-mail: gsavelsbergh@fbw.vu.nl

DOI: 10.1016/S0079-6123(07)64011-4

aim of this chapter is to provide an overview with respect to both the development of perception and action during infancy. The key point is that both need each other to develop. In the final section, we discuss the possibility of a largely independent but interacting developmental trajectories for vision for perception and vision for action. This is compatible with recent theories in neurosciences that emphasize that perception and action are mediated by two interacting, but neuro-anatomical and functionally separate visual systems (Milner and Goodale, 1995; Glover, 2004). Analogously, we argue that although vision for perception and action might already be dissociated early in development, adaptive behavior (and development) also necessitates that both are linked.

Constraint-led perspective for understanding action development

Scientific views on how infants learn to move adaptively in the world have changed dramatically during the last 3 or 2 decades. Traditionally, developmental change was attributed to some kind of general process of maturation of the central nervous system. Achievements or milestones in motor development, such as grasping, sitting, crawling, and walking were believed to occur at a predetermined age (e.g., McGraw, 1943; Gesell and Amatruda, 1945). The observed movement patterns were thought to emerge in a fixed genetically determined order; for instance, in cephalo-to-caudal or proximal-to-distal sequences. The gradually increasing cortical control over lower reflexes resulted in movement patterns that became more and more coordinated. It was also believed that variations in the emergence of the milestones signaled error or deviant development. By contrast, many recent observations have revealed that development is not a single process of gradual change (Thelen and Smith, 1994). Rather, it is much better characterized as a non-linear process at different levels of movement organization (e.g., Thelen and Smith, 1994; Savelsbergh et al., 1999a). In the so-called "constraint-led perspective" (Newell, 1986), no single factor has priority in shaping development. Instead, it advocates that at different times different factors may influence developmental change. Research from this theoretical perspective is influenced by the early neuro-physiological work of Bernstein (1967), the ecological approach to perception and action of the Gibsons (Gibson, 1979, 1988), and the dynamic systems theory as formulated by Kelso, Turvey, and Thelen (Kugler et al., 1982; Newell, 1986; Kugler and Turvey, 1987; Turvey, 1990; Thelen and Smith, 1994; Kelso, 1995).

In the constraint-led perspective, the development of movement coordination is brought about by changes in the constraints imposed upon the organism–environment system. Newell (1986) proposed three categories of constraints: organismic (e.g., maturity of the central nervous system, anthropometrics), task (e.g., the properties of the target object), and environmental (e.g., laws of nature such as gravity). These different types of constraint do not operate in isolation, but interact with each other, leading to a task-specific organization of the coordination pattern. Hence, development of coordination does not uniquely originate from maturation of the central nervous system, as argued by the traditional neuro-maturational theories, but emerges from an interaction between the three types of constraint.

Within this theoretical framework, a particular constraint may act, at a certain developmental time, as a rate-limiting factor in the emergence and mastering of new actions. A classic illustration of a rate-limiting factor is provided by the early research of Thelen on the development of leg movements (Thelen et al., 1984). When 8-week-old infants are held upright, the stepping movements observed at a younger age "disappear." This disappearance was traditionally explained by cortical inhibition. Thelen, however, showed that when these infants were lying supine, they performed leg movements that were kinematically similar to the stepping movements observed at a younger age. Further, she neatly demonstrated that when the legs of these infants are submerged in water (in the upright position), the stepping pattern reemerged. If the disappearance were due to cortical inhibition, then the cortical centers would inhibit movements solely in the upright posture and not in the supine posture. Such a scenario is highly unlikely.

Instead, Thelen explained the disappearance of newborn stepping movements as the consequence of a disproportionate growth of leg muscles and fat tissue. Specifically, infants ~2 months of age acquire fat at a greater rate than muscle mass, which leads to relatively less muscle force. At some critical ratio, stepping disappears from the infant's movement repertoire in the upright orientation, but not in the biomechanically less demanding supine orientation. Thus, the occurrence or disappearance of stepping movements is dependent upon the interaction between organismic constraints (body proportions) and environmental constraints (orientation to the gravity vector); some ratio of an increase in fat and muscle tissue acts as a rate-limiting factor.

We made similar observations with respect to the development of reaching (Savelsbergh and Van der Kamp, 1994). We manipulated the orientation of infants with respect to gravity by reclining an infant seat to 0°, 60°, or 90°. Dependent upon the infants' age, biomechanical demands did or did not play an important role in the occurrence of reaching. Body orientation affected reaching only among the 3- to 4-month-olds and not in older infants. In a similar, longitudinal study conducted by Carvalho et al. (2007), infants' reaching was recorded for 3 months from 4 months onwards in three different body orientations. Again, there was a significant difference in reaching between the body orientations at 4 months of age; the number of reaches increased in the seated position in comparison to supine position, but the total movement and deceleration times were shorter, suggesting more ballistic reaches. By 5–6 months, the differences between body orientations disappeared. These findings suggest that at 5–6 months of age infants are able to adapt to the different constraints and changed the kinematical parameters of the reach movements accordingly.

The final example illustrates how brain development, in this case associated with inter-hemispheric specialization, is only one of the multiple constraints that is responsible for the motor behavior. Infants do not extend the arms across the body midline to reach for (contra-)laterally positioned objects before ~4–5 months of age. Previously, it has been assumed that the development of midline crossing is uniquely determined by the degree of laterality and hemispheric specialization (Provine and Westerman, 1979, see also Morange and Bloch, 1996). Van Hof et al. (2002) examined infants longitudinally at 12, 18, and 26 weeks of age while reaching for two balls (3 and 8 cm in diameter) at three positions with respect to the body midline of the infant (i.e., ipsilateral, midline, and contralateral). With age, the infants increasingly adopted a uni- or bimanual reaching mode in accordance to the size of the ball. At the same time, the number of reaches during which the body midline was crossed increased. We showed that the developmental changes in midline crossing were associated with changes in the development of bimanual reaching; the majority of the initial midline crosses were part of a bimanual reach to one of the larger balls. It is the need to grasp a large ball with two hands that induces midline crossing when the ball is positioned laterally. Hence, the development of midline crossings cannot solely be attributed to organismic constraints; it entails interaction with task constraints (e.g., object size).

In sum, there is clear empirical support for the contention that new motor behaviors emerge from the interactions of task, environmental, and organismic constraints. One of these multiple constraints may act as rate-limiting factors. A relatively minor change (e.g., fat/muscle ratio change; change in object size) leads to changes in the pattern of coordination, both in real time and on a developmental time scale.

Visual constraints in the development of action

As we argued above, new motor behaviors emerge from the interaction between several types of constraint. An important prerequisite for development is therefore that infants gather information about themselves (i.e., organismic constraint) and their surroundings (task and environmental constraints). An experiment by Newell and coworkers (Newell et al., 1989) illustrates this for the development of grasping. They presented 4- to 8-month-old infants four objects of different size (i.e., 1.25, 2.50, 2.54, and 8.5 cm in diameter). The majority of grasping behavior could be classified in five grips, while theoretically there are 1023 possible combinations of the 10 fingers. Presumably, the size of the infants' hand together with the size of the objects limited the number of possible grip configurations. Clearly, this can only be achieved if infants are capable extracting veridical information about the object and hand size. Both the haptic and the visual system are important in this respect (Newell et al., 1989). In what follows, we will focus on the later.

Take for instance binocular vision. Young infants are not sensitive to binocular information until ~5 months of age. They must use monocular sources of information to gather information about object size. Improvement in reaching, and grasping during the later part of the first half-year of age might therefore be related to the ability to detect binocular information (Van Hof et al., 2006). In other words, the developing visual system (Kugler and Turvey, 1987; Warren, 1990; Van der Kamp et al., 1996) might prove to be an important constraint, not only for the development of perception, but also for the development of action. Which potential constraints imposed by the developing visual system do we know from research?

Habituation, preferential looking, and visual evoked potential studies have shown that the sensitivity to different types of visual information follows different developmental trajectories. Roughly, there is a sequence of monocular kinetic information to binocular information to monocular static information (Yonas and Granrud, 1985; Atkinson, 2000; Kellman and Arterberry, 2000). Already in the first week after birth, infants detect optical expansion. Sensitivity for binocular information, such as vergence and disparity, develops between 11 and 13 weeks, while monocular pictorial information is involved from 7 months. It must be noted, however, that this data is primarily derived from studies that assessed infants' perception of the world. Whether this developmental sequence is also valid for the visual control of movement remains to be validated. In the next section therefore, we discuss how changing involvement of visual information can account for changes in infants' control of movements involving moving objects.

Adaptive movement behavior

Infants' reach for moving objects because they are attractive to them and draw their attention. Rosander and Von Hofsten (2002) (Von Hofsten and Rosander, 1996) showed that very soon after birth, i.e., from 3 weeks onwards, eye movements are coupled to an oscillating display (a drum). Infants are able to track objects that move right in front them, and try to get them. Most infants are successful around 4 or 5 months of age, at least when objects move at low speed. This ability indicates that infants detect and use visual information to control the spatio-temporal accuracy of their arm movements. The temporal characteristics of infants' eye blinking to stimuli that approach head-on suggest that the detection and use of visual information in order to control action exists at an even younger age (e.g., Yonas, 1981; Kayed and Van der Meer, 2000). In the remainder of this section, we review this development of the control of timing of avoidance (e.g., blinking) and interception behavior (e.g., catching) in more detail.

Avoidance behaviors

The optical projection of a head-on approaching object on the retina expands symmetrically; the closer the object the bigger the projection. More precisely, the size of the optical angle subtended by the object is inversely related to the distance between the object and the point of observation (i.e., eye). During the approach of the object, the size of the optic angle increases exponentially. This rapid increase is called looming; it specifies that collision is imminent (Gibson, 1966, 1979; Fig. 1).

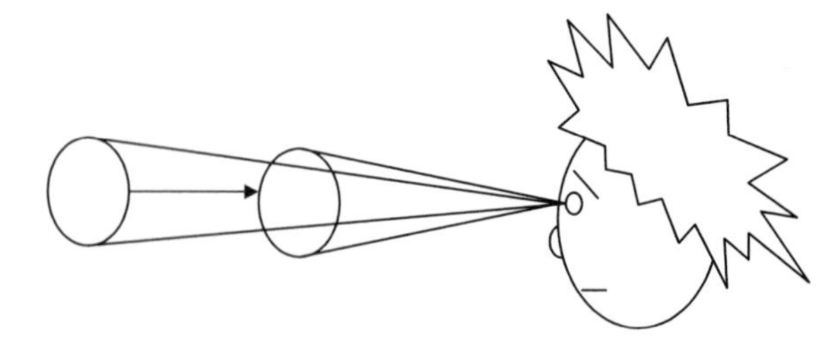

Fig. 1. The optical size of an object (i.e., visual angle) increases with approach.

The majority of studies that examined avoidance behavior used a shadow-casting device, in which an object moves to the light source resulting in an expanding shadow on a screen. Research has shown convincingly that looming elicits avoidance-type behavior in humans and in several other species as well (Schiff et al., 1962; Schiff, 1965); adults blink their eyes, dodge out of its path, or attempt to block the object's approach.

Schiff and coworkers (Schiff et al., 1962; Schiff, 1965) were the first to use "looming" in order to elicit an avoidance response in several animal species. Bower et al. (1970) reported that human infants as young as 6-days-old exhibited avoidance behaviors, like head retraction, raising of the arms, and eye blinking in response to optically expanding shadow patterns. Moreover, the intensity or likelihood of the responses was not different for a virtual or for a real object (Ball and Tronick, 1971), despite that a real approaching object also provides additional information such as air displacement. Thus, very young human infants can use the optical information contained in the expanding pattern to perceive impending collision. In the late seventies, Yonas and coworkers argued that infants' responses to looming were still open to various interpretations (Yonas et al., 1977). They showed that the head and arm movements could be interpreted more properly as instances of tracking behavior. Infants would move their heads because they visually track the rising top contour of the expansion pattern. However, young infants do seem to be aware of collision information available in the expanding optical pattern. Eye blinking at the moment the virtual object would collide may be a case in point. It occurs more frequently to displays specifying approach than to displays specifying a non-expanding rising contour, making an interpretation in terms of tracking less likely (Yonas et al., 1977). Even a 3- to 4-week-old infant blinks in the presence of an approaching object (Nañez, 1988), but its frequency increases dramatically during the first months.

In sum, eye blinking at the moment of virtual collision suggests that the infant perceives that collision is imminent, i.e., that the situation affords some kind of avoidance behavior.

Interception behaviors

For a successful interception, both spatial and temporal information about when and where to intercept the object is mandatory (e.g., Savelsbergh et al., 1991, 1993; Caljouw et al., 2004a, b, 2006). For many years, it was thought that the timing of actions, like reaching for and catching of a moving object, is an advanced skill of which young infants are not capable (Kay, 1969). Infants' first actions were considered to be not much more than simple reflexes or at best fixed action patterns (DiFranco et al., 1978; White et al., 1964, see also Savelsbergh et al., 2006). It was Claes von Hofsten who demonstrated that infants' actions are goal-directed and have some rudimentary spatial and temporal coordination. He found that when a newborn fixated a toy, their reaching movements terminated closer to the object compared to when the object was not fixated (Von Hofsten, 1982). This occurred even though the toy was slowly moving, indicating that infants' first arm movements are not just random or reflexive, but are (partly) under visual control. Although the newborns do not succeed in real interceptions, catching a moving object is an ability that develops early in life. Strikingly, as soon as they make their first successful reaches for stationary objects at ~18 weeks of age, infants are also capable to catch moving ones (Von Hofsten and Lindhagen, 1979; Von Hofsten, 1980, 1982; Out et al., 2001; Van Hof et al., 2005). From 4 to 5 months onwards, infants' readiness to reach for fast moving objects increases, as does the success of their grasps. By 9 months of age, infants achieve a 50 ms precision in catching a target moving at 120 cm/s (Von Hofsten, 1983). Reaching for a moving target requires that the arm movement is continuously adjusted in relation to the perceived (future) interception point, rather than directing the hand to the current object position at the moment of reach onset. There is ample evidence that the arm movements for moving objects are prospectively controlled (Von Hofsten and Lindhagen, 1979; Von Hofsten, 1980, 1983, 1993; Van der Meer et al., 1994; Berthenthal, 1996; Robin et al., 1996; Von Hofsten et al., 1998; Out et al., 2001).

Affordances and action capabilities

Nowadays, we assume that infants are quite capable to learn to detect relevant information from the environment, at least in particular situations, but this assumption has not always been accepted, and strictly speaking it is a relatively new idea. Just like the classical inferential approach that a meaningful image of the outside world can only be achieved by enrichment of the incoming bare sensations through cognitive inferences, the world of the newborn was long characterized as a blooming, buzzing confusion. Because infants seem to do so little during their first months after birth, it was long thought that the young infant was nearly blind and that it is this "handicap" that hinders the young infant from learning about the world. But research over the past 3–4 decades has radically restructured the ideas about the origins of space and motion perception. The old but durable view that infant development begins with sensory systems providing meaningless, disorganized sensations is no longer tenable. Although human infants do have a smaller perceptual repertoire than adults, infants can visually perceive their environment with objects and events occurring in it. It even seems that looking is the major means of information gathering in the relatively immobile newborn. By actively exploring the environment, by means of moving the eye, head, and hand, information can be detected and "affordances" perceived (Gibson, 1988; Van der Kamp and Savelsbergh, 1994; Savelsbergh et al., 1999b). It were the Gibsons (1979, 1988) that coined the concept of affordances.

The affordances of the environment are what it offers the animal, what it provides or furnishes, either for good or ill. The verb "to afford" is found in the dictionary, but the noun "affordance" is not. I have made it up. I mean by it something that refers to both the environment and the animal in a way that no existing term does. It implies the complementarity of the animal and the environment (Gibson, 1979, p. 127).

Gibson argued that information in the environment is not static in time and space, but specifies events, places, and objects. An infant has to learn to pick up and select appropriate information, not how to interpret or construct meaningful perception from stimuli. They discover that the environment consists of things that provide possibilities for action. In young infants, the urge to explore the affordances of the environment is impressive. This exploration is not a blind random search; it is specific and adapted to the properties of the environment (Gibson, 1988; Palmer, 1989). The discovery of affordances therefore depends both on available sources of information and on the infant's capabilities for action. Eleanore Gibson (1988) outlined the development of exploratory behavior into three overlapping phases. The first phase extends from birth to ~4 months. Although the limited mobility and the poor visual acuity constrain infants' scope of exploration to their immediate field of vision, the discovery of object properties is facilitated by visual attention to certain events, in particular motion. Stationary objects are less attended to. In phase two, which begins around the fifth month, infants start to attend to objects. This is facilitated by reaching, increasing visual acuity, and availability of stereoscopic information for depth. Finally, the infants' mastering of locomotor actions leads to the discovery of even broader range of affordances. The works of Adolph (e.g., Adolph et al., 1993; Adolph, 1997) has deepened our understanding of this phase. For example, she encouraged walkers and crawlers to ascend and descend a sloping walkway of 10°, 20°, 30°, and 40°. The findings showed a relation between the exploratory activities and the infants' locomotion abilities. Walkers switched from walking to sliding on descending trials. They touched and hesitated more before descending 10- and 20-degree slopes and explored alternative means for descent by testing different sliding positions before leaving the platform. By contrast, crawlers hesitated most before descending 30- and 40-degree slopes and did not test alternative sliding positions (Adolph et al., 1993). These findings show that not only do the affordances of the environment change with a growing action repertoire, but also that exploration is specific for the infant's mode of action.

The perception of affordances, however, points two ways. It not only refers to the possibilities that the environment offers for action but also equally important for the perception of affordances is the perceiver's ability to realize, and hence to control the action. Learning to move and to coordinate one's actions involve learning to detect and use appropriate sources of information.

Information and the development of action

Information and avoidance behaviors

Caroll and Gibson (1981) reported that 3-month-old infants pushed their heads backward when faced with impending collision with an approaching solid, looming panel but leaned forward when faced with an approaching panel with a window cut out. The solid panel and the window were identical in size to equate visual expansion rate. The observations suggest that infants recognized the functional consequences of contact with these objects, or in other words, differentiated the affordances of the two panels (see also Yonas et al., 1978; Yonas, 1981; Yonas and Granrud, 1985; Nañez and Yonas, 1994; Li and Schmuckler, 1996). Further work suggested that the velocity and size of the object may have an effect on the avoidance behavior but conclusive proof is lacking. As we have argued in the section entitled "Visual constraints in the development of action" looming affords avoidance behavior, at least from 1 month of age. A prime example is eye blinking, which is considered fully established at ~6 months of age (Yonas et al., 1977).

Recently, Kayed and Van der Meer (2000, 2007) have examined the changes in the use of visual information that are associated with the development of eye blinking. They measured the temporal characteristics of infants' eye blinking. Their experiment employed the early shadow-caster paradigm. Infants were seated in front of a large screen on which a virtual approaching object was projected. The projection simulated objects that approached with different velocities either with or without a constant acceleration. The involvement of three optical variables in the timing of eye blinking was examined: the optical angle, its absolute rate of change (looming), and its relative rate of change, which is denoted tau. Lee demonstrated that the pattern of optical expansion, brought about by the relative approach between the actor and the environmental structure of interest contains predictive information about the time-to-contact of an object if it moves with constant velocity (Lee, 1976; see also Savelsbergh et al., 1991). This information is the inverse of the relative rate of expansion of the closed optical contour of a surface. In other words, tau informs about the time it will take before an approaching object reaches the point of observation.

In order to distinguish between the three optical variables (i.e., optical angle, looming, and tau) in the control of eye blink, Kayed and Van der Meer (2000, 2007) determined for each blink, the value of each of these optical variables at the moment the infant blinked. It was assumed that the variable that elicits the blink is kept constant over task constraints (i.e., ball speed) (Michaels et al., 2001; Benguigui et al., 2003; Caljouw et al., 2004a). The findings showed that the 5- to 7-month-old infants geared their blinks to both the optical angle and tau. However, infants who used the optical angle were significantly younger than the infants that used tau. In other words, the infants of different age tended to use different visual information. Importantly, the use of an angle strategy led to difficulties in the case of accelerative collision courses. The infants who kept the optical angle constant across the different approach conditions were too late on the fastest accelerating approach, and blinked after the object would have hit them. The older infants used the more sophisticated tau-strategy by selecting a more useful optical variable that allowed them to deal not only with constant velocity, but also with the fastest accelerative collision courses. During development, infants attune or converge to visual information that specifies the property that results in more adaptive behavior (Jacobs, 2002). In other words, it appears that the informational basis of eye blinking changes with age.

Information and interception behaviors

A few studies suggest that after infants have acquired some rudimentary catching skills, their perception whether or not a moving object can be caught also becomes more accurate. Some researchers reported that infants are less likely to reach for balls that approach at high speeds (Von Hofsten, 1983; Van der Meer et al., 1994; Out et al., 2001). Moreover, in those instances that infants do try to get a fast moving object, they are less likely to be successful (i.e., the hand touching the object). If true, these observations would suggest that infants are able to perceive the catchability of a moving object. Analogously to eye blinking, this change might be associated with a shift in the information-based control of the catching movement. In this respect, Van der Meer et al. (1994) found that from 32 weeks, infants initiated the catch at a fixed or threshold time (e.g., constant time-to-contact) before the object reached at the interception point, independent of the object's speed. By contrast, younger infants initiated their arm movements when the object was at a fixed distance from the interception point (see also Out et al., 2001). Although in the latter case the infants may have tried to compensate by making faster arm movements (Out et al., 2001), the tendency to initiate the arm movements at a certain distance from interception resulted in more failures because eventually the arm can not be moved any faster (Van der Meer et al., 1994, 1995).

It is not unlikely that the tendency to initiate the catch at either a constant time or a constant distance from interception reflects the use of different sources of optical information. For instance, optical angle can provide information about the distance of a head-on approaching object (provided that object size is constant), whereas tau provides information about the time (provided that object velocity is constant). Looming is more complicated, but is more correlated with the time than with distance of the object. Van Hof (2006) (Van Hof et al., 2007) studied perception and action in 3- to 9-month-old infants that were presented with balls that approached at different velocities. In particular, she aimed to uncover whether the information-based regulation of catching movements changes with age, and if so, whether such age-related differences in the visual control of movement are associated with changes in the perception of what the moving objects afford for action. The infants were seated in a baby chair while a ball could approach with constant velocity by means of a specially developed ball transport apparatus (Fig. 2; see Van Hof, 2006; Van Hof et al., 2006, 2007). Infants were presented with a wide range of ball velocities according to the staircase procedure (Adolph, 1997; Van Hof et al., 2005), which ensures that the ball velocities are tailored to each individual infant's catching skill. Each trial infant was classified as either a success (i.e., the ball was contacted or grasped), a failure (i.e., an attempt to get the ball that did not result in a contact), or a refusal (i.e., no attempt was made).

A dramatic change in catching proficiency was found at ~6–7 months of age. The 3- to 5-month-olds made only a few successful interceptions. They had discovered that a moving object afforded catching or making contact, but tried to get the ball regardless of the velocity at which it traveled. In addition, these infants also refused frequently to reach for balls that moved slowly and that they would have caught if only they had tried. Perception of the affordance of catchability was rather inaccurate. By contrast, the 6- to 7-months-old were more proficient catchers; a higher proportion of their attempts resulted in a successful catch, however, they still showed a considerable amount of failures, in particular for the fast moving balls. And although they reached for almost every ball that they could get (i.e., the lower speeds), they also tried balls that moved too fast to contact or

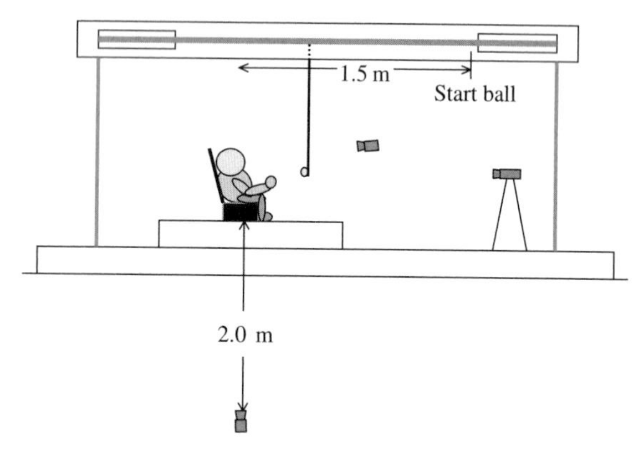

Fig. 2. The ball apparatus and set-up used for controlling the approach velocity of the objects.

grasp. This indicates that the perception of catchabilty was improved but was still not very accurate. Clearly, the 8- to 9-month-old infants were both the best catchers and the best perceivers. They showed only a few failures, and they consistently refused to reach for the fast moving balls they could not get. The latter indicates that by 8–9 months of age the infants did take their *action capabilities* into account when deciding whether or not to try to intercept the moving ball.

In subsequent analyses, Van Hof (2006) assessed for every successful catch the optical variable that may have been used to time the onset of the catching movement, and whether the exploitation of different optical variables would be associated with accuracy in the perception of the affordance of catchability. Von Hof adopted a method introduced by Michaels et al. (2001), who examined the information-based regulation of punching falling balls (see also Benguigui et al., 2003; Caljouw et al., 2004a, b). She assessed whether the timing of catch onset was consistent with the use of a criterion value of one of the three above described monocular optical variables, i.e., optical angle, looming, or tau. The following assumptions were made in this analysis: (1) the infant uses the same variable in each and every attempt, (2) the catch is initiated when the exploited variable reaches a threshold value, and (3) the visuo-motor delay between the moment the threshold value of the optic variable is detected and the moment the arm starts moving is constant across trials (see Van Hof, 2006).

Van Hof (2006) found that infants whose timing pattern was compatible with the use of the optical angle variable were significantly younger than the infants who appeared to control the catch on the basis of the more useful variables like looming or tau. Hence, the successful catches of the younger infants appeared to be based on optical angle and as a consequence were more susceptible to object speed. The faster the balls moved the later the catch was initiated. This typically resulted in insufficient time to fully execute the action, and hence catching failures, in particular for the fast moving balls. It may have been these types of error that enforced infants of ~6–7 months to shift attention to alternative variables like looming and tau (i.e., the absolute and relative rate of change of the optical angle). Looming and especially tau covary closely with the time remaining before the ball reaches the point of interception and are therefore more useful variables in case object velocity varies from trial to trial. The use of these variables would substantially increase the likelihood of making a successful interception, even when balls arrive with high speeds. In fact, Van Hof (2006) found that contrary to the youngest infants, the majority of infants between 6 and 9 months of age exploited either looming or tau.

Van Hof (2006) further demonstrated that the accuracy of infants' perception of whether or not the moving ball can be caught (as indicated by the proportion of success, failure, and refusal) was tightly related (in developmental time) to the degree to which the infants mastered the control of catching action (as indicated by the exploited optic variable). That is, the most dramatic changes for the accuracy of perception as well as the control of action occurred at ~6–7 months of age; the age at which the majority of the infants started to exploit the more useful optic variables (i.e., looming and tau) coincided with the age at which infants considerably improved the accuracy by which they perceived the affordance of catchability. Importantly, these interpretations were corroborated by the findings on the individual level. Infants who used optical angle were significantly less accurate (e.g., they had a lower proportion of refusals and a higher proportion of failures for the high ball speeds) than infants who either used looming or tau. Hence, developmental changes in the perception of affordance perception were closely correlated with developmental changes in the visual control of catching.

The study of Van Hof (2006) (see also Van Hof et al., 2007) illustrates that changes in the accuracy of perception may provide a scaffold for changes in the control of action, which in turn may lead to further changes in the accuracy of perception (see also Gibson, 1988). Hence, the developmental relation between perception and action is not simply causal; the key point is that both need each other to change. This underlines the ecological viewpoint that perception and action are mutually constraining. "Perception and action are interdependent: Perception obtains information for action, and action has consequences that inform perception..." (Gibson, 1997, p. 25).

Reaching for an understanding action development

This chapter describes how visual information constraints the development of action and perception. In this final section, we briefly speculate how this might be related to brain development. We use the two-visual system model of Milner and Goodale (1995) as an avenue into the brain.

Milner and Goodale (1995) (Goodale and Milner, 1992) have proposed that there exists a functional and structural dissociation in two visual pathways in the adult human brain. The ventral system uses visual information for perception of the environment, while the dorsal system is thought to exploit visual information for the control of action. Take for instance a moving object. The ventral system would be involved in obtaining knowledge about the properties of moving objects, such as their color or speed, and perhaps whether they are catchable. By contrast, the dorsal system uses visual information to adjust movement speed to the motion of the ball. By extrapolating the two-visual system model to infancy, several authors have hypothesized that the roles of vision in perception and vision in action may follow separate developmental trajectories (Van der Kamp and Savelsbergh, 2000, 2002; Berthier et al., 2001; Newman et al., 2001; Van der Kamp et al., 2003). If this conjecture is correct, then findings with respect to the development of perception (e.g., as derived through habituation and preferential looking studies) cannot automatically be generalized to the development of action, i.e., the information-based regulation of movement. As an example, Von Hofsten (1982) has found that newborns' arm movements are adapted to the movement direction of the passing object, indicating that newborns use optical information about movement direction of the object to guide their arm movements. By contrast, habituation and preferential looking studies conducted by Wattam-Bell (1992, 1996) indicated that only from 10 weeks onwards infants begin to discriminate the direction of slowly moving stimuli, and that it is not before 4 months that infants' perception of directional information gets more robust. These and other examples (Van der Kamp and Savelsbergh, 2000) illustrate that when trying to understand what and how visual constraints give rise to developmental change in the infants' actions, researchers have to study the actions themselves and should be careful in taking the findings from perception studies as their point of departure.

Summary and conclusions

We reviewed studies that examined the role of visual constraints in the development of avoidance and interception behaviors. In line with the ecological approach to perception and action, we argued that development change in both avoidance and interception behaviors can be understood as attunement or convergence to more useful optic variables. In other words, for a given action (e.g., catching) the coupling between movement and optic variables (e.g., movement onset and optic angle) changes with age (e.g., movement onset and tau). There is some evidence that similar processes of change in coupling between movement and optic variables may have to occur for each new action system as it emerges (Bremner, 2000). Exploration is essential in establishing new couplings and/or selecting the most appropriate coupling for a given action (Gibson, 1988; Savelsbergh and Van der Kamp, 1993).

The actualization of a perceived affordance thus implies the establishment of a coupling between movement and optic variables. An important, but largely unanswered question in this respect concerns the degree to which specificity of learning holds (Adolph, 2000; Savelsbergh and Van der Kamp, 2000). If an infant is able to detect and use a particular optical variable, would she or he be able to also detect and use the same variable in a similar but slightly different action? For instance, in development of eye blinking, we have seen that between 5 and 7 months of age infants converged on the variable tau, which specifies the time remaining before the object reaches the infant. In the development of catching, infants of similar age also attune to variables that covary with time. However, research suggests that this not only involves the variable tau but also looming. Clearly, there are similarities in the development of eye blinking and catching and it would be premature making strong claims on whether a coupling to an optic variable generalize from one action to the other, or whether learning is domain specific. Domain specificity, however, would neatly fit with recent models that suggest that action, and thus coupling between movement and optic variables, is modularly organized (Milner and Goodale, 1995; Atkinson, 2000) and may well follow independent developmental trajectories (Atkinson, 2000; Berthier et al., 2001).

References

Adolph, K.E. (1997) Learning in the development of infant locomotion. Monographs of the Society for Research in Child Development, 62 (3, Serial No. 251).

Adolph, K.E. (2000) Specificity of learning: why infant fall over a veritable cliff. Psychol. Sci., 11: 290–295.

Adolph, K.E., Eppler, M.A. and Gibson, E.J. (1993) Crawling versus walking infants' perception of affordances for locomotion over sloping surfaces. Child Dev., 64: 1158–1174.

Atkinson, J. (2000) The Developing Visual Brain. Oxford University Press, Oxford, UK.

Ball, W. and Tronick, E. (1971) Infants' responses to impending collision: optical and real. Science, 171: 812–820.

Benguigui, N., Ripoll, H. and Broderick, M.P. (2003) Time-to-contact estimation of accelerated stimuli is based on first-order information. J. Exp. Psychol. Hum. Percept. Perform., 29: 1083–1101.

Bernstein, N. (1967) The Coordination and Regulation of Movement. Pergamon Press, New York.

Berthenthal, B.I. (1996) Origins and early development of perception, action, and representation. Ann. Rev. Psychol., 47: 431–459.

Berthier, N.E., Bertenthal, B.I., Seaks, J.D., Sylvia, M.R., Johnson, R.L. and Clifton, R.K. (2001) Using object knowledge in visual tracking and reaching. Infancy, 2: 257–284.

Bower, T.G.R., Brougthon, J.M. and Moore, M.K. (1970) Infant responses to approaching objects: an indicator of response to distal variables. Percept. Psychophys., 9: 193–196.

Bremner, J.G. (2000) Developmental relationships between perception and action in infancy. Infant Behav. Dev., 23: 567–582.

Caljouw, S.R., Van der Kamp, J. and Savelsbergh, G.J.P. (2004a) Catching optical information for the regulation of timing. Exp. Brain Res., 155: 427–438.

Caljouw, S.R., Van der Kamp, J. and Savelsbergh, G.J.P. (2004b) Timing goal directed hitting: impact requirements change the information movement coupling. Exp. Brain Res., 155: 135–144.

Caljouw, S., Van der Kamp, J. and Savelsbergh, G.J.P. (2006) The impact of task constraints on the planning and control of interceptive movements. Neurosci. Lett., 392: 84–89.

Caroll, J.J. and Gibson, E.J. (1981) Differentiation of an aperture from an obstacle under conditions of motion by three-month-old infants. Paper presented at the meetings of the Society for Research in Child Development, Boston, MA.

Carvalho, R.P., Tudella, E. and Savelsbergh, G.J.P. (2007) Spatio-temporal parameters in infant's reaching movements are influenced by body orientation. Infant Behav. Dev., 30: 26–35.

DiFranco, D., Muir, D.W. and Dodwell, D.C. (1978) Reaching in young infants. Perception, 7: 385–392.

Gesell, A. and Amatruda, C.S. (1945) The Embryology of Behavior. Harper, New York.

Gibson, J.J. (1966) The Senses Considered as Perceptual Systems. Houghton-Mifflin, Boston, MA.

Gibson, J.J. (1979) The Ecological Approach to Visual Perception. Houghton-Mifflin, Boston, MA.

Gibson, E.J. (1997) An ecological psychologist's prolegomena for perceptual development: a functional approach. In: Dent-Read C. and Zukow-Goldring P. (Eds.), Evolving Explanations of Development. American Psychological Association, Washington, DC, pp. 23–45.

Gibson, E.J. (1988) Exploratory behavior in the development of perceiving, acting and the acquiring of knowledge. Ann. Rev. Psychol., 39: 1–41.

Glover, S. (2004) Separate visual representations in the planning and control of action. Behav. Brain Sci., 27: 3–78.

Goodale, M.A. and Milner, A.D. (1992) Separate visual pathways for perception and action. Trends Neurosci., 15: 20–25.

Jacobs, D.M. (2002) On Perceiving, Acting and Learning: Toward an Ecological Approach Anchored in Convergence. Digital Printing Partners Utrecht BV, Utrecht, The Netherlands.

Kay, H. (1969) The development of motor skills from birth to adolescence. In: Bilodeau E.A. (Ed.), Principles of Skill Acquisition. Academic Press, New York.

Kayed, N.S. and Van der Meer, A. (2000) Timing strategies used in defensive blinking to optical collisions in five- to seven-month-old infants. Infant Behav. Dev., 23: 253–270.

Kayed, N.S. and Van der Meer, A. (2007) Infants' timing strategies to optical collisions: a longitudinal study. Infant Behav. Dev., 30: 50–59.

Kellman, P.J. and Arterberry, M.E. (2000) The Cradle of Knowledge: Development of Perception in Infancy. The MIT Press, Cambridge, MA.

Kelso, J.A.S. (1995) Dynamic Patterns: The Self-Organization of Brain and Behavior. MIT press, Cambridge.

Kugler, P.N., Kelso, J.A.S. and Turvey, M.T. (1982) On the control and coordination of naturally developing systems. In: Kelso J.A.S. and Clark J.E. (Eds.), The Development of Movement Control and Coordination. Wiley, New York, pp. 5–78.

Kugler, P.N. and Turvey, M.T. (1987) Information, Natural Law, and the Self-Assembly of Rhythmic Movements. Erlbaum, Hillsdale, NJ.

Lee, D.N. (1976) A theory of visual control of braking based on information about time-to-collision. Perception, 5: 437–459.

Li, N.S. and Schmuckler, M.A. (1996) Looming responses to obstacles and apertures: the role of accretion and deletion of background texture. Poster presented at the 10th Biennial International Conference for Infant Studies, April 17–21, 1996, Providence, RI.

McGraw, M. (1943) The Neuromuscular Maturation of the Human Infant. Hafner, New York.

Michaels, C.F., Zeinstra, E.B. and Oudejans, R.R.D. (2001) Information and action in punching a falling ball. Q. J. Exp. Psychol., 54A: 69–93.

Milner, A.D. and Goodale, M.A. (1995) The Visual Brain in Action. Oxford University Press, Oxford, UK.

Morange, F. and Bloch, H. (1996) Lateralization of the approach movement and the prehension movement in infants from 4 to 7 months. Early Dev. Parent., 5: 81–92.

Nañez, J.E. (1988) Perception of impending collision in 3- to 6-week-old human infants. Infant Behav. Dev., 11: 447–463.

Nañez, J.E. and Yonas, A. (1994) Effects of luminance and texture motion on infant defensive reactions to optical collision. Infant Behav. Dev., 17: 165–174.

Newell, K.M. (1986) Constraints on the development of coordination. In: Wade M. and Whiting H.T.A. (Eds.), Motor Development in Children: Aspects of Coordination and Control. Martinus Nijhoff, Dordrecht, The Netherlands, pp. 341–360.

Newell, K.M., Scully, D.M., McDonald, P.V. and Baillargeon, R. (1989) Task constraints and infant grip configurations. Dev. Psychobiol., 22: 817–832.

Newman, C., Atkinson, J. and Braddick, O. (2001) The development of reaching and looking preferences in infants to objects of different sizes. Dev. Psychol., 37: 561–572.

Out, L., Savelsbergh, G.J.P. and Van Soest, A.J. (2001) Interceptive timing in early infancy. J. Hum. Mov. Stud., 40: 185–206.

Palmer, C.F. (1989) The discriminating nature of infants' exploratory actions. Dev. Psychol., 25: 885–893.

Provine, R.R. and Westerman, J.A. (1979) Crossing the midline: limits of early eye-hand behavior. Child Dev., 50: 437–441.

Robin, D.J., Berthier, N.E. and Clifton, R.K. (1996) Infants' predictive reaching for moving objects in the dark. Dev. Psychol., 32: 824–835.

Rosander, K. and Von Hofsten, C. (2002) Development of gaze tracking of small and large objects. Exp. Brain Res., 146: 257–264.

Savelsbergh, G.J.P. and Van der Kamp, J. (1993) The coordination of infants' reaching, grasping, catching and posture: a natural physical approach. In: Savelsbergh G.J.P. (Ed.), The Development of Coordination in Infancy. Elsevier Science Publishers, Amsterdam, pp. 289–317.

Savelsbergh, G.J.P. and Van der Kamp, J. (1994) The effect of body orientation to gravity on early infant reaching. J. Exp. Child Psychol., 58: 510–528.

Savelsbergh, G.J.P. and Van der Kamp, J. (2000) Information in learning to co-ordinate and control movements: is there a need for specificity of practice? Int. J. Sport Psychol., 31: 467–484.

Savelsbergh, G.J.P., Van der Maas, H. and Van Geert, P.C.L. (1999a) Non Linear Analyses of Developmental Processes. Amsterdam, Elsevier.

Savelsbergh, G.J.P., Van Hof, P., Caljouw, S.R., Ledebt, A. and Van der Kamp, J. (2006) No single factor has priority in action development: a tribute to 'Esther Thelen's legacy. J. Integr. Neurosci., 5: 493–504.

Savelsbergh, G.J.P., Whiting, H.T.A. and Bootsma, R.J. (1991) Grasping tau. J. Exp. Psychol. Hum. Percept. Perform., 17: 315–322.

Savelsbergh, G.J.P., Whiting, H.T.A., Pijpers, J.R. and Van Santvoord, A.M.M. (1993) The visual guidance of catching. Exp. Brain Res., 93: 146–156.

Savelsbergh, G.J.P., Wimmers, R.H., Van der Kamp, J. and Davids, K. (1999b) The development of motor control and coordination. In: Kalverboer A.F., Genta M.L. and Hopkins B. (Eds.), Current Issues in Developmental Psychology: Biopsychological Perspective. Kluwer Academic Publishers, Amsterdam, pp. 107–136.

Schiff, W. (1965) Perception of impending collision: a study of visually directed avoidant behavior. Psychol. Monogr., 79: 1–26.

Schiff, W., Caviness, J.A. and Gibson, J.J. (1962) Persistent fear responses in rhesus monkeys to the optical stimulus of "looming." Science, 136: 982–983.

Thelen, E., Fischer, D.M. and Ridley-Johnson, R. (1984) The relationship between physical growth and newborn reflex. Infant Behav. Dev., 7: 479–493.

Thelen, E. and Smith, L.B. (1994) A Dynamic Systems Approach to the Development of Cognition and Action. MIT Press, Cambridge.

Turvey, M.T. (1990) Coordination. Am. Psychol., 45: 938–953.

Van der Kamp, J., Oudejans, R. and Savelsbergh, G.J.P. (2003) The development and learning of the visual control of movement: an ecological perspective. Infant Behav. Dev., 26: 495–515.

Van der Kamp, J. and Savelsbergh, G.J.P. (1994) Exploring exploration in the development of action. Res. Clin. Cent. Child Dev. Rep., 16: 131–139.

Van der Kamp, J. and Savelsbergh, G.J.P. (2000) Action and perception in infancy. Infant Behav. Dev., 23: 238–245.

Van der Kamp, J. and Savelsbergh, G.J.P. (2002) On the development of the two visual systems. Behav. Brain Sci., 25: 120.

Van der Kamp, J., Vereijken, B. and Savelsbergh, G.J.P. (1996) Physical and informational constraints in the coordination and control of human movement. Corp. Psyche Soc., 3: 102–118.

Van der Meer, A.L.M., Van der Weel, F.R. and Lee, D.N. (1994) Prospective control in catching by infants. Perception, 23: 287–302.

Van der Meer, A.L.M., Van der Weel, F.R., Lee, D.N., Laing, I.A. and Lin, J.P. (1995) Development of prospective control of catching moving objects in preterm infants. Dev. Med. Child Neurol., 37: 145–158.

Van Hof, P. (2006) Perception-action coupling in infancy. Ph.D. dissertation, Faculty of Human Movement Sciences, VU University, Amsterdam.

Van Hof, P., Van der Kamp, J., Caljouw, S.R. and Savelsbergh, G.J.P. (2005) The confluence of intrinsic and extrinsic constraints on 3- to 9-month-old infants' catching behavior. Infant Behav. Dev., 28: 179–193.

Van Hof, P., Van der Kamp, J. and Savelsbergh, G.J.P. (2002) The relation of unimanual and bimanual reaching to crossing the midline. Child Dev., 73: 1353–1362.

Van Hof, P., Van der Kamp, J. and Savelsbergh, G.J.P. (2006) Three to eight months infants catching under monocular and binocular vision. Hum. Mov. Sci., 25: 18–36.

Van Hof, P., Van der Kamp, J. and Savelsbergh, G.J.P. (2007) The relation between infants' perception of catchableness and the control of catching (submitted).

Von Hofsten, C. (1980) Predictive reaching for moving objects by human infants. J. Exp. Child Psychol., 36: 369–382.

Von Hofsten, C. (1982) Eye-hand coordination in the newborn. Dev. Psychol., 18: 450–461.

Von Hofsten, C. (1983) Catching skills in infancy. J. Exp. Psychol. Hum. Percept. Perform., 9: 75–85.

Von Hofsten, C. (1993) Prospective control: a basic aspect of action development. Hum. Dev., 36: 253–270.

Von Hofsten, C. and Lindhagen, K. (1979) Observations on the development of reaching for moving objects. J. Exp. Child Psychol., 28: 158–173.

Von Hofsten, C. and Rosander, K. (1996) The development of gaze control and predictive tracking in young infants. Vision Res., 36: 81–96.

Von Hofsten, C., Vishton, P., Spelke, E.S., Feng, Q. and Rosander, K. (1998) Predictive action in infancy: tracking and reaching for moving objects. Cognition, 67: 255–285.

Warren, W.H. (1990) The perception-action coupling. In: Bloch H. and Berthenthal B.I. (Eds.), Sensory-Motor Organizations and Development in Infancy and Early Childhood. Kluwer Academic Publishers, Dordrecht, The Netherlands, pp. 23–37.

Wattam-Bell, J.R.B. (1992) The development of maximum displacement limits for discrimination of motion direction in infancy. Vision Res., 32: 621–630.

Wattam-Bell, J.R.B. (1996) Infants' discrimination of absolute direction of motion. Invest. Ophthalmol. Vis. Sci., 37: S917.

White, B.L., Castle, P. and Held, R. (1964) Observations on the development of visually-directed reaching. Child Dev., 35: 349–364.

Yonas, A. (1981) Infants' responses to optical information for collision. In: Aslin R.N., Alberts J.R. and Peterson M.R. (Eds.), Development of Perception: Vol. 2. The Visual System. New York, Academic Press, pp. 313–334.

Yonas, A., Bechtold, A.G., Frankel, D., Gordon, R.F., McRoberts, G., Norcia., A. and Sternfels, S. (1977) Development of sensitivity to information for impending collision. Percept. Psychophys., 21: 97–104.

Yonas, A. and Granrud, C.A. (1985) Reaching as infants' spatial perception. In: Gottlieb G. and Krasnegor N.A. (Eds.), Measurement of Audition and Vision in the First Year of Postnatal Life: A Methodological Overview. Ablex Publishing Corp, Norwood, NJ, pp. 301–322.

Yonas, A., Pettersen, L. and Lockman, J.J. (1978) Young infants' sensitivity to optical information for collision. Can. J. Psychol., 33: 268–276.

C. von Hofsten & K. Rosander (Eds.)
Progress in Brain Research, Vol. 164
ISSN 0079-6123

CHAPTER 12

Object and event representation in toddlers

Rachel Keen[1,*] and Kristin Shutts[2]

[1]*Department of Psychology, Tobin Hall, University of Massachusetts, Amherst, MA 01003, USA*
[2]*Department of Psychology, William James Hall, Harvard University, Cambridge, MA 02138, USA*

Abstract: Mental representation of absent objects and events is a major cognitive achievement. Research is presented that explores how toddlers (2- to 3-year-old children) search for hidden objects and understand out-of-sight events. Younger children fail to use visually obvious cues, such as a barrier that blocks a moving object's path. Spatiotemporal information provided by movement cues directly connected to the hidden object is more helpful. A key problem for toddlers appears to be difficulty in representing a spatial array involving events with multiple elements.

Keywords: object search; reasoning; cognitive development; toddlers

Representation of absent objects is a hallmark of cognition because it frees the organism from reliance on simply what is in sight at the moment. Tasks requiring search for hidden objects have been used to explore cognitive development since Piaget made it a critical feature of testing infants' representation of out-of-sight objects (Piaget, 1954). While it is easy to replicate Piaget's finding that infants will not search for a desired toy that has disappeared under a cloth, it has been hard to interpret this puzzling behavior. In his chapter in the Handbook of Child Psychology (1983), Paul Harris proposed that the reason for infants' failure in the Piagetian search task is not that they lack object permanence, but rather they do not know where to search for the hidden object. Eventually the infant overcomes this problem and knows to lift the cloth to find the toy.

Knowing where to search for a hidden object depends on one's ability to use the cues that indicate where the hidden object is. For the 10-month-old in Piaget's search task, the cue is the place of disappearance. The infant sees the toy disappear under a cover, so under the cover is the obvious place to look. For the toddler, more complex cues can be used. In our lab we have tested toddlers' understanding of motion cues of disappearance and reappearance from behind multiple hiding sites, and the contact mechanical cue of a barrier. The movement cues are appreciated at a younger age, but the barrier cue is not used until around 3 years of age. What are the constraints that prevent a child from using readily available cues? Why does a child have difficulty in applying relevant knowledge to solve a problem? We will review research from our lab and seek some resolution to these problems in this chapter.

First, let us describe the task. The child is seated in front of an apparatus that features a ramp and an opaque screen that can be placed to hide a large section of the ramp. Four doors cut into the screen can be opened to reveal a hiding place, and a barrier can be placed perpendicular to the ramp behind any of these doors. The barrier protrudes above the screen by several centimeters and stands

*Corresponding author. Tel.: +1 413 545 2655;
Fax: +1 413 545 0996; E-mail: rachelkeen@virginia.edu

DOI: 10.1016/S0079-6123(07)64012-6

out because it is painted a different color from the screen and the ramp. The final element of the apparatus is a ball, which the experimenter holds at the top of the ramp in plain view of the child, and then releases to roll behind the screen and stop at the barrier. The ball can then be retrieved by opening the door by the barrier (see Fig. 1 for a view of a child successfully locating the ball). The actual motor action to retrieve the ball is trivial for a 2-year-old, and experience is given to children before trials start to make sure they know how to open all the doors and find a toy.

Multiple studies in multiple labs have established that 2-year-olds do not know where to search for the ball, apparently ignoring the visible portion of the barrier that should serve to remind them that "the ball stops here". Berthier et al. (2000) presented children with the horizontal version of this task described above and illustrated in Fig. 1, and Hood et al. (2000) presented children with a 2-choice vertical version of the task. The direction of the ball's movement does not appear to make a difference, but the simpler 2-choice task was solved by 2.5-year-olds and the 4-choice task was not solved until 3 years of age.

To understand the toddler's difficulty with this task, we need to examine the underlying components leading to the solution. Previous analyses of the task have yielded many possibilities including (1) understanding the task as a search for a hidden object; (2) understanding the role of the barrier;

Fig. 1. View of the apparatus used in Berthier et al. (2000). The child has opened door 3 and found the ball resting against the barrier. Copyright 2000 by the American Psychological Association. Adapted with permission.

(3) keeping track of an object's hidden movement; (4) attending to the visible portion of the barrier as a cue; (5) predicting the ball's location; (6) using knowledge of the ball's location to orchestrate the appropriate motor response; and (7) representing the spatial integration of the critical but hidden features of ball, barrier, and door (Keen, 2003, 2005; Keen and Berthier, 2004). While these analyses outlined possible problems that toddlers may have with the task, which is rich with cognitive pitfalls, no clear answers have been reached. Through a series of recent studies we have manipulated the available visual information and the response the child makes, in order to arrive at a more comprehensive understanding of toddlers' problem solving.

Beginning with the simplest possibility for failure, did the toddlers exhibit object permanence and understand they were to find the ball? Certainly they did. They always searched eagerly and they stopped opening doors if they found the ball. This was a game they appeared to understand and enjoy. They were willing to play for a dozen or so trials with no more reward than occasionally finding the ball and handing it back to the experimenter for the next round. Children in all studies reported here were thoroughly familiarized with the apparatus and the game before test trials began. They were shown how the ball rolled down the ramp and stopped at the barrier without the screen present. When the screen was then placed in front of the ramp, all four doors were opened so the child could see the ball rolling down the ramp, passing behind the doors and stopping at the barrier, fully visible through one of the doors. Finally, children were pretested on their ability to open the doors by having the experimenter pull down one door, place a toy on the ramp, close the door, and ask the child to find the toy. They were able to find the toy in this simpler situation, presumably because it did not move after it was hidden.

A second possible problem, also easy to rule out, is that toddlers did not understand the physical law of solidity, namely that the solid barrier would stop the solid ball. Mash et al. (2003) gave 2-year-olds full visual access to the ball's movement by first rolling the ball so that it came to rest against the wall before lowering the screen. When the screen was lowered to conceal ramp and ball, the top of the wall showed above the screen, serving as a continual reminder of the ball's location as in the original study (Berthier et al., 2000). Seeing where the ball stopped actually benefited the children very little. In the original study no 2-year-old performed above chance; in Mash et al. (2003) only 2 out of 18 children were above chance. The 2.5-year-old children fared a bit better; 7 out of 18 were above chance compared to 3 out of 16 in the original study. In this study children did not have to remember the bottom of the barrier as a physical obstacle on the ramp and reason how that would stop the ball. Although this would seem to be a major challenge in selecting the correct door, removing this aspect of the problem did not help 2-year-olds, although it may have been a major block for a few 2.5-year-olds. These data are compelling in that they rule out two possible reasons for toddlers' failure to search correctly: the necessity to reason about the solid barrier stopping the ball, and losing track of the ball because of hidden movement.

An important distinction to remember is that simply knowing that the barrier stopped the ball's progress is not sufficient to infer the ball's location. In Mash et al. (2003) after children saw the screen lowered, the problem of where to search came down to either using the top of the barrier as a cue or remembering which of the four doors covered the ball. We now know that it is easy for a 2-year-old to lose track of where they last saw the ball disappear. In Mash et al. (2003) and in Butler et al. (2002) we scored children's eye movements as they tracked the ball's movement down the ramp. In the former study children saw the ball's complete trajectory whereas in the latter study they saw an interrupted trajectory. Butler et al. (2002) used a transparent screen so the ball was visible as it rolled between doors (see Fig. 2). The door of last disappearance marked the ball's hidden location. In both studies merely tracking the ball until it stopped (Mash et al., 2003) or disappeared behind a door and failed to reappear (Butler et al., 2002) did not ensure a correct response. Two-year-olds needed to both track the ball and hold their gaze on the point of last disappearance until they opened a door. If they shifted their gaze elsewhere

Fig. 2. View of the apparatus used in Butler et al. (2002). The transparent screen allows the ball to be seen rolling between doors 2 and 3. Copyright 2002 by the American Psychological Association. Adapted with permission.

before responding, they usually lost track of the correct door and opened another one. Thus, a child might track the ball down to door 4, look somewhere else, and come back to open door 2, despite having seen it roll past this door. Because children tended to frequently break their gaze, they had a high error rate overall. Children at 2.5 years of age seemed able to keep track of the location better. When they tracked the ball to the point of last disappearance, they opened the correct door ~85% of the time (Butler et al., 2002).

In both Mash et al. (2003) and Butler et al. (2002) children could direct their search based solely on spatiotemporal information provided by the ball's movement. Were they using the barrier cue at all? To answer this question an eye-tracking system was employed that measured where children's point of gaze was directed during the entire dynamic event (Kloos et al., 2006). On the assumption that looking at the barrier at some point during the hiding event would signify recognition of its role, we scored whether children fixated the barrier at least once between onset of the ball's movement and opening a door. Even with this generous interpretation of the meaning of eye fixation on the barrier, we failed to find evidence to support the view that 2-year-olds used the barrier as a cue to the ball's location. They fixated the barrier on only 20% of trials and when they did, performance was not improved. Correct search was present on only 25% of trials when children had looked at the barrier. Clearly, a look toward the barrier before opening a door was not guiding their response (Kloos et al., 2006). The clear screen with four opaque doors from Butler et al. (2002) was used in this study, and overall performance of 36% correct was comparable to Butler et al.'s (2002) 39% correct. Thus, the eye data supported the view that although 2-year-olds were using the spatiotemporal information available through the clear screen somewhat, they were not attending to the barrier at all. On the other hand, we know that 3-year-olds, who solve this search task readily, look frequently at the barrier. They also make use of spatiotemporal information, and weigh these two cues appropriately when they are in conflict (Haddad et al., submitted).

Younger children's failure to use the barrier cue may be because it is more abstract than the spatiotemporal cue. Santos (2004) found that rhesus could use spatiotemporal information but not contact mechanical information of the barrier when searching for a hidden food item. She hypothesized that the spatiotemporal cue was more basic and would be used earlier in development than the more complex cue of the barrier. Data from all of our studies presented here and from Hood et al. (2000, 2003) bear this out. Neither search behavior nor eye gaze indicated that 2-year-olds were using the barrier to mark the ball's hiding place. When reasoning about the world, both macaques and young children attended to the direct information about an object's movement, but failed to use more indirect signals.

One obvious reason that toddlers ignore the barrier in favor of motion cues is that movement is extraordinarily captivating. A wealth of research attests to infants' attention to and understanding of objects' motion when undergoing occlusion (e.g., Wynn, 1996; Spelke, 1988; Gredebäck and von Hofsten, 2004; Johnson et al., 2004). The attractiveness of motion may have overshadowed the stationary wall. In a series of studies we sought to enhance the presence of the wall by a number of means. We used a hard wooden ball that clattered down a bumpy track to hit the wall with an audible clunk; we wrapped the wall in flashing red lights; we extended the wall over the screen so that its edge was visible all the way down to the ramp; and finally we substituted a human arm and hand to catch the ball (Keen et al., submitted). None of these manipulations increased correct responding by more than a few percentage points. One could conclude that our imagination was limited, or we were unlucky and did not hit on just the right means to draw attention to the barrier. We think a more conservative conclusion is that getting the child to notice the barrier is not the problem. Remember, even when they happened to look at the wall during the hiding event, performance did not improve (Kloos et al., 2006).

The search task used in all the studies presented here was modeled after studies testing very young infants' knowledge of solidity and continuity (Spelke et al., 1992). Initially both Bruce Hood and we saw the toddlers' failure to search correctly as disagreeing with the infant data (see discussions in Berthier et al., 2000; Hood et al., 2000). At the same time both sets of authors independently suggested that task differences might explain the success of infants and the failure of toddlers to recognize the importance of physical solidity. One key distinction was that the infancy research used preferential looking whereas the toddler research used manual search. A critical difference between these two paradigms is that the latter response requires prediction of the ball's location whereas the former does not. Preferential looking studies rely on the child's ability to recognize an anomalous visual display: the object is not in the right place. Several authors have pointed out that this after-the-fact recognition of something odd does not require the child to predict the object's exact location in advance (Diamond, 1998; Haith, 1998; Meltzoff and Moore, 1998). One would assume that 2-year-old children would respond the same as infants if the "door" task were converted into a looking-time task, but this had to be tested.

Using the 4-door, horizontal version of this task, Hood et al. (2003) tested the same children first with looking-time as the response, then with manual search. For the looking-time procedure Hood et al. (2003) had an experimenter roll the ball that disappeared behind an opaque screen with the barrier showing above one of the doors. The experimenter then simultaneously opened two doors on either side of the barrier. The object was revealed on the correct side of the barrier for half of the test trials and on the incorrect side (as though it had rolled through the solid barrier) for half the trials. Children in this study were 2.5 and 3.0 years of age. As in previous research with infants, toddlers looked longer when the object was on the wrong side of the barrier than when it was in the expected location. Nevertheless, when given the opportunity to search for the object in the usual procedure, 2.5-year-olds failed.

The findings of Hood et al. (2003) pointed to a clear dissociation between looking and searching in children's understanding of a hidden object's location. In a follow-up study Mash et al. (2006) corrected a design problem and changed the procedure to test whether children could predict the ball's location as well as react to an anomalous visual display. The design problem concerned the fact that Hood et al. (2003) always rolled the object from the same direction. This resulted in the toy resting on the same side of the barrier numerous times throughout familiarization of the apparatus and on half of the test trials. When the toy was on the wrong side of the barrier, it was not only anomalous but also novel. Confounding perceptual novelty with the cognitive anomaly can be serious because of the well-known preference for novelty (for review, see Haith and Benson, 1998). Mash et al. (2006) redesigned the ramp so that it could be rotated in either direction. During familiarization and test trials children saw the ball equally often on the right and left of the barrier (see Fig. 3 for a photograph of the apparatus in both positions). A final difference between our study and Hood et al.'s (2003) was age of children. We tested 2-year-olds rather than 2.5- and 3.0-year-olds because the younger group had shown virtually no success on the opaque screen version of this task (Berthier et al., 2000; Mash et al., 2003; Keen et al., in press).

The beginning of our procedure was the same as in the manual search version of the task. The experimenter placed the barrier on the ramp, lowered the screen, and released the ball, which rolled out of sight behind the screen (see Fig. 3). At this point the experimenter manipulated a hand puppet that traversed up and down the screen before opening a door. On 10 standard trials the puppet pulled down the correct door (i.e., on the side nearest to the barrier from the direction the ball had rolled)

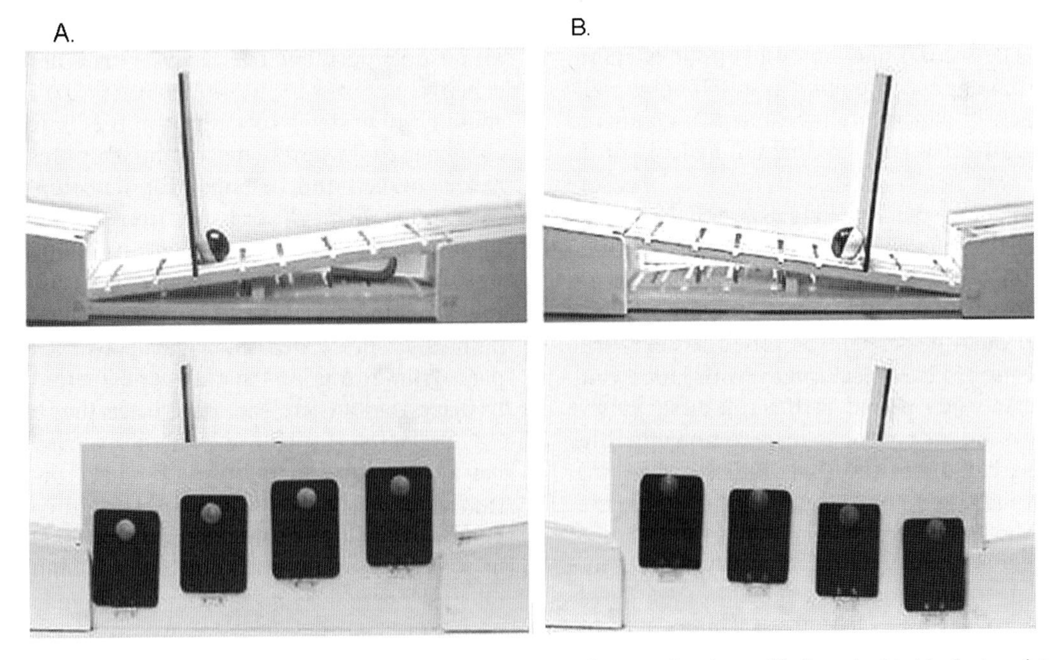

Fig. 3. View of the rotating ramp apparatus used in Mash et al. (2006). The ramp, barrier, and ball are depicted in the top pictures and the apparatus with the door panel in place is depicted in the bottom pictures. (A) Ramp inclined to the right. (B) Ramp inclined to the left. Copyright 2006 by the American Psychological Association. Adapted with permission.

and found the ball. These standard trials were important to maintain the notion that the barrier would stop the ball; we did not want children to regard this as a magic show put on by the puppet in which balls could be found randomly behind doors. There were 4 critical test trials interspersed throughout the standard trials, for a total of 14 trials. On all test trials the puppet opened a door and found no ball on the ramp. On consistent trials, the puppet opened an incorrect door, which should be no surprise if the child realized in advance that it was incorrect. When the screen was lifted the child saw the ball beside the barrier in the correct spot. On inconsistent trials, the puppet opened the correct door to reveal no ball; this would surprise the child if he/she had predicted the ball's location. When the screen was lifted, the ball was seen at a location beyond the barrier.

Given this two-step revelation about the ball's whereabouts, it was possible to devise two measures: (1) looking time when the puppet opened a door and found no ball; and (2) looking time when the screen was lifted to reveal the ball in either the correct or incorrect location. The former measure has the potential to tap whether the child predicted the ball would be in a specific location beside the barrier. The second measure is like that of infant looking-time procedures and also Hood et al.'s (2003) task in which looking time was compared when the object was in an anomalous location to when it was not. Analysis of the first measure found that toddlers looked longer when the puppet opened the correct door to find no ball than when it opened the incorrect door. Looking time did not differ between the two trial types for the second measure when the screen was raised to reveal the ball's location. These results agree with Hood et al. (2003) that when placed in the role of a passive observer, toddlers appear to have some knowledge about the ball's location. Even more compelling, our data indicate that children also appear to predict the ball's specific location by looking longer when it was not in the exact spot where it was expected to be. Why did they also not look longer when the screen was raised to reveal the ball beyond the barrier? We speculated that it was because the view of the ball beyond the barrier simply confirmed what they already knew — that the ball was not in the expected location. The second event was not unexpected, given that they understood and had already responded to the first event (Mash et al., 2006).

Must the child be a passive observer in order to display knowledge of the ball's hidden location? We attempted to devise a task that allowed the child to actively predict where the ball would stop, but without searching manually for it. Using the same rotating ramp and barrier as in Mash et al. (2006), we invited children to place a toy doll, Lorie, on the ramp where she could "catch" the ball (Kloos and Keen, 2005, Experiment 2). The game was explained in terms of Lorie wanting to stand on the ramp and catch the ball. To demonstrate the task, the experimenter placed the barrier on the ramp at an intermediate position. Lorie was then placed on the ramp at three positions in succession: in front of the barrier so the rolling ball knocked her over, right against the barrier where she "caught" the ball, and behind the barrier so the ball stopped at the barrier, leaving Lorie empty-handed. The experimenter commented on what happened each time after the ball was rolled. The children then had six familiarization trials with the ramp tilted in the same direction as the demonstration and the barrier placed in all possible positions over trials. For each trial the experimenter said, "I'm going to roll the ball. Where should Lorie stand to catch the ball?" After the child placed Lorie on the ramp, the experimenter rolled the ball and described the outcome ("Lorie caught the ball", "Lorie was too far away", or "Lorie got knocked over"). For test trials the ramp was tilted in the opposite direction and the ball was rolled just once, with no barrier present. The experimenter proceeded immediately to six trials of placing the barrier on the ramp, and invited children to place Lorie where she could catch the ball.

Children performed at a high level (87% correct) during the familiarization phase, indicating they understood the task; however, their placement of the doll against the correct side of the barrier may have been in imitation of the experimenter's demonstration rather than prediction of where the ball would stop. After the ramp was rotated, children placed Lorie correctly on 49% of trials. An examination of individual children's performance showed the widest possible spread of ability in these 2-year-olds' capacity to predict the ball's future position from a new direction. Two children out of 17 tested had perfect performance, and another 2 maintained high performance at 80% correct, with another 2 near 70%. The rest were at 50% or below and one child had none correct (see Fig. 5 in Kloos and Keen, 2005). Their most frequent error was to put her next to the barrier on the wrong side, clearly a holdover from their placement on familiarization trials.

In a second procedure we tested these same children on placing Lorie behind a screen to catch the ball. Without changing the direction of the ramp the barrier was first placed on the open ramp and the screen was lowered with the barrier showing above it. The children were asked to pull down a door and place Lorie on the ramp so she could catch the ball. After Lorie was placed and the ball was rolled, the experimenter lifted the screen to show the outcome. The purpose of this follow-up procedure was to see the effect of removing the child's view of a critical element in determining placement, the point where the bottom of the barrier met the ramp. Removal of this view resulted in plummeting performance, 32% correct, which was not better than chance.

In some ways, the percentage correct cannot be compared across these two procedures because chance is .25 when the child chooses among four doors, but chance cannot be calculated when Lorie is simply placed on the open ramp because she could be placed at a large number of incorrect locations and only one specific space was correct. We scored the doll's placement as correct if it was within 4 cm of the appropriate side of the barrier. One must appreciate that 49% correct is a high number because Lorie was a small doll (3 cm wide and 7 cm high) and she could be placed anywhere on the 75 cm long ramp. Kloos and Keen (2005) concluded that 2-year-olds could use their knowledge about the barrier to predict where the ball would stop with fair success, but only if they could see all critical elements. In the condition with no screen, children could see the ball, the barrier, and the intersection of ramp and barrier while they were planning where to place the doll within the array. This same task but with a screen occluding the ramp and bottom portion of the barrier appears to wipe out their capacity to imagine the ball's future location, in both the Lorie study and in all the studies reported here when they were simply opening a door to find the ball.

We hypothesized that the key to prediction for the 2-year-old is having the elements visually available within the spatial array. Having to imagine the arrangement of critical elements in the spatial array makes prediction difficult or impossible for children of this age. This analysis of the 2-year-old's problem fits with the looking-time data. In both the Lorie study and the looking-time procedure children do not have to imagine a hidden spatial layout. In Hood et al. (2003) children recognized that the ball was on the wrong side of the barrier. In Mash et al. (2006) they responded to the empty ramp next to the barrier after the puppet opened the door. In this case they responded to the absence of an expected object in an exact location, but note that they could see that exact location rather than having to imagine it.

The results discussed thus far seem to suggest that toddlers' capacity for object representation differs radically from that of adults and older children. Surely no adult (and very few 3-year-olds!) would fail to use the barrier wall to successfully locate the ball in the door task. Previous research suggests however, that even adults can at times be "blind" to the positions of objects that occlude ones they are attempting to track (Scholl and Pylyshyn, 1999). Inspired by literature on constraints on adults' attentive tracking of objects, we conducted a series of experiments to test the hypothesis that young children approach the door apparatus as an object-tracking task (Shutts et al., 2006). Research with adults has shown that attention is directed to whole objects and spreads continuously within an object from points nearer to points farther (e.g., Egly et al., 1994; see Scholl, 2001). We reasoned that if 2-year-olds approached the door apparatus as an object-tracking task, this signature profile should be revealed in their search performance.

We hypothesized that children might perform better on the door task if they could see part of the object they needed to track, because their attention might spread from the visible part of the object to its hidden body. To this end, we attached an antenna with a pompom to a toy car, rolled it down the ramp, and asked children to search for it. In one experiment the antenna was tall and therefore visible above the occluding screen, about equal in height to the barrier wall. In another condition the antenna was short and visible through clear windows we created in the doors. Children in the first experiment performed slightly above chance, but children in the second experiment were near ceiling. A control experiment eliminated the possibility that children in the second experiment were simply reaching for the door with something visible in the window, rather than searching for the specific object to be tracked. This pattern of results is consistent with the idea that children's attention was directed toward the visible part of the object (i.e., the pompom and antenna), and spread in a gradient-like fashion within the object (Egly et al., 1994).

As noted above, if children's attentive tracking resembles adults', then objects further away from the focus of attention may be overlooked. Was the problem with the wall all along due to the screen being too high? Was the top of the wall too far away from its hidden base on the ramp and the ball to serve as a cue? A final experiment tested whether the wall could serve as a cue to the object's location if it showed prominently through the window. Children saw the wall placed on the ramp, and it was positioned to show through the middle of the window, just as the pompom had. Then the car without a pompom was rolled and stopped by the wall out of sight in the same procedure as before. Two-year-olds' performance dropped to chance, just as when the wall was more distant above the screen. The failure to use a nearby cue underscored the need for spatiotemporal information directly connected to the object.

Every experiment that moved 2-year-olds' performance above chance featured spatiotemporal information directly connected to the hidden object. When children saw the ball moving between doors behind a transparent screen, careful tracking and maintenance of gaze helped them choose the correct door (Butler et al., 2002; Kloos et al., 2006). The Shutts et al. (2006) series of experiments similarly showed the power of spatiotemporal information that directly connected the cue with the object's location. During all familiarization procedures children were shown that the barrier stopped the object. Yet knowing that the stationary barrier stops the moving object is not sufficient to support reasoning that connects this information with where to find the ball.

It may be that the spatial layout is too complex for 2-year-olds to imagine the juxtaposition of ramp, barrier, object, and door (Keen, 2005). In this regard, the "door task" joins other search tasks in which toddlers show a remarkable inability to reason about a hidden object's location. DeLoache (1986) found that 2-year-olds could locate a hidden object only when the visual cue was part of the hiding place. When the distinctive cue was indirect, i.e., adjacent but not part of the hiding place, performance dropped. In a unique test of understanding hidden movement, Hood (1995) dropped balls down twisted opaque tubes connected to cups at the bottom. He found that 2-year-olds could not use the path of the tube to predict which cup a ball would land in. Rather they made a "gravity error", choosing the cup directly underneath the position where the ball was dropped, ignoring the path of the tube.

In summary, 2-year-olds have great difficulty following an object's hidden movements when no direct spatiotemporal information is available. Reasoning about contact mechanical effects such as one object blocking another's progress, and combining this information with a hidden spatial layout containing the ball's location requires a level of cognitive development not reached by most 2-year-olds. In tasks of passive observation or active participation toddlers appear to represent complex events, but only if they see the whole spatial array where a hidden past event took place (Hood et al., 2003; Mash et al., 2006), or where a future event will take place (Kloos and Keen, 2005). Children under 3 years of age need the support of viewing the critical elements involved in a contact mechanical event in order to engage in reasoning about an object's specific location. Three-year-olds show a major cognitive advance when they can predict the effect of a barrier in a moving object's path (Berthier et al., 2000; Hood et al., 2000) or evaluate conflicting spatiotemporal and contact mechanical information (Haddad et al., submitted). In a thorough review of the literature Newcombe and Huttenlocher (2000) covered location coding, landmark use, place learning, and motor learning as important processes in understanding space and objects. Our research adds to this body of knowledge by laying out a developmental course for using spatiotemporal and contact mechanical information in young children.

Acknowledgements

Support for this work was provided in part by National Institutes of Health grant R37 HD27714 to Rachel Keen (formerly Rachel Keen Clifton), and by a Mind, Brain, & Behavior fellowship from Harvard University to Kristin Shutts.

References

Berthier, N.E., DeBlois, S., Poirer, C.R., Novak, M.A. and Clifton, R.K. (2000) Where's the ball? Two- and three-year-olds reason about unseen events. Dev. Psychol., 36: 394–401.

Butler, S.C., Berthier, N.E. and Clifton, R.K. (2002) Two-year-olds' search strategies and visual tracking in a hidden displacement task. Dev. Psychol., 38: 581–590.

DeLoache, J.S. (1986) Memory in very young children: exploitation of cues to the location of a hidden object. Cognit. Dev., 1: 123–137.

Diamond, A. (1998) Understanding the A-not-B error: working memory vs. reinforced response, or active trace vs. latent trace. Dev. Sci., 1: 185–189.

Egly, R., Driver, J. and Rafal, R.D. (1994) Shifting visual attention between objects and locations: evidence from normal and parietal lesion subjects. J. Exp. Psychol. Gen., 123: 161–177.

Gredebäck, G. and von Hofsten, C. (2004) Infants' evolving representations of object motion during occlusion: a longitudinal study of 6- to 12-month-old infants. Infancy, 6: 165–184.

Haddad, J., Kloos, H. and Keen, R. (submitted) Conflicting cues in a dynamic search task are reflected in children's eye movements and search errors.

Haith, M.M. (1998) Who put the cog in infant cognition: is rich interpretation too costly? Infant Behav. Dev., 21, 167–179.

Haith, M.M. and Benson, J.B. (1998) Infant cognition. In: Damon, W. (Ed.), Handbook of Child Psychology, Vol. 2. Wiley, Hoboken, NJ, pp. 199–254.

Harris, P.L. (1983) Infant cognition. In: Haith, M.M. and Campos, J.J. (Eds.), Handbook of Child Psychology, Vol. 2. Wiley, New York, pp. 690–782.

Hood, B. (1995) Gravity rules for 2–4 year-olds? Cognit. Dev., 10, 577–598.

Hood, B., Carey, S. and Prasada, S. (2000) Predicting the outcomes of physical events: two-year-olds fail to reveal knowledge of solidity and support. Child Dev., 71: 1540–1554.

Hood, B., Cole-Davies, B. and Dias, M. (2003) Looking and search measures of object knowledge in pre-school children. Dev. Psychol., 39: 61–70.

Johnson, S.P., Slemmer, J.A. and Amso, D. (2004) Where infants look determines how they see: eye movements and object perception performance in 3-month-olds. Infancy, 6: 185–201.

Keen, R. (2003) Representation of objects and events: why do infants look so smart and toddlers look so dumb? Curr. Dir. Psychol. Sci., 12: 79–83.

Keen, R. (2005) Using perceptual representations to guide reaching and looking. In: Reiser, J., Lockman, J. and Nelson, C. (Eds.), Action as an Organizer of Learning and Development: Minnesota Symposia on Child Psychology, Vol. 33. Erlbaum, Mahwah, NJ, pp. 301–322.

Keen, R. and Berthier, N. (2004) Continuities and discontinuities in infants' representation of objects and events. In: Kail, R. (Ed.), Advances in Child Development and Behavior, Vol. 32. Academic Press, New York, pp. 243–279.

Keen, R., Berthier, N., Sylvia, M., Butler, S., Prunty, P. and Baker, R. (in press) Toddlers use of cues in a search task. Infant. Child Dev. (in press).

Kloos, H., Haddad, J. and Keen, R. (2006) Use of visual cues during search in 24-month-olds: evidence from point of gaze measures. Infant Behav. Dev., 29: 243–250.

Kloos, H. and Keen, R. (2005) An exploration of toddlers' problems in a search task. Infancy, 7: 7–34.

Mash, C., Keen, R. and Berthier, N.E. (2003) Visual access and attention in two-year-olds' event reasoning and object search. Infancy, 4: 371–388.

Mash, C., Novak, E., Berthier, N.E. and Keen, R. (2006) What do two-year-olds understand about hidden-object events? Dev. Psychol., 42: 263–271.

Meltzoff, A.N. and Moore, M.K. (1998) Object representation, identity, and the paradox of early permanence: steps toward a new framework. Infant Behav. Dev., 21: 201–235.

Newcombe, N.S. and Huttenlocher, J. (2000) Making Space: the Development of Spatial Representation and Reasoning. MIT Press, Cambridge.

Piaget, J. (1954) The construction of reality in the child. Ballantine, New York.

Santos, L.R. (2004) 'Core knowledges': a dissociation between spatiotemporal knowledge and contact-mechanics in a non-human primate? Dev. Sci., 7: 167–174.

Scholl, B.J. (2001) Objects and attention: the state of the art. Cognition, 80: 1–46.

Scholl, B.J. and Pylyshyn, Z.W. (1999) Tracking multiple items through occlusion: clues to visual objecthood. Cognit. Psychol., 38: 259–290.

Shutts, K., Keen, R. and Spelke, E.S. (2006) Object boundaries influence toddlers' performance in a search task. Dev. Sci., 9: 97–107.

Spelke, E.S. (1988) Where perceiving ends and thinking begins: the apprehension of objects in infancy. In: Yonas A. (Ed.), Perceptual Development in Infancy: Minnesota Symposium on Child Psychology. Lawrence Erlbaum Associates, Hillsdale, NJ, pp. 197–234.

Spelke, E.S., Breinlinger, K., Macomber, J. and Jacobson, K. (1992) Origins of knowledge. Psychol. Rev., 99: 605–632.

Wynn, K. (1996) Infants' individuation and enumeration of actions. Psychol. Sci., 7: 164–169.

C. von Hofsten & K. Rosander (Eds.)
Progress in Brain Research, Vol. 164
ISSN 0079-6123

CHAPTER 13

Learning and development in infant locomotion

Sarah E. Berger[1,*] and Karen E. Adolph[2]

[1]*College of Staten Island, Graduate Center of the City University of New York, Department of Psychology, 2800 Victory Blvd, 4S-221A, Staten Island, NY 10314, USA*
[2]*New York University, Department of Psychology, 6 Washington Place, Room 410, New York, NY 10003, USA*

Abstract: The traditional study of infant locomotion focuses on what movements look like at various points in development, and how infants acquire sufficient strength and balance to move. We describe a new view of locomotor development that focuses on infants' ability to adapt their locomotor decisions to variations in the environment and changes in their bodily propensities. In the first section of the chapter, we argue that perception of affordances lies at the heart of adaptive locomotion. Perceiving affordances for balance and locomotion allows infants to select and modify their ongoing movements appropriately. In the second section, we describe alternative solutions that infants devise for coping with challenging locomotor situations, and various ways that new strategies enter their repertoire of behaviors. In the third section, we document the reciprocal developmental relationship between adaptive locomotion and cognition. Limits and advances in means–ends problem solving and cognitive capacity affect infants' ability to navigate a cluttered environment, while locomotor development offers infants new opportunities for learning.

Keywords: infant; locomotion; affordance; perception-action; crawling; walking; means–ends problem solving

How infants move

Traditional view

For more than 100 years, researchers have posed the question of how infants move (e.g., Burnside, 1927; Ames, 1937; Trettien, 1900). Locomotion is one of the great coups of infancy. The advent of independent mobility marks the transition from helpless newborn to independent agent. How do infants triumph over gravity so as to pull their bodies from the ground and power up to go somewhere?

The traditional answer has two parts. The first part addresses what movements look like at various points in development. In an attempt to minimize extraneous environmental factors and varied task demands, researchers encourage infants to crawl and walk repeatedly over a flat, uniform surface at a steady pace. A century of clever and increasingly sophisticated technologies has produced precise documentation of the timing and coordination of infants' locomotor movements, the trajectories of their limbs, and the forces they produce. Based on thousands of observations of hundreds of children, researchers in the 1930s concluded that the normative development of locomotion was stage-like, identifying, for example, nearly two dozen stages of prone

*Corresponding author. Tel.: 718 982 4148;
E-mail: sberger@mail.csi.cuny.edu

DOI: 10.1016/S0079-6123(07)64013-8

progression, from "passive kneeling" (Stage 1) to crawling on hands and knees (Stage 19) and beyond (Gesell, 1933, 1946; Gesell and Thompson, 1938), and multiple stages in the development of upright locomotion, from newborns' stepping movements to independent walking (Shirley, 1931; McGraw, 1932, 1935, 1945).

The second part of the traditional answer to the question of how infants move concerns how infants acquire the strength and balance to hoist themselves up and around. The original answer was informed and constrained by researchers' reliance on data from infants' locomotion over uniform paths and by the assumption that locomotion unfolds in stages. Gesell, McGraw, and the other early pioneers took the presumed invariance of the locomotor stages as evidence that neural maturation was the driving force of development and that each stage reflected the underlying status of infants' neuromuscular system (Halverson, 1931; Shirley, 1931; Gesell and Thompson, 1934; Gesell and Ames, 1940; McGraw, 1945). On this account, maturation of neural structures and circuitry allows infants intentional control of their limb movements (McGraw, 1935, 1945; Forssberg, 1985), and affects strength and balance by increasing the efficiency and speed with which perceptual information and motor signals are integrated and processed (Zelazo et al., 1989; Zelazo, 1998).

Many modern researchers have shared the traditional emphasis on describing developmental changes in infants' crawling and walking movements over flat, uniform ground, and in understanding the origins of these locomotor patterns. Rather than reifying developmental changes into stages, modern researchers describe the shape of the developmental trajectories in crawling and walking and the relations between various measures of locomotor proficiency (e.g., Bril and Breniere, 1989, 1992; Adolph et al., 1998; Adolph et al., 2003). Rather than relying solely on changes in the brain to explain the developmental origins of locomotion, modern researchers also stress the role of peripheral, biomechanical factors, and the role of experience (for review, see Adolph and Berger, 2006).

New view

Ironically, the abundance of laboratory data collected over the past 100 years with infants crawling and walking at a steady pace over short, straight paths on flat ground cannot speak to the variety and novelty of infants' everyday experiences with balance and locomotion. Outside the laboratory, infants travel along circuitous paths for large distances over variable ground surfaces with different walking speeds (Adolph, 2002; Garciaguirre and Adolph, 2006). In the course of a normal day, infants travel over nearly a dozen different indoor and outdoor surfaces varying in friction, rigidity, and texture. They visit nearly every room in their homes and they engage in balance and locomotion in the context of varied activities. Locomotor experience is distributed in bouts of activity, occasionally over large distances, but most frequently involving only a few steps. Movement is interspersed with periods of quiet stance, when infants stop to rest, play with objects, or interact with caregivers. Sometimes bouts end abruptly when infants fall, and sometimes bouts end more smoothly when infants' attention is captured by new features of the environment.

Thus, there is a very different interpretation of the question about how infants move. This new view is concerned with the adaptiveness of infants' locomotor decisions: How do infants select the appropriate locomotor method for getting from one place to another, modify their ongoing movements appropriately, and discover alternative methods of locomotion to cope with new situations? At the same time that infants acquire new locomotor skills, their bodies and environments are undergoing dramatic developmental change. Infants' body growth is episodic: They can wake up to find themselves 0.5–1.65 cm taller (Lampl et al., 1992). Their bodies become less top-heavy and more cylindrical. With their hands freed from supporting functions, walking infants can vary the location of their center of mass just by picking up a toy. More mature body dimensions and proficient locomotor skills provide infants with increased access to the environment and spur caregivers to allow them greater latitude to encounter novel features of the environment on their own (e.g., removing the gates barring infants from stairs). In short, the new interpretation for the question of how infants move concerns how infants cope with a variable body in a changeable world.

Part of the new answer lies in new methodological approaches to the study of locomotor development. Rather than testing infants on flat, uniform ground, researchers have designed new paradigms that challenge infants with variable features of the environment (e.g., Gibson and Walk, 1960; Gibson et al., 1987; Schmuckler, 1996; Joh and Adolph, 2006). For example, researchers have varied the slant, friction, and rigidity of the ground surface, created gaps in the path, terminated the path in a cliff, blocked the path with overhead or underfoot barriers, varied the height of stair risers and pedestals, varied the width of bridges spanning a precipice, and varied the substance and extent of handrails used for manual support (for reviews, see Adolph et al., 1993b; Adolph, 1997, 2002, 2005; Adolph and Berger, 2005, 2006). The focus is on the strategies infants use to cope with the various obstacles and the adaptiveness of their locomotor decisions. Measures of movement kinematics and forces are used in the service of addressing how well suited infants' movements are to the current environmental constraints.

Chapter overview

This chapter describes examples of the new view of locomotor development. We describe the development of infants' ability to detect and exploit possibilities for locomotion, using slopes, gaps, bridges, and stairs as case studies. Most infants have never navigated slopes, gaps, and bridges on their own (caregivers prevent infants from approaching a precipice), so that these tasks provide a way to observe infants' ability to cope with a novel, potentially risky situation. We focus on how infants gather perceptual information and how they use that information to gauge the possibility or impossibility of crawling or walking. We also discuss how infants learn to solve complex locomotor problems that require cognitive skills for planning and constructing movement strategies. If infants perceive crawling or walking to be possible, they must decide whether to modify their ongoing movements. If crawling or walking is deemed impossible, infants must devise an alternative solution and implement it.

Perceiving affordances for balance and locomotion

Distinguishing possible from impossible actions

The ability to detect affordances lies at the heart of adaptive locomotion. Affordances are possibilities for motor action. Actions are possible only when there is a fit between infants' physical capabilities and the behaviorally relevant features of the environment (Gibson, 1979; Warren, 1984). For example, walking is possible only when infants have sufficient strength, postural control, and endurance relative to the distance they have to walk, obstacles along the way, and the slant, rigidity, and texture of the ground surface. In fact, the body and environment are so intimately connected for supporting motor actions that changes in a single factor on either side of the affordance relationship alter the probability of successful performance. On the body side of the affordance relationship, for instance, as the weight of a load increases, walking becomes impossible. As balance control increases with development, infants can navigate steeper slopes. Reciprocally, on the environment side of the affordance relationship, as the degree of slant increases, walking becomes impossible. With increase in surface traction, infants can walk over steeper slopes. The ability to detect affordances means that infants can take into account both the constraints of their bodies and of the environment when carrying out their movements.

Investigation into infants' perception of affordances has required the design of new laboratory paradigms. The tasks must be novel to control for infants' previous experience and to ensure that infants' behaviors reflect their spontaneous perception of the possibilities afforded by the experimental apparatus. For example, most infants have never descended a steep slope or playground slide on their own. Thus, in a series of studies, infants encountered slopes in the middle of an otherwise flat path (e.g., Adolph et al., 1993a; Adolph, 1995, 1997; Adolph and Avolio, 2000; Mondschein et al., 2000). As shown in Fig. 1, flat starting and landing platforms flanked an adjustable slope. By pressing a remote switch or cranking a jack, the slope could be adjusted to any degree of slant between 0° (flat) and 90° (sheer drop-off). Parents stood at the bottom of the slope encouraging their infants to descend. An experimenter followed alongside infants to ensure their safety. The degree of slant varied from trial to trial, so that some slopes were perfectly safe, some were impossibly risky, and a narrow band of slopes had a shifting probability of success. The primary outcome measures were whether and how infants decided to descend safe and risky slopes to reach the goal at the bottom.

After 10 weeks or so of crawling or walking experience, infants geared their locomotor decisions to the possibilities for action. On trials when slopes were within their abilities, infants crawled or walked to the bottom of the walkway. However, when slopes were beyond their abilities, infants refused to crawl or walk down. After 20 weeks of crawling or walking experience, infants' perception of affordances matched the actual possibilities for action within a few degrees of accuracy. That is, infants' rate of attempts closely matched the conditional probability of success. Information-gathering behaviors became more efficient as the accuracy of infants' motor decisions improved (Adolph, 1995, 1997). Infants looked down the slope as they approached it over the starting platform. They stopped at the edge and generated haptic information from touching. Crawlers put both hands on the slopes and rocked back and forth over their wrists. Walkers stood at the brink of the slopes and rocked back and forth around their ankles. Latency, looking, and touching increased on risky slopes.

Infants' ability to detect affordances is especially impressive because the possibilities for action change from week to week. Affordances were

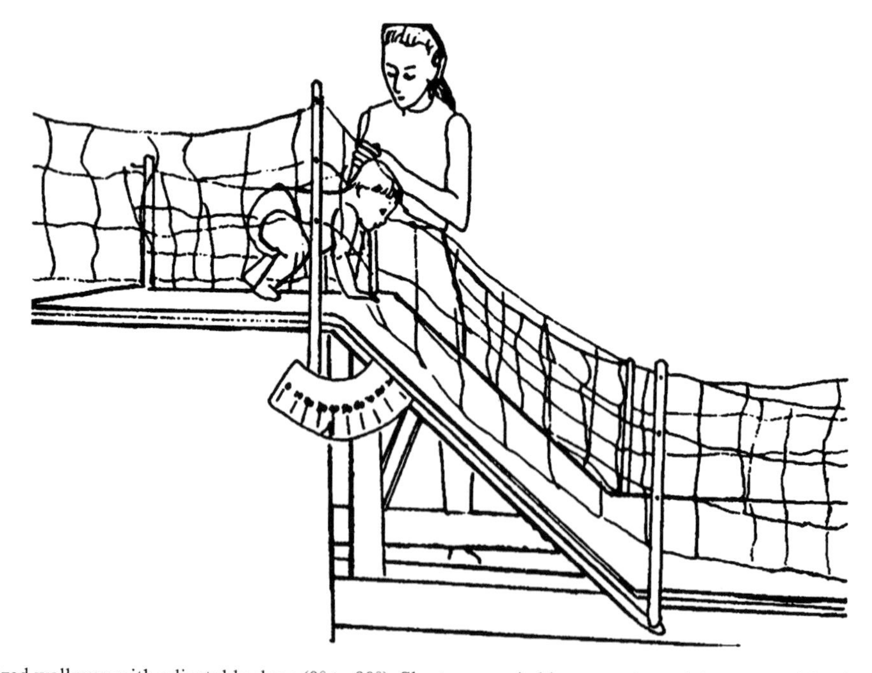

Fig. 1. Mechanized walkway with adjustable slope (0° to 90°). Slant was varied by operating a drive screw with a hydraulic lift. Infants began each trial on the starting platform and traversed the sloping middle section while an experimenter (shown) followed alongside for safety. Parents (not shown) waited at the far end of the walkway. Adapted with permission from the Adolph (1997), Blackwell Publishers.

related to naturally occurring changes in infants' body dimensions, level of locomotor skill on flat ground, and duration of everyday locomotor experience: More maturely proportioned, proficient, and experienced crawlers and walkers, could crawl and walk down steeper slopes. Thus, detecting affordances required infants to relate changes in the degree of slant on each trial to their current bodily propensities. A risky slope one week might be perfectly safe the next week after crawling or walking skill improved. A safe slope for crawling might be risky for walking.

In fact, experienced walking infants were so sensitive to both terms of the affordance relationship that they could simultaneously update their assessment of affordances based on experimental manipulation of their body dimensions and walking proficiency, and to variations in the slope of the ground surface. Fourteen-month-old walking infants were outfitted in a vest with shoulder packs that could be filled with different loads (Adolph and Avolio, 2000). The weight of the shoulder packs changed at the start of each trial. On some trials, the packs were filled with lead-weights (25% of infants' body weight). On other trials, the shoulder packs were filled with feather-weight Polyfil to serve as a control for the heavy condition. The lead-weight packs experimentally changed infants' body proportions to make them more top-heavy and immaturely proportioned (Fig. 2). As a consequence, the heavy loads also decreased infants' walking proficiency on flat ground and slopes. Infants recalibrated their perception of affordances from trial to trial: They treated the same degree of slope differently depending on whether they were wearing feather- or lead-weight packs. Infants were more likely to attempt to descend a given degree of slant wearing the feather-weight packs than when they were wearing the heavy packs.

Experienced infants can detect affordances for stance as well as for locomotion. In stance, the base of support is stationary, but typically the rest of the body is moving — head, arms, and torso move while in sitting and standing positions. We turn to look at something, lift our arms to reach for something, nod, cough, stretch, and bend. Each movement changes the location of infants' center of mass and creates new destabilizing torque. If infants reach or lean too far in any

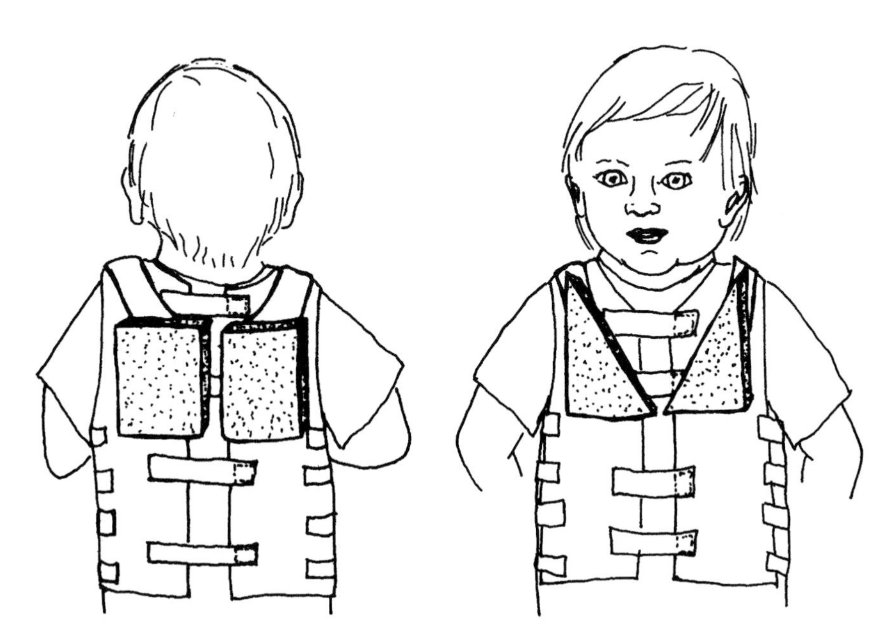

Fig. 2. Velcro vest with removable shoulder packs. Adapted with permission from J. Lockman, J. Reiser, & C. A. Nelson (Eds.), Action as an organizer of perception and cognition during learning and development: Symposium on Child Development (Vol. 33), by K. E. Adolph, "Learning to learn in the development of action," pp. 91–122, 2005, Lawrence Erlbaum Associates.

direction, they will lose balance and fall. To test infants' perception of affordances for maintaining balance in stance, 9.5-month-olds were perched in a sitting position at the edge of an adjustable gap (Adolph, 2000). The gaps apparatus was composed of a fixed starting platform and a movable landing platform. The landing platform slid back and forth along a calibrated track to create gaps 0–90 cm wide between the two platforms. A lure offered at infants' chest height from the landing platform encouraged them to lean forward and stretch an arm out to reach (Fig. 3). Caregivers also offered encouragement from the far side of the landing platform. An experimenter stood nearby infants to rescue them if they fell into the precipice.

Although the gap situation was new, infants were experienced sitters (on average, 15 weeks of experience with independent sitting) and they could detect precisely how far forward they could reach and lean without falling into the gap. On safe gap widths within their ability, infants leaned forward to retrieve the lure. But, on risky gap widths beyond their ability, infants refused to reach. Infants occasionally extended and retracted their arm without touching the lure, as if to explore the limits of the reaching space.

Occasionally, affordances for locomotion are expanded when features of the environment (a handrail or ladder) or extensions of the body (a walking stick or crutch) are used as a means, or tool, for mobility. Tools can make otherwise

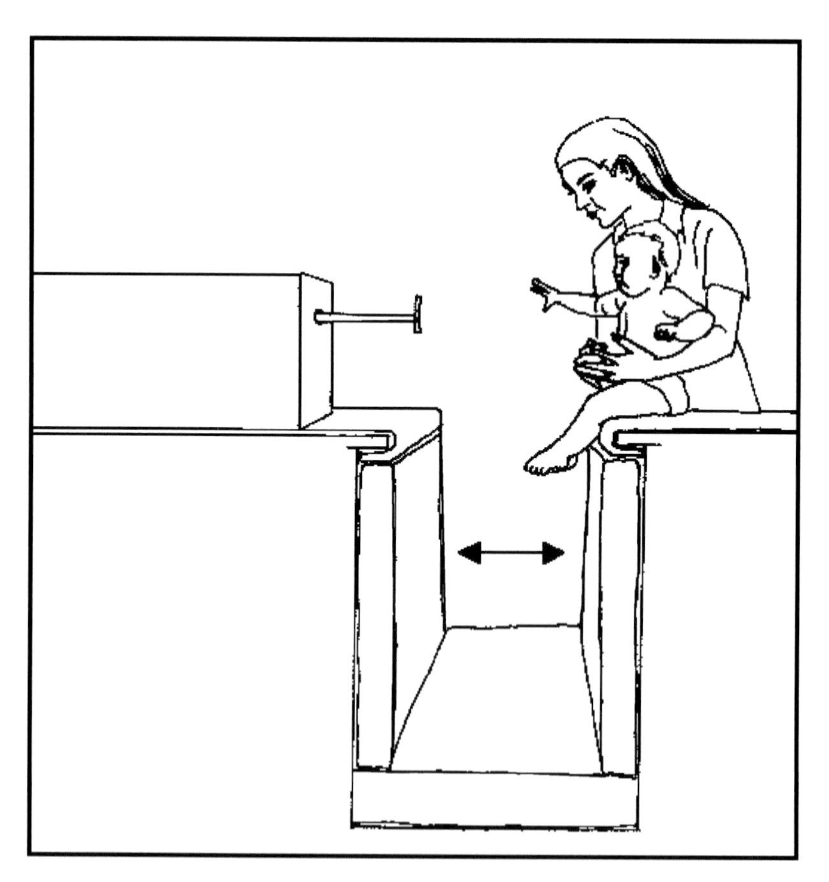

Fig. 3. Illustration of the adjustable gaps walkway for testing sitting infants' perception of affordances. Gap width varied from 0 to 90 cm by pulling the landing platform along a calibrated track. A sitting infant leaning forward over a gap in the surface of support. An experimenter (shown) walked alongside for safety. Parents (not shown) waited at the far end of the walkway. Adapted with permission from Adolph (2000), Blackwell Publishers.

impossible actions possible. Handrails and walking sticks, for example, can be used to augment balance in a treacherous situation. Ladders and climbing/sliding poles can make distant targets accessible. Typically, researchers consider infant tool use only in terms of hand-held implements such as rakes, spoons, and hammers (e.g., Chen and Siegler, 2000; Lockman, 2000; McCarty et al., 2001; Achard and von Hofsten, 2002; McCarty and Keen, 2005). However, recent work shows that experienced walking infants exhibit whole body tool use to expand affordances for locomotion.

By the time infants are 16 months old, their perception of affordances for locomotion is so sophisticated that they can use a handrail as a means for augmenting their balance while crossing narrow bridges (Berger and Adolph, 2003). Infants were encouraged to walk over wide and narrow bridges (12–72 cm) spanning a deep precipice between two flat platforms (Fig. 4). On half of the trials, a sturdy, wooden handrail ran the length of the bridge. On the other half of the trials, the handrail was absent. Parents stood at the far side of the precipice, offering infants toys as a lure, and an experimenter walked alongside infants for safety. Overall, infants were highly accurate in judging their ability to walk over bridges (94% of trials were successful). Regardless of whether the handrail was available, infants crossed the wide bridges (48–72 cm) and avoided the narrowest bridge (12 cm). On intermediate bridges (18–36 cm), infants walked more often when the handrail was available than when it was absent. Use of the handrail varied with bridge width. On wide bridges, the infants ignored the handrail; they ran straight across without touching the rail or they briefly tapped it with one hand in passing. On the 36-cm bridge, they used the handrail by facing forward and sliding one hand along the rail. On the 18- and 24-cm bridges, they turned sideways to face the rail and clutched on with both hands.

A follow-up study showed that infants took the material substance of the handrail into account when gauging its use as a tool for augmenting balance (Berger et al., 2005). Handrails were presented on every trial, but on some of the trials, the handrail was sturdy wood and on other trials the handrail was wobbly foam or rubber (Fig. 5). On intermediate bridges, infants walked more often when the handrail was sturdy and could support their weight than when the handrails were wobbly. These findings suggest that infants perceived the wooden handrail as a structure in the environment, separate from themselves and from the bridges, that could afford bridge crossing by augmenting balance.

Modifying ongoing movements

In a busy, natural environment — or while crossing slopes, bridges, and other situations designed for laboratory experiments — detecting affordances for locomotion requires more than deciding whether to crawl and walk. In addition, infants must modify ongoing crawling and walking movements to suit the current constraints on balance and locomotion. Task constraints, such as the degree of slant or the bridge width, can change from trial to trial. The situation might demand slowing down or speeding up, alterations in step length, turning the body, stooping over, lifting the legs, and so on. How is this accomplished? Experienced crawling and walking infants can plan modifications prospectively, before stepping over the brink of a slope or onto a bridge, concurrently, by monitoring and updating locomotion from step to step, and reactively, in response to a loss of balance.

Prospectivity has practical advantages. Planning gait modifications prior to encountering an obstacle allows infants to prevent themselves from falling, rather than having to recover as they begin to fall. For example, experienced crawling and walking infants take smaller, slower steps to cross steep slopes and narrow bridges (Adolph, 1997; Berger and Adolph, 2003; Berger et al., 2005). The evidence that gait modifications are prospective rather than reactive is that infants begin to curtail their step length and velocity on the starting platform, and turn their bodies to grasp the handrail, in anticipation of stepping onto the obstacle (Gill-Alvarez and Adolph, 2005; Berger et al., 2005). Visual information for the upcoming threat

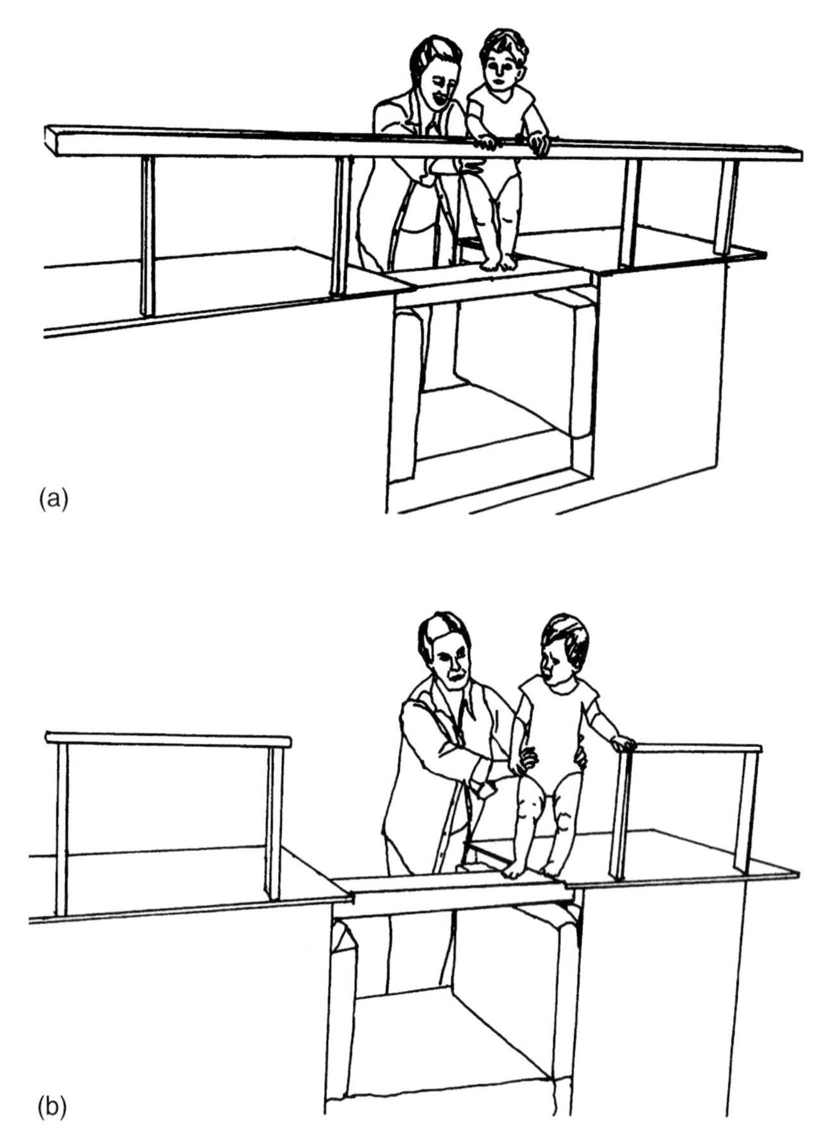

Fig. 4. Walkway with adjustable bridge widths and removable wooden handrail. (a) On handrail trials, a handrail spanned the length of the bridge and rested on permanent support posts. (b) On non-handrail trials, the permanent support posts remained on the starting and finishing platforms. The drop-off beneath the bridge was lined with foam padding. Infants began each trial on the starting platform. Parents (not shown) stood at the far end of the finishing platform offering encouragement. An experimenter (shown) followed alongside infants to ensure their safety. An assistant (not shown) adjusted the bridge widths by fitting them into a grooved slot, and switched the handrails by placing them on the support posts. Adapted with permission from Berger and Adolph (2003), the American Psychological Association.

to balance is sufficient to prompt the behavioral changes. Experienced crawling and walking infants do not have to feel themselves lose balance in order to modify ongoing movements.

Prospective gait modifications are exquisitely attuned to small changes in slant and bridge width. Across experiments, on shallow slopes well within their abilities, infants crawled or walked normally

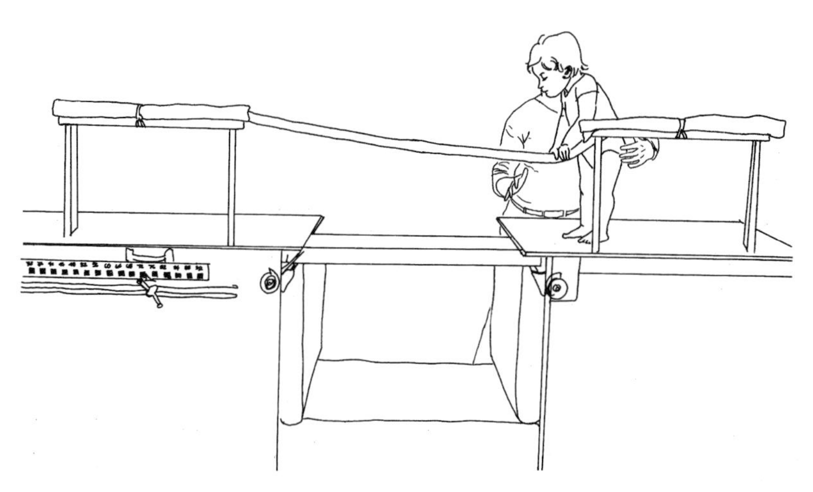

Fig. 5. Walkway with adjustable bridge widths and removable wobbly (foam or latex) handrail (shown with 20-cm bridge). Parents (not shown) stood at the far end of the finishing platform offering encouragement. An experimenter (shown) followed alongside infants to ensure their safety. Adapted with permission from Berger et al. (2005), Blackwell Publishers.

with large, fast steps. But, as slant increased and slopes challenged the limits of their abilities, infants modified their gait by taking smaller steps and more time to reach the goal. Walkers maintained a more stiffly upright posture and crawlers braced themselves by stiffly extending their arms. These gait modifications and postural adjustments served to brake infants' forward momentum from step to step, and thus allowed them to crawl and walk safely down a steeper range of slopes.

Narrow bridges posed a different problem: Rather than fighting gravitational forces as they moved along the bridge, infants had to fit their bodies into a limited space. The narrowest bridges were narrower than the width of infants' shoulders. On wide bridges infants faced forward and walked as usual, but on narrow bridges they modified their gait and posture. Some infants took more than 25 baby steps and longer than 30 s to travel 74 cm over the narrower bridges (Berger et al., 2005). In addition, infants turned their bodies sideways. Rather than using an alternating gait, where each foot traveled a farther distance than the last, they inched along with the trailing leg never passing the leading leg.

Even after infants have committed to crawling or walking over an obstacle, they continue to monitor and update their motor plans. Gait modifications can occur concurrently while taking advantage of the affordance. For example, on intermediate slopes, crawling infants sometimes lowered their chests from a high hands-and-knees position to a lower position with legs splayed wide, momentarily resting their chests on the surface to regain balance (Adolph, 1997). While walking over bridges, infants sometimes began with their bodies facing forward. Then, mid-bridge, they turned their bodies sideways so that they could grab the rail with two hands instead of one, and so that their bodies would fit better on the bridge (Berger and Adolph, 2003). On challenging bridges, infants inched along and then occasionally took a giant step when they neared the end of the bridge as if to facilitate a quicker exit to the landing platform. Updating in the midst of exploiting an affordance generally involved a switch from a less efficient, more precarious gait pattern to a more efficient, more stable strategy. Changes implemented on one trial frequently carried over to the next.

Reactive adjustments are less efficient (infants fall more frequently) than prospective and concurrent gait modifications. However, in situations where access to information about an upcoming perturbation is lacking, infants have no way to prepare for the obstacle, and reactive adjustments are the only option. For example, with the sudden addition of a load, the body can only react to the new distribution of forces. To assess reactive gait modifications and postural adjustments to load carriage, 14-month-old infants were challenged with the sudden addition of symmetrical and asymmetrical loads (15% of body weight) as they walked over flat ground (Garciaguirre et al., in press). The loads were distributed on the front, side, and back of infants' bodies and symmetrically across infants' chests and backs. The distribution of the load varied from trial to trial, so that infants could only determine the location of the load as they felt their bodies begin to tip.

Depending on the load condition, infants displayed 2–5 times as many trials with gait disruptions compared with a featherweight baseline condition. Infants coped with the lead-weight loads by decreasing their step length and walking speed and spending a greater proportion of the gait cycle with both feet on the ground and a smaller proportion of the gait cycle with one foot in the air. The gait modifications were reactive — infants felt themselves pulled off balance and then changed their walking patterns. Although adults show reactive modifications in posture by leaning in the opposite direction of the weights (think of leaning to the left while carrying a heavy suitcase in your right hand, or leaning forward while wearing a heavy backpack), infants reacted by leaning with the loads: They leaned forward while carrying the load in the front, leaned to the right while carrying the load on their right side, and so on. Apparently, infants' reactive gait modifications merely accommodate to the loads, whereas adults' reactive modifications compensate for the loads.

New solutions

To infants, the everyday environment is sometimes like a playground, pregnant with possibilities for playful forms of mobility, and sometimes like an obstacle course, rife with challenges. Crawling and walking are the most studied forms of infant locomotion, and sitting and standing are the most studied forms of stance, but these forms are not infants' sole means of mobility and balance. In situations that offer or demand alternative forms of locomotion, infants jettison their typical crawling and walking postures and search for new solutions. Infants' use of sliding, scooting, backing, and countless other alternative methods of locomotion not only inform on their perception of affordances, but in addition, provide insights into the acquisition of new behavioral forms.

Multiple solutions

Generally, infants find multiple and variable solutions for navigating challenging terrain, rather than a single, fixed approach. For example, successful navigation during descent, such as on a slope, staircase, or drop-off, requires strength and balance to fight gravity as the body is simultaneously lowered and moved forward. Both crawling and walking are difficult during descent because infants' body weight must be supported on a bent limb (an arm in crawling and a leg in walking), requiring more muscle strength than on a fully extended limb during ascent or on flat ground. Thus, on slopes and stairs, walking infants sometimes augmented their balance by holding an available handrail or grabbing onto the experimenter for extra support (Adolph, 1997; Berger, 2004). On steep slopes and stairs, infants abandoned their typical crawling and walking methods in favor of alternative strategies that lowered their center of mass and provided more balance control (Fig. 6). On some trials, infants slid down on their bellies, Superman-style, with arms extended straight in front on their heads and legs extended straight behind. On other trials, infants slid down on their bottoms in a sitting position. On still other trials, infants slid down backward, facing away from the landing platform (Adolph, 1997). Similarly, walking infants descended a steep staircase in a sitting position, scooting down from riser to riser, and in a backing position, crawling or sliding down with their heads pointing away from the floor (Berger, 2004).

Experienced walking infants frequently revert to crawling when balance is threatened. Walkers switch from upright to crawling to descend steep slopes on their hands and knees (Adolph et al., 1993a; Adolph, 1995, 1997), to cross a squishy waterbed (Gibson et al., 1987), and to cross

Fig. 6. Illustration of infants' typical methods and alternative strategies for descending slopes. Adapted with permission from the Adolph (1997), Blackwell Publishers.

narrow bridges (Berger and Adolph, 2003; Berger et al., 2005). Similarly, experienced hands-and-knees crawlers frequently revert to belly crawling for descending steep slopes (Adolph, 1997). The earlier developing form of locomotion appears in infants' repertoires as an alternative strategy before they exhibit new strategies such as sliding down slopes in sitting and backing positions (Adolph, 1997).

Researchers often find themselves outwitted by infants' ingenious solutions to experimental challenges. Indeed, infants may inadvertently benefit from their ignorance about the conventional uses of an apparatus, so that they perceive affordances where adult researchers see none. The standard visual cliff apparatus, for example, is composed of a large, rectangular glass table bisected by a narrow starting board to form two square sides (Gibson and Walk, 1960). On the "shallow" side, a textured surface abutting the glass provides visual information for solid ground. On the "deep" side, the textured surface lies on the floor far below the safety glass, providing visual information for an abrupt drop-off. A raised wooden wall along the outer edges of the apparatus prevents infants from falling to the floor. Infants are placed on the starting board near one of the long sides (due to the size of the apparatus, the experimenter cannot reach the middle of the starting board), and caregivers stand at the far diagonal corner, first on the shallow end and then on the deep end. On the shallow side, experienced crawling infants go straight to their caregivers in a beeline to the diagonal corner. On the deep side, they sometimes avoid crossing by remaining on the starting board. However, infants sometimes use an ingenious detour strategy: They cruise along the wooden walls, so that their hands are always in contact with a solid surface, and their faces are pointing out rather than looking down at the floor beneath the safety glass (Campos et al., 1978; Witherington et al., 2005). In addition, infants "backed" over the safety glass (turned around and crawled backwards to their caregivers), using the same sort of strategy that would be effective if the drop-off were real rather than illusory.

In the "wobbly handrails" experiment, the wobbly rail conditions were intended to be equivalent to the no-handrail conditions in the previous set of studies; the foam and rubber handrails were designed to give way when the infants leaned their weight on them, as indeed they did (Berger et al., 2005). Nonetheless, infants discovered new solutions for crossing narrow bridges with only a wobbly handrail for support (Fig. 7). Most frequently observed was a "hunchback" strategy, where infants walked sideways, stooped over, pressing down on the wobbly handrail. Infants also used a "snowshoe" strategy, where infants walked forward while resting their entire arm on the handrail to prevent it from dipping too deeply, just as snowshoes prevent walkers from plunging

Fig. 7. Illustration of infants' alternative strategies for using the "wobbly" handrails. Adapted with permission from Berger et al. (2005), Blackwell Publishers.

through the snow; a "mountain-climbing" strategy, where infants leaned backward while walking forward and pulled up on the handrail like a rope, dragging themselves along hand-over-hand; a "windsurfing" strategy, where infants walked sideways along the bridge, leaned backward and pulled up on the handrail as high as it would go with both hands; and a "drunken" strategy, where infants faced forward and leaned sideways against the railing, sliding their torsos along the handrail. Infants' alternative strategies were hilarious to observe, but also critical functionally. Infants who adopted one of the alternative handrail strategies were more likely to cross narrow bridges successfully than infants who did not.

Variety of means does not merely reflect differences between infants. Although any single alternative would suffice for navigating an obstacle, many infants use multiple strategies (Siegler et al., 1996). Individual infants use multiple alternative strategies on different trials within the same test session, and multiple means on the same trial. For example, across ages, most infants treated the same degree of slope differently within the same experimental session by using different alternative sliding strategies on different trials for the same increment (Adolph, 1997). While descending stairs, almost all infants (95%) used more than one descent strategy on stairs within the test session and almost all (90%) used more than one descent strategy within the same trial. For example, infants started down the stairs in a sitting position, but partway down switched to backing (Berger, 2004). While using a wobbly handrail to cross narrow bridges, infants often used several of the alternative strategies for crossing (e.g., started with "hunchback" and switched to "snowshoe") in the same trial (Berger et al., 2005).

Situations where infants cannot find a viable alternative tend to be most frustrating, and happily were most rare in laboratory studies. On trials where infants avoided traversal, they did not simply freeze at the brink of the obstacle. However, sometimes their search for alternatives was interrupted by bouts of displacement activities, as if they needed a break from the problem, and sometimes they found a way to escape from the situation entirely. For example, when infants were encouraged to descend a short, very steep, 36° slope and a long, shallow 10° slope, each slanting from a small, 55-cm high platform (Fig. 8), infants occasionally avoided the whole situation by

Fig. 8. Short, steep, 36° slope and long, shallow 10° slope. Infants started each trial on a platform and an experimenter (shown) walked alongside for safety. Parents (not shown) called to their infants from the bottom of each slope. Adapted with permission from Eppler et al. (1996), Elsevier.

scooting down from the backside of the platform rather than descending to their parents at the bottom (Eppler et al., 1996). Similarly, infants sometimes scooted off the backside of the starting platform (54-cm high) rather than descend a small staircase (Berger, 2004) and detoured off the side of the starting platform rather than cross a large, deformable foam pit (Joh and Adolph, 2006). In situations that did not allow a detour (the starting platform was too high, or escape was blocked by a wall), infants coped with the frustration of facing an insurmountable obstacle with displacement behaviors. When facing steep slopes or a rippling waterbed, frustrated infants turned their attention to the overhead ceiling lights, a crumb on the carpet, or their own belly buttons and diapers (Gibson et al., 1987; Adolph, 1997). Nine-month-old sitting infants' response to impossibly wide gaps was a most emphatic rejection: They pivoted away from the gap as if to shut out the sight of their caregivers offering the attractive lure (Adolph, 2000). They sat facing the back wall with their arms crossed until trials thankfully came to an end.

The variety of alternative strategies that infants exhibit for using handrails, descending slopes, and crossing the visual cliff and detour behaviors that remove infants from a frustrating task provide evidence that infants detect an array of affordances — rather than a single solution — for balance and locomotion. Thus, as infants approach an obstacle in their typical crawling or walking posture, they decide whether to continue with their typical method of locomotion, whether to modify their typical method, or whether an alternative strategy is required (Adolph et al., 2000). In the latter case, they must figure out an appropriate alternative by drawing on a pre-existing strategy in their repertoire, constructing a new strategy on the fly, or by trying various strategies until finding one that works.

Explicit instruction

How do infants add new skills to their repertoire of locomotor strategies? One possibility is explicit instruction. Parents might teach infants to execute a new form of locomotion by physically moving the appropriate parts of infants' bodies into the appropriate configurations, modeling the target behavior for infants, and using words and gestures to explain to infants what to do. One situation where infants learn new strategies by explicit instruction is descending stairs.

Infants encounter stairs in their own homes, in the homes of friends and relatives, and outdoors in parks and playgrounds. In a study of 732 families investigating the circumstances surrounding how and when infants learn to navigate stairs (Berger et al., 2007), many parents (58%) reported that they explicitly attempted to teach their infants safe strategies for descending stairs. Parents' most frequent teaching method was hands-on training. They turned infants' bodies at the top landing and moved their arms and legs onto each riser to show them how to get down. Less frequently, parents modeled stair descent by getting down on all fours on the stairs and crawling backward. Parents also gave verbal instructions, such as "turn around and back down" and encouragement such as "good job." Most parents reported using several of these teaching strategies in combination.

Because of the risks associated with descending stairs, most parents preferred that their infants descend in a backing position on their hands and knees because they deemed it safer than attempting to walk, scooting down sitting, and crawling/sliding headfirst. Parents' preference for backing may account for why so many infants were taught to descend stairs, rather than left to discover a strategy on their own. Backing is the most cognitively demanding descent strategy because it requires that infants coordinate several steps into the proper sequence: First, infants must switch from upright to a prone position, then they must pivot their bodies 180° so that they are facing away from the bottom landing, and finally, infants must move their bodies backwards. The initial detour may be especially problematic. Turning away from their destination is difficult for young infants because they tend to be visually and motorically "captured" by a goal (McGraw, 1935; Lockman, 1984; Diamond, 1990; Lockman and Adams, 2001). Although moving backward is not difficult motorically, it is difficult perceptually and cognitively. Infants must give up sight of the goal and deliberately move toward a goal that is only represented, not seen. Thus, it is reasonable that parents determined that their young infants should be helped through the process.

Explicit instruction, however, has limitations. Generally, parents teach their infants the backing strategy on a particular staircase in the home or playground. Learning in the narrow training environment may not transfer when infants encounter a new staircase or a different descent problem where the backing strategy would be useful. Infants who were taught to descend stairs at home using the backing strategy were no more likely to display the backing strategy on laboratory slopes and stairs than infants who had not been taught to back (Adolph, 1995, 1997).

Constructing new solutions

A second way that infants acquire alternative locomotor strategies into their repertoires is by a process of exploration and construction. As they faced impossibly steep slopes and narrow bridges, experienced infants rarely rested quietly. Instead, they concertedly sought out information, weighed whether a particular posture afforded traversal, and continued to try out different combinations of postures and strategies until a solution was found. They shifted their bodies from one position to another, from standing to sitting to squatting to backing, and so on (Adolph et al., 1993a; Adolph, 1995, 1997). They grasped the handrails and support posts in one body configuration and then tested a second or third configuration. These shifts in position had the quality of a concerted means–ends exploration geared toward a goal.

Sometimes the trials ended with infants back in their original position, no closer to the goal. Sometimes trials ended with infants assuming an appropriate crossing position. Frequently, a position assumed on one trial that was not fully executed was retried on a subsequent trial or trials and executed successfully. For some infants, a new strategy might appear full-blown. For example, one experienced crawler pivoted several times in circles on his belly and then shot himself backward down a steep slope (Adolph, 1997). For other infants, the new strategy might appear piece meal, with a 180° pivot on one trial, and the push backward on a subsequent trial.

Sometimes infants took such delight in a newly acquired alternative strategy that they overgeneralized the need for the strategy and used it on increments that could be easily navigated by crawling and walking. In particular, infants enjoyed sliding down slopes. After their first discovery of a new sliding position, they might attempt to cross shallow slopes or even flat ground by backing or scooting down in a sitting position.

Serendipitous discovery

Learning new methods for traversing challenging surfaces was not always intentional. A third route for alternative strategies to enter infants' repertoires was serendipity. Like the process of exaptation in evolution, sometimes infants started out using their typical method of locomotion, but unexpectedly found themselves implementing an alternative. In other words, the biomechanics of the situation may have transformed their intended form of locomotion into a useful alternative.

On slopes, backing and sitting entered some infants' repertoires by the serendipitous route (McGraw, 1935; Adolph, 1997). When experienced crawlers attempted to crawl down steep slopes, they kept their arms stiffly extended. With their legs flexed tightly under their hips, their torsos would be turned sideways due to gravity. Infants would find themselves sliding down sideways. On some trials, infants' bodies would get turned so far around that their feet would be pointing toward the landing platform. Evidence that infants were surprised at this turn of events is that they cried out, "uh oh" and "oh no," crawled back up to the starting platform, and looked down the slope once again in consternation (Adolph, 1997). In time, infants who had several accidental backing trials learned that it was easier to back down a steep slope than to crawl down forward, because it was harder to fight gravity with the heaviest part of their body, their head, going in front. Eventually, infants whose initial backing trials were serendipitous, assumed the backing position before they went over the brink.

Similarly, when crawlers attempted headfirst descent of steep slopes with their knees splayed, rather than glued together, they sometimes ended up with their legs in a straddle-split position. With their stiffly extended arms pushing backward to keep their heads up, infants would accidentally find themselves sliding down in a sitting position. The same infants whose initial sitting trials were serendipitous eventually assumed a sitting position on the starting platform before going over the brink.

The array of strategies for using wobbly handrails may also have been serendipitous (Berger et al., 2005). Infants may have intended to use the wobbly handrails as they would use sturdy ones, but as the wobbly handrails gave way and they found themselves in unusual positions, they would "go with the flow." For example, the "hunchback" strategy may have emerged when infants put their weight on the handrail, as they would with any normal handrail, and it sank to the mid-thigh level. Instead of giving up, as older children and adults might have done, infants simply leaned with the handrail and walked across the bridge sideways, hunched over because that was what the wobbly handrail afforded. Although they may not have begun the trial with the intention of walking hunched over, after finding themselves in that position, they may have accommodated to the circumstances.

Learning by doing

A final possibility for the acquisition of alternative strategies may be learning by doing. Like diving into a pool to test the water, infants may simply throw themselves into the situation with no intended course of action, and then learn by observing the consequences. Learning by doing may be infants' preferred process of exploration in situations where the consequences of falling are negligible. A prime example is situations involving ascent.

Once infants develop the strength in their arms and legs to hoist their entire weight and keep balance on one limb, a new world of ascending opportunities opens up for them. Now infants can climb up any object onto which they can get a foothold. As any parent who has had to babyproof a home can attest, almost any vertical surface can afford the possibility for climbing. Some humorous examples found in the motor development literature include fences, trees, furniture, cribs, bathtubs, and toy boxes (McGraw, 1935; Valsiner and Mackie, 1985; Trettien, 1900). Of course, the most frequently encountered space for climbing is the staircase. Parents report that infants almost always figured out stair ascent on their own (Berger et al., 2007).

In the laboratory, where tasks involving ascent can be observed directly, learning by doing seems to be infants' primary means for acquiring new ascent strategies. On slopes, for example, at every age and level of experience, crawlers and walkers flung themselves at the hill on every trial and then observed the consequences (Adolph et al., 1993a; McGraw, 1935; Adolph, 1995, 1997). In fact, QJ;infants' attempts appeared to be more geared QJ;toward acquiring alternative strategies rather than discriminating affordances for crawling or walking. Attempt rates were always higher for ascent compared with descent, even on impossibly steep slopes. Infants attempted ascent repeatedly on the same trial, happily struggling to get up after sliding back down for the umpteenth time.

Sometimes, infants acquired new gait modifications through learning by doing. Crawlers powering up the slope, sliding back down, and powering up again, switched from hands and knees to using their toes to grip the surface. Walkers learned to lean forward into the slope, after pitching backward on repeated attempts. Sometimes, infants acquired alternative strategies from learning by doing. As walkers flung themselves at the slope, they learned to take a running start, then switch to climbing the remainder on hands and feet. Infants in both crawling and walking groups learned methods for gripping the top of the slope with their forearm and swinging a leg over the brink.

Cognition in motion

On the new view of locomotor development, perception is integral. Traveling over a uniform path at a steady pace encourages infants to execute the same movements over and over. In contrast, navigating through a cluttered environment while adapting to changing goals and varying destinations requires infants to change their movements from step to step. At every point in the development of stance and mobility, infants must use perceptual information to detect the changing constraints on balance and propulsion. To exploit the available affordances, infants must perceive them. And, to perceive the available affordances, infants must turn up the relevant perceptual information by their own spontaneous exploratory movements.

In addition to perception, adaptive mobility involves cognition. Experimental paradigms where infants approach obstacles such as slopes, stairs, bridges, and gaps approximate real-life problems that infants encounter as they wend their way through the maze of household objects, climb over furniture, and scale the equipment at the playground. The problem-solving skills required for coping with such obstacles can involve higher-level cognitive processes. Although means–ends problem solving is typically studied in the context of manual tasks, locomotion can also require infants to discover and coordinate various means to achieve a goal, as when infants identify an alternative sliding position for descending slopes, recognize that a sturdy handrail can augment their balance on narrow bridges, and invent strategies for making a wobbly handrail suffice.

One hallmark of means–ends problem-solving is tool use. The experiments in which walking infants used handrails as tools for keeping balance illustrate how the acquisition of new locomotor means is rooted in the development of both cognitive and motor abilities (Berger and Adolph, 2003; Berger et al., 2005). Successful tool-use requires infants to master three steps and sequence them into the appropriate order: First, infants must recognize the existence of a gap between their own ability and typical means for achieving a goal. Second, infants must understand that an environmental support — a tool — is available that can serve as alternative means for reaching the goal. Third, infants must implement the tool by modifying their typical means.

In the bridges and handrails studies, the first step in the sequence was based on infants' perception of affordances (or lack of them) for crossing narrow bridges. The third step was rooted in infants' ability to modify their typical walking gait by turning sideways and holding the rail. It was the second step that reflected true means–ends problem-solving. Envisioning a handrail as an alternative means for crossing a narrow bridge meant that infants coordinated several pieces of information about the handrail, the bridge and their own bodies. Thus, developmental changes in cognition can affect infants' motor abilities.

Reciprocally, developmental changes in infants' locomotor skills can affect their cognitive abilities. Both cognitive and motor demands compete for infants' finite attentional resources. There is a limit to how much infants can attend to at one time. When infants' attention is overtaxed by a demanding motor challenge, there is a trade-off between cognition and action — part of performance is sacrificed so that the other part of the task can be carried out (Boudreau and Bushnell, 2000; Keen et al., 2003). As infants' crawling and walking proficiency increases, they can devote less attention to keeping balance and, in turn, can allocate more resources to higher-level cognitive processes.

At 12 months, for example, infants no longer exhibit the A-not-B error. After watching an experimenter move a target to a new location, they can inhibit the prepotent response of reaching for the object in the same location where they had retrieved it several times on previous trials. By 12 months, reaching is highly practiced and demands little in terms of cognitive resources. Walking, however, is a different story. Upright balance is still precarious in challenging situations, and infants must devote their attentional resources to solving the motor problems. Thus, in a locomotor version of the A-not-B task, 13-month-old infants repeatedly moved to one location, A, and then were encouraged to go to a second location, B (Berger, 2004). On the B trial, infants inhibited the urge to return to the A location when motor demands were low (walking over flat ground to reach the goal), but the same infants perseverated when motor demands were high (descending a small staircase to the goal). What matters is not the task per se, but rather how much cognitive capacity is required to perform it. Twelve-month-old infants no longer perseverate on a manual search task where they reach for a hidden object because they are expert reachers (e.g., Diamond et al., 1997). In contrast, 13-month-olds do perseverate on a locomotor task on which they are novice stair climbers (Berger, 2004).

Conclusion

These days, it is popular to resurrect Piaget's old idea that motor skill acquisition influences the development of perception and cognition. We have no doubt that Piaget was on the right track. New motor skills provide infants with new opportunities for learning about the self and the environment, and the relations between the two. Stance and mobility take so many different forms in infancy — from sitting on a caregiver's knee to slithering under the coffee table — that the only common denominator may be the involvement of the whole body. The whole body is involved in detecting affordances for locomotion, and finding the means to create new affordances with a tool.

The reciprocal developmental relationship is also true: Developments in perception and cognition can facilitate development of locomotor action. Perception and cognition are integral to independent mobility. For infants, getting down a slope or stair or across a bridge is an opportunity to perceive changing affordances and an invitation to means–ends problem solving. We can best see the relations between perception, cognition, and action when we observe infants engaged in real locomotion in all its manifold forms, through a cluttered environment, with variable goals.

References

Achard, B. and von Hofsten, C. (2002) Development of the infants' ability to retrieve food through a slit. Infant Child Dev., 11: 43–56.

Adolph, K.E. (1995) A psychophysical assessment of toddlers' ability to cope with slopes. J. Exp. Psychol. Hum. Percept. Perform, 21: 734–750.

Adolph, K.E. (1997) Learning in the development of infant locomotion. Monogr. Soc. Res. Child. Dev., 62 (3, Serial No. 251).

Adolph, K.E. (2000) Specificity of learning: why infants fall over a veritable cliff. Psychol. Sci., 11: 290–295.

Adolph, K.E. (2002) Learning to keep balance. In: Kail R. (Ed.), Advances in Child Development and Behavior, Vol. 30. Elsevier Science, Amsterdam, pp. 1–30.

Adolph, K.E. (2005) Learning to learn in the development of action. In: Lockman J. and Reiser J. (Eds.), Action as an organizer of learning and development: The 32nd Minnesota Symposium on Child Development. Lawrence Erlbaum Associates, Hillsdale, NJ, pp. 91–122.

Adolph, K.E. and Avolio, A.M. (2000) Walking infants adapt locomotion to changing body dimensions. J. Exp. Psychol. Hum. Percept. Perform., 26: 1148–1166.

Adolph, K.E. and Berger, S.E. (2005) Physical and motor development. In: Bornstein M.H. and Lamb M.E. (Eds.), Developmental science: an advanced textbook (5th ed.). Lawrence Erlbaum Associates, Mahwah, NJ, pp. 223–281.

Adolph, K.E. and Berger, S.E. (2006) Motor development. In: Kuhn, D. and Siegler, R.S. (Eds.), Handbook of Child Psychology, 6th ed., Vol. 2. Cognition, Perception, and Language. Wiley, New York, pp. 161–213.

Adolph, K.E., Eppler, M.A. and Gibson, E.J. (1993a) Crawling versus walking infants' perception of affordances for locomotion over sloping surfaces. Child. Dev., 64: 1158–1174.

Adolph, K.E., Eppler, M.A. and Gibson, E.J. (1993b) Development of perception of affordances. In: Rovee-Collier C.K. and Lipsitt L.P. (Eds.), Advances in Infancy Research, Vol. 8. Ablex, Norwood, NJ, pp. 51–98.

Adolph, K.E., Eppler, M.A., Marin, L., Weise, I.B. and Clearfield, M.W. (2000) Exploration in the service of prospective control. Infant. Behav. Dev., 23: 441–460.

Adolph, K.E., Vereijken, B. and Denny, M.A. (1998) Learning to crawl. Child. Dev., 69: 1299–1312.

Adolph, K.E., Vereijken, B. and Shrout, P.E. (2003) What changes in infant walking and why. Child. Dev., 74: 474–497.

Ames, L.B. (1937) The sequential patterning of prone progression in the human infant. Genet. Psychol. Monogr., 19: 409–460.

Berger, S.E. (2004) Demands on finite cognitive capacity cause infants' perseverative errors. Infancy, 5: 217–238.

Berger, S.E. and Adolph, K.E. (2003) Infants use handrails as tools in a locomotor task. Dev. Psychol., 39: 594–605.

Berger, S.E., Adolph, K.E. and Lobo, S.A. (2005) Out of the toolbox: toddlers differentiate wobbly and wooden handrails. Child. Dev., 76: 1294–1307.

Berger, S.E., Theuring, C.F. and Adolph, K.E. (2007) How and when infants learn to climb stairs. Infant. Behav. Dev., 30: 36–49.

Boudreau, J.P. and Bushnell, E.W. (2000) Spilling thoughts: configuring attentional resources in infants' goal directed actions. Infant Behav. Dev., 23: 543–566.

Bril, B. and Breniere, Y. (1989) Steady-state velocity and temporal structure of gait during the first six months of autonomous walking. Hum. Move Sci., 8: 99–122.

Bril, B. and Breniere, Y. (1992) Postural requirements and progression velocity in young walkers. J. Mot. Behav., 24: 105–116.

Burnside, L.H. (1927) Coordination in the locomotion of infants. Genet. Psychol. Monogr., 2: 279–372.

Campos, J.J., Hiatt, S., Ramsay, D., Henderson, C. and Svejda, M. (1978) The emergence of fear on the visual cliff. In: Lewis M. and Rosenblum L. (Eds.), The Development of Affect. Plenum, New York, pp. 149–182.

Chen, Z. and Siegler, R.S. (2000) Across the great divide: bridging the gap between understanding of toddlers' and older children's thinking. Monogr. Soc. Res. Child. Dev., 65 (2, Serial No. 261).

Diamond, A. (1990) Developmental time course in human infants and infant monkeys, and the neural bases of inhibitory control in reaching. In: Diamond A. (Ed.), The development and neural bases of higher cognitive functions. The New York Academy of Sciences, New York, pp. 637–676.

Diamond, A., Prevor, M.B., Callender, G. and Druin, D.P. (1997) Prefrontal cortex cognitive deficits in children treated early and continuously for PKU. Monogr. Soc. Res. Child. Dev., 62 (3, Serial No. 252).

Eppler, M.A., Adolph, K.E. and Weiner, T. (1996) The developmental relationship between infants' exploration and action on slanted surfaces. Infant Behav. Dev., 19: 259–264.

Forssberg, H. (1985) Ontogeny of human locomotor control. I. Infant stepping, supported locomotion, and transition to independent locomotion. Exp. Brain Res., 57: 480–493.

Garciaguirre, J.S. and Adolph, K.E. (2006, June) Infants' everyday locomotor experience: A walking and falling marathon. Poster presented at the meeting of the International Conference on Infant Studies. Kyoto, Japan.

Garciaguirre, J.S., Adolph, K.E. and Shrout, P.E. (in press) Baby carriage: Infants walking with loads. Child. Dev.

Gesell, A. (1933) Maturation and the patterning of behavior. In: Murchison C. (Ed.), A Handbook of Child Psychology (2nd ed.). Clark University Press, Worcester, MA, pp. 209–235.

Gesell, A. (1946) The ontogenesis of infant behavior. In: Carmichael L. (Ed.), Manual of Child Psychology. John Wiley, New York, pp. 295–331.

Gesell, A. and Ames, L.B. (1940) The ontogenetic organization of prone behavior in human infancy. J. Genet. Psychol., 56: 247–263.

Gesell, A. and Thompson, H. (1934) Infant behavior: its genesis and growth. Greenwood Press, New York.

Gesell, A. and Thompson, H. (1938) The psychology of early growth including norms of infant behavior and a method of genetic analysis. Macmillan, New York.

Gibson, E.J., Riccio, G., Schmuckler, M.A., Stoffregen, T.A., Rosenberg, D. and Taormina, J. (1987) Detection of the traversability of surfaces by crawling and walking infants. J. Exp. Psychol. Hum. Percept. Perform., 13: 533–544.

Gibson, E.J. and Walk, R.D. (1960) The "visual cliff." Sci. Am., 202: 64–71.

Gibson, J.J. (1979) The Ecological Approach to Visual Perception. Houghton Mifflin Company, Boston.

Gill-Alvarez, S.V. and Adolph, K.E. (2005, November) Emergence of flexibility: How infants learn a stepping strategy. Poster presented to the International Society of Developmental Psychobiology, Washington, DC.

Halverson, H.M. (1931) An experimental study of prehension in infants by means of systematic cinema records. Genet. Psychol. Monogr., 10: 107–283.

Joh, A.S. and Adolph, K.E. (2006) Learning from falling. Child. Dev., 77: 89–102.

Keen, R., Carrico, R.L., Sylvia, M.R. and Berthier, N.E. (2003) How infants use perceptual information to guide action. Dev. Sci., 6: 221–231.

Lampl, M., Veldhuis, J.D. and Johnson, M.L. (1992) Saltation and stasis: a model of human growth. Science, 258: 801–803.

Lockman, J.J. (1984) The development of detour ability during infancy. Child. Dev., 55: 482–491.

Lockman, J.J. (2000) A perception-action perspective on tool use development. Child. Dev., 71: 137–144.

Lockman, J.J. and Adams, C.D. (2001) Going around transparent and grid-like barriers: detour ability as a perception-action skill. Dev. Sci., 4: 463–471.

McCarty, M.E., Clifton, R.K. and Collard, R.R. (2001) The beginnings of tool use by infants and toddlers. Infancy, 2: 233–256.

McCarty, M.E. and Keen, R. (2005) Facilitating problem-solving performance among 9- and 12-month-old infants. J. Cogn. Dev., 2: 209–228.

McGraw, M.B. (1932) From reflex to muscular control in the assumption of an erect posture and ambulation in the human infant. Child. Dev., 3: 291–297.

McGraw, M.B. (1935) Growth: A Study of Johnny and Jimmy. Appleton-Century Co., New York.

McGraw, M.B. (1945) The Neuromuscular Maturation of the Human Infant. Columbia University Press, New York.

Mondschein, E.R., Adolph, K.E. and Tamis-LeMonda, C.S. (2000) Gender bias in mothers' expectations about infant crawling. J. Exp. Child. Psychol., 77: 304–316.

Schmuckler, M.A. (1996) Development of visually guided locomotion: Barrier crossing by toddlers. Ecol. Psychol., 8: 209–236.

Shirley, M.M. (1931) The First Two Years: A Study of Twenty-Five Babies. Greenwood Press, Westport, CT.

Siegler, R.S., Adolph, K.E. and Lemaire, P. (1996) Strategy choice across the life span. In: Reder L. (Ed.), Implicit Memory and Metacognition. Erlbaum, Mahwah, NJ, pp. 79–121.

Trettien, A.W. (1900) Creeping and walking. Am. J. Psychol., 12: 1–57.

Valsiner, J. and Mackie, C. (1985) Toddlers at home: canalization of climbing skills through culturally organized physical environments. In: Garling T. and Valsiner J. (Eds.), Children Within Environments: Toward a Psychology of Accident Prevention. Plenum, New York, pp. 165–190.

Warren, W.H. (1984) Perceiving affordances: visual guidance of stair climbing. J. Exp. Psychol. Hum. Percept. Perform., 10: 683–703.

Witherington, D.C., Campos, J.J., Anderson, D.I., Lejeune, L. and Seah, E. (2005) Avoidance of heights on the visual cliff in newly walking infants. Infancy, 7: 285–298.

Zelazo, P.R. (1998) McGraw and the development of unaided walking. Dev. Rev., 18: 449–471.

Zelazo, P.R., Weiss, M.J. and Leonard, E. (1989) The development of unaided walking: the acquisition of higher order control. In: Zelazo P.R. and Barr R.G. (Eds.), Challenges to Developmental Paradigms. Lawrence Erlbaum Associates, Hillsdale, NJ, pp. 139–165.

C. von Hofsten & K. Rosander (Eds.)
Progress in Brain Research, Vol. 164
ISSN 0079-6123

CHAPTER 14

Core systems in human cognition

Katherine D. Kinzler and Elizabeth S. Spelke*

Department of Psychology, Harvard University, 1130 William James Hall, Cambridge, MA 02138, USA

Abstract: Research on human infants, adult nonhuman primates, and children and adults in diverse cultures provides converging evidence for four systems at the foundations of human knowledge. These systems are domain specific and serve to represent both entities in the perceptible world (inanimate manipulable objects and animate agents) and entities that are more abstract (numbers and geometrical forms). Human cognition may be based, as well, on a fifth system for representing social partners and for categorizing the social world into groups. Research on infants and children may contribute both to understanding of these systems and to attempts to overcome misconceptions that they may foster.

Keywords: core knowledge; infants; cognitive development; cognition

How is the human mind organized, and how does it grow? Does all human development depend on a single, general-purpose learning system? At the opposite extreme, are humans endowed with a large collection of special-purpose cognitive systems and predispositions? Research on human infants, non-human primates, and human children and adults in different cultures provides evidence against both these extremes. Instead, we believe that humans are endowed with a small number of separable systems that stand at the foundation of all our beliefs and values. New, flexible skills, concepts, and systems of knowledge build on these core foundations.

More specifically, research provides evidence for four core systems (Spelke, 2003) and hints of a fifth one. The four systems serve to represent inanimate objects and their mechanical interactions, agents and their goal-directed actions, sets and their numerical relationships of ordering, addition, and subtraction, and places in the spatial layout and their geometric relationships. The fifth system serves to identify members of one's own social group in relation to members of other groups, and to guide social interactions with in- and out-group members. Each system centers on a set of principles that pick out the entities in its domain and support inferences about their interrelationships and behavior. Each system, moreover, is characterized by a set of signature limits that allow for its identification across tasks, ages, species, and human cultures.

Objects

The core system of object representation centers on a set of principles governing object motion: *cohesion* (objects move as connected and bounded wholes), *continuity* (objects move on connected, unobstructed paths), and *contact* (objects influence each others' motion when and only when they touch) (Leslie and Keeble, 1987; Spelke, 1990; Aguiar and Baillargeon, 1999). These principles allow infants of a variety of species, including

*Corresponding author. Tel.: 617-495-3876; Fax: 617-384-7944; E-mail: spelke@wjh.harvard.edu

DOI: 10.1016/S0079-6123(07)64014-X

humans, to perceive the boundaries and shapes of objects that are visible or partly out of view, and to predict when objects will move and where they will come to rest. Some of these abilities are observed in the absence of any visual experience, in newborn human infants or newly hatched chicks (Regolin and Vallortigara, 1995; Lea et al., 1996; Valenza et al., 2006). Moreover, research with older infants suggests that a single system underlies infants' object representations. For instance, 5-month-old infants do not have more specific cognitive systems for representing and reasoning about subcategories of objects such as foods, animals, or artifacts (Shutts, 2006), or systems for reasoning about inanimate, non-object entities such as sand piles (Huntley-Fenner et al., 2002; Rosenberg and Carey, 2006; Shutts, 2006). Finally, infants are able to represent only a small number of objects at a time (about three: Feigenson and Carey, 2003). These findings provide evidence that a single system, with signature limits, underlies infants' reasoning about the inanimate world.

Investigators of cognitive processes in human adults have discovered evidence that the same system governs adults' processes of object-directed attention, which accord with the cohesion, continuity, and contact principles and encompass up to three or four separately moving objects at any given time (e.g., Scholl and Pylyshyn, 1999; Scholl et al., 2001; vanMarle and Scholl, 2003; Marino and Scholl, 2005). Of course, adult humans also have developed more specific knowledge of subdomains of objects such as foods and tools (e.g., Keil et al., 1998; Lavin and Hall, 2001; Santos et al., 2001). When attentional resources are stretched, however, the properties that mark these finer distinctions often fail to guide object representations, whereas core properties continue to do so (Leslie et al., 1998).

If core object representations are constant over human development, then they should be universal across human cultures. Recent studies of remote Amazonian groups support that suggestion. For example, the Pirahã are a highly isolated tribe whose language lacks most number of words and other syntactic devices (Everett, 2005). Nevertheless, the Pirahã distinguish objects from non-object entities (Everett, 2005), and they track objects with the signature set-size limit (Gordon, 2004).

Agents

A second core system represents agents and their actions. Unlike the case of objects, spatio-temporal principles do not govern infants' representations of agents, who need not be cohesive (Vishton et al., 1998), continuous in their paths of motion (Kuhlmeier et al., 2004; although see Saxe et al., 2005), or subject to the constraint of action only on contact (Spelke et al., 1995). Instead, infants represent agents' actions as directed towards goals (Woodward, 1999) through means that are efficient (Gergely and Csibra, 2003). Infants expect agents to interact with other agents, both contingently (Watson, 1972; Johnson et al., 2001) and reciprocally (Meltzoff and Moore, 1977). Although agents need not have faces with eyes (Johnson et al., 1998; Gergely and Csibra, 2003), when they do, even human newborns (Farroni et al., 2004) and newly hatched chicks (Agrillo, Regolin and Vallortigara, 2004) use their gaze direction to interpret their actions, as do older infants (Hood, Willen and Driver, 1998; Csibra and Gergely, 2006). In contrast, infants do not interpret the motions of inanimate objects as goal-directed (Woodward, 1998).

Research on human adults provides evidence for the same system of agent representations. Representations of goal-directed, efficient actions and of reciprocal interactions guide adults' intuitive moral reasoning (Cushman et al., in press; Trivers, 1971). Together, these findings provide evidence for a core system of agent representation that persists over human development, characterized by goal-directedness, efficiency, contingency, reciprocity, and gaze direction.

Number

The core number system shows its own distinctive signature limits. Three competing sets of principles have been proposed to characterize this system (Meck and Church, 1983; Church and Broadbent,

1990; Dehaene and Changeux, 1993). Although their relative merits are still debated (see Gallistel and Gelman, 1992; Izard and Dehaene, in press), there is broad agreement concerning three central properties of core number representations. First, number representations are imprecise, and their imprecision grows linearly with increasing cardinal value. Under a broad range of background assumptions, this "scalar variability" produces a ratio limit to the discriminability of sets with different cardinal values (Izard, 2006). Second, number representations apply to diverse entities encountered through multiple sensory modalities, including arrays of objects, sequences of sounds, and perceived or produced sequences of actions. Third, number representations can be compared and combined by operations of addition and subtraction.

Number representations with these properties have now been found in human infants, children, and adults. Infants discriminate between large numbers of objects, actions, and sounds when continuous quantities are controlled, and their discrimination shows a ratio limit (Xu and Spelke, 2000; Lipton and Spelke, 2003, 2004; Brannon et al., 2004; Wood and Spelke, 2005; Xu et al., 2005). Infants also can add and subtract large numbers of objects (McCrink and Wynn, 2004). In adults and children, cross-modal numerical comparisons are as accurate as comparisons within a single modality (Barth et al., 2003, 2005), and addition of two arrays in different modalities is as accurate as addition within a single modality (Barth et al., in prep.).

Because core representations of number are present throughout development, they should also be present in all cultures, independently of formal education in mathematics. Studies of a second remote Amazonian group with no verbal counting routine, no words for exact numbers beyond "three," and little formal instruction, support this prediction. The Mundurukú discriminate between large numbers with a ratio limit on precision (Pica et al., 2004). Mundurukú adults who have received no instruction in mathematics can perform approximate addition and subtraction on large approximate numerosities (Pica et al., 2004).

Geometry

The fourth core system captures the geometry of the environment: the distance, angle, and sense relations among extended surfaces. This system fails to represent non-geometric properties of the surface layout such as color or odor, or geometric properties of movable objects. When young children or non-human animals are disoriented, they reorient themselves in accord with layout geometry (Cheng, 1986; Hermer and Spelke, 1996; see Cheng and Newcombe, 2005, for review). Children fail, in contrast, to orient themselves in accord with the geometry of an array of objects (Gouteux and Spelke, 2001), and they fail to use the geometry of an array to locate an object when they are oriented and the array moves (Lourenco et al., 2005). Under some circumstances, disoriented children fail to locate objects in relation to distinctive landmark objects and surfaces, such as a colored wall (Wang et al., 1999; Lee, Shusterman and Spelke, 2006). When such children do use landmarks, their search appears to depend on two distinct processes: a reorientation process that is sensitive only to geometry and an associative process that links local regions of the layout to specific objects (Lee et al., 2006).

Human adults show more extensive use of landmarks, but they too rely primarily on surface geometry when they are disoriented (Hermer-Vazquez et al., 1999; Newcombe, 2005). Recent studies of the Mundurukú again suggest that sensitivity to geometry is universal: children and adults with little or no formal education extract and use geometric information in pictures as well as in extended surface layouts (Dehaene et al., 2006).

In summary, research on human infants, children, and adults across very different cultural environments suggests that the human mind is not a single, general-purpose device. Humans learn some things readily, and others with greater difficulty, by exercising more specific cognitive systems with signature properties and limits. The human mind also does not appear to be composed of hundreds or thousands of special-purpose cognitive devices. Rather, the mind is more likely built on a small number of core systems, including the four systems just described.

US vs. Them

Recently, we have begun to investigate a fifth candidate core system, for identifying and reasoning about potential social partners and social group members. Research in evolutionary psychology suggests that people are predisposed to form and attend to coalitions (Cosmides et al., 2003) whose members show cooperation, reciprocity, and group cohesion. An extensive literature in social psychology confirms this predisposition to categorize the self and others into groups. Any minimal grouping, based on race, ethnicity, nationality, religion, or arbitrary assignment, tends to produce a preference for the in-group, or *us*, over the out-group, or *them*. This preference is found in both adults and children alike, who show parallel biases toward and against individuals based on their race (e.g., Baron and Banaji, 2006), gender (Gelman et al., 1986; Miller et al., 2006), or ethnicity.

Studies of infants suggest that these tendencies emerge early in development. Three-month-old infants show a visual preference for members of their own race (Kelly et al., 2005, Bar-Haim et al., 2006). This preference is influenced by infants' experience and depends both on the race of the infant's family members and the predominance of that race in the larger community (Bar-Haim et al., 2006). Race may not be the most powerful or reliable cue to social group membership, however, because contact with perceptibly different races rarely would have occurred in the environments in which humans evolved (Kurzban, Tooby and Cosmides, 2001; Cosmides et al., 2003). A better source of information for group membership might come from the language that people speak, and especially from the accent with which they speak it.

Until recently in human history, languages varied markedly across human groups, even groups living in quite close proximity (e.g., Braudel, 1988). From birth, moreover, infants show a preference for the sound of their native language over a foreign language (Mehler et al., 1988; Moon et al., 1993). We have asked, therefore, whether infants use language to categorize unfamiliar people, and whether they prefer people who speak their native language.

In one series of studies (Kinzler and Spelke, 2005), 6-month-old infants viewed films of the faces of two women who were bilingual speakers of English and Spanish. After the women spoke to the infants in alternation, one in English and the other in Spanish, the two women were presented side by side, smiling without speaking. Although each woman had spoken Spanish to half the infants and English to the others, infants tended to look longer at the woman who had spoken to them in English, their native language.

Further studies revealed that this preference extends to older ages and guides behaviors that are more directly social. For example, 12-month-old infants in Boston were presented with bilingual speakers of English and French who spoke to them in alternation, while each offering two different foods. When later given a choice between the two foods, infants reached preferentially for the food offered by the American speaker (McKee, 2006).

Further experiments reveal the same preference for speakers of the native language in older children from diverse cultures. When 6-year-old English-speaking children in the United States or Xhosa-speaking children in South Africa are shown pictures of two children, one speaking their native language and the other speaking a foreign language (French), the children preferentially select the native speaker as a friend. Variations in accent are sufficient to evoke this preference, both in infants and in children (Kinzler et al., in prep.).

These findings suggest that the sound of the native language provides powerful information for social group membership early in development. Together with the studies of infants' sensitivity to race, they raise the possibility of a fifth core system that serves to distinguish potential members of one's own social group from members of other groups.

Beyond Core Knowledge

Core systems for representing objects, actions, numbers, places, and social partners may provide some of the foundations for uniquely human cognitive achievements, including the acquisition of language and other symbol systems such as maps, the development of intuitive reasoning about physical and biological phenomena, and the development of cognitive skills through formal instruction. A core system for representing potential social partners may be especially useful, as it could guide infants' and children's "cultural learning" (Tomasello, 1999): their acquisition of skills and behaviors that sustain life within a particular human group. In all these cases, core knowledge systems may support and advance human cognitive development, because the principles on which they are based are veridical and adaptive at the scales at which humans and other animals perceive and act on the world.

Nevertheless, core systems of representation also can lead humans into cognitive errors. At the smallest and largest scales that science can probe, the core principles of cohesion, continuity, and contact do not apply. Mathematicians have discovered numbers and geometries beyond the reach of the core domains. Adults and children are prone to errors in reasoning about properties of object mechanics, non-Euclidean geometry, or numbers that violate the principles of core knowledge (e.g., McCloskey, 1983; Gelman, 1991; Randall, 2005).

The most serious errors, however, may spring from the system for identifying and reasoning about the members of one's own social group. A predisposition for dividing the social world into *us* versus *them* may have evolved for the purpose of detecting suitable social partners, but it can cause mischief in modern, multi-cultural societies. It even may support the ravages of discord, violence, and warfare among individuals, groups, and nations. For example, recent world history provides examples of linguicide paired with genocide of the Kurds in Turkey (Phillipson and Skutnabb-Kangas, 1994), and of forced language policies initiating anti-Apartheid riots in South Africa (Sparks, 1996). A preference for one's native language group influences contemporary politics in more subtle ways, as well, such as in debates concerning bilingual education.

Despite these examples, we believe that the strongest message, from human history and cognitive science alike, is that core conceptions can be overcome. The history of science and mathematics provides numerous examples of fundamental conceptual changes that have occurred as thinkers attempted to surmount the limitations of their systems of intuitive reasoning. Despite the pull of core conceptions of Euclidean geometry and object mechanics, cosmologists and particle physicists can test whether space is non-Euclidean and has higher dimensions (e.g., Randall, 2005) and they can use conceptions of massless, discontinuously moving particles to make predictions of astonishing precision (Hawking, 2002). Even preschool children change their conceptions of numbers when they learn to count (Spelke, 2000; Carey, 2001), and they change their conceptions of agents when they learn about biological processes such as eating (Carey, 1985, 2001).

These examples of conceptual change may be useful in thinking about ways to alleviate social conflicts. If core conceptions of social groups fuel such conflicts, they too should be open to change, because understanding of human cognitive development yields insight into its malleability. As the world shrinks in size and different social groups come increasingly into contact, studies of the development of social group preferences may yield valuable insights into the ways in which intergroup conflicts can be moderated or neutralized.

Acknowledgments

We thank Kristin Shutts and Talee Ziv for advice and collaboration. Supported by NIH grant HD23103 to ESS.

References

Agrillo, C., Regolin, L. and Vallortigara, G. (2004) Can young chicks take into account the observer's perspective? 27th Annual Meeting of the European Conference on Visual Perception (ECVP), Budapest.

Aguiar, A. and Baillargeon, R. (1999) 2.5-Month-old infants' reasoning about when objects should and should not be occluded. Cogn. Psychol., 39: 116–157.

Bar-Haim, Y., Ziv, T., Lamy, D. and Hodes, R. (2006) Nature and nurture in own-race face processing. Psychol. Sci., 17: 159–163.

Baron, A. and Banaji, M. (2006) The development of implicit attitudes: evidence of race evaluations from ages 6 and 10 and adulthood. Psychol. Sci., 17: 53–58.

Barth, H., Beckmann, L. and Spelke, E.S. (in prep.) Nonsymbolic approximate arithmetic in children: Abstract addition prior to instruction. (Manuscript under review.)

Barth, H., Kanwisher, N. and Spelke, E. (2003) The construction of large number representations in adults. Cognition, 86: 201–221.

Barth, H., La Mont, K., Lipton, J. and Spelke, E.S. (2005) Abstract number and arithmetic in young children. Proc. Natl. Acad. Sci., 39: 14117–14121.

Brannon, E., Abbott, S. and Lutz, D. (2004) Number bias for the discrimination of large visual sets in infancy. Cognition, 93: B59–B68.

Braudel, F. (1988) The Identity of France (Reynolds, S., Trans.). Collins (original work published 1986), London.

Carey, S. (1985) Conceptual Change in Childhood. Bradford Books, MIT Press, Cambridge, MA.

Carey, S. (2001) Evolutionary and ontogenetic foundations of arithmetic. Mind Lang., 16: 37–55.

Cheng, K. (1986) A purely geometric module in the rat's spatial representation. Cognition, 23: 149–178.

Cheng, K. and Newcombe, N. (2005) Is there a geometric module for spatial orientation? Squaring theory and evidence. Psychon. Bull. Rev., 12: 1–23.

Church, R. and Broadbent, H. (1990) Alternative representations of time, number, and rate. Cognition, 37: 55–81.

Cosmides, L., Tooby, J. and Kurzban, R. (2003) Perceptions of race. Trends Cogn. Sci., 7: 173–179.

Csibra, G. and Gergely, G. (2006) Social learning and social cognition: The case for pedagogy. In: Munakata Y. and Johnson M. H. (Eds.), Processes of Change in Brain and Cognitive Development, Attention and Performance XXI, Oxford University Press, Oxford, pp. 249–274.

Cushman, F., Young, L. and Hauser, M.D. (2006) The role of conscious reasoning and intuition in moral judgments: testing three principles of harm. Psychol. Sci., 17: 1082–1089.

Dehaene, S. and Changeux, J. (1993) Development of elementary numerical abilities: a neuronal model. J. Cogn. Neurosci., 5: 390–407.

Dehaene, S., Izard, V., Pica, P. and Spelke, E.S. (2006) Core knowledge of geometry in an Amazonian indigene group. Science, 311: 381–384.

Everett, D. (2005) Cultural constraints on grammar and cognition in Pirahã: another look at the design features of human language. Curr. Anthropol., 46: 621–634.

Farroni, T., Massaccesi, S., Pividori, D. and Johnson, M. (2004) Gaze following in newborns. Infancy, 5: 39–60.

Feigenson, L. and Carey, S. (2003) Tracking individuals via object-files: evidence from infants' manual search. Dev. Sci., 6: 568–584.

Gallistel, C. and Gelman, R. (1992) Preverbal and verbal counting and computation. Cognition, 44: 43–74.

Gelman, R. (1991) Epigenetic foundations of knowledge structures: initial and transcendent constructions. In: Carey S. and Gelman R. (Eds.), The Epigenesis of Mind: Essays on Biology and Cognition. Lawrence Erlbaum Associates, Inc., Hillsdale, NJ, pp. 293–322.

Gelman, S.A., Collman, P.C. and Maccoby, E.E. (1986) Inferring properties from categories versus inferring categories from properties: the case of gender. Child Dev., 57: 396–404.

Gergely, G. and Csibra, G. (2003) Teleological reasoning in infancy: the naïve theory of rational action. Trends Cogn. Sci., 7(7): 287–292.

Gordon, P. (2004) Numerical cognition without words: evidence from Amazonia. Science, 306: 496–499.

Gouteux, S. and Spelke, E. (2001) Children's use of geometry and landmarks to reorient in an open space. Cognition, 81: 119–148.

Hawking, S.W. (2002) On the shoulders of gaints: the great works of physics and astronomy. Running press book publishers, Philadelphia, PA.

Hermer, L. and Spelke, E. (1996) Modularity and development: the case of spatial reorientation. Cognition, 61: 195–232.

Hermer-Vazquez, L., Spelke, E. and Katsnelson, A. (1999) Sources of flexibility in human cognition: dual-task studies of space and language. Cogn. Psychol., 39: 3–36.

Hood, B.M., Willen, J.D. and Driver, J. (1998) Adults' eyes trigger shifts of visual attention in human infants. Psych. Sci., 9: 131–134.

Huntley-Fenner, G., Carey, S. and Solimando, A. (2002) Objects are individuals but stuff doesn't count: perceived rigidity and cohesiveness influence infants' representations of small groups of discrete entities. Cognition, 85: 203–221.

Izard, V. (2006) Interactions entre les representations numeriques verbales et non-verbales: études théoriques et expérimentales. Unpublished doctoral dissertation, Université Paris VI, Paris.

Izard, V. and Dehaene, S. (in press) Calibrating the number line. Cognition.

Johnson, S., Booth, A. and O'Hearn, K. (2001) Inferring the goals of a nonhuman agent. Cogn. Dev., 16: 637–656.

Johnson, S., Slaughter, V. and Carey, S. (1998) Whose gaze will infants follow? The elicitation of gaze-following in 12-month-olds. Dev. Sci., 1: 233–238.

Keil, F., Smith, W., Simons, D. and Levin, D. (1998) Two dogmas of conceptual empiricism: implications for hybrid models of the structure of knowledge. Cognition, 65: 103–135.

Kelly, D., Quinn, P., Slater, A.M., Lee, K., Gibson, A., Smieth, M., Ge, L. and Pascalis, O. (2005) Three-month-olds, but not newborns, prefer own-race faces. Dev. Sci., 8: F31–F36.

Kinzler, K.D. and Spelke, E.S. (2005) The effect of language on infants' preference for faces. Poster presented at the Biennial Meeting of the Society for Research in Child Development, Atlanta, GA.

Kinzler, K.D., Dupoux, E. and Spelke, E.S. The native language of social cognition. Manuscript under review.

Kuhlmeier, V., Bloom, P. and Wynn, K. (2004) Do 5-month-old infants see humans as material objects? Cognition, 94: 95–103.

Kurzban, R., Tooby, J. and Cosmides, L. (2001) Can race be erased? Coalitional computation and social categorization. Proceedings of the National Academy of Sciences, 98: 15387–15392.

Lavin, T.A. and Hall, D.G. (2001) Domain effects in lexical development: learning words for foods and toys. Cogn. Dev., 16: 929–950.

Lea, S., Slater, A. and Ryan, C. (1996) Perception of object unity in chicks: a comparison with the human infant. Infant Behav. Dev., 19: 501–504.

Lee, S.A., Shusterman, A. and Spelke, E.S. (2006) Reorientation and landmark-guided search by young children: evidence for two systems. Psychol. Sci., 17: 577–582.

Leslie, A. and Keeble, S. (1987) Do six-month-old infants perceive causality? Cognition, 25: 265–288.

Leslie, A., Xu, F., Tremoulet, P. and Scholl, B. (1998) Indexing and the object concept: developing "what" and "where" systems. Trends Cogn. Sci., 2: 10–18.

Lipton, J. and Spelke, E. (2003) Origins of number sense: large-number discrimination in human infants. Psychol. Sci., 14: 396–401.

Lipton, J. and Spelke, E. (2004) Discrimination of large and small numerosities by human infants. Infancy, 5: 271–290.

Lourenco, S., Huttenlocher, J. and Vasilyeva, M. (2005) Toddlers' representations of space. Psychol. Sci., 16: 255–259.

Marino, A. and Scholl, B. (2005) The role of closure in defining the "objects" of object-based attention. Percept. Psychophys., 67: 1140–1149.

van Marle, K. and Scholl, B. (2003) Attentive tracking of objects versus substances. Psychol. Sci., 14: 498–504.

McCloskey, M. (1983) Intuitive physics. Sci. Am., 248: 122–130.

McCrink, K. and Wynn, K. (2004) Large-number addition and subtraction by 9-month-old infants. Psychol. Sci., 15: 776–781.

McKee, C. (2006) The effect of social information on infants' food preferences. Unpublished honors thesis, Harvard University, April 2006 (K. Shutts, K. Kinzler and E. Spelke, advisors).

Meck, W.H. and Church, R.M. (1983) A mode control model of counting and timing processes. J. Exp. Psychol. Anim. Behav. Process., 9: 320–334.

Mehler, J., Jusczyk, P., Lambertz, G. and Halsted, N. (1988) A precursor of language acquisition in young infants. Cognition, 29: 143–178.

Meltzoff, A. and Moore, M. (1977) Imitation of facial and manual gestures by human neonates. Science, 198: 75–78.

Miller, C.F., Trautner, H.M. and Ruble, D.N. (2006) The role of gender stereotypes in children's preferences and behavior. In: Balter, L. and Tamis-LeMonda, C.S. (Eds.), Child Psychology: A Handbook of Contemporary Issues, (2nd ed.). Psychology Press, New York, pp. 293–323.

Moon, C., Cooper, R. and Fifer, W. (1993) Two-day-olds prefer their native language. Infant Behav. Dev., 16: 495–500.

Newcombe, N. (2005) Evidence for and against a geometric module: the roles of language and action. In: Rieser J., Lockman J. and Nelson C. (Eds.), Action as an Organizer of Learning and Development. University of Minnesota Press, Minneapolis, MN, pp. 221–241.

Phillipson, R. and Skutnabb-Kangas, T. (1994) In: Asher, R.E. (Ed.), Linguicide: The Encyclopedia of Language and Linguistics.

Pica, P., Lemer, C., Izard, V. and Dehaene, S. (2004) Exact and approximate arithmetic in an Amazonian indigene group. Science, 306: 499–503.

Randall, L. (2005) Warped Passages. Ecco Press, New York.

Regolin, L. and Vallortigara, G. (1995) Perception of partly occluded objects by young chicks. Percept. Psychophys., 57: 971–976.

Rosenberg, R. and Carey, S. (2006) Infants' indexing of objects vs. non-cohesive entities. Poster presented at the Biennial Meeting of the International Society for Infant Studies.

Santos, L., Hauser, M. and Spelke, E. (2001) Recognition and categorization of biologically significant objects by rhesus monkeys (*Macaca mulatta*): the domain of food. Cognition, 82: 127–155.

Saxe, R., Tenenbaum, J. and Carey, S. (2005) Secret agents: 10- and 12-month-old infants' inferences about hidden causes. Psychol. Sci., 16: 995–1001.

Scholl, B. and Pylyshyn, Z. (1999) Tracking multiple items through occlusion: clues to visual objecthood. Cogn. Psychol., 2: 259–290.

Scholl, B., Pylyshyn, Z. and Feldman, J. (2001) What is a visual object? Evidence from target merging in multiple object tracking. Cognition, 80: 159–177.

Shutts, K. (2006) Properties of infants' learning about objects. Unpublished doctoral dissertation, Harvard University.

Sparks, A. (1996) Tomorrow is Another Country. University of Chicago Press, Chicago, IL.

Spelke, E.S. (1990) Principles of object perception. Cogn. Sci., 14: 29–56.

Spelke, E.S. (2000) Core knowledge. Am. Psychol., 55: 1233–1243.

Spelke, E.S. (2003) Core knowledge. In: Kanwisher, N. and Duncan, J. (Eds.), Attention and Performance, Vol. 20: Functional Neuroimaging of Visual Cognition. MIT Press, Cambridge, MA.

Spelke, E., Phillips, A. and Woodward, A. (1995) Infants' Knowledge of Object Motion and Human Action. Causal Cognition: A Multidisciplinary Debate. Clarendon Press/Oxford University Press, Oxford, pp. 44–78.

Tomasello, M. (1999) The cultural origins of human cognition. Harvard University Press, Cambridge, MA.

Trivers, R.L. (1971) The evolution of reciprocal altruism. Q. Rev. Biol., 46: 35–57.

Valenza, E., Leo, I., Gava, L. and Simion, F. (2006) Perceptual completion in newborn human infants. Child Dev., 77: 1810–1821.

Vishton, P.M., Stulac, S.N. and Calhoun, E.K. (1998) Using young infants' tendency to reach for object boundaries to explore perception of connectedness: rectangles, ovals, and faces. Paper presented at the International Conference on Infant Studies, Atlanta, GA.

Wang, R., Hermer, L. and Spelke, E. (1999) Mechanisms of reorientation and object localization by children: a comparison with rats. Behav. Neurosci., 113: 475–485.

Watson, J.S. (1972) Smiling, cooing, and "The Game." Merrill-Palmer Q., 18: 323–339.

Wood, J. and Spelke, E. (2005) Infants' enumeration of actions: numerical discrimination and its signature limits. Dev. Sci., 8: 173–181.

Woodward, A. (1998) Infants selectively encode the goal object of an actor's reach. Cognition, 69: 1–34.

Woodward, A. (1999) Infants' ability to distinguish between purposeful and non-purposeful behaviors. Infant Behav. Dev., 22(2): 145–160.

Xu, F. and Spelke, E. (2000) Large number discrimination in 6-month-old infants. Cognition, 74: B1–B11.

Xu, F., Spelke, E. and Goddard, S. (2005) Number sense in human infants. Dev. Sci., 8: 88–101.

C. von Hofsten & K. Rosander (Eds.)
Progress in Brain Research, Vol. 164
ISSN 0079-6123

CHAPTER 15

Taking an action perspective on infant's object representations

Gustaf Gredebäck[1] and Claes von Hofsten[2,*]

[1]*Department of Psychology, University of Oslo, Postboks 1094, Blindern, 0317 Oslo, Norway*
[2]*Department of Psychology, Uppsala University, Box 1225, 751 42 Uppsala, Sweden*

Abstract: At around 4 months of age, infants predict the reappearance of temporary occluded objects. Younger infants have not demonstrated such an ability, but they still benefit from experience; decreasing their reactive saccade latencies over successive passages from the earliest age tested (7 weeks of age). We argue that prediction is not an all or none process that infants either lack or possess. Instead, the ability to predict the reappearance of an occluded object is dependent on numerous simultaneous factors, including the occlusion duration, the manner in which the object disappears, and previous experiences with similar events. Furthermore, we claim that infants' understanding of how occluded objects move is based on prior experiences with similar events. Initially, infants extrapolate occluded object motion, because they have massive experience with such motion. But infants also have the ability to rapidly adjust to novel trajectories that violate their initial expectations. All of these findings support a constructivist view of infants object representations.

Keywords: occlusion; representation; action; eye tracking; infant

Taking an action perspective on infant's object representations

As we move around in the environment, objects constantly disappear and reappear from behind one another due to occlusion. Despite this, we as adults manage to maintain a uniform view of the world by compensating for object translations and by representing those objects that are temporarily out of sight. This enables us to predict future events and makes us ready to interact with the environment in a goal directed manner.

Organizing actions towards objects that are temporarily out of view poses specific problems to the perceptual-cognitive system. In order to effectively act towards the future reappearance of a moving object, we must represent that object and be able to estimate both where and when it will reappear. This knowledge is essential for our ability to smoothly carry out action plans despite the fact that objects go in and out of view. Developing stable object representations signify a major improvement of an infants' capability to interact with the environment.

The development of children's understanding of object permanence has been debated with vigour since it was first discussed by Piaget (1954). He considered the development of object permanence to be extremely important. With the establishment

*Corresponding author. Tel: +46 18 471 2133;
Fax: +46 18 471 2123; E-mail: claes.von_hofsten@psyk.uu.se

DOI: 10.1016/S0079-6123(07)64015-1

of object permanence the child goes from living in a fractionated world with no continuity to a world where objects have permanent existence and unique identity. He claimed that infants do not possess an adult-like ability to represent temporarily occluded objects as permanently existing objects until they understand the sequential displacements of a hidden object at the end of the second year of life. At the same time he noted that infants begin to show signs of object representation already during the stage of 'secondary circular reactions', that is, between 4 and 8 months of age but only within the same modality. At this age infants will briefly look for an object that has disappeared but they will not try to retrieve it. From around 12 months of age, infants retrieve hidden objects. If, however, the object is hidden at the same place several times and then hidden at a different place, the infants will reach for it at the previous hiding locations (A not B). It has also been reported that infants in this situation will look at the correct hiding place but reach for the previous one (Mareschal, 2000). Obviously, the relationship between object representation and action is relatively complex.

Piaget's object permanence task is confounded in one important respect. When the object is hidden, the child has to search for it. Failing to do so might reflect an inability to represent the hidden object (out of sight — out of mind) but it might also be caused by inability to formulate an action plan for retrieving the object, that is, a means-ends problem. In order to disambiguate the task, later research has simply presented objects that moves out of sight behind an occluder and observed how the child reacts to those events. This can be done either by measuring their ability to predict where and when the object will reappear or by measuring how their looking times change when some aspect of the events are changed.

Most of this work has focused on how much infants look at occlusion events in which the spatiotemporal continuity has been violated in some way (for related reviews using this methodology see Spelke, 1994; Mareschal, 2000; Baillargeon, 2004). This has been done by making the object reappear at an unexpected location, not reappear at all, reappear at an unexpected time, or by changing the identity of the object during occlusion. Infants looking durations at these various events are coded online (or later from videotapes) by trained observers. The amount of looking is analyzed, whether it declines when the event is presented several times or whether looking is increased when something happens that is not predictable from the previous events. If the infants look longer at those stimuli, it is concluded that the discrepancy has violated the infants' expectancy. For instance, Baillargeon and associates (Baillargeon et al., 1990; Baillargeon and deVos, 1991; Aguiar and Baillargeon, 1999) habituated infants to a tall and a short rabbit moving behind a solid screen. This screen was then replaced by one with a gap in the top. The tall rabbit should have appeared in the gap but did not. Infants from 2.5 month of age looked longer at the tall rabbit event suggesting that they had expected the tall rabbit to appear in the gap.

These studies indicate that the infants are somehow aware of the motion of a temporary occluded moving object but not exactly how it moves or when it will reappear. For instance it is not clear whether the infants expected the tall rabbit to appear at a specific time or not. The infants might have looked longer because they perceived the identity of the object to be changed. Another problem with this paradigm is that it does not address questions related to the micro organization of looking; only the duration is recorded. In many experiments only one data point is collected per subject. In addition, because this method does not record how infants' goal directed responses relate to occurring events, these studies are unable to inform us of the strengths of infants' knowledge; if these representations are strong enough to guide action.

Measuring infants' actions as they interact with the environment represent a different approach to understanding infants' early perceptual-cognitive development. In this paradigm infants are required to organize their actions towards moving objects that become temporarily occluded. Infant's behavioural responses are recorded and related to the spatial-temporal dynamics of the moving object. With this technique we are able to provide a detailed description of how infant's actions relate to events as they occur. This gives us the opportunity to look at how infants' representations, and expectations of when and where an occluded object will reappear, change over time.

This chapter will attempt to review those studies that have looked at how infants come to organize their own actions towards objects that are temporarily occluded. We will both examine when infants come to represent occluded objects and attempt to define those variables that limit (or enhance) infants object representations.

Methodological questions

Several different behaviours have been used as indicators for infants' ability to represent the spatiotemporal continuity of occluded objects and predict their reappearance. Eye movements are of primary interest but are tricky to measure. They can be coded by human observers from video recordings but this method is very time consuming and crude. More direct, precise, and reliable measurements of where gaze is directed at each point in time are needed. It is possible to measure eye movements with electrooculogram (EOG) which gives very high resolution in time (>200 Hz), but as infants rarely move just the eyes, the movements of the head need to be measured as well in order to know where gaze is directed. A new generation of eye trackers measure the reflection of infrared light sources on the cornea relative to the centre of the pupil (usually 50 Hz). For some of these eye trackers, no equipment is applied to the subject who just sits in front of the apparatus. With appropriate calibration, the measurement of cornea reflection provides precise estimates of where gaze is directed in the visual field.

Using gaze tracking as an indicator of predictive behaviour, when tracking occluded objects, relies on the following considerations. While the object is visible, infants from 2 to 3 months of age tracks it at least partially with smooth pursuit (von Hofsten and Rosander, 1997). When the object disappears behind an occluder the eyes are no longer able to sustain its smooth movements (Leigh and Zee, 1999). In order to continue tracking the object the observer has to shift gaze across the occluder in one or more saccades. An example of such behaviour can be observed in Fig. 1. The smooth tracking is visible prior to and following the occlusion in Fig. 1B. During the actual occlusion this infant made a saccade from the disappearance edge to the reappearance edge. The timing of this saccade (when the saccade was initiated or when it terminates at the reappearance location) provides information of when the infant expected the object to reappear (for the development of saccade latencies see Gredebäck et al., 2006). The location where the saccade terminates provides information of

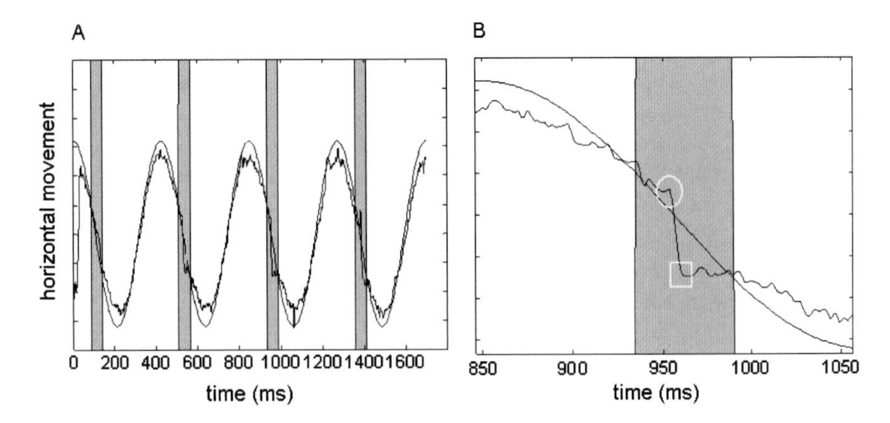

Fig. 1. (A) An object moving with constant velocity on a circular trajectory that is partly occluded (dark grey areas). (B) Enlargement of a single occlusion passage. The circle represent when the saccade is initiated and the square represents the termination of the saccade. Only horizontal eye movements are displayed.

where the object is expected to reappear. Both of these measures are frequently reported in the following text.

The measurements of arm movements is needed for drawing conclusions about infants ability to direct manual actions towards an occluded moving object. In some studies, video has been used but it is also possible to use more automatic motion capture devices where positions are defined by reflecting markers or light emitting diodes. If the infant reaches for the area where the occluded object will appear before the object emerges from behind the occluder, then infants are said to be predictive. That is, the infant has then demonstrated an ability to represent the spatiotemporal properties of the occluded object and the ability to predict how it is going to move in the future. The same logic can be applied to infants' head movements. Moving ones head to fixate the reappearance location ahead of time ensures that the infant fixate the object as it emerges, thereby allowing vision to guide a future reach to the attended object.

At what age do infants start to represent occluded objects?

A series of early reports were performed by Nelson (1971, 1974) and Meichler and Gratch (1980). In these studies 5- and 9-month-old infants were presented with a toy train that moved around on a track and, at one point, past through a tunnel. Infants watched these events and the experimenter recorded the infants' eye movements with a standard video camera. In summary, the videos of the infants' looking at this event gave no indication that 5-month-old infants anticipated the reappearance of the train from the tunnel. Nine-month olds consistently moved their gaze to the reappearance location at the other end of the tunnel and anticipated the emergence of the train there.

In recent years the technology available to measure infant's eye movements have advanced greatly. Numerous studies have taken advantage of the high temporal and spatial resolution provided by state of the art eye tracking technology. One such early eye tracking study was performed by van der Meer et al. (1994). They investigated 4–12 month-old infants' abilities to predicatively track and reach for an occluded toy which moved on a horizontal plane while measuring the infants' eye movements. Infants first started to reach for the toy at 5 months of age. At this age, infants' reaches were reactively launched at the sight of the reappearing toy. However, at the same age, they moved gaze to the reappearance point ahead of time. Not until infants were 8-month-old did they plan the reaching for the object while it was still occluded. This indicates that anticipatory tracking emerges prior to anticipatory reaching; the former exists from at least 5 months of age.

Recently, Johnson et al. (2003) presented 4-month-old infants with objects that became occluded at the centre of the trajectory. These stimuli were presented on a computer monitor and horizontal and vertical eye movements of one eye was recorded using an ASL 504 eye tracker (accuracy 0.5 visual degrees, sampling rate 50 Hz). Four-month-old infants who had previously been presented with fully visible trajectories (without the occluder) were more likely to predict the reappearance of the object than infants who had not been presented with such learning trials. At 6 months of age infants did not demonstrate the same benefits from seeing non-occluded trials. According to the authors these results demonstrate that 4-month-old infants do not possess robust object representations but that 6-month olds do.

In an attempt to trace the development of predictive looking in the occlusion situation, Rosander and von Hofsten (2004) measured head and eye movements of 7-, 9-, 12-, 17-, and 21-week-old infants as they tracked a real object (a happy face) that oscillated on a horizontal trajectory in front of them. Four different conditions were included in this study. The velocity of the object was either constant or sinusoidally modulated. In the former case the object always moved with the same speed and turned abruptly at the endpoints and in the latter case the object accelerated as it moved towards the centre of the trajectory and decelerated before each turn in a smooth fashion. In addition, the object became occluded for 0.3 s at the centre of its trajectory or for 0.6 s at one of the trajectory end points. Trial duration was 20 s which included five cycles of motion. If the occluder covered the centre of the screen each trial included 10 occlusion events and if the occluder covered the end point each trial included 5 occlusion events. In the latter case, the object reappeared on the same side as where it disappeared.

The level of performance in the central occluder condition improved rapidly over age. The youngest infants were purely reactive. It appeared as if the occluder edge itself became the focus of attention after object disappearance. It was found that the gaze of 7- and 9-week-old infants remained at the occluder edge almost 1 s after the object had reappeared on the other side of the occluder. Thus, in many cases the object had already reversed direction of motion and was approaching the occluder again before the infants re-focused their gaze on the object. The relative inability to quickly regain tracking had more or less disappeared for the 12-week olds. At that age, infants moved gaze to the reappearance point as soon as the object became visible (that is after ~0.5 s). Furthermore, the 12-week olds showed signs of being able to represent the moving object after having seen several occlusions. The mean gaze lag at reappearance for the last cycle of the trial with the triangular motion was predictive (see Fig. 2). The fact that also the younger infants became more aware of the reappearing object with experience over a trial suggests that they acquired some kind of representation of the occluded object.

The infants had an increasing tendency with age to extrapolate the occluded motion to the other side of the occluder when it was placed over one of the end points of the trajectory. For the 21-week olds, this tendency was dependent on the motion function used for the oscillation. When the object moved with constant velocity (triangular motion), the subjects made more false gaze shifts to the other side of the occluder. In this condition, there is no way to determine from a single occlusion event whether the object is going to continue or reverse its motion behind the occluder.

To summarize, these studies are all groundbreaking in their own right. The early studies by Nelson (1971, 1974) were the first to measure gaze tracking during occlusion and to demonstrate the importance of learning in occlusion events. The first study to look at eye–hand interaction during occlusion in infancy was provided by van der Meer et al. (1994). At the same time Johnson et al. (2003) and Rosander and von Hofsten (2004) pinpoint the immense importance of previous experiences. Johnson et al. focused on prior experiences with non-occluded objects whereas Rosander and von Hofsten provided a unique illustration that development does not consist of multiple hierarchal knowledge categories. Instead development of object representations is a continuous process that begins as early as 7 weeks-of-age and continuous far beyond 5 months-of-age.

As such, all fail-proof statements about when infants come to represent and predict occluded objects must be regarded with scepticism. Instead the effects of each study that report on the emergence of object representations must be seen in the context of prior experiences (both with fully visible and occluded trials). It should be noted, however, that each of these reports demonstrated a similar onset of object representations at 4 months of age. This is valid even for the study by Johnson et al. (2003); they reported an increase in predictive tracking with prior experience at 4 months of age. This learning appears to be a fundamental component of object representations and should be interpreted in support of the notion that 4-month olds have developed such an ability. To date no study has reported on consistent predictive responses at an earlier age.

Mapping out the psychometric space

Trajectory parameters

Clearly the learning effects described above are not the only component that defines if infants will display mature object representations and have the ability to predict the reappearance location of occluded objects. The ability to represent an occluded object is also dependent on the velocity and amplitude of the moving object and on the duration of the current occlusion event (to name a few contributing factors). The fact that different parameters of the ongoing object motion (independent of previous experiences) is important for infants abilities to predict the reappearance location of

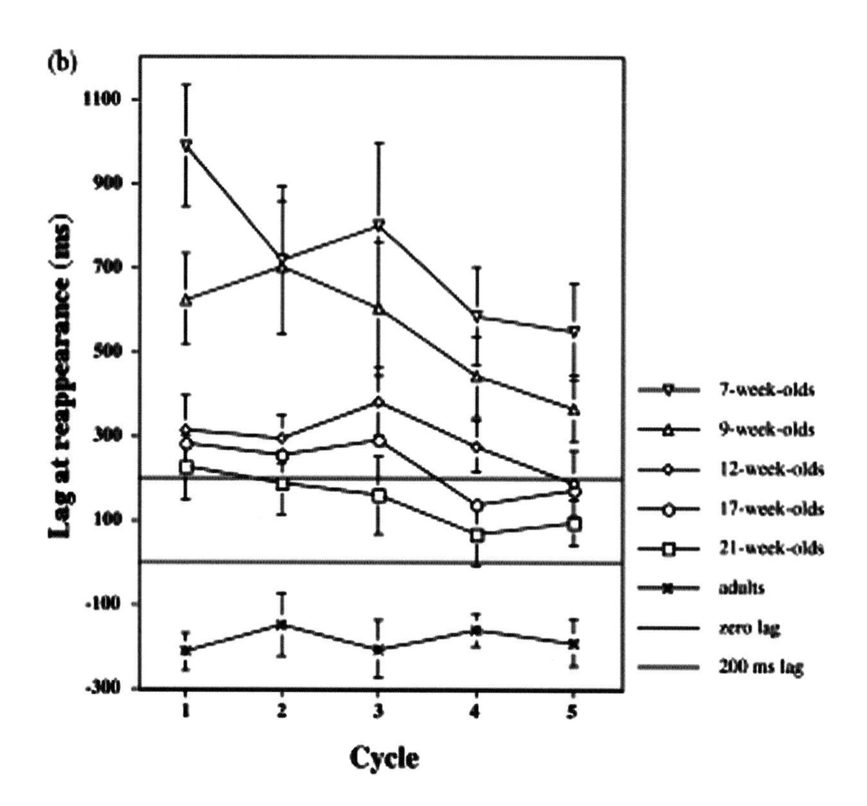
(b)
Lag at reappearance (ms)
1100
900
700
500
300
100
-100
-300
1
2
3
4
5
Cycle
7-week-olds
9-week-olds
12-week-olds
17-week-olds
21-week-olds
adults
zero lag
200 ms lag

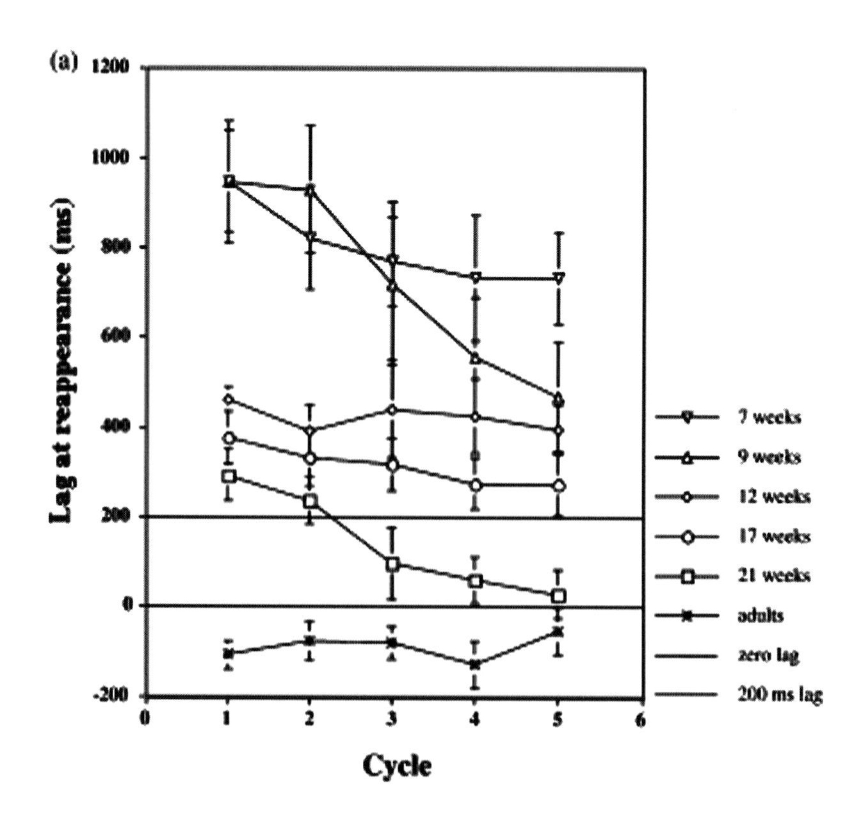
(a)
Lag at reappearance (ms)
1200
1000
800
600
400
200
0
-200
0
1
2
3
4
5
6
Cycle
7 weeks
9 weeks
12 weeks
17 weeks
21 weeks
adults
zero lag
200 ms lag

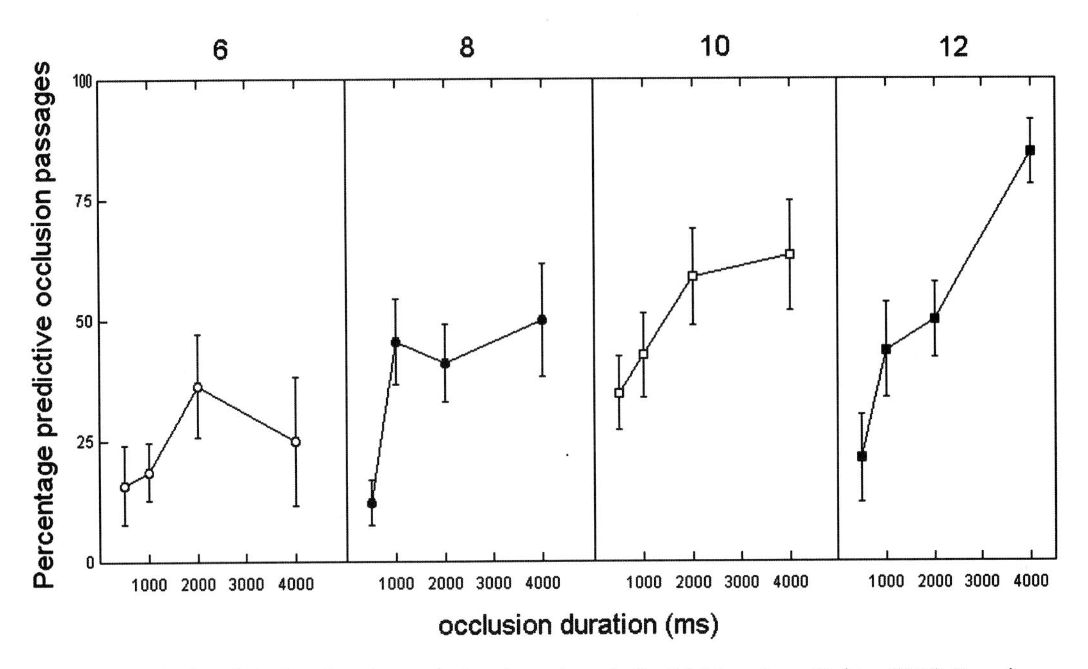

Fig. 3. Percentage predictive trials plotted against occlusion size and age in Gredebäck and von Hofsten (2004). Error bars represent standard error. Note that low occlusion durations equal high velocities (500 ms = 20°/s and 4000 ms = 2.5°/s).

occluded objects is nicely illustrated by two studies performed by Gredebäck and von Hofsten (Gredebäck et al., 2002; Gredebäck and von Hofsten, 2004).

In these studies 6–12 month old infants and adults were presented with an object that moved on a circular trajectory and became occluded once every lap. The study by Gredebäck and von Hofsten (2004), for example, presented such circular trajectories to a group of infants that was followed longitudinally from 6 to 12 months of age. In this study the size of the occluder always remained the same (20%) but velocities of the moving object varied (2.5–20°/s); resulting in four occlusion durations ranging from 500 to 4000 ms. Both studies randomized the presentation order of the different occlusion event and used an ASL 504 eye tracker to measure gaze direction.

The combined experience from these studies is that infants often failed to predict the reappearance of the target (for proportion of successful predictions see Fig. 3), even at 12 months of age (adults performed perfectly). Surely, the between trial randomization lowered the overall performance level and the circular trajectory probably made it more difficult to represent the trajectory of the target. However, the finding illustrated in Fig. 3 is that infants' performance at each age was highly influenced by the velocity (and/or occlusion duration) of the target. The 12-month-old group, for example, ranged in performance from $<20\%$ to $>80\%$ predictions dependent on the stimuli used. Unfortunately, these studies cannot disentangle if the occlusion duration or the velocity of the target is the driving factor behind this change (since they co-vary).

However, a recent study by von Hofsten et al. (2007) presented 4-month-old infants with a series of sinusoidal horizontal trajectories (randomized between trials). The design systematically varied

◀

Fig. 2. The average time differences and SE between object and gaze reappearance at each cycle of the centrally occluded trials. Separate graphs are shown for the sinusoidal (a) and the triangular motion (b). Each data point is the average of one occluder passage in each direction for all subjects in a specific age group. The upper line corresponds to the minimum time required for adults to program a saccade to an unexpected event (200 ms). Adapted with permission from Rosander and von Hofsten (2004).

occluder width, amplitude of the motion, and velocity of the moving object independently of each other. This was done in order to understand which variables contributed to infants' ability to represent and predict the objects reappearance during occlusion. The results demonstrated that infant's performance could not be explained by occluder edge salience, occluder duration on previous trials, or simply the passage of time. They rather geared their proactive gaze shifts over the occluder to a combination of occluder width, oscillation frequency, and motion amplitude that resulted in a rather close fit between the latency of the proactive gaze shifts and occlusion duration. Instead of having explicit knowledge of the relationship between these variables, infants could simply maintain a representation of the object motion and its velocity while the object is occluded. The results of von Hofsten et al. (2007) strongly supported this hypothesis. This can be seen in Fig. 4. It is as if the infants tracked an imagined object in their 'minds eye'. If object motion is represented in this way during occlusion, the effects of occluder width, oscillation frequency, as well as motion amplitude can all be explained.

In summary, numerous variables associated with the ongoing occlusion event determine how well an infant will be able to predict the objects reappearance. Even 12-month-old infants often fail to predict the reappearing object if the velocity is high and the trajectory circular. The final study described above (von Hofsten et al, 2007) made it abundantly clear that object representations are dependent on numerous simultaneous factors associated with the ongoing occlusion event. These findings clearly demonstrate the importance of mapping out the multidimensional psychometric space that governs object representation and the ability to perform an accurate prediction.

What stimulus information defines occlusion?

In the study by Gredebäck and von Hofsten (2004), we argued that infant's difficulties with high velocities could not result from an inability to track fast moving objects. Quite the contrary, we found that infants track (gaze and smooth pursuit) similar fast non-occluded motion with higher accuracy (timing and gain) than slower motion (Gredebäck et al., 2005; Grönqvist et al., 2006). Instead we argued that these difficulties can be related to the duration (clarity) of the gradual disappearance of the object behind the occluder (Gibson and Pick, 1980). In Gredebäck and von Hofsten (2004) the slow moving objects (long occlusion durations) included a slow and clear deletion event. As the velocity of the object increased the duration of the deletion event diminished, making it more and more difficult for the infants to perceive and classify the current events as an occlusion.

To test the hypothesis that infants object representations are influenced by the manner in which the object disappears behind the occluder Gredebäck et al. (in prep.) presented 5- and 7-month-old infants with a ball that moved back and forth along a horizontal path. Gaze were measured with a Tobii eye tracker (accuracy 0.5°, sampling rate 50 Hz). As the object reached the occluder, the ball either became deleted (Fig. 5A) or shrunk (Fig. 5B). It should be noted that the ball reappeared in the same manner as it disappeared in each condition.

The results demonstrate that infants at both 5 and 7 months of age make more predictions in response to the normal deletion condition (~50% predictions at 5 months and ~80% predictions at 7 months) than in response to the shrinking condition (~20% predictions at 5 months and ~50% predictions at 7 months). This suggests that the manner in which the ball became occluded strongly effected infant's representations, in addition to an overall increase in predictive tracking with increased age. Figure 6 include each data point (combined over the two conditions) collected at the two ages. This figure clearly demonstrates that infants track the target and make a saccade over the middle of the occluder.

Another way to manipulate the information pertained in the occlusion event is to turn off the light for the duration of occlusion. With this manipulation it is possible to vary what infants see during the occlusion event at the same time as one maintains both occlusion durations and identical pre- and post-occlusion trajectories. Such studies were performed by von Hofsten et al. (2000) and

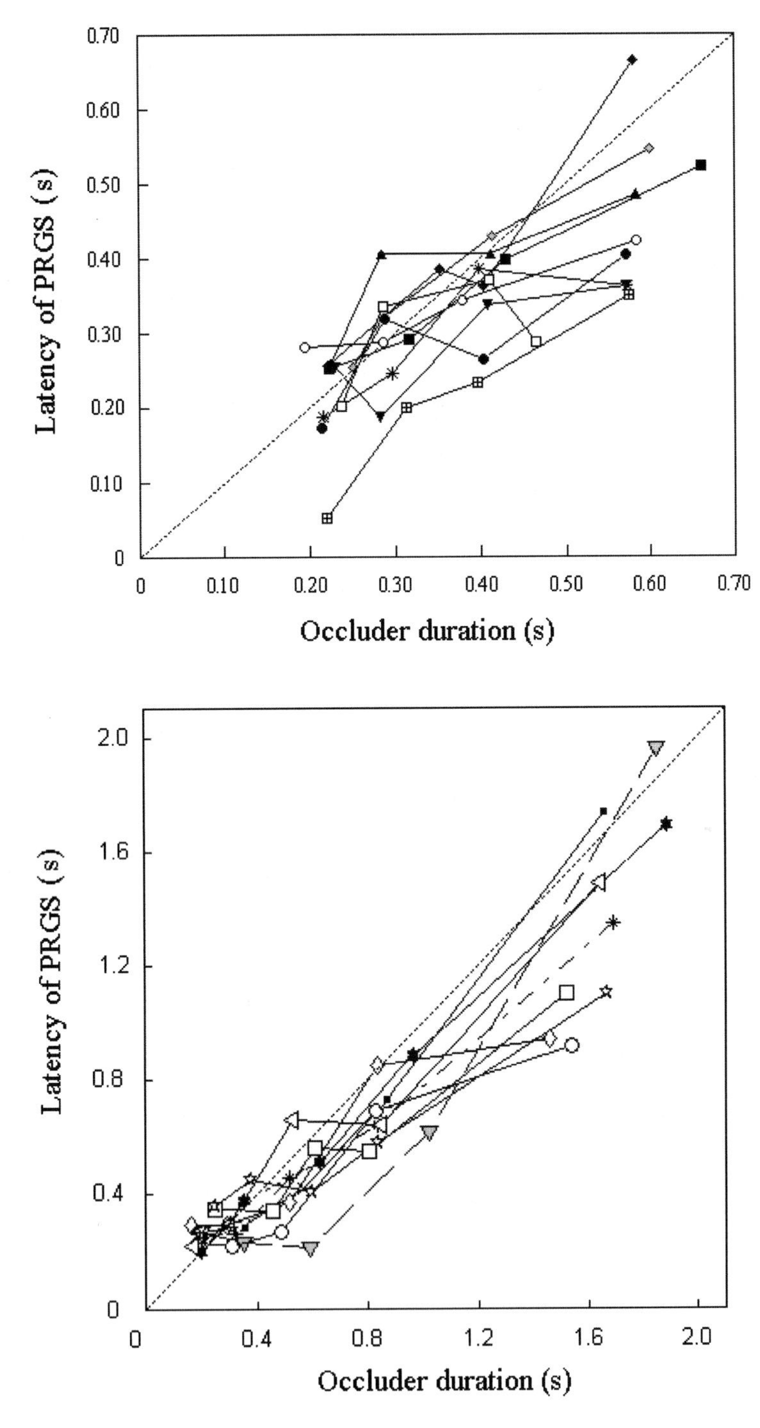

Fig. 4. (a) The relationship between occlusion duration and proactive gaze shifts (PRGS) for individual subjects in Experiment 1 of von Hofsten et al. (in press) that included occlusion durations of from 0.22 to 0.61 s. (b) The relationship between occlusion duration and proactive gaze shifts (PRGS) for individual subjects in Experiment 2 that included occlusion durations from 0.2 to 1.66 s. The dashed line in both figures shows the hypothetical relationship with saccade latency equal to occlusion duration.

Fig. 5. (A) The deletion condition, (B) the shrinking condition, (C, D) the areas where the ball successively disappear and reappear.

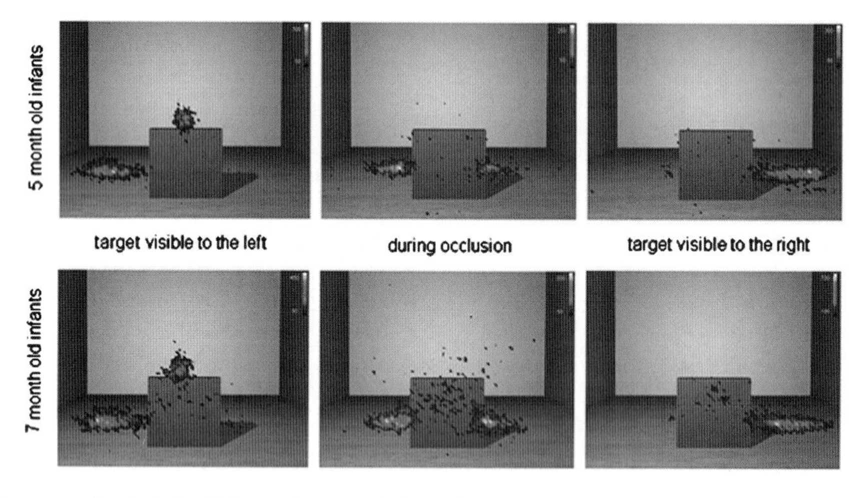

Fig. 6. Colour histograms that include all data points recorded (combined over the two conditions) at each age. Data from each age group is divided in to three pictures dependent on the location of the ball. In between each stimulus infants were presented with an attention-grabbing movie at the centre of the screen. Initial fixations at this location before infants moved their gaze to the ball and started tracking is visible at the centre of the screen when the target is visible to the left.

Jonsson and von Hofsten (2003). Jonsson and von Hofsten (2003) measured 6-month-old infant's head tracking and reaching during occlusion and blackout. During these events a target moved on a straight horizontal path in front of the infants. Either the object was fully visible during the entire trajectory or it became invisible during a period just prior to the optimal reaching space. Three different occlusion durations were used in combination with the two modes of non-visibility (occlusion vs. blackout). In both conditions the object was occluded for 400, 800, or 1200 ms. Infants' head tracking was more inhibited by blackout than by a visible occluder but the opposite effect was observed during reaching. No consistent effects of occlusion duration were observed during blackout. During occlusions, however, the head led at first target reappearance (predictions) and the size of the mean lead increased with prolonged duration of non-visibility.

In summary, these studies add another factor that limits object representations; namely the stimulus information that defines occlusion. The study by Gredebäck et al. (in prep.) demonstrates that a clear deletion event allows infants to classify the stimulus as an occlusion event, and this will in turn, strengthens infant's representations and promote predictions. The study by Jonsson and von Hofsten (2003) demonstrated that the manner in which an object is obstructed from view (occlusion or blackout) also influence the way in which infants are able to deal with the object in its visual absence. Head tracking is more disrupted by competitive visual stimuli (the occluder) and is less disrupted by blackout. Clearly infants' actions on objects that are temporarily out of view are not only influenced by the structure of the stimuli but also by the manner in which it disappears from view.

How specific are object representations?

Several studies indicate that infants' ability to represent occluded objects in the context of reaching is much inferior to their ability to represent them in the context of looking (Spelke and von Hofsten, 2001; Jonsson and von Hofsten, 2003; Hespos et al., in prep.). Spelke and von Hofsten (2001) and Jonsson and von Hofsten (2003) found that predictive reaching for occluded objects were almost totally absent in 6-month-old infants. At the same time they did not seem to have problems with tracking them with their head (von Hofsten et al., 2000; Jonsson and von Hofsten, 2003).

Hespos et al. (submitted) recorded the predictive reaching of 6- and 9-month-old infants who viewed an object that moved in a straight line and, on some trials, was briefly occluded before it entered the reaching space. While there was an increase in the overall number of reaches with increasing age, there were significantly fewer predictive reaches during the occlusion trials than during the visible trials and this pattern showed no age-related change. In a second experiment, Hespos et al. developed a reaching task for adults modelled on the tasks used to assess predictive reaching in infants. Like infants, the adults were most accurate when the target was continuously visible and significantly less accurate when the target was briefly occluded. These findings suggest that the nature and limits to object representations are similar for infants and adults.

Following Shinskey and Munakata (2003) and Scholl (2001), Spelke and von Hofsten (2001) suggested that young infants represent both visible and hidden objects, and their object representations depend on the same mechanisms as those used to represent and attentively track objects in adults (Scholl, 2001). More specifically, the object representations of infants and adults have three properties. First, these representations are more precise, at all ages, when objects are visible than when they are hidden. Second, representations of different objects are competitive; the more objects one attends to, the less precise will be one's representation of each object. Third, precise representations are required for reaching: to reach for an object, one must know where it is, how big it is, what shape it is, and how it is moving. In contrast, less precise representations suffice to determine that a hidden object exists behind an occluder in a scene that one observes but does not manipulate.

Spelke and von Hofsten (2001) proposed that object representations change over human development in just one respect: They become increasingly precise. Just as infants' sensory and perceptual capacities become more accurate with age (e.g. Kellman and Arterberry, 1998), so does their capacity to represent objects. While infants may reliably predict the reappearance of an occluded moving object moving on a linear path from 4 months of age, the ability to predict where and when the moving object will reappear from behind an occluder is problematic to children beyond their first birthday (Gredebäck and von Hofsten, 2004). Both visible and occluded objects are therefore represented with increasing precision as infants grow.

These properties suffice to account for all the reviewed findings. Object representations are more precise in the dark than in the presence of a visible occluder, because the occluder competes with the hidden object for attention, decreasing the precision of both object representations. When a young infant participates in a preferential looking experiment involving an occluded object, she can draw on her imprecise representation of the object to determine that it exists behind the occluder, and in addition identify gross properties of the object such as its approximate location (e.g. Baillargeon and Graber, 1988) and the orientation of its principal axis (Hespos and Rochat, 1997). Nevertheless, a young infant is likely to fail to represent the exact shape, size, or location of an occluded object, because his or her representation is less precise than that of an older child. When a young infant is presented with an occluded object in a reaching experiment, this same imprecise representation is not sufficient to guide object-directed reaching. The differing precision required by many preferential looking experiments vs. many reaching experiments therefore can account for their different outcomes.

Can infants learn new rules of object motion?

We know that infants can extrapolate linear horizontal trajectories from at least 4 months of age (see the discussion on the emergence of object representations above) and that infant's actual performance on a given trial is dependent on the structure of the perceived events, their previous experiences, and the manner in which the object disappears. We also know that infants from at least 6 months of age can extrapolate circular trajectories (Gredebäck et al., 2002; Gredebäck and von Hofsten, 2004). In these studies (reviewed above) both infants' and adults' predictive saccades terminated along the curvature of the circular trajectory (at the reappearance edge of the occluder).

So, we conclude from these studies that infants extrapolate a number of different naturally occurring trajectories. However, what is still unknown from the above-mentioned studies is whether infants can construct new rules of novel trajectories or if infants are solely governed by pre-existing knowledge of how objects naturally move. This question, whether the ontogenetic origin of infants object representations emerge from innate knowledge structures (nativism) or if this knowledge emerge in an interaction with the environment (constructivism) have recently been the focus of much research.

The first two studies to address this issue (while relating the infants' predictions to the actual reappearance location of the object) were performed by von Hofsten and Spelke (von Hofsten et al., 2000; Spelke and von Hofsten, 2001). In these studies the authors measure 6-month-old infants' predictive reaching and head tracking during an occlusion task.

In both studies infants were seated in front of a vertical surface on which a toy moved on linear paths. Half of all trials started with the target moving from the upper edges of the screen, moving downwards on a diagonal path (linear trials). During other trials the toy started moving in the same manner but changed direction at the centre of the screen; continuing downwards but reversing the horizontal direction (non-linear trials). At the intersection between these trajectories (the centre of the screen) the toy moved behind an occluder (see Fig. 7). This event prevented the infants from perceiving whether the toy moved on a straight or turning trajectory. To predict the reappearance of the object, the infants had to turn their head either to the lower right or left side of the occluder (occlusion durations were 400 and 900 ms). Spelke and von Hofsten (2001) contrasted these occlusion events with fully visible trials.

During the first occlusion event infants did not anticipate the reappearance of the toy. However, with experience infants rapidly predicted the reappearance on linear trials (after three trials). Even non-linear trials were anticipated, but learning was slower. These studies demonstrate that 6-month-old infants' are better equipped to learn about linear trajectories than they are to learn about non-linear trajectories. This finding was interpreted in support of the nativist view; suggesting that infants have a pre-existing notion that objects naturally move on linear trajectories (e.g. inertia) and that infants use this knowledge to extrapolate the pre-occlusion trajectory.

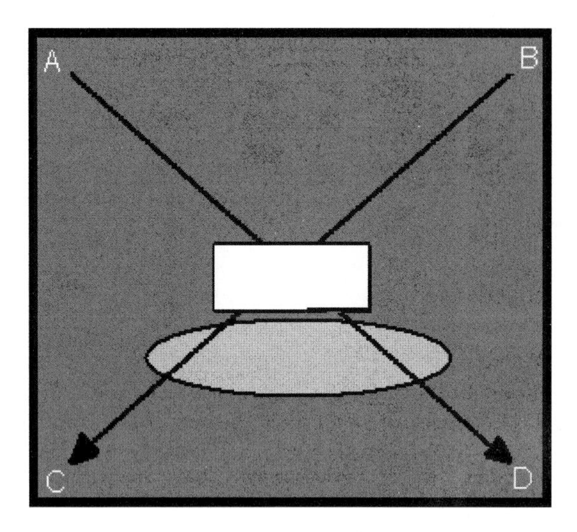

Fig. 7. Arrows and letters indicate the four trajectories used by Spelke and von Hofsten (2001) (A→D, B→C, A→C, B→D). The white square indicate the approximate location of the occluder while the light grew ellipse represent the optimal reaching space of infants.

In retrospect, these papers (von Hofsten et al., 2000; Spelke and von Hofsten, 2001) demonstrate something different altogether. The studies suggest that infants have multiple strategies available to solve an occlusion task. Infants can extrapolate the pre-occlusion trajectory but they also have the ability to learn how to predict novel (non-linear) trajectories. As such, these studies do not inform us about the ontology of infants object representations but illustrate the diversity of recourses available to an infant when faced with an occlusion event.

To better understand the nature of these two forms of prediction, Kochukhova and Gredebäck (in press) presented infants with movies in which a ball rolled back and forth between two endpoints. The middle of the trajectory was covered by a round occluder. Eye movements were measured with a Tobii eye tracker. Experiment 1 compared infants' ability to extrapolate the current pre-occlusion trajectory with their ability to base predictions on recent experiences of novel object motions. In the first (linear) condition infants were presented with multiple linear trajectories. These could be extrapolated but infants were unable to rely on memories of previous events to solve the occlusion task (since each session included multiple trajectories with different directions of motion). In the second (non-linear) condition infants were presented with multiple identical trajectories that turned 90° behind the occluder. These trajectories could not be extrapolated but infants were able to rely on previous experience to predict where the target would reappear.

In the linear condition infants performed at asymptote (~2/3 accurate predictions) from the first occlusion passage and performance did not change over the session. In the non-linear condition all infants initially failed to make accurate prediction. Performance, however, reached an asymptote after two occlusion passages. This initial experiment demonstrates that infants have an initial assumption that objects will continue along the linear extension of the pre-occlusion trajectory. But the results also demonstrate that infants can change their predictions if another source of information is more reliable.

In a second experiment the learning effect observed in response to the non-linear trajectories were replicated and extended. Here infants were presented with the same set of non-linear trajectories on three different occasions; a first session as soon as they arrive in the lab, a second session after a 15 min break, and a third session 24 h later. The results can be observed in Fig. 8.

First of all, infants quickly learned to predict the correct reappearance location of the ball. However, after a 15 min break infants had completely forgotten where the ball reappeared. Infants required a second session to consolidate their experience and form a stable memory of where the ball would reappear. After this second session infants were able to maintain a representation of the trajectory for at least 24 h.

This final study demonstrates that infants' initial assumptions are consistent with a linear extension of the pre-occlusion trajectory. But, more importantly, the study demonstrates that infant can acquire new knowledge after only a few presentations and have the ability to maintain this information over time. We suggest that these different approaches to solving an occlusion task (extrapolations and memories of previous events) are not governed by separate mechanisms. Instead we interpret these

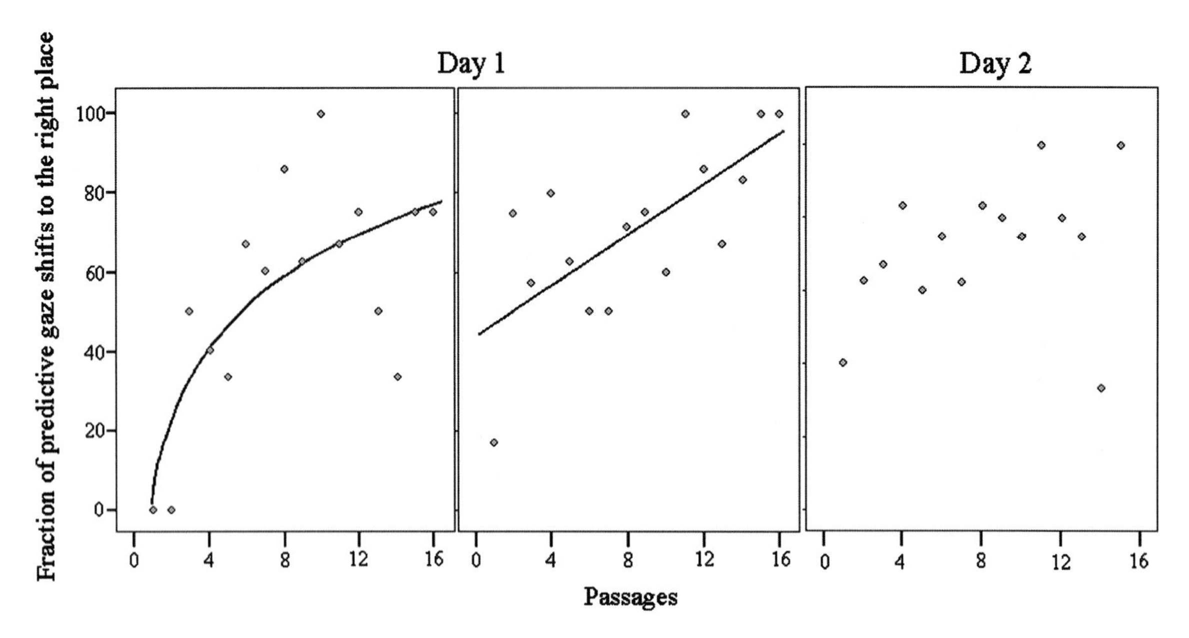

Fig. 8. Percentage of predictive occlusion passages that appear in the correct reappearance location in each of the three sessions of Experiment 2 of Kochukhova and Gredebäck (in press). Each dot represents the average percent accurate predictions on that occlusion passage. Lines depict the regression line with most explained variance; no significant changes were observed during the second day.

findings in support of the constructivist view, suggesting that both steam from the infants' own experience with the environment. Infants learn to predict non-linear trajectories in the lab but have most likely had enough experience with linear (and curvilinear) trajectories in the real world to help them formulate a valid hypothesis about how objects naturally move. From this perspective the current results appear almost trivial; infants are initially more proficient with extrapolation since this is the only trajectory (of the two presented) that infants have had any real experience with (prior to the study). After a number of presentations of non-linear trajectories infants learn to predict these with equal proficiency.

What does prediction really mean?

One noteworthy aspect of measuring anticipatory gaze shifts during occlusions is that predictions occur on only about half of all presented trials in infancy. Despite this, we claim that infants from at least 4 months of age can represent occluded objects. In Rosander and von Hofsten (2004) and in von Hofsten et al. (2007) the 4-month-old infants moved gaze over the occluder ahead of time in 47% of the trials and in Johnson et al. (2003) in 29–46% depending on condition and age. Similar levels of anticipatory gaze shifts have been observed at 6- (Kochukhova and Gredebäck, in press) and 12-month-old infants (Gredebäck and von Hofsten, 2004). If infants track the spatiotemporal contiguity of the occluded object why do they not make accurate predictions on every trial?

First of all, there is no way to ask infants to pay attention to a specific aspect of the visual scene. As an obvious effect thereof infants will on occasion disrupt tracking and look at some non-task related aspect of their visual scene. If infants shift their attention away from the moving object during occlusion then some of these trials will be undistinguishable from a reactive trial (in which the infant only fixated the moving object after it has reappeared from behind the occluder). It is therefore likely that the above-mentioned studies underestimate infants' performance to some degree.

In addition to voluntary changes in attention, infants' ability to actually represent the occluded object is dependent on the relative salience of each aspects of the visual field. As mentioned above, the different elements of the visual field (visible and hidden) compete with each other for available recourses. When a moving object is occluded the relative saliency of visible stimuli (like the occluder) increases. According to this logic, infants might have a general ability to represent non-visible objects but the actual performance on a given trial is easily disrupted.

We have described a number of studies that demonstrate the diversity of infants' performance and the highly variable results obtained through small changes in the psychometric space that make up the visual scene and the occlusion event. Each of these components (e.g. the occluder width, the way the object disappears, and the amplitude of the trajectory) independently influence the relative representational strength of the occluded object and its surroundings. Each helps build up and/or degrade object representations in a non-linear fashion.

Myths about eye tracking and occlusion

This chapter has reviewed a number of studies that measured infants' abilities to predict the reappearance of occluded objects. All of these studies rely on the assumption that predictions are synonymous with (or at least related to) infants' abilities to represent the occluded object and/or its spatio-temporal dynamics. There are, however, a few rival interpretations of these findings. Interpretations that questions the link between prediction and representations, especially when infant's eye movements are used as a dependent measure. The following paragraphs will introduce these alternative interpretations and address why they are unable to account for the obtained results.

Could predictive gaze shifts be the result of random looking?

Is it possible that infants stop tracking at the occluder edge when the object disappears, wait there for a while and then shift gaze anyway in a random fashion. Some of those spontaneous gaze shifts might arrive to the reappearance side of the occluder before the object reappears there. Such random tracking would provide a number of false predictions. Three of the above-mentioned studies clearly demonstrate that this is not the case.

The study by Gredebäck and von Hofsten (2004) presented infants with four different occlusion durations. In this study infants scaled their proactive gaze shifts over the occluder to the actual occlusion duration. More gaze shifts were made after 400 ms in response to a 500 ms occlusion event than in response to a 1000 ms occlusion event, and a similar relationship existed for each of the four-occlusion durations. If gaze moved at random, then the same number of gaze shift would end up on the reappearance side of the occluder independent of the actual occlusion duration. In a similar vain, von Hofsten et al. (2007) demonstrated that the proportion of gaze shifts to the reappearance edge ahead of time showed no relationship with occlusion duration in either of the two experiments. Again, the proportion of gaze shifts ending up at the reappearance side of the occluder would increase with prolonged occlusion durations if infants gaze shifts were launched and directed at random.

A third example comes from the study by Kochukhova and Gredebäck (in press). In this study, the number of gaze shifts during occlusion to each side of the occluder was compared. Infants were only judged to have the ability to predict the actual reappearance location if they made more gaze shifts to this location compared to the alternative reappearance locations along the occluder edge. Their ability to move to the correct location was dependent on the trajectory being presented and on their previous experience with similar events. If infants had moved their gaze at random, then each side of the occluder would be fixated to an equal degree and none of these effects would be significant.

Could predictive gaze shifts be the result of occluder salience?

This alternative account suggests that the salience of the occluder's reappearance edge determine whether infants make predictive saccades across the occluder. If this was the case then stimuli with greater visual salience would attract attention to a higher degree and result in earlier gaze shifts. As contrast sensitivity decreases with increasing eccentricity in the visual field, it is possible that gaze shifts in the presence of a wide occluder will have a longer latency, not because the subject expects the object to reappear later, but because the visual salience of the exiting occluder edge is then lower. One argument against the visual salience hypothesis comes from the reactive saccades in the study by von Hofsten et al (2007). Reactive saccades are by definition elicited by the detection of the reappearing object in the periphery of the visual field. In this study von Hofsten et al. found that the effect of occluder width on reactive saccade latency was small (0.45 s for the narrow and 0.54 for the wide) in comparison to the difference in the latency of the proactive saccades (0.33 s for the narrow and 0.79 for the wide occluder). It is therefore unlikely that it is the visual salience of the exiting occluder edge that determines the difference in saccade latency for the different occluder widths. One can, of course, argue that a non-salient stimulus in the periphery of the visual field like the occluder edge will take longer to detect than a salient one like the reappearing object. However, the latency of proactive saccades for the narrow occluder was shorter than the reaction time to the salient reappearing object in the same condition. Finally and most importantly, visual salience could not be the only determinant of the proactive saccades. The effects of oscillation frequency and motion amplitude were found to be just as important. Motion amplitude and oscillation frequency refer to variables that are not visually present during occlusion and therefore it is inevitable that information from the seen pre-occlusion motion is preserved during occlusion.

Could predictive gaze shifts be the result of conditioning?

This alternative account of the studies reviewed above suggests that predictive saccades are the result of the simple contingency between disappearance and reappearance locations. The hypothesis is derived from operant conditioning and does not involve any representational abilities. At least three of the above-mentioned studies clearly demonstrate that this is not the case. The strongest evidence against this alternative hypothesis comes from the study by Kochukhova and Gredebäck (in press). In the first experiment of this study infants were presented with a numerous linear trajectories with different disappearing and reappearing position. Each trajectory was randomly selected from a set of linear trajectories leaving no room for conditioning of location. Despite this, infants performed at asymptote from the very first trial. The fact that infants predicted the linear trajectory the first time they saw the stimuli clearly indicates that conditioning cannot account for infants' predictions.

The same conclusion can be drawn from the study by Gredebäck and von Hofsten (2004). In their study infants were presented with four different (randomized) occlusion durations and that made conditioning of occlusion duration near impossible.

A third example comes from von Hofsten et al. (2007). They measured whether the previous occlusion duration had an impact of the latency of infants' saccade across the occluder on the current trial. No such factor emerged in the analysis instead infants performance was guided by parameters of the current occlusion event.

Summary

The reviewed research demonstrates that infants' actions are directed to the reappearance of occluded objects from a very early age. At around 4 months of age, infants overcome the temporary occlusion of an object they track by shifting gaze ahead of time to the position where it reappears. Before this age, infants have not demonstrated an ability to predict the reappearance of occluded objects but they still benefit from experience; decreasing their reactive saccade latencies over successive passages from the earliest age tested (7 weeks of age). Occlusion is not only problematic to young infants; they appear to challenge even the adult mind.

We also demonstrate that prediction is not an all or none process that infants either lack or possess. Instead each infant's abilities to predict the reappearance of an occluded object are dependent on numerous simultaneous factors. These include parameters of the current occlusion event (e.g. occlusion duration and the manner in which the object disappears) and previous experiences with similar events (both within the current trial and more long-term experience that predate the experimental session). This illustrate that infant's abilities to predict the motion of an occluded object is determined, in part by their own representational abilities, but also by the dynamics of the current occlusion event, and the relative representational strengths of visible and occluded objects.

We have argued that infants' understanding of how occluded objects move is based on prior experiences with similar events. The functioning of basic biological processes like those related to the perception of object velocity and accretion/deletion at an occluder edge are necessary for allowing the infant to be aware of the object when it is out of sight. We propose that these principles are acquired through an interaction with the environment. Infants will initially extrapolate the trajectories of occluded objects because they have massive experience with linear (and curvilinear) trajectories. But infants also have the ability to rapidly adjust to novel trajectories that violate their initial expectations. All of these findings support a constructivist view of infants' object representations.

Acknowledgements

This paper was supported by Swedish Research Council (421-2006-1794) and the Norwegian Directorate for Children, Youth, and Family Affairs (06/34707).

References

Aguiar, A. and Baillargeon, R. (1999) 2.5 month-old infant's reasoning about when objects should and should not be occluded. Cognit. Psychol., 39: 116–157.

Baillargeon, R. (2004) Infants' reasoning about hidden objects: evidence for event-general and event-specific expectations. Dev. Sci., 7(4): 391–414.

Baillargeon, R. and Graber, M. (1988) Evidence of Location memory in 8-month-old infants in a nonsearch AB task. Dev. Psych., 24: 502–511.

Baillargeon, R., Graber, M., Devos, J. and Black, J. (1990) Why do young infants fail to search for hidden objects? Cognition, 36(3): 255–284.

Baillargeon, R. and deVos, J. (1991) Object permanence in young infants: further evidence. Child Dev., 62(6): 1227–1246.

Gibson, E.J. and Pick, A.D. (1980) An echological approach to perceptual learning and development. In: Reed, E. and Johes (Eds.), Reasons for realism: The selected essays of James J. Gibson. Earlbaum, New York.

Gredebäck, G. and von Hofsten, C. (2004) Infants' evolving representation of moving objects between 6 and 12 months of age. Infancy, 6(2): 165–184.

Gredebäck, G., von Hofsten, C. and Boudreau, J.P. (2002) Infants' visual tracking of continuous circular motion under conditions of occlusion and non-occlusion. IBAD, 25: 161–182.

Gredebäck, G., von Hofsten, C., Karlsson, J. and Aus, K. (2005) The development of two-dimensional tracking: a longitudinal study of circular pursuit. Exp. Brain Res., 163(2): 204–213.

Gredebäck, G., Örnkloo, H. and von Hofsten, C. (2006) The development of reactive saccade latencies. Exp. Brain Res., 173(1): 159–164.

Grönqvist, H., Gredebäck, G. and von Hofsten, C. (2006) Developmental assymetries between horizontal and vertical tracking. Vis. Res., 46(11): 1754–1761.

Hespos, S.J. and Rochat, P. (1997) Dynamic mental representation in infancy. Cognition, 64: 153–188.

von Hofsten, C., Fenq, Q. and Spelke, E.S. (2000) Object representation and predictive action in infancy. Dev. Sci., 3(2): 193–205.

von Hofsten, C., Kochukhova, O. and Rosander, K. (2007) Predictive tracking over occlusions by 4-month-old infants. Dev. Sci., in press.

von Hofsten, C. and Rosander, R. (1997) Development of smooth pursuit tracking in young infants. Vision Res., 37(13): 1799–1810.

Johnson, S.P., Amso, D. and Slemmer, J.A. (2003) Development of object concepts in infancy: evidence for early learning in an eye-tracking paradigm. PNAS, 100(18): 10568–10573.

Jonsson, B. and von Hofsten, C. (2003) Infants ability to track and reach for temporarily occluded objects. Dev. Sci., 6: 88–101.

Kellman, P.J. and Arterberry, M. (1998) The Cradle of Knowledge: Development of Perception in Infancy. MIT Press, Cambridge, MA.

Kochukhova, O. and Gredebäck, G. (in press). Learning about occlusion: initial assumptions and rapid adjustments. Cognition, in press.

Leigh, R.J. and Zee, D.S. (1999) The Neurology of Eye Movements (3rd ed.). Oxford University Press, New York, NY.

Mareschal, D. (2000) Object knowledge in infancy: current controversies and approaches. TICS, 4(11): 408–416.

van der Meer, A.L.H., van der Weel, F.R. and Lee, D.N. (1994) Prospective control in catching by infants. Perception, 23: 287–302.

Meichler, M. and Gratch, G. (1980) De 5 month-olds show object conception in Piaget's sense? IBAD, 3: 256–281.

Nelson, K.E. (1971) Accommodation of visual tracking in human infants to object movement patterns. J. Exp. Child Psychol., 12: 182–196.

Nelson, K.E. (1974) Infants' short-term progress towards one component of object performance. Merril-Palmer Q., 20: 3–8.

Piaget, J. (1954) The Origin of Intelligence in Children. Routledge, New York.

Rosander, K. and von Hofsten, C. (2004) Infants' emerging ability to represent occluded object motion. Cognition, 91(1): 1–22.

Scholl, B. (2001) Objects and attention: The state of the art. Cognition, 80: 1–46.

Shinskey, J.L. and Munakata, Y. (2003) Are infants in the dark about hidden objects? Dev. Sci. 6: 273–282.

Spelke, E.S. (1994) Initial knowledge: six questions. Cognition, 50: 431–445.

Spelke, E.S. and von Hofsten, C. (2001) Predictive reaching for occluded objects by 6 month old infants. J. Cognit. Dev., 2: 261–282.

SECTION IV

The Development of Action and Social Cognition

C. von Hofsten & K. Rosander (Eds.)
Progress in Brain Research, Vol. 164
ISSN 0079-6123

CHAPTER 16

Infants' perception and production of intentional actions

Petra Hauf*

Department of Psychology, St. Francis Xavier University, PO Box 5000, Antigonish, NS, B2G 2W5, Canada

Abstract: Human beings act and interact with their social environment. Thus, it is important not only to understand other individuals' actions, but also to control one's own actions. To understand intentional actions one needs to detect goals in the perceived actions of others as well as to control one's own movements in order to achieve these goals through action production. After a short review of recent studies on the development of action understanding during the first years of life, the role of action effects for action understanding is discussed. In a series of experiments the exchange between action perception and action production is demonstrated, its implications for understanding intentional actions are highlighted.

Keywords: action understanding; infants; action perception; action production; like-me; imitation

Introduction

Activity is part of our everyday life. We explore our world with all our senses, and process a variety of different stimuli and impressions. We represent and reflect on what we perceive and we plan our own productive life on this basis. As human beings we act on the environment, particularly on the social environment. But, there is an important difference between actions that are directed toward objects and social actions. In social interaction we relate our own actions to the actions of other people. Social interaction emphasizes that one's own actions affect the behaviour of the person towards whom they are directed and actions of others affect our own behaviour (e.g., Hauf and Försterling, 2007). Thus, it is important to understand other persons' actions, goals, and desires. But it is equally important to be able to perform one's own actions, to achieve goals, and to express desires. Understanding other people's actions and knowing how those actions can be linked to our own actions is crucial to our social interaction. Actions are organized around goals; they differ from other bodily movements in their intentional character. The ability to detect goals is of capital importance for action understanding, that is, it is important to identify the goals of perceived actions as well as to control certain movements to achieve one's own goals. Furthermore, the ability to predict other people's action goals as well as to predict the environmental outcome and the consequences of one's own actions is of crucial significance. This is the case for children as well as for adults. The last decade has seen an increase of investigations into infants' development of action understanding. These studies mainly focus on infants' developing ability to distinguish between the goal of an action and the movements that are necessary to achieve a certain goal. Thereby, action understanding typically refers to paradigms where infants perceive

*Corresponding author. Tel.: +1(902) 867 5041; Fax: +1 (902) 867 5189; E-mail: phauf@stfx.ca

DOI: 10.1016/S0079-6123(07)64016-3

actions performed by others, whereas studies on action control mainly focus on infants' ability to perform actions on their own. The present paper gives an overview of how infants in their first years of life perceive and understand intentional actions performed by others, of how infants at this age produce intentional actions on their own, and of how the development of these two aspects is linked to each other. The role of action effects is demonstrated in relation to action perception and action production.

Understanding of intentional actions

It is a core assumption in the literature on the understanding of intentional actions that interpreting the actions of other persons as goal-directed is a precursor of intentional understanding (Meltzoff, 1995, 2007a; Carpenter et al., 1998a; Tomasello, 1999a). To understand that other persons have goals in mind while acting is an important step towards a complete 'theory of mind' which enables people to interact with their social partners in a rational way. Children do not obtain a sophisticated 'theory of mind' level — like adults have — before the age of 4 to 6 years (Wellmann et al., 2004, 2007; Aschersleben et al., 2007). This level includes the understanding of goal-directed and intentional actions, and beyond this, the understanding that perceivable behaviour is driven by mental states like emotions, beliefs and desires.

The following sections will focus on intentional actions starting with an overview of how the understanding of goal-directed actions performed by others develops during infancy and of how infants develop such goal-directed actions on their own.

Perception of intentional actions

Recent studies indicate that infants at the age of 5 to 6 months start to understand other persons' intentional actions. With her habituation studies, Woodward (1998, 1999) showed that infants at this age start to understand adult reaching and grasping movements as goal-directed. To demonstrate this, infants watched an actor reach for and grasp one of two toys sitting side-by-side on a stage during habituation trials. In subsequent test trials, the position of the two toys was reversed and the actor grasps either the same toy at a new location or the new toy at the same location. Infants looked longer at events in which the actor reached for the new toy (new goal event) than at events in which the actor reached to the new location (new path event) indicating that infants represent certain single actions as directed at goals, rather than as purely physical trajectories. Interestingly, these results were not found if the grasping action was performed by a mechanical claw (Woodward, 1998) or if an unfamiliar movement — like a touch with the back of the hand — was presented (Woodward, 1999). In these cases infants looked equally long at the new path event and at the new goal event, indicating that the infants did not understand these actions as goal-directed. At first glance, these findings support the notion that infants' understanding of goal-directedness is restricted to familiarity and to experience with human actions. However, recent studies suggest that 6-month-old infants understand even unfamiliar human actions as goal-directed if the presented action leads to a salient action effect (e.g., Hofer et al., 2007; Jovanovic et al., 2007). In general, however, it is still questioned whether these findings could be interpreted as demonstrating infants' understanding of goal-directed actions rather than object-oriented actions.

Nevertheless, there are a few studies indicating that infants at the age of 9–12 months are, indeed, able to interpret perceived actions as goal-directed. These studies mainly focus on purposeful and accidental actions in relation to their successful and/or unsuccessful achievement of a certain goal. For example, Jovanovic and Schwarzer (2007) presented infants with a habituation video in which a hand entered a screen, moved to the contra lateral side, grasped, and lifted an object. On the way to that object, the hand passed another object, touched it accidentally and pushed it over. Importantly, the hand continued on its path until it reached the other object and lifted it. The 9-month-olds looked longer at the display when the object involved in an intentional action was changed than when the object involved in an accidental action was changed, indicating that only the grasping action was interpreted as goal-directed. Behne et al. (2005) investigated young infants' ability to distinguish between an actor's goal and the result of an action. In this study, infants experienced an adult handing them toys. Sometimes, however, the transaction failed either because the adult was in various ways unwilling to give the toy (e.g., she teased the child) or because the adult was unable to give it (e.g., she accidentally dropped it). Infants aged 9 months and older responded differently and appropriately when the experimenter was unwilling to give them a toy (with impatience) compared to when she was trying to give them a toy (with patience), even though they did not actually receive the toy in either case. The ability to predict other people's action goals is important in social interaction. Falck-Ytter et al. (2006) showed that 12-month-old, but not 6-month-old, infants predict others' manual actions. Like adults, the infants moved their gaze predictably to the goal of an object-related grasping action.

Taken together, the reported findings provide evidence that infants begin to understand goal-directed actions at around (6 to) 9 months of age. Detecting the goal of an action helps one to interact with other social agents. Typically, action understanding is investigated with experimental settings presenting short, well-defined action steps. In contrast, everyday actions are often a complex flow of motion in which pauses are rare and only occasionally coincide with boundaries between intentional actions. Nevertheless, adults easily parse continuous everyday actions by detecting the boundaries, extracting important action units, and constructing meaningful representations of these actions. Baldwin et al. (2001) found that 10- to 11-month-old infants who had become familiar with a continuous sequence of everyday intentional actions were subsequently disinterested when motion was artificially suspended at points where intentions were fulfilled compared to points where intentions were not yet fulfilled. The findings indicate that infants at this age parse observed action sequences at junctures coinciding with boundaries between intentions. Parsing ongoing behaviour is likely a precondition to the understanding of intentional actions and action effects help to mark the completion of intentions. I will come back to the role of action effects for infants' action understanding later in this paper, but first, a brief overview on the production of intentional actions will be given.

Production of intentional actions

From very early on, infants produce actions in order to bring about desired action effects. There is ample evidence that infants learn the contingencies between self-performed movements and environmental outcomes that follow these movements (for a review, see Rovee-Collier, 1987). For example, newborns learn to suck in a certain frequency in order to hear their mother's voice (DeCasper and Fifer, 1980) or to turn their heads in anticipation of a bottle (Papousek, 1967). Two- to five-month-old infants learn the coherence between leg kicks and the contingent movements of a mobile (e.g., Rovee and Rovee, 1969; Rovee-Collier and Shyi, 1993). The conjugate reinforcement paradigm allowed infants to produce actions on objects before they would typically do so on their own (Rovee-Collier and Hayne, 2000). In this paradigm, infants lie in a crib with one end of a ribbon tied to their ankle and the other end tied to an overhead mobile. Infants quickly recognize the contingency between their foot kicks and the movements of the mobile, leading to a sharp increase of their leg kicks rate (for related findings, see Rochat and Morgan, 1998). Overall, these studies are mainly concerned with infants' capacity to learn about the contingencies between self-performed actions and their effects. Essentially, the capability to associate actions with action effects seems to be an important requirement for the development of intentional actions. A further precondition for the production of intentional actions is that infants have to be able to differentiate means (movements) from ends (goals) in their own behaviour. In a typical task testing this capacity, infants are required to pull a supporting cloth in order to achieve an interesting toy, which is placed on this cloth out of their reach. Infants typically pass this task at ages 8–9 months (Goubet et al., 2006). The ability to use a cloth as a supporter to obtain an object, or to remove an obstacle to reach an object, has been interpreted as a sign of understanding the differentiation of means and ends (Piaget, 1952; Uzgiris and Hunt, 1975), which seems to be a further precondition for the production of goal-directed actions.

Production of perceived actions

It is important to note that infants do not only learn about actions by analogy to self-produced actions. They also learn about their actions through observing others' actions. Investigations of infants' understanding of intentional actions are spread throughout the field of infant imitation. In order to imitate, infants have to acquire action knowledge by observation and they have to transfer this knowledge to their own actions. Thus, imitation is related to action perception and action production. It has been shown, that usually imitative behaviour starts around 9 months of age (e.g., Meltzoff, 1988; Barr et al., 1996; Tomasello, 1999a; Heimann and Nilheim, 2004; Elsner et al., 2007). In typical studies about action understanding, infants watch a model performing various new actions with new objects with or without salient effects. After a delay — which ranges from several minutes to several months — infants are allowed to play with the objects. The question is whether infants perform the actions they had seen previously in the modelling phase. For example, Barr et al. (1996) investigated the imitation of a three-step action sequence in 6- to 24-month-olds. Infants watched a model removing a mitten from a puppet's hand, shaking the mitten (ringing a bell inside), and replacing it on the puppet's hand. After a 24-h delay 12-, 18-, and 24-month-old infants exhibited clear evidence of imitation; the older infants (18- and 24-month-olds) reproduced two of the target actions whereas the 12-month-olds tended to imitate only one step. The imitation capacities of the 6-month-olds seemed to be much more limited. They removed the mitten only when the imitation phase followed immediately after the modelling phase (without delay) or when the target actions were demonstrated more often (Barr et al., 1996).

For a long time, most imitation studies focused on the memory aspect involved in imitative behaviour. Meltzoff (1988) demonstrated long-term memory recall of novel actions in 14-month-olds. Bauer and Mandler (1989) showed that young children's recall of complex actions is better with action sequences where action steps have to be performed in a specific order (enabling sequences) than with action sequences where action steps can be performed in any order (arbitrary sequences). Only recently has research started to focus on action understanding rather than action memory. Several studies have demonstrated that infants develop a sophisticated understanding of actions performed by others during their second year of life (for an overview, see Hauf, 2007a). At this age they start to copy actions in terms of goals rather than in terms of surface features. For example, in Meltzoff's (1988) study an adult put a flat box on the table and then touched it with his head in order to illuminate it. In two control conditions the experimenter either presented the effect (light on) without demonstrating the target action (head on the box) or there was no demonstration at all (baseline condition). Most of the 14-month-old infants who watched the demonstration of the target action touched the box with their own foreheads whereas none of the infants in the control conditions did so. This finding has been extended by Gergely et al. (2002) who showed that infants perform this head-touch movement under certain conditions and not others. In the 'Hands-free' condition the model pretended to be chilly and wrapped a blanket around her shoulders. After this she placed her hands visibly free onto the table before demonstrating the head-touch movement. In the 'Hands-occupied' condition, however, the model's hands were visibly occupied: again she pretended to be chilly and wrapped a blanket around her shoulders but now holding it with both hands while performing the 'head-action'. In this condition infants were much less likely to imitate the 'head-action'. Instead, they illuminated the box by touching it with their hand, indicating that they assumed 'illuminating the box' as the goal of the observed action (see Williamson and Markman, 2006 for related work).

The fact that infants do not just copy the surface features of a perceived action, but also take into account the environmental circumstances, has been shown in further studies. Meltzoff (1995) showed 18-month-old infants an unsuccessful act. The adult tried to pull apart a dumbbell-shaped object but his hand slipped off the object several times as he pulled. Thus, the goal was not achieved. The infants who observed the model trying, but failing to achieve the goal, produced about as many target actions as infants who observed a successful demonstration, indicating that they read through the literal body movements to the underlying goal of the action. Evidently, young children can understand our goals even if we fail to fulfil them (for related work, see Bellagamba and Tomasello, 1999). In a recent study (Meltzoff, 2007a), the standard unsuccessful-attempt display was shown to 18-month-olds. After this demonstration a trick toy was handed to them. Whenever infants picked it up and attempted to pull it apart, their hands slipped off the ends. Now, their own behaviour matched the perceived surface behaviour of the model. Nevertheless, the infants did not end their behaviour. They repeatedly tried again to pull the object apart and appealed to the adult for help by looking and vocalizing. Obviously, they were trying to achieve the adult's goals, not his literal body movements. Infants also differentially imitate intentional and accidental actions. Carpenter et al. (1998a) showed 14- through 18-month-olds an adult performing a series of two-step actions on objects that made interesting results occur. Actions that were vocally marked as intentional (There!) were imitated almost twice as often as actions that were vocally marked as accidental (Woops!) indicating that they distinguished between purposeful and unpurposeful actions. Researchers using looking-time techniques support infants are able to make the differentiation at the age of 9 months (Behne et al., 2005; Jovanovic and Schwarzer, 2007).

Focussing on the goal not only helps to understand the intentions of actions performed by others, it also triggers infants to produce their actions in line with the inferred goal. To analyse what 12- and 18-month-old infants interpret as being the goal of a perceived action, Carpenter et al. (2005) showed infants an adult making a beeping noise while hopping a toy mouse across a mat. In one condition, the adult ended by placing the mouse in a toy house, whereas in another condition no house was present at the final location. In the first case, infants usually simply put the mouse in the house (ignoring the hopping motion and sound effects). Most likely, infants interpreted the adult's action in terms of the final goal (putting the mouse into the house) and so ignored the behavioural means. In the latter case, infants copied the adult's action (both hopping motion and sound effects) indicating that they saw the action itself as the adult's only goal. In both cases infants were achieving the final goal of the perceived action, which could either be the end state (put the mouse in the house) or the movement (make the mouse hopping) depending on the context.

Taken together, there is evidence that infants at the age of about 9 months begin to understand other persons' goal-directed actions. There is further evidence that around 8 to 9 months of age they are able to perform intentional action on their own. They begin to differentiate between movements and goals of their own self-performed actions. At the same age, they start to transfer action knowledge acquired while observing a model to their own action production. They start to imitate goal-directed actions. Thus, they learn about themselves and their capacities through observing others' actions as well as they learn about actions by analogy to self-produced actions. I will come back to the linkage between perceived and performed actions later, but first, I will discuss the role of action effects on infants' understanding of goal-directed actions.

Role of action effects for the understanding of intentional actions

Although there are a number of studies on the role of action effects in adult action control (see e.g., Prinz, 1997; Hommel et al., 2001; Nattkemper and Ziessler, 2004), only recently has research focused on the role of action effects in infants' understanding of intentional actions. The importance of action effects for infants is obvious. As outlined above, infants learn very early to produce actions in order to achieve desired action effects (e.g.,

DeCasper and Fifer, 1980; Rovee-Collier, 1987; Rochat and Morgan, 1998; Rovee-Collier and Hayne, 2000). They are sensitive to contingencies in social interactions during their first months of life (e.g., Gergely and Watson, 1999; Bigelow and DeCoste, 2003; Bigelow and Rochat, 2006; Soussignan et al., 2006). Contingencies allow infants to detect the effectiveness of their actions in producing external effects. Research about the effect of social interaction on infants' social and emotional development has emphasized the importance of self-performed actions and their external effects. But, contingencies play a crucial role in infants' development of action understanding as well. The following sections will focus on infants' ability to detect action-effect contingencies while observing others' actions, and infants' ability to use this knowledge for their own action production.

The role of action effects in the perception of intentional actions

Only recently, research has started to focus on the role of action effects in infants' understanding of goal-directed actions. For example, Klein et al. (2007), using a looking paradigm, investigated whether 9-month-old infants learn specific object-related action-effect contingencies by observation. Infants watched a video display presenting a person either performing the same action (shaking) on two objects or performing two different actions (shaking and rolling) on the same object. In the first condition, shaking one object elicited an effect whereas shaking the other object did not. In the second condition, producing one specific action elicited an effect whereas producing the other action did not. As a result, both conditions presented only one specific action-effect combination during familiarization. In subsequent test trials, this specific relation was violated. Infants looked longer at the new action-effect contingencies, indicating that they remembered the previously perceived action-effect combinations and detected the changed contingencies. This was the case for one action with two objects as well as for two actions with one object (for related findings in 10-month-olds see Perone and Oakes, 2006).

Additional evidence for the role of action effects in infants' action perception is derived from studies using habituation paradigms. Woodward (1998, 1999) has provided evidence for the understanding of goal-directed actions at 6 months of age, but only for the highly familiar action of grasping. She found no evidence for goal attribution before 12 months of age for a number of unfamiliar action types. However, recent studies, applying a modified version of her paradigm, showed that if a perceptually salient change of state in the goal object was provided, even novel and unfamiliar actions can be understood as goal-directed by 6- to 10-months of age (Király, et al., 2003; Hofer et al., 2007; Jovanovic et al., 2007). For example, Jovanovic and colleagues introduced an action effect consisting of pushing the object smoothly to a new location at the back of the stage after dropping the back of the hand onto the object. Even 6-month-olds looked longer at an object change than at a path change when they watched a human hand performing an unfamiliar action resulting in a change of the object's location (for a replication using video presentation instead of live actions, see Hofer et al., 2007). Furthermore, 9- and 12-month-olds are able to interpret actions performed by a mechanical device as goal-directed if infants perceived a claw grasping and transporting objects to the back of a stage as adding a salient action effect. However, 9-month-olds interpreted this action display as goal-directed only when they additionally received an information phase showing infants how a human held and operated the claw prior to habituation (Hofer et al., 2005).

Action-effect contingencies help infants to understand actions performed by others. But do infants use the action-effect knowledge achieved by observation to guide their own action production?

The role of action effects in the production of intentional actions

Infants recognize contingencies between self-produced movements and related external outcomes early in life. By exploring contingencies, they also learn to associate the effects of their actions with the resultant actions. Infants between 9 and 16 months learn by exploration which actions on a specific object elicit interesting effects. Furthermore, they infer that acting on a similar looking object will lead to similar results (Baldwin et al., 1993; Elsner and Aschersleben, 2003). They are able to detect action-effect contingencies while observing actions performed by others. The following section will focus on the age at which infants start to use action-effect knowledge acquired by observation in order to perform their own intentional actions.

In a longitudinal study, Carpenter et al. (1998b) examined, among other abilities, the imitation of action tasks between 9 and 15 months of age. In order to distinguish between mimicking (reproduction of behaviour without understanding the goal) and imitation (reproduction of behaviour and goal), an action-effect delay was introduced following infants' action reproduction. The 10- and 12-month-olds produced the target actions, but only the older infants tended to look expectantly at the end result following their action. These findings suggest that 12-month-olds encoded the perceived action-effect contingency, and that they expected their own actions to be effective in the same way. In line with this, Elsner and Aschersleben (2003) investigated whether infants transfer observed action-effect relations to their own action production. Infants watched a model performing two target actions (pulling and pressing) on an object, producing a salient effect with each action (sound and light effects). In the test conditions, the action-effect relations remained the same or were reversed. The results showed that 15- but not 12-month-olds detected whether their own actions caused the same effects as the model's actions did, indicated by appropriately modifying their own behaviour. Such detection requires that they retained which action led to which effect.

Further evidence for the impact of perceived action-effect combinations on action production derives from a study by Hauf et al. (2004). A three-step action sequence was demonstrated to 12- and 18-month-old infants. The experimenter took a cylinder off a barrier (1st step), shook it three times (2nd step), and put it back onto the barrier in front of a toy bear (3rd step). Different action-effect combinations were introduced during the demonstration: either the second or the third, or none of the action steps elicited an interesting action effect (sound effect). Following the demonstration, infants initially and with shorter latency produced the action step that was combined with the effect compared to an action step that did not elicit such an effect. This indicates that the action-effect relation helped infants to define the goal leading to a selective production of different action steps (for a replication using video presentation instead of live demonstration, see Klein et al., 2006).

The reported findings suggest that infants start around 12 months of age — but not earlier — to use perceived action-effect relations to control their own action production. Up to now it remained unknown whether younger infants did not encode the specific action-effect relations or whether they failed to transform the observed relations into their own action production. A recent study by Hauf and Aschersleben (2007) suggests that even 9-month-olds are able to use action-effect relations learned by observation to guide their own behaviour, provided the task is not demanding. Infants watched an experimenter pressing two buttons. Pressing one button elicited interesting effects (light and sound effects) whereas pressing the other button did not. The 9-month-old infants initially and with shorter latency produced the target action that had elicited an effect, indicating that they had learned the action-effect relations by observation and used them for own action production. Supporting evidence is provided by Klein et al. (2007). In addition to their looking study reported in the previous section, the actions were demonstrated live in front of the infants during the reaching and imitation study. After having watched an experimenter shaking two objects alternately, with one of the two making a sound, infants were allowed to reach for the objects. Most of the infants chose the object that had elicited the sound effect, indicating that they used the perceived action-effect relations to select the object. To ensure that infants did not choose that object just because it was more salient, infants were also presented with two different actions (shaking and rolling), directed at one object. An experimenter shook and rolled a ball alternately, with only one of the two actions producing a sound. Following this, infants were allowed to act on the ball on their own. Overall, infants initially performed the target action that was combined with an effect, indicating that they used the observed relation between action and effect for their own action production.

Taken together there is growing evidence on the role of action effects in the understanding of goal-directed actions in infants. Focusing on actions and their effects helps infants to infer the goal of actions, and thus to understand intentional actions. During their first year of life, infants are able to understand goal-directed actions performed by others. In addition, recent studies emphasize infants' competence in using this understanding to guide their own production of goal-directed actions.

The interplay of action perception and action production

The findings presented on infants' action perception and action production — supplemented by the role of action effects — demonstrate that young infants understand goal-directed actions of others and they produce goal-directed actions on their own. Infant imitation particularly emphasizes the intimate link between both action perception and action production. Even though there is an agreement in the literature that the capacities to understand one's own and others' actions are linked, there is an ongoing debate about how these two aspects of action control are related to each other and how this relatedness develops during infancy. Do infants first understand their own actions and following this come to understand the actions of others or is it the other way around, that is, do infants first understand the actions of others and following this come to understand their own actions (see e.g., Hauf and Prinz, 2005; Hauf, 2007a)?

The concept of functional equivalence of self-performed actions and perceived actions performed by others offers a powerful framework for action understanding. This idea is reflected in the common coding approach introduced by Prinz (1990, 1997), which emphasizes the functional relationship between action perception and action production. The core assumption of this approach is that perceived and to-be-produced actions share common representational resources, which are used by the perceptual system in order to perceive actions and by the motor system in order to produce actions. Empirical support for this notion derives from rather different domains in adult cognitive psychology (for an overview, see Hommel et al., 2001; Prinz, 2002). Furthermore, neuroscientists have demonstrated that the understanding of other people's actions seems to be conducted by the same neural system by which we understand our own actions. For example, so-called 'Mirror neurons' in the premotor cortex of the monkey brain are activated both when an action is observed and when it is produced (e.g., Rizzolatti et al., 2002; Gallese, 2003). Related findings in humans reveal common brain regions subserving both the perception and the production of actions (e.g., Decety, 1996; Meltzoff and Decety, 2003; Hamilton et al., 2004; Iacoboni, 2005).

Based on the idea of functional equivalence, the common coding approach was recently extended to infancy research. It is assumed that even young infants have representations of actions that are used both by the perceptual system to perceive and understand goal-directed actions of others and by the motor system to produce goal-directed actions. As the common coding approach suggests a functional equivalence of perceived and to-be-produced actions, it facilitates a bidirectional influence of action perception and action production. The idea of common representations is also held by Meltzoff (2002, 2007a), who assumes that infant imitation involves a goal-directed matching process. When infants observe other persons acting, they specify the goal and/or the action visually. When they act on their own afterwards, they compare the feedback of the self-produced action to the representation of the observed action. Such a comparison is possible because the perception and the production of actions are coded in a common framework, the so-called *supramodal act space*. Again, the equivalence between self and other supports bidirectional learning effects:

infants understanding of others' actions is influenced by producing similar actions themselves and infants learn about themselves by observing the actions of others (see Braas and Heyes, 2005, for a recent review about the role of shared representations of perceived and produced actions for imitation). In spite of these theoretical approaches, studies investigating the bidirectional interchange of action perception and action production in infants are still rare. The following section will report recent findings concerning this issue.

Indications for bidirectional interchange of action perception and action production

As outlined above, during their first year of life infants are able to understand goal-directed actions of others as well as to produce their own goal-directed actions. Although the reported findings indicate that action perception and action production are intimately linked to each other early in life, they do not provide evidence for a bidirectional interchange. There are only a few recent studies that support the notion of functional equivalence.

Sommerville and Woodward (2005) used a means–end task to investigate 10-month-olds' ability to complete a cloth-pulling action to reach a toy. Interestingly, only those infants who succeeded in this task understood a similar sequence presented in a visual version indicated by different looking time. Longo and Bertenthal (2006) investigated whether watching an overt reaching action is sufficient to elicit the Piagetian A-not-B error. The 9-month-old infants perseverated only following observation of actions they were themselves able to produce, again indicating a shared representation of perceived and produced actions. Recent studies on imitative play in infants support the idea that infants represent the actions of others and their own actions in commensurate terms. Infants' ability to imitate perceived actions and to recognize that their own actions are imitated by another person suggests that produced and perceived actions share common representations. A preference for imitative contingency compared to other contingent responses is found very early in development. Two-month-olds are more imitative to mothers who imitate them than to mothers who do not imitate them (Nadel et al., 2004). Three-month-olds react more to an imitative mother (Field et al., 1985). Infants gaze and smile more to an imitative adult than to an adult who is timely contingent only (Meltzoff, 1990). Regarding object-related actions, Meltzoff and Moore (1999) were able to show that 9-month-olds tend to look and smile more toward an imitating adult compared to a contingently acting, but not imitating, adult facing them. Moreover, infants systematically produced testing behaviours oriented preferentially toward the imitating actor. In doing so they modulated their own actions on the object while looking at the adult, verifying whether she is changing her imitative behaviour in an appropriate way. Recently, Agnetta and Rochat (2004) replicated and extended these findings. They demonstrated that 9-month-olds differentiate between an imitating and a contingent but not imitating adult; infants preferentially looked at the imitating adult and tried to reengage this person during a still-face period. Given that infants identify the imitative behaviour (action perception) and change their own behaviour (action production) to test the imitative capacities of the adult co-partner, it could be assumed that perceived and to-be-produced actions are influencing each other in a bidirectional manner.

Although these findings indicate a close relation between infants' production of goal-directed actions and their ability to detect such goals in the actions of others, it is crucial that research about a bidirectional interchange include action perception and action production at the same time. Until now there are only a few studies that have addressed this issue by means of the same experimental paradigm. These studies rely on rather simple actions that are useful for action perception and action production tasks. For example, Sommerville et al. (2005) applied a simple object grasp to assess the possible reciprocal relation between action perception and action production. Three-month-old infants took part in a production task and a perception task. During the 'production task', infants were wearing mittens with palms that stuck to the edges of toys, which allowed them to easily pick up the toys. The infants would not have been able to grasp these toys without the mittens. During the 'perception task' the same infants watched an actor reach for and grasp one of two toys arranged side-by-side on a stage in several habituation trials. In following test trials, the position of the two toys was reversed and the actor grasped either the same toy at the new location or the new toy at the same location. All infants participated in both tasks; reach-first infants started with the production task followed by the perception task whereas this order was reversed for watch-first infants. In the case of a bidirectional interchange of action production and action perception, the reach-first infants should manage the perception task better than the watch-first infants *and* the watch-first infants should handle the production task better than the reach-first infants. But this was the case only to some extent. Reach-first infants understood the relation between actor and goal during the perception task whereas the watch-first infants did not, indicating an influence of action production on action perception. Admittedly, watch-first infants and reach-first infants did not differ during the production task. Thus, there was no evidence that action perception influenced subsequent action production. Although the precondition of using the same tasks for investigating the bidirectional exchange was satisfied, the results were not equated in both directions. Whereas the self-produced actions influenced the understanding of a perceived goal-directed action that followed, this was not the case the other way round; the observation of goal-directed actions did not influence subsequent production of goal-directed actions (Sommerville et al., 2005).

At first glance, these results support the notion that infants first have to be able to produce an action in order to understand this action performed by others. But it is also possible that there is indeed a bidirectional influence between action perception and action production, which is difficult to measure in young infants. Accordingly, the studies reported in the following section focus again on the bidirectional interplay by using a combination of looking tasks and imitation tasks.

Studies on bidirectional interplay of action perception and action production

The following studies used the combination of an action-production task and an action-perception task in order to investigate the bidirectional influence. One series of experiments investigated how self-produced actions influenced infants' interest in actions performed by others. In contrast, another series of experiments examined how the perception of actions performed by others influenced infants' own actions that followed. Both studies included the same perception and production tasks, but the sequence of these tasks was varied.

From action production to action perception

In this study infants started to play with a toy (car or ribbons) for a 90 s interval. During this time the 7-, 9-, and 11-month-olds had enough time to elaborate all features of the toys as well as to figure out possible actions related to the toys (action production task). Immediately after this, infants watched two short videos on two video screens simultaneously. Both videos showed the same two adults sitting at a table face to face and alternately performing an action with a toy for a 90 s interval. In one video, the two adults were alternately sliding the car whereas in the other video the two adults were waving the ribbons by taking turns (action perception task; Fig. 1). Depending on the toy infants had used during own action production task, either the car video or the ribbons video could be referred to as being the same-toy video or the different-toy video.

A comparison of the mean looking time data during the perception task demonstrated that 9- and 11-month-olds looked longer at the same-toy video, whereas the 7-month-olds did not show any preference. These findings indicate that the production of goal-directed actions influenced the subsequent perception of goal-directed actions performed by others (Hauf et al., 2007). Apparently, infants' own action production increased their interest in actions of others with the same object. Follow-up studies showed that this influence was restricted to object-related actions. There was no preference for the same-toy video anymore, if the videos showed the

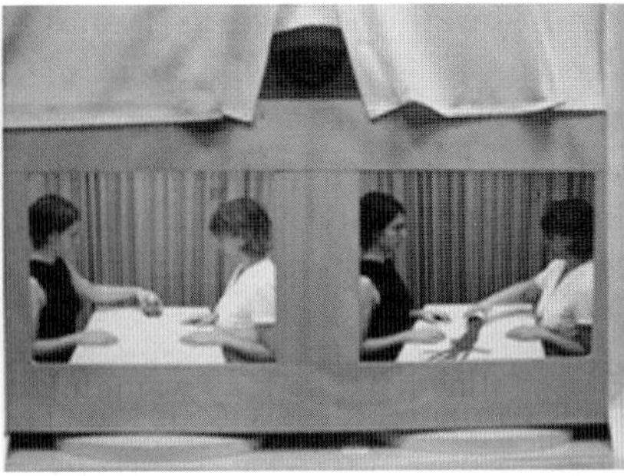

Fig. 1. From action production to action perception: First infants were playing with one toy (left) and subsequently watched two videos simultaneously (right). Each video presented two adults either sliding a car or waving the ribbons (for more details, see Hauf et al., 2007).

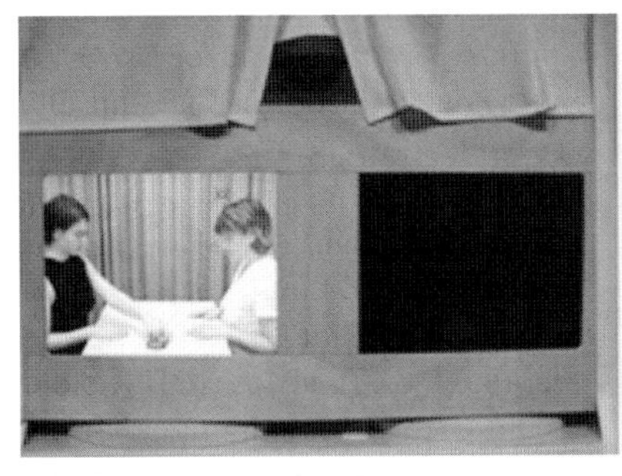
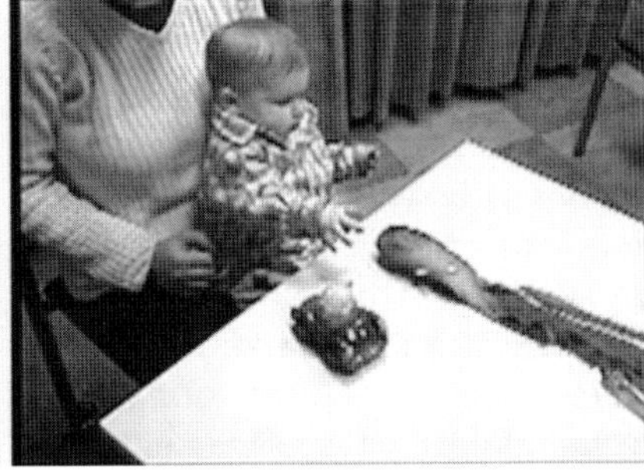

Fig. 2. From action perception to action production: First infants watched a video presenting two adults performing an action, e.g., sliding the car (left). Subsequently infants were seated at a table with two toys given simultaneously within their reach (right; for more details, see Hauf, 2007b).

two adults only looking at the toys, without performing any action and if the videos showed only the toys, without acting persons, respectively. Infants' action production enhanced their interest in actions with objects, but not in watching objects or persons per se (Hauf et al., 2007).

In order to support the idea of a bidirectional interchange of action perception and action production, it also has to be demonstrated that action perception influences action production. How infants' perception of actions performed by others influences their subsequent production of own actions has been investigated with the same tasks.

From action perception to action production

In this study 7-, 9-, and 11-month-old infants first watched a video on a screen, which showed two adults acting upon a toy by taking turns for a 90 s interval (action perception task). The presented videos were identical to those used before. Half of the infants watched the video presenting two adults sliding the car and half of the infants watched the video presenting the two adults waving the ribbons. Following this, the infants were seated at a table and both toys were presented within reach simultaneously (action production task; Fig. 2). Thus, the infants had the possibility to choose and to act either with the same toy they had watched in the video before, or with the new one.

All infants were highly attentive while watching the videos. Nevertheless, a comparison of the mean acting time during the production task yielded no difference at all. The infants of all age groups played equally long with the toy that they had watched during the perception task as with the new toy. Furthermore, only a few infants produced the demonstrated target actions (sliding the car or waving the ribbons). These findings indicate that the perception of goal-directed actions performed by others does not influence the subsequent production of goal-directed actions by the infants (Hauf, 2007b).

In summary, it could be said that the impact of action production on action perception was demonstrated, but not the impact of action perception on action production. This is in line with recent findings emphasizing that the transfer of knowledge from self to others seems to be easier for infants than the transfer of knowledge from others to self. But, does this indeed mean that infants first have to be able to produce a goal-directed action before they are able to understand a similar action of others? Taking all the different aspects into consideration, one has to be very careful in drawing this conclusion. For example, it is surprising that the infants in Hauf's (2007b) study rarely imitated the demonstrated actions (sliding a car or waving ribbons), which were easy to perform. It is well known from imitation literature that infants at the investigated age are able to imitate more complex actions (Meltzoff, 1988; Barr et al., 1996; Heimann and Meltzoff, 1996; Elsner et al., 2007). In a strict sense there is one major difference between the Hauf's study and typical imitation studies that has to be taken into consideration; typical imitation studies demonstrate an action in front of the infants, hand the same object to the infant and thereby encourage the infants to produce the observed target action. In the current study, infants watched two adults performing the same action, imitating each other without demonstrating the action directly to the watching infants. After that the infants obtained two objects. Probably, infants did not interpret this situation as a request to repeat what they have watched before. Maybe having two toys available for playing was just too exciting. Perhaps, they were not focusing on what they had watched just before. This argument seems to be reasonable and in line with recent studies on infants' agentive experience. Sommerville et al. (2005) demonstrated facilitation of action understanding based upon agentive experience but did not provide evidence for an influence of action perception on action production. But it is possible that the perceptual enrichment session provided by the perception task was less rich than that provided by the action task. Due to the 'sticky mittens', the 3-month-olds were able to 'grasp' objects — probably for the first time in their life — providing them with a tremendous new experience that easily could overwrite the impact of what they had watched just before.

Based upon this line of argument, follow-up studies by Hauf (2007b) investigated infants' capacity to imitate with two distinct tasks: (1) Infants watched a video presenting two adults acting upon one toy alternately (same video as used in the former study). Following this perception task, only the same toy was handed out to the infants during the subsequent production task. Note, that infants had no choice anymore and thus were encouraged to play with the same toy they had watched in the video just before. Now the question was whether infants would imitate the perceived target actions. During this production task most infants produced the demonstrated target action (sliding the car or waving the ribbons) indicating an influence of perceived goal-directed actions on self-performed goal-directed actions. Interestingly, age-related analysis revealed significant imitation in 9- and 11-month-olds but not in 7-month-olds. Note that infants in the control condition — that were neither exposed to the toys nor to any action demonstration — did not produce the target action during the production task. (2) Infants watched a video presenting two adults acting upon one toy alternately while ignoring the other toy. This video demonstrated not only two adults producing a target action (sliding the car or waving the ribbons) but also additionally the two adults selecting one toy over the other to play with. The information presented in the video was much more demanding — selection and imitation instead of imitation only. The question was whether infants would select the same toy as the adults and whether the infants would imitate the perceived target action. During the following production task most 9- and 11-month-old infants selected the same toy as the two adults. The infants touched the same toy at first and acted reliably longer upon it. This was not the case in 7-month-olds.

However, even though the 9- and 11-month-olds selected the same toy and acted upon it for a longer time, they only rarely performed the corresponding target action (for more details, see Hauf, 2007b).

Taken together it could be assumed that the interplay of action production and action perception is indeed bidirectional. Through a sequence variation of the same tasks, we were able to demonstrate an influence of action production on action perception (Hauf et al., 2007) as well as an influence of action perception on action production (Hauf, 2007b) during the first year of life. But there is one important issue to note. Even the current studies failed to demonstrate the bidirectional interplay if just the sequence of action production and action perception was reversed (for similar findings, see Sommerville et al., 2005). The impact of perceived actions on performed actions could only be demonstrated when the demands of the perception task matched the demands of the production task. This was the case when the perception task included the demonstration of an action with one toy and the production task offered only one toy or when the perception task demonstrated the selection of one toy and the production task allowed selecting one toy as well. Obviously, further research is necessary to examine the bidirectionality of action perception and action production.

Conclusions

Research on infants' understanding of intentional actions demonstrates that infants at the age of 6–9 months start to understand actions performed by others (e.g., Woodward, 1998, 1999; Behne et al., 2005; Hofer et al., 2007; Jovanovic and Schwarzer, 2007; Jovanovic et al., 2007). Furthermore, it is well documented that infants at around 9 months perform intentional actions themselves (e.g., Uzgiris and Hunt, 1975; Willats, 1999). Beyond this, studies on infant imitation illustrate infants' capacity to transfer action knowledge acquired by observation to their own action production with or without delay (e.g., Meltzoff, 1988; Barr et al., 1996; Tomasello, 1999b; Elsner et al., 2007). Recent studies also provide growing evidence that infants during their second year of life start to copy actions in terms of goals rather than in term of surface features (e.g., Meltzoff, 1988, 1995, 2007a; Carpenter et al., 1998a, 2005; Gergely et al., 2002). Infants are learning efficiently and flexibly by observation. Early in life, they start to focus on goals while watching others' actions. By emphasizing the goal of an action, action effects help infants to infer the goal of a perceived action performed by others (e.g., Király et al., 2003; Hofer et al., 2005, 2007; Jovanovic et al., 2007; Klein et al., 2007) as well as to produce their own goal-directed actions (e.g., Carpenter et al., 1998a, b; Gergely et al., 2002; Hauf et al., 2004; Klein et al., 2006). These studies demonstrate the impact of action perception on action production. Recent studies on agentive experience of infants also reveal the influence of action production on action perception (e.g., Sommerville et al., 2005; Hauf et al., 2007). Clearly, there is evidence that action perception and action production are tightly linked to each other (e.g., Sommerville and Woodward, 2005; Longo and Bertenthal, 2006) and are influencing each other in a bidirectional manner (Agnetta and Rochat, 2004; Meltzoff, 2007a; Hauf, 2007b). Current findings indicate that infants at about 9 months of age have reached a level of functional equivalence between self-produced actions and actions they perceive performed by others, supporting the postulated bidirectional interchange. Infants improve their perception of others' actions through agentive experience *and* they learn about the possibilities of own action production through observing others' actions.

Besides the substantial evidence about shared representations of perceived and to-be-produced actions in adults (for an overview, see Hommel et al., 2001), research now provides evidence that such a functional equivalence holds true in infants as well. Having achieved functional equivalence means that representations of knowledge about one's own actions are similar to representations of knowledge about others' actions. This knowledge can be used by the perceptual system to perceive and understand actions performed by others and by the motor system to produce actions.

Nevertheless, the question how infants attain functional equivalence remains still an unsolved issue. Related research on imitation has focused on the question of how a visual representation of an action is transformed into corresponding motor output. Different theoretical approaches have emphasized either specialized processes like active intermodal mapping (Meltzoff, 2002, 2007a) or processes that are linked more generally to the organization of motor control (Prinz, 1997, 2002) and the priming of motor representations by observation (Heyes, 2001). These approaches differ with respect to the underlying mechanisms proposed for explaining imitative behaviour. But they also share the idea of a direct linkage between the representation of perceived actions and the representation of self-produced actions. Recent findings support that the automatic activation of motor representations by movement observation is used to reproduce observed behaviour (for an overview, see Brass and Heyes, 2005). Beyond this, action observation activates motor representations of the same kind that guide to-be-produced actions (Decety et al., 2002) indicating a bidirectional interchange. Obviously, the motor system and the perceptual system play important roles in the bidirectional interchange in adults and in infants once they have reached functional equivalence. It seems to be reasonable that these systems are involved in the process of establishing functional equivalence as well.

Bidirectional interplay requires cross-modal coordination — information about one's own actions has to be mapped onto the perception of others' actions. Cross-modal coordination develops early in life. For example, cross-modal recognition of shape from hand to eyes is evident in human newborns. Neonates are able to process and encode the shape information of objects by manual exploration, which enables them to distinguish subsequently presented visual objects (Streri and Gentaz, 2003, 2004). At about 5 months of age, infants perceive the boundaries of tactilely presented objects more clearly when the infants actively produced this motion themselves compared to the same motion pattern passively produced by others. This finding by Streri et al. (1993) emphasized that tactile perception is enhanced by active production of surface motion. Infants at this age are able to coordinate information about the tactile and visual features of objects and they use this cross-modal object representation to interpret visually presented physical events (Schweinle and Wilcox, 2004). The capacity of cross-modal coordination is increasing during the first year of life and is well established around 9 months of age (e.g., Bushnell, 1982; Gibson and Walker, 1984; Kellman and Arterberry, 1998). Then infants are able to use tactile experiences for visual object recognition as well as visual experiences for tactile object discrimination. Obviously, the motor system and the perceptual system are linked to each other intimately from early on. This close connection seems to be one important prerequisite for the bidirectional interplay of action perception and action production. Further research has to focus on the impact of the motor system and of the perceptual system on the development of action understanding (von Hofsten, 2004, 2007).

At least as important as the question of how infants achieve functional equivalence of both systems is the question of whether the bidirectional influence of action production and action perception emerges at the same time or whether one direction develops earlier than the other. Based upon the 'supramodal representation' idea (Meltzoff, 2007a) or the common coding perspective (Prinz, 1997) this question is not easy to answer. Both theoretical frameworks allow an influence in all directions at any time and have to be examined with regard to the question of how the bidirectional influence develops during the first months of life. Again it has to be considered that the interchange of action perception and action production is tightly linked to the capacities of the perceptual system and the motor system. Of course both systems develop along different time courses. It seems to be reasonable that the impact of both systems for action perception and action production is not always the same level or the same amount. For example, before infants are able to relate observed actions to correspondent motor patterns, they intensively observe others' actions. Even if they match this action knowledge to their self-produced actions, their action production is constrained by motor development. And even if they are able to produce a goal-directed action and are interested to learn more about goals related to this action, their perception of such actions performed by others may be constrained by environmental aspects.

The development of intentional actions is dependant on various factors. There seems to be no doubt that the perceptual system and motor system are of significant relevance. The achievement of functional equivalence between action perception and action production seems to be an important milestone in the development of action understanding. Infants, like adults, seem to understand others' actions by mapping them onto their own motor representations of the same actions (Meltzoff, 2007a, b; von Hofsten, 2007). Even though it is still a challenge to determine how this equivalence is attained, it is equally important to figure out whether the input of the motor system can increase perceptual capacities and/or whether the input of the perceptual system is able to facilitate subsequent action production. Whatever enhances the understanding of goal-directed and intentional actions boosts the competence in social communication. If we assume that the understanding of one's own and others' actions is a precursor of the understanding of one's own and others' goals, intentions and desires, then it should also be assumed that the mechanisms playing a role in action understanding are playing a similar role in understanding others' desires and mental states.

Acknowledgements

I appreciate the possibility to collect the reported data at the Max Planck Institute for Human Cognitive and Brain Sciences in Munich, Germany. I thank Ann Bigelow for value comments on an earlier draft, and Claes von Hofsten and Kerstin Rosander for the motivation they provided on this project.

References

Agnetta, B. and Rochat, P. (2004) Imitative games by 9-, 14-, and 18-month-old infants. Infancy, 6(1): 1–36.

Aschersleben, G., Hofer, T. and Jovanovic, B. (2007) The link between infant attention to goal-directed action and later theory of mind abilities. Submitted manuscript.

Baldwin, D.A., Baird, J.A., Saylor, M.M. and Clark, M.A. (2001) Infants parse dynamic action. Child Dev., 72(3): 708–717.

Baldwin, D.A., Markman, E.M. and Melartin, R.L. (1993) Infants' ability to draw interferences about nonobvious object properties: Evidence from exploratory play. Child Dev., 64(3): 711–728.

Barr, R., Dowden, A. and Hayne, H. (1996) Developmental changes in deferred imitation by 6- to 24-month-old infants. Infant Behav. Dev., 19: 159–170.

Bauer, P.J. and Mandler, J.M. (1989) One thing follows another: effects of temporal structure on 1- to 2-year-olds' recall of events. Dev. Psychol., 25(2): 197–206.

Behne, T., Carpenter, M., Call, J. and Tomasello, M. (2005) Unwilling versus unable: infants' understanding of intentional actions. Dev. Psychol., 41(2): 328–337.

Bellagamba, F. and Tomassello, M. (1999) Re-enacting intended acts: Comparing 12- and 18-month olds. Infant Behav., 22(2): 277–282.

Bigelow, A.E. and DeCoste, C. (2003) Sensitivity to social contingency from mothers and strangers in 2-, 4-, and 6-month-old infants. Infancy, 4(1): 111–140.

Bigelow, A.E. and Rochat, P. (2006) Two-month-old infants' sensitivity to social contingency in mother-infant and stranger-infant interaction. Infancy, 9(3): 313–325.

Brass, M. and Heyes, C. (2005) Imitation: is cognitive neuroscience solving the correspondence problem? Trends Cogn. Sci., 9: 489–495.

Bushnell, E.W. (1982) Visual-tactual knowledge in 8-, 9.5-, and 11-month-old infants. Infant Behav. Dev., 5: 63–75.

Carpenter, M., Akthar, N. and Tomasello, M. (1998a) Fourteen- through 18-month-old infants differentially imitate intentional and accidental actions. Infant Behav. Dev., 21: 315–330.

Carpenter, M., Call, J. and Tomasello, M. (2005) Twelve- and 18-month-olds copy actions in terms of goals. Dev. Sci., 8(1): F13–F20.

Carpenter, M., Nagell, K. and Tomasello, M. (1998b) Social cognition, joint attention, and communicative competence from 9 to 15 months of age. Monogr. Soc. Res. Child Dev., 63(4): 1–174.

DeCasper, A.J. and Fifer, W.P. (1980) Of human bonding: newborns prefer their mothers' voices. Science, 208: 1174–1176.

Decety, J. (1996) Do imagined and executed actions share the same neural substrate? Cogn. Brain Res., 3: 87–93.

Decety, J., Chaminade, T., Grèzes, J. and Meltzoff, A.N. (2002) A PET exploration of the neural mechanisms involved in reciprocal imitation. Neuroimage, 15: 265–272.

Elsner, B. and Aschersleben, G. (2003) Do I get what you get? Learning about the effects of self-performed and observed actions in infancy. Conscious. Cogn., 12: 732–751.

Elsner, B., Hauf, P. and Aschersleben, G. (2007) Imitating step by step: A detailed analysis of 9- to 15-month-olds' reproduction of a three-step action sequence. Infant Behav. Dev., 30(2): 325–335.

Falck-Ytter, T., Gredebäck, G. and von Hofsten, C. (2006) Infants predict other people's action goals. Nat. Neurosci., 9: 878–879.

Field, T., Guy, L. and Umbel, V. (1985) Infants' responses to mothers' imitative behaviors. Infant Ment. Health J., 6: 40–44.

Gallese, V. (2003) The manifold nature of interpersonal relations: the quest for a common mechanism. Philos. Trans. B Biol. Sci., 358: 517–528.

Gergely, G., Bekkering, H. and Kiraly, I. (2002) Rational imitation in preverbal infants. Nature, 415: 755.

Gergely, G. and Watson, J. (1999) Early socio-emotional development: contingency perception and the social-biofeedback model. In: Rochat P. (Ed.), Early Social Cognition: Understanding Others in the First Months of Life. Lawrence Erlbaum Associates, Mahwah, NJ, pp. 101–136.

Gibson, E.J. and Walker, A.S. (1984) Development of knowledge of visual-tactual affordances of substance. Child Dev., 55: 453–460.

Goubet, N., Rochat, P., Maire-Leblond, C. and Poss, S. (2006) Learning from others in 9- to 18-month-old infants. Infant Child Dev., 15(2): 161–177.

Hamilton, A., Wolpert, D. and Frith, U. (2004) Your own action influences how you perceive another person's action. Curr. Biol., 14: 493–498.

Hauf, P. (2007a) The interchange of self-performed actions and perceived actions in infants. In: Striano T. and Reid V. (Eds.), Social Cognition: Development, Neuroscience and Autism. Blackwell, Oxford.

Hauf, P. (2007b) Baby see — Baby do! What infants learn from other persons' actions. Submitted manuscript.

Hauf, P. and Aschersleben, G. (2007) Action-effect anticipation in infant action control. Psychol. Res., paper in press.

Hauf, P., Aschersleben, G. and Prinz, W. (2007) Baby do — Baby see! How action production influences action perception in infants. Cogn. Dev., 22: 16–32.

Hauf, P., Elsner, B. and Aschersleben, G. (2004) The role of action effects in infants' action control. Psychol. Res., 68: 115–125.

Hauf, P. and Försterling, F. (Eds.). (2007) Making Minds: The Shaping of Human Minds Through Social Context. John Benjamins, Amsterdam.

Hauf, P. and Prinz, W. (2005) The understanding of own and others' actions during infancy: "You-like-me" or "Me-like-you"? Interact. Stud., 6(3): 429–445.

Heimann, M. and Meltzoff, A.N. (1996) Deferred imitation in 9- to 14-month-old infants: a longitudinal study of a Swedish sample. Br. J. Dev. Psychol., 14: 55–64.

Heimann, M. and Nilheim, K. (2004) 6-months olds and delayed actions: An early sign of an emerging explicit memory? Cognitie Creier Comportament, 8(3–4): 249–254.

Heyes, C. (2001) Causes and consequences of imitation. Trends Cogn. Sci., 5: 253–261.

Hofer, T., Hauf, P. and Aschersleben, G. (2005) Infant's perception of goal-directed actions performed by a mechanical device. Infant Behav. Dev., 28(4): 466–480.

Hofer, T., Hauf, P. and Aschersleben, G. (2007) Infants' perception of goal-directed actions on video. Br. J. Dev. Psychol., paper in press.

Hommel, B., Müsseler, J., Aschersleben, G. and Prinz, W. (2001) The theory of event coding: a framework for perception and action planning. Behav. Brain Sci., 24: 849–937.

Iacoboni, M. (2005) Neural mechanisms of imitation. Curr. Opin. Neurobiol., 15: 632–637.

Jovanovic, B., Kiraly, I., Elsner, B., Gergely, G., Prinz, W. and Aschersleben, G. (2007) The role of effects for infant's perception of action goals. Submitted manuscript.

Jovanovic, B. and Schwarzer, G. (2007) Infant perception of the relative relevance of different manual actions. Eur. J. Dev. Psychol., 4(1): 111–125.

Kellman, P. and Arterberry, M. (1998) The Cradle of Knowledge. Development of Perception in Infancy. MIT Press, Cambridge.

Király, I., Jovanovic, B., Prinz, W., Aschersleben, G. and Gergely, G. (2003) The early origins of goal attribution in infancy. Conscious. Cogn., 12(4): 752–769.

Klein, A., Hauf, P. and Aschersleben, G. (2006) A comparison of televised model and live model in infant action control: how crucial are action effects? Infant Behav. Dev., 29: 535–544.

Klein, A., Hauf, P. and Aschersleben, G. (2007) Rattling while shaken and rolled: The observation of action-effect contingencies affects 9-month-olds' action perception and action control. Submitted manuscript.

Longo, M.R. and Bertenthal, B.I. (2006) Common coding of observation and execution of action in 9-month-old infants. Infancy, 10(1): 43–59.

Meltzoff, A.N. (1988) Infant imitation after a 1-week delay: long-term memory for novel acts and multiple stimuli. Dev. Psychol., 24(4): 470–476.

Meltzoff, A.N. (1990) Foundations for developing a concept of self: the role of imitation in relating self to other and the value of social mirroring, social modeling, and self practice in infancy. In: Cicchetti D. and Beeghly M. (Eds.), The Self in Transition: Infancy to Childhood. University of Chicago Press, Chicago, IL, pp. 139–164.

Meltzoff, A.N. (1995) Understanding the intentions of others: re-enactment of intended acts by 18-month-old children. Dev. Psychol., 31: 838–850.

Meltzoff, A.N. (2002) Elements of a developmental theory of imitation. In: Meltzoff A.N. and Prinz W. (Eds.), The Imitative mind. Development, Evolution and Brain Bases. Cambridge University Press, New York, pp. 19–41.

Meltzoff, A.N. (2007a) The 'Like Me' framework for recognizing and becoming an intentional agent. Acta Psychol., 26–43.

Meltzoff, A.N. (2007b) 'Like me': a foundation for social cognition. Dev. Sci., 10(1): 126–134.

Meltzoff, A.N. and Decety, J. (2003) What imitation tells us about social cognition: a rapprochement between developmental psychology and cognitive neuroscience. Philos. Trans. B Biol. Sci., 358: 491–500.

Meltzoff, A.N. and Moore, M.K. (1999) Persons and representation: why infant imitation is important for theories of human development. In: Nadel J. (Ed.), Imitation in Infancy: Cambridge Studies in Cognitive Perceptual Development. Cambridge University Press, New York, pp. 9–35.

Nadel, J., Revel, A., Andry, P. and Gaussier, P. (2004) Toward communication: first imitations in infants, low functioning children with autism and robots. Interact. Stud., 5(1): 45–74.

Nattkemper, D. and Ziessler, M. (2004) Cognitive control of action: the role of action effects. Psychol. Res., 68(2–3): 71–198.

Papousek, H. (1967) Experimental studies of appetitional behavior in human newborns and infants. In: Stevenson H.W., Hess E.H. and Rheingold H.L. (Eds.), Early behaviour: comparative and developmental approaches. Wiley, New York, NY, pp. 249–277.

Perone, S. and Oakes, L.M. (2006) It clicks when it is rolled and it squeaks when it is squeezed: what 10-month-old infants learn about object function. Child Dev., 77(6): 1608–1622.

Piaget, J. (1952) The Origins of Intelligence. Basic Books, New York.

Prinz, W. (1990) A common coding approach to perception and action. In: Neumann O. and Prinz W. (Eds.), Relationships Between Perception and Action: Current Approaches. Springer, Berlin, pp. 167–201.

Prinz, W. (1997) Perception and action planning. Eur. J. Cogn. Psychol., 9(2): 129–154.

Prinz, W. (2002) Experimental approaches to imitation. In: Meltzoff A.N. and Prinz W. (Eds.), The Imitative Mind: Development, Evolution, and Brain bases. Cambridge University Press, Cambridge, pp. 143–162.

Rizzolatti, G., Fadiga, L., Fogassi, L. and Gallese, V. (2002) From mirror neurons to imitation, facts, and speculations. In: Meltzoff A.N. and Prinz W. (Eds.), The Imitative Mind: Development, Evolution, and Brain bases. Cambridge University Press, Cambridge, pp. 247–266.

Rochat, P. and Morgan, R. (1998) Two functional orientations of self-exploration in infancy. Br. J. Dev. Psychol., 16: 139–154.

Rovee, C.K. and Rovee, D.T. (1969) Conjugate reinforcement of infants' exploratory behaviour. J. Exp. Child Psychol., 8: 33–39.

Rovee-Collier, C. (1987) Learning and memory in infancy. In: Osofsky J.D. (Ed.), Handbook of Infant Development. Wiley, New York, NY, pp. 98–148.

Rovee-Collier, C. and Hayne, H. (2000) Memory in infancy and early childhood. In: Tulving E. and Craik F. (Eds.), The Oxford Handbook of Memory. Oxford University Press, New York, NY, pp. 267–282.

Rovee-Collier, C. and Shyi, G. (1993) A functional and cognitive analysis of infant long-term retention. In: Howe M.L., Brainerd C.J. and Reyna V.F. (Eds.), Development of Long-Term Retention. Springer, New York.

Schweinle, A. and Wilcox, T. (2004) Intermodal perception and physical reasoning in young infants. Infant Behav. Dev., 27(2): 246–265.

Sommerville, J.A. and Woodward, A.L. (2005) Pulling out the intentional structure of action: the relation between action processing and action production in infancy. Cognition, 95: 1–30.

Sommerville, J.A., Woodward, A.L. and Needham, A. (2005) Action experience alters 3-month-old infants' perception of others' actions. Cognition, 96: B1–B11.

Soussignan, R., Nadel, J., Canet, P. and Gerardin, P. (2006) Sensitivity to social contingency and positive emotion in 2-month-olds. Infancy, 10(2): 123–144.

Streri, A. and Gentaz, E. (2003) Cross-modal recognition of shape from hand to eyes in human newborns. Somatosens. Mot. Res., 20(1): 13–18.

Streri, A. and Gentaz, E. (2004) Cross-modal recognition of shape from hand to eyes and handedness in human newborns. Neuropsychologia, 42: 1365–1369.

Streri, A., Spelke, E. and Rameix, E. (1993) Modality-specific and amodal aspects of object perception in infancy: the case of active touch. Cognition, 47: 251–279.

Tomasello, M. (1999a) The Cultural Origins of Human Cognition. Harvard University Press, Cambridge, MA.

Tomasello, M. (1999b) Having intentions, understanding intentions, and understanding communicative intentions. In: Zelazo P.D., Astington J.W. and Olson D.R. (Eds.), Developing Theories of Intention: Social Understanding and Self-Control. Erlbaum, Mahwah, NJ, pp. 63–75.

Uzgiris, I.C. and Hunt, J.M. (1975) Assessment in Infancy: Ordinal Scales of Psychological Development. University of Illinois Press, Chicago, IL.

Von Hofsten, C. (2004) An action perspective on motor development. Trends Cogn. Sci., 8(6): 266–272.

Von Hofsten, C. (2007) Action in development. Dev. Sci., 10(1): 54–60.

Wellmann, H.M., Lopez-Duran, S., LaBounty, J. and Hamilton, B., (2007) Infant attention to intentional action predicts preschool theory of mind. Submitted manuscript.

Wellmann, H.M., Phillips, A., Dunphy-Lelii, S. and Lalonde, N. (2004) Infant social attention predicts pre-school social cognition. Dev. Sci., 7: 283–288.

Willats, P. (1999) Development of means-end behaviour in young infants: pulling a support to retrieve a distant object. Dev. Psych., 35(3): 651–666.

Williamson, R.A. and Markman, E.M. (2006) Precision of imitation as a function of preschoolers' understanding of the goal of the demonstration. Dev. Psychol., 42: 723–731.

Woodward, A.L. (1998) Infants selectively encode the goal of objects of an actor's reach. Cognition, 69: 1–34.

Woodward, A.L. (1999) Infants' ability to distinguish between purposeful and non-purposeful behaviors. Infant Behav. Dev., 22: 145–160.

C. von Hofsten & K. Rosander (Eds.)
Progress in Brain Research, Vol. 164
ISSN 0079-6123

CHAPTER 17

The role of behavioral cues in understanding goal-directed actions in infancy

Szilvia Biro[1,*], Gergely Csibra[2] and György Gergely[3]

[1]*Department of Psychology, Leiden University, Wassenaarseweg 52, 2333 AK Leiden, The Netherlands*
[2]*School of Psychology, Birkbeck College, University of London, Malet Street, London WC1E 7HX, UK*
[3]*Institute for Psychology of the Hungarian Academy of Sciences, 18–22 Victor Hugo Street, 1132 Budapest, Hungary*

Abstract: Infants show very early sensitivity to a variety of behavioral cues (such as self-propulsion, equifinal movement, free variability, and situational adjustment of behavior) that can be exploited when identifying, predicting, and interpreting goal-directed actions of intentional agents. We compare and contrast recent alternative models concerning the role that different types of behavioral cues play in human infants' early understanding of animacy, agency, and intentional action. We present new experimental evidence from violation of expectation studies to evaluate these alternative models on the nature of early development of understanding goal-directedness by human infants. Our results support the view that, while infants initially do not restrict goal attribution to behaviors of agents exhibiting self-propelled motion, they quickly develop such expectations.

Keywords: infancy; goal attribution; agency; animacy; action interpretation; intentionality

Introduction

Understanding actions in terms of the goals they are designed to achieve is a fundamental human faculty that emerges early in development (Biro and Hommel, 2007; Csibra and Gergely, 2007). How do infants decide whether a certain behavior is an "action" worthy of goal attribution? What classes of entities invoke infants' interpretation of the behavior of the entity in terms of goals? The central interest of this paper is the nature of the process involved in the identification of the entities that pursue goals. Categorization of entities is generally thought to be based on certain types of cues or features available through perception. Three types of observable cues have been hypothesized to form the basis of identifying the scope of goal attribution in infants (Table 1).

"Featural and/or biomechanical cues" that correspond to the appearance and biomechanical movement properties of humans constitute the first type of cue. It has been proposed that specific cues, like human features such as eyes, a face, hands or body (e.g., Woodward, 1998; Legerstee et al., 2000), or biomechanical bodily movements such as facial expressions or manual acts (Meltzoff and Moore, 1997), can identify the class of entities whose behavior could be interpreted in psychological terms. The category that these cues are assumed to indicate is HUMAN.

The second type of cues are the "Self-propulsion movement cues," which involve behavioral changes in the entity that occur without external

*Corresponding author. Tel.: +31 (0) 71 527-3920;
Fax: +31 (0) 71 5273783; E-mail: sbiro@fsw.LeidenUniv.nl

DOI: 10.1016/S0079-6123(07)64017-5

Table 1. Three types of cues for identifying entities and behaviors that are to be interpreted as goal-directed

Types	Featural cues	Self-propelled movements	Context-sensitive behavioral changes
Cues	• Human surface features (such as eyes, face, hand, body) • Biomechanical movements (such as facial expressions, manual acts)	• Self-generated movement (starting to move by itself, changing speed or direction abruptly, non-rigid transformation) • Independent • Irregular • Unpredictable motion patterns	• Equifinal • Consistently adjusted • Efficient • Predictable motion patterns
Categorization/interpretation	HUMAN	AGENT	GOAL-DIRECTED ACTION

impact. Observing an entity starting to move by itself, abruptly changing its speed and direction, undergoing non-rigid transformation, or generally being capable of independent, irregular, unpredictable movements are all indication of self-propulsion and have been suggested to form the basis for identifying entities in the domain of naïve psychology (e.g., Premack, 1990; Mandler, 1992; Leslie, 1994; Gergely et al., 1995; Tremoulet and Feldman, 2000). The presence of these cues is hypothesized to trigger the categorization of an entity as AGENT.

While the first two cues identify *entities* that are normally engaged in goal-directed actions, the third type of cue single out *behaviors* that are likely to indicate goal-directed actions. These cues can be called "Context-sensitive behavioral changes." This type of cues are characterized by equifinality of the behavior (Heider, 1958), consistent adjustment to changes of the environment (Mandler, 1992; Csibra et al., 1999; Gergely and Csibra, 2003), and performing optimal and efficient actions toward certain states. These cues are assumed to indicate GOAL-DIRECTEDNESS. (See Table 1 for a list of characteristics of the three types of cues.)

The main theoretical questions of this paper are concerned with the relationship between the categorization of an entity and the interpretation of its behavior as goal-directed action. Does the categorization of an entity with one type of cue automatically and necessarily imply the attribution of further properties? For example, if infants have categorized an entity as HUMAN and/or AGENT, do they automatically interpret its actions as GOAL-DIRECTED? Or if infants have identified an action of an entity as GOAL-DIRECTED, would that imply that they consider the entity as an AGENT or HUMAN? What is the underlying mechanism of these possible inferential links? Are they based on pre-wired connections, learnt by registering statistical associations between the observable cues, or are they mediated by automatic simulation processes? Several theoretical accounts have been proposed concerning the relationship between the categorization of an entity and the interpretation of its actions as goal-directed. We will briefly review these theoretical approaches and then outline our position on these issues.

Theoretical accounts on the relationship between HUMAN and AGENT categories and GOAL-DIRECTED ACTION interpretation

"Human first" — cue-based approach

This approach claims that only those entities that can be categorized as HUMAN by featural and/or biomechanical cues belong to the domain of naïve psychology. If the observed entity does not display these cues, infants will not categorize it as human and will not proceed with attributing goals to its actions. Thus, the cue-based categorization of an entity as HUMAN is proposed to be a precondition for interpreting the behavior of the entity as GOAL-DIRECTED. Meltzoff and Moore (1997) proposed that the mechanism to infer further properties of human actions is based on innate simulation processes that are present at birth. These processes allow infants to automatically map observed actions to their own experience and then to project goals and intentions to others based on the relationship between their own acts and underlying mental states. Woodward (1998, 1999) also claimed the primacy of cue-based categorization of entities as HUMAN on the basis of featural cues in evaluating actions as GOAL-DIRECTED. However, she argued that infants learn *gradually* which human actions can be considered as goal-directed through experience with their own actions.

"Agent first" — cue-based approach

The second group of theoretical approaches assumes that the cue-based pre-categorization of the observed entity as an AGENT — on the basis of "Self-propulsion movement cues" — is a precondition for considering the action of the entity as GOAL-DIRECTED. Agents form a broader category than people and are not tied to any specific appearance. Agents are generally defined as a class of objects that have the capacity to be the causal source of events. Agency and its relation to other ontological properties have, however, been conceptualized in different ways in various theoretical accounts. For example, in Leslie's modular theory of agency (Leslie, 1994) self-propulsion defines the first subsystem, in which objects are categorized as mechanical agents with internal and renewable source of energy. The second component of this system — which receives its input form the first subsystem — deals with the actional properties of agents, such as attaining goals and reacting to the environment. Thus, "Context-sensitive cues" such as the repeated equifinal outcome of the agent's action is a useful behavioral cue at this level as it can be teleologically interpreted as the goal state. In contrast, other theories have proposed that the detection of self-propelled movement (Premack, 1990; Baron-Cohen, 1994) or contingent reactivity (Johnson et al., 1998) automatically triggers the interpretation of an entity as an intentional agent. Once an entity has been categorized as an intentional agent, infants can use further "Context-sensitive behavioral cues" to attribute various psychological properties such as goals, perception, or attention to the entity. All of these theories assume that the link from the category AGENT to GOAL-DIRECTED ACTION interpretation is "pre-wired."

"No primacy" — principle-based approach

We propose — as a third, alternative approach — that the interpretation of an action as GOAL-DIRECTED is initially independent from both HUMAN and AGENT categories (see also Csibra et al., 1999). In other words, an entity does not need to have been identified as HUMAN or AGENT in order to allow infants to evaluate its action in terms of goals. Furthermore, this approach claims that the basis of goal attribution is not cue-based but principle-based: the selection of entities belonging to a given domain is not dependent on specific perceptual cues but on the successful application of the central principle that is used to reason about the behavior of entities in that domain (cf. Carey and Spelke, 1994 for a contrast of cue-based vs. principle-based models). We have suggested (e.g., Csibra and Gergely, 1998) that a goal-directed entity can be identified by the successful application of a psychological interpretational system that represents the observed behavior of entities in teleological terms. This interpretational system is assumed to establish a specific explanatory relation among three representational elements: the actions, the goal state, and the constraints of physical reality. However, this representational structure forms a teleological representation only if it satisfies the "principle of rational action," which states that an action can be explained by a goal state if it appears as the most efficient action toward the goal state that is available within the constraints of reality. Therefore, we propose that certain "Context-sensitive behavioral cues" can inform about GOAL-DIRECTEDNESS inasmuch as they can be derived from the principle of rational action itself (see Csibra et al., 1999, 2003; Király et al., 2003). Finally, we believe that categorizing an entity as HUMAN or AGENT and interpreting its action as GOAL-DIRECTED are initially independent from each other, and infants create links between these concepts by gradual learning of the statistical correlations between the observable cues that indicate the three properties.

Our aim in this paper is to contrast these theoretical approaches. First we will review some empirical studies that have investigated the role of the three types of cues in categorizing and reasoning about entities, paying particular attention to what these studies revealed about the inferential link between them. Following this we will report two experiments that tested our hypotheses on the possible links between AGENCY categorization and GOAL-DIRECTED action interpretation.

Research on the inferential link between the categorization of HUMAN or AGENT and GOAL-DIRECTED action interpretation

There is no doubt that from early on infants show sensitivity to the three types of cues described above (see Table 1). Several studies have found that infants pay special attention to human featural cues such as faces, eyes, eye direction, facial expressions (e.g., Farroni et al., 2002, 2005), etc. Three-month-olds have been found to be sensitive to human biomechanical motion: they can differentiate between point-light displays of a human walk and random movement patterns (e.g., Bertenthal, 1993). Similar sensitivity was also shown to Self-propulsion and Context-sensitive behavioral cues. Five-month-olds can distinguish between movements that are self-propelled and movements that are caused by another object (Kaufman, 1997), and can discriminate between non-rigid and rigid transformations of objects (Gibson et al., 1978). Nine-month-old infants' sensitivity to the temporal and spatial characteristics of "causation at a distance" has also been demonstrated (Schlottman and Surian, 1999). As early as 2–3 months of age, infants also show sensitivity to the contingency between observed events and their own movements, or between the movements of two entities (Watson, 1979; Rovee-Collier and Sullivan, 1980; Rochat et al., 1997). What type of inferences can infants make about entities on the basis of these cues?

From HUMAN to AGENT

Some studies provide evidence for infants' ability to infer that an entity that had been categorized as human is capable of self-propelled movement or to show other characteristics of agents. For example, from 7 months of age, infants are surprised if an inanimate object starts to move without external force but not when people move by themselves (Golinkoff et al., 1984; Poulin-Dubois and Shultz, 1990; Spelke et al., 1995; Poulin-Dubois et al., 1996). Infants have also been shown to vary their own behavior (vocalization vs. engaging in manual search) in a "hide and seek" game depending on whether the entity is a human or an inanimate object (Legerstee, 1994). Pauen and Trauble (under review) demonstrated that when 7-month-olds observe a ball and an animal-like furry creature moving together in an ambiguous causal structure, they expect the animal to be able to move by itself and not the ball. Note that these findings cannot tell us whether the link from HUMAN to AGENCY is learnt or pre-wired.

From AGENT to AGENT/HUMAN

Another group of studies has used self-propulsion as the independent variable and tested if infants can use this cue to make further inferences about other agentive properties of the entity. Kaufman (1997) showed that 5-month-olds inferred that self-moving objects can reverse their trajectories while inert objects cannot. Luo and Baillargeon (2000) demonstrated that 6-month-old infants expect inert objects to be displaced when hit while self-propelled objects can "resist." Kotovsky and Baillargeon (2000) found that 7.5-month-old infants know that, in the case of an inert object, a contact with another object is necessary to make the object move. Note that in the above studies both self-propelled and inert objects had identical non-human features, hence infants' inferences were solely based on observable behavioral cues. Saxe et al. (2005) found that only if an object previously showed no evidence of self-propulsion do 10- and 12-month-old infants infer a hidden cause when they see the object move. (In this study, however, available featural cues might have also influenced this inference because the self-propelled object was a furry puppet with a face). These findings suggest that infants do not simply perceptually distinguish self-moving from inert objects but also categorize them as "mechanical agents" (Leslie, 1994). When infants witness an entity displaying one type of behavioral cue for agency (e.g., self-propulsion), they also expect the entity to exhibit other behavioral features that are typical of mechanical agents. Finally, there is a further study that might have provided evidence for the "AGENT to HUMAN" type of inference. Molina et al. (2004) demonstrated that 6-month-old infants inferred the identity of a hidden entity (person vs. ball) on the basis of the type of action (speaking vs. shaking) another person performed to set the objects in motion. Thus, infants inferred that if an entity could move without external physical contact then it is a human.

From HUMAN to GOAL-DIRECTED ACTION

Several studies investigated the role of human featural cues in infants' ability to attribute goals. Meltzoff (1995), for example, showed that 18-month-old infants can successfully infer the goal and reenact a human model's intended act — but not that of a mechanical device — by observing the model's failed attempts to achieve the goal. However, Johnson et al. (2001) found that the model does not need to be a real person: 15-month-olds can imitate the intended action of a non-human agent as long as it has a face and exhibits behavioral agency cues such as self-propulsion and contingent reactivity.

Much recent research on goal attribution applied Amanda Woodward's influential paradigm to investigate infants' emerging ability to interpret action in terms of goals. In Woodward's original visual habituation experiment (1998), 6- and 9-month-old infants were presented with an action in which a hand repeatedly reached toward and grasped one of two toys sitting on a stage. After the infants were habituated to this event, the experimenter swapped the two toys behind a screen. In the test phase, the hand either grasped the same toy as before, which, however, was at a new location, or the other toy at the same location. Looking times for these two events were markedly different: both 6- and 9-month-old infants looked longer if the hand grasped the new toy at the old location than if it grasped the old toy at the new location. This result indicates that infants associated the grasping hand with the grasped object rather than with its location, i.e, they expected the hand to reach toward the same toy. In another version of this study, inanimate novel objects (such as a rod, a mechanical claw, or a flat occluder) replaced the human hand, and in subsequent studies (Woodward, 1999; Woodward et al., 2001) unfamiliar hand actions (the back of the experimenter's hand dropped on the target toy, or a gloved hand grasped the target toy) were used to approach the target toy. Woodward found that in these conditions 6- and 9-month-old infants looked equally long at the two test events (or longer in the old toy/new location event), suggesting that the infants did not selectively encode the goal object in the case of inanimate actors or unfamiliar hand actions.

These findings have been interpreted to show that young infants restrict goal-directed interpretation to human actions and, in fact, only to actions that are already familiar to them, such as grasping. However, Király et al. (2003) and Jovanovic et al. (under review) demonstrated that if the unfamiliar "dropping the back of the hand" action was accompanied with a salient action effect (the hand transferred the toy to another location) infants as young as 6 months were able to encode the goal of the action. Similarly, Biro and Leslie (2007) (Experiment 1) found the same positive result when "equifinal variations" were added as a behavioral cue to another unfamiliar action (a poking hand). These latter studies therefore suggest that the familiarity of the action is not a precondition for goal attribution, it is rather the availability of behavioral cues that determines whether infants are able to interpret an action as goal-directed. In the case of human actions, however, it is problematic to determine the exact role of different behavioral cues (i.e., Self-propulsion cues vs. Context-sensitive cues). Infants have ample experience with their own or others' observed hand actions which could allow them to associate various behavioral cues (such as self-propulsion, variability, adjustments, mechanical effects) with hands and rely on these associations when evaluating the goal-directedness of a hand action even when these cues are not present (for a similar argument concerning the basis of goal attribution in the case of the familiar grasping action, see Király et al., 2003; Biro and Leslie, 2007).

From AGENT to GOAL-DIRECTED ACTION

Several studies looked at the role of behavioral cues in goal attribution in the absence of human features. One group of these studies introduced behavioral cues to the inanimate actors in Woodward's paradigm. Luo and Baillargeon (2005) demonstrated that 5.5-month-old infants could interpret the action of a self-propelled box as goal-directed. A study by Shimizu and Johnson (2004) also found that infants can consider inanimate actions as goal-directed. However, in their study the infants could encode the goal of a novel, faceless, oval-shaped object only if, besides self-propulsion, it also exhibited contingent reactivity and a rotating movement along its axis. This axial turn gave the impression (at least for adults) that the object was making a "choice" by turning away from the non-target toy, i.e., it indicated that it can adjust its behavior for goal attainment. Biro and Leslie (2007) suggested that there is a developmental trend between 6 and 12 months in the extent to which infants rely on behavioral cues for attributing goals to actions of inanimate objects in the Woodward paradigm. They showed that the older age groups can rely on the simultaneous presence of two cues — self-propulsion and a salient action effect — to consider the action of a wooden rod as goal-directed. For the youngest infants, however, these two cues were not sufficient. Only when equifinal variations of the action in goal attainment were also provided did this group of infants attribute a goal.

In another group of studies 6-, 9-, and 12-month-old infants were shown animations of geometric figures. Gergely et al. (1995) (and Csibra et al., 1999) conducted a visual habituation study in which infants observed a small circle repeatedly approach and make contact with a large circle by "jumping over" a rectangular figure separating them in the habituation phase. Adults typically interpret this behavior as a goal-directed action, where the goal state is reaching or contacting the large circle. In the test trials the rectangular figure (the "obstacle") was removed. Infants saw either a novel action (the small circle approached the large circle in a straight line) that was the most efficient action toward the goal in the changed circumstances, or the already familiar jumping action which, however, was no longer the most efficient action to achieve the same goal state. Nine- and twelve-month-old infants showed less recovery of attention to the novel straight-line action, which indicates that they interpreted the action in the habituation event as goal-directed and predicted the most efficient action to achieve the inferred goal in the changed situation.

The actor exhibited several behavioral cues in these studies: self-propulsion (it started to move by itself and it changed direction), contingent reactivity and nonrigid surface movements (the two circles "contracted" and "expanded" in a contingent "turn-taking manner"), equifinal variations in goal attainment (the positions of circles on the left vs. right side of the screen were varied), and situational adjustment of behavior in goal approach (the circle changed its path to jump over the obstacle). In the control condition of this study, infants saw the same event as in the habituation phase except that the rectangular figure (the "obstacle" in the experimental condition) was placed *behind* the small circle and so it did not block the small circle's direct approach route to get to the target circle (i.e., the block was not an "obstacle" anymore). In spite of this, the small circle performed the same jumping behavior (this time, however, jumping over nothing) as in the experimental condition to get to the target circle and so its target approach could not be interpreted as the most efficient action available to achieve its goal. Infants in this condition did not look differently in the two test events, i.e., they did not have any expectations about the small circle's future action.

Note that in the control condition the same behavioral cues were present — including those that indicated agency — as in the experimental condition, however, these cues themselves did not allow infants to interpret the action in terms of goals in the control condition.

Kamewari et al. (2005) also used the Gergely paradigm with 6.5-month-old infants in three different conditions involving real actors: a person, a humanoid robot, or a moving box. They found that infants attributed goal in the first two conditions, but not in the third. Csibra (submitted), however, hypothesized that even though the unfamiliar box has shown self-propulsion, this — in itself — may not have been a sufficient cue for young infants to attribute the final outcome state as the goal of the box's action, as the box had not exhibited any form of behavioral variability across trials in its manner of target approach. He, therefore, replicated the self-propelled box condition but added equifinal variations in goal approach (making the box go around sometimes on the right side while sometimes on the left side of the obstacle to get to the goal object) and found that under these conditions even 6-month-olds interpreted the action of an inanimate and unfamiliar object as goal-directed.

Summary

Two conclusions can be drawn from the research reviewed above. First, the categorization of HUMAN on the basis of featural cues does not seem to be an obligatory precondition for identifying the entity as an AGENT or for considering an entity's action as GOAL-DIRECTED. This has been shown by studies that demonstrated that infants are willing to interpret actions of entities in terms of goals or other agentive or psychological properties that display no human features. Second, there is converging evidence that the categorization of an entity as an AGENT is not sufficient for interpreting the entity's action GOAL-DIRECTED. Thus, the hypothesis that agency cues would directly trigger intentional action interpretation is not supported by the empirical evidence either. The studies we reviewed above, however, cannot tell if categorizing an entity as an AGENT is necessary for giving GOAL-DIRECTED interpretation to its behavior. Although many "Self-propulsion movement cues" and "Context-sensitive behavioral cues" have been used either in solo or in combination with others in these studies, no clear picture has emerged regarding the respective roles of particular behavioral cues in identifying goal-directedness. (This is mainly because the aim of these studies was not to systematically tease these cues apart, but to show that infants can make use of behavioral cues without featural cues in identifying goal-directed actions.) Thus, the "Agent first" (cue-based) and the "No primacy" (principle-based) theoretical approaches have not yet been contrasted with regard to the question whether or not infants must first embrace a prior ontological commitment that the entity in question is an agent for applying a goal-directed interpretation to its behavior. In the next section we will address this question by taking a closer look at the nature of the possible link between the categorization of an entity as an AGENT and interpreting an entity's behavior as GOAL-DIRECTED.

Three possible functional relations between AGENCY and GOAL-DIRECTEDNESS

Both the "Self-propulsion movement cues" that indicate an AGENT and the "Context-sensitive cues" that indicate GOAL-DIRECTEDNESS have been hypothesized to form the basis of perception of animacy in adults as well as in infants (see Scholl and Tremoulet, 2000, for a review). It seems, however, that these two types of behavioral cues sometimes contradict each other and so it would not be easy to build an animacy detector that could rely on both aspects of animacy simultaneously. Let us illustrate this computational problem by a study of Rochat et al. (1997). In that study infants were presented with two discs on a computer screen which were engaged in a "chase" event: one of them, the "chasee" changed its direction and speed unpredictably, while the other, the "chaser" followed it persistently, at a constant speed in a "heat-seeking" fashion. Young infants were able to discriminate this event from another one in which the two disks were performing the same individual movements but this time unrelated to each other's behavior. This suggests that they were sensitive to the presence of the cues embedded in the disks' behavior. One can, however, raise the question: which of the two objects did they consider animate? From the point of view of a "Self-propulsion cues" analysis, the chasee would be categorized as animate because it demonstrated self-generated, independent behavior, while the chaser could be considered as an inanimate object, a perfectly predictable, physical heat-seeking device. However, if infants were looking for "Context sensitive cues" to find an animate entity, the chaser would be considered animate because it adjusted its path consistently to the chasee's path in accord with the goal of catching it up. This example illustrates that coordinating these two aspects of animacy would pose a considerable computational problem for an automatic animacy-detector system.

Nevertheless, since as our adult intuition suggests that both "Self-propulsion" and "Context-sensitive behavior" cues may indicate objects that we treat equally as members of the unitary category of "animate" objects, it is plausible to assume the existence of some *functional* relation between them. This hypothesis can take different forms according to the nature and relative strength attributed to this functional link.

1. The "*Mandatory link*" *hypothesis*: Agent categorization *is required* for goal-directed interpretation of actions. According to this view, goals are attributed to objects *only if* the cues of "self-propulsion" are explicitly present.
2. The "*Exclusive link*" *hypothesis*: Categorization of an entity as non-agent *precludes* goal-directed interpretation of its actions. In this view, goals would not be attributed to objects that are explicitly perceived as being "non-self-propelled."
3. The "*Probabilistic link*" *hypothesis*: Agent categorization *biases* toward goal-directed action interpretation. This view holds that goals will be more likely to be attributed to objects that are perceived "self-propelled."

The "Mandatory link" hypothesis

The first version of this hypothesis, which assumes a mandatory relation between AGENCY and GOAL-DIRECTEDNESS, has already been tested (Csibra et al., 1999, Experiment 2). In this experiment, we sought to determine whether the presence of cues of "self-propulsion" is a necessary precondition for goal attribution in infants. To do so, we created a computer-animation that did not contain *any* information about the source of the motion of the object but still displayed signs of "goal-directedness" in terms of optimal adjustment of its behavior to the relevant changing aspects of the situation across trials. The object repeatedly "flew in" from outside of the screen (the source of the origin of its movement being thus always occluded from view). It always followed an inert, parabolic trajectory — as if driven only by gravitational forces — flying just over a rectangular "obstacle" in the middle of the screen and ending up at the same final spatial position making contact with a stationary target object, which was positioned at the other side of the screen. However, across trials the relative height of the trajectory of the flying object changed to optimally match the varying height of the "obstacle" in the middle, thus managing to fly just over it. This way, the variable behavior of the flying object could be considered to exemplify an efficient action to achieve the same final target position across trials under variable situational constraints. This aspect of the events may have provided sufficient basis for attributing the identical spatial end position reached across trials as the goal of the action.

Whether or not infants made this goal attribution was tested by the same logic that we used in our previous study (Gergely et al., 1995). During the test phase, we removed the obstacle and presented two alternative events to the infants: either the "Old Action" test event, which consisted of the same — already familiar — medium-high parabolic flight trajectory leading to the same end state as had been observed during habituation trials, or the "New Action" test event, which consisted of a novel straight line horizontal approach route to the target position that had become available only now that the "obstacle" has been removed. If infants attribute the final position as the goal of the object's actions, they should see the straight-line approach as being the most efficient action toward that goal in the new situation, hence being compatible with their previous goal-directed interpretation. In contrast, given the removal of the "obstacle" and the subsequent availability of a more efficient alternative new pathway to the goal, the sight of the "Old Action" test event should be considered unjustified and incompatible with the previously attributed goal.

Looking patterns of both 9- and 12-month-olds suggested that this was, indeed, the case. Both age groups looked significantly longer at the familiar "Old Action" event, and an appropriate control condition demonstrated that this could not have been due to an inherent preference toward watching the "Old Action" event over the "New Action" event. These results suggest that positive perceptual evidence indicating that an object's movement is self-propelled is not a necessary precondition for attributing a goal to its behavior. Rather, it seems that infants can recognize the "pure reason" (Csibra et al., 1999) manifested in the observable pattern of justifiable variability of the object's actions in relation to the changing constraints of the situation, which, in itself, may provide sufficient information for goal attribution.

One could object, however, that although the object's observable behavior in these events did not display any direct sign of self-propulsion, it did not exhibit positive evidence to the contrary either: the object's behavior after all *could* have been actually self-propelled for all we know. On this ground, one could still argue that there is a mandatory connection between self-propulsion and goal-directedness if self-propulsion were considered the default interpretation for moving objects whose source of motion cannot be identified. This assumption would make the cues for self-propulsion less valuable because they would just trigger the same interpretation of the event that would be applied to it anyway (unless positive evidence to the contrary is available). In fact, this assumption would change the hypothesis that "self-propelled objects pursue goals" into another hypothesis to the effect that "non-self-propelled objects do not pursue goals." This new form of the hypothesis, which we labeled above as the "Exclusive link hypothesis," leads to the prediction that although the absence of evidence for self-propulsion does not prevent goal-attribution, the presence of positive evidence for *non*-self-propulsion should prevent it. This prediction was tested in Experiment 1.

Experiment 1: the "Exclusive link" hypothesis

Infants were presented with habituation events similar to the previous study (Csibra et al., 1999) with the exception that we provided direct evidence for the object being non-self-propelled. The object again approached its final target position through an inert parabolic pathway whose height was optimally adjusted to match the relative height of the obstacle in the middle that varied its size across habituation events. However, in this condition the object was always visibly launched to its flight trajectory by the direct impact of another moving object hitting it in a Michottean fashion. After habituation, the obstacle was again removed and infants were presented with two alternative test-events: the old "jumping" action vs. the new straight-line approach to target. The moving object was in both cases directly launched by the visible impact of another moving object hitting it. The "non-self-propelled objects do not pursue goals" hypothesis predicts that since the presence of direct evidence for non-self-propulsion must have prevented goal attribution during the habituation events, infants would look equally long at the two test events.

Methods

Participants. Forty-two 12-month-old infants participated in this experiment (27 males and 15 females, mean age: 372.74 days, SD: 12.89 days, range: 352–406 days). An additional 10 infants were also tested but were excluded from the data analysis due to fussiness (7) or short looking time in the test event (3). All of the subjects were healthy, full-term infants who were recruited through advertisements in local magazines.

Stimuli. The stimuli were computer-animated visual events. In the habituation event (see Fig. 1A) first a small red circle on the left side of the screen and a large blue circle on the right side of the screen

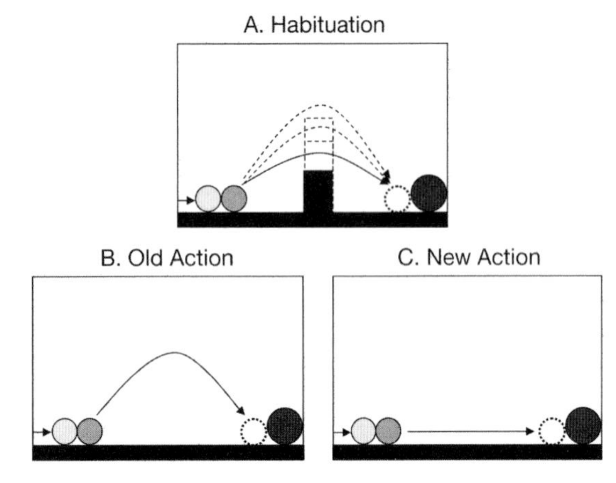

Fig. 1. Habituation event (A), Old Action test event (B), and New Action test event (C) of Experiment 1. Note that in all events the small circle's movement across the screen appeared as if it were launched by the impact of the other one.

appeared with a black rectangular column positioned in between them. The height of the column was randomly varied over trials, being either small, medium, or large. The event started when a third yellow circle entered the screen horizontally from the left side and contacted the small red circle. Upon contact the yellow circle stopped and the small red circle immediately started to move. The small red circle's movement appeared as if it were launched by the impact of the yellow one. It followed a parabolic pathway as if "flying over" the rectangular figure in the middle (the "obstacle"), and then it landed and stopped at the position adjacent to the large blue circle. The rectangular column appeared to form an "obstacle" separating the large circle and the small circle. The height of the small red circle's parabolic pathway was always adjusted to optimally match the variable height of the "obstacle" in such a way that it always just managed to pass over it without colliding with it.

During the test events, the "obstacle" was no longer present and so the small circle's path to the target circle was unobstructed. In the Old Action test event (Fig. 1B) the behaviors of the red and yellow circles were identical to those in the habituation event, i.e., the small red circle was set in motion by the moving yellow circle and then "flew" to the position adjacent to the large blue circle along a parabolic trajectory that corresponded to the average (medium) height of its trajectory during the habituation events. In the New Action test event (Fig. 1C), however, the small red circle approached the same end-point through a novel pathway taking the shortest straight-line route that has now become available.

Apparatus. The infants sat in their parent's lap in a darkened experimental room looking at an 18×24 cm monitor placed at eye level from a distance of 1.2–1.4 m. A video camera focusing on the baby's face was mounted above the monitor peeping through the opening of a black curtain, which allowed the experimenter to monitor the infant's eye fixations. The experimenter controlled the stimulus presentation and registered the looking times by operating a keyboard of a computer.

Procedure. At the beginning of each trial, the experimenter drew the infant's attention to the display by presenting colored flashes on the monitor. When the baby looked at the screen, the experimenter pressed a key that started the presentation of the stimulus event, which was then repeated continuously until the infant looked away for more than 2 s. When the infant looked away, the experimenter released the key on the keyboard, and if she did not press it again within 2 s indicating that the infant looked back again, the computer program stopped the stimulus display and registered the looking time for the trial. When the infant looked at the screen again, the next trial was started. A trial had to last at least 2 s to be treated as valid, i.e., if the infant looked at the event for less than 2 s, the trial was ignored. The computer program calculated the average fixation time for the first three habituation trials and compared this value on-line with the running average of the last three fixation times. We used a habituation criterion that required that the average fixation time for the last three trials be less than half of the average looking times for the first three habituation trials. Thus, the minimal number of habituation trials was six.

After the habituation criterion was reached, a 30 s long break was introduced during which the parent, who was sitting on a swivel chair, was asked to turn with her baby away from the monitor. When they turned back and the test trials

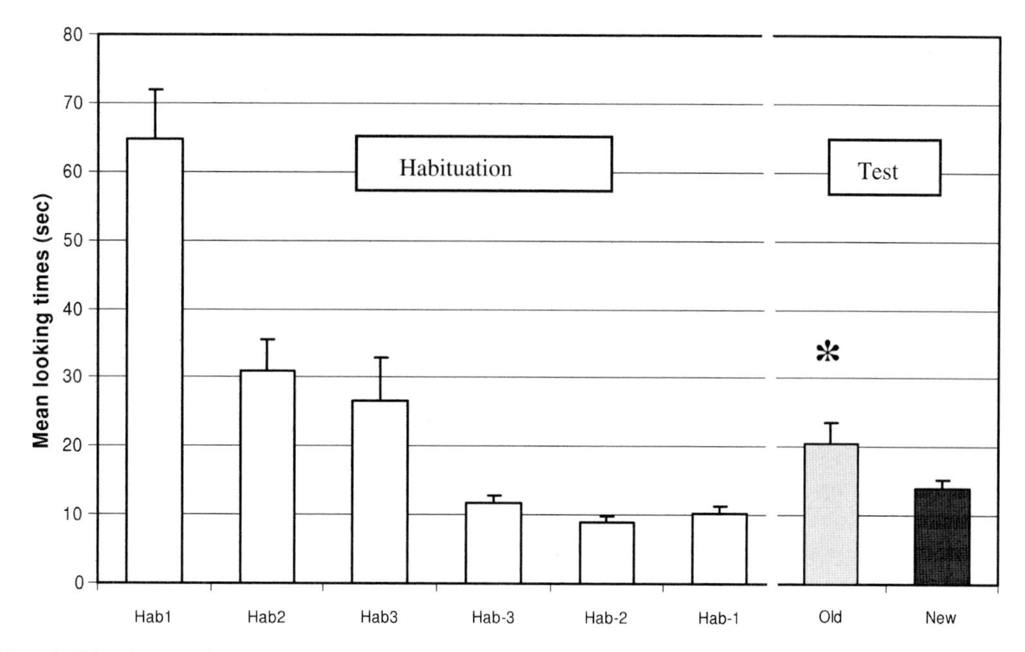

Fig. 2. Mean looking times and standard errors in the first and last three habituation trials, and in the two test events in Experiment 1 (*$p<0.05$).

started, we instructed the parents to close their eyes so that they could not inadvertently bias their child's reaction to the test displays. The test trials were delivered in the same way as the habituation trials. Each infant watched two test trials: an Old Action and a New Action event. For half of the infants the first test trial was an Old Action display followed by a New Action event, while the other half received the same stimuli in the opposite order. The experimenter was blind to the order in which the two test stimuli were presented. In order to ensure that the dishabituation scores reflected the infants' reaction to the nature of the stimulus event, we had to make sure that they had a chance to identify which kind of event was presented to them. Therefore, since the difference between the two test events could not have been detected earlier than 2.5 s within the trial, we excluded from the analysis all the infants who watched either of the test trials for less than 2.5 s and so had no opportunity to observe the full-event structure.

Results

The average number of completed habituation trials was 6.48. The mean looking times during the test phase were analyzed by ANOVAs using event type (old vs. new) as a within-subject factor, and order (old first vs. new first) as a between-subject factor. The analysis revealed a main effect of event type ($F(1,40) = 4.61$, $p<0.05$), indicating that the infants looked at the Old Action test event significantly longer than at the New Action test event. No effects of order were found. The mean looking times for habituation and test events are depicted in Fig. 2.

Discussion

In Experiment 1 we tested whether the perception of an object's movement as non-self-propelled would prevent 12-month-olds from attributing a goal to its behavior. Our findings demonstrate that this was not the case. The only informational basis for goal-attribution in this study was provided by the consistent pattern of adjustments observable in the height of the trajectory of the different goal-approaches performed by the object across trials that always optimally matched the variation in the height of the "obstacle" that separated the object from its target. These adjustments were justifiable in so far as they ensured efficient goal approach across the variable situational conditions. Thus, justifiable adjustment of goal approach in itself proved to be a sufficient cue for goal attribution. In sum, we can conclude that "self-propulsion" is not a necessary prerequisite for goal attribution and its absence does not necessarily inhibit the interpretation of behaviors in terms of goals.

Does this conclusion imply that infants do *not* rely on cues of "self-propulsion" at all when assessing which objects may pursue goals and which may not? Not necessarily. Our evidence demonstrates only that there is no *exclusive link* relating AGENCY to GOAL-DIRECTEDNESS, but it still leaves open the possibility that there is a *probabilistic* relation between them. Perhaps behaviors of self-propelled objects are *more likely* to be interpreted in terms of goals than behaviors of non-self-propelled objects, especially in otherwise ambiguous events which lack any other behavioral cues (such as consistent adjustments) that would positively indicate goal-directedness.

Experiment 2: the "Probabilistic link" hypothesis

One way to investigate whether infants are more likely to attribute goals to self-propelled objects than to externally driven objects is to present such objects engaging in a behavior that is otherwise ambiguous as to whether it is goal-directed or not. In our next study, therefore, we presented such ambiguous object behavior during habituation events. A moving object repeatedly approached a stationary target object at the other side of the screen via the simplest, shortest, straight horizontal pathway leading to it. Note that although this behavior is consistent with the goal of reaching the target and although it qualifies as the most efficient goal-approach available in the given situation, the object's target approach presents no Context-sensitive behavioral cues (such as justifiable adjustment of movement) that would independently indicate whether it is goal-directed or not. Note also that this behavior is equally consistent with a physical causal interpretation that would assume that some kind of external force had been imputed to this object making it roll on a straight-line path until it bumped into the other object in its way.

One can test whether infants interpret this ambiguous behavior in a teleological or causal manner by presenting them with a situation in which a new object (an obstacle) is placed between the horizontally moving and the stationary target object. If infants attributed the goal of reaching the target to the moving object's behavior during the preceding habituation events, in the test event they should expect the ball to adjust its behavior by jumping over the obstacle to reach its goal object via such a new (and justified) detour action. If, however, they interpreted the habituation event in purely causal physical terms, they should expect the ball to repeat its previous straight-line movement during the test event and to bump into the obstacle as a result. We also included a third, control test in which the ball did perform a jumping action while approaching the target ("hopping") without, however, any obstacle blocking its approach toward the target. This "hopping" event would not be justified by either the causal or the teleological interpretation of the object's previous behavior.

Two groups of 12-month-olds were presented with such events, which, however, were embedded in different movement initiating contexts. One group saw that the — originally stationary — object started its movement as soon as it was contacted by another object rolling into it from the side ("Direct Launching") resulting in the impression that it was externally propelled. The other group saw the same initial object contact event with the only difference that a half-a-second pause was introduced between the launcher making contact with the ball and the initiation of the latter's movement ("Delayed Launching"). This delay resulted in the impression that the ball's movement was an internally caused self-propelled behavior.

Now, we have two hypotheses on the table that would predict different response patterns for the test events in this study. According to the first hypothesis, it is the perceived type of the source of motion (self-propelled: "Delayed Launching" vs. externally caused: "Direct Launching") that would determine what kind of interpretation the infant assigns to the object's behavior. On the one hand, this hypothesis predicts that during habituation the sight of a self-propelled object (in the "Delayed Launching" group) taking the most efficient straight-line path available to reach the target object would invoke a teleological interpretation that would assign "contacting the target object" as the goal of the self-propelled object's action. For the test events in which an obstacle appears in the middle of the screen the hypothesized goal assignment would predict longer looking times for the "bumping into the obstacle" event (in which no behavioral adjustment that would have been necessary to achieve the goal occurred) and shorter looking times for the "jumping over the obstacle" action (in which the ball adjusted its behavior to the new situation and achieved its goal of contacting the target circle). On the other hand, this hypothesis predicts that the sight of an externally propelled object (in the "Direct Launching" group) during habituation would invoke a causal interpretation of the event that would produce the opposite pattern of looking times. Additionally, both groups should show violation of expectation (therefore, should look longer) when seeing the object "hopping" without the presence of an obstacle.

In contrast, according to the alternative hypothesis (which is certainly in line with our previous results), the perceived source of motion (self-propelled vs. externally caused) should not have an influence on the interpretation that infants give to the object's behavior. This view then predicts that both groups should show the same looking pattern in the test events.

Methods

Participants. Eighty-three 12-month-old infants participated in this experiment (38 males and 45 females, mean age: 368.49 days, SD: 7.51 days, range: 352–388 days). An additional 13 infants were also tested but were excluded from the data analysis due to fussiness (9) or failing to reach habituation criteria within 15 trials (4). Half of the participants were assigned to the Direct Launching group, while the other half took part in the Delayed Launching group. All of the subjects were healthy, full-term infants who were recruited through advertisements in local magazines.

Stimuli. The stimuli were computer-animated visual events. In the habituation event (see Fig. 3A) first a large blue circle and a small red circle appeared on right and left side of the screen, respectively. The habituation event started when a third small yellow circle entered the screen from the left side and contacted the small stationary red circle. In the Direct Launching group, the yellow circle stopped upon contact and the small red circle immediately started to move toward the right side of the screen on a horizontal path with a constant speed until it reached and made contact with the large blue circle at the other side of the screen. The Direct Launching event resulted in the visual impression as if the small red circle's movement were externally induced by the force impact exerted through the colliding yellow circle. In the Delayed Launching group the habituation event was identical to the one in the Direct Launching group except that the small red circle started to move only 0.5 s after the yellow circle contacted it. The small red circle's movement appeared as if it were self-propelled, internally caused behavior.

In all test events (Fig. 3B–D), the large blue circle and the small red circle appeared in the same starting positions as during habituation events (on the right and left side of the screen, respectively). In the "Jumping" and "Bumping" test events a black rectangular column also appeared between the two circles. All the test events started when the third small yellow circle entered the screen from the left side and contacted the small red circle the same way as it did in the habituation events. In the Direct Launching group, the yellow circle stopped upon contact and the small red circle immediately started to move. In the "Jumping" test event (Fig. 3B), the red circle took a parabolic pathway and "jumped over" the rectangular column landing at the position adjacent to the large blue circle making contact with it. In the "Bumping" test event (Fig. 3C), the red circle followed a horizontal path and "bumped into" the obstacle stopping at the position adjacent to the rectangular column. The red circle in the "Hopping" test event (Fig. 3D) followed the very same parabolic pathway as did

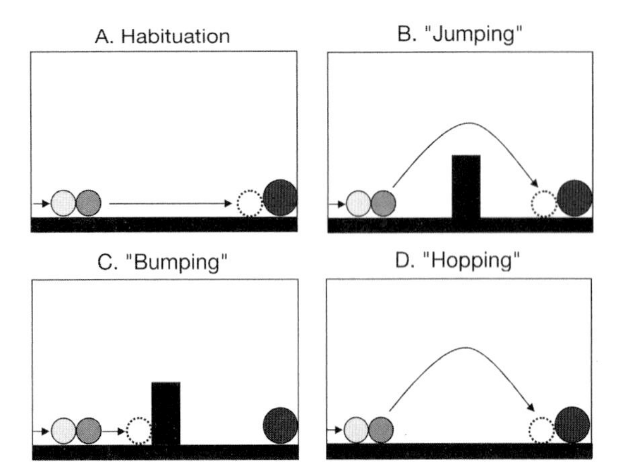

Fig. 3. Habituation event (A), Jumping test event (B), Bumping test event (C), and Hopping test event (D) of Experiment 2. Note that in the Direct Launching group the circle started to move immediately after the other one, coming from outside the screen, contacted it, while in the Delayed Launching group it started to move with a delay of 0.5 s.

the red circle in the "Jumping" test event. However, given the absence of the rectangular column in this test event, there was no "obstacle" to "jump over" and so the red circle's "hopping action" took place over an empty area and appeared unmotivated. The test events in the Delayed Launching group were identical to those in the Direct Launching group except that the small red circle started to move only 0.5 s after the yellow circle contacted it.

Apparatus. The experimental apparatus was identical to the one used in Experiment 1.

Procedure. The procedure was the same as in Experiment 1 with the following difference. Each infant watched three test trials: a Jumping, a Bumping, and a Hopping test event. The order of the test events was counterbalanced which resulted in six order groups.

Results

The average number of completed habituation trials was 7.39. There was no difference in the average number of habituation trials or in the average length of looking during the habituation trials between the Delayed and Direct Launching groups. The looking times during the test phase were first analyzed by a three-way ANOVA in which the event type (Jumping, Bumping, Hopping) was the within-subject factor and the condition (Direct vs. Delayed Launching) and order of the test events were the between-subject factors. This analysis yielded an interaction effect between event type and condition ($F(2,142) = 3.60$, $p < 0.05$), and between even type and order ($F(2,142) = 2.76$, $p < 0.05$). Two-way (event type X order) ANOVAs for each condition were carried out next. In the Direct Launching condition no main effects or interaction was found significant. In the Delayed Launching condition an event type main effect was found ($F(2,72) = 6.50$, $p < 0.05$). Pairwise comparisons revealed that infants looked significantly longer in the Hopping test event than in the Jumping ($p < 0.02$) or the Bumping test events ($p < 0.007$). Looking times in the Jumping and Bumping test events, however, did not differ significantly. In addition, an interaction effect was also found between the event type and order factors in the Delayed Launching condition ($F(2,72) = 2.42$, $p < 0.05$). This interaction was explored with an ANOVA and pairwise comparisons showed that infants in two order groups looked significantly longer in the Hopping test event than in the other two test events, while in the other order groups no difference was found in the looking times between the test trials. The mean looking times for habituation and test events are depicted in Fig. 4.

Discussion

These results confirmed neither of the two hypotheses entirely. Nevertheless, these looking patterns allow us to draw three interesting conclusions. First, the different looking patterns of the two groups (Direct vs. Delayed Launching) suggest that the infants *did* utilize the information about the type of source of motion to interpret the object's behaviors. This result, we believe, clearly demonstrates that 12-month-old infants not only register whether an object is self-propelled or not but also interpret its behavior differently.

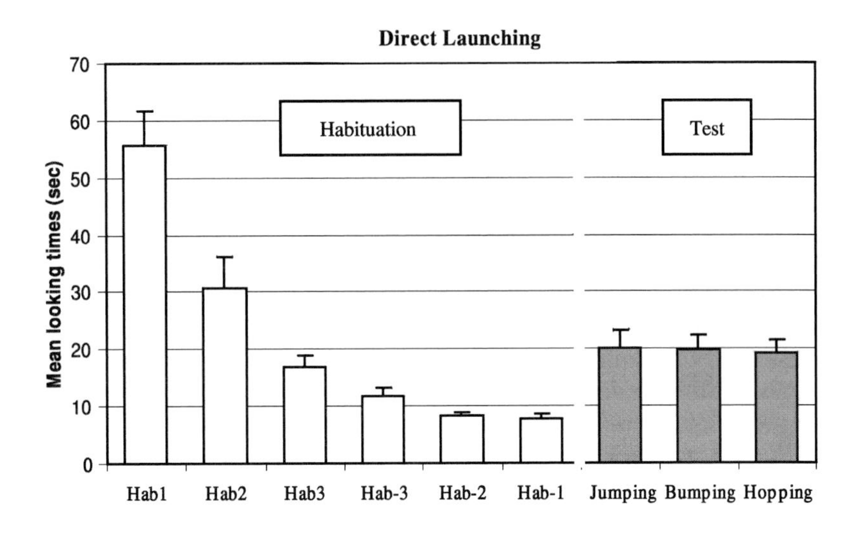

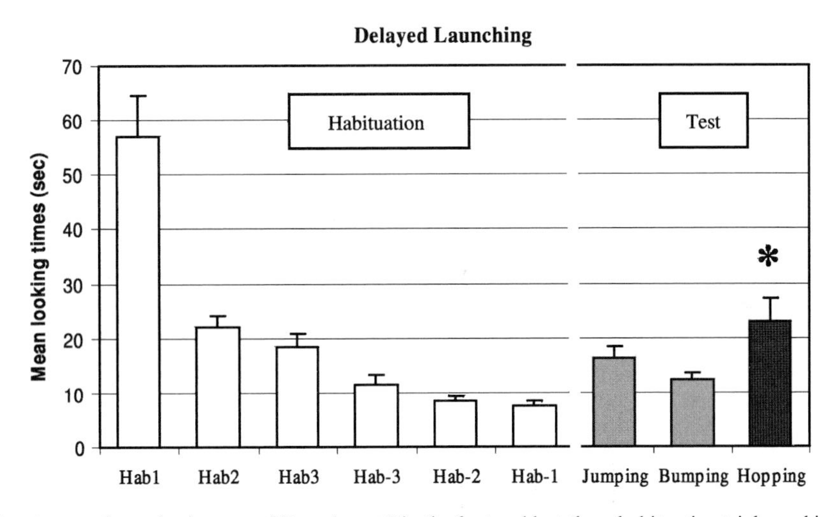

Fig. 4. Mean looking times and standard errors of Experiment 2 in the first and last three habituation trials, and in the three test events by the two groups (Direct and Delayed Launching) ($*p<0.05$).

Second, the most surprising aspect of the findings is the lack of difference in looking times in the "Direct Launching" group. The tacit assumption behind the intuitions about self-propulsion that many researchers seem to share is that infants are only willing to apply a non-physical (i.e., "intentional") interpretation to observed behavior as a kind of "last resort," i.e., only when their causal physical understanding fails to explain the behavior in question. This implies a kind of primacy of physical interpretations, and would suggest that infants should have no difficulty in inferring that an externally launched object will not change its path just because it is blocked by an obstacle. Our infants did *not* make this inference, however. Instead, they seemed to have developed no specific expectations about the subsequent behavior of an externally propelled object. There might be several explanations for this result, but the fact that the other ("Delayed Launching") group *did* differentiate among the test events suggests that causal physical reasoning does not enjoy the kind of primacy in the mind of young infants that is commonly assumed.

Third, the looking times in the "Delayed Launching" group also showed an unexpected pattern. The infants seem to have considered the unjustified hopping action inconsistent with the habituation events, but they accepted the two alternative outcomes (Jumping and Bumping) in the presence of the obstacle equally. While, clearly, we did not predict this result, in retrospect it seems to fit our teleological model in an interesting (though admittedly post-hoc) way. Remember that, lacking cues of behavioral variation that could guide goal assignment, the habituation event in the study was necessarily ambiguous, even if one is inclined to interpret it as a goal-directed action. In fact, the goal of the straight-moving action could equally be either (1) to approach and contact the target object (as we originally hypothesized), or (2) to *move away* from the "intruding" yellow object that first came up close, then established and maintained physical contact with the red object, which the latter eventually "tried to escape from" by moving away toward the other side. (Note that the intentional ascriptions provided here as "reasons" are purely for illustrative purposes and are clearly not necessary for the infant to make in order to attribute the well-formed goal of "moving away from the yellow object" to teleologically explain its observed behavior.) Note that the actions depicted by the two test events that elicited shorter looking times are consistent with these two alternative goals, respectively. Therefore, this looking pattern may reflect genuine goal attribution, but the attribution of two different goals.

It should be pointed out that the hypothesized ambiguity of goal-attribution might be manifested in these looking patterns in two different ways. On the one hand, it is possible that the infants could not decide during habituation which goal to attribute, but they were happy to accept either of them during the test phase. On the other hand, it is also possible that, while observing the habituation events, some infants came to the conclusion that the goal of the red object's action was "to go to the blue object" on the right, while other infants reasoned that its goal was rather "to move away" from the close proximity that the little yellow ball coming from the left has just established with it, and the average looking times, in fact, reflected the differential reactions of these two sub-groups.

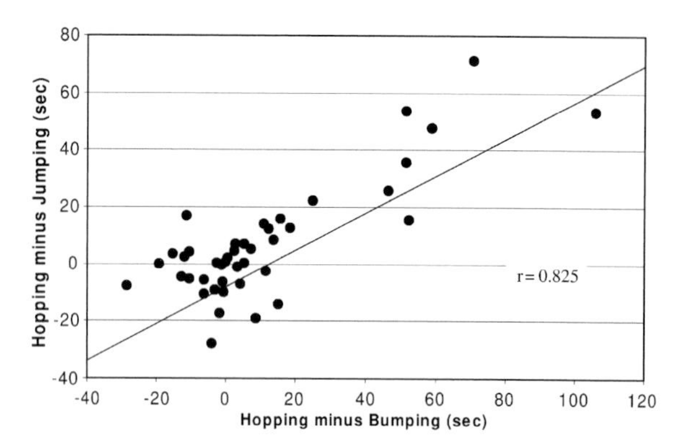

Fig. 5. The association between the difference in looking times between Hopping and Jumping (vertical axis) and Hopping and Bumping (horizontal axis) test events in the Delayed Launching condition for each participant.

There is a way to empirically contrast these two alternatives and to derive differential predictions from them about the clustering of the looking time data for two of the test events. While the first option predicts a positive correlation between the looking times elicited by the "Jumping" and the "Bumping" action test events, the second one predicts a negative correlation for the same two test events: the infants who attributed one specific goal should be surprised to see a behavior consistent with the other, and vice versa. We calculated the difference in looking times between these actions and the "Hopping" test event for each infant and correlated these measures with each other (see Fig. 5). The result ($r = 0.825$, $p < 0.0001$) showed a very strong positive correlation, suggesting that the infants did *not* commit themselves to one goal during habituation.

Conclusions

We have described three types of observable cues assumed to be utilized by infants in determining the scope of goal attribution to various behaviors, and investigated the potential inferential links between the concepts indicated by these cues. In particular, three possible functional links between AGENCY and GOAL-DIRECTEDNESS were hypothesized, differing from each other in the relative strength they attribute to this link. The "Mandatory link" hypothesis claimed that categorization as an agent is *required* for the identification of a goal-directed entity. The second, "Exclusive link" hypothesis stated that categorization of an entity as non-agent *precludes* the identification of a goal-directed entity. Finally, the third, "Probabilistic link" hypothesis assumed that agent categorization *biases* toward the identification of a goal-directed entity.

To test these hypotheses we have reported experiments in which we varied the availability of behavioral cues for self-propulsion that are hypothesized to perceptually identify an AGENT, and Context-sensitive cues, which are assumed to indicate a GOAL-DIRECTED entity. We found no support for the first two hypotheses. Twelve-month-old infants were able to interpret an observed action of an entity as goal-directed and predict its most efficient action in a changed situation even when the cue of self-propulsion was not present (Csibra et al., 1999) or when the entity was explicitly shown as non-self-propelled (Experiment 1). These findings indicate that infants do not need to perceive cues of self-propulsion to interpret behaviors as goal-directed actions. In other words, the categorization as AGENT is not a necessary prerequisite for identification of a GOAL-DIRECTED entity. In addition, perceiving that an entity is *not* self-propelled does not prevent infants from attributing a goal to its action. In other words, the categorization as non-agent does not preclude the identification as a goal-directed entity. The third, "Probabilistic link" hypothesis was, however, partly supported by our findings. In Experiment 2, we found that in an ambiguous situation (in which Context-sensitive cues are not available) perceiving self-propulsion does not by itself allow attributing specific goals, but makes it more likely that 12-month-olds will interpret the ambiguous action as goal-directed. Thus, the categorization of an entity as an AGENT can bias the interpretation of its behavior as GOAL-DIRECTED.

What are the implications of these findings for the theoretical approaches that have been proposed concerning the identification of the scope of naïve psychological reasoning? Regarding the question of the direction of possible inherent inferential links, both our current findings and those that we have reviewed suggest that none of these categories enjoy primacy over the others. In particular, we have demonstrated that prior commitment to conceiving an entity as an agent is neither sufficient nor necessary for applying goal-directed interpretation to the entity's behaviors. Therefore, the "Agent-first" and the "Human-first" views are not confirmed by evidence while the "No-primacy" account is corroborated.

Another related question on which theoretical approaches are divided is whether the nature of the selection process involved in the identification of the entities and events that belong to the domain of naïve psychology is *cue-based* or *principle-based.* We argued that the selection of events to be interpreted as GOAL-DIRECTED is not based on specific human featural or behavioral cues such as self-propulsion or context-sensitive behavioral changes, but on the successful application of the rationality principle. Our present findings confirm this position. Note that although Context-sensitive cues such as contingent behavioral adjustments have been suggested to trigger goal attribution, our previous studies (Gergely et al., 1995; Csibra et al., 1999) demonstrated that the presence of this cue in itself is not sufficient for the interpretation of an event as goal-directed. For example, in the experiment that provided evidence against the "Mandatory link" hypothesis (Csibra et al., 1999) behavioral adjustment was present in both the experimental and the control conditions: the height of the jumping action was contingently adjusted to the height of the rectangular figure. In the control condition, however, the rectangular figure did not form an obstacle between the two circles but was "hanging in the air," leaving the more efficient straight-line approach available. Thus, the jumping action — even though it was contingently varying with the situational changes — was not efficient and could not satisfy the principle of rational action. In the test phase of the control condition, infants did not display expectations about the circle's action in the changed situation, which suggests that they did not interpret its action as goal-directed during habituation. Context-sensitive cues therefore do not elicit goal attribution by themselves, although they can be considered as principle-driven cues that can trigger the *search* for a teleological interpretation. We thus propose that the same principle — namely the principle of rational action — is applied both for identifying entities that are potentially engaged in goal-directed behavior and for reasoning about their behavior. A similar principle-based single-knowledge system was proposed and argued for in the domain of physical objects (e.g., Carey and Spelke, 1994).

Experiment 2 suggested that the infants assume a probabilistic link between AGENT categorization and the identification of a GOAL-DIRECTED action. It is an empirical question whether this probabilistic relation is pre-wired (and acts as a "prior" providing statistical biases and constraints on Bayesian inferences), or learnt by discovering statistical associations between the corresponding cues in the world. By 12 months of age infants are likely to have accumulated sufficient amount of relevant evidence in the domain of action perception that could provide them ample informational basis from which to extract and represent such statistical associations. Such associations may be employed in probabilistic inferences, or can generate biases and constraints, to direct infants' teleological interpretations especially under conditions of underdetermination or ambiguity of the input.

Some findings by Biro and Leslie (2004) and Sommerville et al. (2005) can be viewed as supporting the assumption that gradual associative learning between observable behavioral and featural cues does take place during the first year. Both studies used similar designs: they introduced a "pre-training session" in which infants had the opportunity to learn to associate behavioral cues indicating goal-directedness with featural cues, and to rely on these associations to attribute a specific goal in the subsequent testing session when only the featural but not the behavioral cues were present.

The discovery of the strong correlation between cues of HUMAN features, cues of AGENCY, and cues of GOAL-DIRECTEDNESS in the world leads to the conjecture that the principles that guide psychological reasoning about goals need to be invoked mostly in the case of human intentional actions. Thus, relying, for example, on human/animate appearance to anticipate goal-directed intentional action from an entity is a useful strategy because most of the time it is justified and economical. Recognizing featural cues as opposed to behavioral cues does not require the observation of an entity over time. Featural cues can be processed fast and thus allow us to make quick predictions of the possible actions of an entity, and be prepared for an appropriate response.

On the one hand, while featural cues can function as predictive cues, they can also lead to incorrect conclusions about the domain that an entity belongs to. Imagine that you are in a forest and notice something hanging on a tree in front of you. It does not move and looks like a small piece of a tree branch like many others that you have passed during your walk. Suddenly this piece of wood falls, hurries behind the tree, and disappears in a hole in the tree trunk. You conclude that this was not a piece of wood after all but some kind of animal that got scared and wanted to hide from you. This example also illustrates that if behavioral cues exhibited by the entity are in conflict with our initial categorization based on featural cues, we change the categorization of the entity without hesitation to match to the behavioral cues, and predict the behavior of the entity on the basis of this new category. Moreover, such an experience has an impact on our future categorization of entities that have similar featural cues: these featural cues will now be associated with a new category.

On the other hand, cues of self-propulsion can also generate statistically based expectations that the entity will engage in goal-directed actions. However, we have no difficulty in refraining from reasoning in psychological terms about falling leaves, dropping faucets, or objects blown by the wind.

In conclusion, we propose that the domain of goal attribution is initially defined only by the applicability of its core principle (the principle of rational action, see Gergely and Csibra, 2003), and its ontology is not restricted to featurally or behaviorally defined entities such as persons or agents. During development, however, this purely defined domain becomes "contaminated" by the learnt associations between goal-directedness, human appearance and self-propelledness, allowing children to take the "teleological stance" toward people and other agents.

Acknowledgement

We wish to express our thanks to the Leverhulme Trust whose grant entitled "F/790 Causal versus Teleological Interpretations of Behaviour in Infancy" has supported the research reported in this paper.

References

Baron-Cohen, S. (1994) How to build a baby that can read minds: cognitive mechanisms in mindreading. Cah. Psychol. Cogn. Curr. Psychol. Cogn., 13: 1–40.

Bertenthal, B.I. (1993) Infants' perception of biomechanical motions: intrinsic image and knowledge-based constraints. In: Granrud C. (Ed.), Visual Perception and Cognition in Infancy: Carnegie-Mellon Symposia on Cognition. Erlbaum, Hillsdale, NJ, pp. 175–214.

Biro, S. and Hommel, B. (2007) Becoming an agent. Acta Psychol., 124: 1–7.

Biro, S. and Leslie, A.M. (2004) Interpreting actions as goal-directed in infancy. Poster presented at the 14th International Conference on Infant Studies (ICIS), Chicago, IL, USA.

Biro, S. and Leslie, A.M. (2007) Infants' perception of goal-directed actions. Development through cues-based bootstrapping. Dev. Sci., 10(3): 379–398.

Carey, S. and Spelke, E. (1994) Domain-specific knowledge and conceptual change. In: Hirschfeld L. and Gelman S. (Eds.), Domain Specificity in Cognition and Culture. Cambridge University Press, New York, pp. 169–200.

Csibra, G. (submitted) Goal attribution to inanimate agents by 6.5-month-old infants.

Csibra, G., Biro, S., Koós, O. and Gergely, G. (2003) One year old infants use teleological representation of actions productively. Cogn. Sci., 27(1): 111–133.

Csibra, G. and Gergely, G. (1998) The teleological origins of mentalistic action explanations: a developmental hypothesis. Dev. Sci., 1: 255–259.

Csibra, G. and Gergely, G. (2007) "Obsessed with goals": functions and mechanisms of teleological interpretation of actions in humans. Acta Psychol., 124: 60–78.

Csibra, G., Gergely, G., Biro, S., Koós, O. and Brockbank, M. (1999) Goal attribution without agency cues: the perception of "pure reason" in infancy. Cognition, 72: 237–267.

Farroni, T., Csibra, G., Simion, F. and Johnson, M.H. (2002) Eye contact detection in humans from birth. Proc. Natl. Acad. Sci. U.S.A., 99: 9602–9605.

Farroni, T., Johnson, M.H., Menon, E., Zulian, L., Faraguna, D. and Csibra, G. (2005) Newborns' preference for face-relevant stimuli: effects of contrast polarity. Proc. Natl. Acad. Sci. U.S.A., 102: 17245–17250.

Gergely, G. and Csibra, G. (2003) Teleological reasoning in infancy: the one-year-olds' naive theory of rational action. Trends Cogn. Sci., 7(7): 287–292.

Gergely, G., Nádasdy, Z., Csibra, G. and Biro, S. (1995) Taking the intentional stance at 12 months of age. Cognition, 56: 165–193.

Gibson, E.J., Owsley, C.J. and Johnston, J. (1978) Perception of invariants by five-month-old infants: differentiation of two types of motion. Dev. Psychol., 14: 407–416.

Golinkoff, R.M., Harding, C.G., Carlson, V. and Sexton, M.E. (1984) The infant's perception of causal events: the distinction between animate and inanimate objects. In: Lipsitt L.P. and Rovee-Collier C. (Eds.), Advances in Infancy Research, Vol. 3. Ablex, Norwood, NJ, pp. 145–151.

Heider, F. (1958) The Psychology of Interpersonal Relations. Wiley, Oxford.

Johnson, S., Slaughter, V. and Carey, S. (1998) Whose gaze will infants follow? The elicitation of gaze-following in 12-month-olds. Dev. Sci., 1(2): 233–238.

Johnson, S.C., Booth, A. and O'Hearn, K. (2001) Inferring the goals of a nonhuman agent. Cogn. Dev., 16: 637–656.

Jovanovic, B., Király, I., Elsner, B., Gergely, G., Prinz, W. and Aschersleben, G. (under review) The role of effects for infants' perception of action goals.

Kamewari, K., Kato, M., Kanda, T., Ishiguro, H. and Hiraki, K. (2005) Six-and-a-half-month-old children positively attribute goals to human action and to humanoid-robot motion. Cogn. Dev., 20: 303–320.

Kaufman, L. (1997) Infants distinguish between inert and self-moving inanimate objects. Paper presented at the Biennial Meeting of SRCD, Washington, DC.

Király, I., Jovanovic, B., Prinz, W., Aschersleben, G. and Gergely, G. (2003) The early origins of goal attribution in infancy. Conscious. Cogn., 12: 752–769.

Kotovsky, L. and Baillargeon, R. (2000) Reasoning about collisions involving inert objects in 7.5-month-old infants. Dev. Sci., 3(3): 344–359.

Legerstee, M. (1994) Patterns of 4-month-old infants responses to hidden silent and sounding people and objects. Early Dev. Parent., 3: 71–80.

Legerstee, M., Joanne, B. and Carolyn, D. (2000) Precursors to the development of intention at 6 months: understanding people and their actions. Dev. Psychol., 36(5): 627–634.

Leslie, A.M. (1994) ToMM, ToBy, and agency: core architecture and domain specificity. In: Hirschfeld L.A. and Gelman S.A. (Eds.), Mapping the Mind: Domain Specificity in Cognition and Culture. Cambridge University Press, New York, NY, USA, pp. 119–148.

Luo, Y. and Baillargeon, R. (2000) Infants' reasoning about inert and self-moving objects: further evidence. Manuscript in preparation.

Luo, Y. and Baillargeon, R. (2005) Can a self-propelled box have a goal? Psychological reasoning in 5-month-old infants. Psychol. Sci., 16(8): 601–608.

Mandler, J.N. (1992) How to build a baby: II. Conceptual primitives. Psychol. Rev., 99(4): 587–604.

Meltzoff, A.N. (1995) Understanding of the intentions of others: re-enactment of intended acts by 18-month-old children. Dev. Psychol., 31: 838–850.

Meltzoff, A.N. and Moore, M.K. (1997) Explaining facial imitation: a theoretical model. Early Dev. Parent., 6: 179–192.

Molina, M., Van de Walle, G.A., Condry, K. and Spelke, E.S. (2004) The animate-inanimate distinction in infancy: developing sensitivity to constraints on human actions. J. Cogn. Dev., 5(4): 399–426.

Pauen, S. and Träuble, B. (under review) How 7-month-olds interpret ambiguous motion events: Domain-specific reasoning in infancy.

Poulin-Dubois, D., Lepage, A. and Ferland, D. (1996) Infants' concept of animacy. Cogn. Dev., 11: 19–36.

Poulin-Dubois, D. and Shultz, T.R. (1990) The infant's concept of agency: the distinction between social and nonsocial objects. J. Genet. Psychol., 151: 77–90.

Premack, D. (1990) The infant's theory of self-propelled objects. Cognition, 36: 1–16.

Rochat, P., Morgan, R. and Carpenter, M. (1997) Young infants' sensitivity to movement information specifying social causality. Cogn. Dev., 12: 441–465.

Rovee-Collier, C., and Sullivan, M.W. (1980) Organization of infant memory. J. Exp. Psychol. Hum. Learn. Mem., 6, 798–787.

SEQ CHAPTER Saxe, R., Tenenbaum, J.B. and Carey, S. (2005) Secret agents: inferences about hidden causes by 10- and 12-month-old infants. Psychol. Sci., 16: 995–1001.

Schlottman, A. and Surian, L. (1999) Do 9-month-olds perceive causation-at-a-distance? Perception, 28: 1105–1114.

Scholl, B. and Tremoulet, P.D. (2000) Perceptual causality and animacy. Trends Cogn. Sci., 4(8): 299–309.

Shimizu, Y.A. and Johnson, S.C. (2004) Infants' attribution of a goal to a morphologically unfamiliar agent. Dev. Sci., 7(4): 425–430.

Sommerville, J.A., Woodward, A.L. and Needham, A. (2005) Action experience alters 3-month-old infants' perception of others' actions. Cognition, 96: B1–B11.

Spelke, E.S., Phillips, A. and Woodward, A.L. (1995) Infants' knowledge of object motion and human action. In: Sperber D., Premack D. and Premack A.J. (Eds.), Causal Cognition: A Multidisciplinary Debate. Clarendon Press, Oxford, pp. 44–78.

Tremoulet, P.D. and Feldman, J. (2000) Perception of animacy from the motion of a single object. Perception, 29(8): 943–951.

Watson, J.S. (1979) Perception of contingency as a determinant of social responsiveness. In: Thoman E. (Ed.), The Origins of Social Responsiveness. Erlbaum, Hillsdale, NJ, pp. 33–64.

Woodward, A. (1998) Infants selectively encode the goal object of an actor's reach. Cognition, 69: 1–34.

Woodward, A. (1999) Infants' ability to distinguish between purposeful and non-purposeful behaviors. Infant Behav. Dev., 22(2): 145–160.

Woodward, A., Sommerville, J.A. and Guajardo, J.J. (2001) How infants make sense of intentional action. In: Malle B., Moses L. and Baldwin D. (Eds.), Intentionality: A Key to Human Understanding. MIT Press, Cambridge, MA, pp. 149–169.

C. von Hofsten & K. Rosander (Eds.)
Progress in Brain Research, Vol. 164
ISSN 0079-6123

CHAPTER 18

Seeing the face through the eyes: a developmental perspective on face expertise

Teodora Gliga* and Gergely Csibra

Centre for Brain and Cognitive Development, School of Psychology, Birkbeck College, University of London, Malet Street, London WC1E 7HX, UK

Abstract: Most people are experts in face recognition. We propose that the special status of this particular body part in telling individuals apart is the result of a developmental process that heavily biases human infants and children to attend towards the eyes of others. We review the evidence supporting this proposal, including neuroimaging results and studies in developmental disorders, like autism. We propose that the most likely explanation of infants' bias towards eyes is the fact that eye gaze serves important communicative functions in humans.

Keywords: face recognition; gaze perception; expertise; development; infancy; amygdala

Humans are experts at face processing

Almost all of us are experts in face processing by the time we reach adulthood. We can recognize thousands of individuals in diverse conditions of distance, luminosity, or orientation (Maurer et al., 2002), a performance not yet attained by any automatic face recognition system (Zhao et al., 2003). Experimental studies have also shown that these perceptual capacities are not applied to all object categories: while we can tell apart new faces that are very similar to each other even if they are presented for a very short time (Lehky, 2000), we cannot do the same for other objects (Bruce et al., 1991). This extraordinary face expertise is thought to reflect specialized perceptual and neuronal mechanisms. For example, for faces, as opposed to other visual stimuli, we encode not only the composing elements (features) but also their relative distances. This type of encoding is referred to as *configural* (as opposed to *featural*) processing, and its impairment can have dramatic consequences on face recognition (for a review see Maurer et al., 2002). Face perception is also thought to engage a selective network of brain areas, most notably within the fusiform face area (FFA) (Kanwisher et al., 1997; Haxby et al., 2001).

While there is still controversy on whether people can develop similar expertise for other classes of objects (McKone and Kanwisher, 2005; Bukach et al., 2006), it is widely accepted that faces are the most common and probably the earlier developing domain of expertise. From a functional point of view, the special status of face perception makes sense because only the recognition of individuals allows us to identify our kin, to keep track of friends and enemies, and to maintain group structure and hierarchy. Acquiring specialized face processing mechanisms, either by natural selection during phylogeny, or by intensive learning during ontogeny, seems to be a necessity.

*Corresponding author. Tel.: (+44) 207 631 6322; Fax: (+44) 207 631 6587; E-mail: t.gliga@bbk.ac.uk

DOI: 10.1016/S0079-6123(07)64018-7

This conclusion appears to be trivial, but it tacitly relies on the intuition that faces provide the main source of perceivable individual differences to be employed in recognition of others. People differ from each other in a number of nonvisual characteristics (e.g., voice, odor), but also in a number of nonfacial visual aspects. We have differently proportioned bodies and body parts (e.g., hands), and we also move in very distinct ways. Why do we then rely so much on the face in telling people apart? One possible reason could be that faces provide the richest cues of identity. Faces can differ in the morphology of their internal elements, and also in the relative distance among these elements and between the elements and the face contour. Faces also differ in complexion, and in eye and hair color. It is possible that human bodies and body parts simply do not offer as much variability as do faces, and that we would not be able to tell apart as many people on the basis of bodily cues as we can discriminate based on faces.

One way to demonstrate that other body parts can also be used for recognition of individuals is to show that humans succeed in determining a person's identity based on nonfacial cues when they are explicitly asked to do so. Addressing this question, researchers have focused on both dynamic and structural properties of bodies. Bodily motion is rich in information, as people can tell someone's identity, gender, or age only from point light displays of his or her movement (Hill and Johnston, 2001). Other kinds of information, like kinship, are probably poorly encoded in movement and, more importantly, motion cues are useless when the person is stationary. When focusing on the structural aspects, studies have shown that the human body configuration is processed in a similar way to the configuration of a human face (Reed et al., 2003), and recruits similar brain mechanisms (Stekelenburg and de Gelder, 2004; Gliga and Dehaene-Lambertz, 2005). Brain areas specialized for processing body structure or motion have been described in the vicinity of face-specific cortical areas (Downing et al., 2001; Astafiev et al., 2004). These areas seem to encode subtle structural differences in a selective manner (for example, discriminating between two different hands but not between two motorcycle parts or noses), just as face-selective areas do (Urgesi et al., 2004). None of these studies, however, addressed directly the question of individual recognition from structural bodily cues. Thus, it is not yet known whether people could use nonfacial body-related information for identification, if necessary.

Another way of comparing facial and bodily information for the purpose of individual recognition is to use automatic systems of artificial intelligence. Automatic detection and identification of people from visual scenes is a very active domain of research, mainly because of increased interest in visual surveillance systems. The level of performance of these systems, when using either faces or bodies, could tell us whether one provides potentially more information than the other. Unfortunately, this question is unanswered for the moment, as most of these systems concentrate on face recognition. It is not clear whether this imbalance represents an objectively verified superiority of faces in individual recognition or just the bias of the creators of these systems, who, being human, intuitively turn to the source that they predominantly use for identifying their conspecifics.

Although there is little evidence that body structure is used for identification by people, the potential to do that seems to exist, at least at a neuronal level, as shown by the above mentioned studies (Stekelenburg and de Gelder, 2004; Gliga and Dehaene-Lambertz, 2005). In parallel, studies have shown that, following a long and intense training, people can develop various kinds of expertise: they can recognize individuals of other species: dogs, birds, sheep (Diamond and Carey, 1986; Tanaka and Taylor, 1991; McNeil and Warrington, 1993), or other classes of objects like cars or artificially created "Greebles" (Gauthier and Tarr, 1997; Gauthier et al., 2000). Thus, the question remains: if it is plausible that people become body experts, why does this expertise not develop? What makes faces so special that they become the primary source of recognition of individuals in humans?

In the following sections, we shall offer a developmental answer to this question. We propose that people become face experts because, from birth, children pay more attention to, and hence get more experience with, faces than with other body parts. This preferential treatment of face, we shall suggest, is due to a special interest in the eyes, which, in humans, function as an important communication device.

The interest in faces develops early

The first signs of a specialization for face processing can be seen very early, during the first days of life. In an influential study, Johnson et al. (1991) showed that, only a few hours after birth, infants follow a face-like schematic pattern more persistently than other patterns. This preference decreased between 1 and 2 months of age, but reappeared later under different presentation conditions. The movement of the face stimulus was not necessary at this later age, and photographs of faces were better at triggering the preference than were schematic faces (Johnson et al., 1991; Mondloch et al., 1999). These results led Morton and Johnson (1991) to propose the CONSPEC–CONLEARN model. The model hypothesized that innate orienting mechanisms are responsible for the initial face following in neonates (the CONSPEC mechanism), which were possibly implemented in subcortical structures (the superior colliculus, the pulvinar, or the amygdala). The subcortical origin of this face preference is supported by the special conditions in which it can be triggered — the stimuli have to be in motion in the periphery or presented in the temporal visual field (Simion et al., 1998). A couple of months later, the visual cortex would start playing more and more of a role in face processing, allowing finer face encoding and thus better face recognition (the CONLEARN mechanism).

This interpretation of the visual properties that drive newborns' face preference has recently been challenged. Macchi Cassia et al. (2004) showed that the exact positioning of the facial elements is not important in triggering a preference, as long as the display is symmetric and there are more elements in its upper part (the "top-heavy" hypothesis). Moreover, a general preference for "top-heavy", compared to "bottom-heavy" stimuli, even in the absence of any resemblance to faces, was found in the newborn. These authors concluded that a nonspecific bias, induced by an upper-visual-field advantage in visual sensitivity, drives the initial face preference. Having more elements in their upper part, faces take advantage of this initial bias by capturing newborns' attention. Interestingly, this bias seems to be still present in adulthood, as the FFA is activated more by "top-heavy" stimuli, similar to those used in the infant studies (Caldara et al., 2006).

However, one can also argue that this bias evolved specifically for face processing. Whether or not a visual mechanism acts as a face-preference bias depends not on a goodness-of-fit function to an ideal face template, but on its efficiency in drawing infants' attention to faces in a natural environment. If a bias toward "top-heavy" stimuli successfully selects faces in the species-typical environment of a human newborn without generating too many false alarms, then it is as domain-relevant as a preference for stimuli matching a face template, and they share a common function. Note also that, as we shall see later, "top-heaviness" in a purely geometrical sense does not explain all aspects of newborns' preferences for face-like patterns (Farroni et al., 2005).

Whatever the exact "filters" that assist newborns in finding faces, these innate (or early functioning) biases have always been seen as being beneficial for the infant. Because they bring and maintain human faces in the infant's central visual field, they could be responsible for increasing his/her visual experience with human faces, and thus boosting the perceptual and cortical specialization for this class of objects. A direct influence on the cortical processes, through direct neuronal connections from the subcortical structures, might also be involved in the initial face tracking. These connections would allow the cortex to receive the relevant visual information before the cortical pathways are fully developed (Johnson, 2005).

The speed with which infants acquire knowledge about face structure contrasts with their human body knowledge. It is well established that infants can quickly learn the visual properties of various objects. They can group together different exemplars of an animal category based on both facial and bodily features (Quinn et al., 2001). Surprisingly, however, infants manifest very little knowledge of the human body (Slaughter et al., 2002). If they do display some knowledge, it is when the face is removed from the body (Gliga and Dehaene-Lambertz, 2005). What is in the face that captures infants' attention from the start?

Face expertise is driven by an early interest in eyes

There is one particular feature of faces that seems to be exceptionally effective in triggering young infants' attention — the eyes, especially when they appear to look directly at the infant. Newborns are shown to prefer faces with eyes open to faces with closed eyes (Batki et al., 2000) and, when the eyes are open, they prefer a direct gaze to an averted one (Farroni et al., 2002). In addition, faces with direct gaze are more thoroughly processed, as suggested by the stronger neuronal responses they evoke (Farroni et al., 2002, 2004a) and by the better recognition they produce (Farroni et al., 2006a).

Based on these and other findings that will be reviewed in the following sections, we propose that it is the interest in eyes, and especially in eye gaze, that triggers newborns' orientation towards faces and infants' continuing fascination with this particular body part. Knowing how plastic the face processing neural networks are in the first months, we expect that the consequences of spending more time exploring the eyes and the surrounding area are long lasting, leading to the preferential use of the face for identification in later life. We believe that there is enough evidence to support this hypothesis, and we shall devote the rest of this chapter to reviewing the relevant findings.

Early interest in eyes

Infants are sensitive to eye gaze from birth. Farroni et al. (2002) have shown that 3-day-old newborns look longer, and reorient more frequently, to a face with direct gaze than to a face with averted gaze when these stimuli are shown side by side. The salience of direct gaze could be partly explained by the unique morphology of the human eye. Human eyes are wider in the horizontal direction, expose much higher proportion of sclera than the eyes of other primates, and the sclera is completely white (Kobayashi and Koshima, 1997). Accordingly, direct gaze in a front view face creates an area of high contrast, which could attract infants' attention (Fig. 1b). Further experimental findings suggest that newborns are indeed sensitive to the specific luminosity contrast that consists of dark regions on a white background. Farroni et al. (2005) demonstrated that the preference for the upright schematic face pattern is lost if a contrast-reversed stimulus is presented (white dots on a dark background), but regained if eye-like elements are used instead of the simple white dots (Fig. 1a). When the head is oriented to one side, direct and averted gaze are hardly distinguishable on the basis of the contrast pattern within the eyes only (Fig. 1c). As expected, in this case no preference is recorded in newborns (Farroni et al., 2006b). Simple sensitivity to a white-dark-white pattern is nevertheless not sufficient to explain the initial advantage of direct gaze because inverting the face disrupts this preference (Farroni et al., 2006b). Thus, a complex interplay between gaze perception and structural face perception drives infant's preferences from the very beginning (see Farroni et al., 2005, for a detailed discussion).

A few months later, the sensitivity to the mutual eye gaze is still preserved but it is further refined. Faces with direct gaze, compared to faces with averted gaze, elicit a stronger neurophysiological response (the N290, the event-related potential component that is the likely precursor of adults' face-specific N170) in 4-month-old infants (Farroni et al., 2002). At this age, the same result was also obtained when a three-quarters view of the face was used (Farroni et al., 2004a). Considering the absence of preference for the direct gaze in newborns when presented on 3/4 view faces, the results obtained in 4-month-old suggests that by this time infants have developed the nontrivial capacity of "reading" gaze direction on the basis of both eye and head orientation. The direct gaze in a frontal-view face is nevertheless still privileged, as only this condition, but not the three-quarters view, evokes a frontal burst of gamma-band

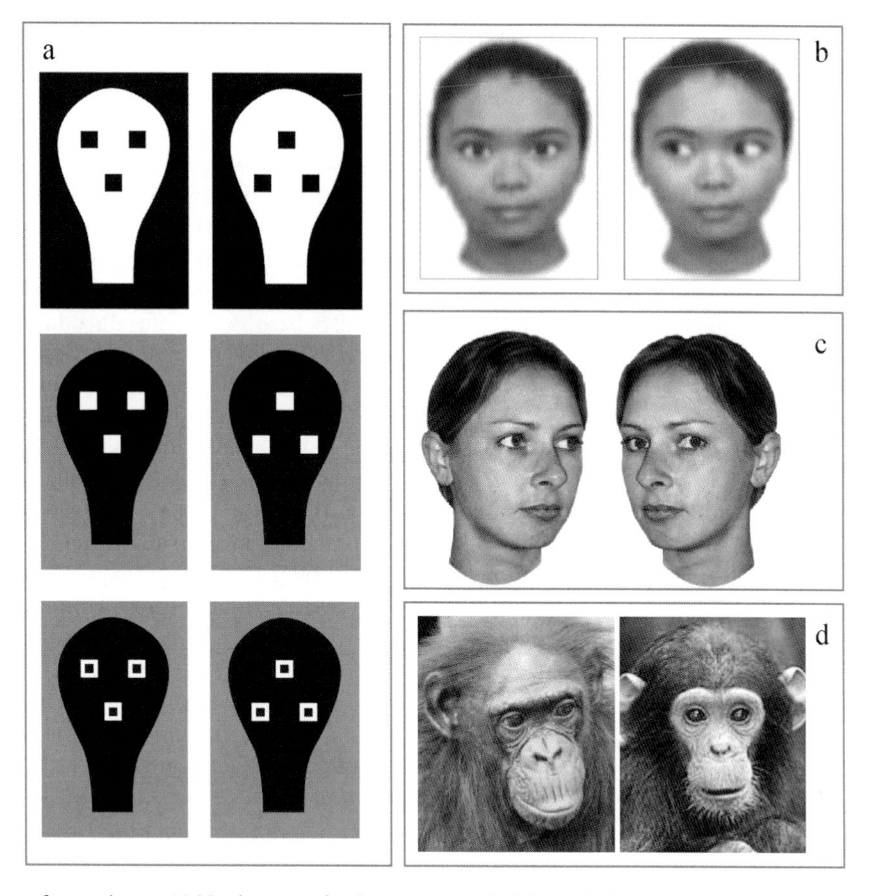

Fig. 1. The saliency of mutual gaze: (a) Newborns prefer the pattern on the left within the top and the bottom pairs, but not within the middle one (Farroni et al., 2005); (b) Mutual and averted gaze in a frontal face, as seen by a 4-months-old infant (Farroni et al., 2002). (c) To distinguish between averted and mutual gaze in 3/4 view faces, gaze direction and head direction have to be integrated (Farroni et al., 2004a). (d) A bonobo face with averted gaze (left) and chimpanzee face with direct gaze (right). The absence of the white sclera makes it difficult to "read" the gaze direction.

oscillatory activity[1] (Grossmann et al., 2007). Frontal activity (especially in the right dorsal medial prefrontal cortex) is associated in adults with processing social signals directed to the self (direct gaze, calling their name, see Kampe et al., 2003). Together these studies suggest that while 4-month-olds already recognize the equivalence between direct gaze configurations, whether they are located in a frontal or averted face, the latter stimulus has not yet gained the same social significance for them as the canonical eye-contact pattern.

Studies around the same age reveal not only excellent gaze discrimination capacities but also a positive emotional response to establishing mutual gaze. Infants remain engaged for longer, and smile more, when their conversation partner is looking at them (Hains and Muir, 1996; Symons et al., 1998). They do not only "enjoy" mutual gaze but also try to elicit it. According to Blass and Camp (2001), in the presence of an actor who looks away or does not interact with them, 12-week-old "infants made considerable efforts, including

[1]Induced gamma-band (>24 Hz, characteristically around 40 Hz) oscillations have been used for studying infants' brain responses to complex visual stimuli (Csibra et al., 2000; Kaufman et al., 2003); for more details, see (Csibra and Johnson, 2007).

smiling, gurgling, and flirting, to establish contact" (p. 771). Direct gaze is an extremely powerful stimulus, which, we believe, attracts infants' attention to faces (1) directly through perceptual salience and (2) indirectly through positive feedback mechanisms mediated by emotional systems. Additional behaviors, like infant-directed speech (motherese), facial expressions and movements produced in contingency with infant's own vocalizations or with his own movements accompany the establishment of a the mutual gaze during mother–infant interactions and further strengthen the saliency of the face (Werker and McLeod, 1989; Cooper and Aslin, 1990; Striano et al., 2005).

It is not only mutual gaze that infants are interested in during their first year of life. When an object is present in the visual scene, 9-month-olds prefer, and show enhanced electrophysiological responses to, gaze shifts towards the object (Senju et al., 2006, 2007). Although this preference makes older infants look towards the object and thus away from the face, such a tendency could also contribute to the developing face expertise by motivating them to search for such object-directed cues on others' face. This argument will be discussed at length in the last section of the paper.

Earlier neuronal specialization for eye processing

Five-month-old infants can discriminate direct gaze from gaze averted by only 5 visual degrees, i.e., when a person is looking at the infant's ear instead of her eyes (Symons et al., 1998). Strikingly, at around the same age, infants' face recognition capacities are by far not as spectacular, as we will see later. Do face-processing mechanisms develop slower than gaze processing mechanisms in the first months of life?

Starting around 4 months, the eyes are the face elements that evoke the strongest event-related potentials (ERPs) response (Gliga and Dehaene-Lambertz, 2007), and this specific signature accompanies eye processing along further development (Taylor et al., 2004). ERP studies comparing the correlates of eye and face processing have brought evidence for an earlier specialization of the eye-processing networks compared to those processing face-general properties. The maturation and/or specialization of neuronal networks is generally associated with a decrease in response latency in different modalities (Taylor and Baldeweg, 2002). This effect could be either due to a general improvement in neuronal conductivity or to more efficient neuronal architectures (Nelson and Luciana, 1998). The shortening of latencies can also be observed for face-induced ERP components. In adults, face and eye perception are associated with a temporal-occipital negativity, at a peak latency of around 170 ms (N170). The generators of the whole-face-evoked and the eye-only-evoked responses are distinct, as suggested by their different topographies on the scalp, and by dipole source localization (McCarthy, 1999; Shibata et al., 2002). As we go back in time from adulthood to early infancy, the latency of the face-evoked "N170" increases by at least 120 ms (Taylor et al., 2004). In contrast, the latency of the eye-evoked response shifts much less (see Fig. 2 and Gliga and Dehaene-Lambertz, 2007). Since these developmental differences are unlikely to be accounted for by general maturation of the brain, which would affect both face- and eye-evoked responses, they require a functional explanation. We suggest that the neural mechanisms responsible for gaze-processing, or at least those involved in detecting mutual gaze, develop earlier than the ones involved in general configural processing, probably because the second task requires more perceptual expertise to accumulate. Thus, by being active early enough in development, eye-orienting mechanisms could bias infants towards developing perceptual face expertise rather than perceptual body expertise.

Direct evidence that eye-processing networks are in operation from the first months of life comes from a study using the repetition-priming paradigm. This paradigm is based on the observation that repetition of stimuli induces diminished activation in the cortical areas that encode their common property. When used with functional magnetic resonance imaging (fMRI), this approach gives access to a finer spatial delineation of the neuronal processes than traditional paradigms (Grill-Spector et al., 1998; Naccache

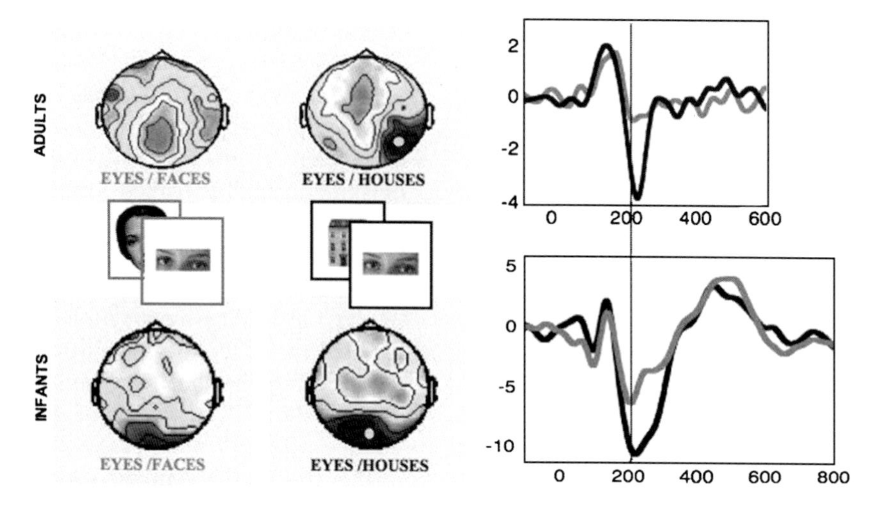

Fig. 2. Targeting eye-selective processing by repetition suppression. Thirty eye images were randomly presented amongst face or house images. The ERPs to eyes were reduced if presented in a face context but not in a nonrelated house context, in both adults and 4-month-old infants. The response suppression started at similar latencies in both ages. The voltage distribution over the scalp in the two conditions (left) is presented along with the curves recorded by the electrodes indicated by a white dot on the voltage maps (light gray — eyes/faces; black — eyes/houses) (Adapted with permission from Gliga and Dehaene-Lambertz, 2007).

and Dehaene, 2001). Repetition-induced ERP effects are similar but occur in the temporal domain. Using this paradigm, Gliga and Dehaene-Lambertz (2007) found a suppression of the eye-evoked responses to the repetition of different human eyes, both in adults and in 3-month-old infants. The latency of the amplitude reduction (~200 ms) was strikingly similar in the two populations (Fig. 2). On the other hand, the same study showed a different pattern of results when structural aspects of faces had to be processed. While adults process human heads in a view-invariant manner, 3-month-old infants do not, as shown by the absence of neuronal adaptation when different face views (front view and profile) are repeated (Gliga and Dehaene-Lambertz, 2007).

Eye detection mechanisms seem to be adult-like at 4 months of age, while processing structural aspects of faces are not. These results add to a number of others showing that "face expertise" (as opposed to "gaze expertise") develops slowly from infancy during childhood and through the teen-age years (Carey, 1992; Taylor et al., 2004). In a recognition memory task where 36 face photographs had to be memorized and retrieved a few minutes later, 6-year-old children performed barely better than chance, while adults' performance was at ceiling (Carey, 1992). The slow improvement of face expertise depends not only on experience with human faces but also on the general improvement of basic visual skills up until adolescence. A recent study by Mondloch et al. (2006) showed that the discrimination among both human and monkey faces improves further between 8 years of age and adulthood.

The studies reviewed in this section indicate that recognizing people by their faces is not an easy task for children, while expertise in eye detection seems to start in early infancy. Because structural face processing is heavily experience-dependent, the emergence of this ability will be susceptible to be influenced by the early developing mechanisms that mediate orienting attention to faces, like the ones that govern eye detection. These studies, however, do not provide a proof for a direct effect of looking for eyes (and especially eyes with direct gaze) on learning about face properties. The next section will offer such evidence.

Learning about faces when looking at the eyes

To demonstrate that infants' interest in eyes and eye gaze has a modulatory effect on face perception, one has to show that facial features are encoded better when the person's eyes are open, and even better when she looks directly at the observer. Such evidence for the role of direct gaze in face recognition has recently been provided. Farroni et al. (2006a) showed that 4–5-month-old infants memorize a stranger's face better if they are presented with direct gaze during the encoding phase. Interestingly, when the experiment started with the presentation of faces with direct gaze, face recognition was better even for the faces with averted gaze presented in a subsequent block. This suggests that direct gaze does not simply induce a local increase in the efficiency of encoding but has a more general and longer lasting attentional effect. In this respect, direct gaze could act similarly to "physiological" learning enhancing factors, like sucrose intake. Blass and Camp (2001) found a strong interdependence between sucrose intake and mutual gaze, showing that infants learn better a stranger's face only when both factors were present.

While the study by Farroni et al. (2006a) shows that mutual gaze has a direct effect on face recognition, it also leaves a lot of questions unanswered. One such question concerns the spatial and temporal extent of these facilitatory effects. Does the direct gaze enhance the encoding of the whole face (thus contributing to the development of configural processing, for example) or would the eyes be privileged? In a habituation paradigm, 4- and 6-month-old infants showed dishabituation when the upper half face changed but not when lower half of the face changed, whereas 8-month-olds detected changes in both conditions (Zauner and Schwarzer, 2003). It is thus possible that the interest in eyes biases further face recognition towards using the upper part of the face, at least in the first months of life. More facial information will be incorporated later on as a result of more exposure or of improved face scanning mechanisms.

There is nevertheless evidence that the eyes are used for recognition more than other face elements even in adulthood. The relative weight given to eyes or any other face part for individual recognition can be estimated by masking various face regions and assessing whether the removal of a region reduces recognition performance drastically. Using this approach, researchers found that, in adults, the head outline, the eyes and the mouth are the most relevant features for face recognition (Schepherd et al., 1981). A new and more precise masking technique (called 'Bubbles') was designed by Gosselin and Schyns (2001) for the same purpose. The procedure involves covering faces with an opaque layer with randomly distributed "holes" in it. The intersection of the stimuli that resulted in successful recognition would highlight the face regions that observers primarily use for identification. The results of a study employing this technique (Vinette et al., 2004) pointed again to the eyes as the primary features used by adult participants when performing a recognition task. Furthermore, these authors were also interested in whether the reliance on eyes was due to the fact that they carry more structural information than other face elements — the question that we asked earlier for faces with respect to the rest of the body. When an automatic template-matcher was presented with the same images as the human observers, it used both the eye region and the outer head contour for recognition. Thus, people seem to be preferentially guided by the eyes even if they could have also used other face regions for identification. The 'Bubbles' technique was recently employed in a study of 7-month-old infants' recognition of familiar and strangers' faces (Humphreys et al., 2006). Consistent with the results obtained with adults, infants manifested a preference for their mother's face only when the eyes were visible in the masked faces.

Another prediction derived from the increased interest in direct gaze on a frontal view face is that the frontal orientation should be the easiest to encode into memory. The opposite prediction could also be made based on the richness of information present in different face orientations. It is generally accepted that the 3/4 view of the face conveys much more identity information than the profile or the frontal-view because it makes both the configuration of the internal elements and the 3D shape properties visible (Baddeley and Woodhead, 1983; Vuilleumier et al., 2005). When these two predictions were contrasted, no advantage of the 3/4 view over the frontal view was found in a recognition task (Liu and Chaudhuri, 2002). While this result does not support one hypothesis over the other, it nevertheless suggests that our face processing system is not only driven by the available information content and optimal bottom-up strategies. Similar studies should also be performed with infants, where we would predict better recognition of front-view faces compared with other orientations.

Further studies will have to explore the impact of the eye preference on learning face properties, and also to address the question of whether the eye preference in early infancy has the potential to shape a life-long face expertise. It is known that the first year of life is a period of increased activity-dependent plasticity in face processing. The variability within the physiognomies of the faces that infants encounter tunes their face representations, allowing them to better recognize those types of faces that they are most likely to come across in the first months of life. Infants gradually become better at processing their own species' faces while losing their ability to tell faces of other species apart (Pascalis et al., 2002). Three-month-old infants brought up amongst Caucasian caregivers are better at discriminating Caucasian than Asian faces (Sangrigoli and De Schonen, 2004), and they prefer to look at their own race faces (Kelly et al., 2005). This own-race preference appears to be a robust phenomenon since it has been replicated and extended under various conditions (Kelly et al., 2005; Bar-Haim et al., 2006), though this advantage disappears if the infants are given a brief "training" with faces of another race (Sangrigoli and De Schonen, 2004). Also, infants have been shown to exhibit an early face gender preference, depending on who their primary caregiver is — the mother or the father (Ramsey-Rennels and Langlois, 2006). Further evidence on the impact of early experience with faces on life-long face processing capacities is provided by studies of children having suffered from congenital cataract during their first year of life. These individuals, years after the cataracts have been removed, are still impaired at face recognition (Le Grand et al., 2003).

We have seen that the interest in eyes facilitates face processing at an age where visual experience actively shapes the perceptual space. Brain imaging studies that target the neural structures involved in eye-gaze processing during development might also shed light on the nature of the link between gaze preference and face expertise. For the time being, no such studies have been carried out in infants. Nevertheless, in the following paragraphs, we will consider a number of studies with adults and children, which, we think, are extremely important for interpreting the findings regarding the early development of face and gaze processing.

The role of the amygdala in face and eye perception

In adults, the amygdala has repeatedly been associated with gaze processing and also with emotion perception from faces and eyes (Adolphs et al., 2002; Adams et al., 2003; Vuilleumier and Pourtois, 2007). Both the fusiform gyrus and the amygdala increase their activity when people look at a face with a direct gaze compared with a face with averted gaze (Kawashima et al., 1999; George et al., 2001, but see Hoffman and Haxby, 2000, for absence of modulation of the fusiform gyrus by the direction of the gaze). Similarly to infants, adults are also faster when having to make judgments about faces with direct gaze, even when their attention is directed away from the gaze information (Macrae et al., 2002; Hood et al., 2003). This effect is not due to increased contrast of the eyes with direct gaze, as it is present, and even stronger, when faces are presented with averted head orientation (Vuilleumier et al., 2005). Moreover, George et al. (2001) did not find an activation difference at occipital areas between the faces with direct and averted gaze, suggesting that direct gaze in adults does not facilitate general visual processing.

More interestingly, an increased coupling between the activation of the amygdala and the fusiform gyrus was also observed in the above studies (Kawashima et al., 1999; George et al., 2001). However, these results cannot inform us about the directionality of this coupling. An involvement of the amygdala in the modulation of fusiform gyrus activation, if shown, would be very interesting from a developmental point of view. Could the amygdala play a role in face processing in infancy? Its selective sensitivity to low spatial frequencies (Vuilleumier et al., 2003) suggests that the poor visual acuity in infancy would not be a limiting factor. Moreover, a recent study (Whalen et al., 2004) showed that a subtle difference in the dark versus white proportion of the eye region, as in the contrast between fearful or surprised and happy facial expressions, can modulate amygdala activity. As we have seen, newborns pay a special attention to high-contrast elements and their contrast polarity within face-like patterns (see Fig. 1a, Farroni et al., 2005). Indeed, the amygdala, together with other subcortical structures, is thought to be involved in enhancing infants' attention to socially relevant stimuli and in modulating the activity of cortical structures that process these stimuli (Johnson, 2005).

The exact role played by this structure in face processing is still under debate. A case study of a patient with amygdala damage, who showed impaired fear recognition but performed as well as controls when she was explicitly instructed to look at the eyes, suggests that the amygdala helps orienting the attention towards the eyes without being necessary for further processing (Adolphs et al., 2005). The amygdala, together with other subcortical and cortical structures, among which are the medial and orbital parts of the prefrontal cortex, is also known to be involved in stimulus-reward learning (Baxter and Murray, 2002). Within this circuit, the amygdala might thus mediate the establishment of the positive feedback loop that makes infants enjoy and seek mutual gaze.

At present, there is no suitable neuroimaging technique to study amygdala activation in human infants. However, one particular condition — autism — offers a unique opportunity to assess the importance of eye gaze and the role of the amygdala during development.

Deficits of gaze and face perception

Another way of testing the potential link between the early tendency of orienting towards eye gaze and the later developing face processing expertise is to look at populations in which the initial interest in eyes may not be present. In this case, one would expect a less developed face expertise or even the development of alternative kinds of expertise for identification of individuals. Children with autism provide a relevant case for this purpose.

Amongst the first signs of autism is the reduced time children spend looking at people's faces or engaging in mutual gaze (Osterling and Dawson, 1994). Based on retrospective analysis of home video recordings, the indifference towards faces and eyes is described as early as the first year of life, persists during the childhood (Dalton et al., 2005), and is still present in adults (Klin et al., 2002). In contrast to typically developing children, children with autism do not show an advantage in detecting faces with direct gaze versus averted gaze (Senju et al., 2005).

Despite a disinterest in faces and in mutual gaze, autistic individuals are not "blind" to eye direction. On the contrary, their physiological measures show a bias for mutual eye gaze, manifested as a stronger ERP or skin conductance response (Kylliainen et al., 2006a, b). Thus, the deficit may lie not in perceiving mutual gaze as a special stimulus but in attributing the correct social relevance (i.e., positive valence) to this stimulus. The lack of orienting to other social signals in autism supports this view. While they can discriminate voices, autistic children orient less to human voice or to infant-directed speech (Ceponiene et al., 2003; Kuhl et al., 2005), even when being called by their name (Lord, 1995; Werner et al., 2000), and are less distressed by a "still-face"[2] situation than typically developing children (Nadel et al., 2000). When asked to judge the emotional status of a face, they are slower than control subjects to detect an emotional expression or a direct gaze, while no such difference was found for neutral faces having averted gaze (Dalton et al., 2005).

[2]When faced by a suddenly unresponsive social partner, young infants typically react by ceasing smiling and by gazing away. This procedure has been used to investigate a broad range of questions about early social and emotional development.

Along with the defective processing of social-communicative cues, mild deficits in face discrimination and recognition have also been described in children with autism. By middle childhood, these children perform worse than mental and chronological aged-matched peers in face recognition tasks (Tantam et al., 1989; Gepner et al., 1996). On the other hand, children with autism are less impaired in recognizing inverted faces than control subjects (Langdell, 1978; Hobson et al., 1988). It is suggested that the superior performance with inverted faces is the result of relying less on holistic face encoding mechanisms and more on local information (Boucher and Lewis, 1992; Davies et al., 1994). In typically developing children, it is the holistic processing that develops slower, being highly dependent on exposure (Carey, 1992). It was also shown that toddlers with autism do not show an advantage for discriminating human faces compared with monkey faces at an age when typically developing children already show superior processing of human faces (Chawarska and Volkmar, 2006).

It is still a matter of debate whether a deficit in face processing mechanisms diminishes the interest of children with autism in the social cues conveyed by faces, or whether it is the indifference towards eyes that gives them less chance to learn about faces, as compared with the normal population (Grelotti et al., 2002). Keeping in line with the main hypothesis of this chapter and with the supportive evidence brought up to this point, we are inclined towards the second causal relationship: that it is the disinterest in eyes and in mutual gaze which lies at the origin of the face recognition deficit in this disorder (but see Behrmann et al., 2006 for an alternative view). As of today, only a few studies have addressed this relationship directly.

Nevertheless those that show that children with autism can become experts in processing other facial features than eyes, or in processing face-like stimuli where the eyes are less salient (e.g., cartoon characters, see Grelotti et al., 2005), speak against the hypothesis of an initial face-processing deficit.

If autism can be characterized by the absence of the positive "eye bias" present in the normal population, it offers us the opportunity to see what alternative cues can be used for identification. Would individual recognition rely more on other face elements or on the body in this disorder? When scanning a face in a dynamic video scene, adults with autism spend less time on the eyes, but also more time on the mouth, than do control individuals (Klin et al., 2002). The same scanning pattern is found in children with autism in a face recognition task (Dalton et al., 2005). This scanning pattern suggests that, in contrast to typical population, individuals with autism would rely more on the lower part of the face for recognition. Indeed, it was shown that children with autism recognize the lower half of faces better, while typically developing children used more the upper part (Langdell, 1978). These findings have recently been replicated by Joseph and Tanaka (2003), who demonstrated not only better recognition performance in children with autism on the basis of the mouth region, but also better performance when the mouth region was upright as opposed to inverted. This suggests that children with autism also rely on holistic processing strategies in face recognition but, because they are not attracted to eyes, they end up focusing on other face regions when encoding identity-specific information. The reliance on body cues for identification was not tested in this population, but we would again expect them to perform better than typically developing individuals.

The brain areas activated by face perception could also shed light on the nature of face processing deficit in autism. One noteworthy discovery is that the neurons of the amygdala in people with autism are smaller and more densely packed than in the normal population (Bauman and Kemper, 2005; Munson et al., 2006). In addition, the amygdala tends to be less active in face processing tasks in this disorder (Baron-Cohen et al., 1999). As we have seen, the amygdala is associated with processing information from the eyes and thought to be involved in orienting the attention towards faces, in infancy. Accordingly, a failure to engage this system may lead to reduced exposure to faces. Baron-Cohen et al. (2000) proposed that a deficit in amygdala function is the basis of failing to engage in eye contact in autism. It is also interesting to note here that a striking similarity has been observed between the reliance on the mouth region in individuals with autism and in a patient with bilateral damage to the amygdala (Adolphs et al., 2005). Atypical functioning is also reported at other areas in the visual stream. The FFA, one of the most important cortical regions for face identity processing, is only weakly activated by faces in individuals with autism, while a network of alternative brain regions prefrontal and primary visual cortices shows activity (Pierce et al., 2001). Dalton et al. (2005) showed that the FFA is activated in autism, but only in those individuals who spend more time with looking at the eyes. The time spent looking at the eyes was positively correlated with the right FFA activity. This result is compatible with the hypothesis that individuals who do not pay special attention to faces will not develop a neuronal specialization for face processing in the fusiform gyrus. This finding is nevertheless also compatible with the opposite hypothesis, according to which impairment in the FFA development could force people with autism to recruit alternative strategies for encoding identity. Looking at the early development of infants at high risk of autism, like the siblings of autistic children, will probably provide us with valuable information regarding the directionality of these effects (see Elsabbagh and Johnson, this volume).

The interest in eyes is driven the social relevance of gaze

We embarked on this chapter with the goal of explaining the selective reliance on faces (in contrast to other equally informative body parts) for the recognition of individuals in humans. We took a developmental perspective and tried to construct an alternative account to what was previously proposed. Our central hypothesis has been that an interest not in facial patterns per se but in human eyes is the driving force of the development of face expertise. We have proposed that because infants are looking for mutual gaze, faces appear more often than other body parts in their central visual field.

Nevertheless, the explanation that infants spend more time looking at faces because the eyes and gaze constitute salient stimuli for them, provides only a partial answer to our question. This answer accounts for what perceptual mechanisms trigger newborns' and infants' bias but it does not explain what functional role this bias serves. What do infants gain by looking at the eyes? Eyes and gaze convey rich information: they have been associated with expressing emotions, mental states, and communicative information. Eyes might even reflect physical health and fitness (subjectively perceived as "beauty") (Rhodes, 2006). Which of these aspects drives infants' tendency to look at the eyes?

Infants' development of understanding of emotional expressions is dependent on their exposure to a normal range of such expressions from their caregivers (de Haan et al., 2004). The straightforward conclusion that infants must look at the eyes in order to extract this information is weakened, nevertheless, by the fact that only some of these expressions are conveyed by the eyes. Expressions like fear, anger, or surprise are accompanied by a clear widening or narrowing of the eyes. While young infants can discriminate between the above expressions, none of them invites them to explore the faces longer (anger has even the opposite effect, Grossmann et al., 2006). It is instead the smiling expression that is preferred by newborns (Farroni et al., 2007). However, when having to judge this expression, adults explore the mouth and not the eye region (Schyns et al., 2002).

While eyes might not help infants understand emotional expressions, the direction of the eye gaze can have an emotional effect on infants. Most episodes of mother–infant interaction are accompanied by direct gaze from the mother (Watson, 1972). A few hypotheses have been put forward on what caregivers and infants gain from such intense and emotionally rich interactions, amongst which the most obvious one is the strengthening of the mother–infant relationship (Watson, 2001). It was suggested that the amount of emotional affect and behavioral contingency during these interactions has an impact on seeking, maintaining, and avoiding contact during social interactions later in life (Volker et al., 1999). The preference for mutual gaze might thus be needed for normal affective development in humans.

Note, however, that filial attachment in many nonhuman species is achieved in the absence of extended face-to-face interactions between parents and infants, which suggest that this end can be achieved without relying on mutual gaze. One further factor that shapes the properties of the mother–infant relationship is bodily contact. Chimpanzee mothers are in constant bodily contact with their offspring in the first 3 months after birth, as the infant clings to the mother and the mother embraces the infant (Matsuzawa, 2006). Human evolution, however, was accompanied by an increasing tendency for physical separation, from which mothers benefited because, having their hands free, they could engage in other activities. It is therefore possible that in exchange for bodily contact, communicative distal contact through facial (eye gaze), gestural and vocal signals between mothers and babies has proliferated. In this context, direct gaze may act as a substitute of bodily contact and will assure infants of being cared for and protected by their mother.

Finally, eyes carry important communicative information, which humans use extensively. In fact, two kinds of communicative information are embedded in eye gaze. Direct gaze (i.e., eye contact) is an ostensive stimulus (Csibra and Gergely, 2006), which signals that the accompanying or subsequent communication is directed to the other party. On the other hand, a gaze shift from direct to averted position may also signal referential information by specifying the direction or location where the referent of the communication can be found. Human newborns are sensitive both to the ostensive (Farroni et al., 2002) and to the referential (Farroni et al., 2004b) aspects of human gaze, and evidence suggests that the latter one depends on the former one (i.e., infants' attention shifts to the direction indicated by others' gaze only after mutual gaze has been established, see Farroni et al., 2003; Senju et al., 2007).

The use of gaze signals in communication is widespread among humans, and it plays an even more important role in infants than in adults. Preverbal infants, who are unable to decode verbal reference, can use gaze information to comprehend nonverbal referential expressions. In fact, a whole body of literature demonstrates that understanding nonverbal reference may be a precondition of efficient word learning in infancy (Tomasello, 2001; Bloom, 2000; Baldwin, 2003). But words are not the only type of knowledge that infants can acquire by referential communication. Valence and function of objects and actions also represent the kinds of information that are much easier to acquire via social than via individual learning. Generally, humans, unlike other animals, tend to acquire a big part of their knowledge through communication. It is thus plausible to assume that human infants' bias towards eyes reflect the functioning of a human-specific social learning system, which evolved to transmit useful cultural knowledge across generations via ostensive communication (Csibra and Gergely, 2006).

Whatever the exact function (or functions) of human infants' obsession with eyes, the evidence we reviewed in this chapter offers an alternative causal story for the development of face expertise in humans. Faces are not the carriers of identity because they are necessarily more informative about individuals or because infants are born with specialized face processing mechanisms, or at least not *only* for the above reasons. Becoming a *face* expert might in addition be a by-product of an increased interest in people as potential sources of valuable information during development. We believe that it is infants' interest in eye-mediated communication that makes them spend more time with looking at human faces, leading to the rapid improvement of face processing skills and eventually to the irreversible perceptual and neuronal specialization for individual recognition by faces.

Acknowledgments

This work was supported by a Pathfinder grant (CALACEI) from the European Commission. We thank to Mayada Elsabbagh, Tobias Grossmann, Atsushi Senju, Victoria Southgate, and Przemyslaw Tomalski for their valuable comments on earlier versions of this paper.

References

Adams Jr., R.B., Gordon, H.L., Baird, A.A., Ambady, N. and Kleck, R.E. (2003) Effects of gaze on amygdala sensitivity to anger and fear faces. Science, 300(5625): 1536.

Adolphs, R., Baron-Cohen, S. and Tranel, D. (2002) Impaired recognition of social emotions following amygdala damage. J. Cogn. Neurosci., 14(2): 1264–1274.

Adolphs, R., Gosselin, F., Buchanan, T.W., Tranel, D., Schyns, P. and Damasio, A.R. (2005) A mechanism for impaired fear recognition after amygdala damage. Nature, 433(7021): 68–72.

Astafiev, S.V., Stanley, C.M., Shulman, G.L. and Corbetta, M. (2004) Extrastriate body area in human occipital cortex responds to the performance of motor actions. Nat. Neurosci., 7(5): 542–548.

Baddeley, A. and Woodhead, M.M. (1983) Improving face recognition ability. In: Llooyd-Bostock S. and Clifford B. (Eds.), Evaluating Witness Evidence. Wiley, Chichester, pp. 125–136.

Baldwin, D. (2003) Infants' ability to consult the speaker for clues to word reference. J. Child Lang., 20(2): 395–418.

Bar-Haim, Y., Ziv, T., Lamy, D. and Hodes, R.M. (2006) Nature and nurture in own-race face processing. Psychol. Sci., 17(2): 159–163.

Baron-Cohen, S., Ring, H.A., Bullmore, E.T., Wheelwright, S., Ashwin, C. and Williams, S.C. (2000) The amygdala theory of autism. Neurosci. Biobehav. Rev., 24(3): 355–364.

Baron-Cohen, S., Ring, H.A., Wheelwright, S., Bullmore, E.T., Brammer, M.J., Simmons, A., et al. (1999) Social intelligence in the normal and autistic brain: an fMRI study. Eur. J. Neurosci., 11(6): 1891–1898.

Batki, A., Baron-Cohen, S., Wheelwright, S., Connellan, J. and Ahluwalia, J. (2000) Is there an innate gaze module? Evidence from human neonates. Infant Behav. Dev., 23: 223–229.

Bauman, M.L. and Kemper, T.L. (2005) Neuroanatomic observations of the brain in autism: a review and future directions. Int. J. Dev. Neurosci., 23(2–3): 183–187.

Baxter, M.G. and Murray, E.A. (2002) The amygdala and reward. Nat. Rev. Neurosci., 3: 563–573.

Behrmann, M., Thomas, C. and Humphreys, K. (2006) Seeing is differently: visual processing in autism. Trends Cogn. Sci., 10(5): 258–278.

Blass, E.M. and Camp, C.A. (2001) The ontogeny of face recognition: eye contact and sweet taste induce face preference in 9-and 12-week-old human infants. Dev. Psychol., 37(6): 762–774.

Bloom, P. (2000) How children learn the meaning of words. MIT Press, Cambridge, MA.

Boucher, J. and Lewis, V. (1992) Unfamiliar face recognition in relatively able autistic children. J. Child Psychol. Psychiatry, 33: 843–859.

Bruce, V., Doyle, T., Dench, N. and Burton, M. (1991) Remembering facial configurations. Cognition, 38(2): 109–144.

Bukach, C.M., Gauthier, I. and Tarr, M.J. (2006) Beyond faces and modularity: the power of an expertise framework. Trends Cogn. Sci., 10(4): 159–166.

Caldara, R., Seghier, M.L., Rossion, B., Lazeyras, F., Michel, C. and Hauert, C.A. (2006) The fusiform face area is tuned for curvilinear patterns with more high-contrasted elements in the upper part. Neuroimage, 31(1): 313–319.

Carey, S. (1992) Becoming a face expert. Philoso. Trans. R. Soc. Lond., 335: 95–103.

Ceponiene, R., Lepisto, T., Shestakova, A., Vanhala, R., Alku, P., Naatanen, R. and Yaguchi, K. (2003) Speech-sound-selective auditory impairment in children with autism: they can perceive but do not attend. Proc. Natl. Acad. Sci. USA, 100: 5567–5572.

Chawarska, K. and Volkmar, F. (2006) Impairments in monkey and human face recognition in 2-years-old toddlers with autism spectrum disorder and developmental delay. Dev. Sci., 10: 266–272.

Cooper, R.P. and Aslin, R.N. (1990) Preference for infant-directed speech in the first month after birth. Child Dev., 61: 1584–1595.

Csibra, G., Davis, G., Spratling, M.W. and Johnson, M.H. (2000) Gamma oscillations and object processing in the infant brain. Science, 290(5496): 1582–1585.

Csibra, G. and Gergely, G. (2006) Social learning and social cogniton: the case for pedagogy. In: Munakata Y. and Johnson M.H. (Eds.), Processes of Change in Brain and Cognitive Development. Attention and Performance XXI. Oxford University Press, Oxford, pp. 249–274.

Csibra, G. and Johnson, M.H. (2007) Investigating event-related oscillations in infancy. In: de Haan M. (Ed.), Infant EEG and Event-Related Potentials. Psychology Press, Hove, England, pp. 289–304.

Dalton, K.M., Nacewicz, B.M., Johnstone, T., Schaefer, H.S., Gernsbacher, M.A., Goldsmith, H.H., et al. (2005) Gaze fixation and the neural circuitry of face processing in autism. Nat. Neurosci., 8(4): 519–526.

Davies, S., Bishop, D., Manstead, A.S.R. and Tantam, D. (1994) Face perception in children with autism and Asperger's syndrome. J. Child Psychol. Psychiatry, 35: 1033–1057.

Diamond, R. and Carey, S. (1986) Why faces are and are not special: an effect of expertise. J. Exp. Psychol. Gen., 115(2): 107–117.

Downing, P.E., Jiang, Y., Shuman, M. and Kanwisher, N. (2001) A cortical area selective for visual processing of the human body. Science, 293(5539): 2470–2473.

Farroni, T., Csibra, G., Simion, F. and Johnson, M.H. (2002) Eye contact detection in humans from birth. Proc. Natl. Acad. Sci. USA, 99(14): 9602–9605.

Farroni, T., Johnson, M.H. and Csibra, G. (2004a) Mechanisms of eye gaze perception during infancy. J. Cogn. Neurosci., 16(8): 1320–1326.

Farroni, T., Johnson, M.H., Menon, E., Zulian, L., Faraguna, D. and Csibra, G. (2005) Newborns' preference for face-relevant stimuli: effects of contrast polarity. Proc. Natl. Acad. Sci. U.S.A., 102(47): 17245–17250.

Farroni, T., Mansfield, E.M., Lai, C. and Johnson, M.H. (2003) Motion and mutual gaze in directing infants' spatial attention. J. Exp. Child Psychol., 85: 199–212.

Farroni, T., Massaccesi, S., Menon, E. and Johnson, M.H. (2006a) Direct gaze modulates face recognition in young infants. Cognition, 102(3): 396–404.

Farroni, T., Menon, E. and Johnson, M.H. (2006b) Factors influencing newborns' preference for faces with eye contact. J. Exp. Child. Psychol., 95(4): 298–308.

Farroni, T., Menon, E., Rigato, S. and Johnson, M.H. (2007) The perception of facial expressions in newborns. Eur. J. Dev. Psychol., 4(1): 2–13.

Farroni, T., Pividori, D., Simion, F., Massaccesi, S. and Johnson, M.H. (2004b) Eye gaze cueing of attention in newborns. Infancy, 5(1): 39–60.

Gauthier, I., Skudlarski, P., Gore, J.C. and Anderson, A.W. (2000) Expertise for cars and birds recruits brain areas involved in face recognition. Nat. Neurosci., 3(2): 191–197.

Gauthier, I. and Tarr, M.J. (1997) Becoming a "Greeble" expert: exploring mechanisms for face recognition. Vision Res., 37(12): 1673–1682.

George, N., Driver, J. and Dolan, R.J. (2001) Seen gaze-direction modulates fusiform activity and its coupling with other brain areas during face processing. Neuroimage, 13(6): 1102–1112.

Gepner, B., de Gelder, B. and de Schonen, S. (1996) Face processing in autistics: Evidence for a generalized deficit? Child Neuropsychology, 2: 123–139.

Gliga, T. and Dehaene-Lambertz, G. (2005) Structural encoding of body and face in human infants and adults. J. Cogn. Neurosci., 17(8): 1328–1340.

Gliga, T. and Dehaene-Lambertz, G. (2007) Development of a view-invariant representation of the human head. Cognition, 102(2): 261–288.

Gosselin, F. and Schyns, P. (2001) Bubbles: a technique to reveal the use of information in recognition. Vision Res., 41: 2261–2271.

Grelotti, D.J., Gauthier, I. and Schultz, R.T. (2002) Social interest and the development of cortical face specialization: what autism teaches us about face processing. Dev. Psychobiol., 40(3): 213–225.

Grelotti, D.J., Klin, A.J., Gauthier, I., Skudlarski, P., Cohen, D.J., Gore, J.C., et al. (2005) fMRI activation of the fusiform gyrus and amygdala to cartoon characters but not to faces in a boy with autism. Neuropsychologia, 43(3): 373–385.

Grill-Spector, K., Kushnir, T., Hendler, T., Edelman, S., Itzchak, Y. and Malach, R. (1998) A sequence of object-processing stages revealed by fMRI in the human occipital lobe. Hum. Brain Mapp., 6(4): 316–328.

Grossmann, T., Johnson M.H., Farroni, T. and Csibra, G. (in press) Social perception in the infant brain: gamma oscillatory activity in response to eye gaze [electronic version]. Social Cogn. Affect. Neurosci.

Grossmann, T., Striano, T. and Friederici, A.D. (2006) Cross-modal binding of emotional information from face and voice in the infant brain. Dev. Sci., 9(3): 309–315.

de Haan, M., Belsky, J., Reid, V.M., Volein, A. and Johnson, M.H. (2004) Maternal personality and infants' neural and visual responsivity to facial expressions of emotion. J. Child Psychol. Psychiatry, 45: 1209–1218.

Hains, S.M. and Muir, D.W. (1996) Infant sensitivity to adult eye direction. Child Dev., 67(5): 1940–1951.

Haxby, J.V., Gobbini, M.I., Furey, M.L., Ishai, A., Schouten, J.L. and Pietrini, P. (2001) Distributed and overlapping representations of faces and objects in ventral temporal cortex. Science, 293(5539): 2425–2430.

Hill, H. and Johnston, A. (2001) Categorizing sex and identity from the biological motion of faces. Curr. Biol., 11: 880–885.

Hobson, R.P., Ouston, J. and Lee, A. (1988) What's in a face? The case of autism. Br. J. Psychol., 79: 441–453.

Hoffman, E.A. and Haxby, J.V. (2000) Distinct representations of eye gaze and identity in the distributed human neural system for face perception. Nat. Neurosci., 3(1): 80–84.

Hood, B.M., Macrae, C.N., Cole-Davies, V. and Dias, M. (2003) Eye remember you: the effects of gaze direction on face recognition in children and adults. Dev. Sci., 6(1): 67–72.

Humphreys, K., Gosselin, F., Schyns, P. and Johnson, M.H. (2006) Using "Bubbles" with babies: a new technique for investigating the informational basis of infant perception. Infant Behav. Dev., 29: 471–475.

Johnson, M.H. (2005) Subcortical face processing. Nat. Rev. Neurosci., 6(10): 766–774.

Johnson, M.H., Dziurawiec, S., Ellis, H. and Morton, J. (1991) Newborns' preferential tracking of face-like stimuli and its subsequent decline. Cognition, 40(1–2): 1–19.

Joseph, R.M. and Tanaka, J. (2003) Holistic and part-based face recognition in children with autism. J. Child Psychol. Psychiatry, 44(4): 529–542.

Kampe, K.K., Frith, C.D. and Frith, U. (2003) "Hey John": signals conveying communicative intention toward the self activate brain regions associated with "mentalizing," regardeless of modality. J. Neurosci., 23: 5258–5263.

Kanwisher, N., McDermott, J. and Chun, M.M. (1997) The fusiform face area: a module in human extrastriate cortex specialized for face perception. J. Neurosci., 17(11): 4302–4311.

Kaufman, J., Csibra, G. and Johnson, M.H. (2003) Representing occluded objects in the human infant brain. Proc. Biol. Sci., 270(Suppl 2): S140–S143.

Kawashima, R., Sugiura, M., Kato, T., Nakamura, A., Hatano, K., Ito, K., et al. (1999) The human amygdala plays an important role in gaze monitoring: a pet study. Brain, 122(4): 779–783.

Kelly, D.J., Quinn, P.C., Slater, A.M., Lee, K., Gibson, A., Smith, M., et al. (2005) Three-month-olds, but not newborns, prefer own-race faces. Dev. Sci., 8(6): F31–F36.

Klin, A., Jones, W., Schultz, R., Volkmar, F. and Cohen, D. (2002) Visual fixation patterns during viewing of naturalistic social situations as predictors of social competence in individuals with autism. Arch. Gen. Psychiatry, 59(9): 809–816.

Kobayashi, H. and Koshima, S. (1997) Unique morphology of the human eye. Nature, 387: 767–768.

Kuhl, P.K., Coffey-Corina, S., Padden, D. and Dawson, G. (2005) Links between social and linguistic processing of speech in preschool children with autism: behavioral and electrophysiological studies. Dev. Sci., 8(1): F9–F20.

Kylliainen, A., Braeutigam, S., Hietanem, J.K., Swithenby, S.J. and Bailey, A.J. (2006a) Face-and gaze-sensitive neural responses in children with autism: a magnetoencephalographic study. Eur. J. Neurosci., 24: 2679–2690.

Kylliainen, A. and Hietanen, J.K. (2006b) Skin conductance responses to another person's gaze in children with autism. J. Autism Dev. Disord., 36(4): 517–525.

Langdell, T. (1978) Recognition of faces: an approach to the study of autism. J. Child Psychol. Psychiatry, 19(3): 255–268.

Le Grand, R., Mondloch, C.J., Maurer, D. and Brent, H.P. (2003) Expert face processing requires visual input to the right hemisphere during infancy. Nat. Neurosci., 6(10): 1108–1112.

Lehky, S.R. (2000) Fine discrimination of faces can be performed rapidly. J. Cogn. Neurosci., 12(5): 848–855.

Liu, C.H. and Chaudhuri, A. (2002) Reassessing the 3/4 view effect in face recognition. Cognition, 83(1): 31–48.

Lord, C. (1995) Follow-up of two-year-olds referred for possible autism. J. Child Psychol. Psychiatry, 36: 1365–1382.

Macchi Cassia, V., Turati, C. and Simion, F. (2004) Can a nonspecific bias toward top-heavy patterns explain newborns' face preference? Psychol. Sci., 15(6), 379–384.

Macrae, C.N., Hood, B.M., Milne, A.B., Rowe, A.C. and Mason, M.F. (2002) Are you looking at me? Eye gaze and person perception. Psychol. Sci., 13(5): 460–464.

Matsuzawa, T. (2006) Evolutionary origins of the human mother–infant relationship. In: Matsuzawa T., Tomonaga M. and Tanaka M. (Eds.), Cognitive Development in Chimpanzees. Springer, Tokyo.

Maurer, D., Grand, R.L. and Mondloch, C.J. (2002) The many faces of configural processing. Trends Cogn. Sci., 6(6): 255–260.

McCarthy, G. (1999) Event-related potentials and functional MRI: a comparison of localization in sensory, perceptual and cognitive tasks. Electroencephalogr Clin Neurophysiol Suppl, 49: 3–12.

McKone, E. and Kanwisher, J.W. (2005) Does the human brain process objects of expertise like faces? A review of the evidence. In: Dehaene S., Duhamel J.R., Hauser M. and Rizzolatti J. (Eds.), From Monkey Brain to Human Brain. MIT Press, Cambridge, MA.

McNeil, J.E. and Warrington, E.K. (1993) Prosopagnosia: a face-specific disorder. Q. J. Exp. Psychol. A., 46(1): 1–10.

Mondloch, C.J., Lewis, T.L., Budreau, D.R., Maurer, D., Dannemiller, J.L., Stephens, B.R., et al. (1999) Face perception during early infancy. Psychol. Sci., 1(5): 419–422.

Mondloch, C.J., Maurer, D. and Ahola, S. (2006) Becoming a face expert. Psychol. Sci., 17(11): 930–934.

Morton, J. and Johnson, M.H. (1991) CONSPEC and CONLERN: a two-process theory of infant face recognition. Psychol. Rev., 98(2): 164–181.

Munson, J., Dawson, G., Abbott, R., Faja, S., Webb, S.J., Friedman, S.D., et al. (2006) Amygdalar volume and behavioral development in autism. Arch. Gen. Psychiatry, 63(6): 686–693.

Naccache, L. and Dehaene, S. (2001) The priming method: imaging unconscious repetition priming reveals an abstract representation of number in the parietal lobes. Cereb. Cortex, 11(10): 966–974.

Nadel, J., Croue, S., Kervella, C., Matlinger, M.J., Canet P., Hudelot, C., Lecuiyer, C. and Martini, M. (2000) Do autistic children have ontological expectations concerning human behavior? Autism, 4(2): 133–145.

Nelson, C.A. and Luciana, M. (1998) The use of event-related potentials in pediatric neuropsychology. In: Coffey C.E. and Brumback R.A. (Eds.), Textbook of Pediatric Neuropsychiatry. American Psychiatric Press.

Osterling, J. and Dawson, G. (1994) Early recognition of children with autism: a study of first birthday home videotapes. J. Autism Dev. Disord., 24: 247–257.

Pascalis, O., de Haan, M. and Nelson, C.A. (2002) Is face processing species-specific during the first year of life? Science, 296(5571): 1321–1323.

Pierce, K., Muller, R.A., Ambrose, J., Allen, G. and Courchesne, E. (2001) Face processing occurs outside the fusiform 'face area' in autism: evidence from functional MRI. Brain, 124(10): 2059–2073.

Quinn, P.C., Eimas, P.D. and Tarr, M.J. (2001) Perceptual categorization of cat and dog silhouettes by 3-to 4-month-old infants. J. Exp. Child Psychol., 79(1): 78–94.

Ramsey-Rennels, J.L. and Langlois, J.H. (2006) Infants' differential processing of female and male faces. Curr. Dir. Psychol. Sci., 15(2): 59–62.

Reed, C.L., Stone, V.E., Bozova, S. and Tanaka, J. (2003) The body-inversion effect. Psychol. Sci., 14(4): 302–308.

Rhodes, G. (2006) The evolutionary psychology of facial beauty. Annu. Rev. Psychol., 57: 199–226.

Sangrigoli, S. and De Schonen, S. (2004) Recognition of own-race and other-race faces by three-month-old infants. J. Child Psychol. Psychiatry, 45(7): 1219–1227.

Schepherd, J.W., Davies, G.M. and Ellis, H.D. (1981) Studies of cue saliency. In: Davies G.M. and Ellis H.D. (Eds.), Perceiving and Remembering Faces. Academic Press, London, pp. 105–133.

Schyns, P., Bonnar, L. and Gosselin, F. (2002) Show me the features! Understanding recognition from the use of visual information. Psychol. Sci., 13(5): 402–410.

Senju, A., Csibra, G. and Johnson, M.H. (2007) Understanding the referential nature of looking: infants' preference for object directed gaze. Manuscript submitted for publication.

Senju, A., Hasegawa, T. and Tojo, Y. (2005) Does perceived direct gaze boost detection in adults and children with and without autism? The stare-in-the-crowd effect revisited. Vis. Cogn., 12: 1474–1496.

Senju, A., Johnson, M.H. and Csibra, G. (2007) The development and neural bases of referential gaze perception. Soc. Neurosci., 1: 220–234.

Shibata, T., Nishijo, H., Tamura, R., Miyamoto, K., Eifuku, S., Endo, S., et al. (2002) Generators of visual evoked potentials for faces and eyes in the human brain as determined by dipole localization. Brain Topogr., 15(1): 51–63.

Simion, F., Valenza, E., Umilta, C. and Dalla Barba, B. (1998) Preferential orienting to faces in newborns: a temporal-nasal asymmetry. J. Exp. Psychol. Hum. Percept. Perform., 24(5): 1399–1405.

Slaughter, V., Heron, M. and Sim, S. (2002) Development of preferences for the human body shape in infancy. Cognition, 85(3): B71–B81.

Stekelenburg, J.J. and de Gelder, B. (2004) The neural correlates of perceiving human bodies: an ERP study on the body-inversion effect. Neuroreport, 15(5): 777–780.

Striano, T., Henning, A. and Stahl, D. (2005) Sensitivity to social contingencies between 1 and 3 months of age. Dev. Sci., 8(6): 509–518.

Symons, L., Hains, S. and Muir, D-. (1998) Look at me: 5-month-old infant's sensitivity to very small deviations in eye-gaze during social interactions. Infant Behav. Dev., 21: 531–536.

Tanaka, J. and Taylor, M. (1991) Object categories and expertise: is the basic level in the eye of the beholder? Cogn. Psychol., 23(3): 457–482.

Tantam, D., Monaghan, L., Nicholson, J. and Stirling, J. (1989) Autistic children's ability to interpret faces: a research note. J. Child Psychol. Psychiatry, 30: 623–630.

Taylor, M. and Baldeweg, T. (2002) Application of EEG, ERP and intracranial recordings to the investigation of cognitive functions in children. Dev. Sci., 5(3): 318–334.

Taylor, M.J., Batty, M. and Itier, R.J. (2004) The faces of development: a review of early face processing over childhood. J. Cogn. Neurosci., 16(8): 1426–1442.

Tomasello, M. (2001) Perceiving intentions and learning words in the second year of life. In: Bowerman M. and Levinson S. (Eds.), Language Acquisition and Conceptual Development. Cambridge University Press, Cambridge.

Urgesi, C., Berlucchi, G. and Aglioti, S.M. (2004) Magnetic stimulation of extrastriate body area impairs visual processing of nonfacial body parts. Curr. Biol., 14(23): 2130–2134.

Vinette, C., Gosselin, F. and Schyns, P.G. (2004) Spatio-temporal dynamics of face recognition in a flash: It's in the eyes! Cogn. Sci., 28, 289–301.

Volker, S., Keller, H., Lohaus, A., Cappenberg, M. and Chasiotis, A. (1999) Maternal interactive behaviour in early infancy and later attachment. Int. J. Behav. Dev., 23(4): 921–936.

Vuilleumier, P., Armony, J.L., Driver, J. and Dolan, R.J. (2003) Distinct spatial frequency sensitivities for processing faces and emotional expressions. Nat. Neurosci., 6(6): 624–631.

Vuilleumier, P., George, N., Lister, V., Armony, J.L. and Driver, J. (2005) Effects of perceived mutual gaze and gender on face processing and recognition memory. Visual Cogn., 12(1): 85–101.

Vuilleumier, P. and Pourtois, G. (2007) Distributed and interactive brain mechanisms during emotion face perception: evidence from functional neuroimaging. Neuropsychologia, 45(1): 174–194.

Watson, J.S. (1972) Smiling, cooing, and "the game." Merrill-Palmer Q., 18: 323–339.

Watson, J.S. (2001) Contingency perception and misperception in infancy: some potential implications for attachement. Bull. Menninger Clin., 65: 296–320.

Werker, J. and McLeod, P.J. (1989) Infants preference for both female and male infant-directed talk: a developmental study of attentional and affective responsiveness. Can. J. Psychol., 43: 320–346.

Werner, E., Dawson, G., Osterling, J. and Dinno, N. (2000) Brief report: recognition of autism spectrum disorder before one year of age: a retrospective study based on home video tapes. J. Autism Dev. Disord., 30: 157–162.

Whalen, P.J., Kagan, J., Cook, R.G., Davis, F.C., Kim, H., Polis, S., et al. (2004) Human amygdala responsivity to masked fearful eye whites. Science, 306(5704): 2061.

Zauner, N. and Schwarzer, G. (2003) Face processing in 8-month-old infants: evidence for configural and analytical processing. Vision Res., 43(26): 2783–2793.

Zhao, W., Chellappa, R., Phillips, P.J. and Rosenfeld, A. (2003) Face recognition: a literature survey. ACM Comput. Surv., 35(4): 399–458.

C. von Hofsten & K. Rosander (Eds.)
Progress in Brain Research, Vol. 164
ISSN 0079-6123

CHAPTER 19

Past and present challenges in theory of mind research in nonhuman primates

Josep Call*

Max Planck Institute for Evolutionary Anthropology, Deutscher Platz 6, D-04103 Leipzig, Germany

Abstract: This paper presents the trajectory of theory of mind research in nonhuman primates, with a special focus on chimpanzees as they have been the most intensely studied species. It analyzes the main developments in the field, the critiques that they raised, the responses that they have generated and the current challenges faced by the field. Currently, the most plausible working hypothesis is that at least chimpanzees know what others can and cannot see. Using tasks with a high ecological validity and mapping out key concepts such association and inference are postulated as fundamental steps to further advance our knowledge in this area.

Keywords: mental attribution; comparative cognition; association; inferences; cognitive evolution

It has been almost 30 years since Premack and Woodruff (1978) posed the question of whether the chimpanzee had a theory of mind. Since then considerable progress has been made that speaks to this question, particularly in the last decade. Obviously, not all questions have been answered yet, notably some of those posed by the critics, but this does not mean that those questions will never be resolved and that no progress has taken place. Scientific progress is made of a mixture of old knowledge, critical analysis, and new insights. Certainly, theory of mind research has had its share of each of these ingredients over the last 30 years as I will show in the next few pages. In order to do so, I will introduce the topic historically, paying special attention to the different routes taken by developmental and comparative scientists, and then move on to review some of the key findings, the critiques of the available evidence, and the replies to those criticisms. Although many of those criticisms have been superseded by new data, others have yet to be resolved and they are some of the most prominent challenges for future research.

A brief history

Although Premack and Woodruff's (1978) paper is commonly regarded as the starting point for the field of theory of mind research, there were some precursors to their work. In developmental psychology, the work by Piaget and Inhelder (1956) on perspective taking and intention attribution is a clear referent. Later on, Flavell and colleagues neo-Piagetian approach significantly advanced our knowledge on topics such as perspective taking and appearance reality in children (see Lempers et al., 1977; Flavell, 1992). With regard to the comparative literature Jolly (1966) and Humphrey (1976) called attention to social problem solving as a breeding ground for intelligence. Kummer (1967)

*Corresponding author. Tel.: +49 341 3550 418/614;
Fax: +49 341 3550 444; E-mail: call@eva.mpg.de

DOI: 10.1016/S0079-6123(07)64019-9

described the complex social tactics in baboons, but it is perhaps the work of Emil Menzel (1974) on deceptive tactics in chimpanzees that is the most important precedent in this area. However, neither developmental psychologists nor primatologists initially established any links with each other. The psychological and ethological literatures developed in parallel until Premack and Woodruff (1978) published their seminal paper on goal attribution by Sarah, a language-trained chimpanzee. Sarah was presented with films depicting a human actor trying to solve various problems (e.g., trying to get an out-of-reach banana) and then she had to select the photograph that depicted the correct alternative. For instance, when Sarah saw a human looking up to an out of reach banana hanging from the ceiling, and she had to pick the photograph in which the human climbed on a box. Sarah was capable of completing video sequences in which a human was depicted trying to solve a variety of problems. Premack and Woodruff argued that Sarah "recognized the videotape as representing a problem, understood the actor's purpose, and chose alternatives compatible with that purpose."

The publication of this paper and the ensuing peer commentary that followed were to have a profound impact on both the comparative and developmental psychology of the following 30 years. However, from the very beginning these two disciplines followed different orientations and methodologies. Developmental psychologists set out to design experiments to assess whether children were able to solve tasks of false belief attribution (Wimmer and Perner, 1983; Baron-Cohen et al., 1985; Perner et al., 1987). These studies typically required children to verbally answer questions posed by the experimenter regarding particular situations in which the protagonist had a false belief about the location of an object or the contents of a box. This experimental approach was supplemented with observational data on the usage of mental terms such as want, know or believe (Bartsch and Wellman, 1995). The initial exclusive reliance on verbal measures prevented the study of certain nonverbal populations, although this shortcoming was later alleviated with the introduction of nonverbal tasks (Clements and Perner, 1994; Call and Tomasello, 1999). Recently, Onishi and Baillargeon (2005) have adapted the looking paradigm to test 15-month-old infants' false-belief attribution. Although the evidence is compelling, there is disagreement about whether this constitutes evidence of false belief attribution (Ruffman and Perner, 2005).

Compared to the research frenzy that Premack and Woodruff's paper unleashed in developmental psychology, the initial reception in the comparative camp was rather cold, at least if one goes by the commentaries that appeared with their paper. For instance, Savage-Rumbaugh et al. (1978) suggested that Sarah may have responded based on perceptual or functional similarity between the films and the alternatives rather than responding to the goals of the actor. Consequently, there was very little experimental work in the next decade, yet some primatologists, who apparently were the only researchers studying nonhuman animals interested by this question, concentrating their efforts on carefully gathering anecdotal reports from various scientists and interpreting the level of mental state attribution shown by various species. Using this method, Whiten and Byrne (1988) suggested that "deception is sometimes sophisticated enough to imply that the agent can mentally represent others' mental states" (p. 211). Although this approach lacks the scientific rigor of systematic observations and experiments, it was a useful heuristic tool. Furthermore, Whiten and Byrne (Byrne and Whiten, 1988; Whiten, 1991) were instrumental in bringing together the psychological and ethological approaches to the study of mental state attribution.

The early 1990s saw a shift in emphasis from observational to experimental studies as a means to study mental state attribution in primates. Povinelli et al. (1990) investigated whether chimpanzees could distinguish between knowledgeable and ignorant experimenters. Two experimenters informed a chimpanzee about the location of food, except that one experimenter had seen where the food was located whereas the other had not. The chimpanzee had to determine which experimenter was a more reliable source of information. In the initial phase of the experimenter, one of the experimenters (the knower) placed food under one of three cups (the chimpanzee could not see which one) while the other experimenter (the guesser) was outside of the room. Once the baiting had been completed the guesser returned to the room and together with knower they pointed to one of the three cups available. The knower always pointed to the correct location while the guesser always pointed to an incorrect location. All four chimpanzees learned to select reliably the cup indicated by the knower after an average of 200 trials. After this initial training, Povinelli et al. (1990) conducted a transfer test in which both experimenters remained in the room facing the baiting event that was carried out by a third experimenter. However, the guesser wore a paper bag over the head so that he was unable to observe the location where the food was deposited. Again, three out of four chimpanzees selected the cup designated by the knower above chance levels after 30 trials. Povinelli et al. (1990) concluded that these results were consistent with mental state attribution in chimpanzees.

Two critiques

Triangulation

Heyes (1993) eloquently criticized previous studies that had been used to bolster the arguments on mental state attribution in nonhuman animals. Her critique encompassed not only those studies based on gathering anecdotes but also extended into the data generated by several experimental approaches. In a nutshell, Heyes (1993) argued that subjects may have used conditional discrimination of observable cues to solve the task, not mental state attribution. Further, she argued that the triangulation method was the best available tool to distinguish between mental state attribution and observable cues.

Triangulation consists of training a conditional discrimination and then administering a transfer test. Recall that this is precisely the method used by Povinelli et al. (1990). Chimpanzees were first trained to select the experimenter that stayed in the room during baiting and later they were tested to see whether chimpanzees would also select the informant that did not wear a bag over the head during baiting in the transfer phase. Povinelli found positive evidence of transfer but Heyes (1993) argued that the critical data to assess transfer were the first few trials, which were not reported in the original paper. On inspection of these data, Povinelli (1994) conceded that there was no evidence of immediate transfer and therefore chimpanzees may have learned to respond to another observable cue during testing.

Recently, Heyes (1998) proposed a way to test mental state attribution via triangulation using a modified version of the Povinelli et al. (1990) setup. Heyes proposed three basic steps. First, chimpanzees should learn to select the cup indicated by the knower much as they had done in the first phase of the Povinelli et al. (1990) experiment. The second step consisted of giving chimpanzees two types of goggles. Blue-rimmed goggles mounted opaque lenses so that subjects wearing them could not see through. In contrast, red-rimmed goggles mounted lenses that were see-through, even though their appearance was identical to the lenses in the blue-rimmed goggles. The idea was that chimpanzees would learn that red goggles were see-through while blue goggles were not. Critically, this information was acquired by the individuals in the absence of competitors and food rewards, for instance, in the course of play. Once subjects had mastered the first two steps then they would receive the third step: the transfer test. Here each human informant wore one of the goggles (blue or red) while baiting took place and then they indicated the food location for the chimpanzee. If chimpanzees selected the cup designated by the experimenter wearing the red-rimmed goggles, this, according to Heyes (1998) would constitute evidence of mental state attribution.

Heyes' (1998) design has some desirable features and some limitations, some of which were already identified in the commentaries that followed her target article. I will defer the discussion of those for the section on "Future directions" because some additional potential limitations of this design have only become apparent after the publication of new data. This new information, which will be presented below, together with previous suggestions can help us to make suggestions on how to improve Heyes' (1998) proposal. For now, let's turn our attention to the second major criticism that the field of mental state attribution in primates has received in recent years.

What chimpanzees know about seeing

The second major critique came from Povinelli and colleagues. Povinelli and Eddy (1996) conducted a series of experiments in which the chimpanzee faced two experimenters and the ape had to request food from one of them. Invariably, one of the experimenters was able to see the chimpanzee while the other was not. Chimpanzees spontaneously discriminated and begged from a human that was facing them compared to one that had the back turned — a finding that had been previously reported in chimpanzees using systematic observations (Tomasello et al., 1994) and experiments in orangutans (Call and Tomasello, 1994). However, chimpanzees failed to discriminate between numerous other combinations including two humans with the back turned but with one looking over her shoulder, one human with a bucket over the head versus a bucket next to the head, or with a blindfold over the eyes versus a blindfold over the mouth. Children presented with some of the same conditions readily distinguished between experimenters that could and could not see them. Nevertheless, if chimpanzees were given enough trials (dozens even hundreds), they were capable of learning to discriminate between the human that could and could not see them. Povinelli and Eddy (1996) concluded that, unlike children, chimpanzees knew very little about what others can and cannot see. Instead, they argued that chimpanzees can learn to use observable cues to decide from whom to beg from and postulated the frontal orientation hypothesis according to which subjects respond to the orientation of the body rather than the face or the eyes. In later writings, Povinelli and Vonk (2003) have questioned the validity of this paradigm as a measure of what individuals can or cannot see. Presumably their cautionary note should also apply to the positive evidence that they reported for children.

A paradigm shift?

These two critiques had a significant impact on the field and interest on theory of mind research in nonhuman primates by the late 1990s had begun to dwindle. Paradoxically, the new revival of theory of mind research that came with the turn of the century was fostered by a third critique. Unlike the previous two critiques, however, this one concentrated on the whole approach that had been used to investigate mental attribution in nonhuman animals, and not on a particular experiment. It is understandable that following the lead of studies on children, comparative psychologists designed experiments with a strong cooperative component (e.g., Povinelli et al., 1990; Povinelli and Eddy, 1996; Call and Tomasello, 1999; Call et al., 2000). Namely, humans indicated the location of hidden food so that chimpanzees could find it or humans helped chimpanzees find food. Such procedures are commonly used in developmental research, but Hare and colleagues (Hare et al., 2000, 2001; Hare, 2001) argued that this manipulation may look rather odd to the eyes of a chimpanzee. Unlike humans, chimpanzees do not share information with others about monopolizable food morsels, they simply eat them. Therefore, Hare et al. (2000) designed an experiment based on competition rather than cooperation by placing a subordinate and a dominant chimpanzee into rooms on opposite sides of a third room in which there were two pieces of food. The subordinate could see a piece of food that the dominant could not see — because it was on her side of a small barrier. Subordinate chimpanzees preferred to approach a piece of food that only they could see much more often than the food that both they and the dominant could see. Several control conditions ruled out the possibility that subordinate chimpanzees were reacting to the behavior of the dominant animal before deciding which piece to approach — a QJ;strategy that can explain the results obtained with capuchin monkeys (Hare et al., 2002) (Fig. 1).

Although Karin-D'Arcy and Povinelli (2002) failed to replicate these results with another group of chimpanzees, Bräuer et al. (in press) did replicate the original results. Furthermore, Bräuer et al. (in press) suggested that Karin-D'Arcy and Povinelli (2002) had been unsuccessful because the competitive arena that the latter used, as well as the distances between the food pieces, were too small compared to the ones originally used by Hare et al. (2000). Indeed, Bräuer et al. (in press) showed that manipulating the distances affected the likelihood

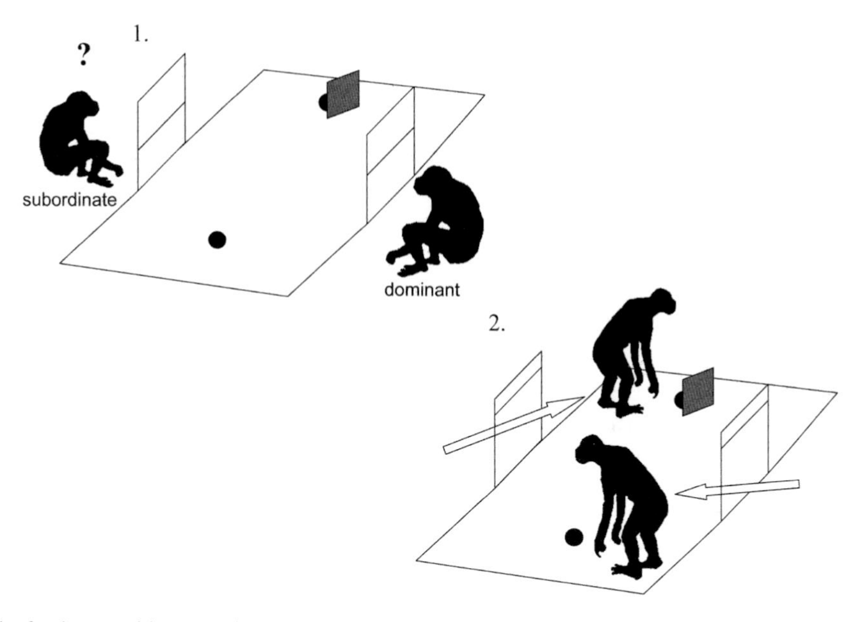

Fig. 1. Setup of the food competition paradigm in chimpanzees. (1) Two pieces of food are located inside a central room. One of the pieces of food is located behind a barrier so that the dominant chimpanzee cannot see it. (2) Subordinate chimpanzees prefer to approach and take food pieces that are hidden from the dominant animal. These preferences are also shown before the dominant animal is released.

of obtaining positive results. The differences between these three sets of studies highlight the importance of taking into account several types of information for decision-making. Thus, subjects take into account not just what other can and cannot see but also the size of the competitive area and the social matrix in which this competition is embedded. This means that lack of evidence on perspective taking may be a consequence of any of these factors overruling the influence of the others.

Leaving aside this methodological consideration, these studies already ruled out some of the alternatives leveled against previous studies. Here subjects did not learn to respond to the correct alternative during the test, they were not responding to cues that were currently present, and there was no evidence that chimpanzees could eavesdrop information about where competitors were going to go. Hare et al. (2000) interpreted these results as evidence that chimpanzees know what others can and cannot see. However, there were some alternative explanations that this initial study did not rule out. In the next section, we explore the most prominent competing hypothesis that has been leveled against the perspective-taking hypothesis.

Replicas, contra-replicas and beyond

Karin D'Arcy and Povinelli (2002) proposed the preferential foraging hypothesis to explain chimpanzees' preference for the food behind the barrier. According to this hypothesis, chimpanzees solved the Hare et al. (2000) tasks because they preferred to forage close to barriers (as opposed to in the open), independently of whether others had any visual access to the food. However, Hare et al. (2001) ruled out this possibility by using two barriers and one piece of food. In experimental trials dominants had not seen the food hidden, or food they had seen hidden was moved to a different location when they were not watching (whereas in control trials they saw the food being hidden or moved). Subordinates always saw the entire baiting procedure and whether the dominant had visual access to the baiting. Subordinates preferentially retrieved and approached the food that dominants had not seen hidden or moved, which suggests that subordinates were sensitive to what dominants had or had not seen during baiting a few moments before. These findings make the peripheral hypothesis moot because unlike the setup in Hare et al. (2000) where there was one piece of food behind the barrier and the other in the open, here the food was always behind a barrier and chimpanzees had to decide whether to take it or not.

In a final experiment, Hare et al. (2001) allowed a dominant A to witness the baiting procedure and then exchanged her with dominant B, who had not seen the baiting. Then B was released to compete with the subordinate animal. The authors compared this condition with one in which the dominant who had seen the baiting was the same individual who competed with the subordinate. Subordinate chimpanzees were more likely to take food from ignorant dominants compared to knowledgeable dominants, suggesting that chimpanzees attribute information to particular individuals who have had visual access to certain events. This means that chimpanzees were also able to take into account past information such as who had seen the baiting, which once again suggests that subjects are not responding to cues that are currently present. Moreover, Hare et al. (2000) had previously shown that the same individual shows a preference for the piece behind the barrier or not depends on whether they are competing against an individual that is higher or lower ranking than themselves. Again this means that individuals have no set preference for foraging behind barriers.

One issue that previous studies based on the competitive paradigm left unanswered is whether subjects were simply responding to the presence/absence of others or whether they had a more refined appreciation of the attentional states of others. In other words, did the face or the eyes play a special role when assessing who can see whom in competitive situations? Povinelli and Eddy's (1996) findings had suggested that chimpanzees did not spontaneously pay attention to such cues. However, recent studies have provided a different picture. Hare et al. (2006) confronted chimpanzees against a human competitor so that they could manipulate the human postural and attention states. Two pieces of food were presented side-by-side in front of the ape and the human could pull the food if he saw the chimpanzee trying to steal it. In one condition the human had his body and face oriented toward one piece. In another condition he had the body oriented toward one piece and the face oriented toward the other piece. Results showed that apes chose to steal the food that the experimenter could not see, regardless of the body orientation. Interestingly, in another experiment with the same basic setup, we found no evidence that subjects were capable of learning to associate arbitrary visual cues (e.g., a colored plastic piece indicated the correct alternative) with the correct response (Melis et al., 2006). Flombaum and Santos (2005) extended these results to rhesus macaques and found that they stole food from humans that were not visually oriented to them in a variety of situations identical to those used by Povinelli and Eddy (1996).

So chimpanzees (and rhesus macaques) avoid the human's face when competing for food. Interestingly, in other studies in which chimpanzees had to gesture to obtain food, they spontaneously gestured in front of the face (e.g., Gómez, 1996; Povinelli et al., 2003; Kaminski et al., 2004; see also Liebal et al., 2004). These findings not only contradict Povinelli and Eddy's (1996) results, but also show that chimpanzees can deploy their behavior flexibly. Chimpanzees seek their partner's face when trying to enlist their help but avoid it when they are trying to steal food from him. This is important because it has been proposed that mental state attribution in nonhuman primates may constitute an encapsulated module — it only works in competitive situations Ghazanfar and Santos (2004). However, we have seen that this is not the case. Chimpanzees, and other primates, can also deploy their perspective-taking skills for exploiting others, for instance, when begging for food or eavesdropping information by following the gaze of others (see Tomasello and Call, 2006; Call, 2007, for recent reviews). The evidence currently available in each of these areas converges, thus suggesting that chimpanzees possess sophisticated perspective-taking skills (Fig. 2).

If perspective taking in chimpanzees is not restricted to competitive situations, it is also not restricted to the visual modality. Some observational

Fig. 2. Chimpanzees follow the gaze of others to distant locations.

data suggest that chimpanzees are also sensitive to what others can and cannot hear (e.g., Hauser, 1990; Boesch and Boesch-Achermann, 2000). Melis et al. (2006) found that chimpanzees trying to steal food from a human who cannot see them but can potentially hear them, preferred to open a silent trap door (as opposed to a noisy one) to get the food without alerting the human. Similarly, rhesus macaques trying to steal food from a human also prefer to open a silent compared to a noisy box (Santos et al., 2006). In contrast, there is no evidence that chimpanzees presented with an auditory version of the Hare et al. (2001) paradigm take advantage of auditory cues produced by others (Bräuer et al., in press). In particular, subordinate chimpanzees did not refrain from approaching food that the experimenter had hidden making noise so that they (and the dominant animal) would be able to locate it. In other words, chimpanzees can use auditory cues to locate hidden food (Call, 2004; Bräuer et al., in press) but they do not appear to take into account that what they have heard, dominant animals may have heard too.

The skeptic could interpret all these findings as evidence that subjects respond to the presence of certain observable cues acting as discriminative stimuli without necessarily inferring any mental states. But again, this interpretation does not fit all the data available. First, we have already seen that chimpanzees can remember things like who has seen what and they make their decisions based on those. More importantly, Melis et al. (2006) have shown that chimpanzees can make decisions about which one of two pieces of food to steal in the absence of overt cues during or before the task. Here chimpanzees chose to reach for the food through an opaque tunnel instead of a transparent one so that the experimenter would not see them trying to get the food until it was too late. Crucially, in this experiment chimpanzees were unable to see the human while reaching for the food, but had to infer whether the human would see the chimpanzee's arm creeping toward the food.

In summary, data accumulated in the last decade has ruled out many of the explanations put forward to explain the findings on perspective taking in primates in the last two decades. Thus, learning to respond to cues during the experiment, reacting to the behavior of others or certain foraging predispositions cannot satisfactorily account for the evidence available today. Moreover, subjects deploy their behavior flexibly in a variety of situations, particularly competitive ones (although not exclusively) and they spontaneously (without training) respond to social stimuli appropriately, even when those are not perceptually available. To some scholars these results represent evidence of perspective taking and suggest that chimpanzees know what others can and cannot see — level 1 perspective taking *sensu* Flavell, 1992. Others argue that chimpanzees have merely associated the presence of certain cues with certain outcomes and learned to use observable cues as discriminative stimuli that control their responses but without attributing mental states to others. This is an important issue to which we turn our attention in the next section.

The 'magic' of complex associations

Very soon after Premack and Woodruff's results were published, it was clear that ruling out associative explanations was going to be a major endeavor for those trying to see whether the chimpanzee had a theory of mind. Thus, in their commentary to Premack and Woodruff's paper, Savage-Rumbaugh et al. (1978) argued that she may have chosen based on associating commonplace situations with certain objects. The fact that Sarah performed better in those situations for which associative procedures were most straightforward (e.g., key with lock) compared to those that were less straightforward supported this alternative. Moreover, Savage-Rumbaugh et al. reported that two language-trained chimpanzees were capable of solving the problems presented to Sarah using a matching to sample paradigm without training. Following the same reasoning, one could argue that the previous studies can all be explained as responding to cues without engaging in mental state attribution. Although this is in principle theoretically possible, researchers have devoted little effort to assess how likely this possibility is. Leaving aside the fact that critics have been quite unsuccessful in pointing out what animals have learned precisely — the list of proposed hypotheses not supported by the data is long and getting longer with each new study (see Tomasello and Call, 2006), it is important to consider associative possibilities carefully. There are two arguments, one empirical and the other theoretical that should raise some doubts about the likelihood of an associative account of the data.

From an empirical standpoint, we can say with confidence that subjects had not learned to respond to the apparatuses and setups used during many of the studies reported above because they had never experienced them before the test took place. Moreover, there was no evidence of learning taking place during the tests. This means that individuals were able to apply whatever socio-cognitive skills they brought to the task to solve a novel problem. Even the interpretation of previous data showing learning effects during transfer as an example to respond to certain observable cues, not attribution of mental states of any kind (e.g., Povinelli et al., 1990; Heyes, 1994) is not so straightforward. If the transfer did not occur in the first trial, both an associative alternative and a mental state attribution alternative are still possible. Learning to respond correctly after the first trial does not rule out mental state attribution. Implicitly, Povinelli (1994) and Heyes (1993) favor the associative account since they make the assumption that chimpanzees would be able to associate very quick stimuli regardless of whether they hold an arbitrary or a perspective-taking relation. There are two reasons, however, that make this assumption problematic.

First, contrary to widespread belief chimpanzees are not so good at associating arbitrary cues with responses (Call, 2006). Chimpanzees require dozens of trials to master simple discrimination, hundreds of trials to master conditional discrimination and even thousands of trials (if they learn at all) for some of the third-order multiple-sign problems that require responding to the color or shape of the stimuli depending on the color of the substrate (Harlow, 1951; Vlamings et al., 2006).

Using situations that approximate the type of social problems reported above does not substantially alter the picture. Chimpanzees required hundreds of trials to reliably distinguish between experimenters that signaled the presence of food from those who did not (Povinelli et al., 1990; Povinelli and Eddy, 1996). Moreover, even if they have learned certain associations, they forgot them if they were not used again (Reaux et al., 1999). It is important to note that these tests were conducted under ideal laboratory conditions in which the interval between the presentation of stimuli and the delivery of the reward and distractions are minimized. Under less controlled conditions, learning may be worse because multiple variables vary and interact and subjects have to select which of those variables are relevant. Furthermore, there are substantial spatiotemporal gaps between the presence of certain cues, the execution of certain responses and the occurrence of rewards. The literature on associative learning is clear in this regard, such gaps profoundly disrupt learning (e.g., Domjan and Burkhard, 1986).

The second problem raised by an associative account is perhaps even more daunting. Stimulus generalization is often invoked as the mechanism by which subjects can apply their knowledge to new stimuli. But stimulus generalization is a concept used to explain shifts along a gradient in a particular perceptual modality, for instance, the change of a sound frequency or a color wavelength. It does not explain how subjects generalize from competing with conspecifics in their outdoor area to the test situation which includes all sorts of novel objects and arrangements and a human competitor, or when an ape has to decide what box to operate (one the is silent or one that is not) when she is trying to steal food from the human. There is surprisingly little in the literature that could support the ability of chimpanzees to make such broad generalizations, yet generalization is continuously invoked as a valid explanation.

From a theoretical standpoint, associative explanations are often favored over mentalistic explanations on the basis of Morgan's Canon. If association based on observable cues can explain a given result, there is no need to postulate mentalistic constructs. One problem arises when one needs to postulate multiple ad hoc associative explanations to explain multiple results while postulating a single mentalistic explanation could account for all the results. Tomasello and Call (2006) have argued that once this decision point has been reached, parsimony dictates that scientists should favor the explanation that invokes fewer presuppositions. Postulating multiple associative ad hoc explanations becomes even more untenable when one has to explain things like chimpanzees seeking or avoiding the face of others depending both on the situation and the presence of certain individuals. Obviously it is theoretically possible to postulate an increasing number of conditional discriminations to account for every single result, but given the difficulty that chimpanzees experience with conditional discrimination as noted above, this "patch approach" to cognition seems a particularly perverse exercise.

There is another argument in favor of the perspective-taking hypothesis. Different paradigms converge in producing similar results. It is therefore sensible to postulate an underlying knowledge that cuts across paradigms. Whiten (1996) has used the concept of intervening variable to make this point. Note that the recourse of finding commonalities across a disparate set of situations is at the core of the triangulation method, which has been advocated as one of the best tools to investigate mental state attribution in nonhuman animals.

Contrary to what it may seem, my goal here is not to close the door to associative learning as a possibility, but to critically examine the probability that a mechanism like conditional discrimination as described in traditional learning theory is responsible for the data available. More importantly, it is necessary to test whether conditional discrimination can generate the observed findings. In the next section, I propose how to refine earlier proposals to test this possibility and outline some theoretical considerations that may pave the road for further work.

Future directions

Refining methods

Heyes' (1998) goggle test is an excellent starting point for further research. It is based on triangulation that is a powerful experimental design and it can potentially rule out association. Moreover, unlike previous tasks it has the potential to resolve whether subjects can apply knowledge gained through their own experience with the physical environment to social problems. In a sense, this could be viewed as a test of simulation or experience projection (Whiten, 1998). However, the goggle test has some limitations both in its design and in the kinds of questions that it can and cannot answer.

From a design standpoint, the test in its current form has three important limitations. First, it is based on a cooperative setup in which the human experimenter informs the chimpanzee about the location of food. I suggest that it would be more profitable to try it on a competitive setup that as we have seen may have a greater power to uncover the socio-cognitive skills. Second, Heyes' task relies heavily on learning to associate an arbitrary cue such as color with a physical property of the glass (see-through or not). This again, may be a problem as apes, unlike humans, are not very adept at associating arbitrary cues (Call, 2006). In fact, color may not be as prominent as other types of information such as position (Haun et al., 2006).

Third, the way to measure the occurrence of transfer is problematic. If subjects perform correctly on the first trial, then it is clear that learning by association cannot explain this result. However, if correct performance is achieved after trial one, it does not necessarily mean an absence of positive transfer but it opens the possibility for associative learning. It is conceivable that chimpanzees may not have only learned to respond correctly, but they have done so based on some perspective-taking ability. In fact, I venture that conditional discrimination of observable cues alone is unlikely to produce those results. Instead, one has to search for the mechanisms responsible for that performance elsewhere. Partly this conclusion may have arisen because unlike humans, chimpanzees experience serious difficulties associating arbitrary cues quickly and transferring those associations across situations. However, they may be able to solve better those problems if there is a causal as opposed to an arbitrary connection structure underlying them (Call, 2006).

Therefore, it is crucial to have a measure of how fast subjects can learn such cues when no information about perspective taking is involved. Thus, besides the condition proposed by Heyes (1998), which we will call *direct condition*, two other groups of subjects should receive two additional conditions. In the *arbitrary condition* (Zentall, 1998), during baiting subjects witness that both experimenters are wearing their respective glasses over their forehead and selecting the individual with the red-rimmed glasses is always rewarded. In the *reverse condition* (Heyes, 1998) the experimenter who is wearing the blue-rimmed (opaque) glasses during baiting is always correct at the time of choice while the experimenter wearing the red-rimmed (see-through) glasses is always incorrect. The prediction of the perspective-taking hypothesis is that subjects should perform best in the direct condition, worse in the reverse condition and intermediate in the arbitrary condition. In contrast, an associative account predicts no difference between conditions.

From a theoretical standpoint, it is important to clarify what this task can and cannot answer. On the one hand, as indicated previously, this task has the potential to rule out certain forms of learning as an explanation for the results. It also can potentially bridge the gap that exists between the knowledge that subjects gather from their own experiences and how they may apply this knowledge to others (see Call, 2003). This is a field that requires attention and there is some research that has begun in this area. Most research has concentrated on attributing mental states to others but less is known about whether individuals know that they themselves have mental states. On the other hand, Heyes' task does not specify the content of the representation that individuals may attribute to others (Slaughter and Mealey, 1998). In other words, this task cannot distinguish between perceptual and epistemic mental states. Subjects may solve this task by reasoning about the visual access that others have, not the knowledge states that visual experience engenders. This means that even with positive results this task is not conclusive evidence for the attribution of epistemic states such as knowledge or beliefs. Although this may be considered a minor point by some authors, others would insist that epistemic states are the hallmark of theory of mind.

Refining concepts

Heyes (1993, 1998) pointed out that much of the debate on nonhuman theory of mind research has revolved around the debate between association versus mentalizing — a debate that in different forms, shapes and intensities can also be found in other areas of animal and human cognition.

Further, Heyes (1993) argued that this was a misleading debate because these two concepts are not directly comparable. She proposed that association should be contrasted to inference (or reasoning) whereas mental constructs should be contrasted to observables. These distinctions are important because they can help us clarify some past misunderstandings and focus our effort on the appropriate levels of analysis — the acquisition mechanism and the content of the represented information. At the level of acquisition mechanism, first we need to specify the features of associative and inferential processes. Currently, we are in a better position to define associative processes. Inferential processes are often invoked when associative processes cannot explain a given observation. However, this default strategy is misleading and some effort should be devoted to specifying and distinguishing the features that characterize inferential processes, particularly contrasting them to associative processes.

At the level of the content of representation, researchers have been quite unsuccessful at pointing out specific stimuli that explain the data available as it has been noted in previous sections (see also Tomasello and Call, 2006). Obviously, even if individuals reason about mental states, they need to use observables to infer those mental states. Thus, using observables as a means to and end (mental state) should be distinguished from using observables as the end point. As indicated above, the variety of observables and situations suggest that individuals may go beyond the information given by those. Nevertheless, it is important to emphasize that going beyond observables does not lead into the attribution of epistemic states (Call, 2001). Reached this point, it is hard to distinguish between the various representations that subjects may create from their partner's behavior. Ruffman and Perner (2005) have suggested that convergent evidence from disparate tasks is one of the best ways to distinguish between perceptual and epistemic states in nonverbal organisms. Similarly, various authors have called for convergent evidence as a way to distinguish mental state attribution from other processes (Premack and Woodruff, 1978; Gallup, 1982; Povinelli et al., 1990). Again, it would be important to devise tests that could directly tease these two options apart, or tease apart observables from more abstract types of information that individuals may have encoded.

One immediate benefit of the previous distinctions is that they help us see that ruling out association does not necessarily make mental state attribution (including false belief attribution) more likely, a point also stressed by Heyes (1993, 1998). It is possible that subjects use inferential processes to solve social problems but the content of representation is anchored in perceptual states, not epistemic states, particularly those dealing with belief attribution. However, very little research has been devoted to refine these distinctions and test them empirically. Over the last few years we have argued that neither associative explanations nor accounts based on meta-representation satisfactorily explain the data currently available. Instead we have proposed a third alternative which we have characterized as insightful, representational, and based on some level of abstraction (Tomasello and Call, 1997; Call, 2001, 2003).

Conclusion

This paper presents the trajectory of theory of mind research in nonhuman primates, with a special focus on chimpanzees as they have been the most intensely studied species. It analyzes the main developments in the field, the critiques that they raised, the responses that they have generated and the current challenges faced by the field. If we have learned anything in the last 30 years is that Premack and Woddruff's question cannot be answered satisfactorily with a single word. Besides the need for clearer distinctions and empirical work that we have highlighted, there is another issue that has contributed, and not always in a constructive manner, to the debate. Theory of mind encompasses different processes including attentional, volitional, and epistemic aspects of the mind that do not necessarily go together (Premack, 1988; Tomasello et al., 2003). Thus, it is conceivable that there may be some organisms that attribute goals and perceptions but not beliefs. Leaving aside those considerations, we should not loose sight of the fact that there has been great progress in this area, particularly in the last 10 years. It is very likely that the next decade will continue to witness important advances in this area. The recent addition of nonprimate species such as canids and corvids in the research program has already started to produce some intriguing and provocative results (Emery and Clayton, 2001; Bräuer et al., 2004; Bugnyar et al., 2004; Dally et al., 2004).

Acknowledgments

This chapter has benefited tremendously from the discussions that I had on this topic with Mike Tomasello and Celia Heyes. Mike Tomasello also provided useful comments on a draft of this chapter. I also thank Nora Tippmann and Juliane Bräuer for help with the drawings in Fig. 1 and Josefine Kalbitz for Fig. 2.

References

Baron-Cohen, S., Leslie, A.M. and Frith, U. (1985) Does the autistic child have a "theory of mind"? Cognition, 21: 37–46.

Bartsch, K. and Wellman, H.M. (1995) Children Talk About the Mind. Oxford University Press, New York.

Boesch, C. and Boesch-Achermann, H. (2000) The Chimpanzees of the Tai Forest. Oxford University Press, New York.

Bräuer, J., Call, J. and Tomasello, M. (2004) Visual perspective taking in dogs (*Canis familiaris*) in the presence of barriers. Appl. Anim. Behav. Sci., 88: 299–317.

Bräuer, J., Call, J. and Tomasello, M. (in press) Chimpanzees really know what others can see in a competitive situation. Anim. Cogn.

Bräuer, J., Call, J. and Tomasello, M. (in press) Chimpanzees do not take into account what others can hear in a competitive situation. Anim. Cogn.

Bugnyar, T., Stowe, M. and Heinrich, B. (2004) Ravens, *Corvus corax*, follow gaze direction of humans around obstacles. Proc. R. Soc. Lond. B Biol. Sci., 271: 1331–1336.

Byrne, R.W. and Whiten, A. (1988) Machiavellian intelligence. Social expertise and the evolution of intellect in monkeys, apes, and humans. Oxford University Press, Oxford.

Call, J. (2001) Chimpanzee social cognition. Trends Cogn. Sci., 5: 369–405.

Call, J. (2003) Beyond learning fixed rules and social cues: abstraction in the social arena. Trans. R. Soc., 358: 1189–1196.

Call, J. (2004) Inferences about the location of food in the great apes (*Pan paniscus*, *Pan troglodytes*, *Gorilla gorilla*, *Pongo pygmaeus*). J. Comp. Psychol., 118: 232–241.

Call, J. (2006) Descartes' two errors: Reasoning and reflection from a comparative perspective. In: Hurley S. and Nudds M. (Eds.), Rational Animals. Oxford University Press, Oxford, pp. 219–234.

Call, J. (2007) Social knowledge in primates. In: Barrett L. and Dunbar R.I.M. (Eds.), Handbook of Evolutionary Psychology. Oxford University Press, Oxford, pp. 71–81.

Call, J., Agnetta, B. and Tomasello, M. (2000) Social cues that chimpanzees do and do not use to find hidden objects. Anim. Cogn., 3: 23–34.

Call, J. and Tomasello, M. (1994) Production and comprehension of referential pointing by orangutans (*Pongo pygmaeus*). J. Comp. Psychol., 108: 307–317.

Call, J. and Tomasello, M. (1999) A nonverbal theory of mind test. The performance of children and apes. Child Dev., 70: 381–395.

Clements, W.A. and Perner, J. (1994) Implicit understanding of belief. Cogn. Dev., 9: 377–395.

Dally, J.M., Emery, N.J. and Clayton, N.S. (2004) Cache protection strategies by western scrub-jays (*Aphelocoma californica*): hiding food in the shade. Biol. Lett., 271: S387–S390.

Domjan, M. and Burkhard, B. (1986) The Principles of Learning and Behavior. Prentice-Hall, Englewood Cliffs, NJ.

Emery, N.J. and Clayton, N.S. (2001) Effects of experience and social context on prospective caching strategies by scrub jays. Nature, 414: 443–446.

Flavell, J.H. (1992) Perspectives on perspective taking. In: Beilin H. and Pufall P.B. (Eds.), Piaget's Theory: Prospects and Possibilities. Lawrence Erlbaum Associates, Hillsdale, NJ, pp. 107–139.

Flombaum, J.I. and Santos, L.R. (2005) Rhesus monkeys attribute perceptions to others. Curr. Biol., 15: 447–452.

Gallup, G.G. (1982) Self-awareness and the emergence of mind in primates. Am. J. Primatol., 2: 237–248.

Ghazanfar, A.A. and Santos, L.R. (2004) Primate brains in the wild: The sensory bases for social interactions. Nat. Rev. Neuros., 5: 603–616.

Gómez, J.C. (1996) Non-human primate theories of (non-human primate) minds: Some issues concerning the origins of mind-reading. In: Carruthers P. and Smith P.K. (Eds.), Theories of Mind. Cambridge University Press, Cambridge, pp. 330–343.

Hare, B. (2001) Can competitive paradigms increase the validity of experiments on primate social cognition? Anim. Cogn., 4: 269–280.

Hare, B., Adessi, E., Call, J., Tomasello, M. and Visalberghi, E. (2002) Do capuchin monkeys, *Cebus apella*, know what conspecifics do and do not see? Anim. Behav., 63: 131–142.

Hare, B., Call, J., Agnetta, B. and Tomasello, M. (2000) Chimpanzees know what conspecifics do and do not see. Anim. Behav., 59: 771–785.

Hare, B., Call, J. and Tomasello, M. (2001) Do chimpanzees know what conspecifics know and do not know? Anim. Behav., 61: 139–151.

Hare, B., Call, J. and Tomasello, M. (2006) Chimpanzees deceive a human competitor by hiding. Cognition, 101: 495–514.

Harlow, H.F. (1951) Primate learning. In: Stone C.P. (Ed.), Comparative Psychology. Prentice-Hall, Englewood Cliffs, NJ, pp. 183–238.

Haun, D.B.M., Call, J., Janzen, G. and Levinson, S.C. (2006) Evolutionary psychology of spatial representations in the Hominidae. Curr. Biol., 16: 1736–1740.

Hauser, M.D. (1990) Do chimpanzee copulatory calls incite male-male competition? Anim. Behav., 39: 596–597.

Heyes, C.M. (1993) Anecdotes, training, traping and triangulating: do animals attribute mental states? Anim. Behav., 46: 177–188.

Heyes, C.M. (1994) Cues, convergence and a curmudgeon: A reply to Povinelli. Anim. Behav., 48: 242–244.

Heyes, C.M. (1998) Theory of mind in nonhuman primates. Behav. Brain Sci., 21: 101–134.

Humphrey, N.K. (1976) The social function of the intellect. In: Bateson P.P.G. and Hinde R.A. (Eds.), Growing Points in Ethology. Cambridge University Press, Cambridge, pp. 7–54.

Jolly, A. (1966) Lemur social behavior and primate intelligence. Science, 153: 501–506.

Kaminski, J., Call, J. and Tomasello, M. (2004) Body orientation and face orientation: two factors controlling apes' begging behavior from humans. Anim. Cogn., 7: 216–223.

Karin-D'Arcy, M.R. and Povinelli, D.J. (2002) Do chimpanzees know what each other see? A closer look. Int. J. Comp. Psychol., 15: 21–54.

Kummer, H. (1967) Tripartite relations in hamadryas baboons. In: Altmann S.A. (Ed.), Social Communication among Primates. The University of Chicago Press, Chicago, IL, pp. 63–71.

Lempers, J.D., Flavell, E.R. and Flavell, J.H. (1977) The development in very young children of tacit knowledge concerning visual perception. Genet. Psychol. Monogr., 95: 3–53.

Liebal, K., Pika, S., Call, J. and Tomasello, M. (2004) To move or not to move: how apes alter the attentional states of humans when begging for food. Inter. Stud., 5: 199–219.

Melis, A.P., Call, J. and Tomasello, M. (2006) Chimpanzees conceal visual and auditory information from others. J. Comp. Psychol., 120: 154–162.

Menzel, E.W. (1974) A group of young chimpanzees in a one-acre field: Leadership and communication. In: Schrier A.M. and Stollnitz F. (Eds.), Behavior of Nonhuman Primates. Academic Press, New York, pp. 83–153.

Onishi, K.H. and Baillargeon, R. (2005) Do 15-month-old infants understand false beliefs? Science, 308: 255–258.

Perner, J., Leekam, S.R. and Wimmer, H. (1987) Three-year-olds' difficulty with false belief: the case for a conceptual deficit. Br. J. Dev. Psychol., 5: 125–137.

Piaget, J. and Inhelder, B. (1956) The child's conception of space. Norton, New York.

Povinelli, D.J. (1994) Comparative studies of animal mental state attribution: a reply to Heyes. Anim. Behav., 48: 239–241.

Povinelli, D.J. and Eddy, T.J. (1996) What young chimpanzees know about seeing. Monogr. Soc. Res. Child Dev., 61: 1–152.

Povinelli, D.J., Nelson, K.E. and Boysen, S.T. (1990) Inferences about guessing and knowing by chimpanzees (*Pan troglodytes*). J. Comp. Psychol., 104: 203–210.

Povinelli, D.J., Theall, L.A., Reaux, J.E. and Dunphy-Lelii, S. (2003) Chimpanzees spontaneously alter the location of their gestures to match the attentional orientation of others. Anim. Behav., 66: 71–79.

Povinelli, D.J. and Vonk, J. (2003) Chimpanzee minds: suspiciously human? Trends Cogn. Sci., 7: 157–160.

Premack, D. (1988) 'Does the chimpanzee have a theory of mind?' Revisited. In: Byrne R.W. and Whiten A. (Eds.), Machiavellian Intelligence. Social Expertise and the Evolution of Intellect in Monkeys, Apes, and Humans. Oxford University Press, Oxford, pp. 160–179.

Premack, D. and Woodruff, G. (1978) Does the chimpanzee have a theory of mind? Behav. Brain Sci., 4: 515–526.

Reaux, J.E., Theall, L.A. and Povinelli, D.J. (1999) A longitudinal investigation of chimpanzees' understanding of visual perception. Child Dev., 70: 275–290.

Ruffman, T. and Perner, J. (2005) Do infants really understand false belief? Trends Cogn. Sci., 9: 462–463.

Santos, L.R., Nissen, A.G. and Ferrugia, J.A. (2006) Rhesus monkeys, *Macaca mulatta*, know what others can and cannot hear. Anim. Behav., 71: 1175–1181.

Savage-Rumbaugh, E.S., Rumbaugh, D. and Boysen, S.T. (1978) Sarah's problems in comprehension. Behav. Brain Sci., 1: 555–557.

Slaughter, V. and Mealey, L. (1998) Seeing is not (necessarily) believing. Behav. Brain Sci., 21: 130.

Tomasello, M. and Call, J. (1997) Primate Cognition. Oxford University Press, New York.

Tomasello, M. and Call, J. (2006) Do chimpanzees know what others see — or only what they are looking at? In: Hurley S. and Nudds M. (Eds.), Rational Animals. Oxford University Press, Oxford, pp. 371–384.

Tomasello, M., Call, J. and Hare, B. (2003) Chimpanzees understand psychological states — the question is which ones and to what extent. Trends Cogn. Sci., 7: 153–156.

Tomasello, M., Call, J., Nagell, K., Olguin, R. and Carpenter, M. (1994) The learning and use of gestural signals by young chimpanzees: a trans-generational study. Primates, 35: 137–154.

Vlamings, P.H.J.M., Uher, J. and Call, J. (2006) How the great apes (*Pan troglodytes*, *Pongo pygmaeus*, *Pan paniscus*, and *Gorilla gorilla*) perform on the reversed contingency task: the effects of food quantity and food visibility. J. Exp. Psychol. Anim. Behav. Process., 32: 60–70.

Whiten, A. (1991) Natural Theories of Mind. Blackwell, Oxford.

Whiten, A. (1996) When does smart behaviour-reading become mind-reading? In: Carruthers P. and Smith P.K. (Eds.), Theories of Theories of Mind. Cambridge University Press, Cambridge, pp. 277–292.

Whiten, A. (1998) Triangulation, intervening variables, and experience projection. Behav. Brain Sci., 21: 132–133.

Whiten, A. and Byrne, R.W. (1988) The manipulation of attention in primate tactical deception. In: Byrne R.W. and Whiten A. (Eds.), Machiavellian intelligence. Social Expertise and the Evolution of Intellect in Monkeys, Apes, and Humans. Oxford University Press, Oxford, pp. 211–223.

Wimmer, H. and Perner, J. (1983) Beliefs about beliefs: representation and constraining function of wrong beliefs in young children's understanding of deception. Cognition, 13: 103–128.

Zentall, T.R. (1998) What can we learn from the absence of evidence? Behav. Brain Sci., 21: 133–134.

C. von Hofsten & K. Rosander (Eds.)
Progress in Brain Research, Vol. 164
ISSN 0079-6123

CHAPTER 20

Infancy and autism: progress, prospects, and challenges

Mayada Elsabbagh* and Mark H. Johnson

Centre for Brain and Cognitive Development, Birkbeck College, University of London, Henry Wellcome Building, WC1E 7HX, London, UK

Abstract: We integrate converging evidence from a variety of research areas in typical and atypical development to motivate a developmental framework for understanding the emergence of autism in infancy and to propose future directions for a recent area of research focusing on infant siblings of children with autism. Explaining the cognitive profile in autism is best achieved through tracing the process through which associated symptoms emerge over development. Understanding this process would shed light on the underlying causes of this multifaceted condition through clarifying how and why a variety of risk factors, single or in combination, exert an impact on the resulting phenotype. We emphasize the importance of integrating theoretical models of typical development in understanding atypical development and argue for the need to develop continuous and individually valid measures for at-risk infants both for predictors and outcomes of autism symptoms.

Keywords: autism; autism spectrum disorder; social orienting; executive function; perception; development; infancy; infant siblings; core deficits

Autism and infancy: an overview

Much research into autism over the past decades has presented a neat developmental story: a form of neurodevelopmental deficit, commonly attributed to brain networks subserving social cognition, leads to decreased attention to, or interest in, the social world. This early lack of attention to or interest in social stimuli interferes with the emergence of critical developmental milestones relevant for social cognition such as shared attention. These cascading influences eventually lead to deficits in theory of mind, and preclude the typical development of socio-communicative skills. Compelling as they are, the elements of such developmental accounts have proved difficult to empirically verify. Because a confirmed diagnosis of autism can only be made from ~3 years of age, our knowledge of the early neural, behavioral, and cognitive profile in autism is very poor, and virtually nothing is known about the underlying causes, or the process through which these deficits emerge. As a result, and despite substantial parallel advances in research on typical development in infancy and research on children and adults with autism, until very recently the two disciplines have rarely been combined.

The purpose of the current chapter is two-fold: First, to review the recent empirical evidence of the neural, cognitive, and behavioral profile of autism in the first few years of life, and second, to situate

*Corresponding author. Tel.: +44 207 631 6231;
Fax: +44 207 631 6587; E-mail: m.elsabbagh@bbk.ac.uk

DOI: 10.1016/S0079-6123(07)64020-5

this fragmentary knowledge within a developmental framework relevant for the study of autism in infancy, specifically through the study of at-risk infant siblings of children diagnosed with autism. We propose that autism in infancy is best understood through a combination of genetic, neural, and cognitive risk factors; the presence and severity of each as well as their interactions over development, give rise to the typical yet considerably variable phenotype observed in autism. This approach is valuable, not only for understanding how symptoms unfold over development, but also for understanding the mechanisms that drive and constrain typical development. In the remainder of this section we will present an overview of how autism is defined and identified. In the next section, we will describe some of the theoretical approaches to the neurocognitive study of this disorder, and review the current evidence relating to them. Finally, in the last section, we will discuss a recent approach focusing on the study of at-risk infants, focusing on the prospects and challenges of this approach.

Autism (or Autism Spectrum Disorder, ASD) is usually defined using a triad of impairment in social interaction, communication, and the presence of restricted and repetitive patterns of behavior (World Health Organization, 1992; American Psychiatric Association, 1994). Some form of autism currently affects 1 in 100 children, with a higher prevalence in males relative to females (Baird et al., 2006).

A confirmed diagnosis of autism is rarely given before 3 years of age; only after this age are the core symptoms sufficiently clear and can developmental delay or other comorbid conditions be clearly ruled out. Deficits or delays in the expected language and social milestones found in typical development, e.g., no babbling by 12 months, no pointing, lack of social smile, or the loss of language or social skills are considered "red flags" for autism (Filipek et al., 1999; Zwaigenbaum, 2001; Sigman et al., 2004), but no tool as yet has provided a reliable and stable diagnosis in the first years of life.

Although the precise causes of autism remain unknown, there is general consensus that it can emerge as a result of interactions among several genetic, prenatal, postnatal, and external environmental factors. Genetic factors appear to play a critical role in autism. The genetic basis of autism have been confirmed through converging lines of evidence from familial aggregation, twin studies, and overlap with other conditions (Autism Genome Project, 2007; Bailey et al., 1995; Newschaffer et al., 2002; Muhle et al., 2004). Nevertheless, there is still neither a clear pattern of inheritance nor a set of specific genes identified. Family aggregation studies indicate that the risk of autism for siblings is increased to 5–10%. One of the most intriguing manifestations of the genetic risk for autism is that it appears to result in an *extended phenotype*. This refers to behavioral and brain characteristics associated with the autism phenotype found not only in affected individuals, but also in their genetic relatives, and to a lesser degree in the general population (Bolton et al., 1994; Pickles et al., 2000; Dawson et al., 2002c). This implies that several characteristics of autism are not themselves uncommon or atypical, but that the number and severity of these symptoms within an individual determine whether they manifest as normative differences in processing styles or in personalities, as opposed to diagnostic symptoms.

Aside from genetic risk, several comorbid conditions that constitute diagnosable medical conditions are associated with autism, including epilepsy, Moebius, Fragile X, and tuberous sclerosis. The study of some of these comorbid conditions has provided some evidence that autism may also be associated with prenatal "accidents." For instance, exposure to teratogens in highly specific time windows may result in an increased risk for autism (Arndt et al., 2005; Miller et al., 2005). However, the precise contribution of such environmental factors remains unclear.

Approaches to the neurocognitive study of autism in infancy

A wide range of theoretical approaches have attempted to specify the "core deficits" in autism in adults and older children. This refers to aspects of cognition and their neural underpinnings that present a unifying and parsimonious characterization and/or causal explanation of autism. Several core deficits have been proposed. These include deficits in social orienting, attention, and motivation; deficits in executive function; and differences in low-level perceptual and attentional processes. Of course, these are not mutually exclusive possibilities, at least as far as characterizing the disorder. However, as we will discuss in more detail in the next section, these characterizations are frequently used to infer underlying causal mechanisms, where other abnormalities are viewed as "secondary" or "peripheral" deficits, resulting from the core deficit. Our focus in this section is to evaluate, against the available evidence, the extent to which these approaches can provide plausible developmental models for the emergence of autism over the first years of life. Before we begin, it is important to address some relevant methodological issues: How is a disorder that is only diagnosed after 3 years studied in infancy?

The fact that a valid and stable diagnosis of autism cannot be obtained prior to 3 years of age, has not been a deterrent in recognizing the importance of this area of research. Consequently, several complementary approaches for investigating autism in the early years have been pursued, yielding valuable results. Some studies have screened for autism in the general population or in high-risk populations, sampling large groups of infants in order to develop appropriate clinical tools (Baron-Cohen et al., 1992; Lord, 1995; Stone et al., 1999; Wetherby et al., 2004). These children are then followed up to the age when some receive a diagnosis and others do not. Studies of this sort are effective in providing some general characteristics of autism early on, but are limited as far as in-depth or experimental assessment, as they require very large samples. Other studies are retrospective, relying on parental descriptions of their children prior to diagnosis (e.g., Wimpory et al., 2000; Werner et al., 2005) or on analysis of home videos of children later diagnosed with autism (reviewed in Palomo et al., 2006). In general, parents appear to be able to provide some valid and specific descriptions of developmental delays and deficits in their children (Filipek et al., 2000), but these studies are inevitably subject to recollection bias. Data from home videos help overcome this problem but are limited by variations in contexts and the lack of experimental control. An alternative approach is to study young children as soon as they are diagnosed, investigating various aspects of perception and cognition (several of these will be presented shortly). Recently, another approach, focusing on the study of infants at risk for autism prior to the onset of overt symptoms, has overcome some of the major challenges faced by these other approaches (Zwaigenbaum et al., 2007). This approach will be discussed in detail in the next section.

In what follows we will present some dominant theoretical approaches to autism (social orienting, executive function, and low-level perceptual and attentional processes). For each putative core deficit, we will briefly describe the neural underpinnings of the target ability in typical adults, its functional development in typical infants, and the model explaining its abnormal functioning in autism. It will become apparent that these approaches often lack a distinction between childhood and adulthood profiles, with the expectation that a core deficit will be present early in life, but manifest slightly differently at different ages. Because our focus is to draw the implications these models have to the development of autism in infancy, we will evaluate each approach against a representative review of the literature focusing on studies of young children (mostly below 6 years) already diagnosed with autism, as well as selective findings from the retrospective approaches mentioned above. Several methodological issues, relevant for the study of special populations, e.g., age of participants, their developmental level, the choice of control groups, etc., are frequently invoked when results across studies do not converge. To allow comparisons across these studies, we will include several of these methodological details. These include sample characteristics as well as information regarding how the groups were matched. They also include brief descriptions of the stimuli and measures used in each study.

Social vs. non-social orienting, attention, and cognition

Autism is primarily defined and diagnosed based on impairment in social interactions and communication. Consequently, it is no surprise that several theoretical approaches have invoked deficits in the neural systems mediating social interactions as characterizing the disorder. In general, these approaches draw on contrasts in the encoding, processing, and interpretation of the social, relative to the physical, world in autism.

In typical adults social stimuli and social contexts are processed by specialized neural systems including the fusiform area, the superior temporal sulcus, and the orbitofrontal cortex (Adolphs, 2003). Further specialization within this "social brain" network has been described. For instance, the fusiform area appears to be selective for faces relative to objects. Detecting the movement of eye gaze in another's face activates the superior temporal sulcus (STS) and the neighboring MT/V5 region (Allison et al., 2000). The basis of these patterns of cortical specialization is currently the subject of debate. Some view specialized areas as "pre-wired" for specific stimuli, e.g., the FFA for faces, whereas others suggest that the specialization is driven by perceptual expertise in a given category of stimuli (Tarr and Gauthier, 2000; Gauthier and Nelson, 2001). An intermediate view is that the typical patterns of cortical specialization for social stimuli emerge during development as a result of interactions and competition between regions that begin with only broadly tuned biases (Johnson, 2001). According to the latter view, some skills may be present at birth, while others become increasingly specialized over time, showing gradual developmental changes leading to adult-like processing patterns.

Johnson and Morton (1991) proposed that a subcortical orienting system (which they termed "conspec") initially biases the newborn to attend toward faces. While this putative system is based on simple low-spatial frequency patterns characteristic of faces, it is sufficient to bias the input to developing cortical visual areas (Johnson, 2005). In response, some of these cortical areas will increase their specialization for faces and related social stimuli. A consequence of this will be increasingly focal patterns of cortical activation with increasing age. Several lines of evidence indicate that infants as young as 2 months of age show activation in similar areas as those found in adults in response to faces. However, the activated network in infants is broader encompassing areas such as the prefrontal cortex (Tzourio-Mazoyer et al., 2002; de Haan et al., 2003; Johnson et al., 2005).

These findings from the adult and infant literature motivated the view of autism as a case of damage to, or lack of specialization of, this social brain network (Klin et al., 2003; Dalton et al., 2005; Dawson et al., 2005; Johnson et al., 2005; Schultz, 2005). Behaviorally, this is manifest in a wide range of impairments or atypical processing of social stimuli and contexts. Not only are social-communicative difficulties diagnostic of the disorder, but they have also been clearly demonstrated through a variety of experimental paradigms and procedures. For instance, one of the most popular views of the core deficit in autism is the marked impairment in inferring other people's beliefs and desires, or theory of mind (Baron-Cohen et al., 1985; Leslie and Thaiss, 1992), even for individuals with average or above average cognitive ability. Similarly, studies examining the neural underpinning of these behavioral differences, have documented atypical activation patterns associated with such tasks (Happé et al., 1996; Baron-Cohen et al., 1999).

Beyond this consistency in the defining features of autism, the functioning of other aspects of the social brain network and the developmental origins of these difficulties is subject of continued debate. This is clearly illustrated in research on face processing in autism. Behavioral studies on the scanning and processing of faces (reviewed in Jemel et al., 2006; Sasson, 2006) are equivocal regarding the nature of the deficits found in autism. On the other hand, there is some consensus that performance of individuals with autism differs from controls insofar as the tasks employed tap into differences in *default* encoding and processing strategies. Individuals with autism scan faces less than typically developing controls, they focus on different regions of the face, and they display greater reliance on featural relative to configural analysis, whereas the opposite strategy is used more by typical individuals. Depending on the task demands, these atypical strategies may or may not produce differences in overall performance levels relative to those found in controls. This implies that individuals with autism may achieve an equivalent level of overall competence on face processing tasks, albeit relying on different underlying strategies. Neuroanatomical studies (reviewed in Schultz, 2005) support this conclusion. Atypical activation patterns in the fusiform and the amygdala are usually found in individuals with autism in response to faces. Furthermore, these neural patterns appear to correlate with atypical behavioral scanning patterns in these individuals (Dalton et al., 2005, 2007).

Based on these adult approaches, developmental models of autism have focused on understanding the precursors of their social difficulties including orienting to social stimuli and events, joint attention, imitation, and social interactions. At the neural level, some developmental models attribute these social difficulties to impairment of mechanisms such as the conspec, which bias infants to orient toward and assign value to socially relevant stimuli. Different models propose alternative explanations of the origins of these difficulties. Some propose that there are impairments in the cortical mechanisms subserving the social brain network, such as the FFA. Other models propose that subcortical impairment in the amygdala leads to reduced attention to (Dawson et al., 2005; Schultz, 2005) or even aversion of socially relevant stimuli such as eye gaze (Dalton et al., 2005). Cascading effects would then follow, precluding the development of social-communicative skills in the typical fashion. With these theoretical proposals in mind, we will now present the available evidence from the early years including studies on social orienting and scanning (Table 1); face, gaze, and emotion processing (Table 2); imitation and understanding of contingency (Table 3); joint attention and communication (Table 4).

Social orienting and scanning

Retrospective studies looking back at the first 2 years of life consistently show less orienting toward social stimuli and a reduced response to name calling from as early as 9 months (Mars et al., 1998; Osterling et al., 2002; Maestro et al., 2005; Werner and Dawson, 2005) or younger (Maestro et al., 2002) in children later diagnosed with autism as compared to those later diagnosed with developmental delay. Consistent with this, experimental studies with already diagnosed children show that they orient less to social stimuli such as people, voices, etc., as compared to non-social stimuli such as objects, sounds, etc., compared to both children of the same age and developmental level (see Table 1 for summary). It is worth noting that these latter studies also report an overall decreased level of orienting to both social and non-social stimuli, but the impairment is greater for the social stimuli. Similarly, in retrospective studies, excessive looking at objects is a reliable risk marker but only when combined with reduced social orienting (Osterling et al., 2002). One possibility is that the increased interest in objects is a consequence of a decreased interest in people.

Compared with numerous studies documenting atypical face scanning in older children and adults, only one study examined this in young children with autism. Anderson et al. (2006) found no differences in visual scanning of social stimuli but they did for landscape images. On the other hand, pupillary responses that index attentional engagement and information processing differentiated the groups when children's faces were viewed, but not when animal faces or landscapes were viewed. Interestingly, scanning of landscapes correlated with diagnostic symptoms in the ASD group; toddlers who fixated more briefly showed less repetitive behaviors (Anderson et al., 2006).

Face and gaze processing

Available studies of face processing in young children with autism point to a pattern of impairment in comparison to both children of the same age and developmentally matched children (see Table 2 for summary). Difficulties in face processing include discrimination as well as understanding of emotion and eye gaze (but see Chawarska et al., 2003). Interestingly, there is evidence for improvement in behavioral performance in slightly older children by ~4 years of age (Chawarska and

Table 1. Social vs. non-social orienting and scanning

Study	Groups	*n* (M:F)	Age (sd)	Matching	Tasks	Measures	Key results
Swettenham et al. (1998)	ASD DD TD	10 17 16	20.7 (1.3) 19.9 (1.4) 20.1 (1.3)	CA	5-min free play session	Looking time, duration, attention shifting to objects or people	Relative to both groups, ASD looked less and switched attention less overall ASD looked less and for shorter duration at people and looked more and for longer at objects. They also shifted attention less between objects and people
Dawson et al. (1998)	ASD DS TD	20 (19:1) 19 (16:3) 20 (19:1)	64.6 65.3 30.9	MA	Social: clapping, name calling Non-social: musical toy, rattle	Head turns	ASD oriented less to all stimuli but the impairment was worse for social stimuli
Dawson et al. (2004a)	ASD DD TD	72 (60:12) 34 (18:16) 39 (30:9)	43.5 (4.3) 44.8 (5.3) 27.1 (8.9)	MA	Social: humming, name calling, snapping fingers, tapping thighs Non-social: timer bleep, phone ring, whistle, car horn	Head turns	ASD oriented less to all stimuli but the impairment was worse for social stimuli
Anderson et al. (2006)	ASD DD TD	9 (8:1) 6 (6:0) 9 (8:1)	49.6 46.3 49.8	CA	Static unimodal images of animal faces, children's faces, landscapes	Scanning and attentional engagement (papillary responses)	Relative to both groups, ASD scanned landscapes less but did not differ in scanning faces ASD showed pupil restriction to internal features of children's faces relative to both groups that showed dilation

Notes: ASD, children with autism; DD, children with developmental delay; DS, children with Down syndrome; TD, typically-developing children; *n* (M:F), number of participants and ratio of males to females in each group; age (sd), mean age in months and standard deviation for each group; CA, group matching based on chronological age; MA, group matching based on mental age.

Table 2. Face, gaze, and emotion processing

Study	Groups	*n* (M:F)	Age (sd)	Matching	Tasks	Measures	Key results
Chawarska and Volkmar (2007) Experiment 1	ASD DD TD	14 (10:4) 16 (12:4) 10 (8:2)	27.9 (4.9) 23.7 (6.7) 19.7 (4.7)	MA	Human and monkey faces	Novelty preference (face recognition)	TD showed enhanced recognition of human but not monkey faces, whereas neither ASD nor DD showed evidence of face recognition regardless of the species
Chawarska and Volkmar (2007) Experiment 2	ASD DD TD	15 (14:1) 10 (8:2) 13 (7:6)	45.9 (6.7) 39.5 (6.7) 38.0 (8.2)	MA	Human and monkey faces	Novelty preference (face recognition)	No group differences. Both human and monkey faces are recognized by all groups
Chawarska et al. (2003)	ASD TD	15 (12:3) 12 (7:5)	28 (6.6) 26 (6.5)	CA	Images of faces where eye gaze is congruent or incongruent with the appearance of a lateral object	Reaction time differences in congruent vs. incongruent trials	No interaction between group (ASD vs. TD) and condition (congruent vs. incongruent)
Johnson et al. (2005)	ASD LD TD	9 13 15	33.4 (5.8) 34.7 (7.4) 36.2 (7.4)	VMA	Images of faces where eye gaze is congruent or incongruent with the appearance of a lateral object	Reaction time differences in congruent vs. incongruent trials	ASD show no difference between congruent and incongruent conditions, whereas TD and LD were faster in the congruent condition
Dawson et al. (2002a)	ASD DD TD	33 (29:4) 17 (11:6) 21 (19:2)	43.5 (4.3) 45.1 (4.9) 45.5 (5.8)	CA	Familiar and unfamiliar faces and objects	Amplitude and latency of ERP	Relative to both groups, ASD did not show a differential response to faces relative to objects but do show P400 and Nc amplitude differences to objects
Gepner et al. (2001)	ASD TD	13 (12:1) 13 (12:1)	69.3 (11.0) 40.7 (13.6)	MA	Still, dynamic, and strobe emotional or non-emotional facial expressions	Accuracy on matching expressions	Mode of presentation affects performance in the emotional condition for TD but not for ASD
Dawson et al. (2004b)	ASD TD	29 (26:3) 22 (19:3)	44.8 (10.0) 43.7 (7.0)	CA	Static photos of an unfamiliar woman posing neutral and fear faces	Amplitude and latency of ERP	Both groups showed faster response to fear TD showed larger N300 and NSW amplitudes to fear compared to neutral whereas ASD showed no difference

Notes: ASD, children with autism or Autism Spectrum Disorder; DD, children with developmental delay; LD, children with language delay; TD, typically developing children; *n* (M:F), number of participants and ratio of males to females in each group; age (sd), mean age in months and standard deviation for each group; CA, group matching based on chronological age; MA, group matching based on mental age; VMA, group matching based on verbal mental age; P400, Nc, N300 and NSW, ERP components sensitive to face processing.

Table 3. Imitation and understanding of contingency

Study	Groups	*n* (M:F)	Age (sd)	Matching	Tasks	Measures	Key results
Rogers et al. (2003)	ASD DD TD	24 (20:4) 20 (9:11) 15 (6:9)	34.2 (3.6) 34.1 (6.5) 21.3 (1.5)	MA	Manual, oral-facial, object-oriented imitation	Imitative behaviors	Relative to both groups, ASD showed overall worse performance ASD were worse on oral-facial and object-oriented imitation relative to manual imitation Imitation problems were unrelated to motor demands
Dawson and Adams (1984)	ASD	15 (12:3)	60.6 (8.2)	Norm	Simultaneous imitation of the child's own actions, a familiar or a novel action, spontaneous imitation, object permanence	Frequency of imitation, social responsiveness	ASD performed well on object permanence but poorly on imitation tasks Those who did poorly on imitation were more socially responsive when the examiner imitated their actions; they were also more likely to imitate familiar relative to novel actions
Escalona et al. (2002)	ASD	20 (12:8)	61.2		Children were randomly assigned to groups where the examiner either imitated their action or was contingently responsive	Behaviors indicating engagement with the adult	Both conditions resulted in increased responsiveness from the child The imitation condition showed additional improvement such as decrease in motor activity and a greater increase in touching the adult

Notes: ASD, children with autism; DD, children with developmental delay; TD, typically developing children; *n* (M:F), number of participants and ratio of males to females in each group; age (sd), mean age in months and standard deviation for each group; MA, group matching based on mental age.

Table 4. Joint attention and communication

Study	Groups	*n* (M:F)	Age (sd)	Matching	Tasks	Measures	Key results
Charman et al. (1998)	ASD PDD DD	8 (7:1) 13 (11:2) 8 (3:5)	21.4 (1.8) 20.5 (1.7) 20.5 (1.0)	MA	Simulated distress by the examiner; spontaneous play; joint attention; goal detection; imitation	Number and quality of target responses	Relative to both groups, ASD did not use gaze declaratively, and showed less target behaviors including empathetic responses, imitation of actions on objects, and pretend play
Stone et al. (1997)	ASD DD/LD	14 14	32.8 (3.5) 31.8 (4.4)	CA & MA	16 situations which elicit requesting or commenting behavior	Number and quality of communicative acts	ASD showed less communicative acts ASD were more likely to directly request objects/actions than to direct the examiner's attention through pointing or eye gaze
Mundy et al. (1986)	ASD DD TD	18 (14:4) 18 18	53.3 50.2 22.2	MA	ESCS; structured and unstructured play	Initiating and responding to social interactions; joint attention; behavior regulation	Relative to both groups, ASD showed less frequency of higher level social interaction and requesting, but they showed the same frequency of lower level ones Relative to both groups, ASD showed deficits in structured but not unstructured play behavior Initiating joint attention was the best predictor of group membership
Dawson et al. (2004a)	ASD DD TD	72 (60:12) 34 (18:16) 39 (30:9)	43.5 (4.3) 44.8 (5.3) 27.1 (8.9)	MA	Simulated distress by the examiner compared to a situation where the examiner displays neutral affect	Attention and reactions	Relative to both groups, ASD looked less at the examiner and showed less concern in the distress relative to the neutral condition
Bacon et al. (1998)	HiASD LoASD LD DD TD	32 (26:6) 51 (47:4) 42 (29:13) 39 (28:11) 29 (18:11)	55.8 (14.3) 62.0 (17.9) 53.9 (12.5) 55.8 (14.6) 55.7 (13.4)		Non-social orienting; simulated distress by the examiner	Orienting; response to distress	Relative to other groups, LoASD showed deficits across all measures, whereas HiASD were moderately impaired in prosocial behaviors Social referencing differentiated the ASD group from all others
Sigman et al. (1992)	ASD DD TD	30 (27:3) 30 (18:12) 30 (27:3)	42.4 (11.2) 41.6 (10.7) 19.83 (8.2)	MA	Simulated distress/fear by the parent or the examiner	Response to distress	Relative to both groups, ASD were significantly less attentive to adults' distress or fear and less likely to show social referencing
Sigman et al. (2003)	ASD DD	22 (20:2) 22 (14:8)	51.3 (11.1) 47.3 (13.1)	MA	Video exposure to an infant crying or playing; interaction with mother or stranger; separation and reunion with the caregiver	Behavioral and cardiac responses	ASD primarily differed in their responses to strangers where they were less interactive and showed less arousal

Table 4 (*continued*)

Study	Groups	*n* (M:F)	Age (sd)	Matching	Tasks	Measures	Key results
Dawson et al. (1990)	ASD TD	16 (13:3) 16 (13:3)	49.6 (12.1)	VMA	Free-play; structured communicative demands; face-to-face interactions	Type of activity; gaze; affective behavior	ASD engaged in non-social activities more than TD ASD were less likely to combine smiles with sustained eye gaze and to smile in response to the mother's smile Mothers of ASD smiled less and were less likely to smile in response to their children's smiles
Kuhl et al. (2005)	ASD TDa TDb	29 (26:3) 29 (22:7) 15 (13:2)	45.3 27.7 48.3	MA & CA	Consonant-vowel syllables /ba/ and /wa/; infant-directed speech and a non-speech analogue	Auditory preference and ERP	Relative to both groups, ASD did not exhibit the typical mismatch negativity response to change in speech syllables Relative to both groups, ASD preferred to listen to non-speech analogues of infant-directed speech, whereas TD showed the opposite pattern Preference for non-speech analogues correlated with ASD symptoms

Notes: ASD, children with autism; HiASD, high-functioning children with autism, LoASD; low-functioning children with autism; PDD, children with pervasive developmental disorder; DD, children with developmental delay; TD, typically developing children; *n* (M:F), number of participants and ratio of males to females in each group; age (sd), mean age in months and standard deviation for each group; CA, group matching based on chronological age; MA, group matching based on mental age; VMA, group matching based on verbal mental age; ESCS, Early Social Communication Scales (Mundy et al., 1986).

Volkmar, 2007). Consistent with these results, a recent review concluded that impairments in face processing appear to be clearer in children relative to adults with autism (Sasson, 2006). However, electrophysiological studies suggest that atypical underlying neural processes mediate performance in both children (Table 2) and adults with autism (Dawson et al., 2004b, 2005).

Imitation and understanding of contingency

As evident from the studies summarized in Table 4, imitation is an area of serious difficulty for young children with autism. However, it appears that certain forms of imitation may pose more difficulty than others. For instance, manual imitation is less problematic for young children with autism than imitation of actions on objects. Interestingly, these studies suggest that imitative responses, and interactions with others in general, may be facilitated in children with autism, if they are based on perfect contingency, rather than on high, but imperfect, contingency patterns preferred by control children. This is consistent with studies of older children (Gergely and Watson, 1999). High, but imperfect, contingency characterizes most social interaction and this may therefore underlie a greater difficulty in dealing with people rather than objects (Dawson and Adams, 1984; Escalona et al., 2002).

Joint attention and communication

Relative to developmentally matched controls, young children with autism demonstrate serious impairments in several tasks of joint attention (summarized in Table 3). Additional data (not reported in Table 3) assessing these abilities using standardized or diagnostic instruments provide similar findings (e.g., Dawson et al., 1998, 2004a, b; Rogers et al., 2003; Wetherby et al., 2004). Further manifestations of social-communicative impairment include difficulties in social referencing, requesting, and noticing or attending to others' distress. Furthermore, retrospective studies in the second year of life also report that children later diagnosed with autism exhibit decreased initiation of social interactions and differences in verbal communication (reviewed in Palomo et al., 2006).

In fact, because the social-communicative difficulties encompass several areas, some studies have attempted to specify which aspects best differentiate children with autism from those with developmental delay. These studies suggest that problems in joint attention are specifically characteristic of young children with autism, relative to children with developmental delay or other comorbid conditions. Moreover, joint attention skills have been frequently found to correlate with difficulties in other areas reported above, including social orienting (Dawson et al., 1998, 2004a), emotion processing (Dawson et al., 2004b), and imitation (Rogers et al., 2003).

In sum, the evidence suggests that autism in early childhood is associated with serious impairments in several social-communicative skills. On the other hand, certain difficulties, namely social orienting and joint attention appear to more specifically characterize the disorder, relative to other conditions. Taken together, despite considerable research into social-communicative abilities in young children with autism and retrospective findings on these abilities, the evidence is limited as far as distinguishing fine-grained hypotheses regarding the origins of these difficulties. Research with older children and adults has suggested that the origin of differences in orienting to, and scanning of, social stimuli is a precursor to the more complex social deficits. However, this claim is difficult to verify given that these problems can also be the result of the social impairment rather than its precursor. By the age of diagnosis and in the few years that follow, several aspects of social-communicative functions are seriously disrupted, making it difficult to differentiate the causes from the consequences. Another challenge for this view is the suggestion that these social difficulties are the product of a qualitatively different form of impairment in non-social abilities. We now turn to review some of these findings, presenting the second general theoretical approach to autism, focusing on difficulties in non-social domains.

Executive function

Executive function is an umbrella term that refers to mechanisms that guide actions in an endogenous goal-driven way, despite conflicting demands from the immediate environment. These include planning, monitoring, working memory, and inhibition. In typical adults, multiple distributed neuroanatomical systems mediate these functions but they tend to be frequently associated with the prefrontal cortex. So-called "executive dysfunction" in autism is commonly manifest in terms of rigidity and preference for sameness, repetitive and restricted interests and behaviors, the latter being among the diagnostic symptoms. There is considerable debate as to whether executive function can be regarded as a core deficit in autism. These impairments tend to not be as universal as those found in social-communication. Executive function deficits tend to be present in various developmental conditions including Attention Deficit Disorder and Tourette Syndrome, but some argue that an atypical profile of executive dysfunction is specific to autism (Ozonoff and Jensen, 1999). Difficulty arises, in part, due to the wide range of tasks and different levels of complexity assessed by these executive function tasks. Performance of individuals with autism on tasks of planning and mental flexibility is more consistently impaired relative to performance on tasks of inhibition and self-monitoring (see Hill, 2004 for a review).

Our understanding of the functional development of the prefrontal cortex, the area frequently associated with executive function, is still limited. However, recent evidence suggests that task-dependent developmental changes in this area begin very early in life and some continue well into adulthood. There is also evidence that this area is important for the development of expertise in other domains including face processing (Johnson et al., 2005) and speech perception (Dehaene-Lambertz et al., 2002).

Developmental models of autism have taken different views on the emergence of executive dysfunction. Some view impairment in the prefrontal cortex as secondary to those found in social-communicative systems in the medial temporal lobe (Griffith et al., 1999; Dawson et al., 2002b). Others suggest the opposite pattern, where certain social deficits such as those in theory of mind are a consequence of executive dysfunction (Russell et al., 1999). Yet another view is that both social-orienting deficits and executive dysfunction originate from early impairments in disengagement of visual attention (Bryson et al., 2004). According to the latter view, the infant's inability to flexibly switch the locus of attention leads to problems in self-regulation as well as a decrease in social orienting. We now evaluate these suggestions against the evidence available from the early years.

Studies on executive function

Little is known regarding restricted and repetitive behaviors in the early years that become characteristic symptoms of older children and adults. Findings from retrospective studies indicate that these may not be prevalent prior to diagnosis, especially when children who are later diagnosed with autism are compared to children later diagnosed with mental retardation (reviewed in Palomo et al., 2006). Complicating this area of research is the fact that repetitive behaviors and preference for sameness manifest in qualitatively different ways in young children relative to adults. Moreover, some of these behaviors are abundant in typical infants and toddlers, e.g., reading the same book, holding onto the same blanket. However, by the age of two, studies using diagnostic instruments report a higher frequency of repetitive sensorimotor behaviors, some of which are atypical, in young children with autism (Lord et al., 2006; Richler et al., 2007).

As far as experimental work is concerned, a handful of studies have evaluated executive functions in young children with autism focusing on a range of skills, including rule-learning, working memory, and response inhibition (see Table 5 for summary). With the exception of one study (McEvoy et al., 1993), the remaining ones found that young children with autism do not perform at the level expected based on their chronological age but their performance is equivalent to children matched on the basis of mental age. This implies that unlike some social-communicative difficulties, performance on these tasks is not specific to the

Table 5. Executive function

Study	Groups	*n* (M:F)	Age (sd)	Matching	Tasks	Measures	Key results
Dawson et al. (2002b)	ASD DD TD	72 (60:12) 34 (18:16) 39 (28:11)	43.5 (4.3) 44.8 (5.3) 27.1 (8.9)	MA	Ventromedial tasks: DNMS; dorsolateral prefrontal tasks: A not B, spatial reversal	Accuracy and number of trials to reach a criterion	No group difference Ventromedial tasks but not dorsolateral prefrontal tasks correlated with joint attention measures
Griffith et al. (1999)	ASD DD	18 (15:3) 17 (10:7)	50.7 (6.7) 50.6 (9.2)	MA	Working memory and response inhibition: A not B, object retrieval, spatial reversal, boxes	Various measures including accuracy	Very few differences where ASD performed better than the DD In both groups performance correlated with mental age A few measures correlated with joint attention in both groups
McEvoy et al. (1993)	ASD DD TD	17 (10:7) 13 (7:6) 16 (10:6)	60.6 (12.9) 50.3 (12.8) 37.9 (20.0)	MA	A not B, DNMS, spatial reversal, alternation	Various measures including accuracy	There was a ceiling effect for the A not B and DNMS tasks, and a floor effect for alternation Relative to both groups, ASD showed more errors on the spatial reversal task In all groups, performance correlated with joint attention
Dawson et al. (2001)	ASD DS TD	20 (19:1) 19 (16:3) 20 (19:1)	64.6 (15.1) 65.3 (16.5) 30.9 (14.4)	MA	Visual paired comparison, DNMS	Accuracy	No group differences

Notes: ASD, children with autism; DD, children with developmental delay; DS, children with Down syndrome; TD, typically developing children; *n* (M:F), number of participants and ratio of males to females in each group; age (sd), mean age in months and standard deviation for each group; MA, group matching based on mental age; A not B, spatial reversal, object retrieval, boxes, alternation, visual paired comparison, and DNMS (delayed non-match to sample) are various tasks assessing executive functions. See cited publications for details.

autism phenotype but is better predicted by mental age. Although the literature evaluating these abilities in older children and adults is far more comprehensive, the studies with younger children with autism indicate that they are not particularly impaired, suggesting that more serious impairments in this area become apparent later in development.

Interestingly, however, some of these studies found that performance on these tasks correlates with measures of social-communicative abilities such as joint attention. More specifically, one study that differentiated several tasks based on their corresponding neuroanatomical systems, found that tasks tapping into the ventromedial cortex were more strongly correlated with measures of joint attention relative to those tapping into the dorsolateral system (Dawson et al., 2002b). This highlights the possibility that some aspects of executive dysfunction relate to social impairment in autism.

In sum, unlike the adult phenotype, executive function in autism appears to be not as seriously affected as social skills. Over development, difficulties in the area become more apparent. In relation to evaluating theoretical models of the emergence of autism, although the evidence is limited, the conclusion that children with autism are not specifically impaired on these tasks seems to suggest that executive function may not be the underlying causal mechanism for social-communicative deficits. However, even though tasks of executive functions do not distinguish young children with autism from those with developmental delay, these tasks still correlate with social skills, suggesting developmental interactions between those systems. Because executive functions and abilities like theory of mind are to some extent similar problem domains, we might expect that these social and non-social skills, which require common computations, are mediated by overlapping neural systems. As a result, the direction of causality is still difficult to establish.

Sensation, perception, and attention

The two approaches described above focusing on social-communicative skills and executive function capture areas of serious difficulties for individuals with autism. However, there are several other characteristics associated with the disorder that are not clearly related to either of these categories. Some of these differences have been broadly described in relation to "central coherence." This refers to the ability to perceptually process the coherent whole of sensory input in order to ascertain meaning (Frith and Happé, 1994). Individuals with autism are often described to have a narrow focus of attention and interest as well as acute perception for details. Interestingly, these differences have been described at multiple levels including sensation, perception, attention, and even style of thought and interests. It is unclear, however, whether these areas are related at all, and whether they share common underlying mechanisms. Unlike the debate over whether executive dysfunction is a primary or secondary deficit in autism, central coherence is usually viewed as a complementary explanation, accounting for aspects of performance that are positive in autism.

Some of the differences in perceptual sensitivity in autism relate to processing of low spatial frequency relative to high spatial frequency visual inputs, associated with the magnocellular and the parvocelluar pathways, respectively (Deruelle et al., 2004; Behrmann et al., 2006). Research in this area is limited even in adults, except for a few studies focusing on motion processing where the results appear to be inconsistent (Spencer et al., 2000; Gepner and Mestre, 2002; Milne et al., 2006). More attention has been paid to aspects of central coherence related to differences in local vs. global processing of visuospatial stimuli. Unlike typically developing individuals, the cognitive processing style in autism appears to be biased toward local rather than global or configural information (Happé and Frith, 2006). This sometimes results in superior performance on tasks that benefit from these abilities (Jolliffe and Baron-Cohen, 1997; Plaisted et al., 1998; Jarrold et al., 2005; Behrmann et al., 2006). In brain imaging studies of adults with autism, increased activation of sensory ventral-occipitotemporal areas and decreased activation in prefrontal areas has been described as reflective of these processing differences in local vs. global information (Ring et al., 1999). Some studies have suggested an association between measures of central coherence and theory of mind (Baron-Cohen and Hammer, 1997; Jarrold et al., 2000), where individuals with a more focal processing style are more likely to have difficulties with theory of mind tasks.

Because central coherence is a theoretical model specific to autism and because it describes several levels of information processing, it is difficult to trace its specific developmental trajectory in infants. Furthermore, we could not find any studies examining these abilities in young children with autism. However, in typical infancy research, a link has been made between early variation in attentional skills in relation to local vs. global processing style, and this has some implications for autism. As typically developing children begin to flexibly scan their environment and switch their attention among different objects, global forms are processed quickly and efficiently. Infants who exhibit a pattern of prolonged look duration, rely more on local elements when processing visual stimuli (Freeseman et al., 1993; Frick and Colombo, 1996). Furthermore, in adults, local directed attention and cue-driven attentional shifting appear to share similar neural substrates in the left superior parietal cortex (Fink et al., 1997). Hence, it is possible that the narrow focus of attention in autism is a developmental consequence of early difficulties in visual disengagement (Landry and Bryson, 2004). Alternatively, it can also result from atypical modulation of early visual processing areas by top-down feedback. More research is required to test whether these differences in information processing found in autism are related to differences in attentional modulation.

Summary

Converging lines of evidence indicate that general deficits as well as specific precursors to some symptoms are present early on in autism. By the time of diagnosis, several co-occurring impairments are clear, and encompass social and non-social domains. After the onset of these symptoms in the early years, different sets of abilities show varying trajectories of development. Some abilities, such as face processing, begin as seriously impaired, but over time, compensatory strategies and atypical neural systems may restore behavioral performance to within the typical range. Other deficits, such as executive dysfunction, may not be evident at younger ages but become clearer over development. In all cases, apart from aspects of behavior that define the disorder itself, substantial variability is seen in the resulting phenotype.

In this section we have briefly reviewed some of the most prominent hypotheses concerning core deficits in autism. Since all of these studies have been conducted with children and adults, it is very difficult to unravel the causal pathways involved and identify the critical primary deficits. Therefore, in the next section we turn our focus to the much smaller literature on infants and toddlers who are at-risk for autism.

Infancy and autism at the crossroads: where do we go from here?

Models of developmental disorders have often focused on fractionating and describing patterns of strength and weaknesses in a given disorder, leaving behind the bulk of theoretical advances in our understanding of the emergent nature of some brain functions during development (Karmiloff-Smith, 1998; Johnson et al., 2005). A recent field of investigation focusing on infant siblings of children with autism has offered promise in contributing to our knowledge of the nature of this developmental condition. In this section, we will argue that integrating theoretical and methodological advances of functional brain development in infancy adds substantial value to this research approach, the results of which will be fruitful to understanding of both typical and atypical development. We begin first with a review of the small but growing literature on infant siblings.

Why infant siblings?

Research on infant siblings of children diagnosed with autism (infant siblings) is motivated by the emergent nature of autism symptoms and the premise that neural plasticity may provide the basis for early intervention, which is likely to lead to improved outcomes. Further, it is hoped that studying infant siblings will reveal the primary deficits in autism before symptoms are compounded by atypical interactions with the social and physical world, and before compensatory strategies and systems cloud the basic processing difficulties.

In fact, the study of siblings offers opportunities to understand not only why autism emerges in some cases and not in others, but also to potentially explain variations associated with the autism phenotype. As mentioned earlier, genetic relatives of individuals with autism including siblings, who do not have an autism diagnosis, share certain characteristics with their affected relatives. This extended phenotype includes aspects of face processing (Dawson et al., 2002c; Dalton et al., 2005), theory of mind (Baron-Cohen and Hammer, 1997), executive function (Ozonoff et al., 1993; Hughes et al., 1997, 1999), and central coherence (Happé et al., 2001). Hence, understanding the precursors of these characteristics in infants can help us understand the underlying mechanisms, which extend to unaffected relatives and potentially to typical development.

Emerging finding from sibling studies

The emerging literature on infant siblings can be broadly divided based on whether or not infant siblings were followed up to the point in time at which some of them received a diagnosis. Studies that have followed up the infants from 4–6 months to the stage of formal diagnosis around 2–3 years provide information regarding the predictors or precursors of autism. In these studies, a few predictor measures have been used longitudinally to examine which factors differentiate infant siblings who develop autism from siblings who do not, or from siblings of unaffected children (Table 6). When overall IQ was used as a predictor, infants later diagnosed with autism showed differences on both verbal and non-verbal measures as early as 14 months (Landa and Garrett-Mayer, 2006). Another battery of risk markers, the Autism Observation Scale for Infants (Zwaigenbaum et al., 2005), focused on precursors of the abilities used in diagnosing older children, including response to name, eye contact, social reciprocity, and imitation. The battery also includes items examining basic visual and motor skills. Here too, infants later diagnosed with autism were clearly differentiated around 12 months on a variety of social and non-social measures. Another approach has focused on specific predictors such as joint attention (Sullivan et al., 2007) or language (Mitchell et al., 2006). Consistent with the other studies, infants later diagnosed with autism show differences in these measures as early as 12 months of age. On the other hand, none of the studies have, as of yet, found reliable predictors of a later diagnosis within the first year.

Other studies have not followed up the infants until the age of 2–3 years, but still found differences in groups of siblings of children with autism relative to groups of infants who do not have affected siblings. These studies cannot provide information about the extent to which performance on certain measures is a predictor of autism but they do provide information regarding the extended autism phenotype that appears to be present in infancy. Three studies of 4–6-month-old infant siblings (Yirmiya et al., 2006; Cassel et al., 2007; Merin et al., 2007) showed subtle differences in the behavior of infant siblings in a task where the mother interacts with the infant, then freezes and displays a neutral, expressionless face, and then finally resumes interaction (still face episode). Two of these studies found that infant siblings showed differences in their affect during the still face episode (Yirmiya et al., 2006; Cassel et al., 2007). The third study did not find the same differences in affect but did report differences in visual fixation during the episode, where infant siblings looked more to the mother's mouth relative to her eyes (Merin et al., 2007). Interestingly, differences between infant siblings and control infants also extend to low-level visual processing. One study found that infant siblings showed enhanced sensitivity to luminance contrasts processed in the magnocellular pathway (McCleery et al., in press). Older siblings above 14 months show somewhat clearer differences in a number of social-communicative measures as well as

Table 6. Studies that followed groups of infant siblings at various ages in early infancy until the stage when some received an autism diagnosis

Study[a]	Groups[b]	n (M:F)	Predictor ages	Predictor measures	Outcome ages	Outcome measures	Key results
Zwaigenbaum et al. (2005)	ASD Sibling TD	19 46 23	6, 12	Risk markers (AOSI); IQ (Mullen); language (CDI); visual attention (disengagement); temperament (IBQ)	24	Diagnosis (ADOS)	At 6 months, ASD differed on only one measure of temperament (activity level) At 12 months, the groups differed on several behavioral markers from the AOSI and showed delayed language, prolonged latency to disengage visual attention, and atypical temperament
Mitchell et al. (2006)	ASD Sibling TD	15 82 49	12, 18	Language (CDI)	24	Diagnosis (ADOS)	Both at 12 and 18 months, the ASD group showed delays in understanding and producing words and gestures
Loh et al. (2007)	ASD Sibling TD	8 (3:5) 9 (3:6) 15 (10:5)	12, 18	Observation of motor behaviors during semi-structured play	36	Diagnosis (DSM-IV, ADOS, ADI)	ASD showed more frequent and atypical stereotyped motor movements and postures
Bryson et al. (2007)	ASD	9 (5:4)	6, 12, 18, 24	Risk markers (AOSI); IQ (Mullen/Bayley)	18, 24, 36	Diagnosis (DSM-IV, ADOS, ADI); IQ (Mullen/Bayley)	Six children showed a serious decrease in IQ between 12 and 24/36 months whereas the remaining 3 were within the average range The earliest signs of autism included social-communicative problems, atypical sensorimotor, and temperamental characteristics
Landa and Garrett-Mayer (2006)	ASD LD Sibling TD	24 11 29 23	6, 14	IQ (Mullen)	24	Diagnosis (ADOS); language (PLS, CDI); IQ (Mullen)	At 6 months the ASD group did not differ on any subscales but at 14 months they differed on all except visual reception At 24 months the ASD group was further differentiated from the LD group based on language scores. The ASD group showed a slower increase in scores over time
Sullivan et al. (2007)	ASD BAP Sibling	16 (14:2) 8 (7:1) 27 (12:15)	14, 24	Various measures of response to joint attention	30/36	Diagnosis (ADOS); IQ (Mullen)	ASD showed subtle differences in joint attention at 14 months, but the differences were clearer and more stable across measures at 24 months Joint attention performance predicted language outcomes for the entire group

Table 6 (*continued*)

Study[a]	Groups[b]	*n* (M:F)	Predictor ages	Predictor measures	Outcome ages	Outcome measures	Key results
Gamliel et al. (2007)	ASD Sibling TD	1 (1:0) 38 (13:25) 39 (14:25)	4, 14	IQ (Bayley)	24, 36, 54	IQ (Kaufman); language (RDLS, CELF-P)	At 62 months, a subgroup of siblings showed difficulties in various cognitive and linguistic domains It was difficult to assess predictors since only one child received a diagnosis At 54 months, the group was within typical levels of functioning with the exception of a few siblings showing some difficulties in language

Notes: ASD, children with autism; LD, children with language delay; sibling, siblings of children with autism who did not themselves receive a diagnosis; BAP, children exhibiting the characteristics of the broader autism phenotype without a diagnosis; TD, typically developing children with no family history of autism; *n* (M:F), number of participants and ratio of males to females in each group; predictor ages, ages in months around which the predictor measures were collected; outcome ages, ages in months around which the outcome measures were collected; AOSI, Autism Observation Scale for Infants (Zwaigenbaum et al., 2005); Mullen, Mullen Scales of Early Learning (Mullen, 1995); Bayley, Bayley Scales of Infant Development (Bayley, 1993); CDI, McArthur Communicative Development Inventory (Fenson et al., 1993); IBQ, Infant Behavior Questionnaire (Rothbart, 1981); ADOS, Autism Diagnostic Observation Schedule (Lord et al., 2000); ADI, Autism Diagnostic Interview (Lord et al., 1994); DSM-IV, clinical diagnosis based on the DSM-IV (APA, 1994); PLS, Preschool Language Scale (Zimmerman et al., 2002); Kaufman, Kaufman Assessment Battery for Children (Kaufman and Kaufman, 1983); RDLS, Reynell Developmental Language Scales (Reynell and Grubber, 1990), CELF-P, Clinical Evaluation of Language Fundamentals-Preschool (Wiig et al., 1992).

[a]Some studies overlap in their samples.

[b]The groups refer to the classification of infants at the time of diagnosis; for studies with different *n* over time, the largest is reported.

cognitive and motor measures (detailed in Table 7). Of course, follow-up data for these studies are necessary to examine the extent to which the group differences between infant siblings and control infants predict any of the characteristics associated with autism later on.

Taken together, and despite the small number of studies to date, some aspects of these data are already revealing. Studies on infant siblings have confirmed that the expression of autism in the early years is subtle, at least as far as overt symptoms or delays in the expected developmental milestones, including some of the precursors to social-communicative difficulties. Infant siblings who do not receive a diagnosis in the second or third year may share some of the early characteristics of siblings who go on to develop autism, suggesting that the presence of specific social or non-social difficulties early on is not sufficient for the disorder to emerge. Although much more research is required before drawing firm conclusions, these new findings can be integrated with findings from other approaches to motivate future directions in this area.

Toward an understanding of the emergent nature of autism

It is appropriate to begin the discussion by addressing what the current understanding of development can contribute toward solving the puzzle of the *emergent* nature of autism: On the one hand, the human brain undergoes substantial and maximal development parentally and postnatally in the early years, with the clear emergence of precursors of many adult skills. Autism, however, frequently associated with serious atypical brain and behavioral organization in childhood onwards, appears to confer only subtle symptoms early in life. Why then are the deficits not so striking in infancy as they are later in childhood and adulthood? There are a number of plausible explanations. First, the core deficits could be specific to social abilities, such as theory of mind, which are acquired late in development. Consistent with certain perspectives, the most sophisticated parts of the social brain network may be silent early in infancy and only later come online. Autism, according to this explanation, is a case of failure of these later developing systems to mature. Alternatively, it is possible that the serious symptoms emerge as a result of interactions among several more subtle brain and behavioral abnormalities early in life. Yet another possibility is that there may be qualitatively different pathways through which infants come to exhibit the autism phenotype. These different pathways may be difficult to detect in group studies. The last two possibilities are consistent with the notion of *probabilistic epigenesis*, where development is the result of bi-directional interactions among brain structure and function (cf., Johnson, 2005). Entertaining these possibilities is essential as these different perspectives would motivate different research questions including what to look for, what to expect, and what theoretical and practical conclusions are drawn. We will now discuss these alternatives in detail, incorporating some existing controversies in research on older children and adults. We begin with the first possibility which is that autism symptoms are subtle in infancy because the core deficit in this disorder relates to later developing skills such as theory of mind.

Is there a unitary core deficit in autism?

In the previous section, we discussed several approaches to autism in older children and adults that seek a core characterization of autism, where other deficits are secondary or peripheral to the primary one. These differing approaches have given rise to several debates including whether the deficit in theory of mind or alternatively in executive function should be viewed as the core one. This debate is similar to a more general one in the area of developmental disorders concerning the utility of fractionating cognitive abilities as a function of the degree of impairment found in each domain (Karmiloff-Smith, 1998).

Several caveats render this frequently recurring debate in developmental disorders less interesting. One of these caveats is related to the behavioral characterization of the autism phenotype. Research studies have increasingly benefited from

Table 7. Studies of the extended phenotype in autism within the first 2 years

Study	Groups	*n* (M:F)	Age (sd)	Measures	Key results
McCleery et al. (in press)	Sib-ASD TD	13 (5:8) 26 (5:8)	6.07 (0.14) 6.07 (0.12)	Contrast sensitivity related to magnocellular (luminance) and parvocellular (chromatic) visual pathways	Sib-ASD showed greater contrast sensitivity related to the magnocellular pathway, but did not differ in their contrast sensitivity related to the parvocallular pathway
Cassel et al. (2007)	Sib-ASD TD	11 19	6.1 (0.3) for overall sample	Emotional reaction during the still face procedure	Sib-ASD smiled less during the overall episode
Merin et al. (2007)	Sib-ASD TD	31 (15:16) 24 (16:8)	6.06 (0.3) 5.9 (0.3)	Emotional reaction and behavioral eye tracking during the still face procedure	More infants in the sibs-ASD group gazed predominantly at the mother's mouth relative to her eyes
Iverson and Wozniak (2007)	Sib-ASD TD	21 (6:15) 18 (8:10)	Around 5, 14	Developmental milestones; production of rhythmic limb movements postural variability	A significant percentage of sib-ASD showed delays including babbling, pointing, first words, and sitting Sib-ASD showed postural instability
Yirmiya et al. (2007)	Sib-ASD TD	30 (8:13) 31 (13:18)	20.2 (3.2) 19.6 (2.8)	Synchrony of interaction; emotional reaction during the still face procedure; response to name	Sib-ASD were less synchronous in their interactions Sib-ASD were less upset during the still face More infants in the sibs-ASD group responded to their name
Toth et al. (2007)	Sib-ASD TD	42 (21:21) 20 (13:7)	20.3 (2.2) 22.4 (2.8)	Immediate and deferred imitation; functional and symbolic play; IQ (Mullen); social communication (CSBS); autism characteristics (ADOS)	No group differences in imitation or play Sib-ASD had lower scores on the receptive language subscale of the Mullen and a lower composite score on the CSBS No group differences on the ADOS except on items of pointing and responsive social smile
Presmanes et al. (2007)	Sib-ASD TD	46 (26:20) 35 (24:11)	15.4 (3.0) 15.5 (2.9)	Different combinations of joint attention cues, e.g., gaze, gaze and head movement; IQ (Mullen); autism characteristics (STAT)	Sib-ASD showed less responding to joint attention for moderately redundant cues but not for highly redundant ones No differences on the Mullen except on the visual reception subscale No differences on the STAT

Notes: Sib-ASD, siblings of children with autism; TD, typically developing children with no family history of autism; *n* (M:F), number of participants and ratio of males to females; age (sd), mean age and standard deviation around which the measures were taken; Mullen, Mullen Scales of Early Learning (Mullen, 1995); CSBS, Communication and Symbolic Behavior Scale (Wetherby et al., 2002); ADOS, Autism Diagnostic Observation Schedule (Lord et al., 2000); STAT, Screening Tool for Autism in 2-year olds (Stone et al., 2004).

relying on more standardized clinical definitions (American Psychiatric Association, 1994; World Health Organization, 1993) and diagnostic tools (Lord et al., 1994, 2000) for autism. The conceptualization of the disorder has become broader, encompassing "non-classic" cases where a diagnosis may or may not be warranted given the individual's overall cognitive and functional ability. In parallel, the diagnostic tools have become more focused on symptoms that are likely to be common to most individuals. These instruments have attempted to reduce variations in clinical judgement, which may lead to highly heterogeneous groups in research studies, by ensuring that the defining characteristics of autism, namely social-communicative symptoms are present in all cases, albeit varying enormously in severity and in comorbid symptoms. In the interest of more homogeneous samples, the definition of autism frequently excludes symptoms which overlap with other disorders including the presence of repetitive and stereotyped behaviors, or those related to overall cognitive or language delay. As a result, autism like several other developmental disorders is diagnosed primarily on the basis of behavioral impairment in a specific area.

While this narrowing of the criteria is motivated by reasonable objectives in terms of specificity of the clinical definition and homogeneity of the samples, from a theoretical point of view, there is doubt on whether it is sensible to look for core deficits in other domains, after having a priori excluded individuals whose impairment on other symptoms is less marked. This does not imply that the social-communicative deficits are the sole characteristics of autism. Associated deficits are clearly present in the majority of cases. These characteristics, however, will be present in variable degrees and may be less striking in some individuals relative to social-communicative impairment. From a scientific point of view, even if certain characteristics of autism are not universal, while others are (by virtue of having defined the disorder as such), the associated impairments still require an explanation, which a single core cognitive deficit falls short of providing.

Beyond these issues of definition of the disorder, several lines of evidence appear to challenge the notion of a single core deficit in autism. Research on other developmental disorders has challenged the notion of selective deficits both at the behavioral and genetic levels. The emerging picture has favored the view that multiple and sometimes subtle deficits across domains lead to atypical patterns of development with some domains being more affected than others (Karmiloff-Smith, 1998). There is also evidence for strong overlap between autism and other developmental disorders such as Specific Language Impairment, implying that common aetiology may lead to qualitatively different phenotypic conditions depending on the presence and severity of other risk factors (Bishop, 2003). Furthermore, population studies indicate that the characteristics of autism tend to cluster differently in the general population making it unlikely that there is a single explanation for the triad (Happe et al., 2006).

Turning to the proposed core deficits from the developmental models of autism presented earlier, there is reasonable evidence that problems in social orienting are present early on. There is also preliminary evidence that impairments in attentional disengagement are expressed between 6 and 12 months of age. Hence, these two mechanisms are potentially viable candidates for a core deficit. However, neither mechanism can be described as directly causal, since (a) both co-occur toward the end of the first year of life, and (b) problems with disengagement occur in various developmental disorders including ones with markedly different social profiles from the one found in autism.

In sum, the search for core deficits at the genetic and cognitive levels has, as yet, proved unsuccessful. This is clearly illustrated by our review of finding in the previous section, where no single account appears to capture all manifestations of the autism profile. From a developmental perspective, even if a cognitive core deficit existed, it would be very hard to detect by later childhood due to the compounding of, and compensatory strategies applied to, different secondary consequences of the original cause. However, it remains possible that there will be a core deficit in the functional development of the brain, and that this could explain the full triad of impairments seen in autism. This core deficit could have immediate multiple causes at the perceptual/cognitive level, as well as a range of developmental consequences.

The emergence of autism as a result of complex interactions

Brain development is a dynamic and self-organizing system (cf., Johnson, 2005). Consequently, the division between primary and secondary functional abnormalities is misleading. For instance neuroanatomical abnormalities in an earlier developing structure will have consequences for later developing structures. Abnormalities in these later developing systems may then exert further atypicality on the earlier developing systems if they provide feedback projections to them. Hence, although the functions mediated by later developing abnormal structures can be described as being caused by the primary abnormalities, these in turn causally modify the earlier structures and their functions through feedback projections. For example, atypical modulation of early visual processing areas by top-down feedback would have far-reaching implications in modifying other perceptual and cognitive processes that rely on these top-down systems. These consideration stress the importance of studying autism at the earliest ages in which symptoms appear, since here we have the most chance to study brain development while the atypical trajectory is only slightly deviant from the typical course, and before compensatory and confounding effects have influenced further development.

In short, no models that involve a single core deficit in autism have yet explained the range of data available, and we should not necessarily assume that there will be a unitary developmental cause of this kind. Instead, it is possible that the functions, which later have been described as core cognitive deficits, arise from widespread albeit subtle impairments in several systems where the typical developmental constraints are altered. A more fruitful approach is to focus on interactions between systems within individuals. The development of autism in infancy is best understood in terms of multiple risk factors, where the presence and the severity of each risk factor as well as their interactions would explain the resulting phenotype.

Considering some concrete examples will clarify this point. An early deficit in orienting to socially relevant stimuli may be necessary but not sufficient for autism to emerge. The deficit in social orienting would result in decreased input from socially relevant stimuli. This deficit, however, would be compounded and amplified by the presence of other difficulties, such as those found in attentional disengagement. A problem with flexibly switching attention between different stimuli would result in "locking" onto certain irrelevant aspects of the diminished input. The infant in this case, would not only receive decreased input from social stimuli, but attentional constraints would impose qualitatively different forms of input, namely focal and irrelevant ones. This would suggest that infants who exhibit a combination of disengagement difficulties with decreased social orienting would be at higher risk for autism than infants who exhibit one of these difficulties in isolation. Several concrete predictions can be further examined in terms of more specific behavioral outcomes. For instance, we might expect the severity of early deficit in social orienting and/or in disengagement to correlate with the severity of impairment in face processing.

Another attractive hypothesis is the extent to which social-orienting deficits are themselves specific to social stimuli. An alternative view is that infants who go on to develop autism differ in their spontaneous preference for the degree of contingency in events, whether social or nonsocial. Consistent with findings from young children with autism (reviewed above in the section titled "Imitation and understanding of contingency"), those infants would show an atypical preference for perfect rather than high, but imperfect, contingency and as a result, they are less attracted to social stimuli because the latter tend to be imperfect. Infancy research with at-risk infants prior to the onset of symptoms would help to disentangle these competing alternatives.

Instead of seeking to affirm that a core deficit is present or not prior to the onset of the symptoms, research focusing on autism in infancy needs to examine interactions among several systems within individuals. Such preliminary integrative accounts of how several risk factors could explain subsequent competence are essential in explaining the resulting adult profiles. Elements of such preliminary developmental accounts can be examined rigorously and verified independently or in combination, but must be integrated theoretically. This would shift our understanding of autism toward a coherent picture of the dynamics through which the adult phenotype is reached. With these considerations in mind, we now turn to a related issue concerning the emergent nature of autism in view of individual variability in early developmental trajectories.

Heterogeneity: merely methodological nuisance?

We have described earlier that the most common and unifying feature of autism is the presence of social-communicative impairment. Beyond this, cases vary enormously in severity and in comorbid symptoms. For instance, two cases meeting an autism classification cut-off may include a child who is nonverbal as well as a child whose language is advanced, albeit peculiar. In fact, the conception of autism as a spectrum suggests that this diagnostic label includes individuals differing greatly in functional impairment. This striking heterogeneity has been considered a daunting issue, not only for autism research but also more generally in developmental disorders. On methodological grounds such individual differences in the resulting profiles are frequently discounted as random noise, where heterogeneity is regarded as an explanation in itself. However, phenotypic expression in developmental disorders is considerably more variable than that seen in typical development, suggesting that not all of the variability is random. In other words, atypical development is inherently more variable than typical development, reflecting different ways of brain and behavioral reorganization. In the previous section we argued that autism symptoms are subtle early on because of the emergent nature of the disorder, where several risk factors interact over time, resulting in the phenotype found in older children and adults. Another possibility, which is not mutually exclusive, is that early symptoms are difficult to detect because they are variable across different individuals. It is possible that a group design fails to capture qualitatively different routes that lead to the autism phenotype.

Several lines of evidence suggest that the early developmental course of autism is in fact heterogeneous. A handful of "classic" cases of autism have been described in infants around the first year of life, where the typical manifestations of autism symptoms are clear (Dawson et al., 2000; Klin et al., 2004). On the other hand, in other infants early signs may be very difficult to detect. Different developmental profiles of children with autism have also been hypothesized (Sigman et al., 2004; Werner et al., 2005; Werner and Dawson, 2005; Palomo et al., 2006). For some, deficits are apparent early on and arrested development continues. In contrast, other children achieve some developmental milestones on the typical developmental schedule but then exhibit loss of some abilities and a developmental regression. Interestingly, heterogeneity in expression of early symptoms appears to map out on less heterogeneous outcomes. An earlier age of recognition correlates with more classic phenotypic autism outcomes, whereas later recognition of early symptoms is associated with more variable manifestations, even if the severity of these symptoms does not differ (Chawarska et al., 2007).

Hence, this variability is likely to be the result of dynamic and probabilistic interactions over development, and the systematic study of these variations can offer important clues toward understanding the emergent nature of autism. This understanding will depend on developing appropriate theoretical and methodological models capable of handling individual differences in phenotypic expression. What are the implications of this for the study of infants at risk for autism? Infancy research has traditionally relied on group methodology. In fact, some classic paradigms such as habituation yield binary looking time measures for which there is little confidence in individual infants' results. In the study of at-risk siblings, it is essential that existing techniques are adapted to boost confidence in individual results, which would, in turn, be fruitful for infancy research in general. Furthermore, as we argued earlier, cross-task correlations within individual infants offer the opportunity to examine individual differences in cognitive profiles. Another possibility is to examine variations in performance within tasks across different infants. The latter requires individual and continuous measures, which would clarify the extent of consistency or variability of each measure across individuals.

Concluding remarks

Recent advances in autism and in infancy research and the recent integration of these directions hold a lot of promise for solving some of the most difficult puzzles in autism. To understand the underlying causes of this condition and the process through which it emerges, any convincing model must encompass not only the symptoms which define the disorder but also other characteristics of the profile.

This is best achieved through making use of existing developmental models to motivate theoretical accounts of how and why risk factors, alone or in combination, exert an impact on the resulting phenotype. It is equally important to understand *protective* factors, as this is likely to offer important clinical implications, if it turns out that simple environmental manipulations might have important consequences in the window of maximal plasticity. A theoretical account of this sort requires the development of continuous and individually valid measurements, both of predictors and outcomes, capable of capturing individual differences. This approach is likely to be productive not only for understanding autism, but could also offer insights into the mechanisms that drive and constrain typical development.

Abbreviations

A not B, Object/spatial retrieval, Boxes, Alternation, Spatial reversal, Visual paired comparison DNMS (delayed non-match to sample)	various tasks assessing executive functions; see cited publications for details
ADI	autism diagnostic interview (standardized instrument for diagnosis)
ADOS	autism diagnostic observation schedule (standardized instrument for diagnosis)
ASD	autism spectrum disorders
Aut	autism
CA	chronological age
CDI	McArthur communicative development inventory (standardized measure of language)
CELF-P	clinical evaluation of language fundamentals - preschool (standardized measure of language)
CSBS	communication and symbolic behavior scales (a measure of communication)
DD	developmental delay
DS	down syndrome
DSM-VI	diagnostic and statistical manual of the American Psychiatric Association
HiAut	high functioning autism
IBQ	infant behavior questionnaire (a measure of temperament)
LD	language delay
LoAut	low functioning autism
MA	mental age
Mullen (Mullen Scales of Early Learning), Bayley (Bayley Scales of Infant Development), Kaufman (Kaufman Assessment Battery for Children) P400, Nc, N300, NSW	various components of event-related potentials in response to face stimuli
PDD	pervasive developmental disorder: not otherwise specified
PLS	preschool language scales (standardized measure of language)

Sib-ASD	sibling of a child diagnosed with an autism spectrum disorder
Sib-Aut	Sibling of a child diagnosed with autism (standardized tasks for measuring general verbal and non-verbal abilities)
STAT	screening tool for autism in two-year-olds (a clinical screening instrument)
STS	superior temporal sulcus

Acknowledgment

We would like to thank Leah Kaminsky for her help with reference materials, and Theodora Gliga, Atsuhi Senju, and Dick Aslin for helpful comments on an earlier version of this manuscript. This work was supported by the UK Medical Research Council Programme Grant G9715587.

References

Adolphs, R. (2003) Cognitive neuroscience of human social behaviour. Nat. Rev. Neurosci., 4: 165–178.

Allison, T., Puce, A. and Mccarthy, G. (2000) Social perception from visual cues: role of the STS region. Trends Cogn. Sci., 4: 267–278.

American Psychiatry Association. (1994) Diagnostic and statistical manual (4th ed). Washington, DC, APA Press.

Anderson, C.J., Colombo, J. and Jill Shaddy, D. (2006) Visual scanning and pupillary responses in young children with Autism Spectrum Disorder. J. Clin. Exp. Neuropsychol., 28: 1238–1256.

Arndt, T.L., Stodgell, C.J. and Rodier, P.M. (2005) The teratology of autism. Int. J. Dev. Neurosci., 23: 189–199.

Autism Genome Project. (2007) Mapping autism risk loci using genetic linkage and chromosomal rearrangements. Nat. Genet., 39: 319–328.

Bacon, A.L., Fein, D., Morris, R., Waterhouse, L. and Allen, D. (1998) The responses of autistic children to the distress of others. J. Autism Dev. Disord., 28: 129–142.

Bailey, A., Le Couteur, A., Gottesman, I., Bolton, P., Simonoff, E., Yuzda, E. and Rutter, M. (1995) Autism as a strongly genetic disorder: evidence from a British twin study. Psychol. Med., 25: 63–77.

Baird, G., Simonoff, E., Pickles, A., Chandler, S., Loucas, T., Meldrum, D. and Charman, T. (2006) Prevalence of disorders of the autism spectrum in a population cohort of children in South Thames: the Special Needs and Autism Project (SNAP). Lancet, 15: 210–215.

Baron-Cohen, S., Allen, J. and Gillberg, C. (1992) Can autism be detected at 18 months? The needle, the haystack, and the CHAT. Br. J. Psychiatry, 161: 839–843.

Baron-Cohen, S. and Hammer, J. (1997) Parents of children with Asperger syndrome: what is the cognitive phenotype? J. Cogn. Neurosci., 9: 548–554.

Baron-Cohen, S., Leslie, A.M. and Frith, U. (1985) Does the autistic child have a "theory of mind"? Cognition, 21: 37–46.

Baron-Cohen, S., Ring, H.A., Wheelwright, S., Bullmore, E.T., Brammer, M.J., Simmons, A. and Williams, S.C. (1999) Social intelligence in the normal and autistic brain: an fMRI study. Eur. J. Neurosci., 11: 1891–1898.

Bayley, N. (1993) Bayley Scales of Infant Development (2nd ed). The Psychological Corporation.

Behrmann, M., Thomas, C. and Humphreys, K. (2006) Seeing it differently: visual processing in autism. Trends Cogn. Sci., 10: 258–264.

Bishop, D.V.M. (2003) Autism and specific language impairment: categorical distinction or continuum? In: Bock G. and Goode J. (Eds.), Autism: Neural Basis and Treatment Possibilities. Wiley, Chichester.

Bolton, P., Macdonald, H., Pickles, A., Rios, P., Goode, S., Crowson, M., Bailey, A. and Rutter, M. (1994) A case-control family history study of autism. J. Child Psychol. Psychiatry, 35: 877–900.

Bryson, S., Landry, R., Czapinski, P., Mcconnell, B., Rombough, V. and Wainwright, A. (2004) Autistic spectrum disorders: causal mechanisms and recent findings on attention and emotion. Int. J. Spec. Educ., 19: 14–22.

Bryson, S.E., Zwaigenbaum, L., Brian, J., Roberts, W., Szatmari, P., Rombough, V. and McDermott, C. (2007) A prospective case series of high-risk infants who developed autism. J. Autism Dev. Disord., 37: 12–24.

Cassel, T.D., Messinger, D.S., Ibanez, L.V., Haltigan, J.D., Acosta, S.I. and Buchman, A.C. (2007) Early social and emotional communication in the infant siblings of children with autism spectrum disorders: an examination of the broad phenotype. J. Autism Dev. Disord., 37: 122–132.

Charman, T., Swettenham, J., Baron-Cohen, S., Cox, A., Baird, G. and Drew, A. (1998) An experimental investigation of social-cognitive abilities in infants with autism: clinical implications. Infant Ment. Health J., 19: 260–275.

Chawarska, K., Klin, A. and Volkmar, F. (2003) Automatic attention cueing through eye movement in 2-year-old children with autism. Child Dev., 74: 1108–1122.

Chawarska, K., Paul, R., Klin, A., Hannigen, S., Dichtel, L.E. and Volkmar, F. (2007) Parental recognition of developmental problems in toddlers with autism spectrum disorders. J. Autism Dev. Disord.

Chawarska, K. and Volkmar, F. (2007) Impairments in monkey and human face recognition in 2-year-old toddlers with Autism Spectrum Disorder and Developmental Delay. Dev. Sci., 10: 266–279.

Dalton, K.M., Nacewicz, B.M., Alexander, A.L. and Davidson, R.J. (2007) Gaze-fixation, brain activation, and amygdala volume in unaffected siblings of individuals with autism. Biol. Psychiatry., 61: 512.

Dalton, K.M., Nacewicz, B.M., Johnstone, T., Schaefer, H.S., Gernsbacher, M.A., Goldsmith, H.H., Alexander, A.L. and Davidson, R.J. (2005) Gaze fixation and the neural circuitry of face processing in autism. Nat. Neurosci., 8: 519–526.

Dawson, G. and Adams, A. (1984) Imitation and social responsiveness in autistic children. J. Abnorm. Child Psychol., 12: 209–225.

Dawson, G., Carver, L., Meltzoff, A.N., Panagiotides, H., Mcpartland, J. and Webb, S.J. (2002a) Neural correlates of face and object recognition in young children with autism spectrum disorder, developmental delay, and typical development. Child Dev., 73: 700–717.

Dawson, G., Hill, D., Spencer, A., Galpert, L. and Watson, L. (1990) Affective exchanges between young autistic children and their mothers. J. Abnorm. Child Psychol., 18: 335–345.

Dawson, G., Meltzoff, A.N., Osterling, J. and Rinaldi, J. (1998) Neuropsychological correlates of early symptoms of autism. Child Dev., 69: 1276–1285.

Dawson, G., Munson, J., Estes, A., Osterling, J., Mcpartland, J., Toth, K., Carver, L. and Abbott, R. (2002b) Neurocognitive function and joint attention ability in young children with autism spectrum disorder versus developmental delay. Child Dev., 73: 345–358.

Dawson, G., Osterling, J., Meltzoff, A.N. and Kuhl, P.K. (2000) Case study of the development of an infant with autism from birth to 2 years of age. J. Appl. Dev. Psychol., 21: 299–313.

Dawson, G., Osterling, J., Rinaldi, J., Carver, L. and Mcpartland, J. (2001) Brief report: recognition memory and stimulus-reward associations. Indirect support for the role of ventromedial prefrontal dysfunction in autism. J. Autism Dev. Disord., 31: 337–341.

Dawson, G., Toth, K., Abbott, R., Osterling, J., Munson, J., Estes, A. and Liaw, J. (2004a) Early social attention impairments in autism: social orienting, joint attention, and attention to distress. Dev. Psychol., 40: 271–283.

Dawson, G., Webb, S., Schellenberg, G.D., Dager, S., Friedman, S., Aylward, E. and Richards, T. (2002c) Defining the broader phenotype of autism: genetic, brain, and behavioral perspectives. Dev. Psychopathol., 14: 581–611.

Dawson, G., Webb, S.J., Carver, L., Panagiotides, H. and Mcpartland, J. (2004b) Young children with autism show atypical brain responses to fearful versus neutral facial expressions of emotion. Dev. Sci., 7: 340–359.

Dawson, G., Webb, S.J., Wijsman, E., Schellenberg, G., Estes, A., Munson, J. and Faja, S. (2005) Neurocognitive and electrophysiological evidence of altered face processing in parents of children with autism: implications for a model of abnormal development of social brain circuitry in autism. Dev. Psychopathol., 17: 679–697.

Dehaene-Lambertz, G., Dehaene, S. and Hertz-Pannier, L. (2002) Functional neuroimaging of speech perception in infants. Science, 298: 2013–2015.

Deruelle, C., Rondan, C., Gepner, B. and Tardif, C. (2004) Spatial frequency and face processing in children with autism and Asperger syndrome. J. Autism Dev. Disord., 34: 199–210.

Escalona, A., Field, T., Nadel, J. and Lundy, B. (2002) Brief report: imitation effects on children with autism. J. Autism Dev. Disord., 32: 141–144.

Fenson, L., Dale, P.S., Reznick, S., Thal, D., Bates, S., Hartung, J.P., Pethick, S. and Reilly, J.S. (1993) MacArthur Communication Development Inventories. San Antonio, TX, Psychological Corporation.

Filipek, P.A., Accardo, P.J., Ashwal, S., Baranek, G.T., Cook Jr., E.H., Dawson, G., Gordon, B., Gravel, J.S., Johnson, C.P., Kallen, R.J., Levy, S.E., Minshew, N.J., Ozonoff, S., Prizant, B.M., Rapin, I., Rogers, S.J., Stone, W.L., Teplin, S.W., Tuchman, R.F. and Volkmar, F.R. (2000) Practice parameter: screening and diagnosis of autism. Report of the Quality Standards Subcommittee of the American Academy of Neurology and the Child Neurology Society. Neurology, 55: 468–479.

Filipek, P.A., Accardo, P.J., Baranek, G.T., Cook Jr., E.H., Dawson, G., Gordon, B., Gravel, J.S., Johnson, C.P., Kallen, R.J., Levy, S.E., Minshew, N.J., Ozonoff, S., Prizant, B.M., Rapin, I., Rogers, S.J., Stone, W.L., Teplin, S., Tuchman, R.F. and Volkmar, F.R. (1999) The screening and diagnosis of autistic spectrum disorders. J. Autism Dev. Disord., 29: 439–484.

Fink, G.R., Halligan, P.W., Marshall, J.C., Frith, C.D., Frackowiak, R.S. and Dolan, R.J. (1997) Neural mechanisms involved in the processing of global and local aspects of hierarchically organized visual stimuli. Brain, 10: 1779–1791.

Freeseman, L.J., Colombo, J. and Coldren, J.T. (1993) Individual differences in infant visual attention: four-month-olds' discrimination and generalization of global and local stimulus properties. Child Dev., 64: 1191–1203.

Frick, J.E. and Colombo, J. (1996) Individual differences in infant visual attention: recognition of degraded visual forms by four-month-olds. Child Dev., 67: 188–204.

Frith, U. and Happe, F. (1994) Autism: beyond "theory of mind." Cognition, 50: 115–132.

Gamliel, I., Yirmiya, N. and Sigman, M. (2007) The development of young siblings of children with autism from 4 to 54 months. J. Autism Dev. Disord., 37: 171–183.

Gauthier, I. and Nelson, C.A. (2001) The development of face expertise. Curr. Opin. Neurobiol., 11: 219–224.

Gepner, B., Deruelle, C. and Grynfeltt, S. (2001) Motion and emotion: a novel approach to the study of face processing by young autistic children. J. Autism Dev. Disord., 31: 37–45.

Gepner, B. and Mestre, D. (2002) Postural reactivity to fast visual motion differentiates autistic from children with Asperger syndrome. J. Autism Dev. Disord., 32: 231–238.

Gergely, G. and Watson, J.S. (1999) Early socio-emotional development: contingency perception and the social-biofeedback. In: Gergely G. and Watson J. (Eds.), Early Social Cognition: Understanding others in the First Months of Life. Lawrence Erlbaum, Mahwah, NJ.

Griffith, E.M., Pennington, B.F., Wehner, E.A. and Rogers, S.J. (1999) Executive functions in young children with autism. Child Dev., 70: 817–832.

de Haan, M., Johnson, M.H. and Halit, H. (2003) Development of face-sensitive event-related potentials during infancy: a review. Int. J. Psychophysiol., 51: 45–58.

Happé, F., Ehlers, S., Fletcher, P., Frith, U., Johansson, M., Gillberg, C., Dolan, R., Frackowiak, R. and Frith, C. (1996) "Theory of mind" in the brain: evidence from a PET scan study of Asperger syndrome. Neuroreport, 8: 197–201.

Happé, F., Briskman, J. and Frith, U. (2001) Exploring the cognitive phenotype of autism: weak "central coherence" in parents and siblings of children with autism: I. Experimental tests. J. Child Psychol. Psychiatry., 42: 299–307.

Happé, F. and Frith, U. (2006) The weak coherence account: detail-focused cognitive style in autism spectrum disorders. J. Autism Dev. Disord., 36: 5–25.

Happé, F., Ronald, A. and Plomin, R. (2006) Time to give up on a single explanation for autism. Nat. Neurosci., 9: 1218–1220.

Hill, E.L. (2004) Executive dysfunction in autism. Trends Cogn. Sci., 8: 26–32.

Hughes, C., Plumet, M.H. and Leboyer, M. (1999) Towards a cognitive phenotype for autism: increased prevalence of executive dysfunction and superior spatial span amongst siblings of children with autism. J. Child Psychol. Psychiatry., 40: 705–718.

Hughes, C., Leboyer, M. and Bouvard, M. (1997) Executive function in parents of children with autism. Psychol. Med., 27: 209–220.

Iverson, J.M. and Wozniak, R.H. (2007) Variation in vocal-motor development in infant siblings of children with autism. J. Autism Dev. Disord., 37: 158–170.

Jarrold, C., Butler, D.W., Cottington, E.M. and Jimenez, F. (2000) Linking theory of mind and central coherence bias in autism and in the general population. Dev. Psychol., 36: 126–138.

Jarrold, C., Gilchrist, I.D. and Bender, A. (2005) Embedded figures detection in autism and typical development: preliminary evidence of a double dissociation in relationships with visual search. Dev. Sci., 8: 344–351.

Jemel, B., Mottron, L. and Dawson, M. (2006) Impaired face processing in autism: fact or artifact? J. Autism Dev. Disord., 36: 91–106.

Johnson, M.H. (2001) Functional brain development in humans. Nat. Rev. Neurosci., 2: 475–483.

Johnson, M.H. (2005) Developmental Cognitive Neuroscience. Blackwell, Oxford.

Johnson, M.H., Griffin, R., Csibra, G., Halit, H., Farroni, T., de haan, M., Tucker, L.A., Baron-Cohen, S. and Richards, J. (2005) The emergence of the social brain network: evidence from typical and atypical development. Dev. Psychopathol., 17: 599–619.

Johnson, M.H. and Morton, J. (1991) Biology and Cognitive Development: The Case of Face Recognition. Blackwell, Oxford.

Jolliffe, T. and Baron-Cohen, S. (1997) Are people with autism and Asperger syndrome faster than normal on the embedded figures test? J. Child Psychol. Psychiatry, 38: 527–534.

Karmiloff-Smith, A. (1998) Development itself is the key to understanding developmental disorders. Trends Cogn. Sci., 2: 389–398.

Kaufman, A.S. and Kaufman, N.L. (1983) Kaufman assessment battery for children (K-ABC). Circle Pines, MN, American Guidance Service.

Klin, A., Chawarska, K., Paul, R., Rubin, E., Morgan, T., Wiesner, L. and Volkmar, F. (2004) Autism in a 15-month-old child. Am. J. Psychiatry, 161: 1981–1988.

Klin, A., Jones, W., Schultz, R. and Volkmar, F. (2003) The enactive mind, or from actions to cognition: lessons from autism. Philos. Trans. R. Soc. Lond. B Biol. Sci., 358: 345–360.

Kuhl, P.K., Coffey-Corina, S., Padden, D. and Dawson, G. (2005) Links between social and linguistic processing of speech in preschool children with autism: behavioral and electrophysiological measures. Dev. Sci., 8: F1–F12.

Landa, R. and Garrett-Mayer, E. (2006) Development in infants with autism spectrum disorders: a prospective study. J. Child Psychol. Psychiatry, 47: 629–638.

Landry, R. and Bryson, S.E. (2004) Impaired disengagement of attention in young children with autism. J. Child Psychol. Psychiatry, 45: 1115–1122.

Leslie, A.M. and Thaiss, L. (1992) Domain specificity in conceptual development: neuropsychological evidence from autism. Cognition, 43: 225–251.

Loh, A., Soman, T., Brian, J., Bryson, S.E., Roberts, W., Szatmari, P., Smith, I.M. and Zwaigenbaum, L. (2007) Stereotyped motor behaviors associated with autism in high-risk infants: a pilot videotape analysis of a sibling sample. J. Autism Dev. Disord., 37: 25–36.

Lord, C. (1995) Follow-up of two-year-olds referred for possible autism. J. Child Psychol. Psychiatry, 36: 1365–1382.

Lord, C., Risi, S., Lambrecht, L., Cook Jr., E.H., Leventhal, B.L., DiLavore, P.C., Pickles, A. and Rutter, M. (2000) The Autism Diagnostic Observation Schedule-Generic: A standard measure of social and communicative deficits associated with the spectrum of autism. J. Autism Dev. Disord., 30: 205–223.

Lord, C., Risi, S., Dilavore, P.S., Shulman, C., Thurm, A. and Pickles, A. (2006) Autism from 2 to 9 years of age. Arch. Gen. Psychiatry, 63: 694–701.

Lord, C., Rutter, M. and Le Couteur, A. (1994) Autism diagnostic interview-revised: A revised version of a diagnostic interview for caregivers of individuals with possible pervasive developmental disorders. J. Autism Dev. Disord., 24: 659–685.

Maestro, S., Muratori, F., Cavallaro, M.C., Pecini, C., Cesari, A., Paziente, A., Stern, D., Golse, B. and Palacio-Espasa, F. (2005) How young children treat objects and people: an empirical study of the first year of life in autism. Child Psychiatry Hum. Dev., 35: 383–396.

Maestro, S., Muratori, F., Cavallaro, M.C., Pei, F., Stern, D., Golse, B. and Palacio-Espasa, F. (2002) Attentional skills during the first 6 months of age in autism spectrum disorder. J. Am. Acad. Child Adolesc. Psychiatry, 41: 1239–1245.

Mars, A.E., Mauk, J.E. and Dowrick, P.W. (1998) Symptoms of pervasive developmental disorders as observed in prediagnostic home videos of infants and toddlers. J. Pediatr., 132: 500–504.

McCleery, J.P., Allman, E., Carver, L.J. and Dobkins, K.R. (in press) Abnormal Magnocellular Pathway Visual Processing in Infants at Risk for Autism. Biol. Psychiatry.

McEvoy, R.E., Rogers, S.J. and Pennington, B.F. (1993) Executive function and social communication deficits in young autistic children. J. Child Psychol. Psychiatry, 34: 563–578.

Merin, N., Young, G.S., Ozonoff, S. and Rogers, S.J. (2007) Visual fixation patterns during reciprocal social interaction distinguish a subgroup of 6-month-old infants at-risk for autism from comparison infants. J. Autism Dev. Disord., 37: 108–121.

Miller, M.T., Stromland, K., Ventura, L., Johansson, M., Bandim, J.M. and Gillberg, C. (2005) Autism associated with conditions characterized by developmental errors in early embryogenesis: a mini review. Int. J. Dev. Neurosci., 23: 201–219.

Milne, E., White, S., Campbell, R., Swettenham, J., Hansen, P. and Ramus, F. (2006) Motion and form coherence detection in autistic spectrum disorder: relationship to motor control and 2:4 digit ratio. J. Autism Dev. Disord., 36: 225–237.

Mitchell, S., Brian, J., Zwaigenbaum, L., Roberts, W., Szatmari, P., Smith, I. and Bryson, S. (2006) Early language and communication development of infants later diagnosed with autism spectrum disorder. J. Dev. Behav. Pediatr., 27: S69–S78.

Mullen, E.M. (1995) Mullen Scales of Early Learning-AGS Edition. AGS Publishing, Circle Pines, MN.

Muhle, R., Trentacoste, S.V. and Rapin, I. (2004) The genetics of autism. Pediatrics, 113: e472–e486.

Mundy, P., Sigman, M., Ungerer, J. and Sherman, T. (1986) Defining the social deficits of autism: the contribution of nonverbal communication measures. J. Child Psychol. Psychiatry, 27: 657–669.

Newschaffer, C.J., Fallin, D. and Lee, N.L. (2002) Heritable and nonheritable risk factors for autism spectrum disorders. Epidemiol. Rev., 24: 137–153.

Osterling, J.A., Dawson, G. and Munson, J.A. (2002) Early recognition of 1-year-old infants with autism spectrum disorder versus mental retardation. Dev. Psychopathol., 14: 239–251.

Ozonoff, S., Rogers, S.J., Farnham, J.M. and Pennington, B.F. (1993) Can standard measures identify subclinical markers of autism? J. Autism Dev. Disord., 23: 429–441.

Ozonoff, S. and Jensen, J. (1999) Brief report: specific executive function profiles in three neurodevelopmental disorders. J. Autism Dev. Disord., 29: 171–177.

Palomo, R., Belinchon, M. and Ozonoff, S. (2006) Autism and family home movies: a comprehensive review. J. Dev. Behav. Pediatr., 27: S59–S68.

Pickles, A., Starr, E., Kazak, S., Bolton, P., Papanikolaou, K., Bailey, A., Goodman, R. and Rutter, M. (2000) Variable expression of the autism broader phenotype: findings from extended pedigrees. J. Child Psychol. Psychiatry, 41: 491–502.

Plaisted, K., O'Riordan, M. and Baron-Cohen, S. (1998) Enhanced discrimination of novel, highly similar stimuli by adults with autism during a perceptual learning task. J. Child Psychol. Psychiatry, 39: 765–775.

Presmanes, A.G., Walden, T.A., Stone, W.L. and Yoder, P.J. (2007) Effects of different attentional cues on responding to joint attention in younger siblings of children with autism spectrum disorders. J. Autism Dev. Disord., 37: 133–144.

Reynell, J.K. and Grubber, C.P. (1990) Reynell developmental language scale. Los Angeles, Western Psychological Association.

Richler, J., Bishop, S.L., Kleinke, J.R. and Lord, C. (2007) Restricted and repetitive behaviors in young children with autism spectrum disorders. J. Autism Dev. Disord., 37: 73–85.

Ring, H.A., Baron-Cohen, S., Wheelwright, S., Williams, S.C., Brammer, M., Andrew, C. and Bullmore, E.T. (1999) Cerebral correlates of preserved cognitive skills in autism: a functional MRI study of embedded figures task performance. Brain, 122(Pt 7): 1305–1315.

Rogers, S.J., Hepburn, S.L., Stackhouse, T. and Wehner, E. (2003) Imitation performance in toddlers with autism and those with other developmental disorders. J. Child Psychol. Psychiatry, 44: 763–781.

Rothbart, M.K. (1981) Measurement of temperament in infancy. Child Development, 52: 569–578.

Russell, J., Saltmarsh, R. and Hill, E. (1999) What do executive factors contribute to the failure on false belief tasks by children with autism? J. Child Psychol. Psychiatry, 40: 859–868.

Sasson, N.J. (2006) The development of face processing in autism. J. Autism Dev. Disord., 36: 381–394.

Schultz, R.T. (2005) Developmental deficits in social perception in autism: the role of the amygdala and fusiform face area. Int. J. Dev. Neurosci., 23: 125–141.

Sigman, M., Dijamco, A., Gratier, M. and Rozga, A. (2004) Early detection of core deficits in autism. Ment. Retard. Dev. Disabil. Res. Rev., 10: 221–233.

Sigman, M., Dissanayake, C., Corona, R. and Espinosa, M. (2003) Social and cardiac responses of young children with autism. Autism, 7: 205–216.

Sigman, M.D., Kasari, C., Kwon, J.H. and Yirmiya, N. (1992) Responses to the negative emotions of others by autistic, mentally retarded, and normal children. Child Dev., 63: 796–807.

Spencer, J., O'Brien, J., Riggs, K., Braddick, O., Atkinson, J. and Wattam-Bell, J. (2000) Motion processing in autism: evidence for a dorsal stream deficiency. Neuroreport, 11: 2765–2767.

Stone, W.L., Lee, E.B., Ashford, L., Brissie, J., Hepburn, S.L., Coonrod, E.E. and Weiss, B.H. (1999) Can autism be diagnosed accurately in children under 3 years? J. Child Psychol. Psychiatry, 40: 219–226.

Stone, W.L., Ousley, O.Y., Yoder, P.J., Hogan, K.L. and Hepburn, S.L. (1997) Nonverbal communication in two- and three-year-old children with autism. J. Autism Dev. Disord., 27: 677–696.

Stone, W.L., Coonrod, E.E., Turner, L.M. and Pozdol, S.L. (2004) Psychometric properties of the STAT for early autism screening. J. Autism Dev. Disord., 34: 691–701.

Sullivan, M., Finelli, J., Marvin, A., Garrett-Mayer, E., Bauman, M. and Landa, R. (2007) Response to joint attention in toddlers at risk for autism spectrum disorder: a prospective study. J. Autism Dev. Disord., 37: 37–48.

Swettenham, J., Baron-Cohen, S., Charman, T., Cox, A., Baird, G., Drew, A., Rees, L. and Wheelwright, S. (1998) The frequency and distribution of spontaneous attention shifts between social and nonsocial stimuli in autistic, typically developing, and nonautistic developmentally delayed infants. J. Child Psychol. Psychiatry, 39: 747–753.

Tarr, M.J. and Gauthier, I. (2000) FFA: a flexible fusiform area for subordinate-level visual processing automatized by expertise. Nat. Neurosci., 3: 764–769.

Toth, K., Dawson, G., Meltzoff, A.N., Greenson, J. and Fein, D. (2007) Early social, imitation, play, and language abilities of young non-autistic siblings of children with autism. J. Autism Dev. Disord., 37: 145–157.

Tzourio-Mazoyer, N., de Schonen, S., Crivello, F., Reutter, B., Aujard, Y. and Mazoyer, B. (2002) Neural correlates of woman face processing by 2-month-old infants. Neuroimage, 15: 454–461.

Werner, E. and Dawson, G. (2005) Validation of the phenomenon of autistic regression using home videotapes. Arch. Gen. Psychiatry, 62: 889–895.

Werner, E., Dawson, G., Munson, J. and Osterling, J. (2005) Variation in early developmental course in autism and its relation with behavioral outcome at 3–4 years of age. J. Autism Dev. Disord., 35: 337–350.

Wetherby, A.M., Allen, L., Cleary, J., Kublin, K. and Goldstein, H. (2002) Validity and reliability of the communication and symbolic behavior scales developmental profile with very young children. J. Speech Lang and Hearing Res., 45: 1202–1218.

Wetherby, A.M., Woods, J., Allen, L., Cleary, J., Dickinson, H. and Lord, C. (2004) Early indicators of autism spectrum disorders in the second year of life. J. Autism Dev. Disord., 34: 473–493.

Wiig, E.H., Secord, W. and Semel, E. (1992) CELF-preschool: Clinical evaluation of language fundamentals – preschool version. San Antonio, Psychological Corporation.

Wimpory, D.C., Hobson, R.P., Williams, J.M. and Nash, S. (2000) Are infants with autism socially engaged? A study of recent retrospective parental reports. J. Autism Dev. Disord., 30: 525–536.

World Health Organization. (1992) The ICD 10 classification of mental and behavioral disorders: Clinical descriptions and diagnostic guidelines. Geneva, Switzerland, World Health Organization.

Yirmiya, N., Gamliel, I., Pilowsky, T., Feldman, R., Baron-Cohen, S. and Sigman, M. (2006) The development of siblings of children with autism at 4 and 14 months: social engagement, communication, and cognition. J. Child Psychol. Psychiatry, 47: 511–523.

Yirmiya, N., Gamliel, I., Shaked, M. and Sigman, M. (2007) Cognitive and verbal abilities of 24- to 36-month-old siblings of children with autism. J. Autism Dev. Disord., 37: 218–229.

Zimmerman, I.L., Steiner, V.G. and Pond, R.E. (2002) Preschool language scaes-4. San Antonio, TX, The Psychological Corporation.

Zwaigenbaum, L. (2001) Autistic spectrum disorders in preschool children. Can. Fam. Physician, 47: 2037–2042.

Zwaigenbaum, L., Bryson, S., Rogers, T., Roberts, W., Brian, J. and Szatmari, P. (2005) Behavioral manifestations of autism in the first year of life. Int. J. Dev. Neurosci., 23: 143–152.

Zwaigenbaum, L., Thurm, A., Stone, W., Baranek, G., Bryson, S., Iverson, J., Kau, A., Klin, A., Lord, C., Landa, R., Rogers, S. and Sigman, M. (2007) Studying the emergence of autism spectrum disorders in high-risk infants: methodological and practical issues. J. Autism Dev. Disord., 37: 466–480.

C. von Hofsten & K. Rosander (Eds.)
Progress in Brain Research, Vol. 164
ISSN 0079-6123

CHAPTER 21

Children–robot interaction: a pilot study in autism therapy

Hideki Kozima[1,*], Cocoro Nakagawa[1] and Yuriko Yasuda[2]

[1]*National Institute of Information and Communications Technology, Hikaridai 3-5, Seika, Soraku, Kyoto 619-0289, Japan*
[2]*Omihachiman-City Day-Care Center for Children with Special Needs, Tsuchida 1313, Omihachiman, Shiga 523-0082, Japan*

Abstract: We present here a pilot study of child–robot interactions, in which we discuss developmental origins of human interpersonal communication. For the past few years, we have been observing 2- to 4-year-old children with autism interacting with Keepon, a creature-like robot that is only capable of expressing its attention (directing its gaze) and emotions (pleasure and excitement). While controlled by a remote experimenter, Keepon interacted with the children with its simple appearance and actions. With a sense of curiosity and security, the children spontaneously approached Keepon and engaged in dyadic interaction with it, which then extended to triadic interactions where they exchanged with adult caregivers pleasure and surprise they found in Keepon. Qualitative and quantitative analysis of these unfolding interactions suggests that autistic children possess the motivation to share mental states with others, which is contrary to the commonly held position that this motivation is impaired in autism. We assume Keepon's minimal expressiveness helped the children understand socially meaningful information, which then activated their intact motivation to share interests and feelings with others. We conclude that simple robots like Keepon would facilitate social interaction and its development in autistic children.

Keywords: human–robot interaction; social interaction; field practice; autism therapy; developmental psychology; interactive robots; minimal design

Introduction

For the past 3 years, at a therapeutic day-care center, we have been observing children (2–4 years old) with developmental disorders, primarily autism, interacting with a simple communication robot. The robot has a small snowman-like body, with which it expresses pleasure (by rocking) and excitement (by bobbing); it has two video cameras in the eyes and a microphone in the nose, with which it looks around for eye contact and further interaction with a child. In longitudinal observations, the children showed various actions in relation to the robot. They often showed vivid facial expressions that even their parents had rarely seen; sometimes they showed caretaking behavior such as putting a cap on the robot and feeding it toy food. Each child showed a different style of interaction that changed over time in a different way, which told us a "story" of his or her personality and developmental profile. We have been providing the therapists and parents with video feedback, which was the story about each child narrated from the first-person perspective of

*Corresponding author. Tel.: +81-774-98-6892;
Fax: +81-774-98-6958; E-mail: xkozima@nict.go.jp

DOI: 10.1016/S0079-6123(07)64021-7

the robot, in order to establish a reference point for sharing and exchanging the caregivers' and specialists' understanding of the children.

Through the field practice, we learned that any therapeutic intervention presupposes that even autistic children possess the motivation for sharing and exchanging mental states with others. This motivates, in turn, therapists and parents to improve not only the quality of the children's social (objective) lives, but also their personal (subjective) lives to enable them to enjoy such interpersonal interaction. Although it is widely believed that this motivation is impaired in autism (Tomasello et al., 2005), we observed in a number of cases that the autistic children established social relationships with the simple robot, which was carefully designed to express its mental states comprehensively. The children spontaneously engaged in dyadic exchange of attention and emotions with it, which then extended to interpersonal interaction with the caregivers to exchange the pleasure and surprise they found in the dyadic interactions. This suggests that "the missing motivation" is not missing in autism.

This paper describes the longitudinal observation of the autistic children interacting with the robot, from which we discuss developmental origins of human interpersonal communication. What we present here includes (1) simple robots with minimal and comprehensive expressiveness facilitate the spontaneous exchange of mental states in autistic children, (2) autistic children therefore possess the motivation for this mental exchange, and (3) the major social difficulties that autistic children generally suffer from would rather stem from the difficulty in sifting out socially meaningful information (e.g., attention and emotions) from the vast incoming perceptual information. In "Designing child–robot interaction", we introduce some of our robots, including Keepon, which we used in the field observations, and their design principles based on recent psychological findings on children's social development. "Interaction in the laboratory" describes how typically developing children interact with the robots, from which we model how social interaction dynamically unfolds as time passes and how such interactions qualitatively change with age. "Field Practices" describes the longitudinal field observation of the autistic children interacting with Keepon; we describe three representative cases that exemplify the emergence of dyadic, triadic, and empathetic interactions. "Discussion" discusses our hypothesis on autistic children's intact motivation for sharing mental states with others.

Designing child–robot interaction

There has been a growing interest in designing interactive robots that human children can naturally and intuitively interact with. This research trend is motivated by the assumption that the underlying mechanism for children's embodied interaction and its development is the fundamental substratum for human social interaction in general. We review here recent engineering and psychological findings, describe our design principle of interactive robots for children, and introduce our robot, Keepon, as an implementation of the design principle.

Interactive robots for children

A number of research projects in the field of embodied interaction have developed interactive robots explicitly for interaction with children. For example, Kismet (Breazeal and Scassellati, 2000) is one of the pioneering examples of "sociable robots". Kismet engages people in natural and intuitive face-to-face interaction, where it exchanges with people a variety of social cues such as gaze direction, facial expression, and vocalization. Kismet emphasized the elicitation of caretaking behavior from people, including children, which facilitates the robot's socially situated learning from those human caregivers.

Another pioneer is the AuRoRa project (Dautenhahn, 1999), which reported that even simple mobile robots gave autistic children a relatively repetitive and predictable environment that encouraged spontaneous interactions, such as playing chasing games, with the robots in a relaxed mood. Billard (2002) developed a doll-like anthropomorphic robot, Robota (Fig. 1, left), for mutual imitative play with autistic children; Robins et al. (2004) intensively analyzed two autistic children playing together with Robota,

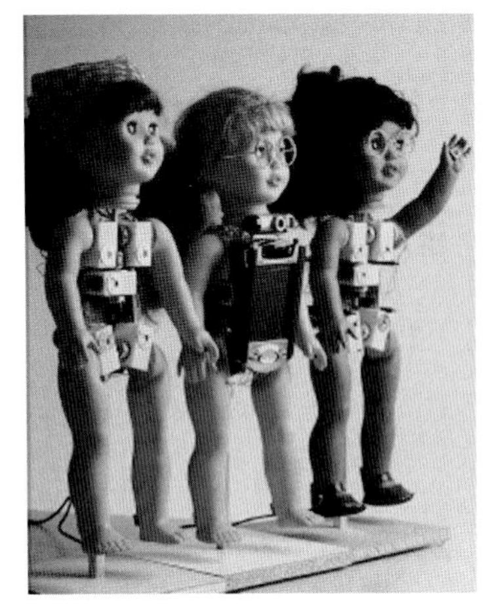

Fig. 1. Some of the robots developed for interaction with children, especially those with autism. From left, Robota (Billard, 2002), Tito (Michaud and Théberge-Turmel, 2002), and Muu (Okada and Goan, 2005).

where they exhibited mutual monitoring and cooperative behavior to derive desirable responses from it.

Scassellati (2005), who has been building and using social robots for the study of children's social development, observed autistic children interacting with a robot with an expressive face and found that they showed positive proto-social behaviors, such as touching, vocalizing at, and smiling at the robot, which were generally rare in their everyday life. Michaud and Théberge-Turmel (2002) devised a number of mobile and interactive robots, including Roball and Tito (Fig. 1, middle), and observed interaction with autistic children in order to explore the design space of child–robot interactions that foster children's self-esteem. Okada and Goan (2005) developed a creature-like robot, Muu (Fig. 1, right), to observe how autistic children spontaneously collaborate with the robot in shared activities, such as arranging colored blocks together.

Autism and PDD (pervasive developmental disorders)

It is notable that several of the research projects previously reviewed have direct or indirect connections with autism, which is a developmental and behavioral syndrome caused by specific and mainly hereditary brain dysfunctions (Frith, 1989). According to the major diagnostic criteria, e.g., DSM-IV (American Psychiatric Association, 1994) and ICD-10 (World Health Organization, 1993), people with autism generally have the following major impairments.

- Social (non-verbal) impairment: difficulty in understanding, as well as indifference to, others' mental states, such as intentions and emotions exhibited by gaze direction, facial expressions, and gestures; difficulty in reciprocal exchange and sharing of interests and activities with others.
- Linguistic (verbal) impairment: difficulty in linguistic communications, especially in pragmatic use of language; delayed or no language development; stereotyped or repetitive speech (echolalia), odd or monotonous prosody, and pronoun reversal.
- Imaginative impairment: difficulty in maintaining a diversity of interest and behavior; stereotyped actions and routines; difficulty in coping with novel situations; absorption in meaningless details, while having difficulty with holistic understanding.

"Autism" (or more precisely, "childhood autism") is a subtype of broader "pervasive developmental disorders (PDD)"; PDD is also referred to as "autism spectrum disorders (ASD)". Roughly speaking, a child will be diagnosed as "autistic", when he or she exhibits all three impairments before the age of 3. Another major subtype is "Asperger's syndrome", in which relatively good language skills (especially syntactic and semantic ones) are demonstrated, but not good at social interaction or imagination. According to a recent statistics (Baird et al., 2006), the prevalence of PDD is estimated to be 116.1 in 10,000, with autism accounting for 38.9 of that figure.

In children with autism or PDD, the above impairments limit the ability to exchange and share mental states with others. This makes it difficult to establish and maintain social relationships, in which they could learn language and other social conventions. Researchers in social robotics therefore have a particular interest in autism to better understand the underlying mechanisms responsible for social interaction and its development.

Eye contact and joint attention

Eye contact and joint attention are fundamental behaviors that maintain child–caregiver interactions (Trevarthen, 2001). A child and a caregiver spatially and temporally correlate their attention and emotions, in which they refer to each other's subjective feelings of pleasure, surprise, or frustration (Dautenhahn, 1997; Zlatev, 2001; Kozima and Ito, 2003). We believe all communication emerges from this mutual reference.

Eye contact is the joint action of two individuals looking into each other's faces, especially the eyes. It serves not only to monitor each other's state of attention (e.g., gaze direction) and of emotion (e.g., facial expressions), but also to temporally synchronize interactions and to establish mutual acknowledgment (Kozima et al., 2004), such as "My partner is aware of me" and "My partner is aware that I am aware of her". It has been widely said that autistic children generally show indifference to others' gaze; however, a brain imaging study (Dalton et al., 2005) revealed that children with autism sometimes emotionally over-react, as evidenced in intensive activation that occurs in the amygdala, to a direct gaze from a non-threatening face of a person. This suggests that autistic children are often emotionally overwhelmed by eye contact with others, and therefore often avert their gaze.

Joint attention is the joint action of two individuals looking at the same target or pointing at it by means of gaze and/or deictic gesture (Butterworth and Jarrett, 1991). In the first stage of joint attention, the caregiver actively follows and guides the child's attention so that the child can easily capture the target; then the child gradually becomes able to follow and guide the partner's attention; finally, the child and caregiver coordinate such initiations and responses, forming spatiotemporal patterns of exchanges of attention. Moreover, in the first stage, children are only capable of coordinating attention with others in relation to a visible object or event; in later stages, invisible psychological entities, such as emotions and concepts, will be incorporated into the target of attention. Joint attention not only provides the interactants with shared perceptual information, but also with a spatial focus for their interaction, thus creating mutual acknowledgment (Kozima et al., 2004), such as "My partner and I are aware of the same target" and "Both of our actions (such as vocalization and facial expressions) relate to the target". Again, it is widely recognized that autistic children rarely engage in joint attention with others; however, when instructed by an experimenter, they often identify the targets of others' attention. This means that they are capable of doing it, but they do not often do it in their daily life.

Keepon, the interactive robot

With inspiration from the psychological study of social development and its deficiency in autism, we first developed an upper-torso humanoid robot, Infanoid (Kozima, 2002), which is 480 mm tall, the approximate size of a 4-year-old human child (Fig. 2). It has 29 actuators and a number of sensors that enables it to express attention (by the direction of its gaze and face, and by pointing), facial expressions (by moving eyebrows and lips), and other hand and body gestures by a set of programs running on PC. Image processing of video streams taken by four CCD cameras (peripheral, foveal × left, right) enables the robot (1) to detect a human face and direct its gaze to the face (eye contact), and (2) to estimate the direction of attention from the

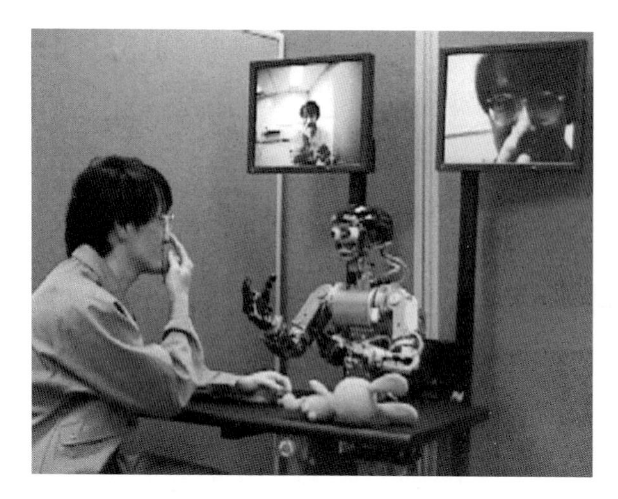

Fig. 2. Infanoid, the upper-torso humanoid robot, engaging in eye contact (left) and joint attention (right) with a human interactant. (The video monitors behind show Infanoid's peripheral and foveal views.)

face and search the attention line for any salient object (joint attention), as illustrated in Fig. 2. With a simple habituation mechanism, which releases attention to a face or an object after a period of fixation, the robot alternates between eye contact and joint attention with a human in the loop.

Through our preliminary experiments in child–robot interaction, we observed that most of the children enjoyed social interactions, where they read the robot's attention and emotions in order to coordinate their behavior. However, we also found that Infanoid conveys overwhelming information to some of the children, especially those under 3 years of age, who generally showed strong embarrassment and anxiety about Infanoid at first sight. This is probably because of its anthropomorphic but strongly mechanistic appearance. The children were often attracted by moving parts, especially hands, fingers, eyes, and eyebrows. Each of the moving parts induces qualitatively different meanings in the children, who then need effortful integration of the separated meanings into a holistic recognition of a social agent.

Having learned from the younger children's difficulty in understanding holistic information from Infanoid's mechanistic appearance, we then built a simple creature-like robot, Keepon (Fig. 3), which has a minimal design for facilitating exchange of attention and emotions with people, especially babies and toddlers, in simple and comprehensive ways.

The simplest robots for our purpose would be one that can display its attention (what the robot is looking at), as exemplified in Fig. 4 (left). The presence of active attention strongly suggests that the robot can perceive the world as we see and hear it. The next simplest one would be a robot that displays not only attentive expression but also emotional expression. Emotive actions strongly suggest that the robot has a "mind" to evaluate what it perceives. However, if we add a number of additional degrees of freedom to the robot's facial expression, this information flood would so overwhelm the children that they would hardly grasp the gestaltic meaning. Therefore, we decided to utilize movement of the robot's body to express some emotional states, such as pleasure (by rocking the body from side-to-side), excitement (by bobbing up and down), and fear (by vibrating) as shown in Fig. 4 (middle). In the final design stage, we made a neck/waist line, as shown in Fig. 4 (left). The neck/waist line provides a clear distinction between the head and the belly, which would give an anthropomorphic (but not overwhelming) impression to the children. The neck/waist line also provides life-like deformation of the body as the robot changes its posture.

As an implementation of the design principle described above, Keepon has a yellow snowman-like body that is 120 mm tall and made of soft silicone rubber. The upper part (the "head") has

Fig. 3. Keepon, the interactive robot, engaging in eye contact (left) and joint attention (right) with a human interactant.

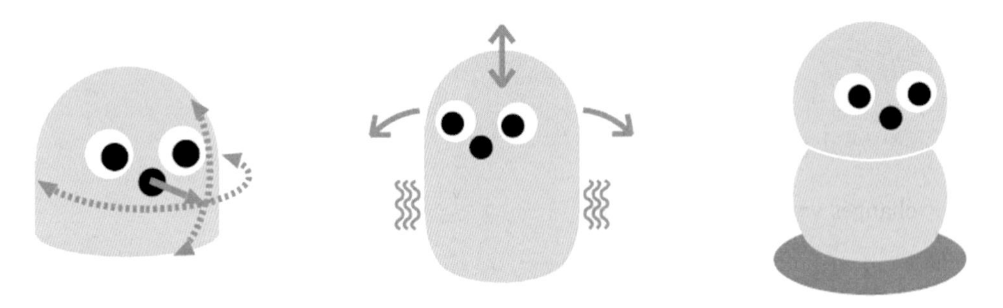

Fig. 4. Designing Keepon's appearance for the interactive functions of expressing attention (left) and emotion (middle), and the final sketch (right).

two eyes, both of which are color CCD cameras with wide-angle lenses (120°, horizontally), and a nose, which is actually a microphone. The lower part (the "belly") contains small gimbals and four wires with which the body is manipulated like a marionette using four DC motors and circuit boards that are encased in the black cylinder (Fig. 5, left). Since the body is made of silicone rubber and its interior is relatively hollow, Keepon's head and belly deform whenever it changes posture or someone touches it (Fig. 5, right).

The simple body has four degrees of freedom: nodding (tilting) $\pm 40°$, shaking (panning) $\pm 180°$, rocking (side-leaning) $\pm 25°$, and bobbing (shrinking down and coming back) with a 15-mm stroke. These four degrees of freedom produce two qualitatively different types of actions:

- Attentive action: Keepon orients toward a certain target in the environment by directing the head (i.e., gaze) up/down and left/right. It appears to perceive the target. This action includes eye contact and joint attention.
- Emotive action: Keepon rocks from left to right and/or bobs its body up and down keeping its attention fixed on a certain target. It gives the impression of expressing Keepon's internal states, which seem to be emotions (e.g., pleasure and excitement) about the target of its attention.

Fig. 5. Keepon's structure: its simple appearance and marionette-like mechanism (left), which drives the deformable body (right).

With these two actions, Keepon can express "what" it perceives and "how" it evaluates the target.

Interaction in the laboratory

Before doing field studies in a therapeutic setting, we carried out preliminary experiments of child–robot interaction using Keepon, in order to verify the effects of our minimal design on the attentive and emotive exchanges with younger children.

Method

Although Keepon can be operated in "automatic" mode, where the robot alternates between eye contact and joint attention in the same way Infanoid does, we intentionally used "manual" mode, where a human operator (or a "wizard", usually at a remote PC) controls the robot's postural orientations, facial/bodily expressions, and vocalizations. The operator watches video from the on-board cameras and listens to sound from the on-board microphone; the operator also watches video from an off-board camera to get a side view of the interaction, which helps the operator to understand what is going on, especially when the on-board cameras are occasionally covered by a child's hand or face.

To perform interactive actions on the robots, the operator uses a mouse to select points of interest on the camera monitor and uses keystrokes that are associated with different emotive actions. The robot thus (1) alternates its gaze between a child's face, the caregiver's face, and sometimes a nearby toy, and (2) produces a positive emotional response (e.g., bobbing its body several times with a "pop, pop, pop" sound) in response to any meaningful action (e.g., eye contact, touch, or vocalization) performed by the child. We thus manually controlled Keepon because Keepon should wait for a child's spontaneous actions, and when the child directs an action, Keepon should respond to it with appropriate timing and in the appropriate manner.

Unfolding interaction with Keepon

We observed 25 typically developing children in three different age groups, i.e., under 1 year ($N = 8$; on average, 9.0 months old), 1–2 years old ($N = 8$; on average, 16.5 months old), and over 2 years old ($N = 9$; on average, 37.3 months old), interacting with Keepon (Fig. 6). Each child, together with the caregiver, was seated in front of Keepon on the floor; there were no particular instruction for them. Interaction continued until the children became tired or bored; on average, each child's dealings lasted ~10–15 min. We found from these observations that children in each age group showed different styles of interaction.

- 0–1-year-olds: interaction was dominated by tactile exploration using the hands and/or

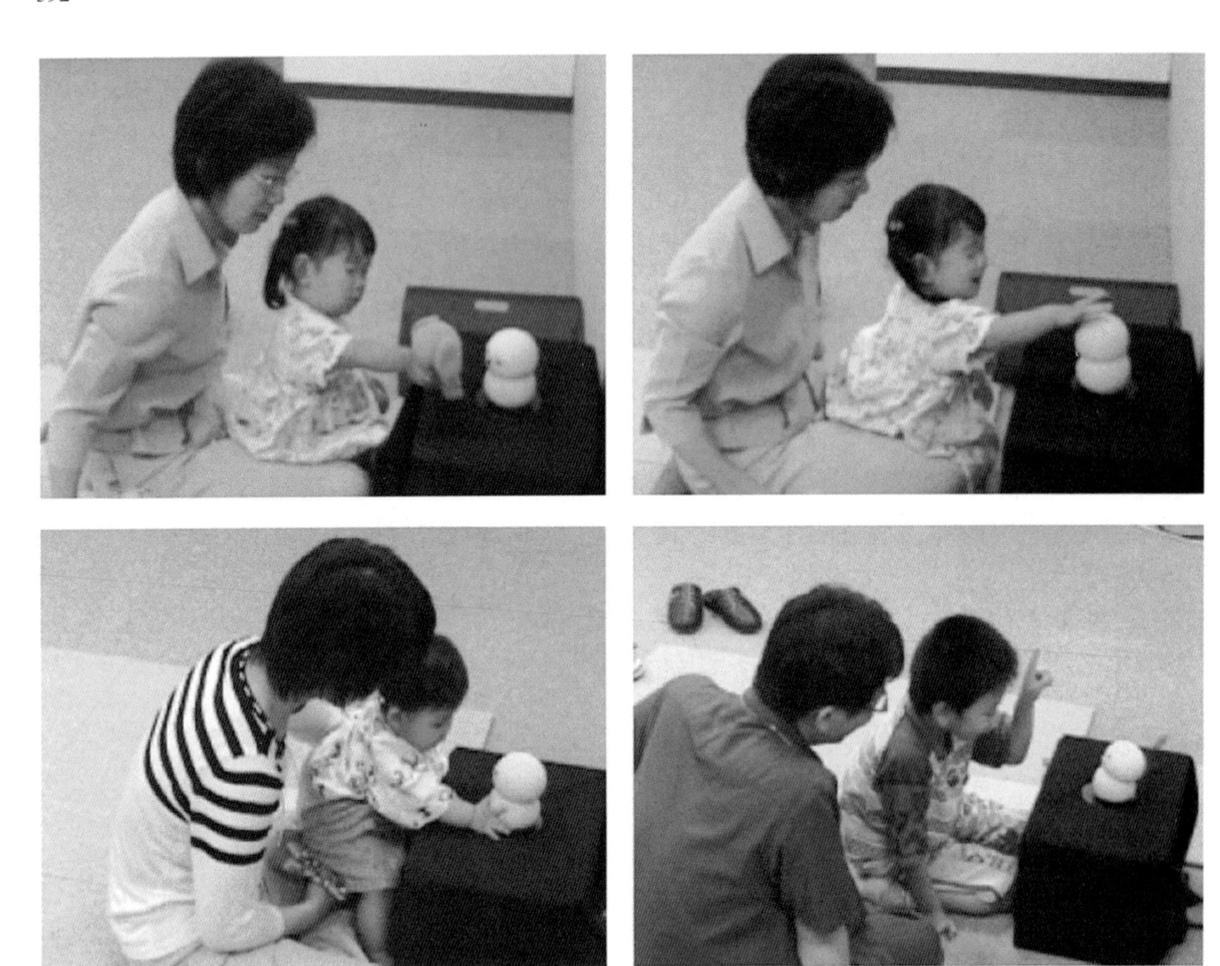

Fig. 6. Interaction between Keepon and typically developing children: a 2-year-old showing a toy to, and soothing the robot (upper), a 6-month-old touching (lower left), and a 5-year-old challenging the robot (lower right).

mouth. The children did not pay much attention to the attentive expressions of the robot, but they exhibited positive responses, such as laughing or bobbing their bodies, to its emotive expressions.

- 1–2-year-olds: the children demonstrated not only tactile exploration, but also awareness of the robot's attentive state, sometimes following its attention. Some mimicked its emotive expressions by rocking and bobbing their own bodies.
- 2+-year-olds: these children first carefully observed the robot's behavior and how caregivers interacted with it. Soon they initiated social exploration by showing it toys, soothing it (by stroking its head), or verbally interacting with it (such as asking questions).

The differences between the interactions of each age group reflect differences in their ontological understanding of Keepon. The 0–1-year-olds recognized the robot as a "moving thing" that induced tactile exploration. The 1–2-year-olds first encountered a "moving thing", but after observing the robot's responses to various environmental disturbances, they recognized that it was an "autonomous system" that possesses attention and emotion as an initiator of its own expressive actions. The 2+-year-olds, through the recognition of a "thing" and a "system", found that the robot's responses, in terms of attention and emotion, had a spatiotemporal relationship with what they had done to it; finally, they recognized it as a "social agent" with whom they could play with exchanging and coordinating their attention and emotions.

Field practices

Our primary research site was a day-care center for children with developmental disorders, especially those with PDD, including childhood autism and Asperger's syndrome. In the playroom at the day-care center, children (mostly 2–4 years old), their parents (usually mothers), and therapists interact with one another, sometimes in an unconstrained manner (i.e., individually or within a nuclear relationship of child/mother/therapist), and sometimes in rather organized group activities (e.g., rhythmic play and storytelling). In these dynamically and diversely unfolding interactive activities, the children's actions are watched, responded to, and gradually situated in the social context of everyday life.

Keepon in the playroom

A wireless version of Keepon was placed in the playroom just as if it were another toy on the floor. In their daily therapeutic sessions (3 h each), seven to eight combinations of child/mother/therapist got together in the playroom, where they sporadically interacted with Keepon. In the morning, before any children showed up, we put Keepon in the playroom; the initial positions of Keepon varied in each session according to how other toys were arranged. During free play (i.e., the first hour), children could play with Keepon at any time. During group activities (i.e., the following 2 h), Keepon was moved to the corner of the playroom so that it did not interfere with the activities; however, if a child became bored or stressed by the group activities, he or she would be allowed to play with Keepon. Between the free play and the group activities, the children had a snack in an adjacent room, during which Keepon was placed in the room.

Throughout the observations, we tele-controlled Keepon from a remote room and recorded live interactions with the children from Keepon's perspective (Fig. 7). In other words, we recorded all the information from the subjective viewpoint of Keepon as the first person of the interaction. Strictly speaking, this subjectivity belongs to the operator; however, the interaction is mediated by the simple actions that Keepon performs, and every action Keepon performs can be reproduced from the log data. Therefore, we may say that Keepon is both subjective (i.e., interacting directly with children) and objective media (through which

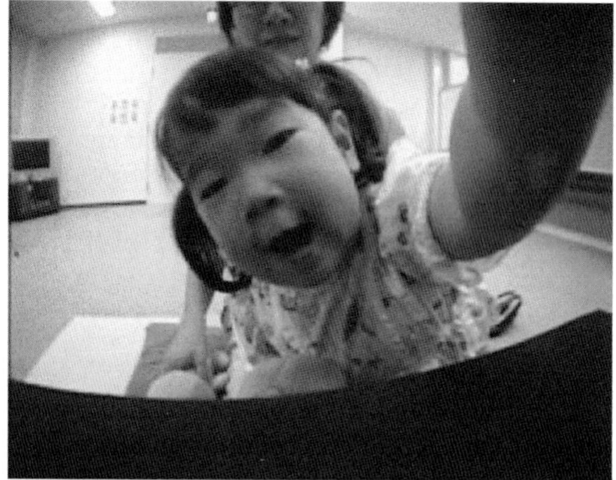

Fig. 7. A child seen from the subjective viewpoint of Keepon as the first person of the interaction. (From preliminary observations in our laboratory.)

anyone can re-experience the interactions), enabling human social communications to be studied.

Case 1 — emergence of dyadic interaction

For the past 3 years (over 100 sessions, or 700 child-sessions in total), we have been longitudinally observing a group of children with PDD (including childhood autism and Asperger's syndrome), Down's syndrome, and other developmental disorders. We observed over 30 children in total; some of the children enrolled in or left the center during this period. We describe here three representative cases, each of which best demonstrates the emergence of dyadic, triadic, and empathetic interactions.

The first case is a 3-year-old girl, M, with autism. At CA 1:11 (chronological age of 1 year and 11 months), her mental age (MA) was estimated at 0:10 by Kyoto Scale of Psychological Development. At CA 3:5, she was diagnosed as autism with moderate mental retardation. Here, we describe how the interaction between M and Keepon unfolded in 15 sessions over 5 months (CA 3:9–4:1), during which she did not exhibit any apparent production of language.

- From Session 1 (hereafter referred to as S1), M exhibited a strong interest in Keepon, but did not get close to it. Through S1–S7, M avoided being looked at directly by Keepon (i.e., gaze aversion); however, M gradually approached it from the side and looked at it in profile.
- In S5, after watching a boy put a paper cylinder on Keepon's head, M went to her therapist and pulled her by the arm to Keepon, non-verbally asking her to do the same thing. When the therapist completed her request, M left Keepon with a look of satisfaction on her face. Through S5–S10, her distance to Keepon gradually decreased to less than 50 cm.
- In the free play of S11, M touched Keepon's head using a xylophone stick. During the group activity, M reached out with her arm to Keepon but did not actually touch it. In the intermission of the group activity, M sat in front of Keepon and touched its belly with her left hand, as if examining its texture or temperature.
- After this first touch in S11, M began acting exploratively with Keepon, such as looking into its eyes, waving her hand at it, and listening to its sound. From S12, M vocalized non-words to Keepon, as if she expected some vocal response from it. In S13, M put a knitted cap on its head, and then asked her mother to do the same thing. In S14, M actually kissed the robot.

We can see that M's persistent curiosity gradually reduced her fear of Keepon. We also see here the emergence of both spontaneous dyadic interactions (Baron-Cohen, 1995; Tomasello, 1999), such as touching Keepon with a xylophone stick, and interpersonally triggered dyadic interaction, such as putting a paper cylinder on its head. The latter especially suggests that M was a good observer of others' behavior, although she seldom imitated others even when instructed. Because the boy's action was mediated by Keepon and an object (e.g., the paper cylinder) that were of interest to M, it would be relatively easy for her to emulate (Tomasello, 1999) the same action and result.

Case 2 — emergence of triadic interaction

The second case is a 3-year-old girl, N, with autism and moderate mental retardation (MA 1:7 at CA 3:1; no apparent language). We observed her interactions with Keepon for 39 sessions, which lasted for ~17 months (CA 3:4–4:8).

- In S1, N gazed at Keepon for a long time. After observing another child playing with Keepon using a toy, N was encouraged to do the same, but did not show any interest in doing that.
- Through S2–S14, N did not pay much attention to Keepon, even when she sat next to it. However, N often glanced at the robot, when she heard sounds coming from it. The first touch was in S10.
- In S15, after observing another child place a cap on Keepon's head, N touched Keepon with her finger.

- In S16 (after a 3-month interval from S15), N came close to Keepon and observed its movements. During snack time, N came up to Keepon again and poked its nose, to which Keepon responded by bobbing, and N showed surprise and a smile; the mothers and therapists in the playroom burst into laughter. During this play, N often looked referentially and smiled at her mother and therapist.
- From S17, N often sat in front of Keepon with her mother; sometimes she touched Keepon to derive a response. From S20, N started exploring Keepon's abilities by walking around it to see if it could follow her.
- During snack time in S33, N came up to Keepon and started an "imitation game". When N performed a movement (bobbing, rocking, or bowing), soon Keepon mimicked her; then N made another, and Keepon mimicked her again. Through S33–S39, N often played this "imitation game" with Keepon, during which she often looked referentially at her mother and therapist.

For the first 5 months (15 sessions), N did not show strong curiosity about Keepon; even when N was held in her carer's arms in front of Keepon, she simply saw Keepon but did not act on it. After a 3-month break, especially in S16 and S33, we witnessed the emergence of triadic interactions (Tomasello, 1999; Trevarthen, 2001), where Keepon or its action functioned as a pivot (or a shared topic) for interpersonal interactions between N and her mother or therapist. In those triadic interactions, which were spontaneously performed in a playful and relaxed mood, it seemed that N wanted to share with the adults the "wonder" she had experienced with Keepon. Within this context, the "wonder" was something that induced smiles, laughter, or other emotive responses in herself and her interaction partner. It is also notable that the "imitation play" first observed in S33 was unidirectional one, in which Keepon was the imitator and N was the model and probably the referee; however, this involved reciprocal turn-taking, which is one of the important components of social interaction.

Case 3 — emergence of empathetic interaction

The third case is a boy, S, with Asperger's syndrome with mild mental retardation (MA/cognition 3:2 and MA/language 4:3 at CA 4:6). Here, we describe the first 15 sessions, which lasted for ~9 months (CA 3:10–4:6).

- In the first encounter, S violently kicked Keepon and knocked it over; then, he showed embarrassment, not knowing how to deal with the novel object.
- From S2, S became gentle with Keepon. Often S scrambled with another child for Keepon (S3 and S6), suggesting his desire to possess the robot. In S5, S showed his drawing of the both of them to Keepon, saying "This is Pingpong [Keepon]; this is S".
- In S8, S asked Keepon, "Is this scary?" showing bizarre facial expressions to the robot. When an adult stranger approached Keepon, S tried to hide it from her, as if he were protecting Keepon.
- In S11 and S16, when another child behaved violently with Keepon, S often hit or pretended to hit the child, as if he were protecting Keepon. During snack time in S14, S was seated next to Keepon. S asked the robot and another child if the snacks were good by saying, "Yummy?".
- In S15, Keepon wore a flu mask. S came up to Keepon and asked "Do you have a cough?" a couple of times. When his therapist came in, S informed her of the presence of the mask, saying, "Here's something".

In the early stages of interaction, we saw a drastic change in S's attitude toward Keepon. S exhibited exceptionally violent behavior toward Keepon in the first encounter. But after S2, S demonstrated exceptionally gentle behavior toward Keepon, trying to monopolize and sometimes to protect it. His therapist suggested that S usually expressed violent behavior toward strangers to whom he did not know how to relate, but he would behave socially after getting used to them. It is noteworthy that S seemed to regard Keepon as a human-like agent that not only perceived the environment and evaluated its emotional content, but also understood language, regardless of S's relatively good cognitive and linguistic capabilities.

Discussion

From the field observations at the therapeutic day-care center, we found a number of observations that are contradictory to the widely accepted knowledge of autism. We discuss here why and how such things happened.

Is the motivation missing?

In Case 1, the autistic girl M first avoided being looked at by Keepon. As suggested from the brain imaging study (Dalton et al., 2005), autistic children would emotionally over-react to someone's direct gaze even when the facial expression was non-threatening. According to her mother and therapist, however, M had never over-reacted to a direct gaze from any person or animal, and she was rather indifferent to others' gazes and presence. However, in the first several sessions, M over-reacted selectively to the direct gaze from Keepon. Meanwhile, M also showed strong curiosity about Keepon, and this curiosity was competing with her fear of the direct gaze. As illustrated in Fig. 8, M slowly but steadily got close to Keepon, exploring what Keepon was and how Keepon would respond to various interventions, using her therapist as a secure base.

We assume that Keepon's simple appearance and actions helped the autistic children, especially M, to sense social contingency, or presence of a "mind", in its attentive and emotive actions. Intuitively speaking, a "mind" would be an internal process that selects a particular target from the environment and displays a particular response to the target. The selection is what we call "attention", and the response comes from what we call "emotion". As we have seen before, Keepon is only capable of expressing attention by orienting its gaze and expressing emotions by rocking and/or bobbing its body. This simplicity and comprehensiveness would open a bypass channel (Fig. 9) through which the children directly perceived Keepon's attention, emotions, and therefore "mind", without being perceptually overwhelmed.

This also suggests that autistic children possess the motivation to share mental states with others, which is contrary to the commonly held position that this motivation is impaired in autism (Tomasello et al., 2005). What we observed in the child–robot interactions was that the autistic children spontaneously engaged in dyadic exchanges of attention and emotions with Keepon. As exemplified in Cases 2 and 3, some of the children then extended it to triadic interaction, where Keepon

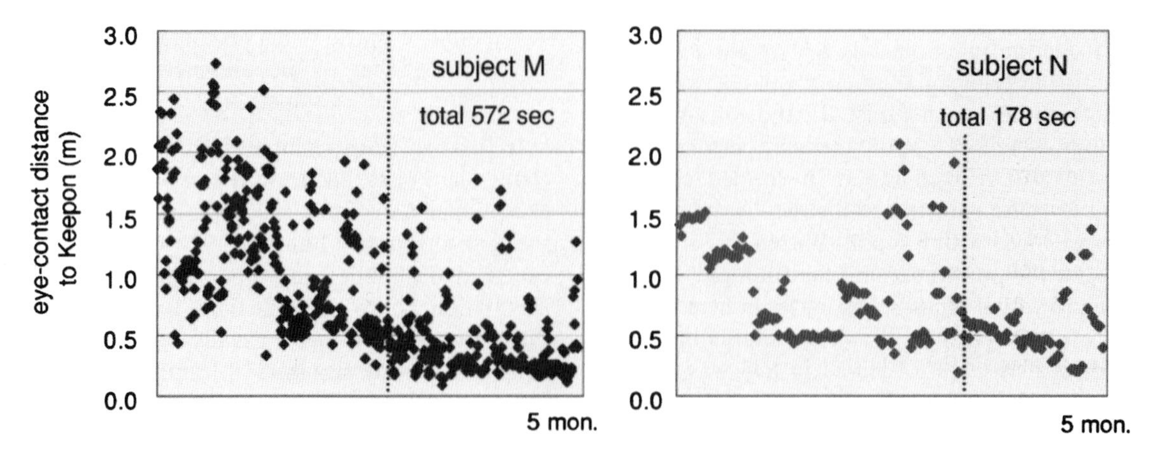

Fig. 8. Gradual decrease in eye contact distance between children and Keepon plotted for the first 5 months of interactions. The distance was geometrically estimated from (1) the distance between eyes, (2) the size of a face, or (3) the body height. The dotted lines represent when "first touch" occurred.

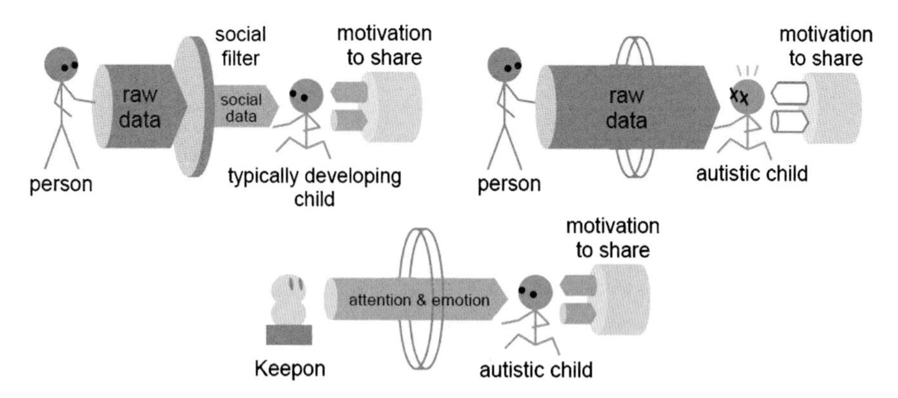

Fig. 9. The social filter in typically developing children (left), its possible dysfunction in autistic children (right), and the bypass channel produced by Keepon (bottom).

functioned as a pivot to exchange with others the pleasure and surprise they found in the dyadic interactions. Their motivation for relating to others would be intact, but difficult to be activated because of their failure in sifting out meaningful information, such as the attention and emotions of interactants, from the perceptual information flooding in (Fig. 9).

Comparing with typically developing children

Now we look into how a group of 27 children in a class of 3-year-olds (at the beginning of the observation; average CA was 4:0 throughout the yearlong period) interacted with Keepon in the playroom of their preschool. In each session, at around 8:30 a.m., one of the teachers brought Keepon to the playroom and put it on the floor with other toys. In the first 90 min, the children arrived at the preschool, gradually formed clusters, and played freely with each other and with Keepon. In the next 90 min, under the guidance of three teachers, the children engaged in various group activities, such as singing songs, playing musical instruments, and doing paper crafts. Keepon was moved as necessary by the teachers so that it did not interfere with the activities; sometimes it sat beside the teacher who was telling a story, or sat on the piano watching the children who were singing or dancing.

Throughout the longitudinal observations (25 3-h sessions) we telecontrolled Keepon's attentional expression, emotional expression, and vocalizations (of simple sounds), by watching and listening to the video from Keepon's on-board cameras and microphone. The children showed various spontaneous interactions with Keepon, individually and in a group; the style of these interactions changed over time. Here are some anecdotes about what Keepon experienced in the playroom.

- In S1, the children demonstrated shyness and embarrassment around Keepon, not knowing well what it was and how they should interact with it. From S2, they gradually initiated various interventions with Keepon — from hitting it to feeding it.
- In S5, a girl NK (hereafter NK/f) put a cap on Keepon. When the cap was gone, a boy YT (hereafter YT/m) put his own cap on Keepon. In S7, when it was lost again, TK/m and NK/f soothed Keepon, saying, "Did you loose your cap?" and "Try to get by without your cap."
- In S6, KT/f played with Keepon in the outdoor playground; a boy in the 4-year-old class approached Keepon and said to KT/f (referring to Keepon), "This is a camera. This is a machine," but KT/f insisted, "No, these are Keepon's eyes!"
- In S8, pointing to an insect cage, SR/f guided Keepon's attention to it. In S9, when NR/m hit Keepon's head several times, HN/f stopped him by saying, "It hurts! It hurts!" During reading time in S11, NK/f and TM/m came up and showed their picture books to Keepon.

- In S13, FS/m and TA/m, strongly hit Keepon's head a couple of times, as if demonstrating their braveness to each other. YT/f and IR/f, observing this, approached Keepon and checked if it had been injured, then YT/f said to Keepon and IR/f, "Boys are all alike. They all hit Keepon," stroking its head gently.
- In S16, being away for a couple of sessions, NK/f came to Keepon and said, "We haven't seen each other for a while," as if easing Keepon's loneliness.
- In S17, YT/f taught Keepon some words — showing it the cap, she said, "Say, Bo-shi," then switched to Keepon's knitted cap and said, "This is a Nitto Bo-shi, which you wear in winter." (Keepon could only respond to her by bobbing up and down with the "pop, pop, pop" sound.)
- Also in S17, after two girls hugged Keepon tightly, other girls found a scar on its head. NK/f pretended to give medicine to Keepon with a spoon, saying, "Good boy, you'll be all right."
- In S19, after playing with Keepon for a while, IZ/m asked other children nearby, "Please take care of Keepon." IZ/m managed to get an OK from KT/f, and then left Keepon.
- In S22, after all the children practiced a song with the teachers, several of them ran to Keepon and asked one by one, "Was it good?" to which Keepon responded by nodding and bobbing to give praise.
- In S23, NZ/m noticed Keepon had a flu mask and asked Keepon, "Caught a cold?" NK/f then put a woolen scarf around Keepon's neck, then NR/m and YS/f asked NK/f, "Is he ill?" and "Does he have a cold?"
- In S25, NK/f gave a toy sled to Keepon. Keepon showed a preference to another toy NK/f was holding. After some negotiation, NK/f brought over another sled and persuaded Keepon, "Now you have the same thing as me."

Especially during free play time (the first 90 min), the children showed a wide range of spontaneous actions, not only dyadic interaction between a particular child and Keepon, but also n-adic one, in which Keepon functioned as a pivot of interpersonal play with peers and sometimes with teachers. Since the children were generally typically developing, they often spoke to Keepon, as if they believed that it had a "mind" or "intelligence". They interpreted Keepon's responses, although they were merely simple gestures and sounds, as having communicative meanings within the interpersonal context. We have never observed this with the autistic children, who rarely interacted with peers. Compared with the experimental setting ("Interaction in the laboratory"), where children became bored after 15-min interactions, it is interesting that children in the preschool never lost interest even after 20 sessions (60 h in total).

Conclusion

We have described in this paper the 3-year-long longitudinal observation of autistic children interacting with a creature-like robot, Keepon, which is only capable of expressing attention by orienting its gaze and expressing emotions by rocking and/or bobbing up and down. From the observed interactions in rather unconstrained situations in their everyday life, we learned the following points.

- The autistic children, who generally have difficulty in interpersonal communication, were able to approach Keepon and gradually establish physical and social contact with Keepon. Since the robot exhibited its attention and emotions in simple and comprehensive ways, the children would understand the social meaning of the robot's actions without getting bored or overwhelmed.
- Some of the children extended their dyadic interactions with Keepon into triadic interpersonal ones, where the robot functioned as a pivot of sharing and exchanging the pleasure and surprise they found in the dyadic interaction with Keepon.

Based on these empirical findings, which are somewhat contrary to the common assumption about autism, it is suggested that (1) simple robots like Keepon, with minimal and comprehensive expressiveness, facilitate in autistic children spontaneous exchange and sharing of mental states with others, (2) autistic children therefore possess the motivation for this exchange and sharing, which is commonly considered to be dysfunctioning in autism, and (3) these imply that autistic children's major social difficulties would come from a dysfunction of the social filter that sifts out socially meaningful information from the vast amount of incoming perceptual information. These hypotheses have to be investigated in further research.

In the therapeutic field, therapists and parents generally believe that autistic children possess the motivation for sharing and exchanging mental states, with which the children would enjoy such interactions with the parents, therapists, and peers. This is exactly what we observed in the therapeutic practices, where each of the children spontaneously enjoyed such interactions in a different style that changed over time, which told us a "story" about his or her individual developmental profile. These unique tendencies cannot be thoroughly explained by a diagnostic label such as "autism" or "Asperger's syndrome". The "story" has been accumulated as video data, which is being utilized by therapists, psychiatrists, and pediatricians at the day-care center to help plan their therapeutic intervention. We also provided the video data to parents with the hope that it may positively influence their own child-care.

Acknowledgments

The authors are grateful to Motoaki Kataoka (Kyoto Women's University), and Ikuko Hasegawa (Hachioji Preschool) for their support for the field study. We are also thankful to Hiroyuki Yano, Daisuke Kosugi, Chizuko Murai, Nobuyuki Kawai, and Yoshio Yano, who collaborated with us in the preliminary experiments on child–robot interactions.

References

American Psychiatric Association. (1994) DSM-IV: Diagnostic and Statistical Manual of Mental Disorders (4th ed.). American Psychiatric Association, Arlington, VA.

Baird, G., Simonoff, E., Pickles, A., Chandler, S., Loucas, T., Meldrum, D. and Charman, T. (2006) Prevalence of disorders of the autism spectrum in a population cohort of children in South Thames: The Special Needs and Autism Project (SNAP). Lancet, 368: 210–215.

Baron-Cohen, S. (1995) Mindblindness: An Essay on Autism and Theory of Mind. MIT Press, Cambridge, MA.

Billard, A. (2002) Play, dreams and imitation in Robota. In: Dautenhahn K., Bond A., Cañamero L. and Edmonds B. (Eds.), Socially Intelligent Agent. Kluwer Academic Publishers, Dordrecht, The Netherlands, pp. 165–173.

Breazeal, C. and Scassellati, B. (2000) Infant-like social interactions between a robot and a human caretaker. Adapt. Behav., 8: 49–74.

Butterworth, G. and Jarrett, N. (1991) What minds have in common is space: spatial mechanisms serving joint visual attention in infancy. British Journal of Developmental Psychology, 9: 55–72.

Dalton, K.M., Nacewicz, B.M., Johnstone, T., Schaefer, H.S., Gernsbacher, M.A., Goldsmith, H.H., Alexander, A.L. and Davidson, R.J. (2005) Gaze fixation and the neural circuitry of face processing in autism. Nat. Neurosci., 8: 519–526.

Dautenhahn, K. (1997) I could be you: the phenomenological dimension of social understanding. Cybernet. Syst. J., 28: 417–453.

Dautenhahn, K. (1999) Robots as social actors: aurora and the case of autism. Proceedings of the International Cognitive Technology Conference, pp. 359–374.

Frith, U. (1989) Autism: Explaining the Enigma. Blackwell, London, UK.

Kozima, H. (2002) Infanoid: A babybot that explores the social environment. In: Dautenhahn K., Bond A., Cañamero L. and Edmonds B. (Eds.), Socially Intelligent Agent. Kluwer Academic Publishers, Dordrecht, The Netherlands, pp. 157–164.

Kozima, H. and Ito, A. (2003) From joint attention to language acquisition. In: Leather J. and van Dam J. (Eds.), Ecology of Language Acquisition. Kluwer Academic Publishers, Dordrecht, The Netherlands, pp. 65–81.

Kozima, H., Nakagawa, C. and Yano, H. (2004) Can a robot empathize with people? Int. J. Artif. Life Robot., 8: 83–88.

Michaud, F. and Théberge-Turmel, C. (2002) Mobile robotic toys and autism. In: Dautenhahn K., Bond A., Cañamero L. and Edmonds B. (Eds.), Socially Intelligent Agent. Kluwer Academic Publishers, Dordrecht, The Netherlands, pp. 125–132.

Okada, M. and Goan, M. (2005) Modeling sociable artificial creatures: findings from the Muu project. Proceedings of the 13th International Conference on Perception and Action, 2005.

Robins, B., Dickerson, P., Stribling, P. and Dautenhahn, K. (2004) Robot-mediated joint attention in children with autism: a case study in robot-human interaction. Interact. Stud., 5: 161–198.

Scassellati, B. (2005) Using social robots to study abnormal social development. Proceedings of the 5th International Workshop on Epigenetic Robotics, p. 1114.

Tomasello, M. (1999) The Cultural Origins of Human Cognition. Harvard University Press, Cambridge, MA.

Tomasello, M., Carpenter, M., Call, J., Behne, T. and Moll, H. (2005) Understanding and sharing intentions: the origins of cultural cognition. Behav. Brain Sci., 28: 675–691.

Trevarthen, C. (2001) Intrinsic motives for companionship in understanding: their origin, development, and significance for infant mental health. Infant Mental Health J., 22: 95–131.

World Health Organization. (1993) The ICD-10 Classification of Mental and Behavioural Disorders: Diagnostic Criteria for Research, World Health Organization, Geneva, Switzerland, 1993.

Zlatev, J. (2001) The epigenesis of meaning in human beings, and possibly in robots. Minds Mach., 11: 155–195.

SECTION V

The Development of Artificial Systems

C. von Hofsten & K. Rosander (Eds.)
Progress in Brain Research, Vol. 164
ISSN 0079-6123

CHAPTER 22

Sensorimotor coordination in a "baby" robot: learning about objects through grasping

Lorenzo Natale[1], Francesco Orabona[2], Giorgio Metta[1,2,*] and Giulio Sandini[1,2]

[1]*Italian Institute of Technology, via Morego 30, 16163, Genoa, Italy*
[2]*LIRA-Lab, DIST, University of Genoa, viale Causa 13, 16145, Genoa, Italy*

Abstract: This paper describes a developmental approach to the design of a humanoid robot. The robot, equipped with initial perceptual and motor competencies, explores the "shape" of its own body before devoting its attention to the external environment. The initial form of sensorimotor coordination consists of a set of explorative motor behaviors coupled to visual routines providing a bottom-up sensory-driven attention system. Subsequently, development leads the robot from the construction of a "body schema" to the exploration of the world of objects. The "body schema" allows controlling the arm and hand to reach and touch objects within the robot's workspace. Eventually, the interaction between the environment and the robot's body is exploited to acquire a visual model of the objects the robot encounters which can then be used to guide a top-down attention system.

Keywords: development; humanoid robotics; body schema; top-down and bottom-up attention

Introduction

In the past few years there has been significant technological advance in computer technology and robotics. Today computers are much more powerful than they used to be and they can be interconnected through fast networks, which allow efficient parallel computation. At the same time digital cameras have higher resolution, better quality and higher frame rate. This notwithstanding, we are still far from achieving the dream of artificial intelligence. Artificial systems (computer programs, expert systems or robots) are not able to face the challenges of the real world. We are still not capable of building devices which are able to cope with the variability of the world where, on the other hand, even the simplest animal can thrive. Likewise there is a growing interest in the scientific community to the study of cognitive systems with the aim of implementing cognitive abilities in artificial systems. The study of cognition is still in the preparadigmatic stage and, indeed, little agreement can be found even in its definition (see Clark (2001) for a review). According to *cognitivism*, cognition is "a computational process carried out on a symbolic representation of the world". Symbols represent the world and can be shared across different entities (artificial or biological); they are a complete characterization of the world in which the entity is located, and as such are independent of the entity itself and its past experience. Somewhat at the other extreme, emergent approaches define cognition as the result of the interaction and co-development between the agent's body and the environment in which it lives

*Corresponding author. Tel.: +39 010 7178-1411;
Fax: +39 010 720321; E-mail: pasa@liralab.it

DOI: 10.1016/S0079-6123(07)64022-9

(Maturana and Varela, 1998; Beer, 2000; Sandini et al., 2004).

Although the definitive answer is still to be found, the observation of biological systems provides hints to plausible solutions. Two aspects look crucial: (i) the existence of a body (embodiment) and (ii) the fact that the internal representation of the world is acquired by acting in the environment. The two requirements are obviously intertwined, as the interaction between the agent and the environment is possible only by means of a physical body. As a consequence, internal representations become function of the particular embodiment and, perhaps more importantly, of the history of experiences of the agent.

Subscribing to the emergent approach implies that internal representations cannot be built into the system "by design"; instead the cognitive system has to be able to create these representations by directly interacting with the environment or, indirectly, with other agents. Through action, the embodiment and environment co-determine the resulting representations.

Motivated by these considerations, this paper proposes a developmental approach to the realization of a number of cognitive abilities in a humanoid robot. Although a fair amount of cognitivism is still present, especially in the realization of the visual system, learning permeates the implementation at various levels. Learning and a certain degree of adaptation is clearly the prerequisite to a fully emergent design, although not yet an end or a definite answer to the understanding of cognitive systems altogether.

We identified the minimum requirements for our robot as having an oculomotor system, an arm, and a hand. Although simplified this configuration suffices in allowing active manipulation of the world via reaching and grasping. The robot follows a developmental route that goes initially through the exploration of its body and terminates into the characterization of external objects (e.g. segmentation) by effect of grasping.

Conceptually this process can be divided in three phases. The first stage is devoted to learning the internal models of the body (we call it "learning the body-map"), which provides basic motor and perceptual skills like gaze control, eye–head coordination and reaching. Based on these abilities the interaction with the external world is investigated in the second phase where the robot discovers properties of objects and ways of handling them (learning to interact). The robot tries simple stereotyped actions like pushing/pulling and grasping of objects, which allow starting the acquisition of information about the entities that populate its environment and simultaneously discover new more efficient ways of interaction (for example different grasp types). Finally, the third stage concerns learning to understand and interpret events; the robot has associated its actions with the resulting perceptual consequences. Interpretation is achieved by inverting this association; perceptions are projected into the corresponding actions that work as a reference frame to give meaning to what happens in the environment.

In our past work we have addressed some of the aspects related to this third phase (Natale et al., 2002; Fitzpatrick et al., 2003). In this paper we focus on the two first phases: learning a body-map and learning to interact.

We show how the robot can acquire an internal model of its hand, which allows the robot to localize it and anticipate its position in the visual scene during action execution. The hand internal model is then used to learn to reach a point in space and to accommodate the position of the hand with respect to the object during grasping. The robot uses these abilities to build a visual model of the objects it grasps. Once an object is grasped, in fact, the robot can move and rotate it to build a statistical model of its visual appearance.

Experimental platform

The experiments reported in this paper were carried out on a robotic platform called Babybot (Fig. 1). The Babybot is an upper torso humanoid robot that consists of a head, an arm and a hand. The head has 5 degrees of freedom, two of which control the neck in the pan and tilt direction, whereas the other three actuate the two eyes to pan independently and tilt on a common axis. The arm is a Unimate PUMA 260, an industrial

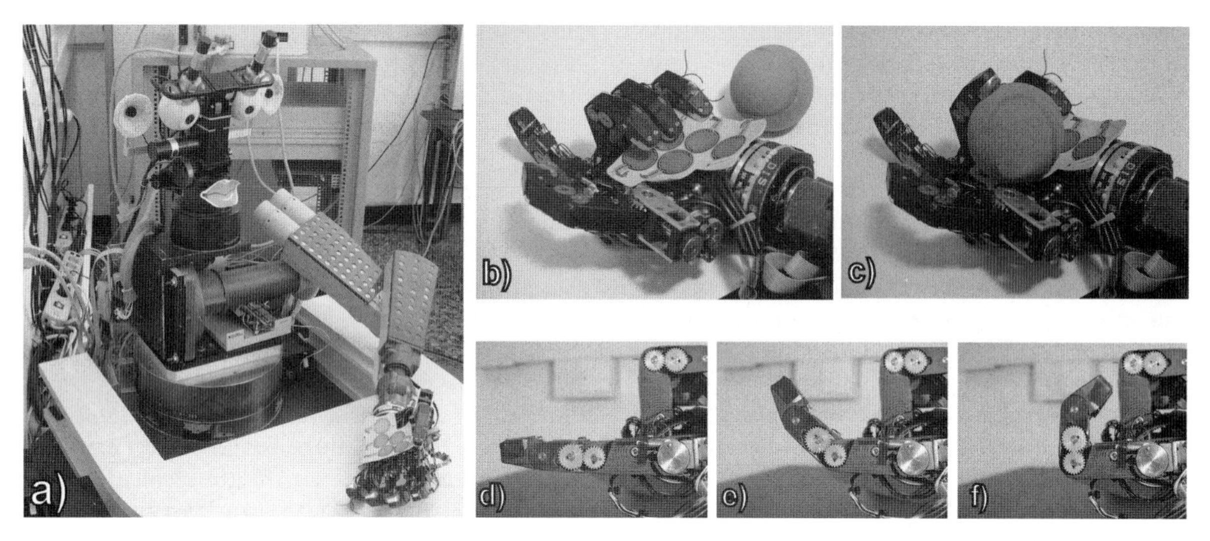

Fig. 1. (a) The experimental setup, the Babybot. Left: details of the hand. (b) and (c): elastic compliance. (d–f): mechanical coupling between phalanges.

manipulator with 6 degrees of freedom; it is mounted horizontally to better mimic the human kinematics. The hand has 5 fingers; each finger has three phalanges, the thumb has an additional degree of freedom, which allows it to perform a rotation toward the palm. Overall the number of joints is 16 but for reasons of space and weight they are controlled by using only six motors. Two motors are connected to the index fingers: they are linked to the first (proximal) and second phalanges. The distal (small) phalange is mechanically coupled to the preceding one so that the two bend together (see Fig. 1). Two motors control the motion of middle, ring and little finger. As in the case of the index finger, the proximal phalanges are actuated by a single motor, while the second and third phalanges are actuated by a second motor. The mechanical coupling between the joints is realized by means of springs to allow a certain degree of adaptation in case of physical contact or impact with solid objects. For example, during a movement of flexion of the fingers toward the palm, if the middle finger were to be blocked by an obstacle the others would continue to bend up to the equilibrium of the torque generated by the motor and that of the spring (Figs. 1b and c). The same would happen in case the distal phalanges had hit the obstacle. The thumb is different as one motor controls the rotation around an axis parallel to the palm and a second motor is connected to the three phalanges, whose independent motion is permitted by elastic coupling as for the other fingers.

The sensory system of the Babybot consists of two cameras and two microphones for visual and auditory feedback. Tactile feedback is provided by 17 force-sensing resistors mounted on the hand, five of which are placed on the palm and the remaining 12 evenly distributed on the thumb, index, middle and ring fingers. A JR3 6-axial force sensor provides torque and force feedback measured at the wrist. Further proprioceptive information is provided by encoders mounted on all motors and by a three-axis gyroscope mounted on the head. More details about the Babybot architecture can be found elsewhere (Natale, 2004).

Visual system

One of the first steps of any visual system is that of locating suitable interest points in the scene ("salient regions" or events) and eventually direct gaze toward these locations. Human beings and many animals do not have a uniform resolution view of the visual world but rather only a series of snapshots acquired through a small high-resolution sensor (e.g. our fovea). This leads to two questions: (i) how to move the eyes efficiently to important locations in the visual scene, and (ii) how to decide what is important and, as a consequence, where to look next.

The literature follows two different approaches in the attempt of accounting for these facts. On the one hand, the space-based attention theory holds that attention is allocated to a region of space, with processing carried out only within a certain spatial window. Attention in this case could be directed to a region of space even in absence of a real target (the most influential evidences for the spatial selection come from the experiments of Posner et al. (1980) and Downing and Pinker (1985).

On the other hand, object-based attention theories argue that attention is directed to an object or a group of objects, and that the attention system processes properties of object(s), rather than regions of space. This object-based theory is supported by growing behavioral and neurophysiological evidence (Egly et al., 1994; Scholl, 2001). In other words, the visual system seems optimized for segmenting complex 3D scenes into representations of (often partly occluded) objects for recognition and action. Indeed, perceivers must interact with objects in the world and not with disembodied locations.

Finally, another classification can be made depending on which cues are actually used in modulating attention. One approach uses bottom-up information including basic features such as color, orientation, motion, depth, and conjunctions of features. A feature or a stimulus catches the attention of the system if it differs from its immediate surrounding in some dimensions and the surround is reasonably homogeneous in those same dimensions. However, higher-level mechanisms are involved as well; a bottom-up stimulus, for example, may be ignored if attention is already focused elsewhere. In this case attention is also influenced by top-down information relevant to a particular task.

In the literature a number of attention models that use the first hypothesis have been proposed (Giefing et al., 1992; Milanese, 1993; Itti et al., 1998); most of them are derived from Treisman's Feature Integration Theory (FIT) (Treisman and Gelade, 1980). This model employs a separate set of low-level feature maps, which are combined together by a spatial attention window operating in a master saliency map. An important alternative model is given by Sun and Fisher (2003), who proposed a combination of object- and feature-based theory (this model, unfortunately, requires hand-segmented images as input for training).

While it is known that the human visual system extracts basic information from images such as lines, edges, local orientation etc., vision not only represents visual features but also the items that such features characterize. But to segment a scene into items, objects, that is to group parts of the visual field as units, the concept of "object" must be known by the system. In particular, there is an intriguing discussion underway in vision science about reference to entities that have come to be known as "proto-objects" or "pre-attentive objects" (Pylyshyn, 2001). These are steps up from mere localized features, and they have some but not all of the characteristics of objects.

The visual attention model we propose starts by considering the first stages of the human visual system, using then a concept of salience based on "proto-objects" defined as blob of uniform color in the images. Then, since the robot can act on the world, it can do something more: once an object is grasped the robot can move and rotate it to build a statistical model of the color blobs, thus effectively constructing a representation of the object in terms of proto-objects and their spatial relationships. This internal representation feeds then back to the attention system of the robot in a top-down way; as an example we show how the latter can be used to direct attention to spot one particular object among others that are visible on a table in front of the robot.

Our approach integrates bottom-up and top-down cues; in particular bottom-up information suggests/identifies possible regions in the image where attention could be directed, whereas top-down information works as a prime for those regions during the visual search task (i.e. when the robot seeks for a known object in the environment).

Log-polar images

Figure 2 shows the block diagram of the first stage of the visual processing of the robot. The input data are a sequence of color log-polar images (Sandini and Tagliasco, 1980). The log-polar transformation models the mapping of the primate visual pathways from the retina to the visual cortex. The idea of employing space-variant vision is derived from the observation that the distribution of the cones, i.e. the photoreceptors of the retina involved in diurnal vision, is not uniform: cones have a higher density in the central region called fovea, while they are sparser in the periphery. Consequently the resolution is higher and uniform in the center while it decreases in the periphery proportionally to the distance from the fovea.

The main advantage of log-polar sensors is computational, as they allow to acquire images with a small number of pixels and yet to maintain a large field of view and high resolution at the center (Sandini and Tagliasco, 1980). Moreover, this particular distribution of the receptors seems to influence the scanpaths of an observer (Wolfe and Gancarz, 1996), so it has to be taken into account to better model the overt visual attention.

The radial symmetry of the distribution of the cones can be approximated by a polar distribution, whereas their projection to the primary visual cortex is well represented by a logarithmic-polar (log-polar) distribution mapped onto an approximately rectangular surface (the cortex). From the mathematical point of view the log-polar mapping can be expressed as a transformation between the polar plane (ρ, θ) (retinal plane), the log-polar plane (ξ, η) (cortical plane) and the Cartesian plane (x, y) (image plane), as follows (Sandini and Tagliasco, 1980):

$$\begin{cases} \eta = q \cdot \theta \\ \xi = \log_a \dfrac{\rho}{\rho_0} \end{cases} \quad (1)$$

where ρ_0 is the radius of the innermost circle, $1/q$ is the minimum angular resolution of the log-polar layout and (ρ, θ) are the polar co-ordinates.

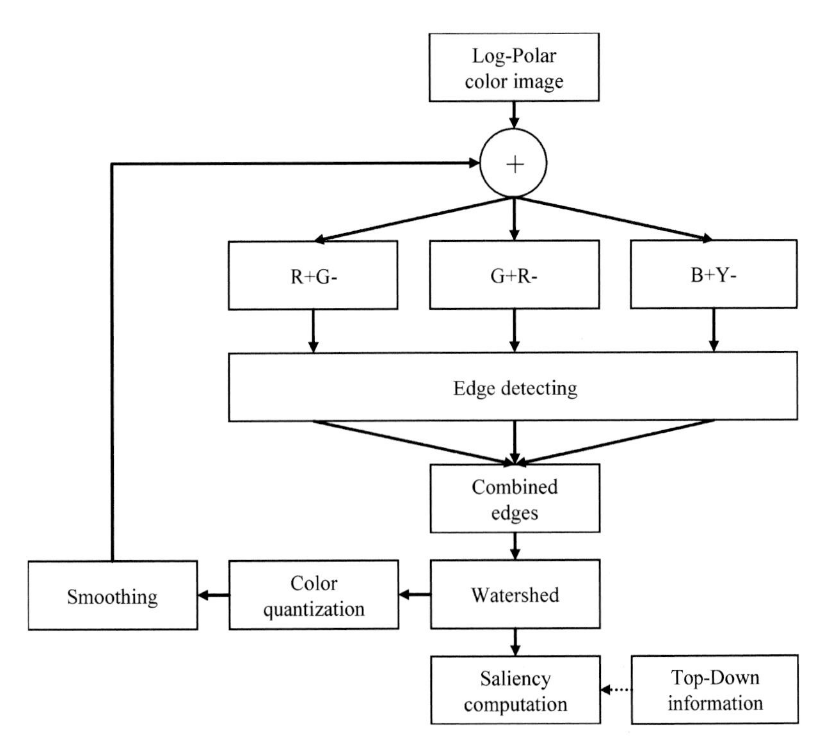

Fig. 2. The visual attention system: block diagram (see text for details).

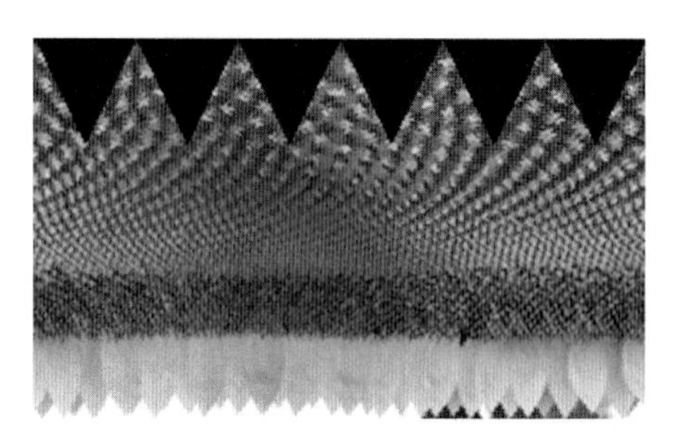

Fig. 3. Log-polar mapping. The original image (left) and the result of the log-polar mapping in the cortical plane (right).

Figure 3 illustrates the log-polar layout by showing a standard rectangular image and its log-polar counterpart. It is worth noting that the flower's petals, that have a polar structure, are mapped horizontally in the log-polar image. Circles, on the other hand, are mapped vertically. Furthermore, the stamens that lie in the center of the image of the flower, occupy about half of the corresponding log-polar image (the cortical magnification).

Visual attention

As a first step the input image is smoothed, by taking the average between the current frame and the output of the color quantization (see later) on the previous frame. Then the red, green, and blue channels of each image are separated, and the yellow channel is calculated as the mean of the red and green one. These four channels are combined to generate three color-opponent channels, similar to those of the retina. Each of these channels, typically indicated as (R^+G^-, G^+R^-, B^+Y^-), has a center-surround receptive field (RF) with spectrally opponent color responses. That is, for example, a red input in the center of a particular RF increases the response of the channel R^+G^-, while a green one in the surrounding decreases its response. The spatial response profile of the RF is expressed by a Difference-of-Gaussians (DoG) function. Each pixel is considered as the center of a RF, so that the output of the RF filtering is simply obtained by a convolution of the whole image with a DoG kernel, generating an output image of the same size of the input. This computation, considering for example the R^+G^- channel, is expressed by:

$$R^+G^-(\mathbf{x}) = a \cdot R(\mathbf{x}) \otimes \gamma_c(\mathbf{x}, \sigma_c) - b \cdot G(\mathbf{x}) \otimes \gamma_s(\mathbf{x}, \sigma_s) \qquad (2)$$

The two Gaussian functions $\gamma_c(\mathbf{x}, \sigma_c)$ and $\gamma_s(\mathbf{x}, \sigma_s)$ are not balanced and the ratio b/a is 1.5, consistent with the study of Smirnakis et al. (1997) Similarly to what happens in the human retina (Billock, 1995) the unbalanced ratio implicitly code the achromatic information. It is worth noting that filtering the log-polar images with a standard space-invariant filter corresponds to a space-variant filtering in the original Cartesian image (von Seelen and Mallot, 1990).

Edges are then extracted on the three channels separately by employing a generalization of the Sobel filter due to Li et al. (2003). The resulting edge maps are combined together to generate a single map as follows:

$$E(\mathbf{x}) = \max\{\mathrm{abs}(E_{\mathrm{RG}}(\mathbf{x})), \mathrm{abs}(E_{\mathrm{GR}}(\mathbf{x})), \mathrm{abs}(E_{\mathrm{BY}}(\mathbf{x}))\} \quad (3)$$

It has to be noted that the log-polar transform has the side effect of sharpening the edges near the fovea due to the already mentioned magnification factor. To compensate for this effect the edge map is multiplied by an exponential function, and normalized to a fixed range (0–255).

It has been speculated, that synchronizations of visual cortical neurons may serve as the carrier for the observed perceptual grouping phenomenon (Eckhorn et al., 1988; Gray et al., 1989). The differences in oscillator phase between spatially neighboring spiking cells could be used in principle to label different objects in the scene. We have used a watershed transform (rainfalling variant) (Vincent and Soille, 1991; Smet and Pires, 2000) on the edge map to simulate the result of this synchronization and to generate the proto-objects. The activation is spread from the center of the image (in the edge map) until all spaces between edges are filled in. As a result the image is segmented into blobs with either uniform color or uniform gradient of color.

Each blob is then tagged with the mean color of the pixels within its internal area (this leads to a sort of quantized image). The result is blurred with a Gaussian filter and stored: it will be averaged with the next frame to obtain a temporal smoothing and reduce the effect of noise. After an initial startup delay of 4–5 frames, the number of blobs and their size stabilizes.

As discussed above, it is known that a feature or stimulus is salient if it differs from its immediate surrounding area. We chose to calculate the bottom-up salience as the Euclidean distance in the color-opponent space between each blob and the average color in a ball surrounding it. The radius of the ball (the spot or focus of attention) is not fixed: it changes with the size of the objects in the scene. In the same way the definition of "immediate surrounding area" should be relative to the size of the focus of attention. For this reason the greater part of the visual attention models in the literature uses a multi-scale approach and filters the salience map with suitable filters, or "blob" detectors (Itti and Koch, 2001). These approaches lack continuity in the choice of the size of the attention focus. We propose instead to vary dynamically the region of interest depending on the size of the blobs. In other words, we compute the salience of each blob in relation to a neighborhood region whose size is proportional to that of the blob itself. In our implementation we use a rectangular region three times the size of the bounding box of the blob. The choice of a rectangular window is not incidental, it was chosen because filters over rectangular regions can be computed efficiently by employing the integral image as in Viola and Jones (2004). Blobs that are too small or too big are discarded from the saliency computation and will not be considered as possible candidates to be part of objects (proto-objects).

The bottom-up saliency is computed as:

$$S_{\text{bottom-up}} = \frac{1}{\sqrt{3}}\sqrt{\left(\langle \mathrm{R^+G^-}\rangle_{\text{blob}} - \langle \mathrm{R^+G^-}\rangle_{\text{surround}}\right)^2 + \left(\langle \mathrm{G^+R^-}\rangle_{\text{blob}} - \langle \mathrm{G^+R^-}\rangle_{\text{surround}}\right)^2} + \sqrt{\left(\langle \mathrm{B^+Y^-}\rangle_{\text{blob}} - \langle \mathrm{B^+Y^-}\rangle_{\text{surround}}\right)^2} \quad (4)$$

where $\langle\rangle$ indicates the average of the pixel values over a certain area (as in the subscripts).

The top-down influence on attention is, at the moment, calculated in relation to the visual search task. When the robot has acquired a model of the object and begins searching for it, it uses the visual information of the object to bias the saliency map. In practice, the top-down saliency map is computed as the distance between the average color of each blob and that of the target:

$$S_{\text{top-down}} = 255 - \frac{1}{\sqrt{3}}\sqrt{\left(\langle R^+G^-\rangle_{\text{blob}} - \langle R^+G^-\rangle_{\text{object}}\right)^2 + \left(\langle G^+R^-\rangle_{\text{blob}} - \langle G^+R^-\rangle_{\text{object}}\right)^2 + \sqrt{\left(\langle B^+Y^-\rangle_{\text{blob}} - \langle B^+Y^-\rangle_{\text{object}}\right)^2}} \quad (5)$$

The total salience is simply estimated as the linear combination of the two terms above:

$$S = \alpha \cdot S_{\text{top-down}} + \beta \cdot S_{\text{bottom-up}} \quad (6)$$

The total salience map S is eventually normalized in the range 0–255, as a consequence the salience of each blob in the image is relative to the most salient one. The target of the next saccade is the center of mass of the most salient blob (this is in agreement with human behavior; Melcher and Kowler, 1999).

As a final note on efficiency, it is worth saying that the use of log-polar images allows to compute the saliency map in real time (15 frames per second on a 2.8 Ghz Pentium IV).

IOR

Local inhibition is transiently activated in the salience map. This prevents the focus of attention to be redirected immediately to a location that was previously attended. Experiments in human psychophysics have demonstrated the existence of such an "inhibition of return" (IOR) coded in an allocentric reference frame (Posner and Cohen, 1984) and in an object-based coordinates (Tipper, 1994).

Our system implements a simple object-based IOR. The robot maintains a list of the last five positions (Wolfe, 2003) it has visited, coded in a body-centered coordinate system. The color information of the relative blobs is also stored in the list, which is updated with a First-In First-Out policy. When the robot moves its gaze — for example by moving the eyes or the head in coordination — it keeps memory of the blobs it has visited earlier. Inhibition occurs only if the blob presents the same color that is stored in the list; in case the object moves or its color changes the location becomes available for fixation.

Learning about the self

Internal models are thought to be available to the brain and responsible for formulating predictions about the world or simulating the body (Wolpert and Miall, 1996). In general the collection of the internal models required to represent the body is called the *body schema*: it involves, for example, the relative positions of the limbs, and their weight and size. In humans and biological systems the internal representation of the body is shaped during development and maintained adapted to the physical modification occurring in life. In artificial agents (where the body does not change with time) adaptation can spare the tedious operation of manually tuning the system's internal models and their calibration. The latter might be required to compensate changes in the visual appearance of the body or drift in the sensors (e.g. the motor encoders).

In infants this sense of the body is acquired during development and emerges a few months after birth (Rochat and Striano, 2000). This is a cause–effect problem because on the one hand the brain uses internal models to recognize the body whereas on the other it has to acquire the body schema and maintain it up to date. To solve it, the brain needs a "bootstrapping" mechanism that allows the identification of the body and, in this way, the acquisition of the internal representation. To distinguish the body from the rest of the world the brain is thought to take advantage of extra information. For example, while a child waves the hand in front of his eyes, his brains "knows" what kind of motion is producing since it has exclusive access to the motor commands it sends to the muscles and the relative proprioceptive feedback (Rochat and Striano, 2000).

In robotics there have been attempts to replicate self-recognition mechanisms. Yoshikawa et al. (2003) exploit the invariance of the body with respect to the external world to train a neural network to segment the arm of the robot. Their idea is that during learning, when the robot moves in the environment, the background changes, whereas the arms remain stationary with respect to the proprioceptive feedback.

Instead, the active behavior of the robot is used by Metta and Fitzpatrick (2003); in this case the robot identifies its body because it moves with respect to the background. Since motion alone is not sufficient to segment out external objects that move in the environment, the system seeks similarities between proprioceptive and visual feedback. Among the others, periodic actions may add robustness because offer the possibility to exploit repeatability (Fitzpatrick and Arsenio, 2004).

Segmentation of the hand

Repeated, self-generated actions were performed by the robot during the learning phase. In particular the robot was programmed to execute periodic movements of the wrist. The resulting motion of the hand was detected by computing the image difference between the current frame and an adaptive model of the background. The period of motion of each pixel in the resulting motion image was then computed with a zero-crossing algorithm; similar information was extracted from the proprioceptive feedback of each motor encoder. As a result, the hand of the robot was segmented by selecting, among the pixels that moved periodically, those whose period matched that of the wrist joints. Conversely non-periodic pixels or pixels moving with different periods were identified as being externally originated and discarded. Figure 4 shows an example of the detection for two different pixels whose motion was (a) correlated and (b) uncorrelated with that of the robot's hand. Low-pass filtering and a threshold were applied after the detection to obtain a dense segmented image (see Fig. 5).

This algorithm forces the robot to stop and wait until the periodic movement of the wrist is

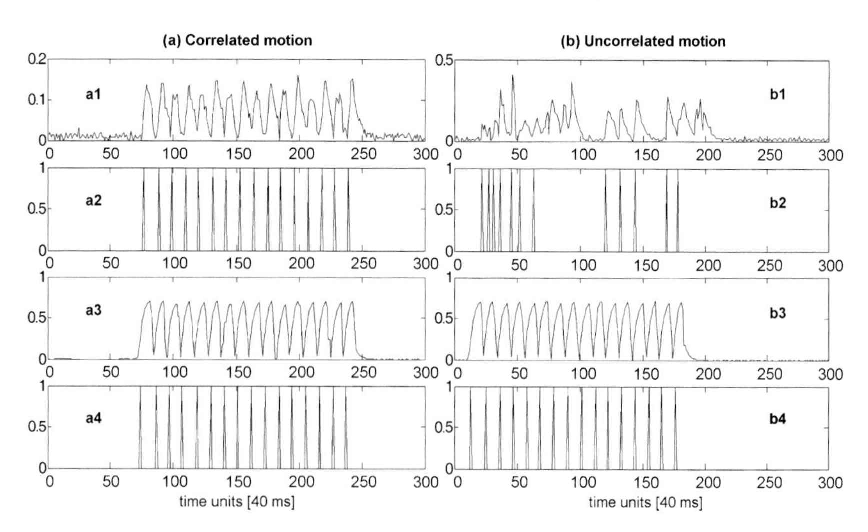

Fig. 4. Correlated versus uncorrelated motion, an example. The plots represents the time course of the variables involved in the detection procedure for two exemplar pixels whose motion matched (a) and did not match (b) that of the hand. (a1) and (b1) show the value of the motion for the pixel (normalized between 0 and 1). The result of the zero-crossing algorithm is reported in (a2) and (b2). The same procedure is replicated for the wrist proprioceptive feedback: (a3) and (b3) show the speed of the joint (normalized arbitrary scale), whereas (a4) and (b4) show the result of the zero-crossing algorithm. Compare (a2) to (a4) and (b2) to (b4).

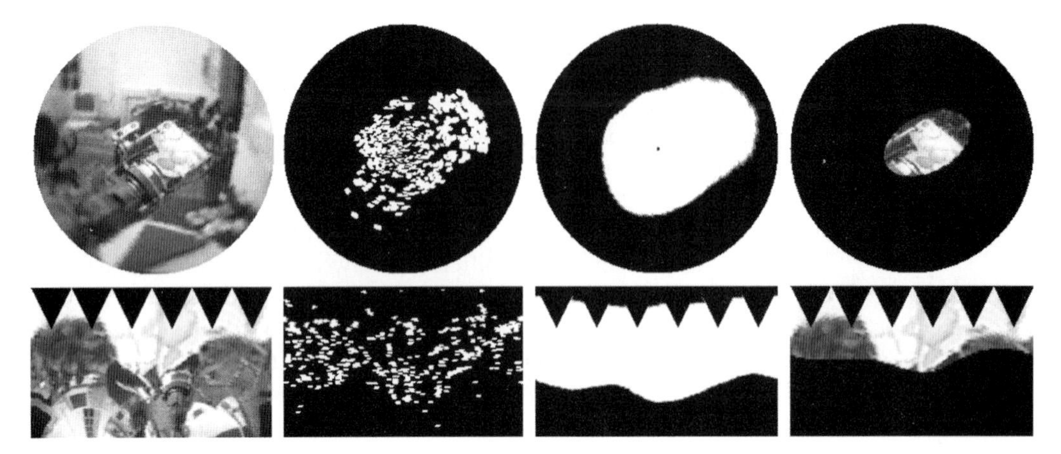

Fig. 5. An example of the detection procedure. From left to right: the original image at the beginning of the procedure, the result of the detection (that is the pixels whose motion was correlated with that of the hand), the result of the low-pass filtering, and the segmentation after the ellipse fitting. Notice that the ellipse tends to collapse towards the center, because the log-polar transformation gives more weight to the pixels close to the fovea.

performed. For this reason it is not useful during action or to drive a feedback control loop; it is instead ideally suited as a bootstrapping mechanism to acquire an internal model of the hand, which can provide faster localization. In practice this was implemented with two neural networks: one trained to compute the position of the hand in the visual field given the current arm and head posture, and another to estimate the hand's shape and orientation (in this case the hand was represented as an ellipse). Indeed, these neural networks can also predict the expected location and the (simplified) appearance of the hand in the visual field given the current posture of the robot (its "felt" position). The approach we followed here to perform the segmentation of the hand is similar to the one of Metta and Fitzpatrick (2003); the main difference with our approach is the use of periodicity that allows the detection of the hand in real time at high resolution. The result is a dense segmentation from which it is possible to derive additional information like shape and orientation.

The hand internal model, expectation and prediction

To gather the training data the robot moved the arm randomly and then waved the hand for a few seconds; for each spatial location the segmentation of the hand was performed as described in the previous section. For each trial the center of mass of the segmented area was computed along with the best fitting ellipse parameters. The complete algorithm is reported in Fig. 6.

The resulting (x, y) coordinates were used to train the first neural network whereas the ellipse parameters (orientation, major and minor axis) constituted the training samples for the second neural networks. It is important to take into account that the position of the hand in the visual field depends both on the posture of the arm and head (other parameters like orientation and size of the hand are less influenced, if not at all). Unfortunately this enlarges the learning space and increases the time required for exploration (to collect the training set) and learning (higher dimensionality). For this reason the position of the hand was projected into an egocentric reference frame before being used to train the neural network. This last operation significantly reduced the dimensionality of the input space of the neural network. When needed, the output of the neural network is projected back to the retinocentric reference frame. Both projections (back and forth from egocentric and retinocentric reference frame) require knowledge of head inverse and direct kinematics. In the experiments reported here they were hardwired in the system, a possible procedure to learn a model

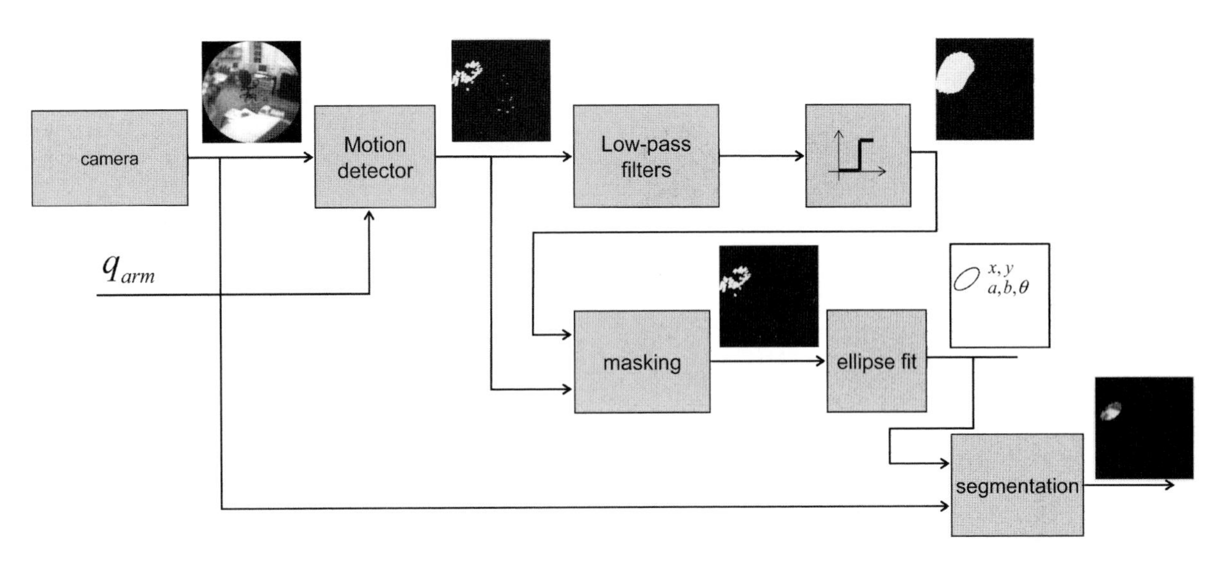

Fig. 6. Detection algorithm, block schema. Images are captured from the camera. The "motion detector" block compares the motion in the image with the proprioceptive feedback from the arm (the wrist). A series of low-pass filters identify the blob, which contains the hand. The blob is used to mask the result of the "motion detector" to remove possible outliers. An ellipse shape is fitted on the remaining pixels and, eventually, the hand is segmented.

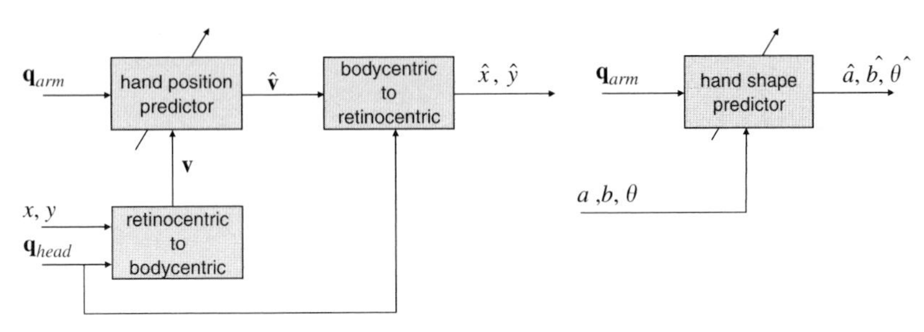

Fig. 7. Left: hand position predictor. Right: hand shape predictor. In the experiments reported in this paper the learning modules were multi-layer perceptrons with a hidden layer and sigmoidal units.

of them is suggested by Arsenio (Fitzpatrick and Arsenio, 2004). Figure 7 reports the block diagrams of the two models.

As learning module we employed a multi-layer perceptron network with sigmoidal units trained with backpropagation; learning was performed online by storing all new samples and performing batch learning every 100 new samples. The learning process was validated by testing the ability of the network to predict new samples; when a new sample was obtained the network was used to predict the output given the input. The resulting output was compared to the current sample and the error computed. The increasing ability of the network to predict new samples proved that learning was effective. Figure 8 (left) reports the plot of the error during an experiment (in this case the error is computed in the image plane to simplify visualization of the results); the total time of this experiment was ~2 h.

At the end of the exploration phase the robot had trained an internal model of the hand by which it could (i) localize its center of mass (ii) estimate its orientation and approximate size. The output of these models is not based on actual visual feedback, but on the mere projection of the proprioceptive information about the hand: they represent the expectation the robot possesses about its body (in this case, the hand).

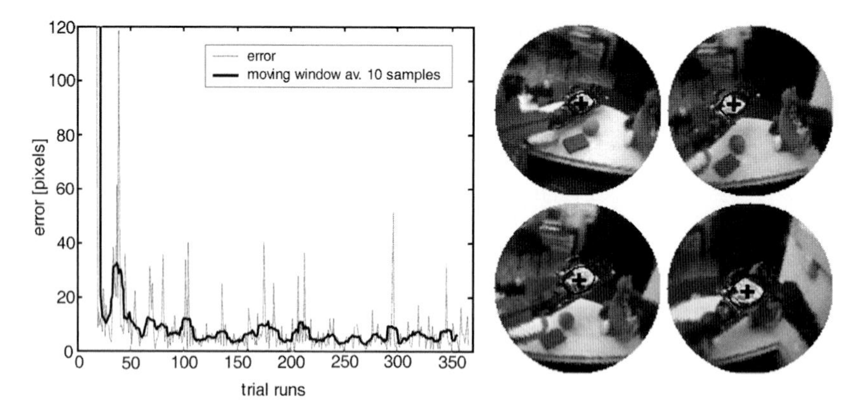

Fig. 8. Hand localization error trend (left). As new examples are presented to the network the performance improves. Example of the localization after learning (right). The cross corresponds to the position of the hand, whereas the ellipse represents its approximate shape and orientation. The size of the network was 20 units in the hidden layer, the total time of this experiment was ~2 h.

These measures were used in numerous ways. The center of mass was employed to close a visual loop to direct gaze towards the hand (see Fig. 8 right). For this task the internal model was addressed with the proprioceptive feedback of the arm. Another possibility was to address the model with the arm motor command (final joint position) to obtain the position of the hand at the end of the movement. In general this model offers a means of computing a prediction of the position, size and orientation of the hand from a given arm configuration or, in other words, of simulating a motor action. In the next section this will be used to learn the reaching map and estimate the visuomotor Jacobian matrix for a reaching task.

Reaching

The solution we propose is based on the use of a direct mapping between the eye–head motor plant and the arm motor plant (Metta et al., 1999). Flanders et al. (1999) suggested that the information about gaze direction might be employed by the brain to establish a reference point for reaching. They analyzed the error when reaching in the dark and showed how this correlates to the error of the gaze (the gaze drifts away from the target in the dark). Accordingly one premise we make is that the position of the fixation point coincides with the object to be reached. In other words, reaching for an object starts by looking at it. Under this assumption, the fixation point can be considered as the "end-effector" of the eye–head system. The position of the eyes with respect to the head, determines uniquely the position of the fixation point in space relative to the shoulder. The arm motor command can be obtained by a transformation of the eye–head motor/positional variables. We called this approach "motor–motor coordination", because the coordinated action is obtained by mapping motor variables into motor variables:

$$\mathbf{q}_{arm} = \mathbf{f}(\mathbf{q}_{head}) \tag{7}$$

where $\mathbf{q}_{head}$ and $\mathbf{q}_{arm}$ are head and arm posture, respectively (joint space).

What is interesting in this approach is not Eq. (7) per se, which, after all, implements the inverse kinematics of the arm, but the mechanisms used to learn it. In fact, this mapping can be easily learnt when the tracking behavior described in the previous section is active. The robot explored the workspace by moving the arm randomly, while simultaneously, it tracked its hand; whenever the eyes fixated the hand a new sample consisting of the arm and head joint angles was acquired and used to train a neural network approximating Eq. (7). In this case learning was performed online by using the Schaal et al. model (Schaal and Atkeson, 1998). The exploration was conducted in two ways. A first movement of the arm was performed by sampling a random uniform distribution within the part of the arm workspace in front of the robot. Small subsequent movements were performed randomly with Gaussian distribution with zero mean and standard deviation equal to 5°. This last step while not strictly required sped up learning by sampling quickly large portions of the arm's workspace: i.e. for small movements of the order of 5° the arm fixation was achieved rapidly and thus a new sample was added to the training set. When a sufficient number of samples were acquired, the robot started using the motor–motor map to actively reach for visually identified objects while learning could continue.

Learning can be further improved by reducing the dimensionality of the input vector $\mathbf{q}_{head}$. In fact, only three variables are needed to code the position of the fixation point; for this purpose we decided to use azimuth, elevation, and distance — in substitution for the five angles of the head joints. This transformation is motivated by practical reasons, but it is also biologically plausible (Lacquaniti and Caminiti, 1998).

Similarly to the previous section, learning was tested by comparing every new sample to the output of the network (see text for details). The graph of the error during an experiment is reported in Fig. 9 (left) for each sample (dotted line) and the moving window average over 20 samples (the total time of the experiment was ~1.5 h). From the first plot it is hard to determine a real increment of performance as several samples at the end of the learning session present relatively large errors. This is due to noise in the training data, which affects not only learning, but also the measure of performance. In particular noise is higher in those configurations of the arm where the hand is closer to the head and the system fails to control the angle of vergence between the eyes. In these situations the error is large because the position of the fixation points varies significantly (from very far to very close). The average error, however, has a distinguishable uniform trend. Figure 9 (right) shows a sequence of images taken from the robot left eye during an exemplar reaching action.

It is worth mentioning that there is no need to separate the exploration/training phase and reaching (exploitation). An initial "reflex" can be employed as substitute for the reaching map at the very beginning; this simple behavior could, for example, populate the robot workspace with three positions (left, center and right). Exploration in this case would still be guaranteed by a random procedure, similar to the one described earlier. This approach was followed in Metta et al. (1999) and Metta (2000).

The reaching problem can also be solved in the image plane. Consider the planar case (i.e. no 3D information is available and one of the arm joints is maintained to a fixed position) and suppose to measure the position of the end point in the image plane $\mathbf{x}_{hand}$. We want to control the arm to reach a target point $\mathbf{x}^*_{hand}$. If the robot is not in a singular configuration we can solve the problem by following a standard visual servoing approach

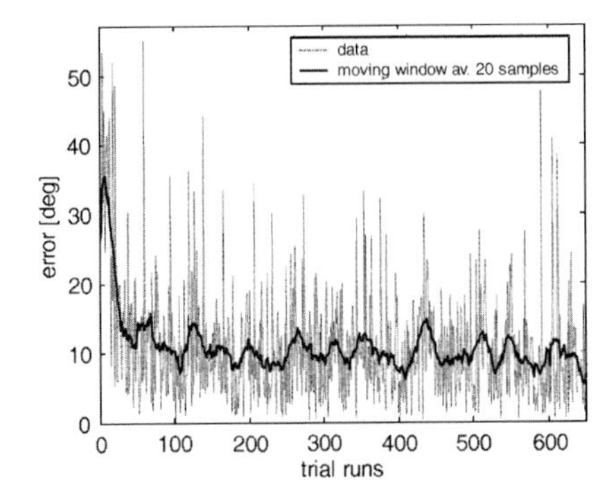

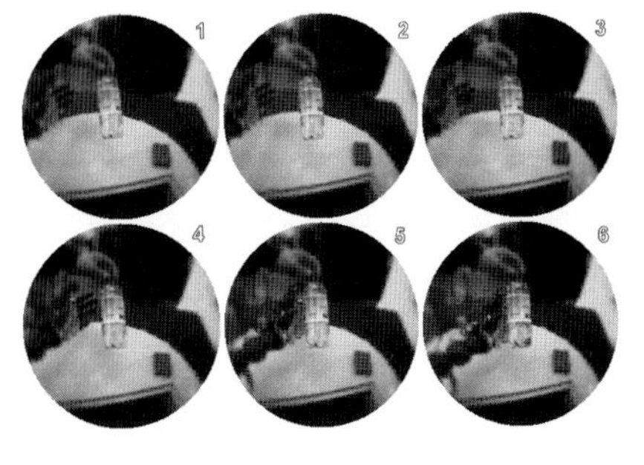

Fig. 9. Reaching error (left). As new examples are gathered and presented to the network the performance increases. This improvement is less remarkable; we believe this is due to noise in the training data that affects not only learning, but also the measure of performance. An exemplar sequence of a reaching action after the learning is reported on the right. The number of units of the network after the learning was 12, the total time required to perform this experiment was ~1.5 h.

(Espiau et al., 1992; Hutchinson et al., 1996):

$$\dot{\mathbf{q}} = -k \cdot \mathbf{J}^{-1}(\mathbf{q}_{\mathrm{arm}}) \cdot \Delta\mathbf{x} \tag{8}$$

where

$$\Delta\mathbf{x} = \mathbf{x}_{\mathrm{hand}} - \mathbf{x}^{*}_{\mathrm{hand}} \tag{9}$$

$k > 0$ is a scalar and $\mathbf{J}(\mathbf{q}_{\mathrm{arm}})$ is the Jacobian of the transformation between the image plane and the arm joint space. $\mathbf{J}^{-1}(\mathbf{q}_{\mathrm{arm}})$ is 2 by 2 matrix whose elements are a non-linear function of the arm joint angles. Given $\mathbf{J}^{-1}(\mathbf{q}_{\mathrm{arm}})$ it is possible to drive the endpoint toward any point in the image plane. At least locally $\mathbf{J}^{-1}$ can be approximated by a constant matrix:

$$J^{-1}(\mathbf{q}_{\mathrm{arm}}) \approx \hat{\mathbf{J}}^{-1}(\mathbf{q}_{\mathrm{arm}}) = \begin{bmatrix} a_{11} & a_{12} \\ a_{21} & a_{22} \end{bmatrix} \tag{10}$$

Since following the procedure described in the previous section the robot has learnt a direct transformation between the arm joint angles and the image plane (see for example Fig. 7), it can now recover the position of the endpoint from a given joint configuration:

$$\mathbf{x}_{\mathrm{hand}} = \mathbf{f}(\mathbf{q}_{\mathrm{arm}}) \tag{11}$$

Indeed, to compute a local approximation of $\mathbf{J}^{-1}$, a random sampling of the arm joint space around a given point $(\bar{\mathbf{x}}, \bar{\mathbf{q}})$ can be performed:

$$\mathbf{q}_{\mathrm{i}} = \bar{\mathbf{q}} + \Delta\mathbf{q}_{\mathrm{i}} \tag{12}$$

with

$$\Delta\mathbf{q}_{\mathrm{i}} = \boldsymbol{\eta}(\mathbf{0}, \boldsymbol{\sigma}) \tag{13}$$

and where $\boldsymbol{\eta}(\mathbf{0},\boldsymbol{\sigma})$ follows a normal distribution of zero mean and standard deviation of 5°.

For each sample, by applying Eq. (11) we obtain a new value $\mathbf{x}_i = \mathbf{x} + \Delta\mathbf{x}_i$ that can be used to estimate $\mathbf{J}^{-1}$ around $\bar{\mathbf{q}}$ with a least squares procedure:

$$\Delta\mathbf{q}_{\mathrm{i}} = \begin{bmatrix} \Delta\mathbf{x}_{\mathrm{i}}^{\mathrm{T}} & \mathbf{0} \\ \mathbf{0} & \Delta\mathbf{x}_{\mathrm{i}}^{\mathrm{T}} \end{bmatrix} \cdot \begin{bmatrix} a_{11} \\ a_{12} \\ a_{21} \\ a_{22} \end{bmatrix} \tag{14}$$

$\hat{\mathbf{J}}^{-1}(\bar{\mathbf{q}})$ can then be used in the closed loop controller to drive the arm toward a specific position in the image plane. However, there is no need to close the loop with the actual visual feedback. By using the map in Eq. (11), in fact, we can substitute the actual visual feedback with the internal simulation provided by the model. From the output of the closed loop controller we can estimate the position of the arm at the next step, by assuming a pure kinematic model of the arm; in this way the procedure can be iterated several times to obtain the joint motor commands required to perform a reaching movement. The flowchart in Fig. 10 explains this procedure.

In principle the inverse Jacobian could be learnt by using the visual feedback of the hand. In practice, however, this is often impractical because

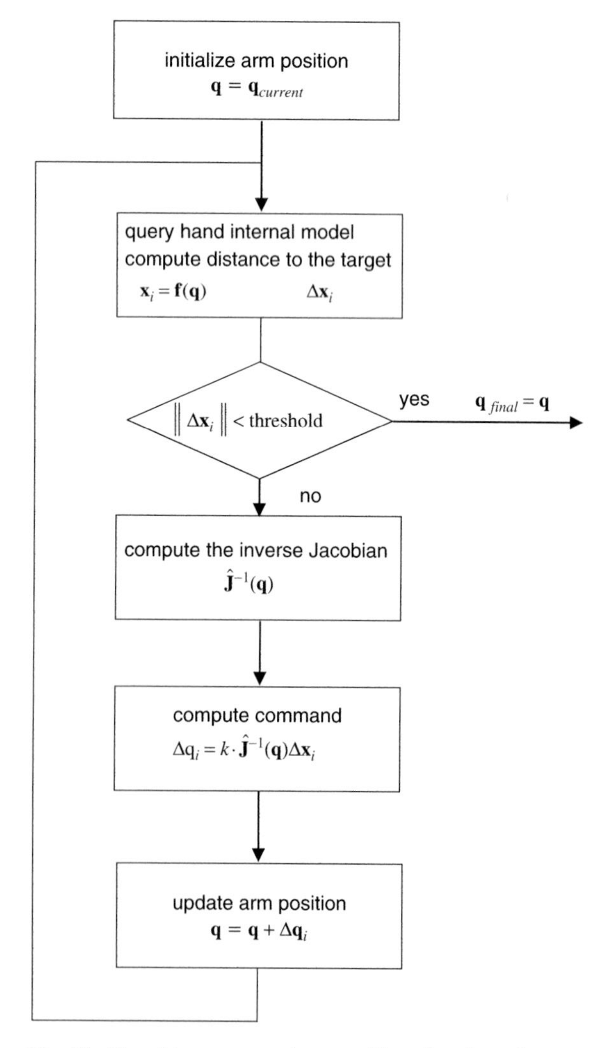

Fig. 10. Closed-loop approach to reaching, flowchart. See text for further details.

continuous visual feedback from the hand is rarely available. The approach we propose here requires only knowledge of the forward kinematics (as estimated in the previous section); the estimation of the inverse Jacobian with the approach we described is fast and can be easily performed online. Note also that the inverse Jacobian could have been computed analytically by taking the first derivative of Eq. (11). By selecting a least square solution, in our case, we added an extra smoothing factor that is beneficial in considering a control application. Also, in theory, our approach is more flexible since it does not require the knowledge of the number of units and structure of the neural network employed to approximate Eq. (11) and can be completely automatic.

The main limitation of this approach is that we do not make use of 3D visual information; while this is a clear limitation of this implementation, the same approach can be easily extended to the full 3D case. The implementation is consistent with the hand internal model, which provides the position of the hand in the image plane of one of the eyes only (left). Since in the Babybot the hand position is uniquely described by 3 degrees of freedom (the first three joints of the Puma arm), this technique was used to control only two of them (arm and forearm). Given the kinematics of the Puma arm this allowed performing movements on the plane defined by the shoulder joint. Another point worth discussing is that the closed loop controller does not use real visual feedback, and, therefore, its accuracy depends on the precision of the hand internal model. To achieve better performances, actual visual feedback might be required.

Let us summarize what we have described in this section. We have introduced two approaches to solving the inverse kinematics of the manipulator. The first method uses a mapping between the posture of the head (whose fixation point implicitly identifies the target) and the arm motor commands; it allows controlling the arm to reach any point fixated by the robot[1]. The second approach uses the hand internal model to compute a piecewise constant approximation of the inverse Jacobian and simulate small movements of the arm in the neighborhood of the desired target. The procedure is iterated several times to compute the motor command required for reaching the target. Reaching in this case is planned in the image plane; however, since the internal model is 2D, the approach is limited to the plane identified by the shoulder. For these reasons, the two methods were mixed in the experiment reported in the next section. The motor–motor mapping is employed to plan a first gross movement to approach the target, whereas the "closed-loop approach" allows a finer positioning of the fingers on the target. This second part of the movement is planned by considering the point of the ellipse at maximum distance from the robot's body (which corresponds to the fingers) as the arm endpoint (Fig. 11). This strategy proved successful because it substantially increased the probability to grasp the objects on the table.

Once the robot has computed the final arm posture, planning of the actual movement is still required. This was done with a simple linear interpolation between the current and final arm configuration. The trajectory was divided in steps that were then effected by the low level controller; to this purpose we employed a low-stiffness PD controller with gravity compensation. The gravity load term for each joint was learnt online as described in Natale (2004).

Learning about objects

In this section we describe a method for building a model of the object the robot grasps. We assume for a moment that the robot has already grasped an object; this can happen because a collaborative human has given the object to the robot (as we describe in the next section) or because the robot has autonomously grasped the object. In this case the robot may spot a region of interest in the visual scene and apply a stereotyped action involving the arm and hand to catch it. Both solutions are valid bootstrapping behaviors for the acquisition of an internal model of the object. When the robot holds the object it can be explored through movements of the arm and rotations of the wrist.

[1]During the learning of the motor–motor map, the robot tracks the palm of the hand.

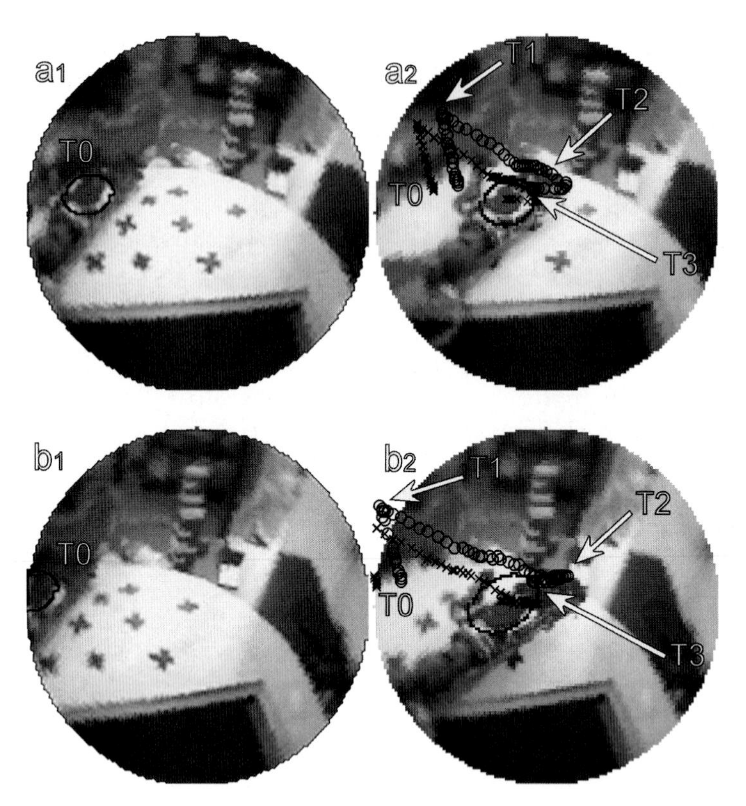

Fig. 11. Arm trajectories for two reaching actions (a) and (b). T0 marks the position of the hand at the beginning of the action. Crosses correspond to the position of the palm; circles show the position of the fingers. The action is divided in three phases. From T0 to T1 arm prepositioning. From T1 to T2, reaching: in this case the motor–motor map is used to move the palm towards the center of the visual field (the target). A small adjustment with the arm Jacobian is performed to position the fingers on the target (T2 to T3).

In short, the idea is to represent objects as collections of blobs generated by the visual attention system and their relative positions (neighboring relations). The model is created statistically by looking at the same object several times from different points of view (see Fig. 12). At the same time the system estimates the probability that each blob belongs to the object by counting the number of times each blob appears during the exploration.

In the following, we use the probabilistic framework proposed by Schiele and Crowley (1996a, b). We want to calculate the probability of the object O given a certain local measurement M. This probability $P(O|M)$ can be calculated using Bayes' formula:

$$P(O|M) = \frac{P(M|O)P(O)}{P(M)} \quad (15)$$

where $P(O)$ is the a priori probability of the object O, $P(M)$ the a priori probability of the local measurement M, and $P(M|O)$ is the probability of the local measurement M when the object O is fixated. In the following experiments we carried out only a single detection experiment, there are consequently only two classes, one representing the object and another representing the background. For lack of better estimations we set $P(O)$ and $P(\sim O)$ to 0.5 (this is equivalent to doing a maximum likelihood estimation).

Since a single blob is not discriminative enough, we considered the probabilities of observing pairs of blobs; the local measurement M becomes the event of observing both a central (i.e. fixated) and surrounding blobs:

$$P(M|O) = P(B_i|B_c \text{ and } (B_i \text{ adiacent } B_c)) \quad (16)$$

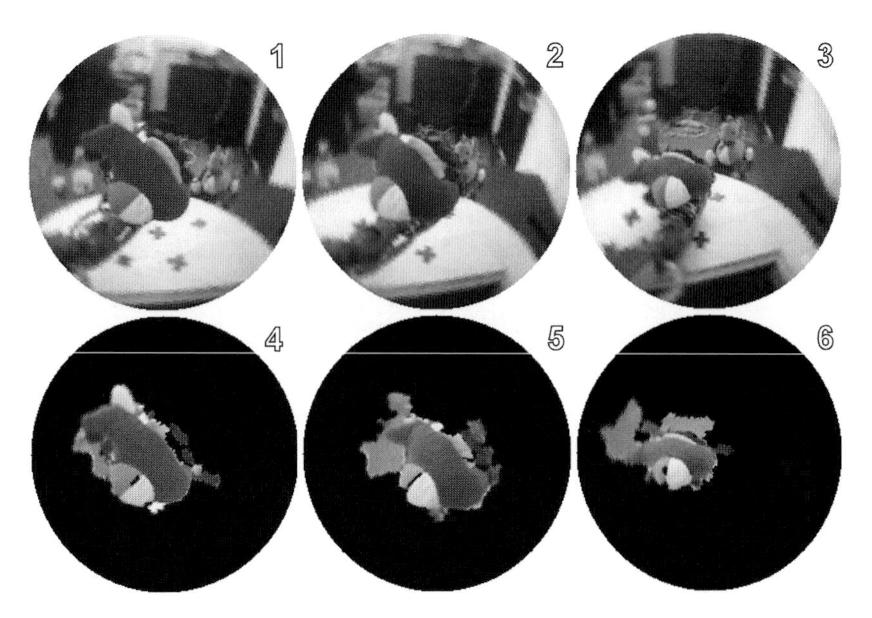

Fig. 12. Object exploration and corresponding blobs (1–3 and 4–6, respectively). The blobs used in training the object model are the central and the adjacent ones. An example of the resulting segmentation is reported in Fig. 13. Notice that fixation is maintained on the object by using the hand localization module (see text).

where B_i is the ith blob surrounding the central blob B_c which belongs to the object O. That is, we exploit the fact the robot is fixating the object and assume B_c to be constant across fixations of the same object — this is guaranteed by the fact the object is being hold by the hand. In practice this corresponds to estimating the probability that all blobs B_i adjacent to B_c (which we take as a reference) belong to the object. Moreover, the color of the central blob B_c will be stored to be used during visual search to bias the salience map. This procedure, although requiring the "active participation" of the robot (through gazing) is less computationally expensive compared to the estimation of all probabilities for all possible pairs of blobs of the fixated object. Estimation of the full joint probabilities would require a larger training set than the one we used in our experiments. For the same reason we assumed statistical independence of the blobs of the objects; under this assumption the total probability $P(M1,\ldots,MN|O)$ can be factorized in the product of the probabilities $P(M_i|O)$. The probabilities $P(M|\sim O)$ are estimated during the exploration phase with the blobs not adjacent to the central blob. An object is detected if the probability $P(O|M1,\ldots,MN)$ is greater than a fixed threshold.

Our requirement was that of building the object model with the shortest possible exploration procedure. Unfortunately, the small training set might give histograms $P(M|*)$ with many empty bins zero counts bins. To overcome this problem a probability smoothing method was used. A popular method of zero smoothing is Lidstone's law of succession (Lidstone, 1920):

$$P(M|O) = \frac{\text{count}(M \wedge O) + \lambda}{\text{count}(O) + v\lambda} \qquad (17)$$

for a v valued problem. With $\lambda = 1$ and a two valued problem ($v = 2$), we obtain the well-known Laplace's law of succession. Following the results of Kohavi et al. (1997) we choose $\lambda = 1/n$ where n is equal to the number of frames utilized during the training. The model of an object is trained in real time; the duration of the training is determined by the time required by the robot to rotate and move the object with the hand (currently $\sim$30 s).

When an object is detected after visual search, a possible figure-ground segmentation is attempted, using the information gathered during the exploration phase. Each blob is segmented from the background if it is adjacent to the central blob and

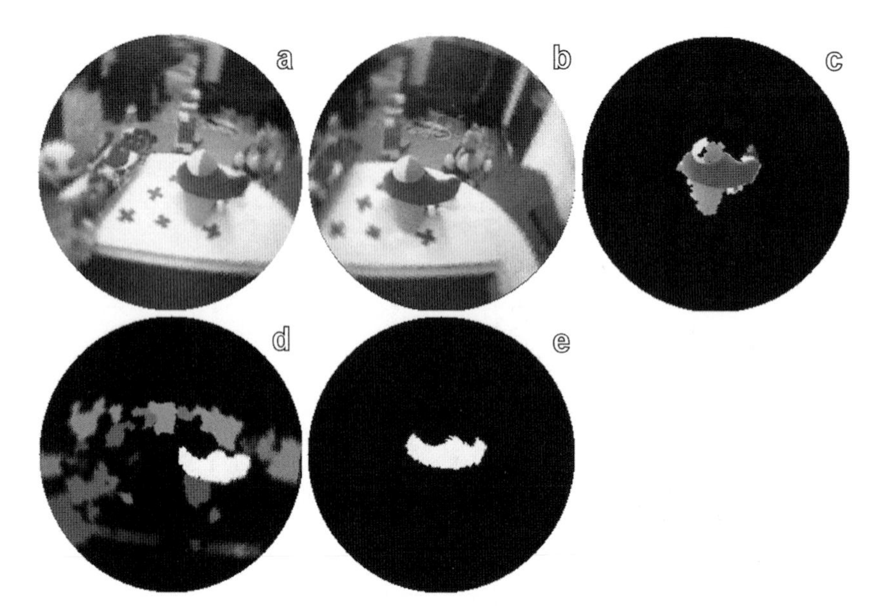

Fig. 13. Visual search. The robot has acquired a model of the airplane toy during an exploration phase (not shown); this information primes the attention system. The blue blob at the center of the airplane is selected and a saccade performed. (a) and (b) show the visual scene before and after the saccade. (d) and (e) show the output of the visual attention system synchronized with (a) and (b), respectively. The result of the segmentation after the saccade is in (c).

if its probability to belong to the object is greater than 0.5. This probability is approximated using the estimated probability as follows:

$$P(B_i \in O | B_c \text{and } (B_i \text{ adiacent } B_c)) \cong P(B_i | B_c \text{and } (B_i \text{ adiacent } B_c)) \quad (18)$$

As an example Fig. 13 shows the result of the segmentation procedure. These results could be further improved by adding some hypothesis about the regularity of the object boundary. However, for the purpose of this paper (object identification for the manipulation task) these refinements were not necessary.

In Table 1, results are shown of using a toy car and a toy airplane as target objects; 50 training sessions were performed for each object. The first column shows the recognition rate, the second the average number of saccades (mean $\pm$ standard deviations) it takes the robot to locate the target in case of successful recognition. The recognition rate of the toy airplane is lower than the one of the toy car because the former is more similar (by virtue of its color and number of blobs) to the background.

Table 1. Performance of the recognition system measured from a set of 50 trials

Object	Recognition rate (%)	Number of saccades when recognized
Toy car	94	3.19 ± 2.17
Toy airplane	88	3.02 ± 2.84

Grasping behavior

The modules described in the previous sections can be integrated to achieve an autonomous grasping behavior. Figure 14 can be used as a reference for the following discussion. The action starts when an object is placed in the robot's hand and the robot detects pressure in the palm (frame 1). This elicits a clutching action of the fingers; the hand follows a preprogrammed trajectory, the fingers bend around the object toward the palm. If the object is of some appropriate size, the intrinsic elasticity of the hand facilitates the action and the grasping of the object. The robot moves the arm to bring the object close to the cameras and begins its exploration. The object is placed in four positions with different orientations and background

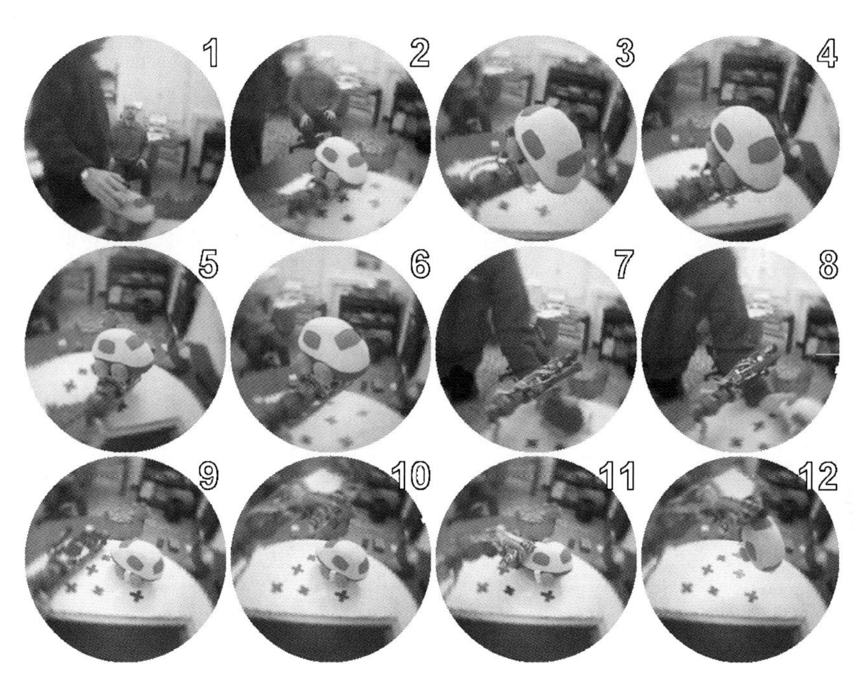

Fig. 14. A sequence of the robot grasping an object. The action starts when an object is placed on the palm (1). The robot grasps the object and moves the eyes to fixate the hand (2). The exploration starts in (3) when the robot brings the object close to the camera. The object is moved in four different positions while maintaining fixation; at the same time the object model is trained (3–6). The robot drops the object and starts searching for it (7). The object is identified and a saccade performed (7–9). The robot eventually grasps the toy (10–12).

(frames between 2 and 6). During the exploration, the robot tracks the hand/object; when the object is stationary and fixation is achieved, a few frames are acquired and the model of the object trained as explained above. At the end of the exploration the object is released (frame 4). At this point the robot has acquired the visual model of the object and starts searching for it in the visual scene. To do this, it selects the blob whose features better match those of the object's main blob and perform a saccade. After the saccade the model of the object is matched against the blob that is being fixated and its surrounding. If the match is not positive the search continues with another blob, otherwise grasping starts (frames 7–9). At the end of the grasp the robot uses haptic information to detect whether it is holding the object or rather the action failed. In this process the weight of the object and its consistence in the hand is checked (the shape of the fingers holding the object). If the action is successful the robot waits for another object, otherwise it performs another trial (search and reach).

It is fair to say that part of the controller was preprogrammed. The hand was controlled with stereotyped motor commands. Three primitives were used: one to close the hand after pressure was detected, and two during the grasping to preshape the hand and actually clasp the object. The robot relied on the elasticity of the hand to achieve the correct grasping. To facilitate grasping, the trajectory of the arm was also programmed beforehand; waypoints relative to the final position of the arm were included in the joint space to approach the object from the top.

Discussion and conclusions

In this paper we have presented a developmental approach to the realization of cognitive abilities in a humanoid robot, which starts from the exploration of the body and unfolds by eventually exploring the external world. The robot starts from a limited set of initial motor and perceptual competencies and autonomously develops more sophisticated ways to interact with the environment. This knowledge is used to begin the exploration of the environment and to build a visual model of the objects that are grasped.

We have presented an implementation of a visual attention system properly taking into account top-down and bottom-up information. The top-down system divides the visual scene into color blobs; each blob is assigned a saliency depending on the ratio between its color and the color of the area surrounding it. The robot actively explores the visual appearance of the objects it grasps: every time an object is placed on the palm a statistical model of the blobs that are part of it is constructed. This information is subsequently fed to the attention system as a bottom-up primer to control the visual search of the same object. Thus, the robot experience allows it to build a representation of the object with which it interacts while, at the same time, modulates the visual attention system. The robot's ability to act is used together with the body internal model to drive the exploration of the environment. This facilitates learning in different ways. Firstly, it helps the robot to focus attention both in space and in time. During the acquisition of the object visual model, in fact, the robot can track the object because it knows the position of the hand from its proprioceptive feedback. The latter is also useful to detect when the acquisition of the model can be initiated because the object does not move and the eyes have acquired a stable fixation on it. Finally, the fact that the object is being held by the hand guarantees the link between different sensory modalities (for example the sight of the object and the kinesthetic information from the hand). The object model makes use of visual information; in Natale et al. (2004) we show how it is possible to build a model of the objects based only on haptic information. In the future we would like to investigate the integration of the two approaches.

We support the enactive view of cognition in showing how much the body and the ability to build the representation of the external world through the interaction between the body and the environment can be useful for an autonomous agent. Even a simple set of behaviors (such as the one initially provided to the robot) is sufficient to begin the exploration of the environment and acquire an internal representation of it. On the other hand it is fair to say that much of the system presented in this paper is still "cognitivist" and more or less carefully handcrafted into the robot. For practical reasons, our implementation lays in between a full emergent and a cognitivist approach although biologically informed choices were made when possible.

We have also shown how this initial body-environment interaction is sufficient to start linking actions with their resulting consequences to form prediction about the behavior of the robot. Very often prospective control is required to plan a successful action. During grasping, for example, the correct timing of preshaping and closure of the fingers is required; the lags in the sensory streams (visual and tactile) typical of artificial and natural systems make feedback control ineffective. To be able to anticipate the impact of the hand with the object, the robot is required to control the timing between preshaping and actual grasping; clearly this cannot be based only on visual and tactile feedback. Prospective control, however, is not only important for action. It gives an agent the possibility to create expectations on which to base the interpretation of the world and the actions performed by others. By means of the interaction with the world the agent builds a model of the behavior of external entities (objects, people, etc.) and the associated sensory feedback. This link can be used afterward to anticipate the consequences of a similar action and, eventually, to compare them with the real feedback. In the same way new situations can be interpreted by matching them against the robot's past experience. For example, the event of a ball that falls on the floor (and the resulting visual and auditory sensations) can be associated to the action of dropping it. Anticipation and predictions enhance the agents' ability to understand and interact with the environment and, for this reason, are important aspects of cognition. The results of this paper represent the first steps into the implementation of cognitive abilities in an artificial system. It is difficult to think, at least from an emergent perspective, of a shortcut that prescinds from sensorimotor coordination in achieving cognitive skills to be used in the real world.

To conclude, we would like to comment on the effort required to build a complete robotic platform on the one hand, and the software architecture on the other. Presently, the Babybot is an integrated robotic platform where it is extremely easy for software modules controlling different subparts (arm, head or hand to mention just a few) to exchange information and coordinate with each other (Metta et al., 2006). This is not very common, as usually in the literature papers report single experiments where the robotic platform is specifically programmed to perform the desired task, but care is not taken to realize a system, which can grow in complexity as new modules are added. The experiment reported in the last section does not only show the integration between the visual attention system and the motor system but also the complexity of the system as a whole. We believe that this is a necessary prerequisite to carry out research in humanoid robotics as the complexity and number of skills increase.

Acknowledgments

This work is funded by the European Commission's Cognition Unit, Directorate-General Information Society, as part of project no. IST-2004-004370: RobotCub — ROBotic Open-architecture Technology for Cognition, Understanding, and Behaviour.

References

Beer, R.D. (2000) Dynamical approaches to cognitive science. Trends Cogn. Sci., 4(3): 91–99.

Billock, V.A. (1995) Cortical simple cells can extract achromatic information from the multiplexed chromatic and achromatic signals in the parvocellular pathway. Vision Res., 35(16): 2359–2369.

Clark, A. (2001) Mindware: An Introduction to the Philosophy of Cognitive Science. Oxford University Press, Oxford, UK.

Downing, C. and Pinker, S. (1985) The spatial structure of visual attention. In: Posner, M. and Marin, O.S.M. (Eds.), Attention and Performance, Vol. XI. Lawrence Erlbaum Associates, Hillsdale, New Jersey, U.S.A, pp. 171–187.

Eckhorn, R., Bauer, R., Jordan, W., Brosch, M., Kruse, M., Munk, W. and Reitboeck, H.J. (1988) Coherent oscillations: a mechanism of feature linking in the visual cortex? Biol. Cybern., 60(8), 121–130.

Egly, R., Driver, J. and Rafal, R. (1994) Shifting visual attention between objects and locations: evidence for normal and parietal subjects. J. Exp. Psychol. Gen., 123(2): 161–177.

Espiau, B., Chaumette, F. and Rives, P. (1992) A new approach to visual servoing in robotics. IEEE Trans. Rob. Autom., 8(3): 313–326.

Fitzpatrick, P. and Arsenio, A. (2004) Feel the beat: using cross-modal rhythm to integrate perception of objects, others and self. In Proceedings of the Fourth International Workshop on Epigenetic Robotics, August, 25–27, 2004, Genoa, Italy.

Fitzpatrick, P., Metta, G., Natale, L., Rao, S. and Sandini, G. (2003) Learning about objects through action: initial steps towards artificial cognition. In Proceedings of the IEEE International Conference on Robotics and Automation, May 12–17, 2003, Taipei, Taiwan.

Flanders, M., Daghestani, L. and Berthoz, A. (1999) Reaching beyond reach. Exp. Brain Res., 126(1): 19–30.

Giefing, G.J., Janssen, H. and Mallot, H.A. (1992) Saccadic object recognition with an active vision system. In Proceedings of the 10th European Conference on Artificial Intelligence, August 1992, Vienna, Austria.

Gray, C.M., König, P., Engel, A.K. and Singer, W. (1989) Oscillatory responses in cat visual cortex exhibit intercolumnar synchronization which reflects global stimulus properties. Nature, 338: 334–336.

Hutchinson, S., Hager, G. and Corke, P. (1996) A tutorial on visual servo control. IEEE Trans. Rob. Autom., 12(5): 651–670.

Itti, L. and Koch, C. (2001) Computational modelling of visual attention. Nat. Rev. Neurosci., 2(3): 194–203.

Itti, L., Koch, C. and Niebur, E. (1998) A model of saliency-based visual attention for rapid scene analysis. IEEE Trans. Pattern Anal. Mach. Intell., 20(11): 1254–1259.

Kohavi, R., Becker, B. and Sommerfield, D. (1997) Improving simple Bayes. In Proceedings of the European Conference on Machine Learning, April 1997, Prague, Czech Republic.

Lacquaniti, F. and Caminiti, R. (1998) Visuo-motor transformations for arm reaching. Eur. J. Neurosci., 10(1): 195–203.

Li, X., Yuan, T., Yu, N. and Yuan, Y. (2003) Adaptive color quantization based on perceptive edge protection. Pattern Recognit. Lett., 24(16): 3165–3176.

Lidstone, G. (1920) Note on the general case of the Bayes-Laplace formula for inductive or a posteriori probabilities. Trans. Fac. Actuaries, 8: 182–192.

Maturana, H.R. and Varela, F.J. (1998) The Tree of Knowledge, the Biological Roots of Human Understanding. Shambhala Publications, Inc., Boston & London.

Melcher, D. and Kowler, E. (1999) Shapes, surfaces and saccades. Vision Res., 39(17): 2929–2946.

Metta, G. (2000) Babyrobot: A Study into Sensori-motor Development. PhD Thesis, University of Genoa, Genoa, Italy.

Metta, G. and Fitzpatrick, P. (2003) Early integration of vision and manipulation. Adapt. Behav., 11(2): 109–128.

Metta, G., Fitzpatrick, P. and Natale, L. (2006) YARP: yet another robot platform. Int. J. Adv. Robot. Syst., 3(1): 43–48 Special issue on Software Development and Integration in Robotics.

Metta, G., Sandini, G. and Konczak, J. (1999) A developmental approach to visually-guided reaching in artificial systems. Neural Netw., 12(10): 1413–1427.

Milanese, R. (1993) Detecting salient regions in an image: From biological evidence to computer implementation. PhD Thesis, University of Geneva, Geneva, Switzerland.

Natale, L. (2004) Linking action to perception in a humanoid robot: a developmental approach to grasping. PhD Thesis, University of Genoa, Genoa, Italy.

Natale, L., Metta, G. and Sandini, G. (2004) Learning haptic representation of objects. In Proceedings of the International Conference on Intelligent Manipulation and Grasping, July 1–2, 2004, Genoa, Italy.

Natale, L., Rao, S. and Sandini, G. (2002) Learning to act on objects. In Proc. of the Second International Workshop, BMCV 2002, November 22–24, 2002, Tubingen, Germany.

Posner, M.I. and Cohen, Y. (1984) Components of visual orienting. In: Bouma, H. and Bouwhuis, D.G. (Eds.), Attention and Performance, Vol. X., Erlbaum, Hillsdale, NJ, pp. 531–556.

Posner, M.I., Snyder, C.R.R. and Davidson, B.J. (1980) Attention and the detection of signals. J. Exp. Psychol. Gen., 109(2): 160–174.

Pylyshyn, Z. (2001) Visual indexes, preconceptual object, and situated vision. Cognition, 80(1–2): 127–158.

Rochat, P. and Striano, T. (2000) Perceived self in infancy. Infant Behav. Dev., 23(3–4): 513–530.

Sandini, G., Metta, G. and Vernon, D. (2004) RobotCub: An open framework for research in embodied cognition. In Proceedings of the IEEE-RAS/RJS International Conference on Humanoid Robotics, November 10–12, 2004, Santa Monica, California, USA.

Sandini, G. and Tagliasco, V. (1980) An anthropomorphic retina-like structure for scene analysis. Comput. Vis. Graphics Image Process., 14(3): 365–372.

Schaal, S. and Atkeson, C.G. (1998) Constructive incremental learning from only local information. Neural Comput., 10(8): 2047–2084.

Schiele, B. and Crowley, J.L. (1996a) Probabilistic object recognition using multidimensional receptive field histograms. In Proceedings of the 13th International Conference on Pattern Recognition, August, 1996, Vienna, Austria.

Schiele, B. and Crowley, J.L. (1996b) Where to look next and what to look for. In Proceedings of the IEEE/RSJ International Conference on Intelligent Robots and Systems, December, 1996, Osaka, Japan.

Scholl, B.J. (2001) Objects and attention: the state of the art. Cognition, 80(1–2): 1–46.

Smet, P.D. and Pires, R. (2000) Implementation and analysis of an optimized rainfalling watershed algorithm. In Proceedings of the IS&T/SPIE's 12th Annual Symposium Electronic Imaging 2000: Science and Technology, 23–28 January, San Jose, California, U.S.A.

Smirnakis, S.M., Berry, M.J., Warland, D.K., Bialek, W. and Meister, M. (1997) Adaptation of retinal processing to image contrast and spatial scale. Nature, 386: 69–73.

Sun, Y. and Fisher, R. (2003) Object-based visual attention for computer vision. Artif. Intell. Med., 146(1): 77–123.

Tipper, S.P. (1994) Object-based and environment-based inhibition of return of visual attention. J. Exp. Psychol. Hum. Percept. Perform., 20(3): 478–499.

Treisman, A.M. and Gelade, G. (1980) A feature integration theory of attention. Appl. Cogn. Psychol., 12(1): 97–136.

Vincent, L. and Soille, P. (1991) Watersheds in digital spaces: an efficient algorithm based on immersion simulations. IEEE Trans. Pattern Anal. Mach. Intell., 13(6): 583–598.

Viola, P. and Jones, M.J. (2004) Robust real-time face detection. Int. J. Comp. Vis., 57(2): 137–154.

von Seelen, W. and Mallot, H.A. (1990) Neural Mapping and Space-Variant Image Processing. In Proceedings of the International Joint Conference on Neural Networks (IJCNN), San Diego, California, U.S.A.

Wolfe, J.M. (2003) Moving towards solutions to some enduring controversies in visual search. Trends Cogn. Sci., 7(2): 70–76.

Wolfe, J.M. and Gancarz, G. (1996) Guided search 3.0. In: Lakshminarayanan V. (Ed.), Basic and Clinical Applications of Vision Science. Kluwer Academic, Dordrecht, Netherlands, pp. 189–192.

Wolpert, DM. and Miall, R.C. (1996) Forward models for physiological motor control. Neural Netw., 9(8): 1265–1279.

Yoshikawa, Y., Hosoda, K. and Asada, M. (2003) Does the invariance in multi-modalities represent the body scheme? — a case study with vision and proprioception. In Proceedings of the 2nd Intelligent Symposium on Adaptive Motion of Animals and Machines, Kyoto, Japan.

C. von Hofsten & K. Rosander (Eds.)
Progress in Brain Research, Vol. 164
ISSN 0079-6123

CHAPTER 23

Emergence and development of embodied cognition: a constructivist approach using robots

Yasuo Kuniyoshi[1,2,*], Yasuaki Yorozu[1], Shinsuke Suzuki[1], Shinji Sangawa[1], Yoshiyuki Ohmura[1], Koji Terada[1] and Akihiko Nagakubo[3]

[1]*Department of Mechano-Informatics, School of Information Science and Technology, The University of Tokyo, 7-3-1 Hongo, Bunkyo-ku, Tokyo, 113-8656, Japan*
[2]*JST ERATO Asada Synergistic Intelligence Systems Project, Japan*
[3]*Intelligent Systems Division, National Institute for Advanced Industrial Science and Technology, 1-1-1 Umezono, Tsukuba, Ibaraki 305-8568, Japan*

Abstract: A constructivist approach to cognition assumes the minimal and the simplest set of initial principles or mechanisms, embeds them in realistic circumstances, and lets the entire system evolve under close observation. This paper presents a line of research along this approach trying to connect embodiment to social cognition. First, we show that a mere physical body, when driven toward some task goal, provides a clear information structure, for action execution and perception. As a mechanism of autonomous exploration of such structure, "embodiment as a coupled chaotic field" is proposed, with experiments showing emergent and adaptive behavior. Scaling up the principles, a simulation of the fetal/neonatal motor development is presented. The musculo-skeletal system, basic nervous system, and the uterus environment are simulated. The neural-body dynamics exhibit spontaneous exploration of a variety of motor patterns. Lastly, a robotic experiment is presented to show that visual-motor self-learning can lead to neonatal imitation.

Keywords: embodiment; actions; symbols; information structure; emergent behavior; learning; imitation; fetus; cortical maps; body schema/image; coupled chaos system

Introduction

How does the human mind emerge, develop, and function? Being able to artificially create and activate it would be the ultimate form of unraveling the mystery. It may even be the most appropriate methodology for cognitive science.

Constructivist approach to cognition

Cognition is not dissociable from body–environment interactions (Pfeifer and Scheier, 1999). Any attempt to do so would destruct its natural function. That is why the traditional reductionist approach does not work for a true understanding of cognition. Then, without dividing the problem and examining each piece, how can we determinate its basic principles?

A *constructivist* approach to cognition (Drescher, 1991; Hashimoto, 1999; Asada et al., 2001)

*Corresponding author. Tel.: +81-3-5841-6276; Fax: +81-3-5841-6276; E-mail: kuniyosh@isi.imi.i.u-tokyo.ac.jp

DOI: 10.1016/S0079-6123(07)64023-0

provides an alternative methodology. It assumes a minimal and the simplest set of initial principles or mechanisms, embeds them in realistic circumstances (i.e., the simulated body and environment), and lets the entire system evolve under close observation. If the resulting behavior is identical to a particular aspect of the target system behavior, then the initial principles or mechanisms are assumed to capture the essence of why the target system behave or evolve the way it does in its natural environment.

The above approach inevitably assumes the notion of emergence, adaptation, development, evolution, or in general, how the behavior or the system changes (expressed as "evolve," in the previous paragraph). This is essentially due to the fundamental assumption that the observed behavior is the result of the complex interaction between the system and its circumstances, and *not directly* specified by or predicted from the description of the initial state.

In the context of cognitive science, where the study of individuals is targeted, the above approach asks for a minimal set of interested features of the initial configuration (*predispositions* (Johnson, 2005)), embeds them in appropriate circumstances such as the general learning mechanism, nervous system structure, body, and environment, and observes the course of its development. By changing the initial features and see how the system develops differently, we will understand the meaning of the initial features. This would be the operational extreme version of the developmental cognitive neuroscience (Johnson, 2005).

Body shapes the brain

Pushing the constructivist approach, we take an extreme position claimed as "body shapes the brain." It assumes that the physical structures and properties of the body and the environment determine the potential information structures that serve as the foundation of cognitive structures, and that during early development, the brain (mainly the neo-cortex) self-organizes to capture and exploit such information structures. Hence we claim that the body serves as the origin of behavioral repertoire, and even symbols and concepts. We assume very little or no "innate" cognitive functions and cortical representations.

This may be regarded as a radical recurrence of *tabula rasa* approach. However, our position is very different. We do assume genetic effects on the anatomy of the body, sensory organs, and nervous system. These are essentially important factors of the "circumstances" (see above) called *embodiment*.

We do *not* assume the variety of very early behavioral and cognitive functions called "reflexes," "instincts," and "innate functions" to exist *a priori* unless they have direct and clear correspondences with specific neuronal circuits determined by structural genes. Moreover, we seriously question the notion of "innateness," as we are now starting to simulate the fetal development. Along with the morphogenesis, the fetuses are already engaged in motor activities (Kisilevsky and Low, 1998), and the neuronal organization depends on signal activities (Crair, 1999).

At the genetic level, according to the modern view of gene network, environmental factors can influence gene expression (Ridley, 2003). Then ontogeny should be viewed as an interactive emergence (Hendriks-Jansen, 1996) process where genes regulate behavior and the behavior influences gene expression.

To summarize, when we view the development "from embryo" (Karmiloff and Karmiloff-Smith, 2001), the notion of innateness must be seriously re-examined. It is trivial that genes play an important role. However, the classical view of development as a genetic clockwork is completely wrong. The modern view is a dynamic, interactive, and emergent process. At the same time, the process has strong structural stability that from time to time, stable patterns are formed. That is why "normal" infants acquire almost the same set of skills in mostly the same order (Thelen and Smith, 1994).

A constructivist approach attempts to reveal the principles governing the above dynamic, interactive, and emergent process. In order to obtain a clear understanding, we start with the minimal initial configuration, and then add mechanisms and principles one by one.

At its early stage, presented in this paper, we start with only a physical body and the environment. Then we test a simple principle of chaos-driven behavior emergence, which is essentially a version of central pattern generator (CPG) activity. Finally, we add very simple neural mechanisms corresponding to spine, medulla, and primary sensory-motor cortices. So far, we try to show how much the system can do without any "innate" or "genetically coded" circuits for particular behavior control (except for the clearly understood stretch reflex).

From robot body to robot brain

In the following sections, we present an overview of our recent studies along the above-discussed approach. The descriptions are limited to convey the mainstream ideas. The readers are advised to refer to each corresponding original paper for technical details.

We start in the next section by showing how a mere physical body, when driven toward some task goal, provides a clear information structure, called "knacks and focuses." They play crucial roles both in execution and visual recognition of the task.

Then, in the third section, we look for an appropriate mechanism for driving the body without *a priori* knowledge to explore and discover the potential information structure provided by the physical body. We propose a novel mechanism called "embodiment as a coupled chaotic field," that serves the above role. The mechanism has a biological correspondent in the central nervous system under certain conditions.

The fourth section presents our ongoing efforts toward simulating the fetal/neonatal motor development. The musculo-skeletal system of the body, the basic nervous system, and the uterus environment are simulated. The chaotic field principle is naturally embedded in the neural-body dynamics, resulting in spontaneous exploration of a variety of motor patterns.

The fifth section presents a model that provides a partial account for how perceptual-motor learning during the fetal period might contribute to the neonatal imitation. This serves as the bridge between the explorative learning of motor repertoire and the initial form of social interactions toward human communication.

"Knacks and focuses" of dynamics actions: information structure emerging from body–environment interaction

In order to investigate the effect of having a human form body, we chose a particular task called "roll-and-rise" action (see Fig. 1). It is quite a dynamic motion with intensive body–environment contacts. This characteristic of the task strongly emphasizes the effect of the physical property of the body such as the outer shape, mass distribution, articulation of the limbs, joint limits, etc.

From a robot control point of view (Kuniyoshi et al., 2004a), the above task is very interesting. The differential equation of the system dynamics cannot be analytically solved, meaning that the standard control theory cannot provide a proper control law. Also, due to the intensive contacts, there is much uncertainty in the behavior of the system. However, humans can learn to do the action reliably within just a few trials.

"Knacks" in dynamic whole body human action

In order to reveal human strategy for roll-and-rise action, we carried out a series of motion measurement and analysis experiments. Subjects are asked to do the task repeatedly and their whole-body

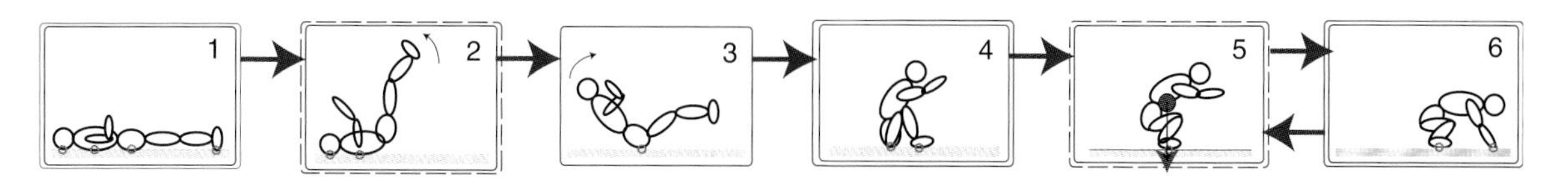

Fig. 1. "Roll-and-rise" action. Starting from a lying flat posture (1), swing up both legs, swing them down (2), and climb on the feet (5) in one action. In case of failing to hold self at the crouching posture, falling forward (6), push back with both hands to recover.

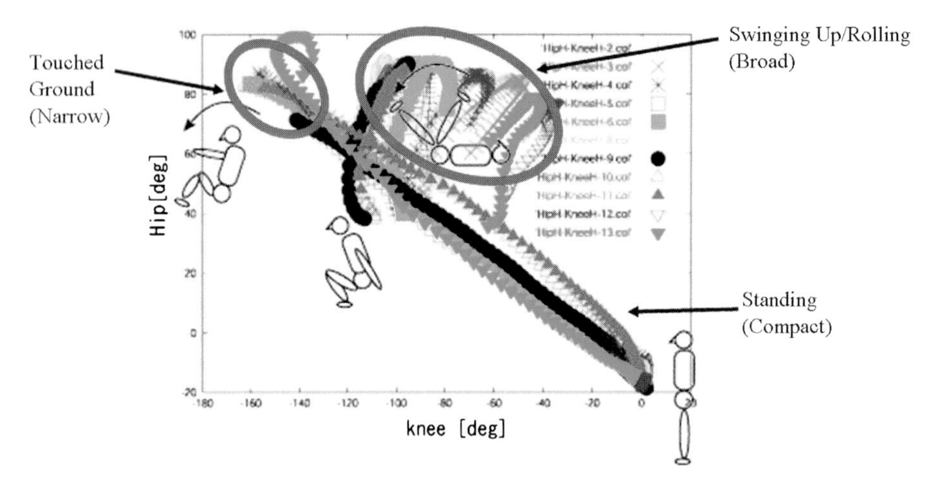

Fig. 2. Non-uniform trajectory structure of roll-and-rise actions. Hip–knee angle trajectories from 11 trials of the same subject are overlaid. Trajectories while swinging up the legs are distributed and those when the feet touched the ground are converged.

motion is measured with an optical motion capture system.

Figure 2 shows hip–knee angle trajectories from 11 trials of the same subject, overlaid on the same plane. A significant feature is that in some parts the trajectories are distributed (top-middle part) and in some other parts (for example, top-left part) they converge. In other words, in some parts trajectories are different between multiple trials, and in other parts they are always the same. Figure 3 shows the statistical data, the average, and variance, which confirm the non-uniform trajectory distribution.

The above data suggests that human strategy for dynamic whole-body actions have non-uniform trajectory bundle structure on the state-space of the task. Figure 4 shows the schematic representation of the "global dynamics" structure on the state-space. The left-most disc (henceforth, "node") denotes the initial lying-flat state and the right-most node denotes the goal state of standing still. There are several possible strategies for standing up, and the lowest path correspond to the dynamic roll-and-rise action. The gray bands denote the trajectory bundles. Note that the width of the gray bands is non-uniform. Around the intermediate nodes (the bottom-middle node, for example) they narrow down, and in between the nodes they broaden.

Although the rigorous analysis of the entire dynamics is very difficult, it is possible to model and analyze the dynamics around particular states. The two color map graphs at the bottom of Fig. 4 shows the results of numerical analyses. The axes of the graphs correspond to the state parameters (e.g., hip and knee angle, in the right graph) at the point of analysis, and the color indicates the success index (i.e., required head angular velocity to reach the goal state, meaning how hard to accomplish the task). The red regions correspond to the successful states; the tiny bit at the top-left corner in the right graph, and the wedge-like region in the left graph. Imagine that the gray bands have 3D volumes and the graphs are the intersections. The red regions correspond to the area where the successful trajectories penetrate the graph planes.

What the analyses tell us is that the bottom-middle node, the moment where the ground contact shifts from the hip to the feet, is the most critical point in the entire action. One has to go through an extremely narrow region at this point in order to reach the goal state. However, at some distance from this point, there is a broad margin. In other words, we only need to make sure to meet the few critical conditions, and can relax and let the natural dynamics (rolling) do the work at other points.

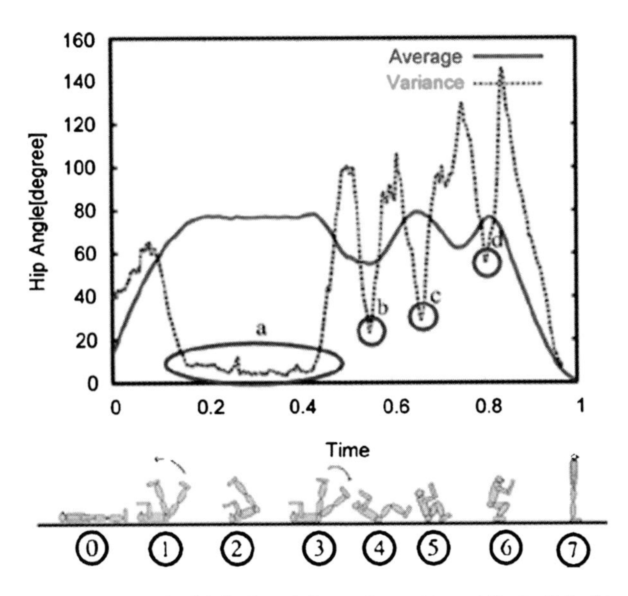

Fig. 3. Statistics of multiple trials. The average (solid line) and the variance (dotted line) of the hip angle trajectories from multiple trials are plotted against the time axis. Note the drastic ups and downs of the variance graph.

Symbols from physical actions

The nodes represent important points of the task, and we call them "knacks." Knowing and paying attention to the knacks is critical to the success of the task. The nodes also define the boundaries and branching points of the dynamics. Because the boundary conditions change at the nodes, different equations govern the dynamics before and after the nodes. Also, since the trajectories get separated by the critical conditions at the nodes into the successful group and non-successful one, the trajectories branch at the nodes.

In summary, the nodes can be regarded as *symbols* representing the action unit boundaries and essential information about the task, which can be communicated to other agents in order to share the skill.

If the nodes convey symbolic information, we should be able to recreate successful actions based on the node information, for a different body such as a humanoid robot.

Figure 5 shows our humanoid robot. We developed the robot with special considerations about the outer shape, joint configuration and limits, and mass and power distribution, to be as close as possible to average adult humans. It can move very fast and exert large enough impulse force to simulate biological motion.

With the aid of numerical simulation and semi-automatic search on computers, we generate motor patterns that conform to the node information, which are then tested on the real robot.

Figure 6 shows a successful case. Within ~2 s (from starting to raise the legs), the robot rose on its feet. The success rate is 100% including the cases where the robot did not stop at the crouching posture and rolled forward hitting the ground with the hands (this is considered as another form of success because it is easy to recover by pushing back by the hands), and 70% excluding such cases (i.e., counting only those where the robot could hold itself at the crouching posture).

"Focuses" in action recognition

It has been shown in the above discussion that roll-and-rise action has a few symbolic nodes. We investigate how they relate to perception (Kuniyoshi et al., 2004b). Our hypothesis is that human observers obtain important information for recognizing the action from these nodes.

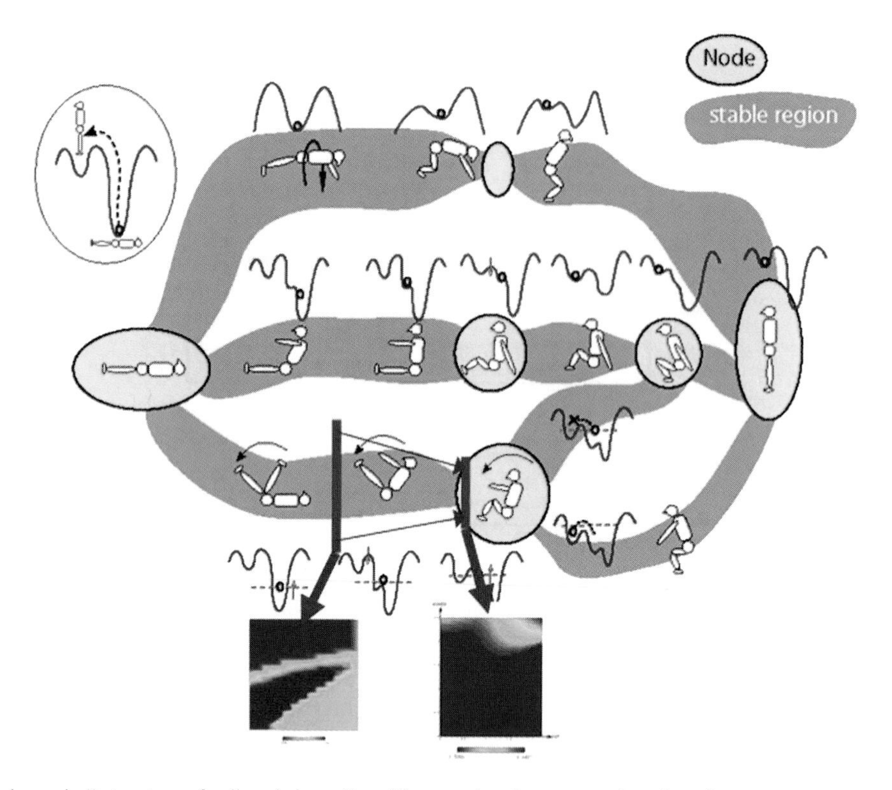

Fig. 4. "Global dynamics" structure of roll-and-rise action. The gray bands represent bundles of many state-space trajectories. They converge at the critical states denoted by the discs. The two color maps at the bottom indicate the regions of successful states computed by analyzing the dynamics of the task; the red wedge-like successful region in the left map narrows down to a tiny red successful region at the top-right corner of the right map. They correspond to the region of successful trajectories at the two cross-sections of the trajectory bundle.

Since the nodes are sparsely located along the time axis, we hypothesize that the amount of information extracted by human observers is not uniformly distributed, but localized along the time axis.

In order to test the above hypothesis, we designed a psychological experiment in which the subjects are asked to watch movie clips of various roll-and-rise performances by human performers (see Fig. 7), including both successes and failures.

The movies are of different temporal length, most of them being cut off before the end, but always starting from the beginning (see Fig. 8). Hence the subjects see the movies up to various moments. Each time a subject sees a movie clip, he/she is asked to guess whether the performance was a success or a failure. If the subject has obtained enough information up to that point, then the probability of a correct guess increases.

The number of successes and failures of the trials were: 31 successes and 17 failures by performer A, and 21 successes and 8 failures by performer B. For each performer, four trials (two successes and two failures) were selected and used for further experiment.

Thirty students, 23 males and 7 females of age between 21 and 24 were recruited as subjects. Prior to the experiment, each subject was given a complete description of the purposes and the procedure of the experiment, time to freely decide about participation, and then asked to sign an agreement. This experiment does not include any deceiving procedure.

Results

The numbers of correct answers for the movie clips with each length are plotted with bars in Fig. 9.

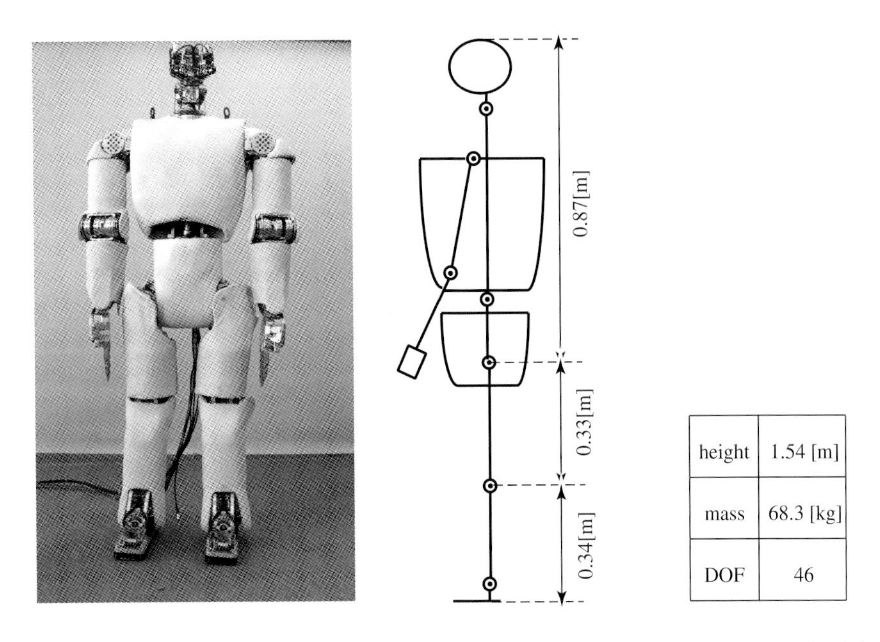

height	1.54 [m]
mass	68.3 [kg]
DOF	46

Fig. 5. Humanoid robot. Front view (left), DOF configuration in sagittal plane (middle), size and weight (right).

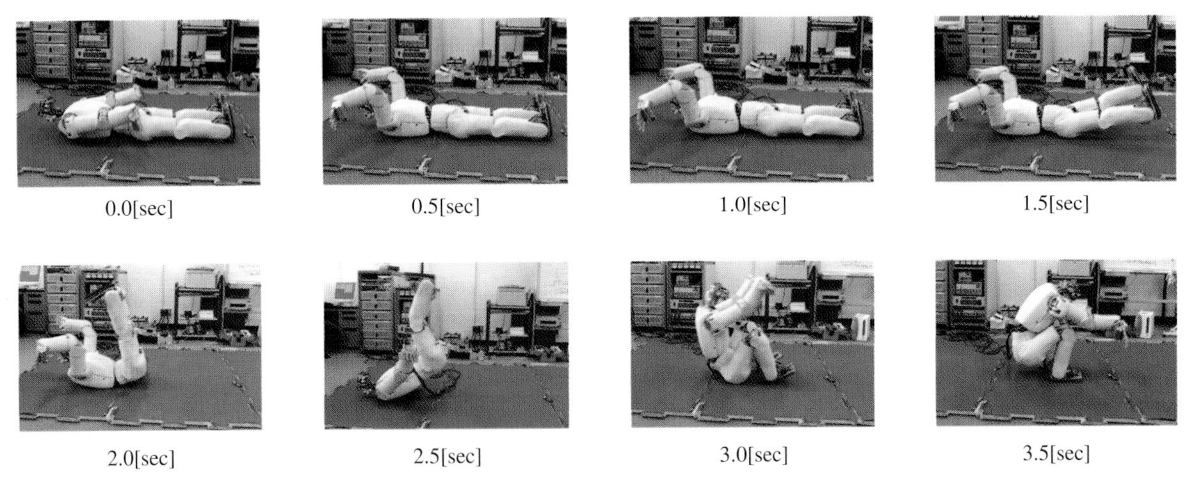

Fig. 6. Successful execution of a roll-and-rise motion by the humanoid robot.

Each graph corresponds to each original performance of roll-and-rise action. The line graphs represent the change rates of the number of correct answers along the time axes.

The rate of change indicates how many subjects changed their guesses to the correct one during each time interval. This indicates the amount of information obtained during the interval from the movie clip that contributed to correct recognition.

If the subjects collect information uniformly along the time axis from the movie clips, we should see linear increase of correct answers, which will take the form of a flat horizontal line graph. However, as shown in Fig. 9, it is strongly nonlinear. Each change rate line graph has one or two

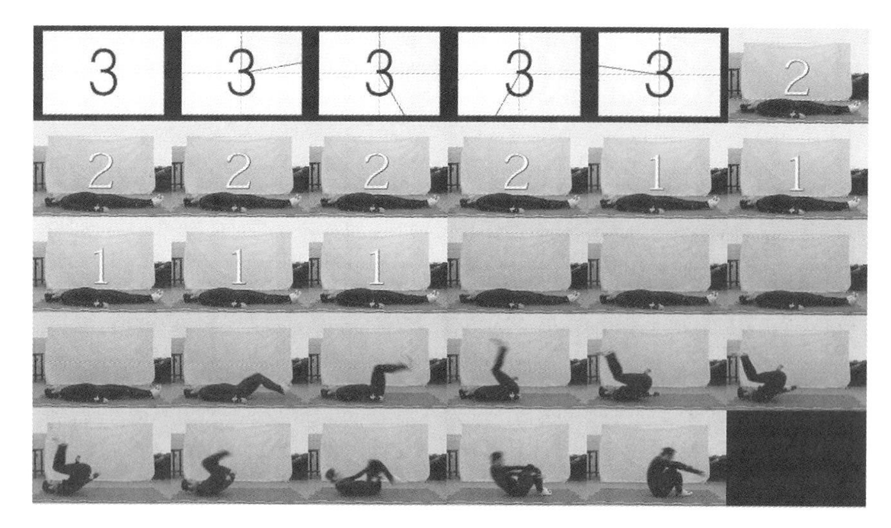

Fig. 7. A movie clip presented to subjects (A 1000 ms example).

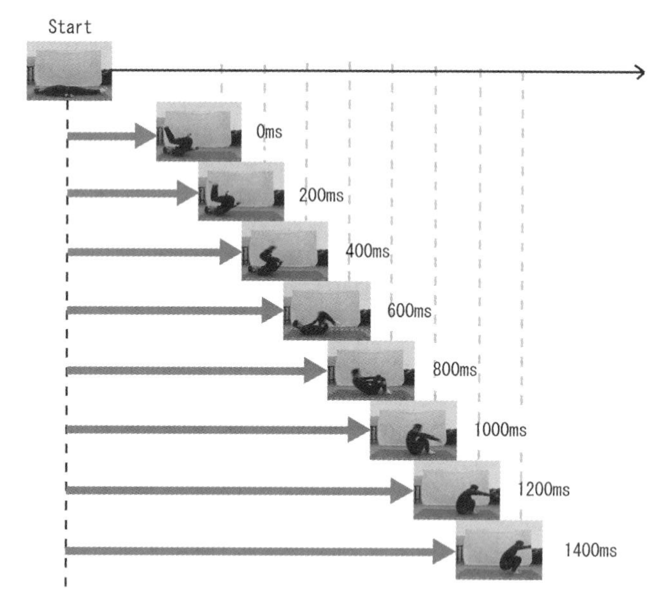

Fig. 8. Generation of eight movie clips with different length.

peaks, indicating that important information for recognizing the success or the failure of the roll-and-rise action is strongly localized in a few points along the time axis. Comparing the peaks of change rate graphs with the contents of the movie clips (shown in Fig. 8), we can identify that at least the strongest peaks correspond to the feet landing timing, i.e., 800–1000 ms. Some data indicate the correspondence to earlier time points but require further experiments with more precise timings.

The above results confirm that the nodes, the states with critical conditions, are important both for execution and recognition of the task. The idea that the nodes represent symbols that can be shared among people is supported. It is important to remember that such nodes emerge from the pure physics of the body–environment interaction.

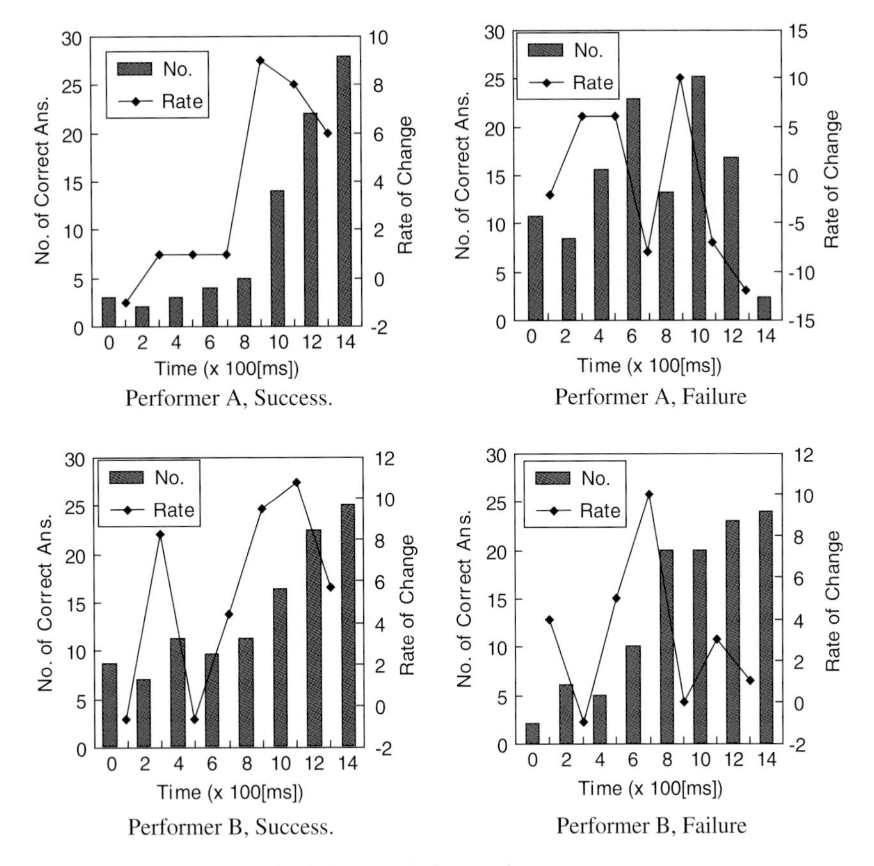

Fig. 9. Temporal change of correctness.

Exploration and emergence of embodied interaction patterns

We have seen in the previous section that the physics of body–environment interaction gives rise to clear information structures that can be exploited for both control and recognition. Then, in this section, we investigate how to autonomously explore for such structures.

It is trivial that unless we drive the body, no such structure as above emerges. But then, how should we drive the body when we do not know in advance what kind of structure emerges? Traditional robot learning methods typically assume an *a priori* set of *primitives* that are atomic unit actions that can be combined in time/space to express arbitrary behavior. However, this method suffers from serious problems: (1) The search space size, i.e., the number of possible combinations, becomes intractably large for dynamic motion of multiple joint body such as a humanoid, and (2) It optimizes for a particular goal or evaluation function, and does not explore broad enough possibilities of actions.

We propose a very different novel model in which a distributed set of chaotic elements are coupled with the multiple-element musculo-skeletal system (Kuniyoshi and Suzuki, 2004; Kuniyoshi et al., 2007). The model assumes no *primitives*, and no evaluation functions. Yet it explores broad possibilities of "what can be done with the given body." Once it discovers some consistent pattern exploiting the natural dynamics of the body (e.g., energy efficient cyclic motion — such as walking), the model spontaneously stabilizes the motion pattern by mutual entrainment. Further more, it has immediate adaptation capability to changing constraints, switching to different/novel motion patterns.

Coupled chaotic system

Coupled map lattice (CML) and globally coupled map (GCM) (Kaneko and Tsuda, 2001) have been investigated in complex systems science for their rich dynamics properties. They follow (1)–(2). CML is a coupled chaotic system with local interaction (1). GCM is one with global interaction (2).

$$x^i_{n+1} = (1-\varepsilon)f(x^i_n) + \frac{\varepsilon}{2}\{f(x^{i+1}_n) + f(x^{i-1}_n)\} \quad (1)$$

$$x^i_{n+1} = (1-\varepsilon)f(x^i_n) + \frac{\varepsilon}{N}\sum_{j=1}^{N} f(x^j_n) \quad (2)$$

Where, x^i_n denotes the internal state of i th element at time n, N the total number of elements, and ε the connection weight between elements. $f(x)$ can be any chaos function. In this paper, we adopt a standard *logistic map* represented as the following.

$$f(x) = 1 - ax^2 \quad (a > 1.4011\ldots) \quad (3)$$

With no interaction between the elements, all of them behave chaotically. But with interaction, depending on the parameters (a; ε), a rich variety of dynamical structures emerge such as ordered phases (with clusters of resonating elements) and partially ordered phases (configuration of the clusters changes with time).

This phenomenon is essentially caused by a competition of two tendencies: (1) A tendency to synchronize each other by the effect of the mean-field, and (2) A tendency to take arbitrarily different values due to the nature of chaos dynamics.

Body and environment as an interaction field of chaotic elements

Figure 10 shows our model of chaos coupling through robotic embodiment.

N chaotic elements are connected with actuators and sensors of the robot body. Each element drives a corresponding actuator based on its current internal state. The effect of N actuators collectively changes the physical state of the body that is constrained by and interacting with the environment. In other words, the output of N chaotic elements are mixed together and transformed by the embodied dynamics. The result is then sensed at each site of the actuator, e.g., in terms of joint angle or muscle length. Each sensor value is input to the corresponding chaotic element. Then each element updates, by chaotic mapping, its internal state from the new sensor value and the previous internal state.

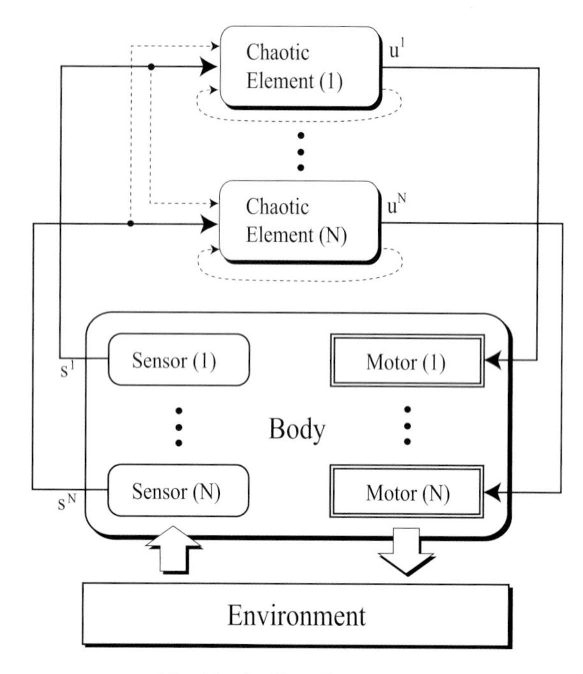

Fig. 10. Outline of our model.

In our model, body–environment interaction dynamics, or *embodiment*, serves as the chaos coupling field, which is nonlinear and time-varying. Theoretically very little is known about such cases, but since the coupling field directly reflects the current body–environment dynamics, we believe that the emergent ordered patterns correspond to useful motor coordination patterns which immediately get reorganized in response to dynamically changing environmental situation.

Experiments with a muscle–joint body

A series of experiments with a simple 1-joint body is carried out in order to investigate the effect of our model.

The structure of the 1-joint body is shown in Fig. 11. It consists of two cylindrical rigid bodies connected with a free joint, and 12 muscle fibers connecting them at equal spacing around the cylinders.

Each muscle fiber is modeled with Hill's characteristic equation (Hill, 1938). The length of each fiber is fed back to the corresponding chaotic element as a sensor value.

Figure 12 shows the behavior of the system. Once the simulation starts, the top cylinder starts to oscillate coherently. In some of the experiments, occasional spontaneous change of oscillation direction is observed.

Note that even for this simple behavior, the 12 muscle fibers must be properly coordinated. And the coordination is emergent.

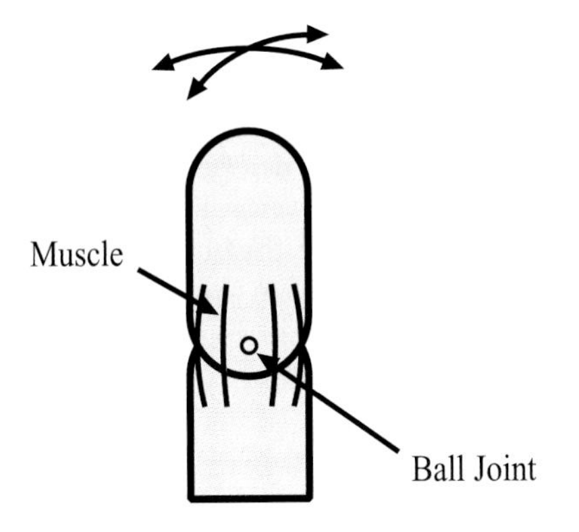

Fig. 11. Appearance of the muscle–joint model.

In a control experiment, all the sensor feedback signals are disabled. The resulting behavior, shown in Fig. 13, is jerky and disordered, with chaotic internal states.

In other experiments (Kuniyoshi and Suzuki, 2004), various modifications to the physical arrangement of the body were tested. The system exhibits different behaviors as if it exploits the new physical features: when some muscle fibers are added in slanted arrangements, the body mostly rotated, etc.

The environment also makes a part of the interaction field for the chaotic elements. In another experiment, we introduced a dynamic obstacle that is raised in the middle of the experiment to interfere with the oscillating body (Fig. 14). Figure 15 shows the result: Initially, the body oscillated coherently in the same way as Fig. 12. Then the obstacle is raised and the body collided with it. After the collision, the body made a complex motion with repeated collisions for a short duration. The internal state is chaotic during the period. But soon after that, within ~3 s, the body began to oscillate coherently in a new collision-free direction. The overall behavior looks as if the system immediately adapted to the change of the environment by a quick exploration.

Experiments with an insect-like robot: emergence of walking

In order to investigate the effects of our model in a more meaningful behavior with more complex

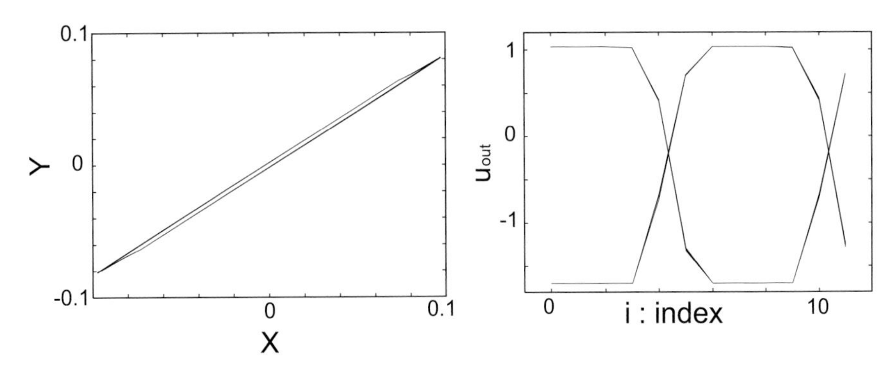

Fig. 12. Experiment with feedback of length sensor. Trajectory of the center of the upper cylinder (left), chaos unit value pattern (right). In the right graph, the output values for each chaotic unit (index i) are connected to form a line graph for each time step, overlaid for the duration of the record. This clean graph with two (alternating over time) lines means that all the units form one cluster with periodic oscillation.

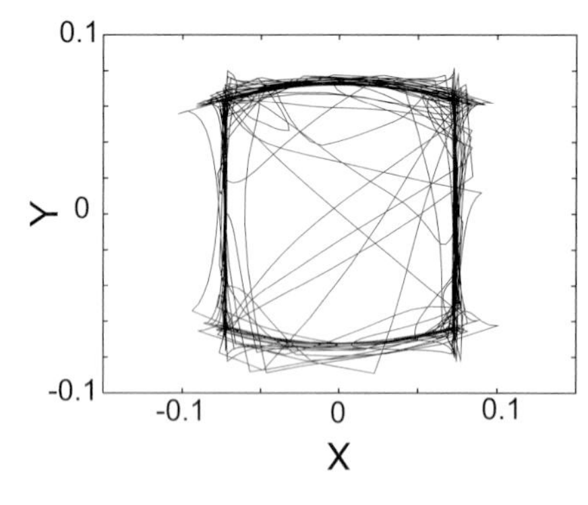

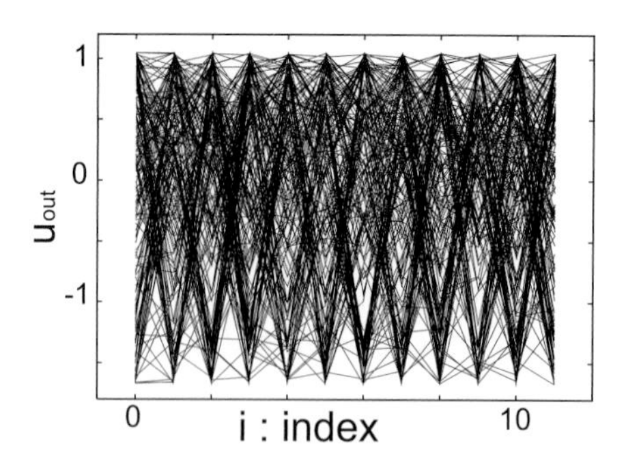

Fig. 13. Experiment with no sensor feedback. Trajectory of the center of mass of the upper link projected on x–y plane (left graph). Cluster plot of the chaotic elements (right). For each element with index i, its motor output u_{out} is overlaid for $n = 10,11,12,\ldots$. The points of all the elements are connected with a line for each time step.

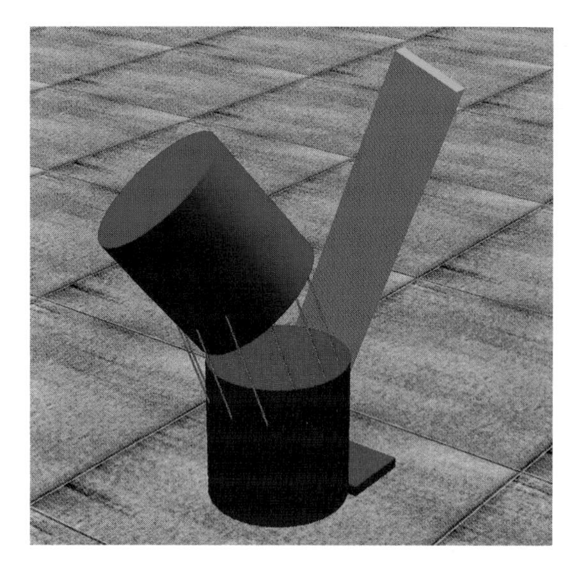

Fig. 14. The muscle–joint model and an obstacle.

interactions with the environment, we defined an insect-like multi-legged robot. The robot has a disc-shaped body with 12 legs attached on its fringe with regular spacing (Fig. 16). Each leg is connected to the body by a rotational joint and two springs, and can swing in the radial direction as shown in the middle of Fig. 16.

When the simulation is run with no sensor feedback, the legs moved in uncoordinated ways, and no order was observed in the motion of the robot. It just kept on randomly struggling around the same spot on the ground.

On the other hand, when the sensor feedback is introduced, after the initial chaotic period (a few seconds), the robot started to move in a certain direction, and then finally showed a stable locomotive behavior with a constant speed in a stable direction (Fig. 17). The locomotive behavior was realized by synchronizing the three or four hind legs and kicking the ground with them.

Emergence of body afforded behaviors

The above results show that our proposed mechanism has the capability of spontaneously exploring and stabilizing various behavior patterns that exploit the intrinsic oscillatory modes of the complex body. Moreover it sometimes spontaneously switches from one pattern to another. Although further capabilities and limitations are yet to be discovered, the proposed mechanism makes a good candidate for explorative learning of *body/environment affordances*, and discovering the embodied information structures.

Simulated baby

The above model correlates with the essential structure of vertebrates, i.e., the spine/medulla circuit and the musculo-skeletal body. It is well established that parts of spine/medulla circuit act as non-linear oscillators, called CPG (central pattern generator). Under certain conditions, a coupled system of non-linear oscillators acts as a coupled chaotic system. Therefore, it is quite plausible that vertebrates exploit the similar principle

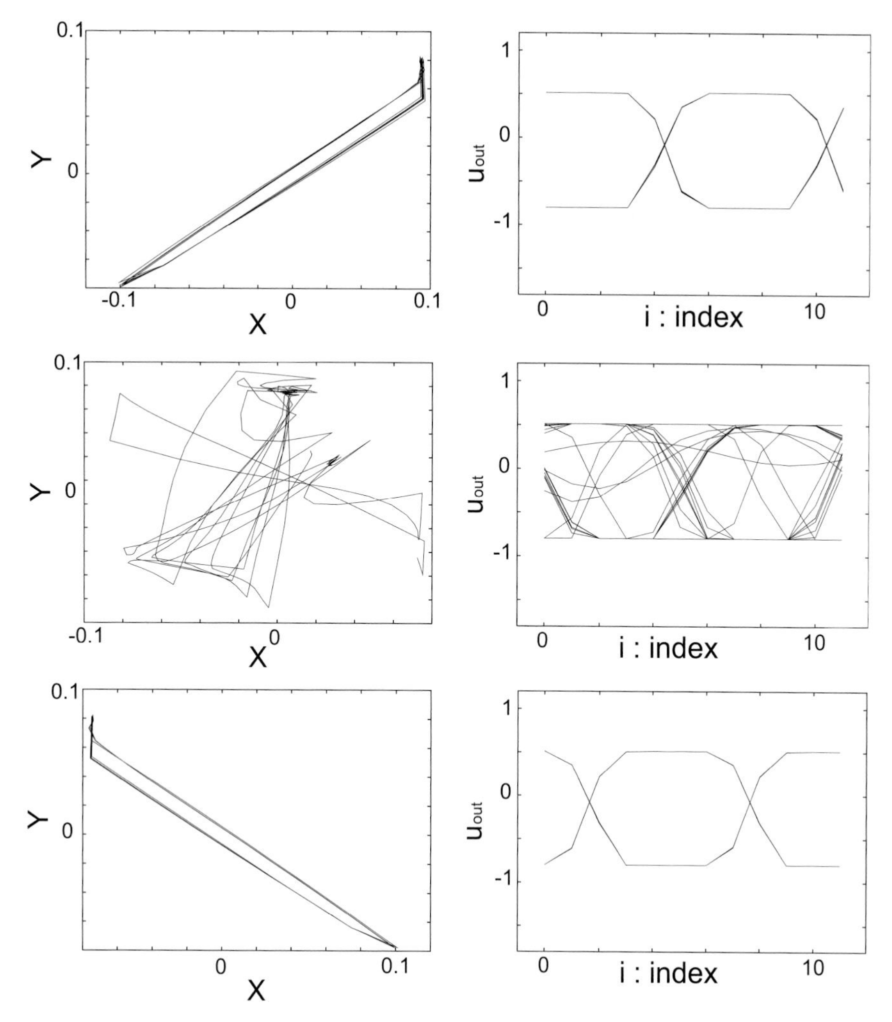

Fig. 15. Experiment with an obstacle. Coherent oscillation before the obstacle is raised (top), complex explorative motion right after the obstacle is raised (middle). Another coherent oscillation emerged after a few seconds (bottom).

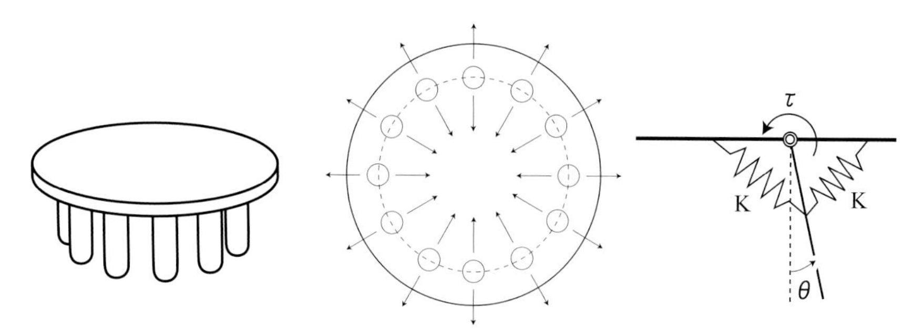

Fig. 16. Appearance of the insect-like robot (left), direction of leg motion (middle), and the mechanism of a leg (right).

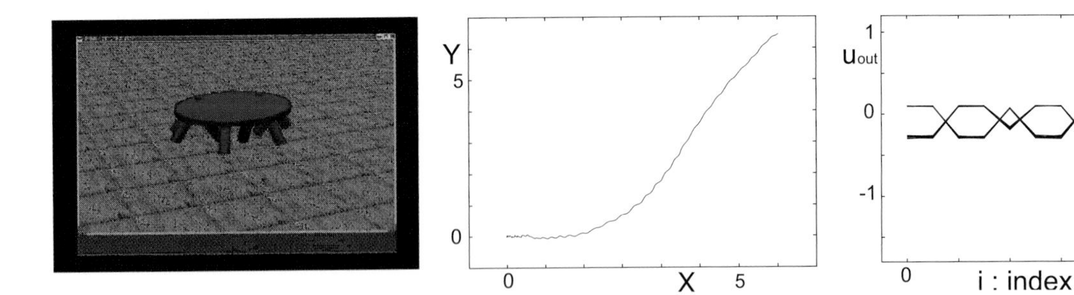

Fig. 17. Experiment with sensor feedback. The walking insect-like robot (left). Trajectory on the ground plane (middle) exhibits self-entrained stable orientation. Internal state (right) is stable and coordinated.

as our model for acquisition and adaptation of motor behavior.

Since our model explores and discovers motion patterns that fit the natural property of the body, it may be a good candidate for simulating the initial mechanism of motor development. It may be able to start with very little predefined knowledge and autonomously acquire appropriate motor primitives.

A human body is so complex, and a systematic search for all possible motion patterns is virtually impossible. However, our model should be able to discover appropriate motions very quickly. Moreover, the cluster emergence in pure CML and GCM are known to scale to thousands of elements. This is a good reason to expect that our model can handle the musculo-skeletal system of a human body.

Baby body

Figure 18 shows the view of our simulated baby (Kuniyoshi and Sangawa, 2006). The outlook is crude as we invest little effort on the quality of graphics. However, the musculo-skeletal system is modeled at a highly detailed level.

Our model has 198 muscles. We omitted the wrist, ankle, fingers, toes, neck, and face. But the body stem and the limbs are quite faithfully modeled. The dimensions, mass, and inertial parameters of all the body parts are defined according to the measurements of real babies.

The proprioceptive sensing organs, i.e., the muscle spindles and Golgi tendon organs, are also modeled as precisely as possible. The muscles are also modeled to match the performance of real babies.

All the physical body parameters are modeled as functions of week age after gestation (in the uterus). So the body can simulate physical growth of the fetal and neonatal periods.

The simulated baby body is placed in two types of simulated environments (Fig. 19): The "fetus" is placed in a simulated uterus with a movable wall (modeled as a nonlinear spring-damper membrane), filled with liquid. The "neonate" is placed on a flat floor surrounded by flat walls (like a playpen).

Cortico-medullar-spinal-musculo-skeletal system: overview

Based on the core model of emergent explorative dynamics above, we propose an integrated model of explorative motor learning with minimum components (Kuniyoshi and Sangawa, 2006). Figure 20 shows the overall structure of the model. It consists of network models of the primary somatosensory (S1) and primary motor (M1) areas of the cerebral cortex, the CPG model of medulla, the efferent (α and γ neurons) and afferent (S0) spinal neurons, and the muscle with sensory organs (Ia: spindle, Ib: tendon).

The CPG model of medulla, together with nonlinear characteristics of the muscle and associated sensory organs, behaves as a high dimensional coupled chaos system.

The cortical areas S1 and M1 are modeled as self-organizing neural networks. The connections between the cortical areas and the medullar-spinal

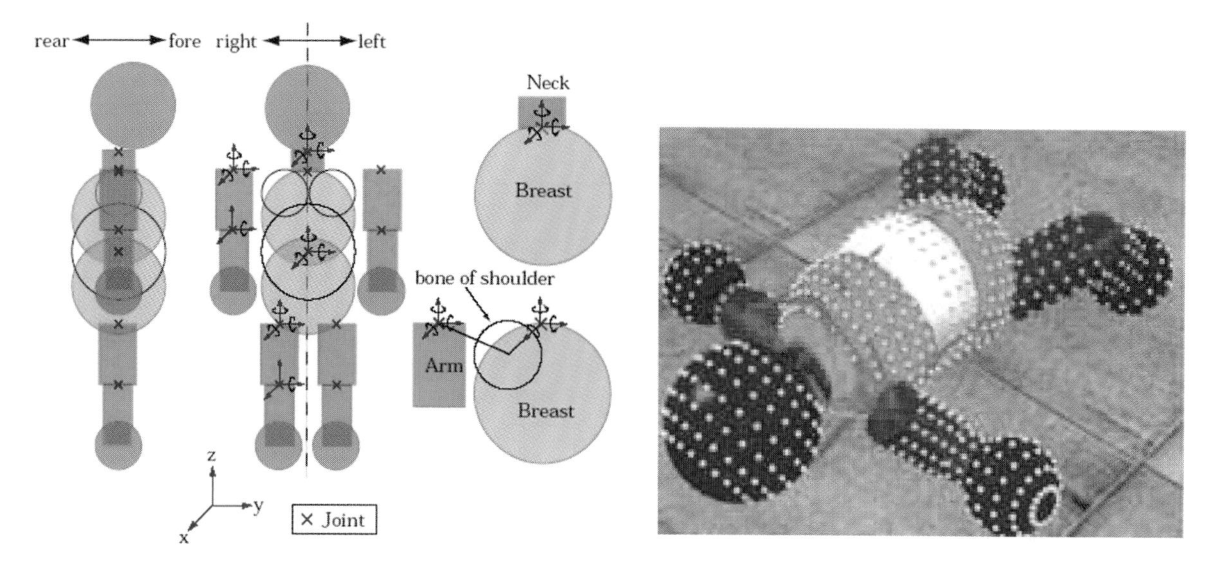

Fig. 18. Simulated baby. White dots over the surface in the left graphics are tactile sensing points.

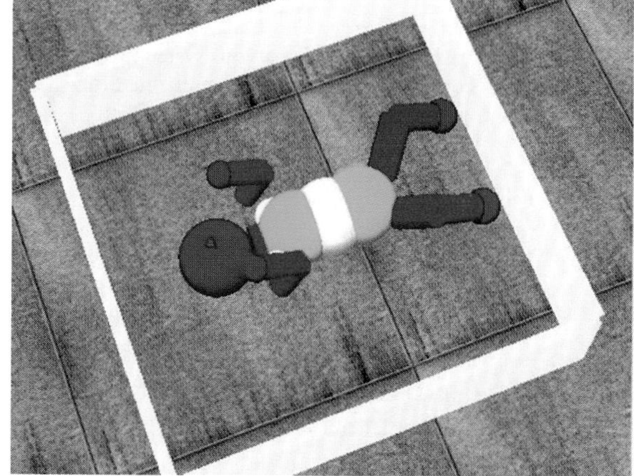

Fig. 19. Simulated environments. A fetus in the uterus (left) and a newborn on a flat bed with a fence (right).

circuit have plasticity modeled with Hebbian learning.

While the overall neural-musculo-skeletal system generates exploratory bodily movements, the neural system learns the sensory-motor patterns. If some coherency emerges in the sensory-motor patterns due to the global neural-musculo-skeletal entrainment, it will be captured as stable maps in the self-organizing cortical area models.

Human cerebrum has left and right hemispheres connected by a bundle of fibers called *corpus callosum*. Each hemisphere controls the other side of the body. Accordingly, the spinal fibers are also grouped into two sides. As shown in Fig. 20, we construct left and right nervous systems. And we connect the left and right cortical models with the model of corpus callosum.

Emergent behavior

We are starting to observe emergence of patterned motions and stable clustering of cortical neurons (Kuniyoshi and Sangawa, 2006).

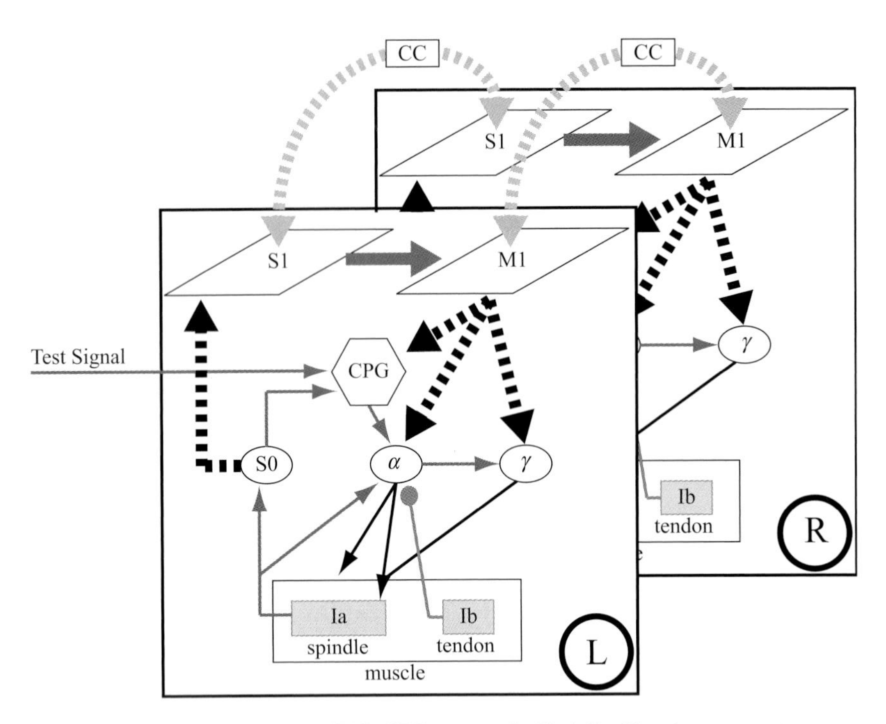

Fig. 20. Lateral organization of the nervous system. CPG: CPG neurons in Medulla, S1: primary somatosensory area, M1: primary motor area, CC: corpus callosum, α: alpha motor neuron, γ: gamma motor neuron, S0: afferent interneuron. Arrow and filled circle represent excitatory and inhibitory connection, respectively. Thick broken lines represent all to all connections with plasticity.

Figure 21 shows a series of motor behavior observed in one of the experiments. In the figure, the infant model first rolls over from the face-up to face-down posture, then started to do crawling-like behavior (but without moving forward), which continued for a while. Throughout the experiment, the model often exhibited mixtures of periodic and aperiodic complex movements.

The precise cause of the above meaningful behavior must wait for a detailed analysis. It is highly plausible to assume that the emergence of meaningful behavior such as crawling owes much to the natural body property: the overall geometry of the skeletal system, natural joint positions due to the natural length settings of the muscles, the intrinsic frequency of the limbs, all contribute to the resulting behavior, and the neural system could explore and get entrained into such consistent oscillatory movement as the crawling-like behavior.

Emergence of imitation from learning about self

So far we proposed a plausible model of autonomous motor development. But how does it relate to "human" cognition that should involve social interaction aspects? Remember that in Section "Knacks and focuses" of dynamics actions: information structure emerging from body–environment interaction," we pointed out that the information structure emerging from embodiment serves dual functions: action execution and action recognition. Thus, learning about self-behavioral patterns implies acquiring capability to recognize other's behavior. In this section, we propose a QJ;basic model that supports this view; the system that learns about self-motor patterns whose internal representation has the (unintended) function of early behavior imitation (Kuniyoshi et al., 2003).

Meltzoff and Moore (1977) discovered that very early neonates exhibit facial and manual imitation.

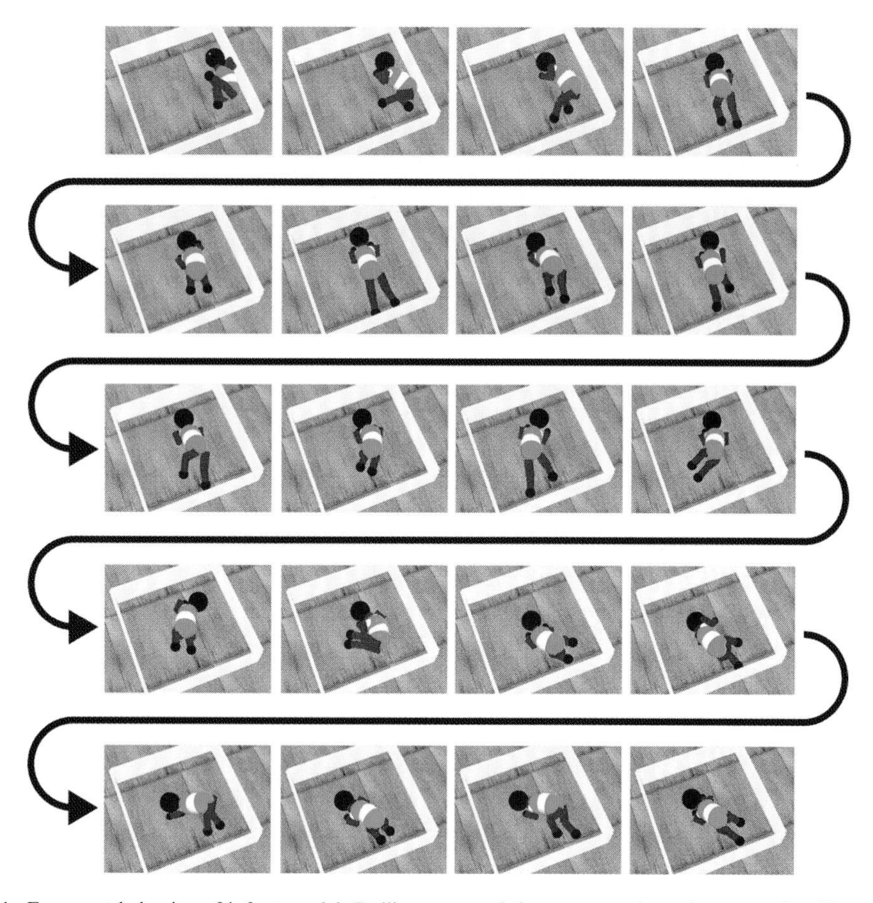

Fig. 21. Emergent behavior of infant model. Rolling over and then starts and continues crawling-like motion.

They pose two alternative explanations: (1) Existence of an innate mechanism which represents the gestural postures of body parts in supra-modal terms, i.e., representations integrating visual, somatosensory, and motor domains. (2) Possibility of creating such representation through self-exploratory "body babbling" during the fetus period. Meltzoff and Moore submit to the first hypothesis, proposing a mechanism called AIM.

We support the second hypothesis above. Our aim is to show that assuming primitive configuration of fetuses and neonate's brain, the result of self-exploratory sensory-motor learning can be immediately reused in imitating a first-seen other human's gestural motion (Fig. 22).

We attempt to achieve this by creating a robot system that exhibits such behavior. Although it is a minimalist model, it works on a real humanoid robot and responds to a real human gesture in the real world. It should serve as an initial model for a synthetic developmental study of imitation, which will clarify necessary modules to be added/acquired for more complex types of imitation.

Overview of the system

Neonatal imitation is typically attributed to some innate visual-motor mapping mechanism. However, recent findings about rich exploratory movements of fetuses and extensive neural development during this period suggest a possibility of self-exploratory learning during the fetal period and its effect on newborn imitation abilities.

As a constructivist approach to this hypothesis, we developed a visuo-motor neural learning

Fig. 22. Scenes from self-learning (left), robot's view during learning (middle), imitating human gesture after learning (right).

system that consists of orientation selective visual movement representation, distributed arm movement representation, and a high dimensional temporal sequence learning mechanism (Fig. 23).

Some evidences suggest excessive cross-modal connections in neonatal brain. So we hypothesize that fetuses/neonates treat visual, proprioceptive, and motor data as "melted together" and extracts information structure in an amodal manner. Based on the hypothesis, we combine the incoming vision and motor feature vectors into a single high-dimensional vector and feed to a uniformly connected large neural network that extracts spatio-temporal patterns from the sensory-motor-time space.

In the experiments, our robot first generates several arm motion patterns and learns the correspondence between the motor commands and the proprioceptive and visual sensor data. This situation simulates a fetus moving its hands and vaguely observing the motion.

During learning, the vision and motor data are fed to the neural network as a temporal sequence of high-dimensional scalar vectors. In our experiment, a 36 dimensional visual feature vector and a 150 dimensional joint angle vector (population coding), spliced together into a 186 dimensional sensory-motor vector, is generated every 33 ms. In order to cope with this hard learning task with high-dimensional and long sequence of data, we adopt a neuron model called "non-monotonic neural net" proposed by Morita (1996). It is basically a dynamic continuous-valued Hopfield network consisting of neurons with a special non-monotonic output function. It learns coherent spatio-temporal patterns as high-dimensional

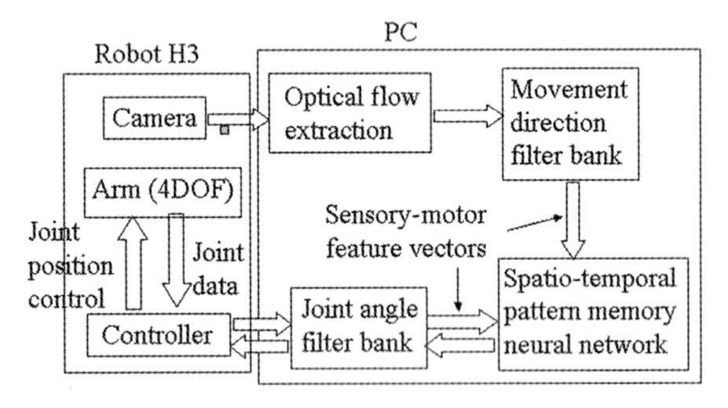

Fig. 23. Configuration of the learning to imitate system.

trajectory attractors in the state-space of the neural network.

After the learning, a human comes in front of the robot, showing arm movements that are similar to the ones in self-learning.

Figure 24 shows some of the joint angle profiles of the arm movement during self-learning phase. It also shows some of the profiles of the generated arm movement in response to given human gestures. The figure shows three trials for each type of arm movement patterns.

In seven out of nine trials shown in Fig. 24, the system successfully recognized first-seen human arm movements as known patterns and generated correct arm movements. However, in Pattern1-Trial2 and Pattern3-Trial2, the system failed to recognize the given patterns in the middle of the sequences. We made 20 trials and the overall success rate of correct imitation was 81%.

Although the robot has never seen or programmed to interpret human arm movement, and the detail of visual stimuli are very different, the robot identifies some of the patterns as similar to those in self-learning, and responds by generating

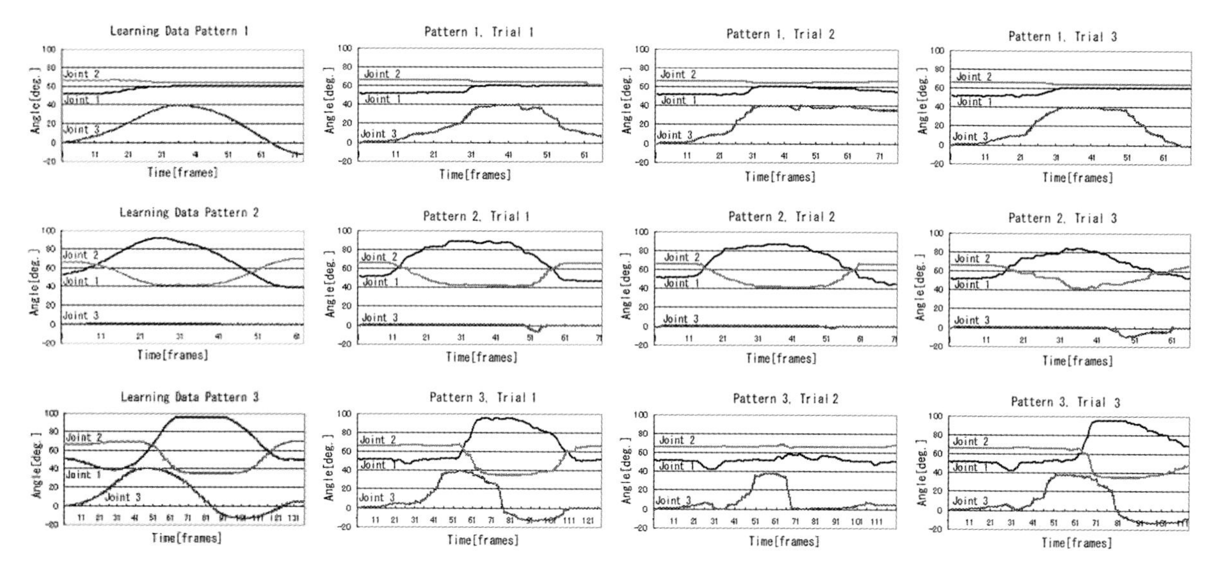

Fig. 24. The joint angle profile of the three arm movement patterns during self-learning (left-most column). Vertical axis indicates the angles (degree). Horizontal axis indicates time (video frames). Each line corresponds to a joint angle trajectory. Pattern 1: Horizontal swing (left), Pattern 2: Vertical swing (middle), Pattern 3: Circular movement (right). The joint angle profile of the generated arm movement in response to the three arm movement patterns given by a human demonstrator (right 3 columns). Vertical axis indicates the angles (degree). Horizontal axis indicates time (video frames). Each line corresponds to a joint angle trajectory. Pattern 1: Horizontal swing (top row), Pattern 2: Vertical swing (middle row), Pattern 3: Circular movement (bottom row). Trial 1 (left column), Trial 2 (middle column), Trial 3 (right column).

the previously learned arm movement. In other words, the robot exhibits early imitation ability based on self-exploratory learning.

There is no self–other distinction. So the robot is responding to the visual stimuli *as if* it were just remembering playing with its own hand and being attracted with the idea.

Cognitive development from embryo: summary and discussions

A constructivist approach to revealing the principles of emergence and development of embodied cognition is presented.

It starts with the minimal initial configuration, embeds the system in interaction with its circumstances, and investigates its behavior to see if some important aspect of the target system is realized. It then augments the initial configuration one by one to further match the resulting behavior to the target.

In our line of research, we started with only a physical body without any nervous system. In a whole body dynamic actions, we showed that pure physics gives rise to a meaningful information structure, "knacks"/"focuses," the flip sides of the same structural feature. They are invariant over perturbations and self–other differences, emerge when the body is driven toward a goal state, define action unit boundaries, and serve as the basis for both execution and perception of actions and for imitation and communication of skills. These are strongly relevant to the notion of actions and prospective control by von Hofsten (1993).

A fundamental problem, though, is the principle of autonomously driving the system. On one hand, in order for the above structure to emerge, we need to drive the system toward some goal. On the other hand, if we apply explicit control toward a particular goal, the system just obeys it without any exploration. Moreover, in order to write a concrete control law, we have to have a complete knowledge of the dynamics of the system, meaning that the system does not have to learn any more. Then, how can the system do an effective exploration without knowing how to reach the goal? Our proposal is to introduce a system of highly flexible pattern generators with strong entrainment property. Such a system can explore the natural modes of motion of the body, getting entrained into the self-generated motion pattern, but also capable of shifting to different modes. This self-entrained pattern may be a primitive form of autonomous goal-directed behavior.

We proposed a novel framework for the above kind of autonomous exploration. The core mechanism is based on a coupled chaotic system, which autonomously explores and generates various coordination patterns of multiple degrees of freedom. The emergent motion patterns exploit and resonate with the body–environment dynamics. Therefore our model is a very good candidate as the initial core mechanism to simulate very early motor development of human babies. It should be important for human babies to acquire motor primitives exploiting the characteristics of body–environment dynamics.

The above model correlates with real human babies because the CPG in spine/medulla can generate high dimensional chaos under certain conditions, and the resulting whole-body movement has the similar property as the general movement (GM) that appears in early motor development of human babies.

We are now constructing and experimenting with a simulated baby. It is designed to be very close to real human babies in terms of musculo-skeletal system. The coupled chaotic system corresponds to the medulla circuit behavior at a certain input signal condition, coupled with the musculo-skeletal system dynamics. When an emergent motion pattern persists for certain time duration, the learning in the cortex model and other neural connections fixates it in the neural connections. This way the system should be able to explore, discover, and learn various motor primitives that fully exploit the natural body–environment dynamics. It is still an open question how to design a mechanism that appropriately integrates the learning and emergence.

The last experiment suggests how self-explorative learning during the fetal period can support neonatal imitation. Since the human form bodies provide a common information structure, and it serves both action execution and perception, the learned information structure naturally serves as the basis for behavioral imitation.

The above steps connect the pure body properties to the early form of social intelligence, i.e., imitation.

Obviously the above results are yet to be conclusive and a lot more work must be devoted. However, they suggest a plausible set of principles of early cognitive development. Among many other open questions, some immediate next steps would be to address the issue of body schema and body image acquisition/exploitation, clarifying what are the essential elements to be added to the initial configuration, or alternatively showing that the current system can acquire them without modification. Body image, or self-perception, would establish broad bi-directional pathways of interaction between embodiment, social cognition, and self-awareness.

Acknowledgments

The present paper summarizes research results from several different projects. They have been supported in parts by Grant-in-Aid for Scientific Research from Japan Society for Promotion of Science, and in other parts by JST ERATOAsada Synergistic Intelligence Project.

Ichiro Tsuda, Rolf Pfeifer, Ruzena Bajcsy, Minoru Asada, Jun Tani, Kazuo Hiraki, Max Lungarella, Alex Pitti, and many others provided valuable suggestions and comments.

References

Asada, M., MacDorman, K.F., Ishiguro, H. and Kuniyoshi, Y. (2001) Cognitive developmental robotics as a new paradigm for the design of humanoid robots. Robot. Auton. Syst., 37: 185–193.

Crair, M.C. (1999) Neuronal activity during development: permissive or instructive? Curr. Opin. Neurobiol., 9: 88–93.

Drescher, G.L. (1991) Made-Up Minds: A Constructivist Approach to Artificial Intelligence. MIT Press, Cambridge, MA, USA.

Hashimoto, T. (1999) Modeling categorization dynamics through conversation by constructive approach. In: Floreano, D., Nicoud, J.-D., Mondada, F., Carbonell, J.G. and Siekmann, J. (Eds.), Advances in Artificial Life: 5th European Conference, Ecal'99, Lausanne, Switzerland, September 1999: Proceedings. Lecture Notes in Computer Science. Springer, pp. 730–734.

Hendriks-Jansen, H. (1996) Catching Ourselves in the Act. MIT Press, Cambridge, MA, USA.

Hill, A.V. (1938) The heat of shortening and the dynamic constants of muscle. Proc. R. Soc. Lond. B, 126: 136–195.

von Hofsten, C. (1993) Prospective control: a basic aspect of action development. Hum. Dev., 36(5): 253–270.

Johnson, M.H. (2005) Developmental Cognitive Neuroscience (2nd ed.). Blackwell, Oxford, England.

Kaneko, K. and Tsuda, I. (2001) Complex Systems: Chaos and Beyond. Springer, Berlin-Heidelberg, Gemany.

Karmiloff, K. and Karmiloff-Smith, A. (2001) Pathways to Language: From Fetus to Adolescent (Developing Child). Harvard University Press, Cambridge, MA, USA.

Kisilevsky, B. and Low, J. (1998) Human fetal behavior: 100 years of study. Dev. Rev., 18: 1–29.

Kuniyoshi, Y., Ohmura, Y., Terada, K. and Nagakubo, A. (2004a) Dynamic roll-and-rise motion by an adult-size humanoid robot. Int. J. Humanoid Robot., 1(3): 497–516.

Kuniyoshi, Y., Ohmura, Y., Terada, K., Nagakubo, A., Eitoku, S. and Yamamoto, T. (2004b) Embodied basis of invariant features in execution and perception of whole body dynamic actions: knacks and focuses of roll-and-rise motion. Robot. Auton. Syst., 48(4): 189–201.

Kuniyoshi, Y. and Sangawa, S. (2006) Early motor development from partially ordered neural-body dynamics: experiments with a cortico-spinal-musculoskeletal model. Biol. Cybern., 95(6): 589–605.

Kuniyoshi, Y. and Suzuki, S. (2004) Dynamic emergence and adaptation of behavior through embodiment as coupled chaotic field. In: Proceedings of IEEE International Conference on Intelligent Robots and Systems, pp. 2042–2049.

Kuniyoshi, Y., Suzuki, S. and Sangawa, S. (December 2007) Emergence, exploration and learning of embodied behavior. In: Thrun, S., Brooks, R. and Durrant-whyte, H. (Eds.), Robotics Research: Results of the 12th International Symposium ISRR. Springer Tracts in Advanced Robotics, Vol. 28. Springer Verlag.

Kuniyoshi, Y., Yorozu, Y., Inaba, M. and Inoue, H. (2003) From visuo-motor self learning to early imitation: a neural architecture for humanoid learning. In: Proceedings of IEEE International Conference on Robotics and Automation, pp. 3132–3139.

Meltzoff, A.N. and Moore, M.K. (1977) Imitation of facial and manual gestures by human neonates. Science, 198: 75–78.

Morita, M. (1996) Memory and learning of sequential patterns by nonmonotone neural networks. Neural Netw., 9: 1477–1489.

Pfeifer, R. and Scheier, C. (1999) Understanding Intelligence. MIT Press, Cambridge, MA, USA.

Ridley, M. (2003) Nature Via Nurture: Genes, Experience, and What Makes Us Human. Harper Collins, New York, NY, USA.

Thelen, E. and Smith, L.B. (1994) A dynamic systems approach to development of cognition and action. MIT Press.

Subject Index